IN
TO MAGNETI(

B. D. CULLITY
University of Notre Dame

INTRODUCTION TO MAGNETIC MATERIALS

ADDISON-WESLEY PUBLISHING COMPANY
Reading, Massachusetts
Menlo Park, California · London · Amsterdam · Don Mills, Ontario · Sydney

This book is in the

Addison-Wesley Series in Metallurgy and Materials

Consulting Editor

MORRIS COHEN

 Printed in the United States of America. Published simultaneously in Canada. Library of Congress Catalog Card No. 71-159665.

ISBN 0-201-01218-9

13 14 15-MA-97 96

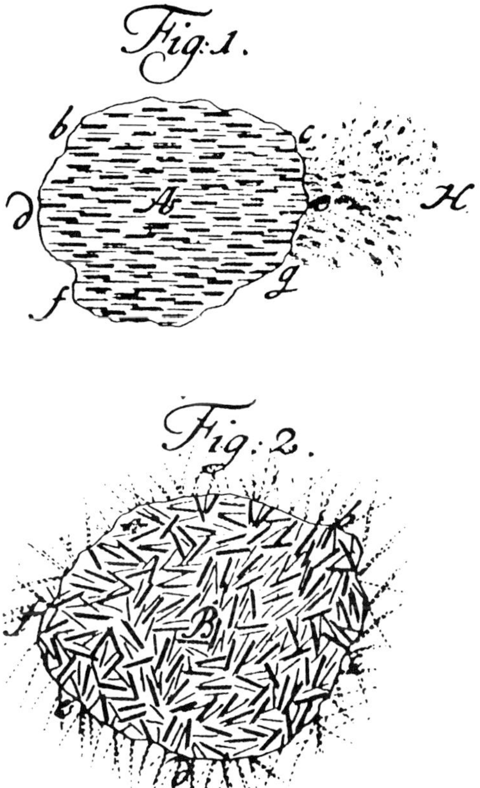

An illustration taken from the *Principia Rerum Naturalium,* a treatise published in 1734 by Emanuel Swedenborg (1688–1772), Swedish scientist, philosopher, and theologian. The drawings indicate his view of the difference between magnetized (Fig. 1) and unmagnetized (Fig. 2) iron. They are in remarkable agreement with the modern domain theory of ferromagnetism, which came almost two centuries later.

The rich diversity of ferromagnetic phenomena, the perennial challenge to skill in experiment and to physical insight in coordinating the results, the vast range of actual and possible applications of ferromagnetic materials, and the fundamental character of the essential theoretical problems raised have all combined to give ferromagnetism a width of interest which contrasts strongly with the apparent narrowness of its subject matter, namely, certain particular properties of a very limited number of substances.

Edmund C. Stoner: "Ferromagnetism,"
Reports on Progress in Physics,
Vol. 11, 1948, The Physical Society,
London

PREFACE

Take a pocket compass, place it on a table, and watch the needle. It will jiggle around, oscillate, and finally come to rest, pointing more or less north. Therein lie two mysteries. The first is the origin of the earth's magnetic field, which directs the needle. The second is the origin of the magnetism of the needle, which allows it to be directed. This book is about the second mystery, and a mystery indeed it is, for although a great deal is known about magnetism in general, and about the magnetism of iron in particular, it is still impossible to predict from first principles that iron is strongly magnetic.

This book is for the beginner. By that I mean a senior or first-year graduate student in engineering, who has had only the usual undergraduate courses in physics and materials science taken by all engineers, or anyone else with a similar background. No knowledge of magnetism itself is assumed.

People who become interested in magnetism usually bring quite different backgrounds to their study of the subject. They are metallurgists and physicists, electrical engineers and chemists, geologists and ceramists. Each one has a different amount of knowledge of such fundamentals as atomic theory, crystallography, electric circuits, and crystal chemistry. I have tried to write understandably for all groups. Thus some portions of the book will be extremely elementary for most readers, but not the same portions for all readers.

Despite the popularity of the *mks* system of units in electricity, the overwhelming majority of magneticians still speak the language of the *cgs* system, both in the laboratory and in the plant. The student must learn that language sooner or later. This book is therefore written in the *cgs* system.

The beginner in magnetism is bewildered by a host of strange units and even stranger measurements. The subject is often presented on too theoretical a level, with the result that the student has no real physical understanding of the various quantities involved, simply because he has no clear idea of how these quantities are measured. For this reason methods of measurement are stressed throughout the book. All of the second chapter is devoted to the most common methods, while more specialized techniques are described in appropriate later chapters.

The book is divided into four parts:

1. Units and measurements.

2. Kinds of magnetism, or the difference, for example, between a ferromagnetic and a paramagnetic.
3. Phenomena in strongly magnetic substances, such as anisotropy and magnetostriction.
4. Commercial magnetic materials and their applications.

The references, selected from the enormous literature of magnetism, are mainly of two kinds, review papers and classic papers, together with other references required to buttress particular statements in the text. In addition, a list of books is given, together with brief indications of the kind of material that each contains.

Magnetism has its roots in antiquity. No one knows when the first lodestone, a natural oxide of iron magnetized by a bolt of lightning, was picked up and found to attract bits of other lodestones or pieces of iron. It was a subject bound to attract the superstitious, and it did. In the sixteenth century Gilbert began to formulate some clear principles.

In the late nineteenth and early twentieth centuries came the really great contributions of Curie, Langevin, and Weiss, made over a span of scarcely more than ten years. For the next forty years the study of magnetism can be said to have languished, except for the work of a few devotees who found in the subject that fascination so eloquently described by the late Professor Stoner in the quotation printed at the beginning of this book.

Then, with the end of World War II, came a great revival of interest, and the study of magnetism has never been livelier than it is today. This renewed interest came mainly from three developments:

1. *A new material.* An entirely new class of magnetic materials, the ferrites, was developed, explained, and put to use.
2. *A new tool.* Neutron diffraction, which enables us to "see" the magnetic moments of individual atoms, has given new depth to the field of magnetochemistry.
3. *A new application.* The rise of computers, in which magnetic devices play an essential role, has spurred research on both old and new magnetic materials.

And all this was aided by a better understanding, gained about the same time, of magnetic domains and how they behave.

In writing this book, two thoughts have occurred to me again and again. The first is that magnetism is peculiarly a hidden subject, in the sense that it is all around us, part of our daily lives, and yet most people, including engineers, are unaware or have forgotten that their lives would be utterly different without magnetism. There would be no electric power as we know it, no electric motors, no radio, no TV. If electricity and magnetism are sister sciences, then magnetism is surely the poor relation. The second point concerns the curious reversal, in the United States, of the usual roles of university and industrial laboratories in the area of magnetic research. While Americans have made sizable contributions to the international pool of knowledge of magnetic materials, virtually all of these contributions have come from industry. This is not true of other countries or other subjects. I do not pretend to know the reason for this imbalance, but it would certainly seem to be time for the universities to do their share.

Most technical books, unless written by an authority in the field, are the result of a collaborative effort, and I have had many collaborators. Many people in industry have given freely from their fund of special knowledge and experience. Many others have kindly given me original photographs. The following have critically read portions of the book or have otherwise helped me with difficult points: Charles W. Allen, Joseph J. Becker, Ami E. Berkowitz, David Cohen, N. F. Fiore, C. D. Graham, Jr., Robert G. Hayes, Eugene W. Henry, Conyers Herring, Gerald L. Jones, Fred E. Luborsky, Walter C. Miller, R. Pauthenet, and E. P. Wohlfarth. To these and all others who have aided in my magnetic education, my best thanks.

Notre Dame, Indiana B. D. C.
February 1972

CONTENTS

Chapter 4 Ferromagnetism

Chapter 5 Antiferromagnetism

Chapter 6 Ferrimagnetism

MAGNETIC PHENOMENA

Chapter 7 Magnetic Anisotropy

Chapter 8 Magnetostriction and the Effects of Stress

Chapter 9 Domains and the Magnetization Process

Chapter 10 Induced Magnetic Anisotropy

Chapter 11 Fine Particles and Thin Films

Appendixes

INTRODUCTION

CHAPTER 1

DEFINITIONS AND UNITS

1.1 INTRODUCTION

The story of magnetism begins with a mineral called magnetite (Fe_3O_4), the first magnetic material known to man. Its early history is obscure, but its power of attracting iron was certainly known for centuries before Christ. Magnetite is widely distributed. In the ancient world the most plentiful deposits occurred in the district of Magnesia, in what is now modern Turkey, and our word *magnet* is derived from a similar Greek word, said to come from the name of this district. It was also known to the Greeks that a piece of iron would itself become magnetic if it were touched, or, better, rubbed with magnetite.

Later on, but at an unknown date, it was found that a properly shaped piece of magnetite, if supported so as to float on water, would turn until it pointed approximately north and south. So would a pivoted iron needle, if previously rubbed with magnetite. Thus was the mariner's compass born. This north-pointing property of magnetite accounts for the old English word *lodestone* for this substance; it means "waystone," because it points the way.

The first truly scientific study of magnetism was made by the Englishman William Gilbert (1540–1603), who published his classic book *On the Magnet* in 1600 [1.1].* He experimented with lodestones and iron magnets, formed a clear picture of the earth's magnetic field, and cleared away many superstitions that had clouded the subject. For more than a century and a half after Gilbert, no discoveries of any fundamental importance were made, although there were many practical improvements in the manufacture of magnets. Thus, in the eighteenth century, compound steel magnets were made, composed of many magnetized steel strips fastened together, which could lift 28 times their own weight of iron [1.2]. This is all the more remarkable when we realize that there was only one way of making magnets at that time: the iron or steel had to be rubbed with a lodestone, or with another magnet which in turn had been rubbed with a lodestone. There was no other way until the first electromagnet was made in 1825, following the great discovery made in 1820 by Hans Christian Oersted (1775–1851) that an electric current produces a magnetic field. Research

* Numbers in square brackets relate to the references at the end of the book. "G" numbers are keyed to the General References.

on magnetic materials can be said to date from the invention of the electromagnet, which made available much more powerful fields than those produced by lodestones, or magnets made from them.

In this book we shall consider basic magnetic quantities and the units in which they are expressed, ways of making magnetic measurements, theories of magnetism, magnetic behavior of materials, and, finally, the properties of commercially important magnetic materials. The study of this subject is complicated by the unfortunate existence of two different systems of units: the *mks* and the *cgs* (electromagnetic or emu) systems. The *mks* system, currently taught in most physics courses, simplifies many of the equations of electricity but introduces no such simplification into the field of magnetism. Perhaps for this reason, the vast majority of workers in magnetism still use the *cgs* (emu) system, both in their laboratories and in their publications. This book is accordingly written in the *cgs* system, simply in order that the reader may learn the current language of this field. The final section of this chapter is devoted to a brief treatment of the *mks* system and to the interconversion of units.

Many of the equations in this introductory chapter and the next are stated without proof, because their derivations can be found in any physics textbook.

1.2 MAGNETIC POLES

Almost everyone as a child has played with magnets and felt the mysterious forces of attraction and repulsion between them. These forces appear to originate in regions called *poles*, located near the ends of the magnet. That end of a pivoted bar magnet which points approximately toward the north geographic pole of the earth is called the north-seeking pole, or, more briefly, the north pole. Since unlike poles attract, and like poles repel, one another, this convention means that there is a region of south polarity near the north geographic pole. The law governing the forces between poles was discovered independently in England in 1750 by John Michell (1724–93) and in France in 1785 by Charles Coulomb (1736–1806). This law states that the force F between two poles is proportional to the product of their pole strengths p_1 and p_2 and inversely proportional to the square of the distance d between them:

$$\boxed{F = \frac{p_1 p_2}{d^2}.} \tag{1.1}$$

Here the proportionality constant has been put equal to 1. If we measure F in dynes and d in centimeters, then this equation becomes the definition of pole strength: *a unit pole, or pole of unit strength, is one which exerts a force of one dyne on another unit pole located at a distance of one centimeter.* (The dyne is in turn defined as that force which gives a mass of 1 g an acceleration of 1 cm/sec^2. The weight of a 1 g mass is 981 dynes.) The unit of pole strength has no name.

Poles always occur in pairs in magnetized bodies, and it is impossible to separate them. (If a bar magnet is cut in two transversely, new poles appear on the

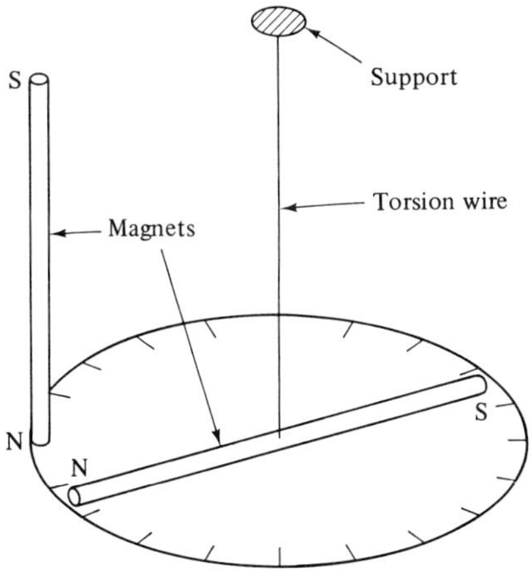

Fig. 1.1 Torsion balance for measuring the forces between poles.

cut surfaces and two magnets result. Despite this fact, at least one manufacturer still receives requests for single-pole magnets and must reluctantly decline to fill such orders.) The experiments on which Eq. (1.1) is based were performed with magnetized needles, so long that the poles at each end could be considered approximately as isolated poles, and the torsion balance sketched in Fig. 1.1. If the stiffness of the torsion-wire suspension is known, the force of repulsion between the two north poles can be calculated from the angle of deviation of the horizontal needle. The arrangement shown minimizes the effects of the two south poles.

A magnetic pole creates a field around it, and it is this magnetic field which produces a force on a second pole nearby. Experiment shows that this force is directly proportional to the product of the pole strength and *field strength* or *field intensity* H:

$$\boxed{F = pH.} \tag{1.2}$$

The proportionality constant has again been put equal to 1, and this equation then defines H: *a field of unit strength is one which exerts a force of one dyne on a unit pole.* (If an unmagnetized piece of iron is brought near a magnet, it will become magnetized, again through the agency of the field created by the magnet. For this reason H is also sometimes called the *magnetizing force*.) A field of unit strength is also said to have an intensity of one *oersted* (Oe). How large is an oersted? The magnetic field of the earth amounts to a few tenths of an oersted, that of a bar magnet near one end is about 5000 Oe, while that of a powerful electromagnet is 20,000–30,000 Oe. Another unit of field strength, used in describing the earth's field, is the *gamma* ($1\gamma = 10^{-5}$ Oe).

A unit pole in a field of one oersted is acted on by a force of one dyne. But a unit pole is also subjected to a force of one dyne when it is one centimeter away

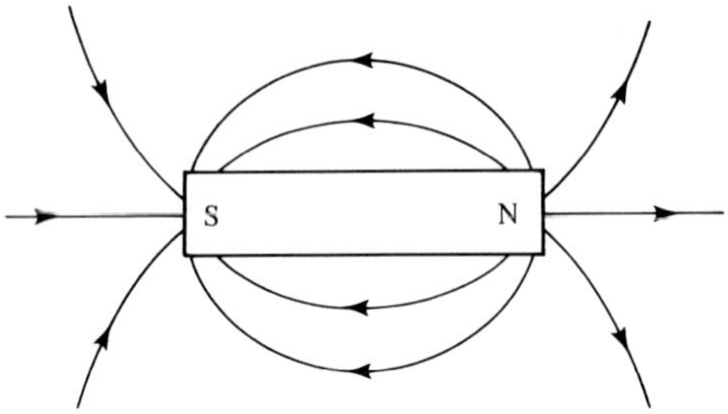

Fig. 1.2 External field of a bar magnet.

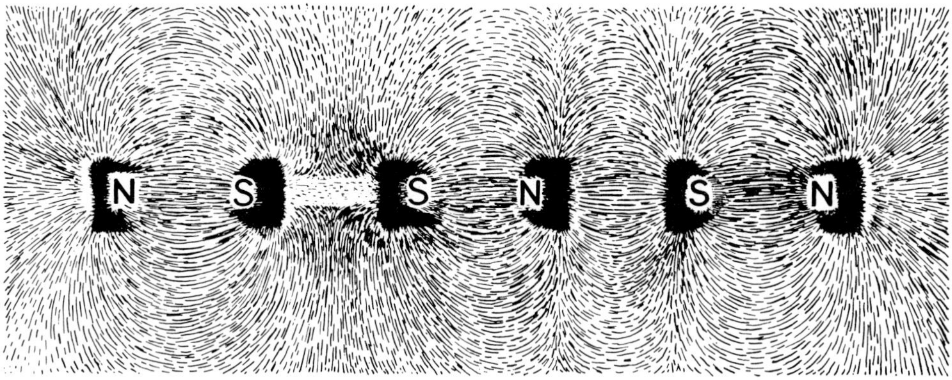

Fig. 1.3 Fields of bar magnets revealed by iron filings. (Courtesy Encyclopedia Brittanica).

from another unit pole. Therefore, the field created by a unit pole must have an intensity of one oersted at a distance of one centimeter from the pole. It also follows from Eqs. (1.1) and (1.2), that this field decreases as the inverse square of the distance d from the pole:

$$H = \frac{p}{d^2}. \tag{1.3}$$

Michael Faraday (1791–1867) has the very fruitful idea of representing a magnetic field by "lines of force." These are directed lines along which a single north pole would move, or to which a small compass needle would be tangent. Evidently, lines of force radiate *outward* from a single north pole. Outside a bar magnet, the lines of force leave the north pole and return at the south pole. (Inside the magnet, the situation is more complicated and will be discussed in Section 2.6) The resulting field (Fig. 1.2) is easily made visible in two dimensions by sprinkling iron filings on a card placed on the magnet. Each filing becomes magnetized and acts like a small compass needle, with its long axis parallel to the lines of force. Figure 1.3 shows the fields of three adjacent magnets delineated in this way.

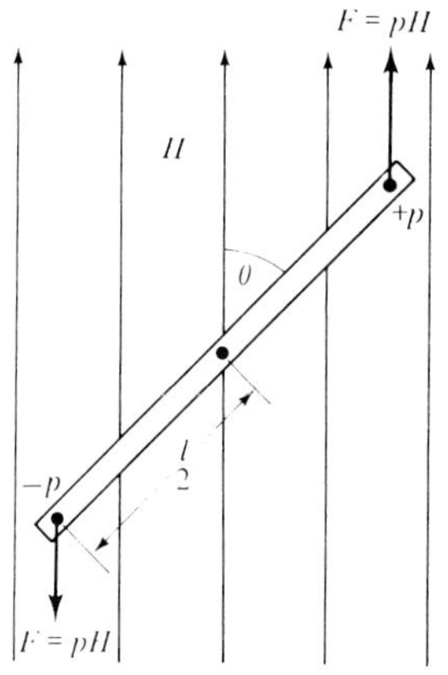

Fig. 1.4 Bar magnet in a uniform field. [Note use of plus and minus signs to designate north and south poles, respectively.]

The notion of lines of force can be made quantitative by defining the field strength H as the number of lines of force passing through unit area perpendicular to the field. A line of force, when referred to this quantitative sense, is called a *maxwell*.* Thus

$$1 \text{ Oe} = 1 \text{ line of force/cm}^2 = 1 \text{ maxwell/cm}^2.$$

Imagine a sphere with a radius of 1 cm centered on a unit pole. Its surface area is 4π cm^2. Since the field strength at this surface is 1 Oe, there must be 4π lines of force passing through it. In general, *$4\pi p$ lines of force issue from a pole of strength p.*

1.3 MAGNETIC MOMENT

Consider a magnet with poles of strength p located near each end and separated by a distance l. Suppose the magnet is placed at an angle θ to a uniform field H (Fig. 1.4). Then a couple acts on the magnet, tending to turn it parallel to the field. The moment of this couple is

$$(pH \sin \theta)(l/2) + (pH \sin \theta)(l/2) = pHl \sin \theta.$$

When $H = 1$ Oe and $\theta = 90°$, the moment is given by

$$\boxed{m = pl,} \tag{1.4}$$

where m is the *magnetic moment* of the magnet. It is the moment of the couple exerted on the magnet when it is at right angles to a uniform field of 1 Oe. (If

* James Clerk Maxwell (1831–1879), Scottish physicist, who developed the classical theory of electromagnetic fields.

the field is nonuniform, a translational force will also act on the magnet. See Sec. 2.9.)

Magnetic moment is an important and fundamental quantity, whether applied to a bar magnet or to the "electronic magnets" we will meet in a later chapter. Magnetic poles, on the other hand, represent a mathematical concept rather than physical reality; they cannot be separated for measurement and are not localized at a point, which means that the distance l between them is indeterminate. Although p and l are uncertain quantities individually, their product is the magnetic moment m, which can be precisely measured. This does not mean that the concept of magnetic poles should be abandoned; on the contrary, this concept is a useful aid to thought and almost indispensable in many problems.

Returning to Fig. 1.4, we note that a magnet not parallel to the field must have a certain potential energy E_p relative to the parallel position. The work done (in ergs) in turning it through an angle $d\theta$ against the field is

$$dE_p = 2(pH \sin \theta)(l/2)\, d\theta = mH \sin \theta \, d\theta.$$

It is conventional to take the zero of energy as the $\theta = 90°$ position. Therefore,

$$E_p = \int_{90°}^{\theta} mH \sin \theta \, d\theta,$$

$$\boxed{E_p = -mH \cos \theta.} \tag{1.5}$$

Thus E_p is $-mH$ when the magnet is parallel to the field, zero when it is at right angles, and $+mH$ when it is antiparallel. Actually, the magnetic moment m is a vector which, for a bar magnet, is drawn from the south pole to the north. In vector notation, Eq. (1.5) becomes

$$E_p = -\mathbf{m} \cdot \mathbf{H}. \tag{1.6}$$

Equation (1.5) or (1.6) is an important relation which we will need frequently in later sections.

Because the energy E_p is in ergs, the units of magnetic moment m are ergs/oersted.

1.4 INTENSITY OF MAGNETIZATION

When a piece of iron is subjected to a magnetic field, it becomes magnetized, and the extent of its magnetism depends on the strength of the field. We therefore need a term to describe the extent to which a body is magnetized.

Consider two bar magnets of the same size and shape, each having the same pole strength p and interpolar distance l. Assume that the poles are located exactly at the ends. (Actually, this assumption is true only in the limit of very long, thin magnets. In most bar magnets the poles are not located exactly at the ends, nor are they localized at points.) If placed side by side, as in Fig. 1.5 (a), the poles add, and the magnetic moment $m = (2p)l = 2pl$, which is double the moment of the individual magnets. If they are placed end to end, as in (b), the

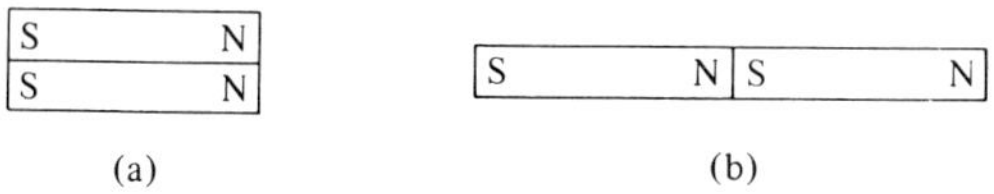

Fig. 1.5 Compound magnets.

adjacent poles annul each other and $m = p(2l) = 2pl$, as before. Evidently, the total magnetic moment is the sum of the magnetic moments of the individual magnets.

In these examples, we have doubled the volume. The *magnetic moment per unit volume* has not changed in this process of combining magnets and is therefore a quantity descriptive of the extent to which the magnets are magnetized. It is called the *intensity of magnetization*, or simply the *magnetization*, and is written M (or I or J by some authors). Since

$$\boxed{M = \frac{m}{v},} \tag{1.7}$$

where v is the volume, we can also write

$$M = \frac{pl}{v} = \frac{p}{v/l} = \frac{p}{a}, \tag{1.8}$$

where a is the cross-sectional area of the magnet. We therefore have an alternative definition of the magnetization M as the *pole strength per unit area of cross section.*

Inasmuch as the units of magnetic moment m are ergs/oersted, the units of magnetization M are ergs/oersted cm^3. However, these units are more often written simply as emu/cm^3, where "emu" is understood to mean the electromagnetic unit of magnetic moment.

Sometimes it is more convenient to refer the extent of magnetization to unit mass rather than unit volume. Thus the *specific magnetization* σ is defined as

$$\sigma = \frac{m}{w} = \frac{m}{v\rho} = \frac{M}{\rho} \text{ emu/g}, \tag{1.9}$$

where w is the mass and ρ the density.

(A few writers use the abbreviation "emu" to mean ergs/oersted cm^3. Then M is in emu and σ in emu-cm^3/g.)

1.5 MAGNETIC DIPOLES

As shown in Appendix 4, the field of a magnet at a distance r from the magnet, of pole strength p and length l, depends only on the moment pl of the magnet and not on the separate values of p and l, provided r is large relative to l. Thus

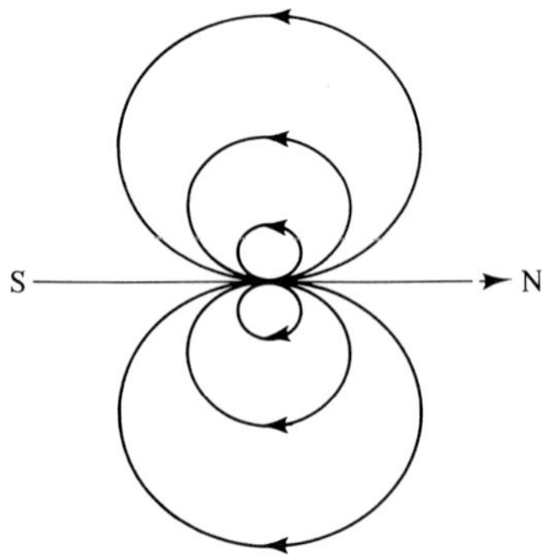

Fig. 1.6 Field of a magnetic dipole.

the field is the same if we halve the length of the magnet and double its pole strength. Continuing this process, we obtain in the limit a very short magnet of finite moment called a *magnetic dipole.* Its field is sketched in Fig. 1.6. We can therefore think of any magnet, as far as its external field is concerned, as being made up of a number of dipoles; the total moment of the magnet is the sum of the moments, called dipole moments, of the constituent dipoles.

1.6 MAGNETIC EFFECTS OF CURRENTS

A current in a straight wire produces a magnetic field which is circular around the wire axis in a plane normal to the axis. Outside the wire the magnitude of this field, at a distance r cm from the wire axis, is given by

$$H = \frac{2i}{10r} \text{ Oe}, \tag{1.10}$$

where i is the current in amperes. (Inside the wire, $H = 2ir/10r_o^2$, where r_o is the wire radius.) The direction of the field is that in which a right-hand screw would rotate if driven in the direction of the current (Fig. 1.7 a). [In Eq. (1.10) and other equations for the magnetic effects of currents, we are using "mixed" practical and *cgs* electromagnetic units. The electromagnetic unit of current equals 10 amperes, which accounts for the factor 10 in these equations.]

If the wire is curved into a circular loop of radius R cm, as in Fig. 1.7(b), then the field at the center along the axis is

$$H = \frac{2\pi i}{10R} \text{ Oe}. \tag{1.11}$$

The field of such a current loop is sketched in (c). Experiment shows that a current loop, suspended in a uniform magnetic field and free to rotate, turns until the plane

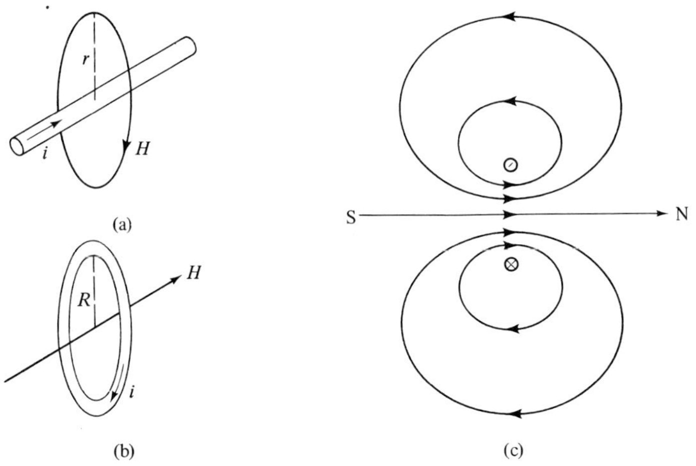

Fig. 1.7 Magnetic fields of currents.

of the loop is normal to the field. It therefore has a magnetic moment, which is given by

$$m \text{ (loop)} = \frac{\pi R^2 i}{10} = \frac{Ai}{10} \text{ ergs/Oe}, \tag{1.12}$$

where A is the area of the loop in cm^2. The direction of m is the same as that of the axial field H due to the loop itself (Fig. 1.7 b).

A helical winding produces a much more uniform field than a single loop. Such a winding is called a *solenoid*, after the Greek word for a tube or pipe. The field along its axis is given by

$$\boxed{H = \frac{4\pi ni}{10L} = \frac{1.257ni}{L} \text{ Oe},} \tag{1.13}$$

where n is the number of turns and L the length of the winding in cm. Note that the field is independent of the solenoid radius as long as the radius is small compared to the length. Inside the solenoid the field is quite uniform, except near the ends, and outside it resembles that of a bar magnet (Fig. 1.8). The magnetic moment of a solenoid is given by

$$m \text{ (solenoid)} = \frac{nAi}{10} \text{ ergs/Oe}, \tag{1.14}$$

where A is the cross-sectional area.

As the diameter of a current loop becomes smaller and smaller, the field of the loop (Fig. 1.7 c) approaches that of a magnetic dipole (Fig. 1.6). Thus it is just

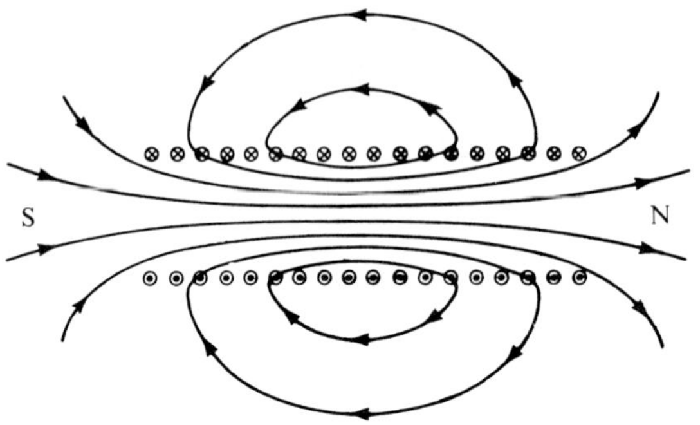

Fig. 1.8 Magnetic field of a solenoid.

as permissible to regard a magnet as being a collection of current loops as a collection of dipoles. In fact, André-Marie Ampère (1775–1836) suggested over a hundred years ago that the magnetism of a body was due to "molecular currents" circulating in it. These were later called Amperian currents. Figure 1.9 (a) shows the current loops on the cross section of a uniformly magnetized bar. At interior points the currents are in opposite directions and cancel one another, leaving the net, uncanceled loop shown in (b). On a short section of the bar these current loops, called *equivalent surface currents*, would appear as in (c). (In the language of poles, this section of the bar would have a north pole at the position indicated.) The similarity to a solenoid is evident. In fact, given the magnetic moment and cross-sectional area of the bar, we can calculate the equivalent surface current in terms of the product ni from Eq. (1.14). However, it must be remembered that, in the case of the solenoid, we are dealing with a real current, called a *conduction current*, whereas the equivalent surface currents, with which we replace the magnetized bar, are entirely imaginary.

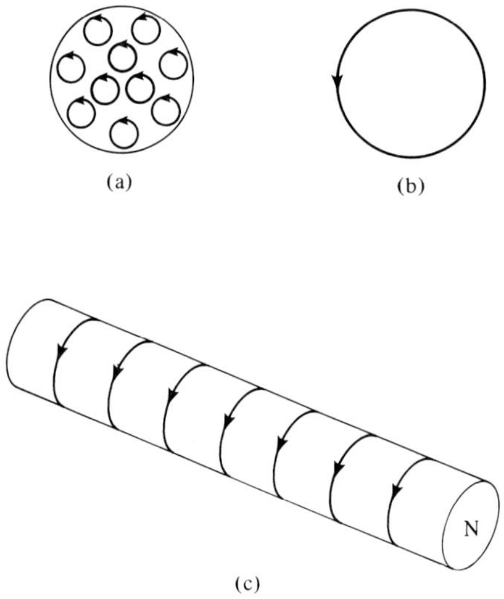

Fig. 1.9 Amperian current loops in a magnetized bar.

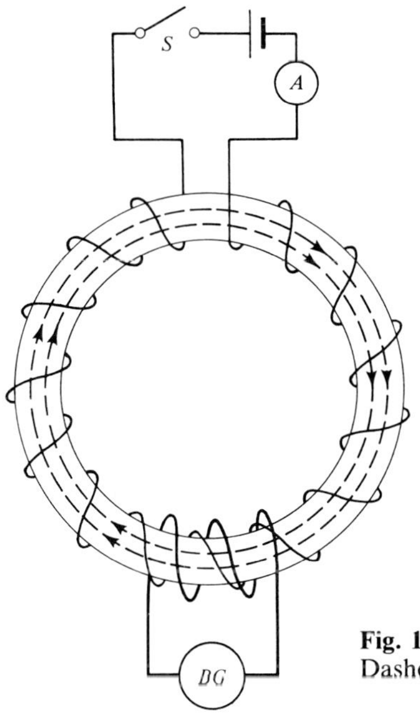

Fig. 1.10 Circuit for magnetization of a ring specimen. Dashed lines indicate flux. (After Sears [1.3]).

It is possible to describe all magnetic phenomena in terms of equivalent surface currents, without any reference to magnetic poles. However, discussion in these terms of, for example, the magnetization of a bar magnet becomes unduly complicated. There the language of poles is simpler and more direct. Neither Amperian currents nor magnetic poles have any physical reality. They are merely descriptive concepts and, of the two, we will choose the pole concept as being more generally useful, noting that the magnetic moment, to which both concepts lead, is a more fundamental quantity than either one.

1.7 VARIETIES OF MAGNETISM

We are now in a position to consider how magnetization can be measured and what the measurement reveals about the magnetic behavior of various kinds of substances. Figure 1.10 shows one method of measurement. The specimen is in the form of a ring,* wound with a large number of closely spaced turns of insulated wire, connected through a switch S and ammeter A to a battery. This winding is called the primary, or magnetizing, winding. It forms an endless

---

* Sometimes called a Rowland ring, after the American physicist H. A. Rowland (1848–1901) who first used this kind of specimen in his early research on magnetic materials. He is better known for the production of ruled diffraction gratings for the study of optical spectra.

solenoid, and the field in it is given by Eq. (1.13); this field is, for all practical purposes, entirely confined to the region within the coil. This arrangement has the advantage that the material of the ring becomes magnetized without the formation of poles, a circumstance which simplifies the interpretation of the measurement. Another winding, called the secondary winding or search coil, is placed on a short section of the ring and connected to a ballistic galvanometer *BG*.

Assume that the ring contains nothing but empty space. If the switch *S* is closed, a current i is established in the primary, producing a field of H oersteds, or H maxwells/cm^2, within the ring. If the cross-sectional area of the ring is A cm^2, then the total number of lines of force in the ring is $HA = \phi$ maxwells, which is called the *flux*. (It follows that H may be referred to as a *flux density*.) The change in flux $\Delta\phi$ through the search coil, from 0 to ϕ, induces an emf in the search coil according to Faraday's law, and this emf causes a deflection of the galvanometer, as explained in detail in the next chapter. If the galvanometer is calibrated, its deflection is a measure of $\Delta\phi$, which in this case is simply ϕ. If the ring contains empty space, it is found experimentally that ϕ (observed), obtained from the galvanometer deflection, is exactly equal to ϕ (current), which is the flux produced by the current in the primary winding, i.e., the product A and H calculated from Eq. (1.13).

However, if there is any material substance in the ring, ϕ (observed) is found to differ from ϕ (current). This means that the substance in the ring has added to, or subtracted from, the number of lines of force due to the field H. The relative magnitudes of these two quantities enable us to classify all substances according to the kind of magnetism they exhibit:

ϕ (observed) $<$ ϕ (current), diamagnetic (for example, Cu, He)
ϕ (observed) $>$ ϕ (current), paramagnetic (for example, Na, Al) or antiferromagnetic (for example, MnO, FeO)
ϕ (observed) $\gg$ ϕ (current), ferromagnetic (for example, Fe, Co, Ni) or ferrimagnetic (for example, Fe_3O_4).

Paramagnetic and antiferromagnetic substances can be distinguished from one another by magnetic measurement only if the measurements extend over a range of temperature. The same is true of ferromagnetic and ferrimagnetic substances.

All pure substances are magnetic to some extent. However, examples of the first three types listed above are so feebly magnetic that they are usually called "nonmagnetic," both by the layman and by the engineer or scientist. The observed flux in a typical paramagnetic, for example, is only about 0.02 percent greater than the flux due to the current. The experimental method outlined above, although useful for illustrative purposes, is not capable of accurately measuring such small differences, and entirely different methods have to be resorted to. In ferromagnetic and ferrimagnetic materials, on the other hand, the observed flux is thousands or even millions of times larger than the flux due to the current.

We can formally understand how the material of the ring causes a change in flux if we consider the fields which actually exist inside the ring. Imagine a very thin, transverse cavity cut out of the material of the ring, as shown in Fig. 1.11. Then H lines/cm^2 cross this gap, due to the current in the magnetizing winding,

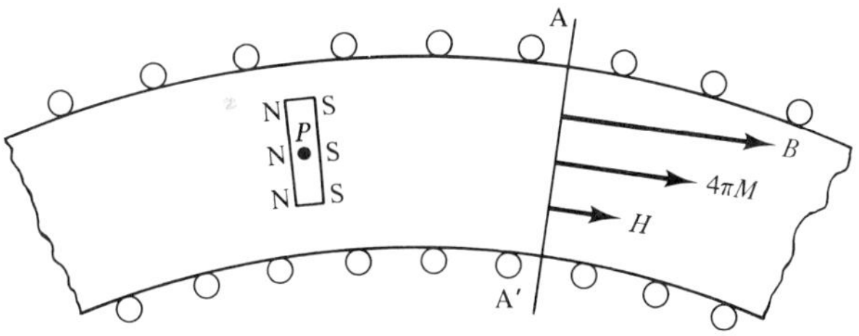

Fig. 1.11 Transverse cavity in portion of Rowland ring.

in accordance with Eq. (1.13). (This flux density is the same, incidentally, *whether or not there is any material in the ring.*) In addition, the applied field H, acting from left to right, magnetizes the material, and north and south poles are produced on the surface of the cavity, just as poles are produced on the ends of a magnetized bar. If the material is ferromagnetic, the north poles will be on the left-hand surface and south poles on the right. If the intensity of magnetization is M, then each square centimeter of the surface of the cavity has a pole strength of M, and $4\pi M$ lines issue from it. These are sometimes called *lines of magnetization.* They add to the *lines of force* due to the applied field H, and the composite group of lines crossing the gap is called *lines of induction.* The number of lines of induction per cm^2 is called the *magnetic induction B*. Therefore,

$$\boxed{B = H + 4\pi M.} \tag{1.15}$$

Because lines of B are always continuous, this expression gives the value of B, not only in the gap, but also in the material on either side of the gap and throughout the ring. Although B, H, and M are actually vectors, they are usually parallel, so that the above equation is normally written in scalar form. These vectors are indicated at the right of Fig. 1.11, for a hypothetical case where B is about three times H. They indicate the values of B, H, and $4\pi M$ at the section AA' or, for that matter, at any other section of the ring.

(The word "induction" is a relic from an earlier age: if an unmagnetized piece of iron were brought near a magnet, then magnetic poles were said to be "induced" in the iron, which was, in consequence, attracted to the magnet. Later the word took on the quantitative sense, defined above, of the total flux density in a material, denoted by B.)

Equation (1.15) is so important that it is instructive to consider another way of deriving it. Equation (1.2) defines a field H in terms of the force exerted on a unit pole placed in that field. The lines of induction B also constitute a field, and we can determine the B field if we can measure, or calculate, the force on a unit pole placed in that field. Our problem then is to calculate the total force on an imaginary unit pole placed at the center P of the gap in the ring. This force will be the sum of that due to the applied field and that due to the north and south poles produced on the sides of the gap by the magnetization of the material. Consider an

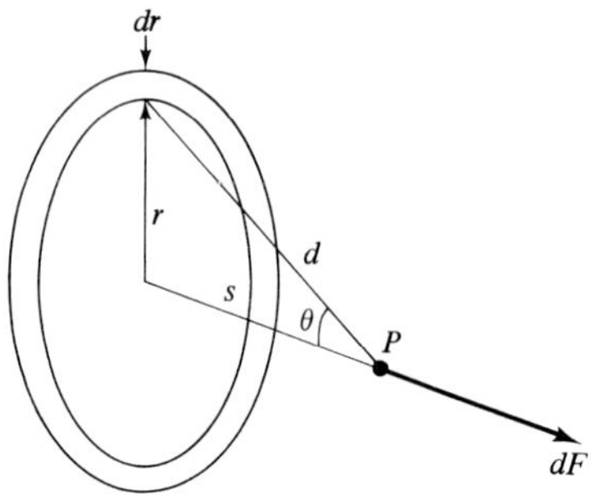

Fig. 1.12 Axial force on unit pole at *P*.

annular ring of radius r and width dr on the north-pole face of the gap (Fig. 1.12). The total pole strength on this ring is $(M)(2\pi r)\,dr$, and the axial force which this exerts on a unit north pole at P is, from Eq. (1.1),

$$dF = \frac{(2\pi Mr\,dr)(1)\cos\theta}{d^2}.$$

But $r = d\sin\theta = (r^2 + s^2)^{1/2}\sin\theta$. When the value for $r\,dr$ found from this relation is inserted into the expression for dF, we obtain

$$dF = 2\pi M\sin\theta\,d\theta.$$

Since P is very near the face of the gap, the total force is obtained by integrating from $\theta = 0$ to $\theta = \pi/2$:

$$F = \int_0^{\pi/2} 2\pi M\sin\theta\,d\theta = 2\pi M.$$

The other face of the gap, covered with south poles, exerts an equal force in the same direction. The force due to the poles on the gap surfaces is therefore $4\pi M$, and that due to the field is H. Thus the total force on a unit pole at P, or the total field in the gap, is $(H + 4\pi M)$.

Although B, H, and M must necessarily have the same units (lines or maxwells/cm^2), quite different names are given to these units. A maxwell per cm^2 is customarily called a *gauss* (G),* when reference is made to B, and an oersted when referred to H, while the preferred unit for M is emu/cm^3. However, even these conventions are not always followed, and both H and M are often stated in terms of

* Carl Friedrich Gauss (1777–1855), German mathematician. Renowned for his genius as a mathematician, he also developed magnetostatic theory, devised a system of electrical and magnetic units, designed instruments for magnetic measurements, and investigated terrestrial magnetism.

gauss. In this book we will write M in emu/cm^3, but $4\pi M$ in gauss, to emphasize that the latter forms a contribution to the total flux density B. Note that this discussion concerns only the *names* of these units. There is no question here of any numerical conversion of one to the other; they are all numerically equal. (It should also be noted that it is not usual to refer, as is done above, to H as a flux density and to HA as a flux, although there would seem to be no logical objection to these designations. Instead, most writers restrict the terms "flux density" and "flux" to B and BA, respectively.)

Remembering our experiment with the Rowland ring, we now see that ϕ (observed) $= BA$, because the galvanometer measures the change in the *total* number of lines enclosed by the search coil. On the other hand, ϕ (current) $= HA$. The difference between them is $4\pi MA$. The magnetization M is zero only for empty space. The magnetization, even for applied fields H of many thousands of oersteds, is very small and negative for diamagnetics, very small and positive for para- and antiferromagnetics, and large and positive for ferro- and ferrimagnetics. The negative value of M for diamagnetic materials implies that south poles are produced on the left side of the gap in Fig. 1.11 and north poles on the right.

Note that $B = H$ in empty space, because there is no matter to magnetize, and M is therefore zero. This circumstance, and others, has led to considerable confusion between the B and H fields and to arguments as to which is the more "fundamental." We will therefore digress briefly to consider this point.

Writers of current physics textbooks take the position that B is the fundamental vector, and they would write Eq. (1.10), for example, as

$$B = 2i/10r \text{ G},$$

for the field in empty space outside a current-carrying wire, in mixed *cgs*-practical units. They can then develop almost the entire subject of electromagnetism in terms of B alone; not until it is a question of describing the effect of a field on magnetizable matter is there any necessity for introducing a second field H. They then, in effect, rewrite the field-current relation as $H = 2i/10r$. They would also hold that only B, and not H, is measurable.

This position is not the only tenable one. In fact, it was almost unheard of a few decades ago and is not the position currently adopted by people who study and work with magnetic materials. These people find it much more logical to think of H as the primary field, primary in the sense that it is the field which makes things happen. The H field can be produced by a current in a wire or by magnetic poles. If at a point exterior to these field-producing devices there is some magnetizable matter, then the H field produces some magnetization M at that point, and the total field at that point becomes B, which is the sum of H and $4\pi M$. Thus H is the cause and B the overall effect. If there is no matter at the point, then the field there may be called H or B.

Which field is measurable? The answer to this question depends only on the kind of field to which the field-measuring device is exposed. In terms of the search coil-galvanometer measurement described above, if the search-coil is wrapped around a piece of matter, then the galvanometer deflection is a measure of a flux change $\Delta(BA)$ through the coil. If the coil encloses only empty space, the measured flux change is $\Delta(HA)$, which here equals $\Delta(BA)$.

Further discussion of the difference between B and H fields may be found in the next chapter, particularly in Section 2.6.

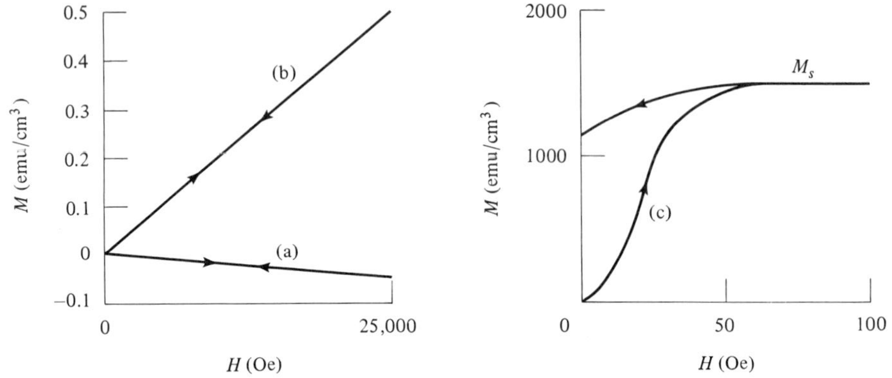

Fig. 1.13 Typical magnetization curves of (a) a diamagnetic, (b) a para- or antiferromagnetic, and (c) a ferro- or ferrimagnetic. (After Schoenberg [1.4]).

The magnetic properties of a material are characterized not only by the magnitude and sign of M but also by the way in which M varies with H. The ratio of these two quantities is called the *susceptibility* κ:

$$\boxed{\kappa = \frac{M}{H}} \quad \text{emu/cm}^3\ \text{Oe}. \tag{1.16}$$

Since M is the magnetic moment/cm^3, κ also refers to unit volume and is sometimes called the volume susceptibility to emphasize this fact. In any case, it is well named, since it indicates how responsive a material is to an applied magnetic field. Various other susceptibilities are defined as follows:

$\chi = \kappa/\rho$ = mass susceptibility (emu/g Oe), where ρ = density,
$\chi_A = \chi A$ = atomic susceptibility (emu/g atom Oe), where A = atomic weight,
$\chi_M = \chi M'$ = molecular susceptibility (emu/g mol Oe), where M' = molecular weight.

Typical curves of M vs H, called *magnetization curves*, are shown in Fig. 1.13 for various kinds of substances. Curves (a) and (b) refer to substances having volume susceptibilities of -2×10^{-6} and $+20 \times 10^{-6}$, respectively. These substances (dia-, para-, or antiferromagnetic) have linear M,H curves under normal circumstances and retain no magnetism when the field is removed. The behavior shown in curve (c), of a typical ferro- or ferrimagnetic is quite different. The magnetization curve is nonlinear, so that κ varies with H and passes through a maximum value of about 40 for the curve shown. Two other phenomena appear:

1. *Saturation.* At large enough values of H, the magnetization M becomes constant at its saturation value of M_s.

2. *Hysteresis*, or irreversibility. After saturation, a decrease in H to zero does not reduce M to zero. Ferro- and ferrimagnetic materials can thus be made into *permanent magnets*. (The word *hysteresis* is Greek, meaning "a coming late," and is today applied to almost any phenomenon in which the effect lags behind the cause. Its first use in science was by Ewing* in 1881, with reference to the magnetic behavior of iron.)

The physicist is mainly interested in the variation of M with H, because it immediately gives the susceptibility and discloses the kind of substance. The engineer, on the other hand, is usually concerned only with ferro- and ferrimagnetic materials and wants to know the total flux density B produced by a given field. He therefore finds the B,H curve, also called a magnetization curve, more useful. The ratio of these two quantities is the *permeability* μ:

$$\boxed{\mu = \frac{B}{H}.} \tag{1.17}$$

Since $B = H + 4\pi M$, we have

$$B/H = 1 + 4\pi(M/H);$$

$$\boxed{\mu = 1 + 4\pi\kappa.} \tag{1.18}$$

Note that μ is *not* the slope dB/dH of the B,H curve, but rather the slope of a line from the origin to a particular point on the curve. Two special values are often mentioned, the initial permeability μ_0 and the maximum permeability μ_m. These are illustrated in Fig. 1.14, which also shows the typical variation of μ with H for a ferro- or ferrimagnetic.

We can now characterize the magnetic behavior of various kinds of substances by their corresponding values of κ and μ:

1. *Empty space*. $\kappa = 0$, since there is no matter to magnetize, and $\mu = 1$.
2. *Diamagnetic*. κ is small and negative, and μ slightly less than 1.
3. *Para- and antiferromagnetic*. κ is small and positive, and μ slightly greater than 1.
4. *Ferro- and ferrimagnetic*. κ and μ are large and positive, and both are functions of H.

Permeability is a very useful property for describing ferro- and ferrimagnetics, since it immediately gives the induction B produced by a given field H. It there-

---

* J. A. Ewing (1855–1935), British educator and engineer. He taught at Tokyo, Dundee, and Cambridge and did research on magnetism, steam engines, and metallurgy. During World War I he organized the cryptography section of the British navy. During his five-year tenure of a professorship at the University of Tokyo (1878–1883), he introduced his students to research on magnetism, and Japanese research in this field has flourished ever since.

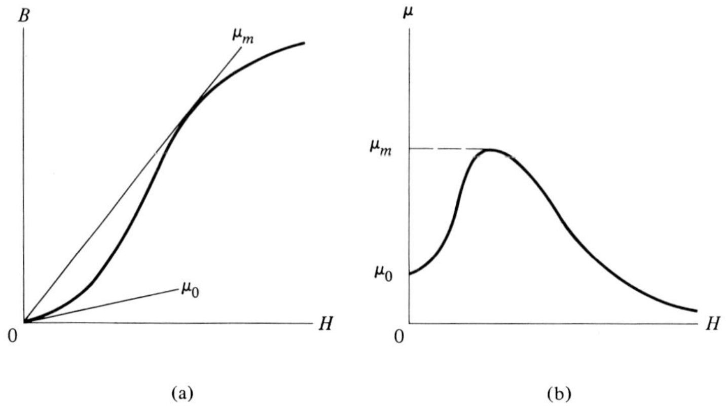

Fig. 1.14 (a) B versus H curve of a ferro- or ferrimagnetic, and (b) corresponding variation of μ with H.

fore discloses how much the flux density due to a current in the magnetizing winding is *multiplied* by the material inside the winding. In terms of the Rowland-ring experiment, $\mu = \phi$ (observed)/ϕ (current).

The permeability of air is 1.000 000 37. The difference between this and the permeability of empty space is negligible, relative to the permeabilities of ferro- and ferrimagnetics, which typically have values of μ of several hundreds or thousands. We can therefore deal with these substances in air as though they existed in a vacuum. In particular, we can say that B equals H in air, with negligible error.

1.8 MAGNETIZATION CURVES AND HYSTERESIS LOOPS

Both ferro- and ferrimagnetic materials differ widely in the ease with which they can be magnetized. If a small applied field suffices to produce saturation, the material is said to be *magnetically soft* (Fig. 1.15 a). Saturation ofanother material, which will in general have a different value of M_s, may require very large fields, as shown by curve (c). Such a material is *magnetically hard.* Sometimes the same material may be either magnetically soft or hard, depending on its physical condition: thus curve (a) might relate to a well-annealed material, and curve (b) to the heavily cold-worked state.

Figure 1.16 shows magnetization curves both in terms of B (full line from the origin in first quadrant) and M (dashed line). Although M is constant after saturation is achieved, B continues to increase with H, because H forms part of B. Equation (1.15) shows that the slope dB/dH is unity beyond the point B_s, called the *saturation induction;* however, the slope of this line hardly ever appears to be unity, because the B and H scales are usually quite different. Continued increase of H beyond saturation will cause μ to approach 1 as H approaches

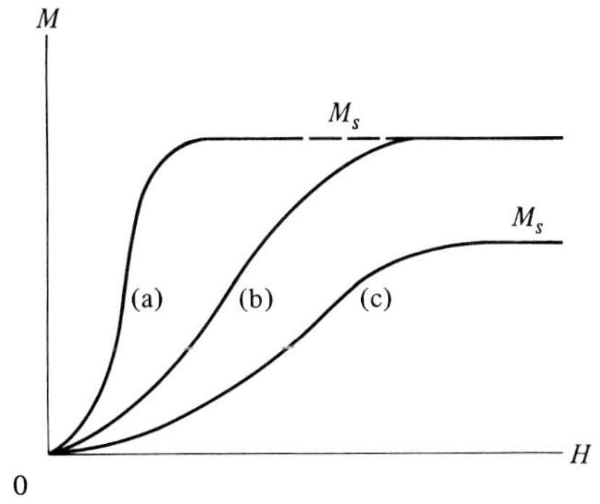

Fig. 1.15 Magnetization curves.

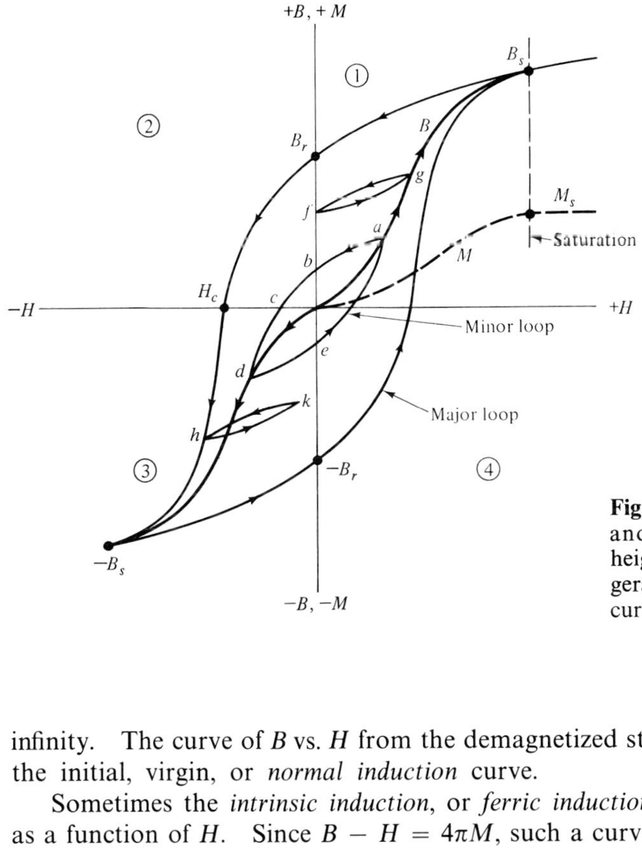

Fig. 1.16 Magnetization curves and hysteresis loops. (The height of the *M* curve is exaggerated relative to that of the *B* curve.)

infinity. The curve of *B* vs. *H* from the demagnetized state to saturation is called the initial, virgin, or *normal induction* curve.

Sometimes the *intrinsic induction*, or *ferric induction*, $B_i = B - H$ is plotted as a function of *H*. Since $B - H = 4\pi M$, such a curve will differ from an *M*,*H* curve only by a factor of 4π applied to the ordinate. B_i measures the number of lines of magnetization/cm^2.

If *H* is reduced to zero after saturation has been reached in the positive direction, the induction in a ring specimen will decrease from B_s to B_r, called the *retentivity* or *residual induction*. If the applied field is then reversed, by reversing the current in the magnetizing winding, the induction will decrease to zero when the negative applied field equals the *coercivity* H_c. This is the reverse field necessary to "coerce"

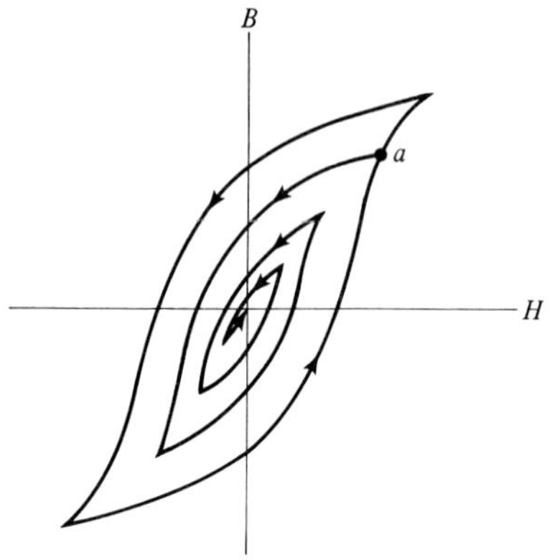

Fig. 1.17 Demagnetization by cycling with a decreasing amplitude.

the material back to zero induction; it is usually written as a positive quantity, the negative sign being understood. At this point, M is still positive and is given by $|H_c/4\pi|$. (The reverse field required to reduce M to zero is called the *intrinsic coercivity* H_{ci}. To emphasize the difference between the two coercivities, some authors write ${}_BH_c$ for the coercivity and ${}_MH_c$ for the intrinsic coercivity.)

If the reversed field is further increased, saturation in the reverse direction will be reached at $-B_s$. If the field is then reduced to zero and applied in the original direction, the induction will follow the curve $-B_s$, $-B_r$, $+B_s$. The loop traced out is known as the *major hysteresis loop*, when both tips represent saturation. It is symmetrical about the origin as a point of inversion, i.e., if the right-hand half of the loop is rotated 180 about the H axis, it will be the mirror image of the left-hand half. The loop quadrants are numbered 1 to 4 as shown in the drawing, since this is the order in which they are usually traversed.

If the process of initial magnetization is interrupted at some intermediate point such as a and the corresponding field is reversed and then reapplied, the induction will travel around the minor hysteresis loop *abcdea*. Here b is called the *remanence* and c the *coercive force*. (Despite the definitions given above, the reader should note that the terms remanence and retentivity, and coercive force and coercivity, are used almost interchangeably by some writers. In particular, the term coercive force is often loosely applied to any field, including H_c, which reduces B to zero, whether the specimen has been previously saturated or not. When "coercive force" is used without any other qualification, it is usually safe to assume that "coercivity" is actually meant.)

There is an infinite number of symmetrical minor hysteresis loops inside the major loop, and their tips all lie on the normal induction curve. There are also many nonsymmetrical minor loops, some of which are shown at *fg* and *hk*.

If a specimen is being cycled on a symmetrical loop, it will always be magnetized in one direction or the other when H is reduced to zero. Demagnetization is accomplished by interrupting the cycling at some point, such as a in Fig. 1.17, followed by further cycling in a field whose amplitude is decreased a little during each cycle. In this way the induction is made to traverse smaller and smaller loops until it finally arrives at the origin. The only other way to demagnetize a ferro- or ferrimagnetic material is to heat it above its *Curie point.* Above this temperature both kinds of material become paramagnetic. If they are then cooled in the absence of any applied field, they will be in the demagnetized condition at room temperature.

1.9 MKS UNITS

In this section we will consider how various magnetic quantities are expressed in *mks* units and how to convert units from one system to the other.

Actually, it would be more correct to speak of the "*mks* system," rather than simply of "*mks* units," because the shift from *cgs* to *mks* units in the teaching of physics has also involved a quite different approach to the subject matter, and not merely a change in the size of the units. Thus, in the *mks* system, B is defined (in terms of the force between two current-carrying wires) before H is defined, a distinction is made between B and H in empty space, and magnetization is treated with emphasis on Amperian currents (equivalent surface currents) rather than magnetic poles. As a result, the *form* of many basic equations is entirely changed.

The following is intended as a summary, with no attempt at logical development. Equations are given the same number as their *cgs* equivalents in this chapter, followed by the suffix *a.*

Coulomb's law of the force between poles becomes

$$F = \frac{p_1 p_2}{4\pi \mu_0 d^2} \text{ newtons,} \tag{1.1a}$$

where d is in meters and μ_0, called the permeability of empty space, is equal to $4\pi \times 10^{-7}$ weber/ampere meter. (This μ_0 is not to be confused with the same symbol in the *cgs* system for the initial permeability of a ferro- or ferrimagnetic substance.) A newton is the force required to give a mass of 1 kg an acceleration of 1 m/sec^2; it is equal to 10^5 dynes.

$$\text{Force on a pole} = F = pH \text{ newtons} \tag{1.2a}$$

$$\text{Field of a pole} = H = \frac{p}{4\pi \mu_0 d^2} \text{ ampere-turns/meter} \tag{1.3a}$$

$$\text{Magnetic moment} = m = pl \text{ weber-meter} \tag{1.4a}$$

$$\text{Potential energy} = E_p = -mH \cos\theta \text{ joules} \tag{1.5a}$$

$$\text{Magnetization} = M = \frac{m}{v} = \frac{p}{a} \text{ weber/meter}^2 \qquad \text{(1.7a, 1.8a)}$$

$$\text{Field of straight wire} = H = i/2\pi r \text{ ampere/meter} \tag{1.10a}$$

$$\text{Field of current loop} = H = i/2R \text{ ampere/meter} \tag{1.11a}$$

$$m \text{ (loop)} = \mu_0 A i \text{ weber-meter} \tag{1.12a}$$

$$\text{Field of solenoid} = H = ni/L \text{ ampere-turns/meter} \tag{1.13a}$$

$$m \text{ (solenoid)} = \mu_0 n A i \text{ weber-meter} \tag{1.14a}$$

In empty space, $B = \mu_0 H$, and the basic equation connecting B, H, and M in a material substance is

$$B = \mu_0 H + M, \tag{1.15a}$$

where B and M are in weber/meter^2 and H is in ampere-turns/meter.

$$\text{Volume susceptibility} = \kappa = M/H \text{ weber/ampere meter} \tag{1.16a}$$

$$\text{Absolute permeability} = \mu = B/H \text{ weber/ampere meter} \tag{1.17a}$$

$$\text{Relative permeability} = \mu_r = \frac{\mu}{\mu_0} = \frac{B}{\mu_0 H}$$

From Eq. (1.15a) we have

$$\frac{B}{H} = \mu_0 + \frac{M}{H},$$

$$\mu = \mu_0 + \kappa, \tag{1.18a}$$

$$\frac{\mu}{\mu_0} = \mu_r = 1 + \frac{\kappa}{\mu_0}.$$

The physical significance of the three terms in Eq. (1.15a) should be noted. A line of induction is called a *weber*, and the total flux density or induction B (in webers/m^2) is made up of (a) the flux density $\mu_0 H$ due to the current in the magnetizing winding, and (b) the density M of the lines of magnetization contributed by the material in the winding. Since there are M lines of magnetization/m^2 and M poles/m^2, there must be one line (one weber) issuing from each unit north pole.

Because of the differences between M, and between H, in the two systems, κ and μ have different physical meanings. κ (*mks*) gives directly the ratio of density of lines of magnetization to density of lines of applied field, whereas κ (*cgs*) is $1/4\pi$ times this ratio. On the other hand, μ (*cgs*) gives directly the ratio of induction flux density to applied field density, whereas μ (*mks*) is μ_0 (or $4\pi \times 10^{-7}$) times this ratio. The relative permeability μ_r (*mks*) and the permeability μ (*cgs*) are numerically equal and physically equivalent. The most important numerical conversions are:

H: 1 ampere-turn/m $= 4\pi \times 10^{-3}$ oersted.

B: 1 weber/meter^2 $= 10^4$ gauss. (A weber/m^2 is sometimes called a *tesla*.)

M: 1 weber/meter^2 $= 10^4/4\pi$ emu/cm^3.

ϕ: 1 weber $= 10^8$ maxwells.

Other conversion factors are given in Appendix 8.

(The reader should also be aware that there is no common agreement as to how the basic relation of Eq. (1.15a) should be written in the *mks* system. Although Eq. (1.15a) is the most common form, three others exist:

$$B = \mu_0 H + \mu_0 M,$$

$$B = H + \mu_0 M,$$

$$B = H + M.$$

These other relations clearly involve other definitions and units for H and/or M. This lack of agreement on the interrelation of B, H, and M is probably the largest single obstacle to the widespread adoption of the *mks* system in magnetism.)

PROBLEMS

1.1 From the defining equation for pole strength p, find its dimensional units in terms of g, cm, and sec.

1.2 Show that magnetization M and field strength H have the same dimensional units. What are they?

1.3 A cylindrical bar magnet 6 in. long and 1 in. in diameter has a magnetic moment of 14,000 ergs/Oe.

a) What is its magnetization?

b) What current would have to be passed through a 100-turn solenoid of the same dimensions to give it the same magnetic moment?

1.4 A paramagnetic bar has the same dimensions as the magnet of the preceding problem and a volume susceptibility of 20×10^{-6} emu/cm^3 Oe. What magnetic moment would it acquire in a field of 20,000 Oe?

1.5 A demagnetized iron ring is placed in the circuit of Fig. 1.10. It has a mean diameter of 6 in., a cross-sectional area of 1/64 in^2, and it is wrapped with a magnetizing winding of 300 turns. When a current of 1.25 amperes is sent through the winding, the ballistic galvanometer indicates a flux change of 2000 maxwells. What is the induction, magnetization, and permeability?

CHAPTER 2

EXPERIMENTAL METHODS

2.1 INTRODUCTION

No clear understanding of magnetism can be attained without a sound knowledge of the way in which magnetic properties are measured. Such a statement, of course, applies to any branch of science, but it seems to be particularly true of magnetism. The beginner is therefore urged to make some simple, quantitative experiments early in his study of the subject. Quite informative measurements on an iron rod, which will vividly demonstrate the difference between B and H, for example, can be made with inexpensive apparatus: a ballistic galvanometer or fluxmeter, an easily made solenoid, some wire, a few switches and resistors, a battery, and an ammeter. Most books on magnetism contain some information on experimental methods. Those by Bates [G.13] and Ewing [G.1] are particularly helpful, and so is an inexpensive pamphlet by Sanford and Cooter [2.1]. Standard texts on electrical measurements, such as that by Vigoureux and Webb [2.14], will also be found useful. Books and papers devoted entirely to magnetic measurements are those of Astbury [G.5], Zijlstra [G.31], McGuire and Flanders [2.23], Oguey [2.24], and the ASTM [2.26].

The experimental study of magnetic materials requires (a) a means of producing the field which will magnetize the material, and (b) a means of measuring the resulting effect on the material. We will therefore first consider ways of producing magnetic fields, by solenoids and by electromagnets. Then we will take up the various methods of measuring the magnetization curve and the hysteresis loop of a strongly magnetic substance, and finally, in the last section of this chapter, the methods of measuring the susceptibility of a weakly magnetic substance. Methods of measuring more specialized magnetic properties (e.g., anisotropy, magnetostriction, core losses, etc.) will be dealt with at the appropriate place in later chapters.

The student who wishes to gain a good understanding of magnetic materials cannot afford to slight the contents of this chapter, even if he is not particularly interested in measurements, because some quite basic magnetic phenomena are first introduced here. For example, the demagnetizing fields discussed in Section 2.6 have an importance not restricted to measurements; these fields can affect the magnetic state and magnetic behavior of many specimens.

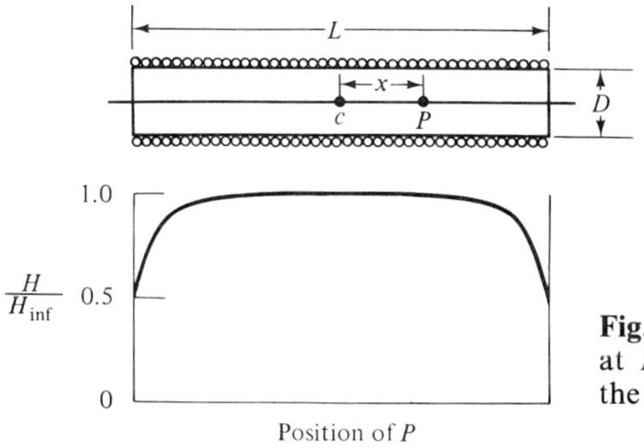

Fig. 2.1 Single-layer solenoid. The axial field at P is expressed as a fraction of the field at the center of an infinitely long solenoid.

2.2 FIELD PRODUCTION BY SOLENOIDS

Solenoids are useful for measurements on specimens of almost any shape, but are particularly suited to rod and wire specimens. They can be designed to produce fields ranging from a fraction of an oersted up to more than 200 kilooersteds (kOe). Two types can be distinguished, normal and superconducting.

Normal solenoids

These are usually made with insulated copper wire, wound on a tube of any electrically insulating material such as plastic. For the dimensions shown in Fig. 2.1, the field H in oersteds at a point P on the axis, distant x cm from the center C, is given by

$$H = \frac{4\pi ni}{10L}\left[\frac{L+2x}{2\sqrt{D^2+(L+2x)^2}} + \frac{L-2x}{2\sqrt{D^2+(L-2x)^2}}\right], \tag{2.1}$$

where n is the number of turns. At the center of the solenoid ($x = 0$), this reduces to

$$H = \frac{4\pi ni}{10L}\left(\frac{L}{\sqrt{D^2+L^2}}\right), \tag{2.2}$$

and, when $L \gg D$, to

$$H = \frac{4\pi ni}{10L} = \frac{1.257ni}{L}. \tag{2.3}$$

In any solenoid, the field decreases as x increases. The field at the end of a long solenoid is just one-half of the field at the center. But the field over the middle half is surprisingly uniform, as shown by Fig. 2.1 and the values in Table 2.1, which are derived from Eq. (2.1). In this table, H_{inf} is the field, given by Eq. (2.3), at the center of an infinitely long solenoid. When the L/D ratio is 20, for example, the field over the middle half is uniform to within 0.15 percent and only about 0.3 percent less than that produced by an infinitely long solenoid.

To increase the field beyond a certain value, it is better to increase n/L, by winding the wire in two or more layers, than to increase the current. Although H is proportional to i, the heat developed in the winding is proportional to i^2R,

Table 2.1 Field Uniformity in Solenoids

L/D	H at center	H at edge of middle half
5	0.9806 H_{inf}	0.9598 H_{inf}
10	0.9950	0.9892
20	0.9987	0.9972
50	0.9996	0.9994

where R is its resistance. Thus doubling the number of layers will double H, R, and the amount of heat; doubling the current will double H, but quadruple the heat. Cooling of the winding becomes necessary for fields larger than about 1000 Oe. This can be accomplished by winding the wire on a water-cooled copper tube or by forming the winding from copper tubing rather than wire; the tubing can then carry both the electric current and the cooling water. Solenoid design is a matter of balancing several conflicting requirements, and the following points should be kept in mind:

1. D is determined by the working space required within the solenoid.
2. The ratio L/D is fixed by the distance over which reasonable field uniformity is required. Because the specimen to be tested must be subjected to a reasonably uniform field, this means that the maximum specimen length effectively determines the ratio L/D. Specimen length is in turn governed by the factors discussed in Section 2.6.
3. For a given L, the field is proportional to the number of ampere-turns ni, and the power required (= rate of heat generation) is proportional to i^2R.
4. For a given current, the voltage required at the power source is proportional to R, which in turn is proportional to n.

Helmholtz coils will produce an almost uniform field over a much larger volume than the usual solenoid. Two thin, parallel coils are placed at a distance apart equal to their common radius r (Fig. 2.2). The field parallel to the axis of the coils, at a point P on the axis, distant x from one coil, is given by

$$H = \frac{\pi ni}{5r}\left[\left\{1 + \frac{x^2}{r^2}\right\}^{-3/2} + \left\{1 + \left(\frac{r - x}{r}\right)^2\right\}^{-3/2}\right] \qquad (2.4)$$

where n is the number of turns in each coil. At the center of the coil system ($x = r/2$):

$$H = \frac{0.899ni}{r}. \qquad (2.5)$$

Scott [2.2] gives equations for both components of the field at points off the axis. For the same current and the same total length of wire in the winding, i.e. same power consumption, Helmholtz coils produce a field which is only a few percent of that produced by a solenoid of length r. They are thus confined mainly to low-field applications.

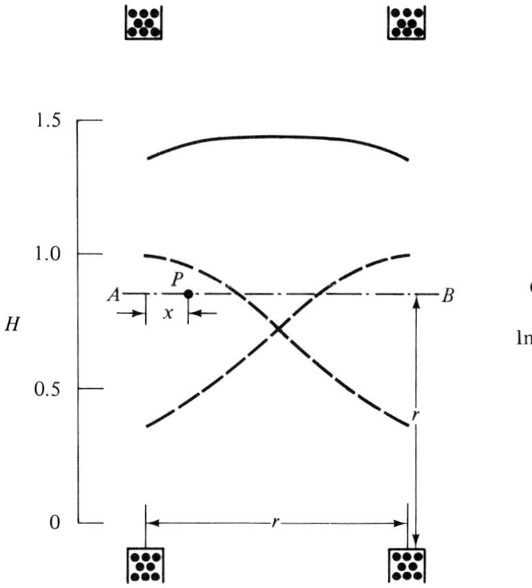

Fig. 2.2 Cross section through a Helmholtz coil system. The axial field at any point P on the axis AB is shown by the solid line. The field produced by each coil is shown by dashed lines, and the unit of field strength is the field produced by one coil at its center.

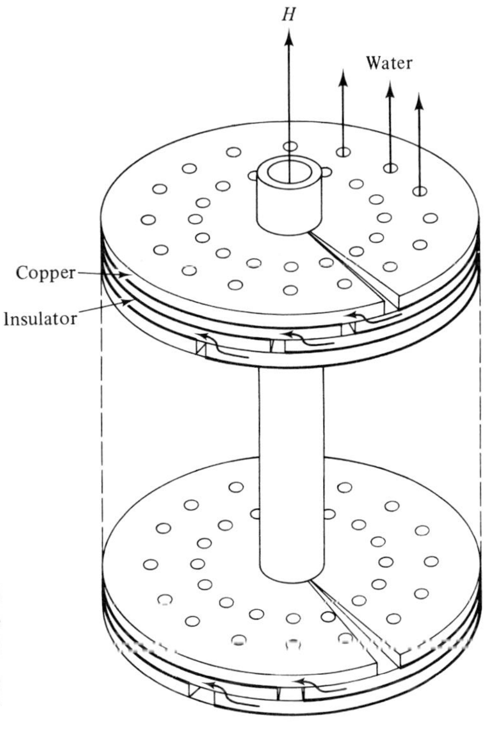

Fig. 2.3 Solenoid of a Bitter magnet (schematic). Arrows show current path from one disc to another.

To produce very high fields by means of normal solenoids requires very large currents, and the design problem then becomes almost entirely one of removing very large amounts of heat. (Note that the process of maintaining a steady magnetic field by means of an electric current is one conducted at absolutely zero efficiency. *All* the input power goes into heat.) Beginning about 1936, Francis Bitter [2.3] began the development of high-field solenoids of a new type. The coil of a *Bitter magnet* is sketched in Fig. 2.3. (Any device which produces a field is commonly referred to as a "magnet," whether or not it contains iron.) The winding is composed, not of wire, but of thin discs of copper. These discs, about 1 ft in diameter and 0.04 in. thick, have a central hole and a narrow radial slot and are insulated from each other by similarly cut sheets of thin insulating material. Each copper disc is rotated about 20° with respect to its neighbor, and the region of overlap provides a conducting path for the current to flow from one disc to the other. The current path through the entire stack of discs, which is about one foot high, is therefore helical, as in a conventional solenoid. The discs are clamped tightly together and enclosed in a case (not shown). Cooling water under high pressure is forced axially through the magnet, through a large number of small holes cut in each disc. A field of about 100 kOe can be

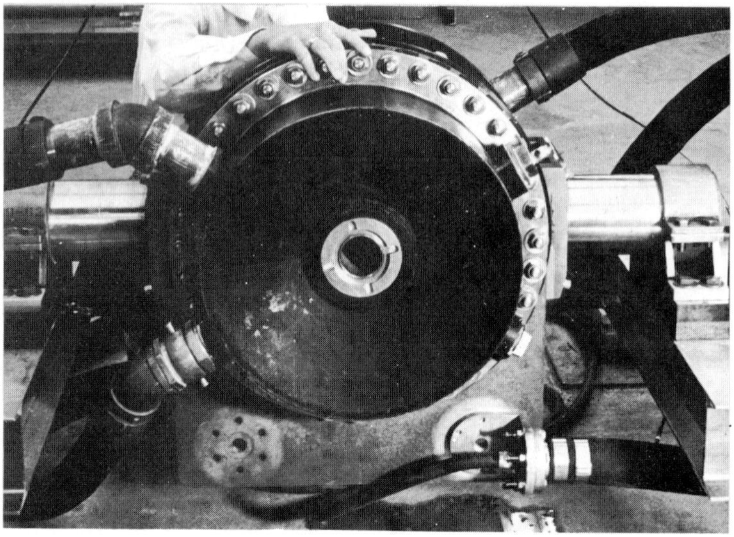

Fig. 2.4 Bitter magnet which produces 110 kOe in a 2-in. bore with 2.5 megawatts input. Two cooling-water hoses can be seen on the right, and two on the left. One of the electrical leads, carrying 10,000 amperes, can be seen at the bottom; the other is behind the magnet. (Courtesy Francis Bitter National Magnet Laboratory)

produced at the center of a magnet with a $1\frac{1}{8}$ in. diameter bore, with the expenditure of 1700 kilowatts of power. To do this requires a very large current (10,000 amperes) and large amounts of cooling water (800 gallons per minute). The large size of the water and electrical connections to a Bitter magnet is apparent in Fig. 2.4. Bitter magnets have been made that produce a field of over 200 kOe.

The *power supply* to a solenoid must provide direct current. Dry cells, lead storage batteries, or alternating current rectifiers are suitable for low fields. High-field devices like Bitter magnets require large motor-generator sets, consisting of an ac motor driving a dc generator. Such installations, together with the necessary water-pumping facilities, are expensive.

Pulsed fields offer a less costly approach to the problem of measurements in high fields. If the measurement can be made quickly, by means of an oscilloscope, for example, then only a transient field is needed. This can be produced by slowly charging a bank of capacitors and then abruptly discharging them through a solenoid. A large pulse of current, lasting only a small fraction of a second, is produced, and the problem of heat removal is thus greatly minimized or even eliminated, depending on the magnitude of the field required. Pulsed fields of several hundred kilo-oersteds have been produced in special water-cooled solenoids, and pulsed fields of moderate strength (10-30 kOe) in conventional, wire-wound, noncooled solenoids are easily obtained [2.4, 2.5]. (One factor which must not be overlooked in the design of high-field coils, whether intended for pulsed or continuous operation, is mechanical rigidity. A magnetic field

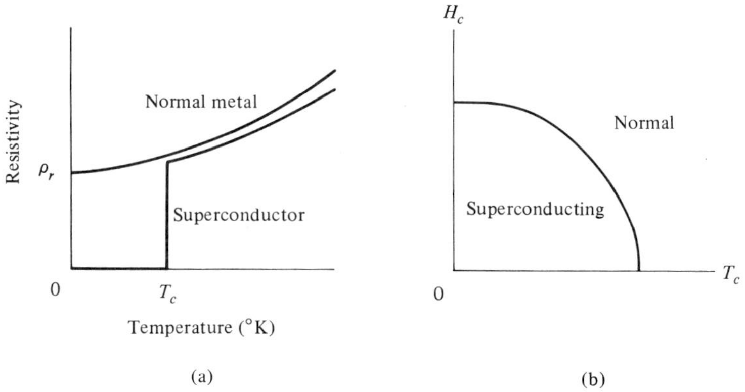

Fig. 2.5 (a) Variation of electrical resistivity with temperature for a normal metal or alloy and a superconductor. (b) Variation of critical temperature with field.

exerts a force on a current-carrying conductor, and in a solenoid developing 100 kOe there is an outward pressure of about 5800 lb/in^2 which tries to distort the coil into a spherical shape. A method of forming sheet metal is based on this effect: a pulsed magnetic field forces the metal into a suitable die.)

Superconducting solenoids

The phenomenon of superconductivity provides a radically different approach to the high-field problem. When a normal metal is cooled to 0°K, its electrical resistivity decreases to a low but finite value ρ_r, called the residual resistivity (Fig. 2.5a). However, the resistivity of some metals and alloys decreases abruptly to *zero* at a critical temperature T_c of a few degress Kelvin; these materials are called superconductors; lead (T_c = 7.2°K) and tin (T_c = 3.7°K) are examples. (The magnetic properties of superconductors are also quite unusual. See Section 3.5). If a current is once started in a circuit formed of a superconductor maintained below T_c, it will persist indefinitely without any power input or heat generation, because the resistance is zero. The attractive possibility at once presents itself of producing very large magnetic fields by making a solenoid of, for example, lead wire and operating it below T_c by immersing the windings in liquid helium (4.2°K). However, soon after the discovery of superconductivity in 1911, it was found that an applied magnetic field decreased T_c and a field of a few hundred oersteds destroyed the superconductivity completely (Fig. 2.5b). Thus, when the field produced by the solenoid itself exceeded a critical value H_c, the normal resistivity of the wire would return, along with the attendant problems of heating and power consumption.

The solution to this problem was not found until 1961 when Kunzler *et al.* [2.6] discovered that the niobium-tin alloy Nb_3Sn remained superconducting even at a field of 88 kOe. It was later found that the critical field of this alloy at 4.2°K,

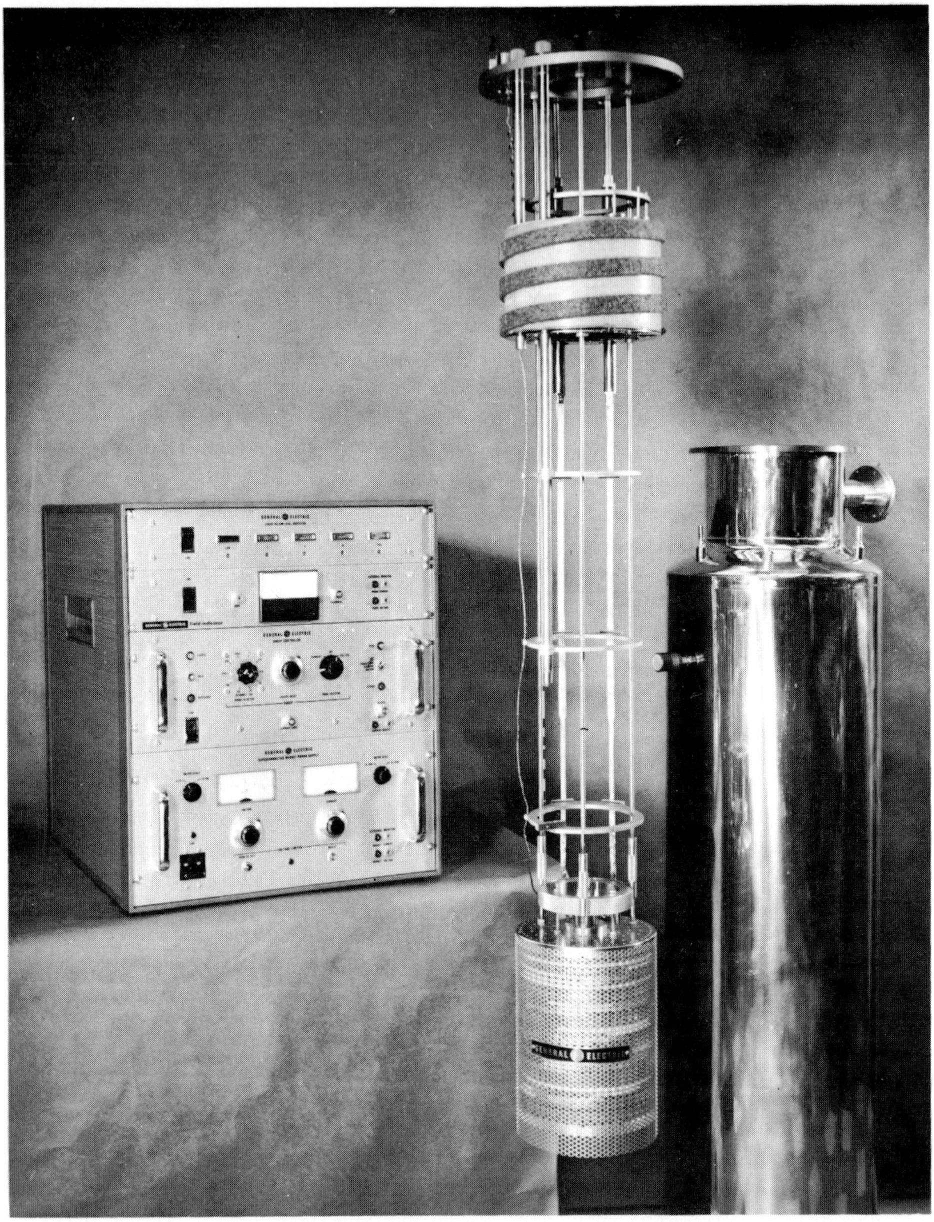

Fig. 2.6 Superconducting solenoid. (Courtesy General Electric Company)

the temperature of liquid helium, is 220 kOe. Nb_3Sn is very brittle, and many metallurgical problems had to be solved before it was successfully made in the form of a composite tape suitable for a solenoid winding. It was later found that Nb-Zr and Nb-Ti alloys, which are reasonably ductile, are superconducting up to fields of the order of 80 kOe at 4.2°K. Superconducting solenoids of all three of these materials are commercially available, and an example is shown in Fig. 2.6. The solenoid itself appears at lower center; it is maintained at 4.2°K when inserted in the liquid-helium dewar at the right; the control cabinet at the left contains the power supply and instruments to indicate the field strength in the solenoid and the helium level in the dewar. The solenoid has a length of 10 in., an outer diameter of 8 in., a bore of 1.25 in., and is wound with Nb_3Sn tape. It produces a field of 143 kOe with a current of 140 amperes.

An ordinary lead storage battery is enough to start the current in a superconducting solenoid, and thereafter no power input is required for the solenoid itself. However, it must be maintained at the temperature of liquid helium, which means that power must be expended to operate a helium liquifier, or liquid helium must be bought. But the power to operate the liquifier, about 5 kilowatts, is negligible in comparison with the more than 1000 kilowatts required for a water-cooled solenoid of the Bitter type. It has been estimated that the cost of a superconducting magnet and cryostat is about one-tenth that of a Bitter magnet and its associated equipment.

The great interest in high magnetic fields extends far beyond studies of their effects on the magnetic properties of materials. They are needed for a wide range of experiments in solid-state physics and biology, for bubble chambers and high-energy particle accelerators. Superconductors are in the operating or planning stage as field sources for such applications. Another area of interest is magnetohydrodynamics, whereby electric power is generated from hot ionized gases (plasmas) heated by conventional fuels. Perhaps the most challenging application is in the area of power generation by controlled nuclear fusion. One problem that must be solved before this can become a reality is the confinement of the plasma by a magnetic field, and superconductors to provide these fields are being considered. Research on high-field superconductors is only in its infancy today, and much can be expected of these materials in the future. Sampson [2.7] has reviewed the state of the art as of 1968.

The search continues for superconducting materials with higher critical fields and higher critical temperatures. The latest discovery [2.8] is a Nb-Al-Ge alloy with $T_c = 20.7°K$ and a critical field at 4.2°K of 410 kOe. These are the highest values so far observed.

In the early 1960's it was thought that superconducting magnets would soon replace Bitter magnets for high-field research. This has not happened. Actually, Bitter magnets can be made to produce higher fields in small volumes than present-day superconductors, and Bitter magnets are being built in laboratories all over the world. The U. S. center for high-field research is the Francis Bitter National Magnet Laboratory, located at the Massachusetts Institute of Technology. It contains both Bitter and superconducting magnets.

2.3 FIELD PRODUCTION BY ELECTROMAGNETS

In the average laboratory the need for fields larger than those obtainable from conventional solenoids is usually met with electromagnets. Most of these produce fields in the range of 0–20 kOe, although some are capable of 30–50 kOe.

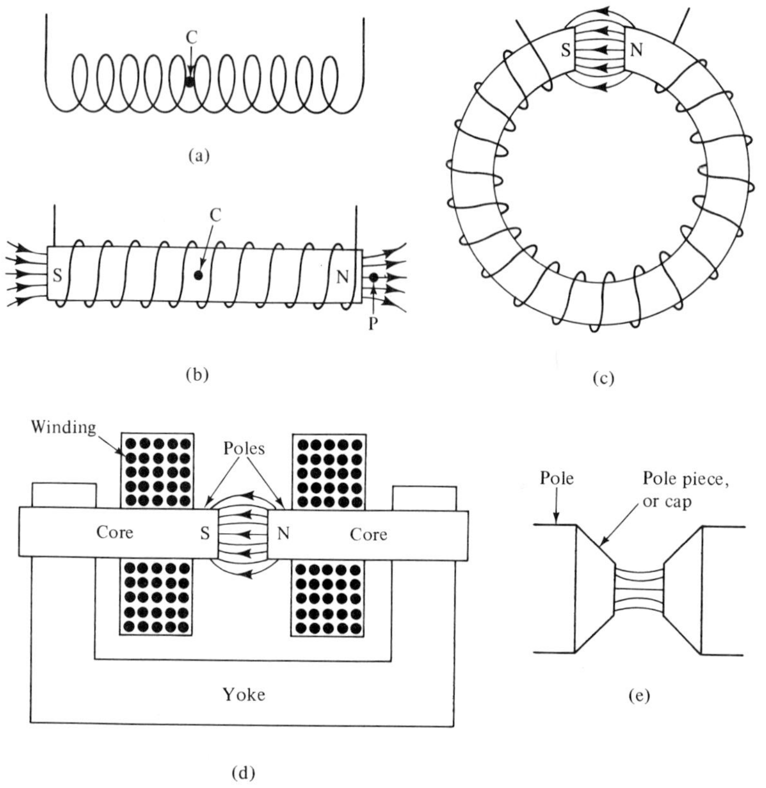

Fig. 2.7 Evolution of the electromagnet.

An electromagnet consists essentially of an iron "core" around which is wrapped a coil of wire carrying a direct current. Figure 2.7 illustrates its development. A simple solenoid is shown in (a); the field H at its center C is proportional to the number of ampere-turns per unit length of its winding, in accordance with Eq. (2.3). If an iron rod is inserted in the coil, as in (b), the field at its center C, inside the iron, is now very much larger, because the field is now given by B, which is the sum of H due to the current and $4\pi M$ due to the iron [Eq. (1.15)]. We cannot, of course, make any use of the field at a point inside the iron. However, the field at the point P, just outside the end of the rod, is also equal to B. Further away, the lines of force diverge and the flux density, or field strength, decreases. The iron has, in effect, multiplied the field due to the current, and the multiplying factor is simply the permeability μ, because $B = \mu H$. Thus, if the permeability is 2000 for $H = 10$ Oe, the field inside and just outside, the iron is 20,000 Oe. In this way quite a large field can be obtained with a relatively low current. Here we have ignored the fact that B inside the rod near one end is much less than B in the center of the rod, as we shall see in Section 2.6. When this effect is taken into account, we find that the current in the coil has to produce a field H many

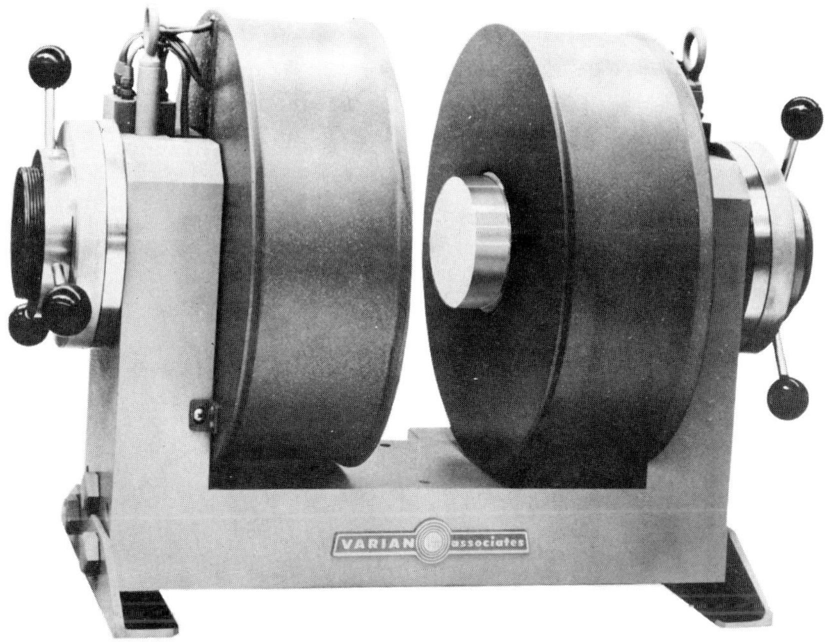

Fig. 2.8 Laboratory electromagnet. The 4-in. diameter cylindrical pole caps shown may be replaced with caps of different shape, if desired. (Courtesy Varian Associates)

times larger than 10 Oe if the flux density B at the end of the rod is to be 20,000 G. Even so, this current will still be very much less than if the iron were absent.

(The rotor of an electric motor or generator is fundamentally a rotating electromagnet, composed of an iron core wound with copper wire. If these machines are to do any useful amount of work, they must operate at high flux densities; without the flux-multiplying power of iron they would be nothing but scientific toys. It is indeed a most remarkable fact of history that, when Faraday and Henry and their followers needed this power, they found it right at hand in iron, so cheap, so common, which later work has shown to be the best element for this purpose in the whole periodic table. One wonders what direction the development of technology would have taken, if only one element had turned out to be strongly magnetic and if that element were, for example, gold.)

The divergence of the lines of force near the ends of a straight iron rod can be reduced by bending the rod into a circle so that the ends nearly touch, as in Figure 2.7 (c). The flux then travels directly from one pole to the other across the air gap. As the current in the winding is increased, the magnetization of the iron increases to its saturation value M_s of 1714 emu/cm^3. The maximum contribution of the iron to the field in the air gap is therefore $4\pi M_s$, or about 21,500 G, if the pole faces are flat. Any further increase is due only to an increased current in the winding. To make this contribution more effective the turns of the winding are brought close to the air gap, and the magnet assumes the final form shown in section in Fig. 2.7 (d). The flux generated by the winding passes

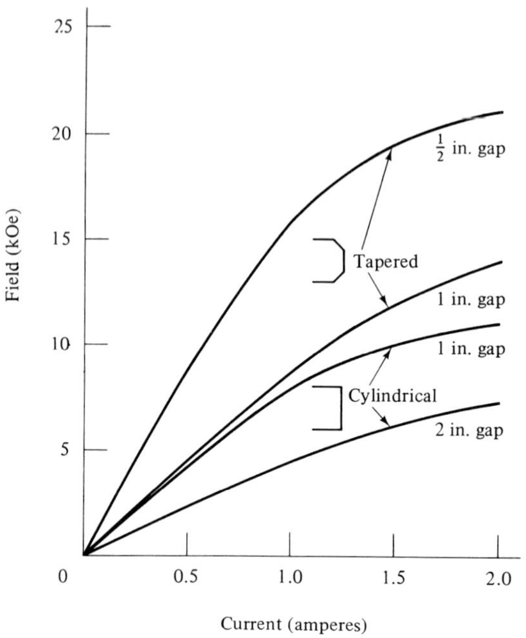

Fig. 2.9 Field in the gap of the magnet of Fig. 2.8 as a function of current in the winding and gap size, for 4-in. cylindrical pole caps and caps tapering from 4 to 2 in. (Courtesy Varian Associates)

through the core into the yoke, and back through the other core. Both core and yoke are made of iron or very low carbon steel, annealed to produce high permeability. The yoke must also be massive enough to resist the strong force of attraction between the two poles. The windings are often water cooled. The size of a magnet is specified in terms of the diameter of its poles, which usually ranges from 4 to 12 in. Figure 2.8 shows a commercial 4-in. magnet.

When a field uniform over a fairly large volume is required, flat pole faces are used. To achieve higher fields, tapered pole pieces (pole caps) can be attached to the poles. The free poles formed on the tapered surfaces contribute to the field at the center of the gap, as suggested by Fig. 2.7 (e). This contribution can be far more than $4\pi M_s$ [G-1, G-31], but this higher field is achieved only in a smaller volume and it is not very uniform. A still further, but moderate, increase in field can be obtained by making the pole pieces of a 50 percent Fe and 50 percent Co alloy, which has a value of $4\pi M_s$ of more than 23,000 G.

In most magnets the size of the air gap (distance between poles) can be adjusted by screwing the cores in or out of the yoke. The larger the gap, the smaller the field, because much of the flux then leaves the volume of the gap proper and forms the "fringing flux" indicated in Figs. 2.7 (c) and (d). Figure 2.9 shows how the field in the gap depends on gap size and the current in the winding for the magnet of Fig. 2.8. The minimum gap size is fixed by the size of the specimen and other experimental apparatus which must be inserted in the gap.

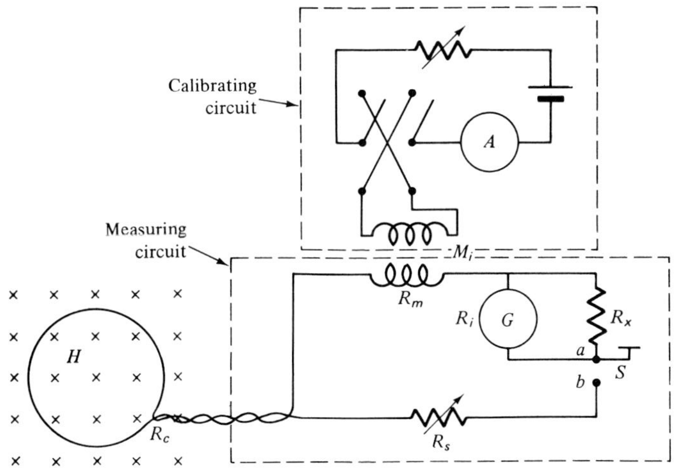

Fig. 2.10 One-turn search coil connected to a ballistic galvanometer G. The lines of the field H, indicated by crosses, are normal to the plane of the drawing. The leads from the search coil are twisted to prevent flux passage between the leads.

2.4 MEASUREMENT OF FIELD STRENGTH

The field H to which a specimen is subjected in a measurement of its magnetic properties must be known, by calculation or measurement. If the field is produced by a solenoid, its intensity can be calculated from the current, number of turns, and length of the winding, because these quantities can all be measured with sufficient accuracy. However, the field in the gap of an electromagnet must be measured; it depends not only on the current in the windings but also on the magnetic properties of the core and yoke, and these are seldom accurately known. It is usual to measure this field for several currents and prepare calibration curves like those of Fig. 2.9.

The two most popular methods of measuring magnetic fields are by means of the current induced in a search coil (ballistic method) or the voltage produced in a semi-conductor through the Hall effect.

Ballistic method

This method is the more important of the two, inasmuch as it can be applied to the measurement of B and M, as well as H. We will therefore consider it in some detail, paying particular attention to the operation of the measuring instruments involved (ballistic galvanometer or fluxmeter). A search coil is first made by winding fine insulated wire on a circular, nonmagnetic form. The plane of the coil is then placed normal to the field to be measured, and the winding is connected to an appropriate measuring circuit (Fig. 2.10) containing a ballistic

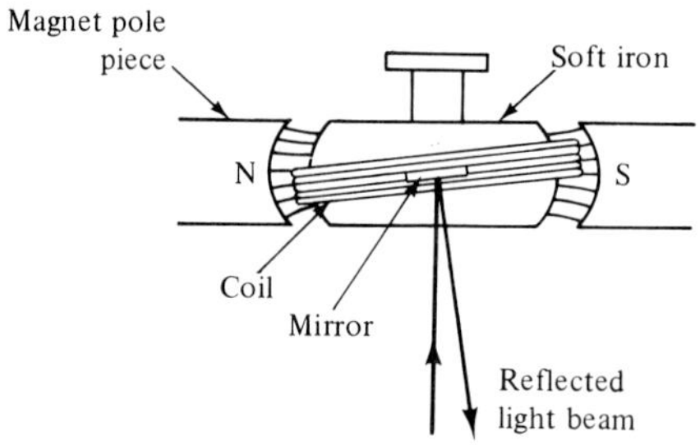

(a)

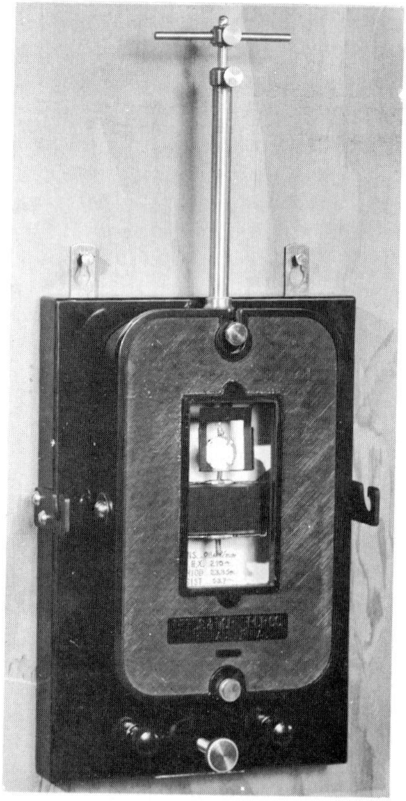

(b)

Fig. 2.11 (a) Plan view of ballistic galvanometer coil. (b) Wall-mounted ballistic galvanometer of moderate sensitivity. (Courtesy of Leeds and Northrup Co.)

galvanometer. By Faraday's law, any change in the total number of lines of force ϕ through the coil will induce an emf e in the coil, proportional to the *rate of change* of flux and the number of turns N, given by

$$\boxed{e = 10^{-8}\, N \frac{d\phi}{dt} \text{ volts.}} \tag{2.6}$$

If the area of the search coil is A cm^2, then the flux through it is $\phi = HA$ maxwells. To measure a steady field, the flux through the coil must be changed by a known amount. For example, the coil may be suddenly moved from the region where H is to be measured to a region where the field is known to be essentially zero, e.g., from the air gap of an electromagnet to a point two or three feet away. Then the flux change $\Delta\phi$ is simply ϕ. Or the coil may be rapidly rotated by 180° about an axis normal to the field, so that $\Delta\phi = 2\phi$. (It is then called a flip coil.)

The emf induced by the change in flux in the search coil produces a current through the galvanometer which at any instant is given by $i = e/R$, where R is the total resistance of the galvanometer circuit. This current causes a deflection d of the galvanometer proportional to the total charge $q = \int i\,dt$ flowing through the instrument, provided the time during which the charge flows is *small* compared to the natural period of swing of the galvanometer. Then

$$i = \frac{e}{R} = \frac{10^{-8}\,NA}{R}\frac{dH}{dt}, \qquad \int i\,dt = \frac{10^{-8}\,NA}{R}\int dH,$$

$$q = \frac{10^{-8}\,NA\,\Delta H}{R}\ \text{coulombs.} \tag{2.7}$$

But

$$q = kd, \tag{2.8}$$

where k is a constant for a particular galvanometer and a particular value of R. Therefore,

$$\Delta H - \left(\frac{10^{8}\ Rk}{NA}\right) d\ \text{Oe}, \tag{2.9}$$

if R is in ohms, A in cm^2, k in coulomb/mm, and d in mm. This equation enables us to measure a field H by causing it to change by some known amount ΔH through a search coil, as mentioned above. The key S in Fig. 2.10 must be depressed during the measurement.

Ballistic galvanometer

This instrument (Fig. 2.11) consists of a rectangular coil of fine wire, through which passes the charge to be measured, suspended by a thin metal ribbon in the air gap of a permanent magnet. The dimensions of the coil and suspension are chosen to make the time required for one complete oscillation (the period) rather long, of the order of 20 sec. If a charge is suddenly made to flow through the coil, the coil receives a sudden angular impulse, rotates through a certain maximum angle (hence the name "ballistic," as in the ballistic pendulum), and then turns back toward the zero position because of the restoring torque in the suspension. The maximum rotation is measured in terms of the deflection d of a beam of light reflected from a mirror on the suspension to a graduated scale, usually located one meter from the mirror. If Eq. (2.8) is to be valid, the charge must pass through the coil before it has moved appreciably from its rest position.

The sensitivity and general behavior of a ballistic galvanometer depend markedly on the resistance of the circuit in which it is used. When the coil is moving, the galvanometer acts like an electrical generator. Motion of the conductors in the coil through the magnetic field of the instrument causes a back emf to be induced in the coil. This emf produces a current in the coil, and the interaction of this current and the magnetic field is such as to resist the motion of the coil. The rotation of the coil is therefore slowed down, or damped. The smaller the resistance of the circuit, the larger the induced current, and the greater the electromagnetic damping.

This effect can be demonstrated with the circuit of Fig. 2.10. Here R_c is the resistance of the search coil and leads, R_s is a variable series resistance, and R_m is the resistance of the secondary of a standard mutual inductance M_i. (The purpose of M_i will be explained

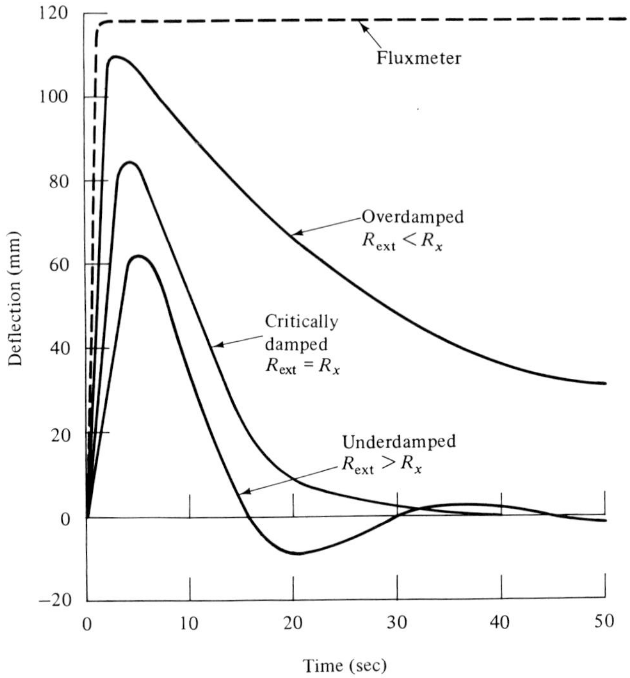

Fig. 2.12 Deflection versus time curves (solid lines) for a ballistic galvanometer of internal resistance R_i of 52 ohms, and external resistances of 25 ohms (overdamped), 270 ohms (critically damped), and 625 ohms (underdamped).

later.) Suppose the key S is depressed, so that the galvanometer is connected to b. Then the resistance of the circuit, external to the galvanometer, is $(R_c + R_m + R_s)$, and we can vary it by changing R_s. If the external resistance is varied, but the flux change through the search coil is kept constant, then the swing of the galvanometer is found to vary as shown in Fig. 2.12. When the resistance is small, so much current flows through the coil during its swing that its motion is very sluggish (overdamped), and it requires a long time to return to zero. On the other hand, when the resistance is large, the motion is underdamped, and the coil will swing rapidly back to zero, overshoot, and oscillate back and forth, again requiring a long time to reach a stable zero. When the resistance has an intermediate value R_x, the critical external damping resistance (CXDR), the coil is critically damped: it approaches zero asymptotically and reaches a stable zero in the least possible time. The value of R_x, which is 270 ohms in this case, is usually specified by the maker of the galvanometer.

Figure 2.12 also shows that the sensitivity, as measured by the maximum deflection, increases as the external resistance decreases, for a constant flux change in the search coil. [The lower the resistance, the greater the charge produced by a given flux change, as shown by Eq. (2.7)]. Equation (2.9) shows that the sensitivity is governed entirely by the total circuit resistance $R = R_c + R_m + R_s + R_i$, where R_i is the internal resistance of the galvanometer. Sensitivity and damping are not independent, however, and it is best to operate at or near critical damping. This may not always be possible. If, for example, the field to be measured is weak, then the number of turns on the search coil must be large, and so R_c is large. If R_c exceeds R_x, the coil is underdamped and much time is wasted in waiting for it to come to rest

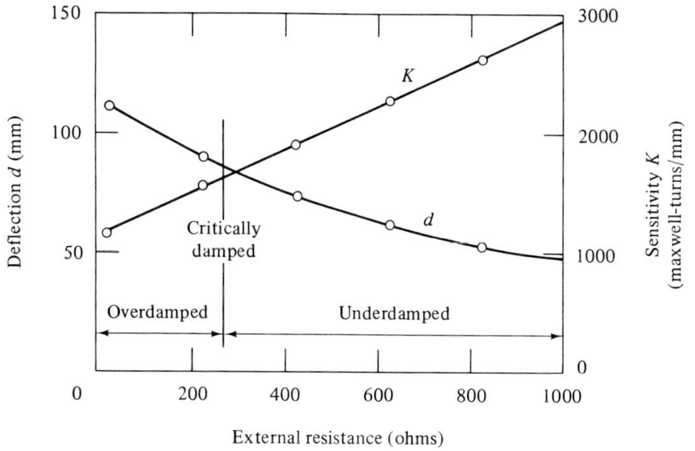

Fig. 2.13 Maximum deflection d and sensitivity K as a function of external resistance for the galvanometer of Fig. 2.12 and the same flux change as in Fig. 2.12.

after each measurement. This is avoided by means of the key S and the critical resistance R_x in Fig. 2.10. S is normally held in position a by a spring. It is depressed to connect the galvanometer to b when making a measurement of the field, and released as soon as the maximum deflection of the galvanometer is noted. Thus the galvanometer is always critically damped during the return swing, whatever the extent of damping during the initial deflection.

There are many measures of galvanometer sensitivity. The value of k in coulombs/mm, which appears in Eq. (2.8), is one such measure. Another, and one more useful in magnetic measurements, is the value of $K = (N\Delta\phi)/d$, or the number of maxwell-turns/mm of deflection. "Maxwell-turns" may seem like an odd unit, until one realizes that a flux change of 500 maxwells (lines) through a 1-turn search coil produces the same effect as a flux change of 1 maxwell through a 500-turn coil. (The product of the flux and the number of turn is often called the flux linkage.) Since $\phi = HA$, Eq. (2.9) can be written

$$\Delta\phi = \frac{10^8 Rkd}{N},$$

or

$$\Delta\phi = \frac{Kd}{N}, \tag{2.10}$$

where

$$K = 10^8 Rk = \frac{N\,\Delta\phi}{d} \quad \text{maxwell-turns/mm.} \tag{2.11}$$

Figure 2.13 shows how K and d vary with external resistance. Note that a galvanometer of high sensitivity is one with a *low* value of K.

In order to measure flux changes with a ballistic galvanometer, the value of K must be known. It may be determined by means of the calibrating circuit shown in the upper part of Fig. 2.10. The primary of the standard mutual inductance M_i is connected through a reversing switch to a battery, ammeter, and variable resistor; the secondary is connected in the search-coil circuit. A mutual inductance consists of two coils so arranged that the flux due to a current in one coil (the primary) passes through all or part of the other coil (the secondary). The two coils are defined as having a mutual inductance M_i of one henry* if an emf of one volt is induced in the secondary when the current in the primary is changing at the rate of one ampere/sec:

$$e = M_i \frac{di}{dt} \quad \text{volts,} \tag{2.12}$$

Comparison of this with Eq. (2.6) suggests another definition: two coils have a mutual inductance of one henry if a current of one ampere in the primary causes a flux linkage $N\phi$ in the secondary of 10^8 maxwell-turns, or

$$M_i i = 10^{-8} N\phi. \tag{2.13}$$

To calibrate the galvanometer, a known current i is set up in the primary, reversed, and the galvanometer deflection d noted. Then

$$M_i \Delta i = 2M_i i_r = 10^{-8} N \Delta \phi.$$

Therefore,

$$K = \frac{N \Delta \phi}{d} = \frac{2 \times 10^8 M_i i_r}{d}, \tag{2.14}$$

where i_r is the current *reversed*. It is best to reverse a number of different currents and plot d versus i_r; the slope of the straight line through these points will give the best value of $2 \times 10^8 M_i/K$, from which K can be determined. A convenient value for M_i is about one millihenry.

A standard mutual inductance, fixed or variable, may be purchased, or one can easily be made. For example, we can take an ordinary solenoid as the primary coil and wind another wire around its central section to form the secondary. If the solenoid has n turns, a length of L cm, and a cross-sectional area of A cm^2, then the flux through it is, from Eq. (1.13),

$$\phi = HA = \frac{4\pi ni A}{10L} \quad \text{maxwells}$$

where i is the current in amperes. Therefore, the mutual inductance is

$$M_i = \frac{10^{-8} N\phi}{i} = \frac{4\pi \times 10^{-9}\, n\, NA}{L} \quad \text{henrys,} \tag{2.15}$$

where N is the number of turns in the secondary.

Because the sensitivity K of a galvanometer depends on the resistance of the circuit in which it is used, the value of K determined in a calibration against a known mutual inductance is valid only when the galvanometer circuit has the same resistance as it had during calibration. For this reason M_i is often left permanently connected in the measuring circuit of Fig. 2.10. If it is removed, R_s must be increased by an amount R_m in order to keep the circuit resistance constant. Overlooking this fact is a frequent source of error. When measurements are made over a range of temperature, it is important to remember that the resistance of the search coil will change with temperature; a different setting of R_s is then required for each temperature in order to keep the circuit resistance constant.

The effective cross-sectional area A of a search coil is not always easy to measure or compute, particularly if the coil has a large number of turns wound in several layers. Actually, the area

---

* Joseph Henry (1797–1878), American physicist. Independently of Faraday, he discovered electromagnetic induction.

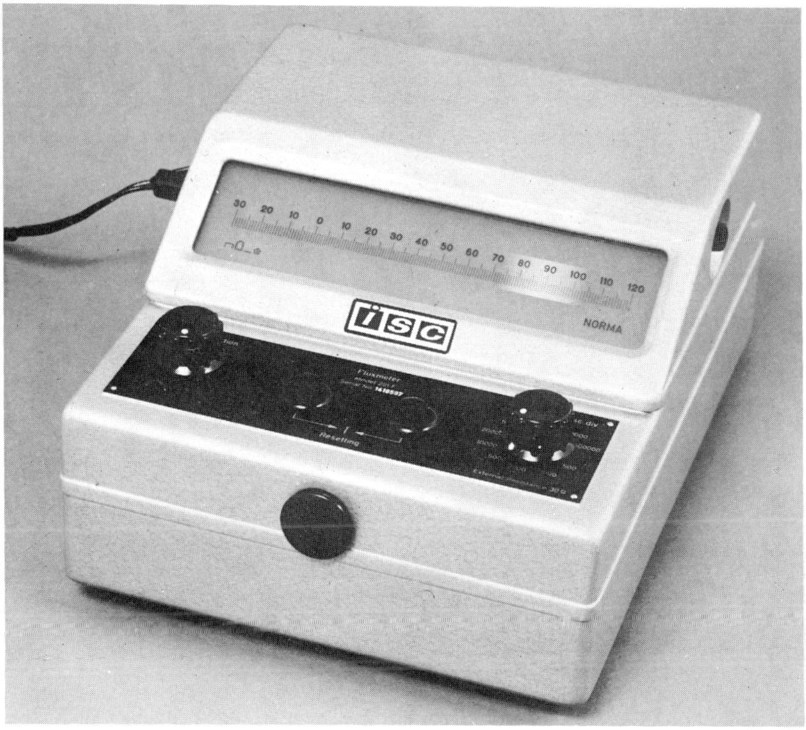

Fig. 2.14 "Norma" fluxmeter. (Courtesy Instrument Systems Corporation).

need not be known, because it always enters the relevant equations (2.9 or 2.10) as the product NA, called the *area-turns* of the coil. This quantity may be determined experimentally, after the galvanometer or fluxmeter has been calibrated, by placing the coil in a known field (for example, at the center of a solenoid), reversing the flux through the solenoid, and noting the galvanometer deflection. Equation (2.10) can then be solved for $NA = Kd/\Delta H$ cm^2.

Fluxmeter

This may be substituted for a ballistic galvanometer for the measurement of flux changes. A fluxmeter, sometimes called a creep galvanometer, is also a moving-coil instrument, but it differs from a ballistic galvanometer in two respects: (1) the torsional stiffness of the suspension is as near zero as the manufacturer can make it [2.9], and (2) it is used in a low-resistance circuit and is, therefore, heavily overdamped. As a result, when a charge passes quickly through a fluxmeter, the coil deflects very rapidly to its final position and remains there, as shown by the dashed line in Fig. 2.12. (There may be a slow drift, but this can usually be eliminated by proper adjustment.) After the deflection is noted, the coil is returned to its zero position by passing a small current through it.

The fluxmeter is a portable instrument, the position of the coil being indicated by a pointer or an internally reflected light beam. In the fluxmeter shown in Fig. 2.14, the beam is multiply reflected over a total distance of one meter and focused on the scale at the front. This is a multiple-range instrument which has various sensitivities K available at the turn of a switch,

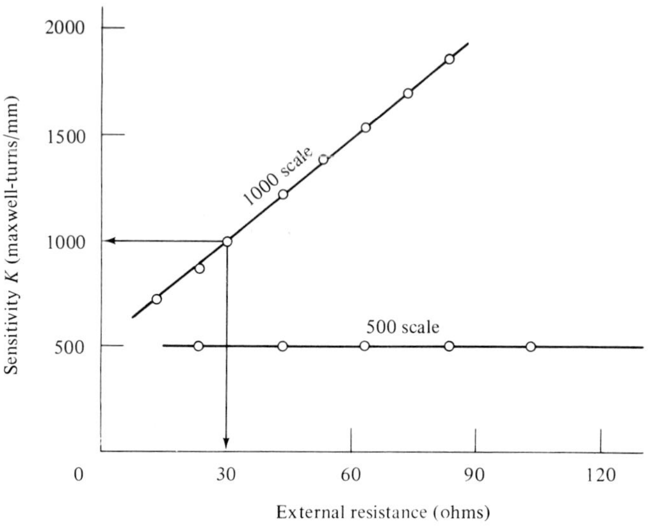

Fig. 2.15 Sensitivity K as a function of external resistance for two scales of the fluxmeter shown in Fig. 2.14. The 1000 scale is calibrated for an external resistance of 30 ohms.

ranging from 500 to 500,000 maxwell-turns/mm in ten steps. These values of K apply only when the external resistance has the value specified by the maker. For the instrument shown, this value is 30 ohms for all scales except the 500 one, where the external resistance may be 0 to 100 ohms without affecting the sensitivity. Figure 2.15 shows the effect of variations in resistance.

A fluxmeter is faster and easier to read than a ballistic galvanometer. It has another, more important, advantage; it can accurately measure slow, as well as fast, changes in flux. Because the suspension imposes essentially no torsional constraint, the fluxmeter coil moves in unison, so to speak, with the change in flux in the search coil. As long as the flux changes, the coil rotates; when the flux becomes constant, the coil stops.

On its most sensitive range, the 500 scale for the instrument of Fig. 2.14, the moving coil of a fluxmeter is connected directly to the search coil. Because this coil moves *during* the change in flux through the search coil, the back emf induced tends to prevent current flow in the forward direction and thus acts as a resistance. (This is not true of the ballistic galvanometer, because all the charge has gone through its coil before it has moved appreciably from its rest position.) This resistance, called the dynamic resistance, exceeds 5000 ohms in the instrument shown. In comparison to this large resistance, variations in search-coil resistance from 0 to 100 ohms are insignificant; thus the sensitivity K is virtually independent of search-coil resistance, as shown in Fig. 2.15. This is not true of the less sensitive scales. These are achieved by a set of internal shunts, selected by the range switch, which divide the voltage generated by the search coil so that only a part of it is applied to the fluxmeter coil; the calibration of this voltage divider requires that the search coil have a fixed resistance. In summary, the variation of sensitivity K with search-coil resistance is inherent in the ballistic galvanometer, but is only the consequence of the internal voltage divider in a multirange fluxmeter. On its most sensitive scale the fluxmeter has a sensitivity independent of search-coil resistance over a rather large range.

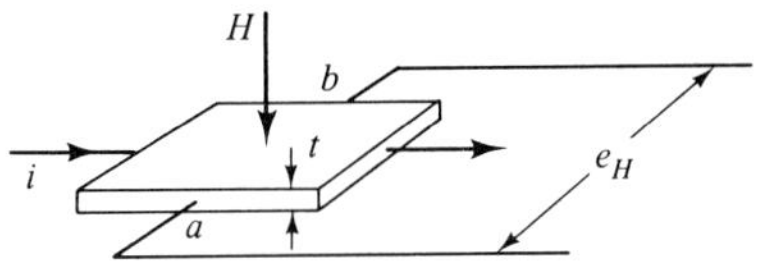

Fig. 2.16 Relation between field, current, and emf in the Hall effect.

On their most sensitive range, fluxmeters have sensitivities varying from about 10,000 to several hundred maxwell turns/div, depending on construction, with recommended search-coil resistances of up to about 100 ohms. Ballistic galvanometers cover a similarly wide range, except that the most sensitive have K less than 100 maxwell-turns/div, and search-coil resistances may be as high as 1000 ohms.

A fluxmeter is calibrated exactly like a ballistic galvanometer. This may be done either to check the calibration furnished by the maker or to determine K for values of external resistance other than that specified by the maker. The measuring circuit of a fluxmeter contains, besides the search coil, a variable series resistance and a key arranged to short circuit the instrument when not in use, as indicated at the right of Fig. 2.17.

Both the ballistic galvanometer and the fluxmeter are voltage integrators. Instruments are also available, called "electronic fluxmeters," in which this integration is performed electronically. Their electrical output is proportional to the change in the field in the search coil and may be fed to an automatic recorder (Section 2.5).

Hall effect

This effect* occurs in any conductor carrying a current in the presence of a transverse magnetic field. If there is a current i in a plate-shaped conductor (Fig. 2.16), then two opposite points a and b will be at the same potential in the absence of a magnetic field. When a field H acts at right angles to the plate, the current path is distorted, and an emf e_H is developed between a and b. The magnitude of this Hall emf is proportional to the product of the current and the field:

$$e_H = \frac{R_H \, i \, H}{t} \text{ volts} \tag{2.16}$$

where t is the thickness of the plate in cm, and R_H, the Hall constant, is a property of the material. The effect occurs both in metals and semiconductors, but is very much larger in the latter. R_H is typically of the order of 10^{-5} volt-cm/ampere-oersted in a semiconductor.

If i is kept constant, then e_H is a measure of field strength, after calibration in a known field. If i is alternating, then e_H is also alternating and can be amplified before measurement. The sensing element, called a Hall probe, is usually the semiconductor InSb in commercial instruments. The probe can be made very small ($\frac{1}{16} \times \frac{1}{16} \times 0.020$ in., for example) and is fixed to the end of a thin rod connected by a light cable to the unit containing the current source, amplifier,

* Discovered in 1879 by the American physicist Edwin Hall (1855–1938) when he was a student under H. A. Rowland at Johns Hopkins University.

and indicating meter. Multirange instruments are available for measurement of fields ranging from fractions of an oersted to tens of kilo-oersteds. Because of the small size of the probe, these instruments are well suited to the measurement of fields in confined regions or of field gradients. Alternating fields may also be measured.

The relation between Hall emf and field is not linear to any better than 1 percent and may become markedly nonlinear at fields greater than 20 kOe. The instrument must be calibrated regularly at zero field and at one or more known fields.

Note that a Hall probe has a continuous voltage output when placed in a constant magnetic field; there is no need to change the flux through the sensing element in order to make a measurement, as there is with a search coil.

Other methods

Magnetic fields may also be measured by:

1. *Rotating coil.* If a search coil is rapidly and continuously rotated in a field, an alternating emf will be generated that is proportional to the field strength. The coil is located at the end of a long shaft. A dc output can be obtained by means of a commutator and brushes.
2. *Moving-magnet magnetometer.* See Section 2.7.
3. *Saturable core.* See Section 13.5.
4. *Nuclear magnetic resonance.* See Section 12.7.
5. *Josephson effect.* This effect consists of a current across a very narrow insulating gap between two superconductors. The magnitude of the current is extraordinarily sensitive to the presence of a magnetic field. As a means of measuring a field this effect is so sensitive (fields of the order of 10^{-9} Oe can be detected) that its use is restricted to certain unusual applications [2.28].

Devices for measuring magnetic fields are sometimes called *gaussmeters*, sometimes *magnetometers*.

2.5 MAGNETIC MEASUREMENTS IN CLOSED CIRCUITS

Lines of magnetic induction B are continuous and form closed loops. The region occupied by these closed loops is called a *magnetic circuit.* Sometimes the flux follows a well-defined path, sometimes not. When the flux path lies entirely within strongly magnetic material, except possibly for a small amount of leakage flux, the circuit is said to be closed. If the flux passes partially through "nonmagnetic" material, the circuit is said to be open.

An important property of a closed and homogeneous magnetic circuit is that the material comprising it can be magnetized without the production of magnetic poles. As we shall see, this circumstance considerably simplifies the determination of the field H which causes the magnetization.

The simplest example of a closed magnetic circuit is a ring magnetized circumferentially, and we will now consider how the normal induction curve and hysteresis loop of a ring specimen can be determined. A search coil of N turns is wound on the ring, and over this is placed a magnetizing winding of n turns, both windings extending completely around the circumference. A current i through this winding subjects the material of the ring to a field H, given by

$$H = \frac{1.257ni}{L} \text{ Oe}, \tag{1.13}$$

where L is now the mean circumference of the ring in centimeters.

The method is ballistic. A sudden change in i produces a sudden change in H, and therefore in M, and therefore in B, because $B = H + 4\pi M$. If the cross-sectional area of the ring is A cm^2, then the flux change $\Delta\phi(= A\,\Delta B)$ produces a deflection d of the ballistic galvanometer or fluxmeter connected to the search coil, and $\Delta\phi$ can be found from d by Eq. (2.10).

The *normal induction curve* may be determined with the circuit of Fig. 2.17, which is designed to have a minimum of switches and resistors. Switch S_2 is needed only for the determination of the hysteresis loop and is kept closed for the present. S_1 thus acts as a simple reversing switch, and R_1 controls the current in the magnetizing winding. The procedure is as follows:

1. Demagnetize the specimen, unless it is definitely known to be demagnetized. Do this by setting H at a high value (at least five times the coercive force), and reduce it slowly to zero, while continuously reversing H by means of S_1.
2. Set H at a low value, say H_1, and reverse it several times. Then zero the fluxmeter, depress S_3, reverse H_1 once again, and note the fluxmeter deflection. The value of ΔB so found must be divided by *two*, in order to obtain the value of B_1.
3. Set H at a higher value H_2 and repeat, so obtaining a set of points on the normal induction curve.

This method gives better results than one in which H is changed from 0 to H_1, from H_1 to H_2, etc., a process in which errors tend to accumulate. The reversal of H_1 in step 2 above carries the material repeatedly around a minor hysteresis loop. It is then said to be in a cyclic state and, after several reversals, the tips of the minor loops will become reproducible.

In carrying out the above procedure, it is important that each successive value of H be larger than the preceding one. If H is increased accidentally from H_1 to H_3, for example, and then reduced to H_2 and cycled at H_2, quite erroneous results will be obtained for the value of B corresponding to H_2. (This incorrect procedure will place the material on a minor loop not centered on the origin, and the reader will find it instructive to sketch the path which this procedure will make the material follow on a B,H diagram.) Following such an accidental increase to H_3, the material must be demagnetized from H_3, after which H can be increased to H_2 for a determination of B.

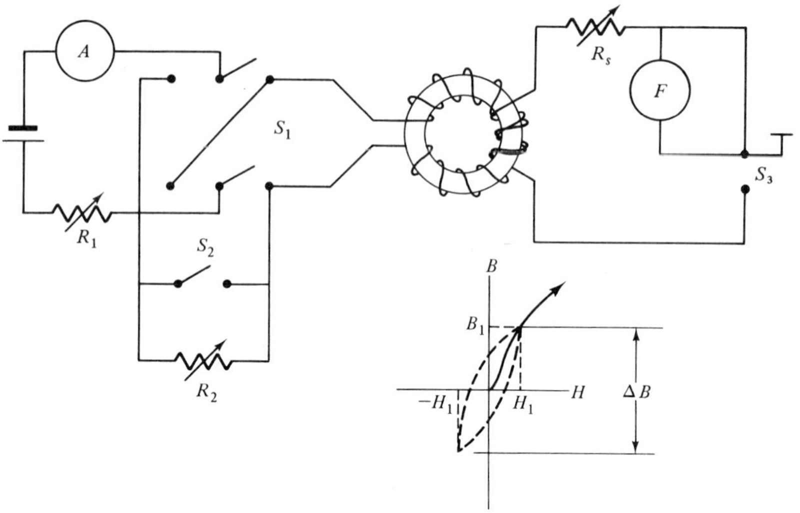

Fig. 2.17 Circuit for magnetic measurements on a ring specimen. A = ammeter, F = fluxmeter.

To determine the *hysteresis loop* of Fig. 2.18 we must operate both switches, S_1 and S_2. When S_1 is thrown to the right, we obtain the maximum value i_m of the magnetizing current, corresponding to the maximum field H_m at the tip of the hysteresis loop, whether S_2 is open or closed. When S_1 is thrown to the left, the current is reversed and equal to or less than i_m, depending on whether S_2 is closed or open, respectively. In the following procedure, the right-hand position of S_1 corresponds to the right-hand (positive) tip of the hysteresis loop:

1. Demagnetize the specimen.
2. Close S_2 and throw S_1 to the right.
3. Adjust R_1 to produce the maximum current i_m desired. R_1 remains constant thereafter.
4. Cycle the specimen several times from tip to tip of the loop by reversing S_1. Reverse once more and note the fluxmeter deflection. This deflection gives a value of ΔB equal to $2B_m$ and establishes the position of the loop tip.
5. With S_1 on the right, open S_2, and then throw S_1 to the left. The field thus changes from H_m to $-H_1$. The fluxmeter deflection gives a value of ΔB which is subtracted, *without division by two*, from B_m to obtain B_1.
6. Close S_2. This changes the field from $-H_1$ to $-H_m$, and brings the specimen to the negative tip of the loop. Throw S_1 to the right to bring the specimen back to the starting point at the positive tip.
7. Decrease R_2 and repeat steps 5 and 6. In this way, points between B_r and $-B_m$ are obtained. B_r itself is found simply by moving S_1 from the right-hand to the open position.

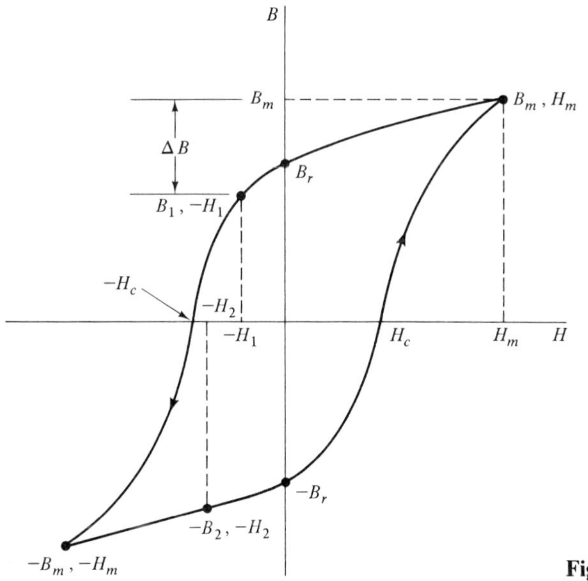

Fig. 2.18 Hysteresis loop.

8. Points between $-B_m$ and $-B_r$ are obtained by operating from the negative tip. With S_2 closed, throw S_1 from right to left. Then open S_2 to change the field from $-H_m$ to $-H_2$, say. The fluxmeter deflection gives a value of ΔB which is measured upward from $-B_m$ to obtain $-B_2$.

In this way, the entire left-hand portion of the loop may be determined. The right-hand portion is drawn in by symmetry, and the coercive force H_c found graphically. Note that the various switching operations must always result in counterclockwise travel of the material around the hysteresis loop. If a mistake in switching is made, throw S_1 to the right, close S_2, and reverse S_1 several times in order to cycle the material around the loop being determined.

If a large number of measurements is to be made by the methods outlined above, faster operation can be achieved by replacing the variable resistors R_1 and R_2 of Fig. 2.17 with a set of fixed resistors, each of which can be separately switched in or out of the circuit by its own switch.

Ring specimens, although free from magnetic poles, have some disadvantages. First, windings must be applied to each specimen to be tested, and this can be time-consuming. Second, some specimen shapes simply cannot be formed into a satisfactory ring. For example, if a wire or rod is bent into a circle, the joint, even though welded, is not in the same magnetic state as the rest of the ring and can lead to erroneous results. Sheet material, on the other hand, is quite satisfactory; rings can be stamped out, and a number of these stacked together to form a composite, laminated ring. (However, sheet is usually magnetically anisotropic; it has different properties in directions at different angles to the direction in which the sheet was originally rolled. Therefore, measurements on rings cut from

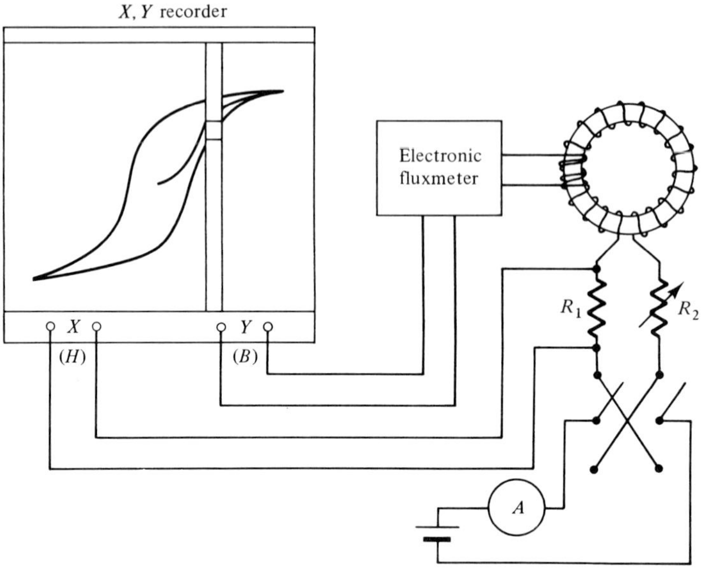

(a) Hysteresigraph

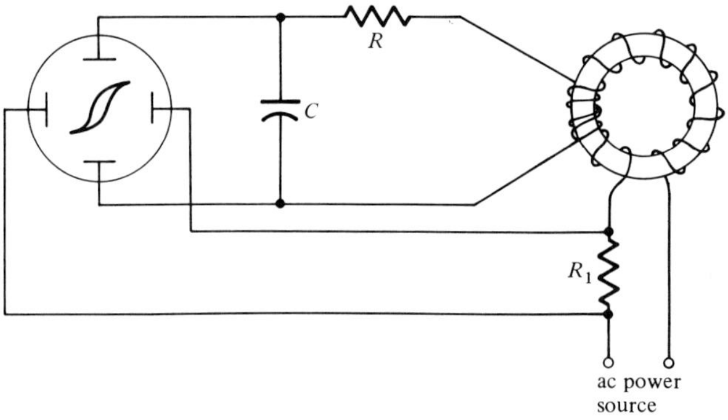

(b) Loop tracer

Fig. 2.19 Circuits for automatic recording of B,H curves and loops.

such sheet reveal only the average properties over the various directions in the sheet.)

Some applications of magnetic materials require that the material be operated in the form of a ring. Here there is no problem, other than that of applying the windings, because the material to be tested is in exactly the form in which it is to be used.

The determination and plotting of B,H curves and loops by the point-by-point methods outlined above can be very time-consuming. It is possible to record the data automatically and quickly, although more elaborate equipment is necessarily required. There are two main methods (Fig. 2.19):

a) *Hysteresigraph.* This method is essentially "static," in that the hysteresis loop is traversed slowly, in a time of the order of one minute. The search coil is connected through an electronic fluxmeter to the $Y(B)$ axis of an X,Y recorder; the terminals of a fixed resistor R_1, in series with the magnetizing winding, are connected to the X (H) axis. The output of the fluxmeter is proportional to ΔB, and the voltage drop across R_1 is proportional to the current in the magnetizing winding and hence to the field H. As H is varied continuously, by means of R_2, the pen of the recorder automatically traces the curve.

b) *Loop tracer.* Often the property of greatest interest is the dynamic behavior of a material, i.e., its response to alternating magnetization at a frequency of the order of 60 hertz (Hz) or higher (1 Hz = 1 cycle/sec). The dynamic hysteresis loop can be observed by connecting the search coil through a resistor R to a capacitor C and then to the vertical-deflection plates of an oscilloscope. This "RC circuit" integrates the emf induced in the search coil; the voltage across the capacitor at any instant is then proportional to the value of ΔB at that instant. The voltage across the resistor R_1 provides the H deflection of the oscilloscope beam. Figure 12.12 shows loops recorded in this way.

These two methods are by no means confined to measurements on ring specimens; they are also applicable to many of the other measurements described in this chapter.

2.6 DEMAGNETIZING FIELDS

Before considering magnetic measurements in open circuits, we must examine the nature of the fields involved. A magnetic field H can be produced either by electric currents or by magnetic poles. If due to currents, the lines of H are continuous and form closed loops; for example, the H lines around a current-carrying conductor are concentric circles. If due to poles, on the other hand, the H lines begin on north poles and end on south poles.

Suppose a bar is magnetized by a field applied from left to right, and subsequently removed. Then a north pole is formed at the right end, and a south pole at the left, as shown in Fig. 2.20 (a). We see that the H lines, radiating out from the N pole and ending at the S pole, constitute a field *inside the magnet* which acts from N to S and which therefore tends to *demagnetize* the magnet. This self-demagnetizing action of a magnetized body is important, not only because of its bearing on magnetic measurements, but also because it is involved in a great deal of magnetic theory. We will therefore consider it in detail.

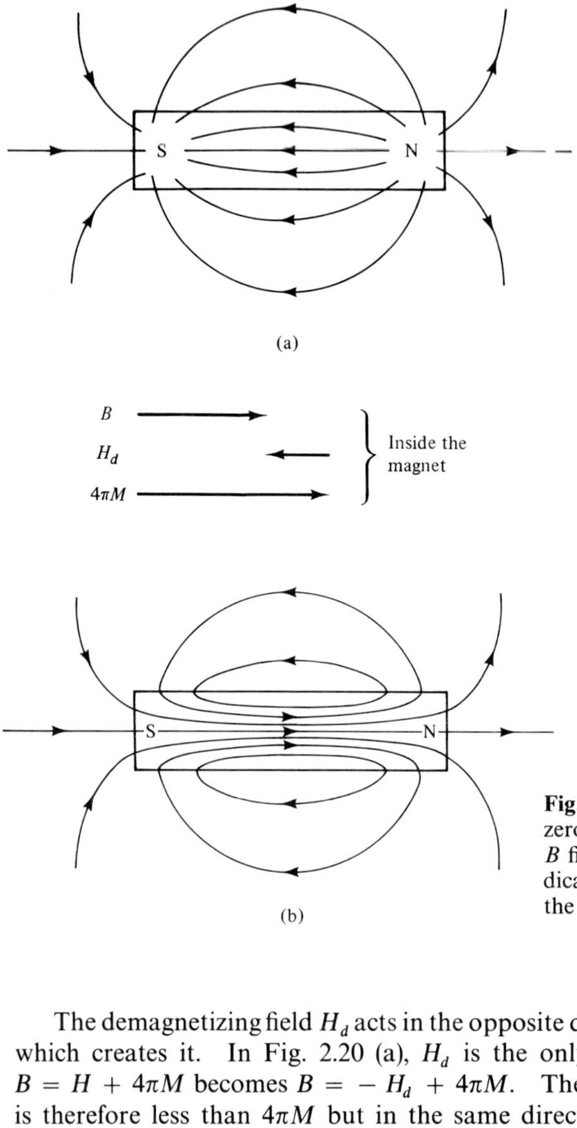

Fig. 2.20 Fields of a bar magnet in zero applied field: (a) H field, and (b) B field. The vectors in the center indicate the values of these quantities at the center of the magnet.

The demagnetizing field H_d acts in the opposite direction to the magnetization M which creates it. In Fig. 2.20 (a), H_d is the only field acting, and the relation $B = H + 4\pi M$ becomes $B = -H_d + 4\pi M$. The induction B inside the magnet is therefore less than $4\pi M$ but in the same direction, because H_d can never exceed $4\pi M$ in magnitude. These vectors are indicated in Fig. 2.20 (b), in which the B field of the magnet is sketched. Note that *lines of B are continuous* and are directed from S to N inside the magnet. Outside the magnet, $B = H$, and the external fields in Figs. 2.20 (a) and (b) are therefore identical. The magnet of Figure 2.20 (b) is in an open magnetic circuit, because part of the flux is in the magnet and part is in air.

As Fig. 2.20 (b) shows, the magnetization of a bar magnet is not uniform: the lines diverge toward the ends, so that the flux density there is less than in the center. This results from the fact that H_d is stronger near the poles, and Fig. 2.21 (a) shows why: the dashed lines show the H field due to each pole separately, and

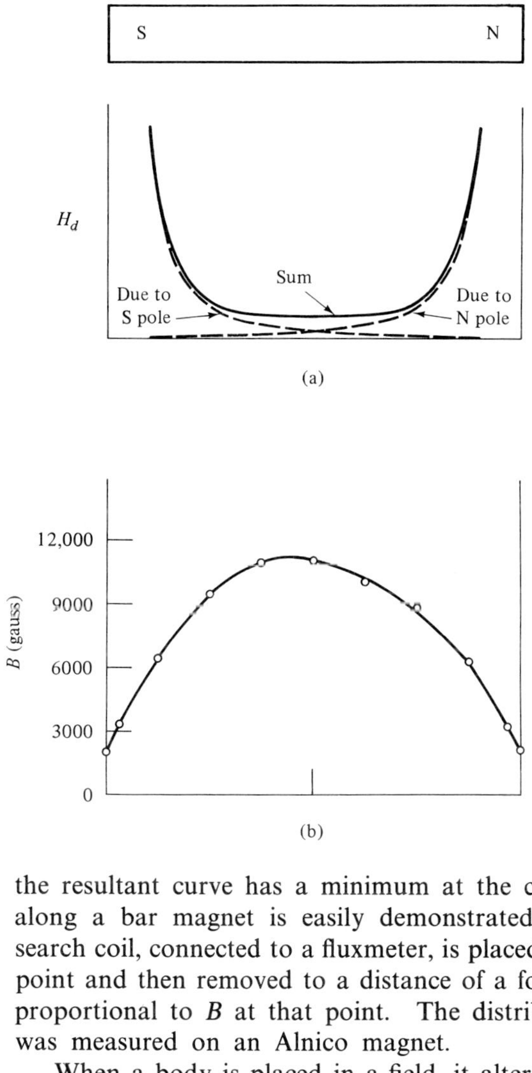

Fig. 2.21 Variation along the length of a bar magnet of (a) the demagnetizing field H_d on the axis (schematic), and (b) the induction B (values measured on a cylindrical magnet, 0.75 in. in diameter and 5.75 in. long).

the resultant curve has a minimum at the center. The variation in induction along a bar magnet is easily demonstrated experimentally. A closely fitting search coil, connected to a fluxmeter, is placed around the magnet at a particular point and then removed to a distance of a foot or so; the resulting deflection is proportional to B at that point. The distribution of B shown in Fig. 2.21 (b) was measured on an Alnico magnet.

When a body is placed in a field, it alters the shape of that field. Thus, in Figure 2.22, suppose that (a) is a uniform field, e.g., the field of a solenoid. It may be regarded as either an H field or a B field. The B field of a magnet in zero applied field is shown in (b). The B field in (c) is the vector sum of the fields in (a) and (b). The flux tends to crowd into the magnet, as though it were more permeable than the surrounding air; this is the origin of the term *permeability* (μ). At points outside the magnet near its center, the field is actually reduced. The same general result is obtained if the body placed in the field is originally unmagnetized, because the field itself will produce magnetization. Figure 2.22 applies to a material like iron, with $\mu > 1$. The opposite effect occurs for a diamagnetic body: the flux tends to avoid the body, so that the flux density is greater outside than inside. [Because lines of B are continuous, the B lines of Fig. 2.22 (c) must

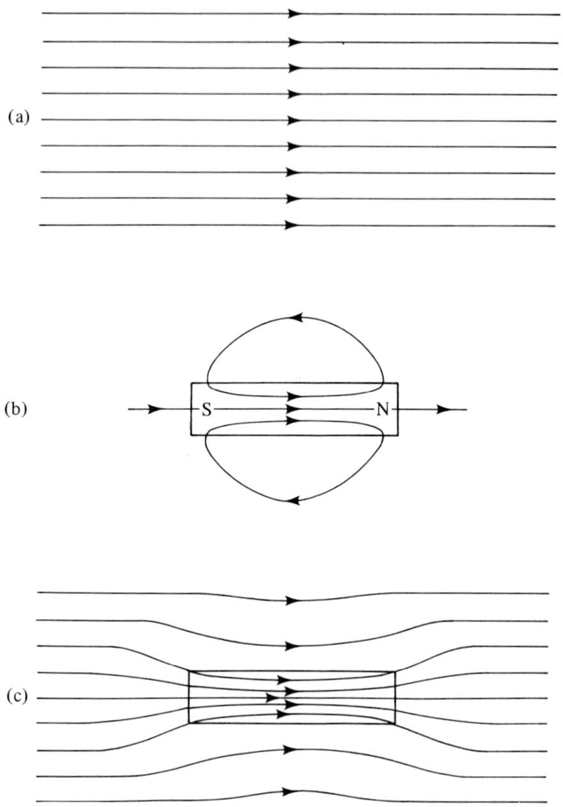

Fig. 2.22 Result of placing a magnetized body, having $\mu > 1$, in an originally uniform field.

close on themselves outside the drawing. If the field in which the body was placed was generated by a solenoid, then the manner in which the lines close is suggested in Fig. 1.8.]

The extent to which a body, originally unmagnetized, disturbs the field in which it is placed depends on its permeability. For strongly magnetic materials (ferro- and ferrimagnetic) the disturbance is considerable; for weakly magnetic materials it is practically negligible. Steel ships produce appreciable disturbance of the earth's magnetic field at a considerable distance from the ship, and the magnetic mines used in warfare take advantage of this fact. As the ship passes, the change in field at the position of the mine is sensed by some kind of magnetometer which then actuates an electrical circuit to detonate the mine [2.10].

If a high-permeability ring is placed in a field, it tends to shield the space inside from the field, somewhat as suggested by Fig. 2.23 (a). The field lines tend to follow the magnetic material around the perimeter and emerge from the other side. The difficulty with this "explanation" of shielding is that it suggests that the part of the ring normal to the field plays a primary role in diverting the flux. Actually, a field normal to the center of a flat plate passes right through, undevi-

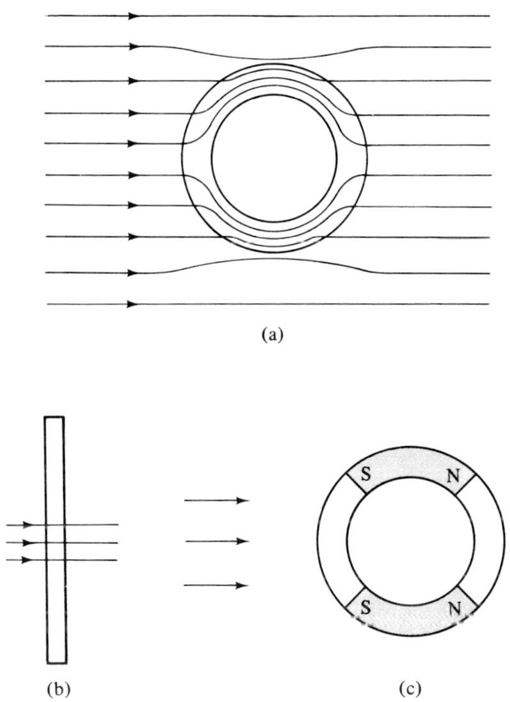

Fig. 2.23 Magnetic shielding.

ated, as shown in (b). It is the sides of the ring parallel to the applied field that have the greatest effect. These become magnetized, with poles as shown in (c), and they reduce the field inside the ring by exactly the same mechanism as that by which the bar magnet of Fig. 2.22 (c) reduces the field in the region adjacent to its center. (The portions of the ring normal to the field acquire little magnetization, because of their very large demagnetizing factor. See below.) Two or more thin concentric rings, separated by air gaps, are more effective than one thick ring. Certain components of some electronic circuits need to be shielded from stray magnetic fields; this is done by enclosing them in a thin sheet of a high-permeability material, usually a Ni-Fe alloy.

Returning to the bar magnet of Fig. 2.20 (b), we might ascribe the nonuniformity of the induction inside the magnet to the fact that lines of B "leak out" of the sides. If we taper the magnet toward each end to make up for this leakage, the induction can be made uniform throughout. It may be shown [2.11], although with some difficulty, that the correct taper to achieve this result is that of an ellipsoid (Fig. 2.24). If an unmagnetized ellipsoid is placed in a uniform magnetic field, it becomes magnetized uniformly throughout; the uniformity of M and B is due to the uniformity of H_d throughout the volume. This uniformity can be achieved only in an ellipsoid. (These statements require qualification for ferro- and ferrimagnetic materials, because they are made up of *domains*. Even an ellipsoidal specimen of such a material cannot be uniformly magnetized, although a condition of uniform M is approached as the domain size becomes small relative to the specimen size. See Sections 4.1 and 7.2.)

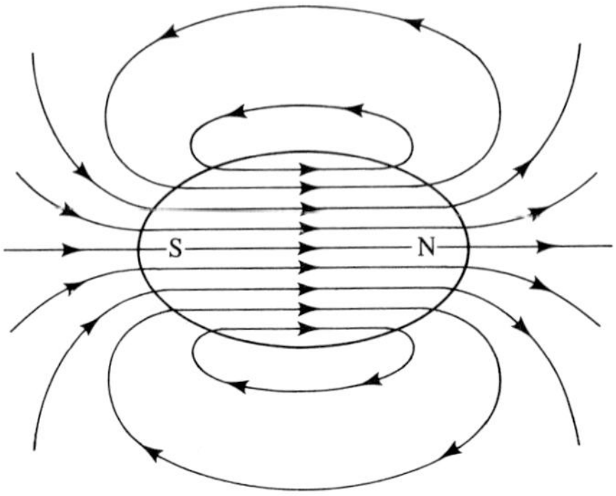

Fig. 2.24 The B field of an ellipsoidal magnet in zero applied field.

The demagnetizing field H_d of a body is proportional to the magnetization which creates it:

$$\boxed{H_d = N_d M,} \tag{2.17}$$

where N_d is the *demagnetizing factor* or coefficient. N_d depends mainly on the shape of the body, and can be calculated exactly only for an ellipsoid. The general case is too complex to treat here, but we will calculate N_d for the special case of a sphere.

Uniform magnetization of a sphere by an applied field H_a will cause north and south poles to appear on the surface, as indicated in Fig. 2.25, and these free poles produce the demagnetizing field H_d. (The term "free poles," often encountered in the literature of magnetism, merely means poles which are not compensated by other poles of opposite sign in the immediate neighborhood; they are not really free, in the sense of being separable from the magnet on which they reside. They have nothing to do with the really free poles, called monopoles, which some theoretical physicists have postulated as existing in very small concentrations among the debris of nuclear disintegrations; an experimental search for these monopoles has so far been unsuccessful [2.25].) Let ρ_s be the pole density on the surface of the sphere and a its radius. To find ρ_s we note that the pole strength per unit area on the equatorial cross-section (Fig. 2.26) is M, if M is the magnetization of the sphere. The number of poles on an annular strip of radius r and width dr on this section is $2\pi r\ M dr$. Since M is uniform throughout the sphere, there will be the same number of poles on an annular strip on the surface, formed by projecting the equatorial strip up to the surface. The area of this surface strip is $(2\pi a \sin\theta)(a\,d\theta)$. Therefore, the surface pole density is

$$\rho_s = \frac{2\pi r\ M\ dr}{2\pi a^2 \sin\theta\ d\theta}.$$

But $r = a \sin\theta$, and $dr = a \cos\theta\ d\theta$. Therefore.

$$\rho_s = \frac{M(a \sin\theta)(a \cos\theta)\,d\theta}{a^2 \sin\theta\ d\theta},$$

$$\rho_s = M \cos\theta. \tag{2.18}$$

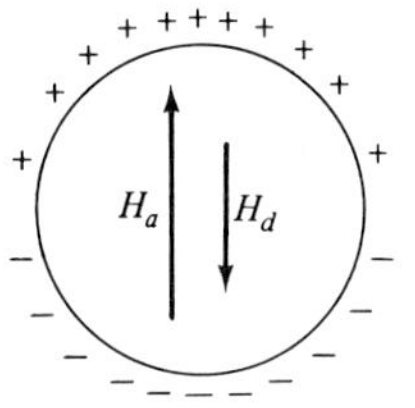

Fig. 2.25 Poles created on surface of magnetized sphere. North and south poles are here represented by plus and minus signs, respectively.

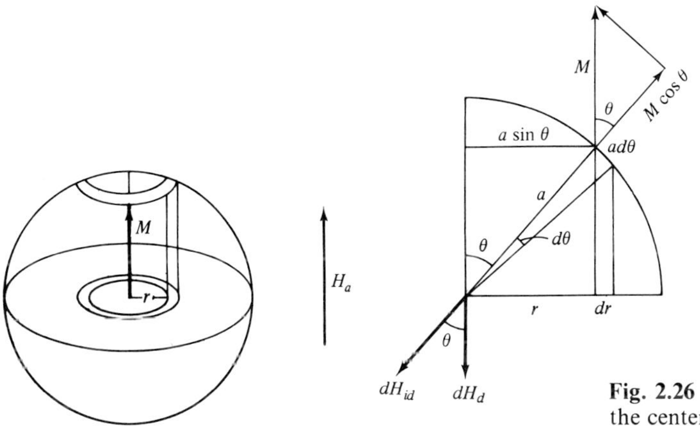

Fig. 2.26 Demagnetizing field at the center of a sphere.

This shows, as might be expected, that the surface density of poles is not uniform: it decreases from a maximum of M at the top and bottom to zero at the sides.

The inclined demagnetizing field dH_{id} at the center of the sphere, due to poles on annular strips at the top *and* bottom of the sphere, is given by the force on a unit pole placed at the sphere center:

$$dH_{id} = \frac{[(2\rho_s)(2\pi a^2 \sin\theta \, d\theta)][1]}{a^2}.$$

The demagnetizing field antiparallel to M is

$$dH_d = dH_{id} \cos\theta.$$

Substituting, we obtain

$$H_d = \int dH_d = 4\pi M \int_0^{90^\circ} \cos^2\theta \sin\theta \, d\theta,$$

$$H_d = \frac{4\pi M}{3}.$$

Therefore, the demagnetizing factor is

$$N_d = \frac{4\pi}{3}. \qquad (2.19)$$

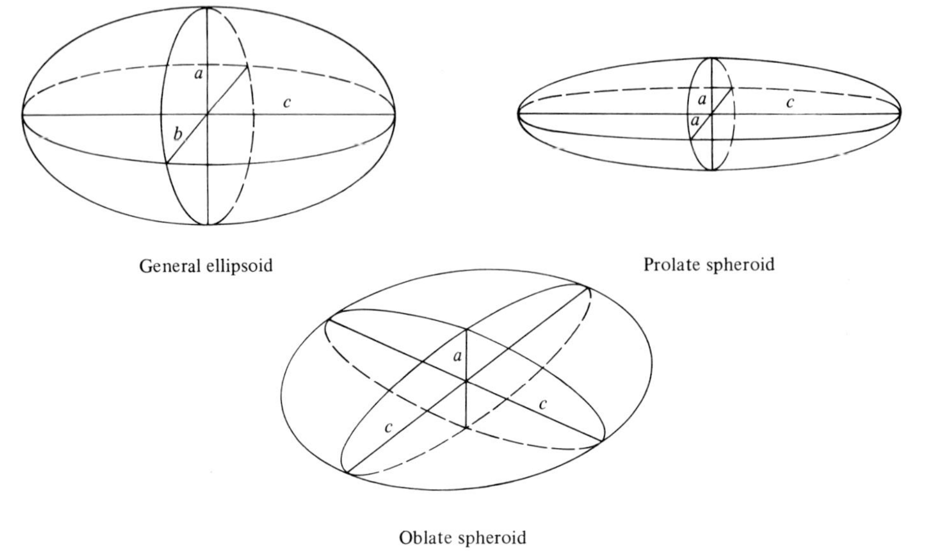

Fig. 2.27 Ellipsoids.

Demagnetizing factors for the general ellipsoid have been given by Stoner [2.12] and Osborn [2.13]. The general ellipsoid has three unequal axes $2a$, $2b$, $2c$, and a section perpendicular to any axis is an ellipse (Fig. 2.27). Of greater practical interest is the ellipsoid of revolution, more properly called a spheroid. A prolate spheroid is formed by rotating an ellipse about its major axis $2c$; then $a = b$, and the resulting solid is rod-shaped. Rotation about the minor axis $2a$ results in the disc-shaped oblate spheroid, with $b = c$. If N_a, N_b, and N_c are particular values of the demagnetizing factor N_d for magnetization along the a, b, or c axis, then $N_a + N_b + N_c = 4\pi$ for any ellipsoid.

Equations for the demagnetizing factors of ellipsoids of revolution have been given by Maxwell [2.11] and equations, tabular data, and graphs by Stoner [2.12]. The results are as follows:

1. *Prolate spheroid*, or rod. $a = b \neq c$. Put $c/a = r$. Then,

$$N_c = \frac{4\pi}{(r^2 - 1)}\left[\frac{r}{\sqrt{r^2 - 1}} \ln\left(r + \sqrt{r^2 - 1}\right) - 1\right], \tag{2.20}$$

$$N_a = N_b = \frac{4\pi - N_c}{2}. \tag{2.21}$$

When r is very large (very long rod), then

$$N_a = N_b \approx 2\pi \tag{2.22}$$

$$N_c \approx \frac{4\pi}{r^2}(\ln 2r - 1). \tag{2.23}$$

N_c approaches zero as r approaches infinity.

(Example: For $r = 10$, $N_a = N_b = 6.16$ and $N_c = 0.256$.)

2. *Oblate spheroid*, or disc. $a \neq b = c$, and $c/a = r$.

$$N_a = \frac{4\pi r^2}{r^2 - 1}\left(1 - \sqrt{\frac{1}{r^2 - 1}}\sin^{-1}\frac{\sqrt{r^2 - 1}}{r}\right), \tag{2.24}$$

$$N_b = N_c = \frac{4\pi - N_a}{2}. \tag{2.25}$$

When r is very large (very thin disc), then

$$N_a \approx 4\pi, \tag{2.26}$$

$$N_b = N_c \approx \pi^2/r. \tag{2.27}$$

N_b and N_c approach zero as r approaches infinity.

(Example: For $r = 10$, $N_a = 10.82$ and $N_b = N_c = 0.88$.)

Specimens often encountered in practice are the cylindrical rod magnetized along its axis and the cylindrical disc magnetized in the plane of the disc. Such specimens are never uniformly magnetized except when completely saturated. The demagnetizing field varies from one point to another in the specimen and cannot be exactly calculated. Instead, H_d must be measured, and different investigations are not usually in good agreement. (For the method of measurement see, for example, Vigoureux and Webb [2.14] and Appendix 11.) It is also found experimentally that the demagnetizing factor depends somewhat on the permeability of the specimen as well as its shape. Values of demagnetizing factors have been selected by Bozorth from a number of investigations and are shown in Fig. 2.28 in the form of $N_d/4\pi$.

(In the *mks* system, if B, H, and M are related by the equation $B = \mu_0 H + M$, then demagnetizing factors are $1/4\pi$ times the values in the *cgs* system. Thus N_d for a sphere is $\frac{1}{3}$ and $N_a + N_b + N_c = 1$ for any ellipsoid.)

Some important conclusions can be drawn from our derivation of the demagnetizing field of a magnetized sphere, because Eq. (2.18) for the surface pole density ρ_s is of general validity. It applies to any body if θ is taken as the angle between M and the normal to the surface at the point where ρ_s is to be evaluated. Note that $M \cos\theta$ is the component of the magnetization normal to the surface inside the body, and that M is zero outside. Therefore, *the pole density produced at a surface equals the discontinuity in the normal component of M at that surface.* If $\mathbf{n}$ is a unit vector normal to the surface, then

$$M \cos\theta = \mathbf{M}\cdot\mathbf{n} = \rho_s. \tag{2.28}$$

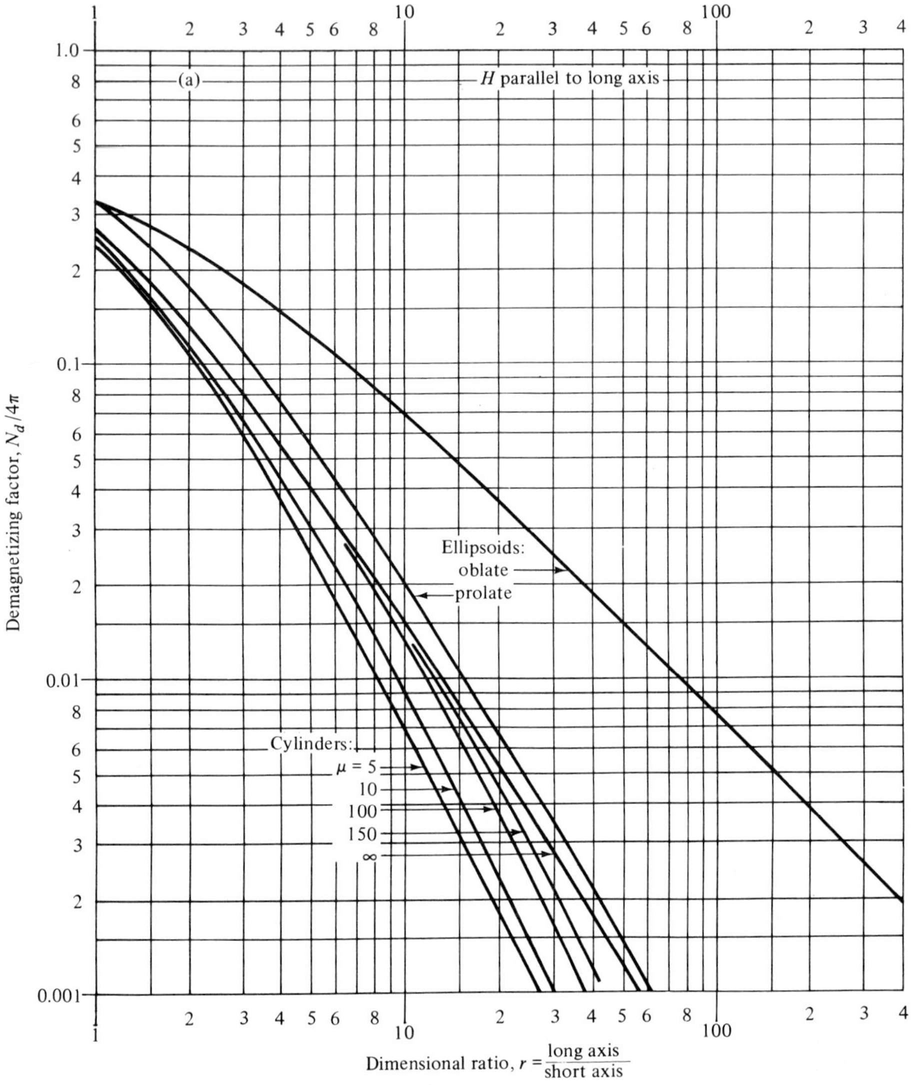

Fig. 2.28 Demagnetizing factors $N_d/4\pi$ of cylinders and ellipsoids of revolution magnetized along the long axis, according to Bozorth [G-4, p. 846].

Note that this agrees with one of the definitions of M as the pole strength per unit area of cross section. (The sign of the pole is given by the following rule: if the normal component of M decreases as a surface is crossed in the direction of M, the polarity of the surface is positive, or north.) Free poles can also be produced at the interface between two bodies magnetized by different amounts and/or in different directions. If $\mathbf{M}_1$ and $\mathbf{M}_2$ are the magnetizations of the two bodies, then the discontinuity in the normal component is

$$\mathbf{M}_1 \cdot \mathbf{n} - \mathbf{M}_2 \cdot \mathbf{n} = \rho_s. \tag{2.29}$$

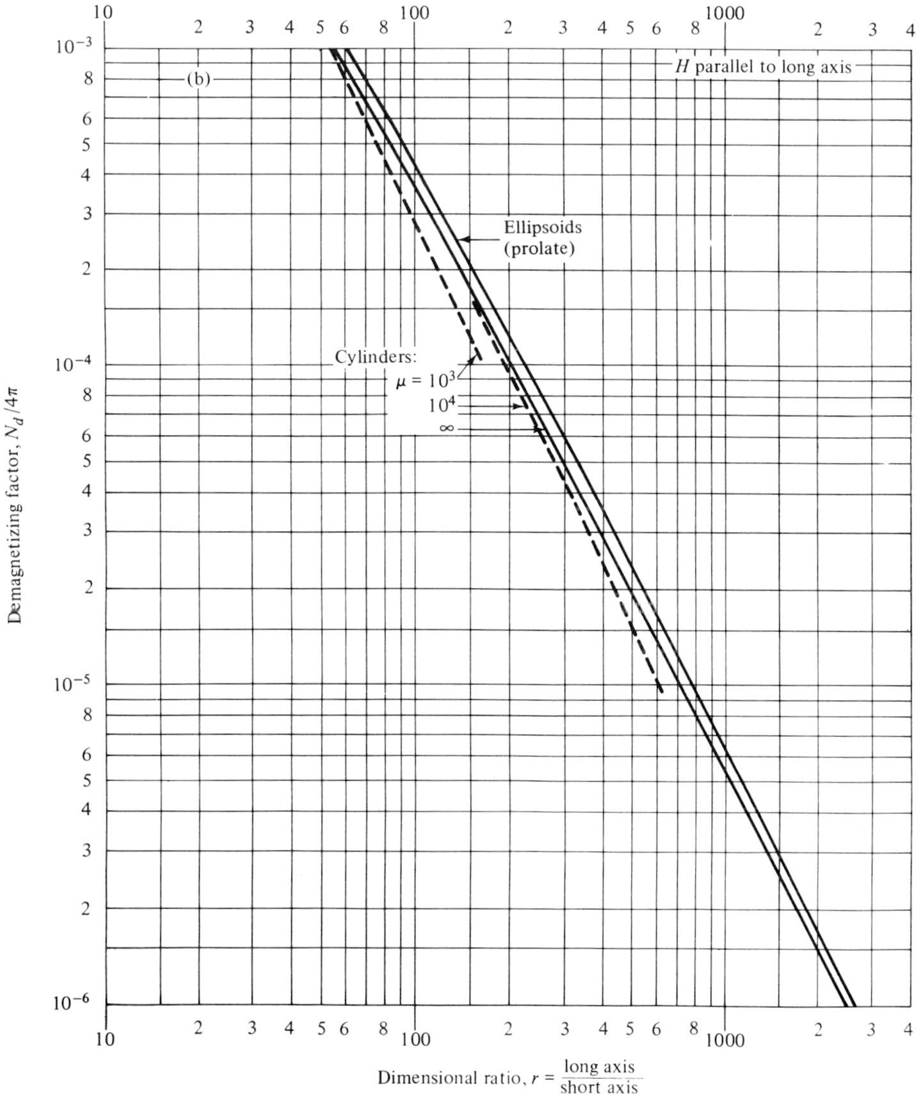

Fig. 2.28 *(continued)*

This is an important principle, which we shall need later.

(We might also note that, at the interface between two bodies or between a body and the surrounding air, certain rules govern the directions of **H** and **B** at the interface:

1. The tangential components of **H** on each side of the interface must be equal.
2. The normal components of **B** on each side of the interface must be equal.

These conditions govern the angles at which the *B* and *H* lines meet the air-body interfaces depicted in Fig. 2.20, for example.)

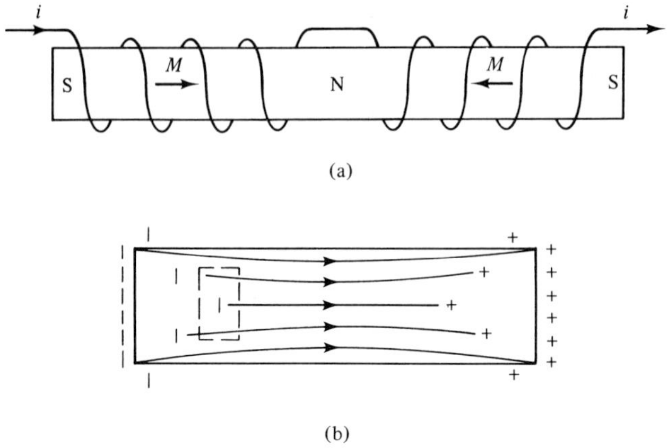

Fig. 2.29 Internal poles.

Free poles may exist, not only at the surface of a body, but also in the *interior*. For example, on a gross scale, if a bar has a winding like that shown in Fig. 2.29 (a), south poles will be produced at each end and a north pole in the center, for a current i in the direction indicated. On a somewhat finer scale, free poles exist inside a cylindrical bar magnet, as very approximately indicated in Fig. 2.29 (b). The condition for the existence of interior poles is *nonuniform magnetization*. An ellipsoidal body can be uniformly magnetized, and it has free poles only on the surface, unless it contains domains. A body of any other shape, such as a cylindrical bar, cannot be uniformly magnetized because the demagnetizing field is not uniform, and it always has interior as well as surface poles. Nonuniformity of magnetization means that there is a net outward flux of M from a small volume element, i.e., the divergence of $\mathbf{M}$ is greater than zero. But if there is a net outward flux of M, there must be free poles in the volume element to supply this flux. Such a volume element is delineated by dashed lines in Fig. 2.29 (b), in which lines of M have also been drawn, going from south to north poles. (For clarity, the lines of M connecting surface poles on the ends have been omitted.) If ρ_v is the volume pole density (pole strength per unit volume), then

$$\text{div } \mathbf{M} = \nabla \cdot \mathbf{M} = \frac{\partial M_x}{\partial x} + \frac{\partial M_y}{\partial y} + \frac{\partial M_z}{\partial z} = -\rho_v. \tag{2.30}$$

On the axis of a bar magnet, $\mathbf{M}$ decreases in magnitude from the center toward each end, as indicated qualitatively by the density of lines in Fig. 2.29 (b). Suppose the axis of the magnet is the x axis, and we assume for simplicity that $\mathbf{M}$ is uniform over any cross section. Then only the term $\partial M_x/\partial x$ need be considered. Between the center of the magnet and the north end $\partial M_x/\partial x$ becomes increasingly negative, which means that ρ_v is positive and that it increases in magnitude toward the end, more or less as depicted in Fig. 2.29 (b). Although the interior pole distributions in Figs. 2.29 (a) and (b) differ in scale, both are rather macroscopic; we shall see in Chapter 9 that interior poles can also be distributed on a microscopic scale.

The general derivations of Eqs. (2.28) and (2.30) may be found in any intermediate-level text on electricity and magnetism.

In summary:

1. Lines of B are always continuous, never terminating.
2. a) If due to currents, lines of H are continuous.
   b) If due to poles, lines of H begin on north poles and end on south poles.
3. At an interface,
   a) the normal component of **B** is continuous,
   b) the tangential component of **H** is continuous, and
   c) the discontinuity in the normal component of **M** equals the surface pole density ρ_s at that interface.
4. The negative divergence of **M** at a point inside a body equals the volume pole density at that point.
5. The magnetization of an ellipsoidal body is uniform, and free poles reside only on the surface, unless the body contains domains. See Section 9.5.
6. The magnetization of a nonellipsoidal body is nonuniform, and free poles exist on the surface and in the interior. (The saturated state constitutes the only exception to this statement. A saturated body of any shape is uniformly magnetized and has poles only on its surface.)

2.7 MAGNETIC MEASUREMENTS IN OPEN CIRCUITS

Measurements of this type are made by the following methods: ballistic, extraction, vibrating-sample, and magnetometer.

Ballistic method

In this method, usually applied to rods, a search coil is wound on the center of the specimen, which in turn is placed in the center of a solenoid (Fig. 2.30) or in the air gap of an electromagnet. The solenoid and search coil are connected to the magnetizing and fluxmeter circuits, respectively, exactly as in Fig. 2.17, and the operations required to obtain the normal induction curve and hysteresis loop are exactly the same as for a ring specimen. B is found from fluxmeter deflections, as before. However, the applied field H_a due to the solenoid must

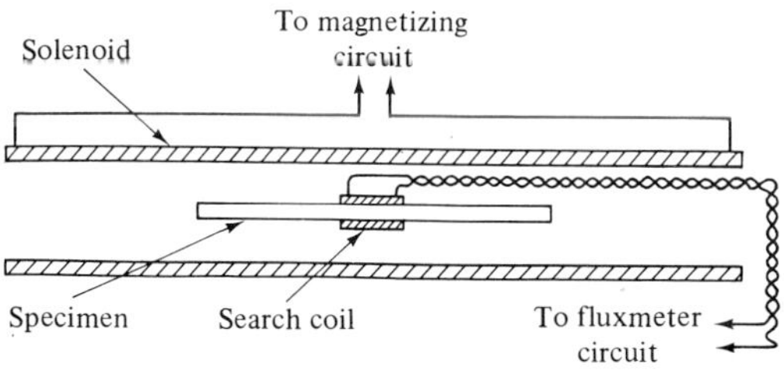

Fig. 2.30 Ballistic methods for rods. Windings are indicated by cross-hatching.

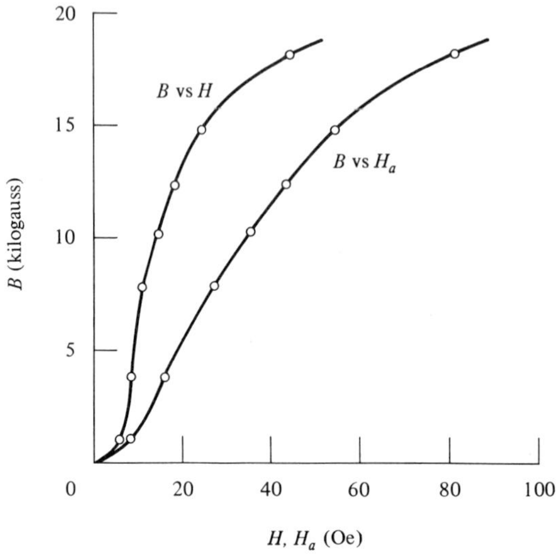

Fig. 2.31 Induction curves of cold-worked iron.

now be corrected for the demagnetizing effect to obtain the true field H acting on the specimen:

$$H = H_a - H_d = H_a - N_d M. \tag{2.31}$$

This value of H applies only to the center of the specimen, because the value of N_d as given, for example, in Fig. 2.28 applies only to the center. (Such values of N_d are accordingly called ballistic demagnetizing factors.) Since B rather than M is measured in this method, H and M are evaluated as follows:

$$B = H + 4\pi M = H_a - N_d M + 4\pi M,$$

$$M = \frac{B - H_a}{4\pi - N_d}, \tag{2.32}$$

$$H = H_a - N_d \left(\frac{B - H_a}{4\pi - N_d} \right) = H_a - \frac{(B - H_a)}{(4\pi/N_d) - 1}. \tag{2.33}$$

These relations are illustrated by the experimental data in Table 2.2, obtained from a rod of commercially pure iron in the cold-worked condition. The rod was 9.44 in. long and 0.27 in. in diameter and hence had a length/diameter ratio of 35.0. From Fig. 2.28 we find $N_d/4\pi = 0.0020$ and $N_d = 0.025$. This leads to the demagnetizing fields H_d listed in the table, and we see that they form a very substantial fraction of the applied fields. The induction is plotted in Fig. 2.31 as a function of both applied and true fields. It is clear that the *apparent permeability*, given by B/H_a, is much less than the *true permeability*, or B/H. It may be

Table 2.2 Magnetization of Iron

H_a (Oe)	B (G)	$(B - H_a)$ (G)	M (emu/cm^3)	H_d (Oe)	H (Oe)
8.1	1,080	1,070	85	2.1	6.0
16.2	3,850	3,830	305	7.7	8.5
26.9	7,910	7,880	628	15.8	11.1
35.0	10,080	10,040	801	20.1	14.9
43.0	12,420	12,380	987	24.8	18.2
53.9	14,860	14,810	1,180	29.6	24.3
80.7	18,220	18,140	1,447	36.3	44.4

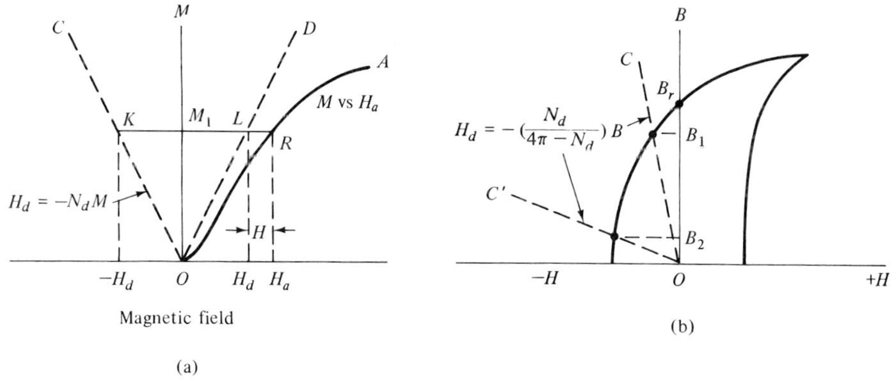

Fig. 2.32 Graphical representation of demagnetizing fields.

shown, by means of relations previously given, that, if μ is large,

$$\frac{1}{\mu_{\text{true}}} = \frac{1}{\mu_{\text{app}}} - \frac{N_d}{4\pi}. \tag{2.34}$$

A curve of B vs. H (or M vs. H) can be obtained from a B,H_a (or M,H_a) curve by a graphical method. In Fig. 2.32 (a) OA is a magnetization curve. The dashed line OC is a plot of the relation $H_d = -N_d M$ and has a slope of $-1/N_d$; it shows the value of the demagnetizing field at any level of M. Thus $H_d = -KM_1$ at $M = M_1$ If we plot OC with an equal but positive slope, it becomes OD. Then $M_1 L = -H_d$, $M_1 R = H_a$, and $LR = H$. Therefore, if OD and OA are sheared to the left by an amount sufficient to make OD vertical, OA will then be a curve of M as a function of the true field H. A curve of B vs. H_a may be similarly transformed; there the demagnetizing-field line OD has a slope of $(4\pi - N_d)/N_d$. The same graphical shear can be applied to a B,H_a hysteresis loop to obtain a true B,H loop.

When the field applied to a specimen on open circuit is reduced to zero, the induction remaining is always *less* than in a ring specimen, because of the demagnetizing field. In Fig. 2.32 (b) the induction in a ring specimen would be B_r, because $H_d = O$ and $H_a = H$. But in an open-circuit specimen, the remanent induction is given by the intersection of the demagnetizing line OC or OC' with the second quadrant of the hysteresis loop. If the specimen is long and narrow, this line (OC) will be steep and the residual induction B_1 will not differ much from that of the ring. If the specimen is short and thick, the line (OC') will be so nearly flat that the residual induction B_2 will be very small.

There are two approaches to the problem of correcting the applied field for the demagnetizing effect. One is to make the specimen in the form of an ellipsoid. Then H_d can be exactly calculated, but at the cost of laborious specimen preparation. The other is to use a cylindrical rod specimen of very large length/diameter ratio. The demagnetizing factor is then so small that any error in it has little effect on the computed value of the true field.

The demagnetizing effect can assume huge proportions in short specimens of magnetically soft materials. For example, suppose a particular material can be brought to its saturation value of $M_s = 1700$ emu/cm^3 by a field of 10 Oe when it is in the form of a ring. If it is in the form of a sphere, H_d will be $(4\pi/3)(1700) = 7120$ Oe at saturation, and the applied field necessary to saturate it will be $7120 + 10 = 7130$ Oe. As shown in Problem 2.14, the M,H_a (or B,H_a) curve will be a straight line almost to saturation, with a slope related to N_d, and many details of the true curve, such as the initial permeability, will be unobservable. In fact, one method of measuring N_d for a short specimen is to determine the slope of the initial linear portion of the M,H_a or B,H_a curve.

If the properties of a rod of a magnetically soft material are being measured by the ballistic method, the effect of the horizontal component of the earth's field, which amounts to two or three tenths of an oersted, should be eliminated by placing the specimen at right angles to this component, as determined by a compass.

The ballistic method is well suited to rod specimens, particularly when the effect of tension or torsion on magnetic properties is to be studied. For a rod of given diameter, the length will be fixed by the requirement of attaining a sufficiently low value of N_d. The rod length will then fix the solenoid length, because the entire rod must be subjected to an essentially uniform field.

If a large number of rods is to be tested, time will be saved by slipping each rod into a single search coil, previously wound on a nonmagnetic form. Since the cross-sectional area A_c of the search coil will usually be larger than the area A_s of the specimen, an *air-flux correction* must be made for the flux in air outside the specimen but inside the search coil:

$$\phi_{\text{observed}} = \phi_{\text{specimen}} + \phi_{\text{air}},$$

$$B_{\text{apparent}}\, A_s = B_{\text{true}}\, A_s + H(A_c - A_s),$$

$$B_{\text{true}} = B_{\text{apparent}} - H\left(\frac{A_c - A_s}{A_s}\right). \tag{2.35}$$

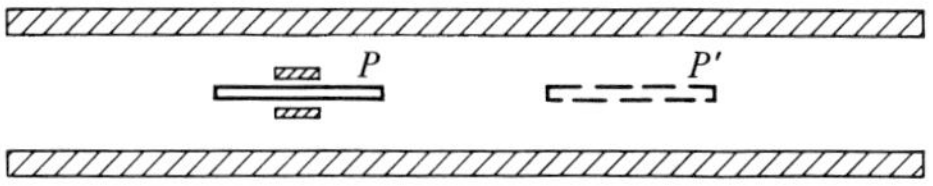

Fig. 2.33 Extraction method with solenoid.

Extraction method

This method is based on the flux change in a search coil when the specimen is removed from the coil. When the solenoid in Fig. 2.33 is producing a magnetic field, the total flux through the search coil is

$$\phi_1 = BA = (H + 4\pi M)A = (H_a - H_d + 4\pi M)A, \tag{2.36}$$

where A is the specimen or search-coil area. (The two are assumed equal here merely to simplify the equations, i.e., the air-flux correction is omitted.) If the specimen is suddenly moved from P to P', then the flux through the coil is

$$\phi_2 = H_a A. \tag{2.37}$$

The fluxmeter will therefore deflect an amount proportional to the flux change

$$\phi_2 - \phi_1 = (N_d - 4\pi)MA. \tag{2.38}$$

The extraction method therefore measures M directly, rather than B. (Although the extraction method is ballistic in nature, the name "ballistic method" is usually reserved for one in which a change in B is produced by a change in H.) The extraction method differs from the ballistic in two respects:

1. M is measured *at* a particular field strength, rather than a change in M due to a change in field.
2. The flux change in the search coil does not involve H. This fact results in greater sensitivity when M is small compared to H, as it is for weakly magnetic substances; M is then determined from the difference between two large numbers in the ballistic method, but directly in the extraction method.

If, as is frequently true, the second position P' of the specimen is not far enough removed from the search coil, part of the external field of the specimen will extend back to the coil, and Eq. (2.37) will be invalid. The apparatus must then be calibrated with a specimen of known magnetization. Still more frequently the specimen is shorter than the search coil, so that Eq. (2.36) is invalid, and it is difficult to calculate the flux through the coil when specimen is inside it; calibration is again necessary.

An important variation of the basic extraction method involves two search coils, one at P and the other at P'. These are connected in "series opposition," i.e., if one is wound clockwise, viewed along their common axis, the other is wound counterclockwise. If the area-turns of both coils are equal, then a change in H_a will induce equal and opposite emfs in the two coils, and the fluxmeter will show no deflection. Thus variations in H_a caused, for example, by variations in the

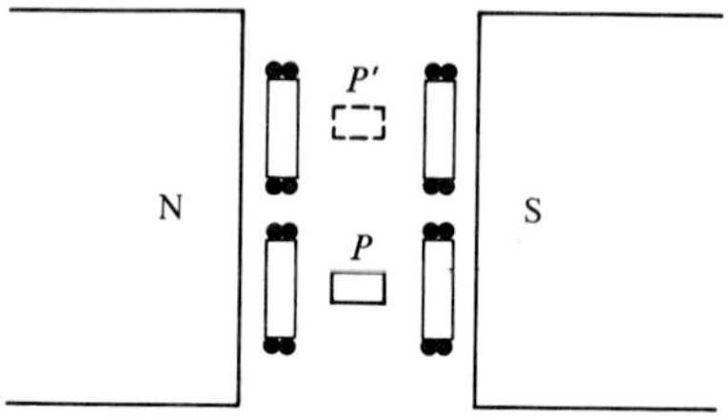

Fig. 2.34 Extraction method with electromagnet.

voltage applied to the solenoid will not influence the results. When the specimen is moved out of one coil and into the other, the galvanometer deflection is twice the amount obtained with one coil.

The extraction method, with either a single or double coil, is often used with an electromagnet as a field source. In Fig. 2.34 the coil axes are parallel to the field but displaced from one another; each coil is also split in two, so that the entire coil system resembles two Helmholtz pairs.

Open-circuit measurements made with an electromagnet must be corrected for the *image effect*. Figure 2.35 (a) shows the exterior field of a short magnetized specimen. If this specimen is now placed in the gap of a nonexcited electromagnet, the lines of force from the specimen swing around to the positions indicated in (b). [These new positions are just those which would result if the lines of force were connected to the "magnetic images" of the specimen in the pole caps, as shown by the dashed lines. An analogous effect occurs in electrostatics, when the lines of force from a point charge shift their position when a conducting plate is placed nearby. The magnetic moment of the magnetic image exactly equals that of the specimen when the permeability μ of the pole caps is infinite, decreases as μ decreases, and vanishes when the pole caps are removed ($\mu = 1$).] The new positions of the lines from the specimen, caused by the presence of the pole caps, increases the flux linkage with the search coil and therefore increases the fluxmeter deflection due to removal of the specimen from the search coil. However, this change in flux linkage is automatically taken care of by the calibration, with a specimen of known M_s, which the method of Fig. 2.34 requires. The troublesome aspect of the image effect is that the number of flux linkages with the coil decreases by several percent as the field due to the electromagnet varies from low values to high, *even when the magnetic moment of the specimen is kept constant.* (The higher the field, the closer the pole caps are to saturation, and the lower their permeability.) This can be shown by means of an artificial specimen, consisting of a small coil of the same shape and size as the real specimen, in which a constant current is maintained in order to produce a constant magnetic moment of known magnitude. For work of the highest precision, corrections for the change in the image effect with field should be obtained by this artificial-specimen method [2.15]. An alternative method is based on the fact that the image effect is more important at small gaps than at large. M,H curves are measured, for various gap sizes, on a specimen of known properties. The shifts required to make the points on the small-gap curve fit on the large-gap curve constitute the corrections.

Precise measurements by the method of Fig. 2.34 require that the position of the specimen be accurately reproducible, both with respect to the coils and, because of the image effect, with respect to the pole pieces. In addition, the standard specimen used to calibrate the apparatus must have the same size and shape as the unknown.

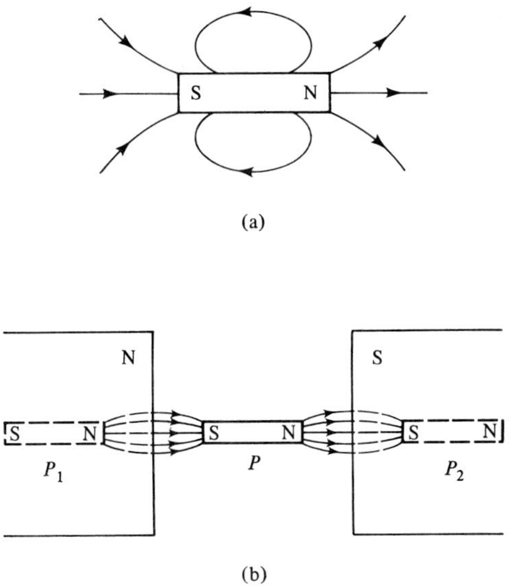

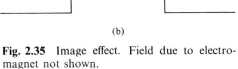

Fig. 2.35 Image effect. Field due to electromagnet not shown.

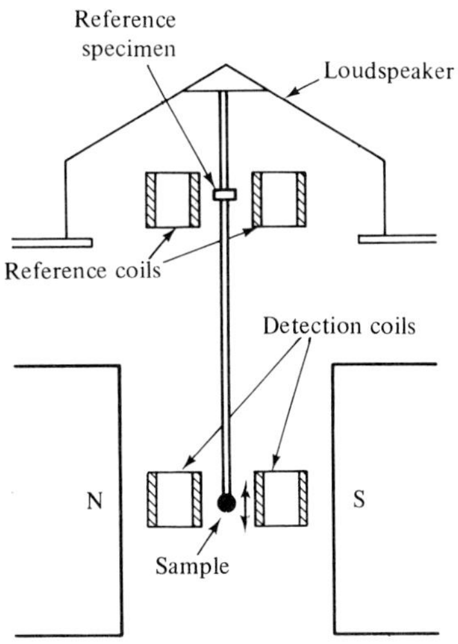

Fig. 2.36 Vibrating-sample magnetometer, after Foner [2.16]

Vibrating-sample method

This method was developed by Foner [2.16] and is based on the flux change in a coil when the sample is vibrated near it. The sample, usually a sphere or small disc, is cemented to the end of a rod, the other end of which is fixed to a loudspeaker cone (Fig. 2.36) or to some other kind of mechanical vibrator. Current through the loudspeaker vibrates the rod and sample at about 80 cycles/sec and with an amplitude of about 0.1 mm in a direction at right angles to the magnetic field. The oscillating magnetic field of the sample induces an alternating emf in the detection coils. The vibrating rod also carries a reference specimen, in the form of a small permanent magnet, near its upper end; the oscillating field of this induces another emf in two reference coils. The voltages from the two sets of coils are compared, and the difference is proportional to the magnetic moment of the sample. This procedure makes the measurement insensitive to changes in, for example, vibration amplitude and frequency. The detection-coil arrangement shown in Fig. 2.36 is only of several possible ones described by Foner. The apparatus must be calibrated with a specimen of known M_s.

This method is very versatile and sensitive. It may be applied to both weakly and strongly magnetic substances, and it can detect a change in magnetic moment of 5×10^{-5} erg/Oe, which corresponds to a change in mass susceptibility of 5×10^{-9} emu/g Oe for a one-gram sample in a field of 10,000 Oe. However, the electronic measuring circuits required make it considerably more expensive than other methods. It is not generally suited to the determination of the magnetization curve or hysteresis loop of a magnetically soft material; the specimen has to be short, and the demagnetizing field is then such a large fraction of the applied field that the true field is uncertain; this objection does not apply to the determination of the value of M_s, because knowledge of the true field at saturation is not necessary.

Magnetometer method

This method, virtually obsolete today, is a very old one and the simplest and most direct of all. It is based on a static measurement of the external field of the specimen by means of the deflection produced by that field in the angular position of a pivoted or suspended magnet. The essential features of the method are just those of the Coulomb-law experiment of Fig. 1.1, if the vertical magnet there be regarded as the specimen. The field applied to the specimen is normally supplied by a solenoid. The magnetometer method is fully described by Ewing [G.1] and by Vigoureux and Webb [2.14]. (The name "magnetometer" is today applied to a large variety of diverse instruments. Some are used for the study of magnetic properties, like the instrument of Fig. 2.36; others are designed for the measurement of the earth's field. The instrument discussed in this section, which is simply called a magnetometer in the older literature, should now probably be referred to more specifically as a suspended-magnet magnetometer.)

In the magnetometer method the deflection of the magnet measures the magnetization of the entire specimen averaged over its volume. The value of the demagnetizing factor N_d appropriate to this method is therefore one averaged over the whole specimen, and magnetometer demagnetizing factors are substantially larger than ballistic ones [G.13, G.15]. Only in ellipsoidal specimens, magnetized uniformly, are the two kinds of factors equal.

The magnetometer method can be made very sensitive. Years ago Rayleigh [2.17] was able to measure the magnetization produced in iron by fields as low as 0.00004 Oe, and a variant of the method is still used in magnetic observatories to monitor, with a precision exceeding 1 part in 10,000, the small daily changes that occur in the magnitude of the earth's field [2.18]. Another current application is in the study of rock magnetism.

However, the sensitivity of the method can make measurements difficult, and it is rarely used today in the study of magnetic materials. Local variations of the "earth's field" can be extremely troublesome. They are caused, for example, by large moving masses of iron nearby (elevators in the building, automobiles outside), by natural currents in the earth, by artificial direct currents (due to arc welding, even hundreds of feet away, and to dc motors), and by the leakage flux from electromagnets (even though located in a different room of the same building). Disturbance by these transient fields can be minimized to some extent by means of astatic magnetometers [2.14, 2.19]. Note that the magnetometer method could be modernized by substituting some other field-measuring device, such as a Hall probe, for the suspended magnet. Although subject to the disturbances of transient fields, this method has the advantage that the measurement is made at a point several inches away from the specimen, which may be, for example, enclosed in a furnace. All things considered, the magnetometer method is best suited to the measurement of very small magnetization or of changes in magnetization with time.

Any kind of work involving weak fields, be it the study of their effects on magnetic materials or the calibration of satellite-borne magnetometers for measuring magnetic fields in space, will be plagued by the ambient field variations mentioned above. The experimenter in search of magnetic peace and quiet then has the choice of building a special laboratory, remotely situated, or of controlling the field in his own laboratory. This control can be achieved by building a magnetically-shielded room [2.27] and/or by generating compensating fields. The latter are produced by an arrangement of three large Helmholtz-coil pairs on three mutually perpendicular axes. A magnetometer at the center senses the field there and through appropriate circuits controls the currents in the Helmholtz coils. In this manner the field

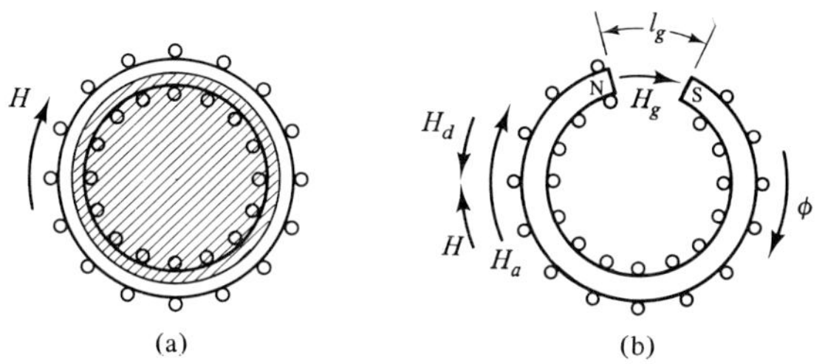

Fig. 2.37 Magnetic circuits: (a) closed, and (b) open.

in the central region can be made to have any required (low) value, including zero, regardless of field changes outside the system.

The four methods of measurement just described fall into two groups. In the first group is the ballistic method, in which the magnetization M is measured by making a small change in M. The second group includes the other three methods (extraction, vibrating-sample, and magnetometer); here M is measured statically, since the measurement does not depend on a change in M but only on its value at the time of measurement.

2.8 MAGNETIC CIRCUITS AND PERMEAMETERS

A permeameter is a device for testing straight bars or rods in a *closed* magnetic circuit. To understand this method requires some general knowledge of the magnetic circuit, which we will now consider. The concept of a magnetic circuit is a very useful one, not merely with respect to materials testing, but also in the design of electric motors and generators and devices containing permanent magnets.

Suppose an iron ring of permeability μ, circumferential length l cm, and cross section A cm^2 is uniformly wound with n turns of wire carrying a current of i amperes (Fig. 2.37 a). Then the field and the flux are given by

$$H = \frac{4\pi ni}{10l} \text{ Oe}, \tag{2.39}$$

$$\phi = BA = \mu HA \text{ maxwells}. \tag{2.40}$$

Combining these equations gives

$$\phi = \mu A\left(\frac{4\pi ni}{10l}\right) = \frac{4\pi ni/10}{l/\mu A}. \tag{2.41}$$

This should be compared with the following equation for the current i in a wire of length l, cross section A, resistance R, resistivity ρ, and conductivity σ $(= 1/\rho)$, when an electromotive force e is acting:

$$i = \frac{e}{R} = \frac{e}{\rho l/A} = \frac{e}{l/\sigma A}. \tag{2.42}$$

Table 2.3 Circuit Analogies

Magnetic	Electric
Flux $= \dfrac{\text{magnetomotive force}}{\text{reluctance}}$	Current $= \dfrac{\text{electromotive force}}{\text{resistance}}$
Flux $= \phi$	Current $= i$
Magnetomotive force $= \dfrac{4\pi ni}{10}$	Electromotive force $= e$
Reluctance $= \dfrac{l}{\mu A}$	Resistance $= R = \dfrac{\rho l}{A} = \dfrac{l}{\sigma A}$
Reluctivity $= \dfrac{1}{\mu}$	Resistivity $= \rho$
Permeance $= \dfrac{\mu A}{l}$	Conductance $= \dfrac{1}{R}$
Permeability $= \mu$	Conductivity $= \sigma = \dfrac{1}{\rho}$

The similarity in form between Eqs. (2.41) and (2.42) suggests the various analogies between magnetic and electric quantities listed in Table 2.3. The most important of these are the *magnetomotive force* (mmf) $= 4\pi\, ni/10$, of which the unit is the *gilbert*, or oersted-centimeter, and the *reluctance* $= l/\mu A$.

A magnetic circuit may consist of various substances, including air, in *series*. We then add reluctances to find the total reluctance of the circuit, just as we add resistances in series:

$$\phi = \frac{\text{mmf}}{(l_1/\mu_1 A_1) + (l_2/\mu_2 A_2) + \cdots}, \tag{2.43}$$

where l_1, l_2, ... are the lengths of the various portions of the circuit, μ_1, μ_2, ... their permeabilities, and A_1, A_2, ... their areas. Similarly, if the circuit elements are in *parallel*, the reciprocal of the total reluctance is equal to the sum of the reciprocals of the individual reluctances.

The open magnetic circuit of Fig. 2.37 (b), consisting of an iron ring with an air gap, may be regarded as a series circuit of iron and air. Because the permeability of air is so small compared to that of iron, the presence of a gap of length l_g greatly increases the reluctance of the circuit. Thus

$$\frac{\text{reluctance with gap}}{\text{reluctance without gap}} = \frac{\dfrac{(l - l_g)}{\mu A} + \dfrac{l_g}{(1)(A)}}{\dfrac{l}{\mu A}} = 1 + (l_g/l)(\mu - 1). \tag{2.44}$$

A value for μ of 5000 is typical of iron near the knee of its magnetization curve. If the ring has a mean diameter of 10 cm and a gap length of 1 cm, then the reluctance of the gapped ring is 160 times that of the complete ring, although the gap amounts to only 3 percent of the circumference. Even if the gap is only 0.05 mm (2×10^{-3} in.), so that it is more in the nature of an imperfect joint than a gap, the reluctance is 1.8 times that of a complete ring. Since the magnetomotive force is proportional to the reluctance (for constant flux), the current in the winding would have to be 160 times as large to produce the same flux in the ring with a 1-cm gap as in the complete ring, and 1.8 times as large for the 0.05 mm gap.

These results may be recast in different language. To say that the current in the winding must be increased to overcome the reluctance of the gap is equivalent to saying that a current increase is required to overcome the demagnetizing field H_d created by the poles formed on either side of the gap. We will then regard the field due to the winding ($=4\pi ni/10l$) as the applied field H_a. If H_a (and ϕ) are clockwise, as in Fig. 2.37 (b), H_d will be counterclockwise. This demagnetizing field can be expressed in terms of a demagnetizing coefficient N_d, which we will find to be directly proportional to the gap width. If H is the true field, then

$$H = \frac{B}{\mu} = \frac{\phi}{\mu A} = \frac{1}{\mu A}\left(\frac{\text{mmf}}{\text{reluctance}}\right) \tag{2.45}$$

$$= \frac{1}{\mu A}\left[\frac{4\pi ni}{10\left(\dfrac{l - l_g}{\mu A} + \dfrac{l_g}{A}\right)}\right] \tag{2.46}$$

$$= \left(\frac{4\pi ni}{10}\right)\left[\frac{1}{l + l_g(\mu - 1)}\right] = \frac{H_a l}{l + l_g(\mu - 1)}. \tag{2.47}$$

But $\mu - 1 = 4\pi\kappa = 4\pi M/H$, where κ is the susceptibility. Substituting and rearranging, we find

$$H = H_a - (4\pi l_g/l)\, M = H_a - N_d M = H_a - H_d. \tag{2.48}$$

The demagnetizing coefficient is thus given by $4\pi l_g/l$. The field in the gap is $H_g = B_g$, which, because of the continuity of lines of B, is equal to the flux density B in the iron, provided that fringing (widening) of the flux in the gap can be neglected.

Returning to the ungapped ring, we may write Eq. (2.39) as

$$Hl = \frac{4\pi ni}{10} = \text{mmf}. \tag{2.49}$$

Since l is the mean circumference of the ring, Hl is the *line integral of H around the circuit*, which we may take as another definition of magnetomotive force:

$$\boxed{\oint H\, dl = \text{mmf} = 4\pi ni/10 = 1.257ni.} \tag{2.50}$$

(It is understood here that H is parallel to l, as is usually true. If not, we must write the integral as $\oint H \cos\theta\, dl$, where θ is the angle between H and dl.) Although

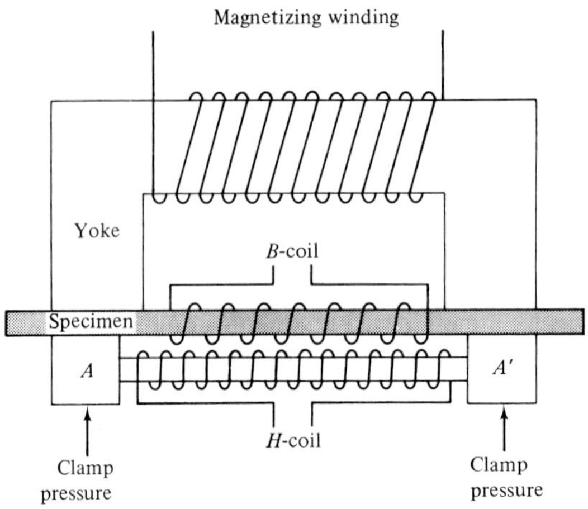

Fig. 2.38 Plan view of Fahy Simplex permeameter (schematic).

we have extracted Eq. (2.50) from a particular case, it is quite generally true and is known as Ampere's law: *the line integral of H around any closed curve equals* $4\pi/10$ *times the total current through the surface enclosed by the curve.* Thus, if l is the boundary of the shaded surface in Fig. 2.37 (a), the total current through this surface is ni, so that $Hl = 1.257\, ni$. Ampere's law often provides a simple means of evaluating magnetic fields. For example, the value of $\oint H\, dl$ around a wire carrying a current i, and at a distance r from the wire, is simply $(H)(2\pi r)$; therefore, $2\pi r H = 4\pi i/10$, or $H = 2i/10r$, in agreement with Eq. (1.10).

Magnetomotive force may also be defined as the work required to take a unit magnetic pole around the circuit. Since the force exerted by a field H on a unit pole is simply H, the work done in moving it a distance dl is Hdl, and we again arrive at $\oint H\, dl$ as the magnetomotive force in the circuit. Pursuing the analogy with electricity still further, we may define the difference in *magnetic potential* V between two points as the work done in bringing a unit magnetic pole from one point to the other against the field, or

$$V_2 - V_1 = -\int_1^2 H\, dl. \tag{2.51}$$

In a circuit, closed or open, composed of permanently magnetized material, flux exists even though the magnetomotive force is zero. Discussion of such circuits will be deferred to Chapter 14.

Although the analogy between magnetic and electric circuits is useful, it cannot be pushed too far. The following differences exist:

1. There is no flow of anything in a magnetic circuit corresponding to the flow of charge in an electric circuit.
2. No such thing as a magnetic insulator exists, except under quite abnormal circumstances (Section 3.5). Thus flux tends to "leak out" of magnetized

bodies instead of confining itself to well-defined paths. This fact alone is responsible for the greater difficulty and lower accuracy of magnetic measurements, compared to electrical.

3. Electrical resistance is independent of the current strength. But magnetic reluctance depends on the flux density, because μ is a function of H, and H determines B.

In a *permeameter* a closed magnetic circuit is formed by attaching a yoke, or yokes, of soft magnetic material to the specimen in order to provide a closed flux path. A magnetizing winding is applied to the specimen or yoke, or both. Therefore, a permeameter is essentially an electromagnet in which the gap is closed by the specimen.

Many types exist, distinguished by the relative arrangement of specimen, yoke, and magnetizing winding; they are fully described by Vigoureux and Webb [2.14] and by Sanford and Cooter [2.1]. We will consider only the Fahy Simplex permeameter shown in Fig. 2.38. The specimen, usually in the form of a flat bar, is clamped between two soft iron blocks, A and A', and a heavy U-shaped yoke. The yoke is made of silicon steel (iron containing 2 or 3 percent Si), a common high-permeability material, laminated to reduce eddy currents. Its cross section is made large, relative to that of the specimen, so that its reluctance will be low. The lower the reluctance, the greater will be the effect of a given number of ampere-turns on the yoke in producing a field through the specimen.

A B-coil is wound on a hollow form into which the specimen can be easily slipped. If a change is made in the current through the magnetizing winding, the field H acting on the specimen will change. The resulting change in B in the specimen is measured ballistically in the usual way, by means of a fluxmeter connected to the B-coil. The only problem then is the measurement of H, and the same problem exists in any permeameter because none of them has a perfectly closed magnetic circuit. In the present permeameter the field in the specimen is not the same as the field through the magnetizing coil because of leakage; therefore, H cannot be calculated from the magnetizing current. Instead, H is measured by means of the H-coil, which consists of a large number of turns on a nonmagnetic cylinder placed near the specimen and between the blocks A and A'. When the magnetizing current is changed, H is measured ballistically by another fluxmeter connected to the H-coil. In effect, the assumptions are made that the magnetic potential difference between the ends of the specimen in contact with the blocks is the same as that between the ends of the H-coil, and that the H-coil accurately measures this difference. This is found to be not quite correct, and rather low accuracy results unless the H-coil is calibrated by means of a specimen with known magnetic properties.

The properties of magnetically soft materials are usually quite sensitive to strain, and this must be kept in mind when clamping a specimen of such material in most forms of permeameter. Too much clamp pressure alters the properties of the specimen, and too little results in a magnetically poor joint and greater leakage.

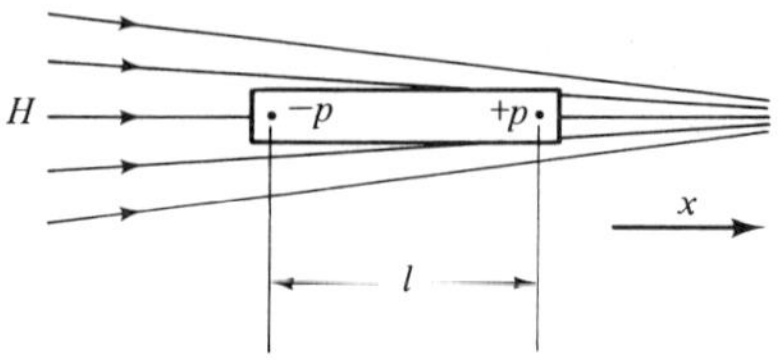

Fig. 2.39 Body in a nonuniform field.

2.9 SUSCEPTIBILITY MEASUREMENTS

The chief property of interest in the case of weakly magnetic substances (dia-, para-, and antiferromagnetic) is their susceptibility. The M,H curve is linear, and measurements at one or two values of H are enough to fix the slope of the curve, which equals the susceptibility. Fields of the order of several kilo-oersteds, at least, are usually necessary in order to produce an easily measurable magnetization, and these fields are usually provided by an electromagnet. Because M is small, the demagnetizing field H_d is small, even for short specimens, and entirely negligible relative to the large applied fields H_a involved. Therefore H can be taken equal to H_a with no appreciable error.

The following methods are available for measuring susceptibility:

1. *Extraction method.* Because the external field of the specimen is small, the search coil or coils must have a large number of turns. This method has insufficient sensitivity when the susceptibility κ is small, and it is normally limited to measurements at room temperature and below, because κ decreases with increasing temperature for para- and antiferromagnetics.
2. *Vibrating-sample method.* (See Section 2.7)
3. *Force methods.* These methods, described below, are based on measurement of the force acting on a body when it is placed in a nonuniform magnetic field. An instrument designed for this purpose is usually called a *magnetic balance.*

A body placed in a *uniform* field will rotate until its long axis is parallel to the field. The field then exerts equal and opposite forces on the two poles so that there is no net force of translation. On the other hand, consider the nonuniform field, increasing from left to right, of Fig. 2.39. In a body of positive κ, such as a paramagnetic, poles of strength p will be produced as shown. Because the field is stronger at the north pole than at the south, there will be a net force to the right, given by

$$\begin{aligned} F_x &= -pH + p\left(H + l\frac{dH}{dx}\right) \\ &= pl\frac{dH}{dx} = m\frac{dH}{dx} = Mv\frac{dH}{dx} \\ &= \kappa v\, H\frac{dH}{dx} = \frac{\kappa v}{2}\frac{dH^2}{dx} \text{ dynes,} \end{aligned} \tag{2.52}$$

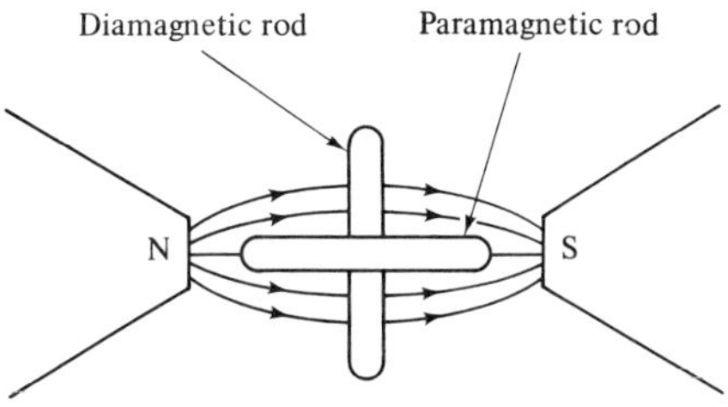

Fig. 2.40 Positions of para- and diamagnetic rods in the field of an electromagnet.

where m is the magnetic moment and v the volume of the body. Thus the body, if free to do so, will move to the right, i.e., into a region of greater field strength. [Note that the body moves in such a way as to increase the number of lines through it, just as the same body, if placed at rest in a previously uniform field, acts to concentrate lines within it as shown in Fig. 2.22 (c).] If the body is diamagnetic (negative κ), its polarity in the field will be reversed and so will the force: it will move toward a region of lower field strength. This statement should be contrasted with the ambiguous remark, sometimes made, that a diamagnetic is "repelled by a field."

(The orientation of a para- or diamagnetic rod in a field depends on the shape of the field, in a manner analyzed in detail by Laufer [2.20]. If the field is uniform, both rods will be parallel to the field. If the field is axially symmetrical, with the field lines concave to the axis, as in Fig. 2.40, the paramagnetic rod will lie parallel to the field (Greek *para* = beside, along) and the diamagnetic rod at right angles (*dia* = through, across). These terms were originated by Faraday. A field of this shape always exists between the tapered pole pieces of an electromagnet or between flat pole pieces if they are widely separated. A much less common field shape is one in which the field lines are convex to the axis; in such a field the rod positions of Fig. 2.40 are reversed.)

If the field H has components H_x, H_y, H_z, then $H^2 = H_x^2 + H_y^2 + H_z^2$, and the force on the body in the x direction is

$$F_x = \frac{\kappa v}{2}\left(\frac{\partial H_x^2}{\partial x} + \frac{\partial H_y^2}{\partial x} + \frac{\partial H_z^2}{\partial x}\right)$$

$$= \kappa v\left(H_x \frac{\partial H_x}{\partial x} + H_y \frac{\partial H_y}{\partial x} + H_z \frac{\partial H_z}{\partial x}\right).$$

It is often necessary to correct for the effect of the medium, usually air, in which the body exists, because the susceptibility κ of the body may not be greatly different from the susceptibility κ_0 of the medium. The force on the body then becomes

$$\boxed{F_x = (\kappa - \kappa_0)\, v\left(H_x \frac{\partial H_x}{\partial x} + H_y \frac{\partial H_y}{\partial x} + H_z \frac{\partial H_z}{\partial x}\right),} \tag{2.53}$$

because motion of the body in the $+x$ direction must be accompanied by motion of an equal volume of the medium in the $-x$ direction.

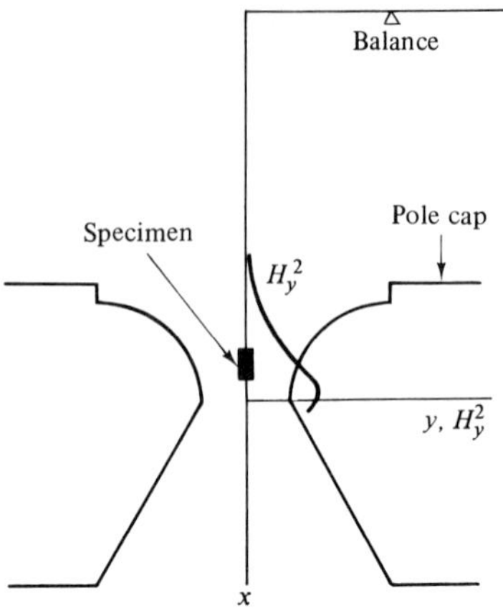

Fig. 2.41 Faraday method. (After Heyding, Taylor, and Hair [2.21]).

The two most important ways of measuring susceptibility are the Faraday method (also called the Curie method) and the Gouy method.

In the *Faraday method* the specimen is small enough so that it can be located in a region where the field gradient is constant throughout the specimen volume. The pole pieces of an electromagnet may be shaped or arranged in various ways to produce a small region of constant field gradient. Figure 2.41 shows one example. The field is predominantly in the y direction and, in the region occupied by the specimen, H_x and H_z and their gradients with x are small. The variation of H_y^2 with x is shown by the curve superimposed on the diagram, and it is seen that $dH_y^2/dx(= 2H_y dH_y/dx)$ is approximately constant over the specimen length. Equation (2.53) therefore reduces to

$$F_x = (\kappa - \kappa_0)\, vH_y \frac{dH_y}{dx} \text{ dynes.} \tag{2.54}$$

F_x is measured by suspending the specimen from one arm of a sensitive balance. If $(\kappa - \kappa_0)$ is positive, there will be an apparent increase in mass Δw when the field is turned on. Then

$$F_x = g\, \Delta w \text{ dynes,} \tag{2.55}$$

where g is the acceleration due to gravity (981 cm/sec^2) and w is in grams. The Faraday method is not a good absolute method because of the difficulty of determining the field and its gradient at the position of the specimen. But it is capable of high precision and can be calibrated by measurements on specimens of known susceptibility, determined, for example, by the Gouy method.

The specimen in the *Gouy method* is in the form of a long rod (Fig. 2.42). It is suspended so that one end is near the center of the gap between parallel magnet pole pieces, where the field H_y is uniform and strong. The other end extends to a region where the field H_{y0} is relatively weak. The field gradient dH_y/dx therefore produces a downward force on the specimen, if the net susceptibility

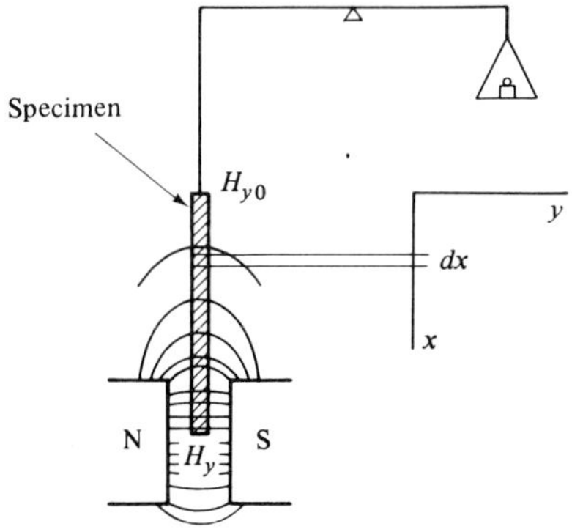

Fig. 2.42 Gouy method.

$(\kappa - \kappa_0)$ of specimen and displaced medium is positive. The force on a small length dx of the rod, of volume dv and cross-sectional area a, is

$$dF_x = \frac{(\kappa - \kappa_0)\,dv}{2}\frac{dH_y^2}{dx} = \frac{(\kappa - \kappa_0)a\;dx}{2}\frac{dH_y^2}{dx}.$$

The force on the whole rod is

$$F_x = \frac{(\kappa - \kappa_0)a}{2}\int_{H_{y0}}^{H_y} dH_y^2 = \frac{(\kappa - \kappa_0)a}{2}(H_y^2 - H_{y0}^2).$$

H_y is of the order of 10,000 Oe, and H_{y0} is less than 100 Oe. Hence H_{y0}^2 is negligible compared to H_y^2, and

$$F_x = \frac{(\kappa - \kappa_0)a\;H_y^2}{2}. \tag{2.56}$$

Therefore, a field gradient does not have to be determined in this method; all that is needed is a measurement of the uniform field in the magnet gap. The only disadvantage of the method is that a fairly large amount, about 10 cm^3, of specimen is required, since the specimen must be 10 to 15 cm long. Solid rods may be measured directly. Powdered materials are enclosed in a glass tube, and the force on the empty tube is measured separately to find the correction for the container.

In this section, statements regarding paramagnetics apply to any material with positive κ. If the material is ferro- or ferrimagnetic, the forces involved will be quite large, and force methods have been used, chiefly by Sucksmith *et al.* [2.22], to measure their magnetization.

As will become apparent in later chapters, susceptibility measurements made only at room temperature are of very limited value. Only by making measurements over an extended temperature range can we obtain information which will permit a description of the true magnetic nature of the specimen. Thus

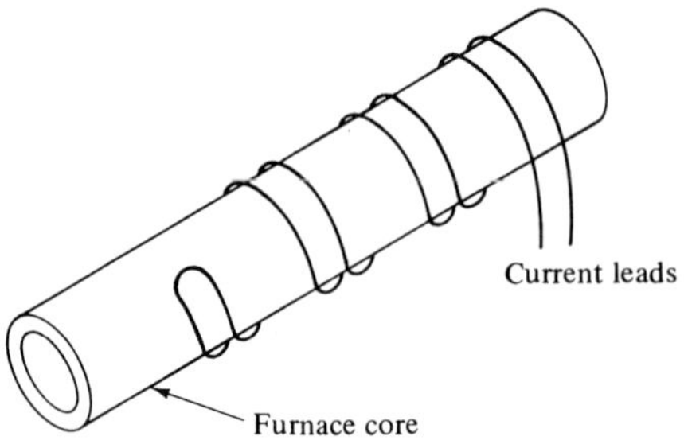

Fig. 2.43 Noninductive, or bifilar, winding [schematic].

furnaces and/or cryostats are needed, and these must be narrow enough to fit in a magnet gap and yet enclose the specimen. Furnaces are usually of the electrical resistance type, water-cooled on the outside. (It should be noted that the usual helically-wound tube furnace constitutes a solenoid, and the heating current produces an appreciable magnetic field. In magnetic experiments it may be important to suppress this field. This can be done by winding the furnace "noninductively," as shown in Fig. 2.43. The wire is simply doubled on itself before it is wound on the furnace core. Currents in adjacent turns are then in opposite directions, and their magnetic fields annul each other. Whether wound inductively or not, furnaces exposed to strong magnetic fields should be powered with direct, not alternating, current. Because of the force on a current-carrying conductor in a magnetic field, windings carrying ac will vibrate at twice the frequency of the ac. This vibration can quickly loosen the refractory cement and cause failure of the furnace.) The liquids in general use for cooling cryostats are liquid helium (normal boiling point = 4.2°K), liquid hydrogen (20.4°K), and liquid nitrogen (77.3°K).

When measurements extend over a range of temperature, the mass susceptibility χ is a more convenient property to deal with than the volume susceptibility κ, because the specimen mass does not change with temperature, whereas its volume does. Since $\kappa v = \chi w = m/H$ = magnetic moment produced by unit field, all that is necessary is to replace κv by χw in equations like (2.54), where w is the mass of the specimen.

PROBLEMS

2.1 A solenoid is 24 in. long and 1 in. in diameter. It is closely wound with one layer of wire 0.025 in. in diameter, which has a resistance of 16 ohms per 1000 ft. For a current of 2 amperes, find (a) the field at the center, (b) the voltage needed, and (c) the power consumed.

2.2 For the solenoid of the previous problem, plot the field on the axis from the center to the end. What is the ratio

a) of the field at the edge of the middle half to the field in the center?

b) of the field at the end to the field in the center?

2.3 A certain power source can deliver a maximum current of i at a voltage E to a single-layer solenoid of length L and diameter D, which is wound with a wire of resistivity ρ. In terms of these quantities, obtain an expression for the wire diameter d that is needed, if the solenoid is to operate at maximum power W.

2.4 If the wire diameter found in the previous problem is doubled, with i constant at its maximum value, what is the effect on H, E, and W?

2.5 What is the ratio of the field at the center of symmetry of two Helmholtz coils to the field at the center of a solenoid, for the same current in each and the same total length of wire? The solenoid has $L/D = 10$, and the distance between the Helmholtz coils is L.

2.6 A mutual inductance is made by winding a secondary coil of 800 turns on the central section of a solenoid of 1350 turns, 37.7 in. length, and 1.38 in. diameter. The resistance of the secondary is 25 ohms. In series with the secondary is a resistance of 245 ohms and a ballistic galvanometer (of internal resistance 52 ohms). In a calibration experiment, the galvanometer deflections produced by reversal of current in the solenoid were as follows:

Current reversed	Deflection
0.20 ampere	35.8 mm
0.40	72.1
0.60	109.8
0.80	146.1

a) What is the sensitivity k in coulombs/mm?
b) What is the sensitivity K in maxwell-turns/mm?

2.7 The normal induction curve and hysteresis loop of a ring specimen are determined with a fluxmeter of 1000 maxwell-turns/mm sensitivity. The ring has a circular cross section of $\frac{1}{4}$ in. diameter, and the mean diameter of the ring is $6\frac{1}{4}$ in. The magnetizing winding has 300 turns and the search coil 25 turns. The current changes Δi (in amperes) and the corresponding fluxmeter deflections d (in millimeters) were as follows:

Induction curve			Hysteresis loop		
Δi		d	Δi		d
−0.05 to	0.05	7.0	0.5 to	0	30.0
−0.10	0.10	20.0	0.5	−0.1	45.0
−0.15	0.15	42.2	0.5	−0.2	67.4
−0.20	0.20	62.0	0.5	−0.3	93.4
−0.25	0.25	75.8	0.5	−0.4	103.5
−0.30	0.30	85.4	0.5	−0.5	110.0
−0.35	0.35	94.0	−0.5	−0.4	3.0
−0.40	0.40	100.8	−0.5	−0.3	7.3
−0.45	0.45	106.0	−0.5	−0.2	13.0
−0.50	0.50	110.0	−0.5	−0.1	20.8

a) Plot the normal induction curve, the hysteresis loop, and the μ versus H curve.
b) What is the value of B_{max}, B_r, H_c, μ_{max}, and M_{max}?

2.8 A magnet is placed in a uniform applied field H_a directed antiparallel to the original magnetization. In the manner of Fig. 2.20, sketch the vector relations inside the magnet and the B field inside and outside the magnet, under the following two conditions:

a) H_a is sufficient to reduce B to zero inside the magnet (coercive-force point).
b) H_a is sufficient to reduce M to zero (intrinsic-coercive-force point).

2.9 Assume that the magnetization of a bar magnet of length l is uniform on any cross section and that it can be represented by $M = c(l^2 - x^2)$, where x is the distance from the center of the magnet and c is a constant. What is the pole strength per unit volume in the interior and per unit area on the ends?

2.10 A field of 500 Oe is incident at right angles on a flat iron plate of infinite extent in two dimensions and permeability 40. What is the induction in the iron?

2.11 An unmagnetized body, in the shape of a prolate spheroid and with a permeability of 5000, is placed in a uniform field with its long axis parallel to the field. What is the ratio, to two significant figures, of the flux density in the body to the flux density in the original field, when the axial ratio of the spheroid is (a) 1 (sphere), (b) 5, (c) 100, and (d) infinity?

2.12 The normal induction curve of an annealed iron rod is measured with a ballistic galvanometer of 1688-maxwell-turns/mm sensitivity, and a solenoid which produces 4.65 Oe/ampere. A 40-turn search coil is wound directly on the specimen, which is 7.00 in. long and 0.1775 in. in diameter. The current reversed i_r in the solenoid and the corresponding galvanometer deflection d were as follows:

i_r (amperes)	d (mm)
1.5	24.0
3.1	49.2
4.9	77.6
8.5	103.7
11.0	107.5
12.7	109.1

a) Plot the B,H_a and B,H curves.
b) What is the true permeability at $B = 10{,}000$ G?

2.13 The normal induction curve of a nickel wire ($M_s = 484$ emu/cm^3), 1 mm in diameter, is to be measured with a fluxmeter of 500 maxwell-turns/mm sensitivity. A value of ΔB equal to $B_s/10$ is to give a fluxmeter deflection of at least 5 mm. How many turns are needed on the search coil? Assume that saturation can be obtained with a true field of 300 Oe.

2.14 The following values of H (in Oe) and M (in emu/cm^3) were measured on a ring specimen of annealed nickel.

H	M	H	M
4.04	49.1	17.48	330.
5.38	91.9	21.5	349.
8.08	196.	26.9	365.
10.77	258.	33.9	379.
13.44	298.	40.4	388.

a) Plot the M,H curve for the ring specimen.
b) Compute and plot the M,H_a curve for the same material in the form of a sphere.
c) What is the reciprocal slope of the initial portion of the latter curve?
d) Show that a M,H_a curve has an initial slope of $1/N_d$ if μ is large.

2.15 A cylindrical silver rod, of susceptibility $\chi = -0.181 \times 10^{-6}$ emu/g Oe, is weighed in a Gouy balance. It is 1.00 cm in diameter, 15.00 cm long, and weighs 123.5 g. The susceptibility κ of the air in which it is weighed is 0.027×10^{-6} emu/cm^3 Oe. The field in the magnet gap is 10,000 Oe, and the field at the upper end of the rod is negligible. What is the apparent change in mass when the field is turned on?

KINDS OF MAGNETISM

CHAPTER 3

DIAMAGNETISM AND PARAMAGNETISM

3.1 INTRODUCTION

By appropriate methods we can measure the gross magnetic properties of any substance and classify that substance as diamagnetic, or paramagnetic, or ferromagnetic, etc. We will now proceed, in Chapters 3 through 6, to examine the internal mechanisms responsible for the observed magnetic behavior of various substances. In these chapters we will be interested only in two structure-insensitive properties, the susceptibility of weakly magnetic substances and the saturation magnetization of strongly magnetic ones. (These properties are said to be structure-insensitive in the sense that they do not depend on details of the fine structure, such as strain, lattice imperfections, or small amounts of impurities.) We will also pay particular attention to the variation of these properties with temperature, because this variation provides an important clue to the magnetic nature of the substance.

The magnetic properties in which we are interested are due entirely to the *electrons* of the atom, which have a magnetic moment by virtue of their motion. The nucleus also has a small magnetic moment, but it is insignificant compared to that of the electrons, and it does not affect the gross magnetic properties.

3.2 MAGNETIC MOMENTS OF ELECTRONS

There are two kinds of electron motion, orbital and spin, and each has a magnetic moment associated with it.

The *orbital motion* of an electron around the nucleus may be likened to a current in a loop of wire having no resistance; both are equivalent to a circulation of charge. The magnetic moment of an electron, due to this motion, may be calculated by an equation similar to Eq. (1.12), namely,

$$\mu = (\text{area of loop})(\text{current in emu}). \tag{3.1}$$

(Use of the symbol μ both for permeability and for an electronic or atomic moment is common and ordinarily does not lead to any ambiguity.) To evaluate μ we must know the size and shape of the orbit and the electron velocity. In the original (1913) Bohr theory of the atom the electron moved with velocity v in a circular orbit of radius r. If e is the charge on the electron in esu and c the velocity of light, then e/c is the charge in emu. The current, or charge passing a given point per unit time, is then $(e/c)(v/2\pi r)$. Therefore,

$$\mu \text{ (orbit)} = \pi r^2 \left(\frac{ev}{2\pi rc} \right) = \frac{evr}{2c}. \tag{3.2}$$

An additional postulate of the theory was that the angular momentum of the electron must be an integral multiple of $h/2\pi$, where h is Planck's constant. Therefore,

$$mvr = nh/2\pi. \tag{3.3}$$

Combining these relations, we have

$$\mu \text{ (orbit)} = eh/4\pi mc \tag{3.4}$$

for the magnetic moment of the electron in the first ($n = 1$) Bohr orbit.

The *spin* of the electron was postulated in 1925 in order to explain certain features of the optical spectra of hot gases, particularly gases subjected to a magnetic field (Zeeman effect), and it later found theoretical confirmation in wave mechanics. Spin is a universal property of electrons in all states of matter at all temperatures. The electron behaves as if it were spinning about its own axis, as well as moving in an orbit about the nucleus, and associated with this spin are definite amounts of magnetic moment and angular momentum. It is found experimentally and theoretically that the magnetic moment due to electron spin is equal to

$$\mu \text{ (spin)} = \frac{eh}{4\pi mc} \tag{3.5}$$

$$= \frac{(4.80 \times 10^{-10} \text{ esu})(6.62 \times 10^{-27} \text{ erg sec})}{4\pi(9.11 \times 10^{-28} \text{ g})(3.00 \times 10^{10} \text{ cm/sec})}$$

$$= 0.927 \times 10^{-20} \text{ erg/Oe}.$$

Thus the magnetic moment due to spin and that due to motion in the first Bohr orbit are exactly equal. Because it is such a fundmental quantity, this amount of magnetic moment is given a special symbol μ_B and a special name, the *Bohr magneton.* Thus,

$$\boxed{\mu_B = \text{Bohr magneton} = eh/4\pi mc = 0.927 \times 10^{-20} \text{ erg/Oe}.} \tag{3.6}$$

It is a natural unit of magnetic moment, just as the electronic charge e is a natural unit of electric charge.

How can the magnetic moment due to spin be understood physically? We may, if we like, imagine an electron as a sphere with its charge distributed over its surface. Rotation of this charge produces a lot of tiny current loops (Fig. 3.1), each of which has a magnetic moment directed along the rotation axis. But if we calculate the resultant moment of all these loops, we obtain the wrong answer,

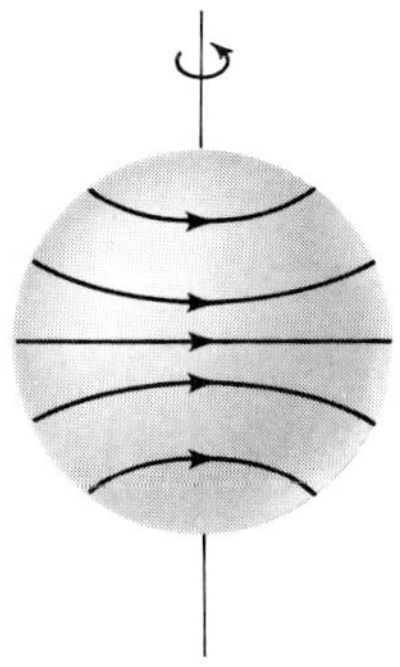

Fig. 3.1 Electron spin.

$5\mu_B/6$ instead of μ_B. Nor does the right answer result from the assumption that the charge is uniformly distributed through the volume of the sphere. Such calculations are fruitless, because we do not know the shape of the electron or the way in which charge is distributed on or in it. The spin of the electron, and its associated magnetic moment, simply has to be accepted as a fact, consistent with wave mechanics and with a large number of diverse kinds of experiments, but with no basis in classical physics. The model of Fig. 3.1 is therefore only an aid to visualization; it has no quantitative significance.

3.3 MAGNETIC MOMENTS OF ATOMS

Atoms contain many electrons, each spinning about its own axis and moving in its own orbit. The magnetic moment associated with each kind of motion is a vector quantity, parallel to the axis of spin and normal to the plane of the orbit, respectively. The magnetic moment of the *atom* is the vector sum of all its electronic moments, and two possibilities arise:

1. The magnetic moments of all the electrons are so oriented that they cancel one another out, and the atom as a whole has no net magnetic moment. This condition leads to diamagnetism.

2. The cancellation of electronic moments is only partial and the atom is left with a net magnetic moment. (Such an atom is often referred to, for brevity, as a "magnetic atom.") Substances composed of atoms of this kind are para-, ferro-, antiferro-, or ferrimagnetic.

To calculate the vector sum of the magnetic moments of all the electrons in any particular atom is a rather complex problem which the reader will find treated in any book on atomic physics. However, this problem is not particularly relevant here because the result applies only to the free atom, such as the atoms in a monatomic gas. The calculation from first principles of the net magnetic moment of an atom in a solid is, in general, not yet possible, and the net moment must be determined experimentally. This knowledge of atomic moments,

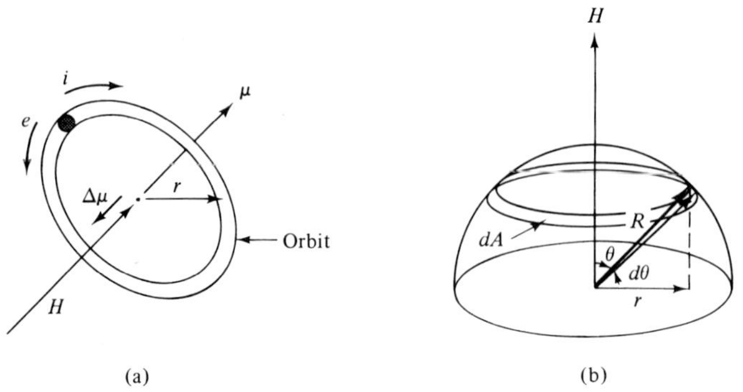

Fig. 3.2 Effect of a field on orbital moments.

obtained experimentally, is of great importance in the continued development of solid-state physics, entirely aside from its relevance to commercially important magnetic materials.

3.4 THEORY OF DIAMAGNETISM

A diamagnetic is a substance that exhibits, so to speak, negative magnetism. Even though it is composed of atoms which have no net magnetic moment, it reacts in a particular way to an applied field. The theory of this effect was first worked out by the French physicist Paul Langevin (1872–1946) in a classic paper [3.1] published in 1905. He refined and made quantitative some ideas which had been earlier advanced by Ampère and by the German physicist Wilhelm Weber (1804–1891).

Consider a particular electron orbit normal to an applied magnetic field, as in Fig. 3.2 (a). Motion of the electron is equivalent to current in a loop, and, as soon as the field is increased from zero, the change in flux through the loop induces an emf $\mathscr{E}$ in the loop, according to Faraday's law (Eq. 2.6):

$$\mathscr{E} = -10^{-8}\frac{d\phi}{dt} = -10^{-8}\frac{d(HA)}{dt} \text{ volts,} \tag{3.7}$$

where A is the area of the loop. As the minus sign indicates, this emf acts in such a way as to oppose the change in flux (Lenz's law), that is, to decrease it. This can be achieved by a decrease in the loop current or, in other words, the electron velocity. The result is a decrease in the magnetic moment of the loop. (The symbol $\mathscr{E}$ is used for emf in this section in order to avoid confusion with the symbol e for electron charge.)

The electron orbit is not only assumed to act as a current loop to which Faraday's law can be applied, but the current loop is assumed to act like a wire without resistance. Thus the change in current produced by $\mathscr{E}$ persists, even though $\mathscr{E}$ is finite only while the applied field is changing from 0 to H. The effect is *not* momentary. The magnetic moment is decreased as long as the field H is acting.

In the orbit shown in Fig. 3.2 (a), the electron is moving counterclockwise, so that the direction of positive current is clockwise. The magnetic moment μ of the loop is thus in the same direction, taken as positive, as that of the applied field H. The effect of H is to change μ by an amount $\Delta\mu$ in the negative direction. Now imagine another orbit adjacent to the one shown, with its plane parallel to the plane of the first, but with an electron moving clockwise. In zero field the moment of this second orbit is in the negative direction, and the change in moment, when a field is applied, is also negative. Thus the moments of these two orbits cancel in zero field, but both orbits produce a negative moment when a field is applied. Similarly, the moments of all the electrons in an atom can be so oriented in space that they all cancel in zero field, and yet each orbit reacts to an applied field by producing a moment antiparallel to that field. As a result each atom acquires a negative moment when the field is on.

We can compute the change in moment of a single orbit as follows. If E is the electric field intensity acting around a circular orbit of length l, then

$$E = \frac{\mathscr{E}}{l} = -10^{-8}\frac{A}{l}\frac{dH}{dt} = -10^{-8}\frac{\pi r^2}{2\pi r}\frac{dH}{dt}$$

$$= -10^{-8}\frac{r}{2}\frac{dH}{dt} \text{ volts/cm.}$$

The force exerted on the electron by this field is

$$F = 10^8\ E(e/c) = ma \text{ dynes.}$$

The resultant acceleration is

$$a = \frac{dv}{dt} = 10^8\frac{Ee}{mc} = -\frac{er}{2mc}\frac{dH}{dt} \text{ cm/sec}^2.$$

Integrating over a change in magnetic field from 0 to H, and assuming that the orbit radius r does not change during the application of the field, we have

$$\int_{v_1}^{v_2} dv = -\frac{er}{2mc}\int_0^H dH,$$

$$v_2 - v_1 = \Delta v = -\frac{erH}{2mc} \text{ cm/sec.}$$

But Eq. (3.2) shows that this change in electron velocity will lead to a change in magnetic moment of

$$\Delta\mu = \frac{er\Delta v}{2c} = -\frac{e^2r^2H}{4mc^2} \text{ erg/Oe.} \tag{3.8}$$

This result applies only when the plane of the orbit is perpendicular to the applied field. In general, it will be inclined to the field, and r in the above equation must then be interpreted as the projection of the orbit radius R on a plane normal to the field. Figure 3.2 (b) shows the radius R of an orbit inclined at an angle θ to the field, and we must now find the average of r^2 when R takes on all possible orientations in the hemisphere shown. This average value is

$$\overline{r^2} = \overline{R^2 \sin^2 \theta} = (R^2 \int \sin^2 \theta \, dA)/A,$$

where $dA (= 2\pi R^2 \sin\theta \, d\theta)$ is an annular element of area. Therefore,

$$\overline{r^2} = [R^2 \int_0^{\pi/2} (\sin^2 \theta)(2\pi R^2 \sin\theta \, d\theta)]/2\pi R^2 = 2R^2/3.$$

Inserting this in Eq. (3.8), we find that the change in moment caused by the field is

$$\Delta\mu = -\frac{e^2 R^2 H}{6mc^2}, \tag{3.9}$$

where R is the radius of an orbit that can take on all possible orientations with respect to the field.

So far we have considered only a single electron. The change in moment of an *atom* containing Z electrons will be given by a sum of Z terms like the one above, or

$$\Delta\mu \text{ (per atom)} = -\frac{e^2 H}{6mc^2} \sum R_n^2, \tag{3.10}$$

where R_n is the radius of the nth orbit. The summation $\sum R_n^2$ may be replaced by $Z\overline{R^2}$, where $\overline{R^2}$ is the average of the squares of the various orbital radii. And to put our result on a volume basis, we note that the number of atoms per unit volume is $N\rho/A$, where N is Avogadro's number, ρ the density, and A the atomic weight. Therefore,

$$\Delta\mu \text{ (per cm}^3) = -(N\rho/A)(e^2 Z\overline{R^2} H/6mc^2). \tag{3.11}$$

But we have assumed that each atom has no net magnetic moment in the absence of a field. Therefore, the moment acquired per unit volume, which is simply the magnetization M, is given by Eq. (3.11). Since the volume susceptibility κ is given by M/H, we have

$$\kappa = -(N\rho/A)(e^2 Z\overline{R^2}/6mc^2) \text{ emu/cm}^3 \text{ Oe}. \tag{3.12}$$

As an example, we may apply this equation to carbon. X-ray diffraction measurements show that the radius of the carbon atom is about 0.7 angstrom (1 Å $= 10^{-8}$ cm). We may therefore take $\overline{R^2}$ as approximately equal to $(0.7 \times 10^{-8})^2$ cm^2. Inserting this value, we have

$$\kappa = -\frac{(6.02 \times 10^{23})(2.22 \text{ g/cm}^3)(4.80 \times 10^{-10} \text{ esu})^2 (6) (0.7 \times 10^{-8} \text{ cm})^2}{(12.01)(6)(9.11 \times 10^{-28} \text{ g})(3.00 \times 10^{10} \text{ cm/sec})^2}$$

$$= -1.5 \times 10^{-6} \text{ emu/cm}^3 \text{ Oe}.$$

The experimental value is -1.1×10^{-6}. The agreement between calculated and experimental values for other diamagnetic substances is generally not that good, but it is at least within an order of magnitude. All in all, this classical theory of Langevin is a good example of the use of a simple atomic model to quantitatively explain the bulk properties of a material.

None of the quantities in Eq. (3.12) varies much with temperature. This agrees with the experimental fact that the susceptibilities of diamagnetic substances, with a few exceptions, are *independent of temperature*.

The *quantum theory* of diamagnetism arrives at the same expression as Eq. (3.12), and then attempts to calculate $\overline{R^2}$ by a calculation of the electron charge distribution within the atom. This can be done, at present, only for the simpler kinds of atoms.

3.5 DIAMAGNETIC SUBSTANCES

Electrons which constitute a closed shell in an atom usually have their spin and orbital moments so oriented that the atom as a whole has no net moment. Thus the monatomic rare gases He, Ne, A, etc., which have closed-shell electronic structures, are all diamagnetic. So are most polyatomic gases, such as H_2, N_2, etc., because the process of molecule formation usually leads to filled electron shells and no net magnetic moment per molecule.

The same argument explains the diamagnetism of ionic solids like NaCl. The process of bonding in this substance involves the transfer of an electron from each Na atom to each Cl atom; the resulting ions, Na^+ and Cl^-, then have closed shells and are both diamagnetic. Covalent bonding by the sharing of electrons also leads to closed shells, and elements like C (diamond), Si, and Ge are diamagnetic.

Almost all organic compounds are diamagnetic, and magnetic measurements have furnished much useful information about the structure of organic molecules. The measured susceptibility is inserted into Eq. (3.12), which is then solved for $\overline{R^2}$, which gives data on the size of the electron orbits. If measurements are made in various directions in a single crystal, some conclusions as to the shape of the molecule can be reached.

But not all gases are diamagnetic, nor are all ionic or covalent solids. Generalizations in this area are dangerous. The reader interested in further details should consult books on *magnetochemistry* [G.8, G.20, G.32], which is a subject devoted to the relation between magnetic properties and the chemical bond. The behavior of the metals is particularly complex; most are paramagnetic, but some diamagnetic; they are discussed in Section 3.8.

The magnetic behavior of superconductors is most unusual. If a metal like lead, normally diamagnetic, or tantalum, normally paramagnetic, is cooled in a magnetic field, it will become superconducting at its critical temperature T_c and, at the same time, all flux will be excluded from it except for a thin surface layer which is only a few hundred angstroms deep (Fig. 3.3). This is called the Meissner effect. Recalling the statement made in Section 2.6 that lines of force tend to

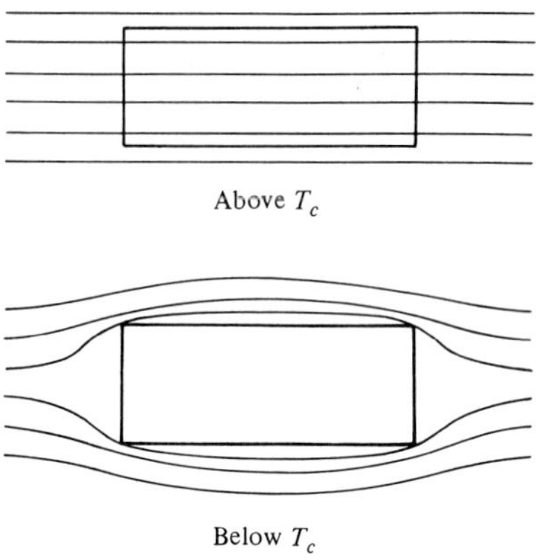

Fig. 3.3 Behavior of a superconductor like lead in a magnetic field, above and below its critical temperature T_c. (The slight diamagnetism of lead above T_c is neglected in this drawing.)

avoid a diamagnetic body, we see that a superconductor is a perfect diamagnet. Since B is zero inside, the permeability μ and $(H + 4\pi M)$ are also zero, so that $M = -H/4\pi$ and $\kappa = -1/4\pi$. The effect disappears when H exceeds a few hundred oersteds in metals like lead or tantalum, called "ideal" (Type I) superconductors. In "nonideal" (Type II) superconductors, like Nb_3Sn, complete flux exclusion persists up to fields of several thousand oersteds, and partial exclusion to still higher fields.

A perfect diamagnetic is a magnetic insulator in the sense that it bars the passage of magnetic flux, just as an electrical insulator bars the passage of electric charge. Cioffi [3.2] has exploited this fact by constructing electromagnets in which the air gap and coils are surrounded by a superconductor. When they are operated in a helium cryostat, leakage and fringing flux are markedly reduced. A magnet so designed and operated is therefore much smaller than a conventional one, for the same field in the air gap.

3.6 CLASSICAL THEORY OF PARAMAGNETISM

The first systematic measurements of the susceptibility of a large number of substances over an extended range of temperature were made by Curie * and reported by him in 1895. He found that the mass susceptibility χ was independent of temperature for diamagnetics, but that it varied inversely with the absolute temperature for paramagnetics:

$$\chi = C/T \tag{3.13}$$

* Pierre Curie (1859–1906) French physicist. He and his wife, Marie (Sklodowska) Curie (1867–1934), later became famous for their research on radioactivity.

This relation is called Curie's law, and C is the Curie constant per gram. It was later shown that the Curie law is only a special case of a more general law,

$$\chi = C/(T - \theta), \tag{3.14}$$

called the Curie-Weiss law. Here θ is a constant, with the dimensions of temperature, for any one substance, and equal to zero for those substances which obey Curie's law.

Curie's measurements on paramagnetics went without theoretical explanation for ten years, until Langevin in 1905 took up the problem in the same paper [3.1] in which he presented his theory of diamagnetism. Qualitatively, his theory of paramagnetism is simple. He assumed a paramagnetic to consist of atoms, or molecules, each of which has the same net magnetic moment μ, because all the spin and orbital moments of the electrons do not cancel out. In the absence of an applied field, these atomic moments point at random and cancel one another, so that the magnetization of the specimen is zero. When a field is applied, there is a tendency for each atomic moment to turn toward the direction of the field; if no opposing force acts, complete alignment of the atomic moments would be produced and the specimen as a whole would acquire a very large moment in the direction of the field. But thermal agitation of the atoms opposes this tendency and tends to keep the atomic moments pointed at random. The result is only partial alignment in the field direction, and therefore a small positive susceptibility. The effect of an increase in temperature is to increase the randomizing effect of thermal agitation and therefore to decrease the susceptibility.

We will now consider the quantitative aspects of the theory in some detail, not because the magnetic properties of paramagnetics are of much practical importance (they are not), but because the theory of paramagnetism leads naturally into the theory of ferro and ferrimagnetism.

We consider a unit volume of material containing n atoms, each having a magnetic moment μ. Let the direction of each moment be represented by a vector, and let all the vectors be drawn through the center of a sphere of unit radius. We wish to find the number dn of moments inclined at an angle between θ and $\theta + d\theta$ to the field H. In the absence of a field the number of μ vectors passing through unit area of the sphere surface is the same at any point on the sphere surface, and dn is proportional simply to the area dA, which is given, as in Fig. 3.2(b), by $2\pi \sin\theta \, d\theta$ for a sphere of unit radius. But when a field is applied, the μ vectors all shift toward the direction of the field. Each atomic moment then has a certain potential energy E_p in the field, given by Eq. (1.5), so that

$$E_p = -\mu H \cos\theta. \tag{3.15}$$

In a state of thermal equilibrium at temperature T, the probability of an atom having an energy E_p is proportional to the Boltzmann factor $e^{-E_p/kT}$ where k is the Boltzmann constant. The number of moments between θ and $\theta + d\theta$ will now be proportional to dA, multiplied by the Boltzmann factor, or

$$dn = K\, dA\, e^{-E_p/kT} = 2\pi K e^{(\mu H \cos\theta)/kT} \sin\theta \, d\theta, \tag{3.16}$$

where K is a proportionality factor, determined by the fact that

$$\int_0^n dn = n.$$

For brevity we put $a = \mu H/kT$. We then have

$$2\pi K \int_0^\pi e^{a \cos \theta} \sin \theta \, d\theta = n. \tag{3.17}$$

The total magnetic moment in the direction of the field acquired by the unit volume under consideration, i.e., the magnetization M, is given by multiplying the number of atoms dn by the contribution $\mu \cos \theta$ of each atom and integrating over the total number:

$$M = \int_0^n \mu \cos \theta \, dn.$$

Substituting Eqs. (3.16) and (3.17) into this expression, we have

$$\begin{aligned} M &= 2\pi K\mu \int_0^\pi e^{a \cos \theta} \sin \theta \cos \theta \, d\theta \\ &= \frac{n\mu \int_0^\pi e^{a \cos \theta} \sin \theta \cos \theta \, d\theta}{\int_0^\pi e^{a \cos \theta} \sin \theta \, d\theta}. \end{aligned}$$

To evaluate these integrals, we put $x = \cos \theta$ and $dx = -\sin \theta \, d\theta$. Therefore,

$$\begin{aligned} M &= \frac{n\mu \int_1^{-1} x e^{ax} \, dx}{\int_1^{-1} e^{ax} \, dx} \\ &= n\mu \left(\frac{e^a + e^{-a}}{e^a - e^{-a}} - \frac{1}{a} \right) = n\mu \left(\coth a - \frac{1}{a} \right). \end{aligned} \tag{3.18}$$

But $n\mu$ is the maximum possible moment which the material can have. It corresponds to perfect alignment of all the atomic magnets parallel to the field, which is a state of complete saturation. Calling this quantity M_0, we have

$$\boxed{\frac{M}{M_0} = \coth a - \frac{1}{a}.} \tag{3.19}$$

The expression on the right is called the Langevin function, usually abbreviated by $L(a)$. Expressed as a series, it is

$$L(a) = \frac{a}{3} - \frac{a^3}{45} + \frac{2a^5}{945} - \cdots \tag{3.20}$$

$L(a)$ as a function of a is plotted in Figure 3.4. At large a it tends to 1, and at small a it has a slope of $\frac{1}{3}$, as may be seen from Eq. (3.20). When a is small, less than about 0.5, $L(a)$ is practically a straight line.

The Langevin theory leads to two conclusions:

1. Saturation will occur if a $(= \mu H/kT)$ is large enough. This makes good physical sense, because large H or low T, or both, is necessary if the aligning tendency of the field is going to overcome the disordering effect of thermal agitation.

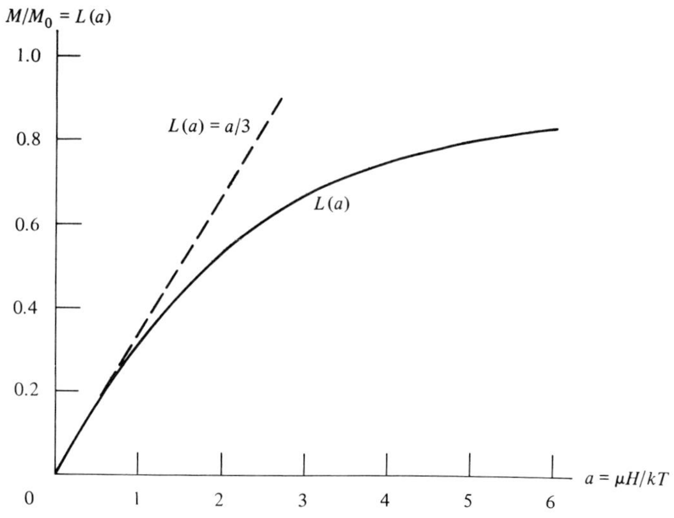

Fig. 3.4 Langevin function.

2. At small a, the magnetization M varies linearly with H. As we shall see presently, a *is* small under "normal" conditions, and linear M,H curves are observed, like that of Fig. 1.13 (b).

The Langevin theory also leads to the Curie law. For small a, $\mathrm{L}(a) = a/3$, and Eq. (3.18) becomes

$$M = \frac{n\mu a}{3} = \frac{n\mu^2 H}{3kT}. \tag{3.21}$$

Therefore,

$$\chi = \frac{\kappa}{\rho} = \frac{M}{\rho H} = \frac{n\mu^2}{3\rho kT}. \tag{3.22}$$

But n, the number of atoms per unit volume, is equal to $N\rho/A$. Therefore,

$$\chi = \frac{N\mu^2}{3AkT} = \frac{C}{T} \text{ emu/g Oe,} \tag{3.23}$$

which is Curie's law, with the Curie constant per gram given by

$$C = \frac{N\mu^2}{3Ak}. \tag{3.24}$$

The net magnetic moment μ per atom may be calculated from experimental data by means of Eq. (3.23). Consider oxygen, for example. It is one of the few gases which are paramagnetic; it obeys the Curie law and has a mass susceptibility of 1.08×10^{-4} emu/g Oe at 20°C. Therefore, writing M' (molecular weight) instead of A (atomic weight) in Eq. (3.23) because the constituents of oxygen are molecules, we have

$$\mu = \left(\frac{3M'kT\chi}{N}\right)^{1/2}$$

$$= \left[\frac{(3)(32)(1.38 \times 10^{-16}\ \text{erg/deg})(293)(1.08 \times 10^{-4}\ \text{emu/g Oe})}{6.02 \times 10^{23}}\right]^{1/2}$$

$$= 2.64 \times 10^{-20} \quad \text{erg/Oe per molecule}$$

$$= \frac{2.64 \times 10^{-20}}{0.927 \times 10^{-20}}\,\mu_B = 2.85\,\mu_B \text{ per molecule.}$$

This value of μ is typical. Even in heavy atoms or molecules containing many electrons, each with orbital and spin moments, most of the moments cancel out and leave a net magnetic moment of only a few Bohr magnetons.

We can now calculate a and justify our assumption that it is small. Typically, H is about 10,000 Oe in susceptibility measurements. Therefore, at room temperature,

$$a = \frac{\mu H}{kT} = \frac{(2.64 \times 10^{-20}\ \text{erg/Oe})(10^4\ \text{Oe})}{(1.38 \times 10^{-16}\ \text{erg/deg})(293)}$$

$$= 0.0065,$$

which is a value small enough so that the Langevin function $L(a)$ can be replaced by $a/3$.

The effect of even very strong fields in aligning the atomic moments of a paramagnetic is very feeble compared to the disordering effect of thermal energy at room temperature. For example, there are $N/32$ oxygen molecules per gram, each with a moment of 2.64×10^{-20} emu. If complete alignment could be achieved, the specific magnetization σ of oxygen would be $(6.02 \times 10^{23}/32)(2.64 \times 10^{-20})$, or 497 emu/g. (This value, incidentally, is more than double that of saturated iron.) On the other hand, the magnetization acquired in a field as strong as 100,000 Oe is only $\sigma = \chi H = (1.08 \times 10^{-4})(10^5) = 10.8$ emu/g, or about 2 percent of the saturation value.

There was nothing in our previous discussion of the diamagnetic effect to indicate that it was restricted to atoms with no net magnetic moment. In fact, it is not; the diamagnetic effect occurs in all atoms, whether or not they have a net moment. A calculation of the susceptibility of a paramagnetic should therefore be corrected by subtracting the diamagnetic contribution from the value given by Eq. (3.23). This correction is usually small (of the order of -0.5×10^{-6} emu/g Oe) and can often be neglected in comparison to the paramagnetic term.

The Langevin theory of paramagnetism, which leads to the Curie law, is based on the assumption that the individual carriers of magnetic moment (atoms or molecules) *do not interact with one another*, but are acted on only by the applied field and thermal agitation. Many paramagnetics, however, do not obey this law; they obey instead the more general Curie-Weiss law,

$$\chi = C/(T - \theta). \tag{3.14}$$

In 1907 Weiss* [3.3] pointed out that this behavior could be understood by postulating that the elementary moments *did* interact with one another. He suggested that this interaction could be expressed in terms of a fictitious internal field which he called the "molecular field" H_m and which acted in addition to the applied field H. The molecular field was thought to be in some way caused by the magnetization of the surrounding material. (If Weiss had advanced his hypothesis some ten years later, he would probably have called it the hypothesis of the atomic field. X-ray diffraction was discovered in 1912, and by about 1917 diffraction experiments had shown that all metals and simple inorganic solids were composed of atoms, not molecules.)

Weiss assumed that the intensity of the molecular field was directly proportional to the magnetization:

$$\boxed{H_m = \gamma M,} \tag{3.25}$$

where γ is called the molecular field constant. Therefore, the total field acting on the material is

$$H_t = H + H_m. \tag{3.26}$$

Curie's law may be written

$$\chi = M/\rho H = C/T.$$

H in this expression must now be replaced by H_t:

$$\frac{M}{\rho(H + \gamma M)} = \frac{C}{T}.$$

Solving for M, we find

$$M = \frac{\rho C H}{T - \rho C \gamma}.$$

Therefore,

$$\chi = \frac{M}{\rho H} = \frac{C}{T - \rho C \gamma} = \frac{C}{T - \theta}. \tag{3.27}$$

---

* Pierre Weiss (1865–1940), French physicist. He deserves to be called the father of modern magnetism, because almost the whole theory of ferromagnetism is due to him, and his ideas also permeate the theory of ferrimagnetism. Most of his work was done at the University of Strasbourg.

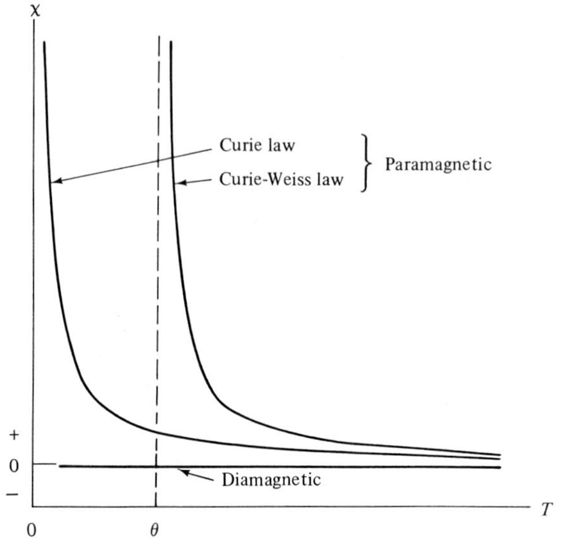

Fig. 3.5 Variation of mass susceptibility with absolute temperature for para- and diamagnetics.

Therefore, θ $(= \rho C\gamma)$ is a measure of the strength of the interaction because it is proportional to the molecular field constant γ. For substances that obey Curie's law, $\theta = \gamma = 0$.

Figure 3.5 shows how χ varies with T for para- and diamagnetics. If we plot $1/\chi$ vs. T for a paramagnetic, a straight line will result; this line will pass through the origin (Curie behavior) or intercept the temperature axis at $T = \theta$ (Curie-Weiss behavior). Data for two paramagnetics which obey the Curie-Weiss law are plotted in this way in Fig. 3.6, and we note that *both positive and negative* values of θ are observed, positive for $MnCl_2$ and negative for $FeSO_4$. Many paramagnetics obey the Curie-Weiss law with small values of θ, of the order of 10°K or less. A positive value of θ, as illustrated in Fig. 3.5, indicates that the molecular field is aiding the applied field and therefore tending to make the elementary magnetic moments parallel to one another and to the applied field. Other things being equal, the susceptibility is then larger than it would be if the molecular field were absent. If θ is negative, the molecular field opposes the applied field and tends to decrease the susceptibility.

It is important to note that the molecular field is in no sense a real field; it is rather a force, which tends to align or disalign the atomic or molecular moments, and the strength of this force depends on the amount of alignment already attained, because the molecular field is proportional to the magnetization. Further discussion of the molecular field will be deferred to the next chapter.

Early in this section it was stated that the effect of an applied field on the atomic or molecular "magnets" was to turn them toward the direction of the field. This

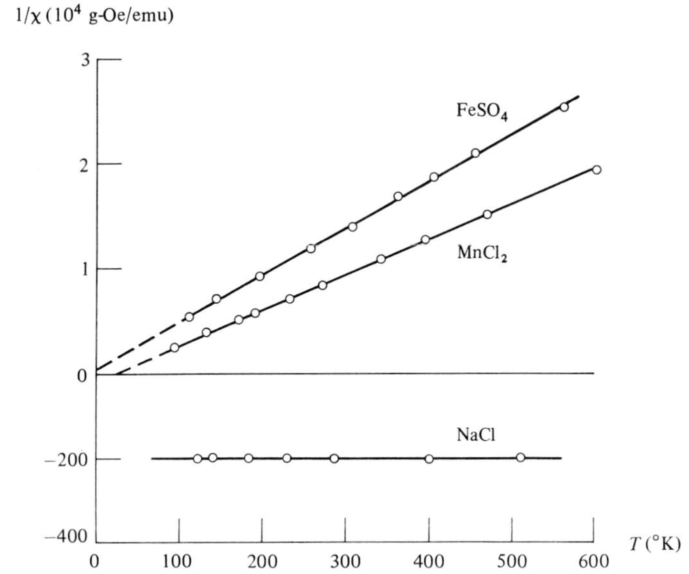

Fig. 3.6 Reciprocal mass susceptibilities of one diamagnetic and two paramagnetic compounds (after Ishiwara and Honda [3.4, 3.5]). Note change in vertical scale at origin.

statement requires qualification, because the effect of the field is not just a simple rotation, like that of a compass needle exposed to a field not along its axis. Instead, there is a precession of the atomic moments about the applied field, because each atom possesses a certain amount of *angular momentum* as well as a magnetic moment. This behavior is analogous to that of a spinning top. If the top in Fig. 3.7(a) is not spinning, it will simply fall over because of the torque exerted by the force of gravity F about its point of support A. But if the top is spinning about its axis, it has a certain angular momentum about that axis; the resultant of the gravitational torque and the angular momentum is a precession of the axis of spin about the vertical, with no change in the angle of inclination θ. In an atom, each electron has angular momentum by virtue of its spin and its orbital motion, and these momenta combine vectorially to give the atom as a whole a definite angular momentum. We might then roughly visualize a magnetic atom as a spinning sphere, as in Fig. 3.7(b), with its magnetic moment vector and angular momentum vector both directed along the axis of spin. A magnetic field H exerts a torque on the atom because of the atom's magnetic moment, and the resultant of this torque and the angular momentum is a precession about H. If the atom were isolated, the only effect of an increase in H would be an increase in the rate of precession, but no change in θ. However, in a specimen containing many atoms, all subjected to thermal agitation, there is an exchange of energy among atoms. When a field is applied, this exchange of energy disturbs the precessional motion enough so that the value of θ for each atom decreases slightly, until the distribution of θ values becomes appropriate to the existing values of field and temperature.

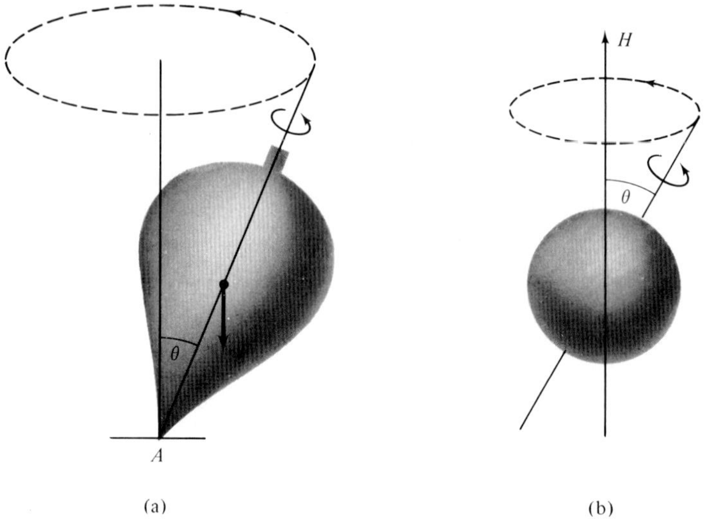

Fig. 3.7 Precession of (a) a spinning top in a gravitational field, and (b) a magnetic atom in a magnetic field.

3.7 QUANTUM THEORY OF PARAMAGNETISM

The main conclusions of the classical theory are modified by quantum mechanics, but not radically so. We will find that quantum theory greatly improves the quantitative agreement between theory and experiment without changing the qualitative features of the classical theory.

The central postulate of quantum mechanics is that *the energy of a system is not continuously variable.* When it changes, it must change by discrete amounts, called quanta of energy. If the energy of a system is a function of an angle, then that angle can undergo only discontinuous changes. This is precisely the case in a paramagnetic substance, where the potential energy of each atomic moment μ in a field H is given by $-\mu H \cos\theta$. In the classical theory, the energy, and hence θ,

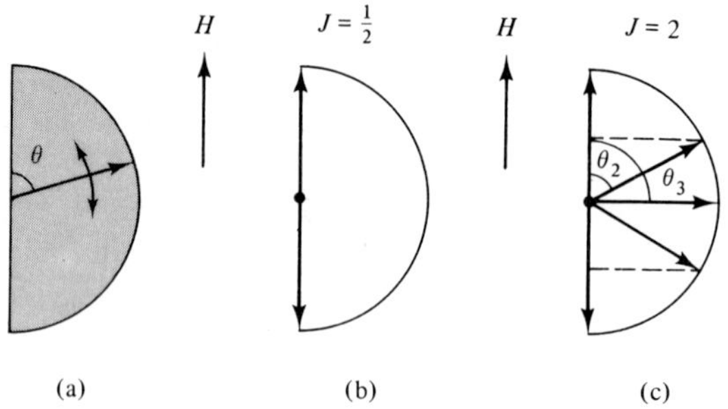

Fig. 3.8 Space quantization. (a) Classical case. (b) and (c) Two quantum possibilities.

was regarded as a continuous variable, and μ could lie at any angle to the field. In quantum theory, θ is restricted to certain definite values $\theta_1, \theta_2, \ldots$, and intermediate values are not allowed. This restriction is called *space quantization*, and is illustrated schematically in Fig. 3.8, where the arrows indicate atomic moments. The classical case is shown in (a), where the moments can have any direction in the shaded area; (b) and (c) illustrate two quantum possibilities, in which the moments are restricted to two and five directions, respectively. The meaning of J is given later.

The rules governing space quantization are usually expressed in terms of angular momentum rather than magnetic moment. We must therefore consider the relation between the two, first for orbital and then for spin moments. The orbital magnetic moment for an electron in the first Bohr orbit is, from Eq. (3.4),

$$\mu_{\text{orbit}} = \frac{eh}{4\pi mc} = \frac{e}{2mc}\left(\frac{h}{2\pi}\right).$$

If we write the corresponding angular momentum $h/2\pi$ as p, we have

$$\mu_{\text{orbit}} = \frac{e}{2mc}(p_{\text{orbit}}). \tag{3.28}$$

The angular momentum due to spin is $sh/2\pi = h/4\pi$, where s is a quantum number equal to $\frac{1}{2}$. Therefore, from Eq. (3.5),

$$\mu_{\text{spin}} = \frac{eh}{4\pi mc} = \frac{e}{mc}\left(\frac{h}{4\pi}\right) = \frac{e}{mc}(p_{\text{spin}}). \tag{3.29}$$

Therefore the ratio of magnetic moment to angular momentum for spin is twice as great as it is for orbital motion. The last two equations can be combined into one general relation between magnetic moment μ and angular momentum p by introducing a quantity g:

$$\mu = g(e/2mc)(p), \tag{3.30}$$

where $g = 1$ for orbital motion and $g = 2$ for spin. The factor g is called the *spectroscopic splitting factor*, or *g factor*.

In an atom composed of many electrons the angular momenta of the variously oriented orbits combine vectorially to give the resultant orbital angular momentum of the *atom*, which is characterized by the quantum number L. Similarly, the individual electron spin momenta combine to give the resultant spin momentum, described by the quantum number S. Finally, the orbital and spin momenta of the atom combine to give the total angular momentum of the atom, described by the quantum number J. Then the net magnetic moment of the atom, usually called the effective moment μ_{eff}, is given in terms of g and J, as we might expect by analogy with Eq. (3.30). The relation is

$$\mu_{\text{eff}} = g(eh/4\pi mc)\sqrt{J(J+1)} \text{ erg/Oe},$$

$$\mu_{\text{eff}} = g\sqrt{J(J+1)}\,\mu_B. \tag{3.31}$$

The moment may be said to be made up of an effective number n_{eff} of Bohr magnetons:

$$n_{\text{eff}} = g\sqrt{J(J+1)}.$$

Because of spatial quantization the effective moment can point only at certain discrete angles $\theta_1, \theta_2, \ldots$ to the field. Rather than specify these angles, we specify instead the possible values of μ_H, the component of μ_{eff} in the direction of the applied field H. These possible values are

$$\mu_H = g\, M_J\, \mu_B, \tag{3.32}$$

where M_J is a quantum number associated with J. For an atom with a total angular momentum J, the allowed values of M_J are

$$J, \quad J-1, \quad J-2, \ldots, \quad -(J-2), \quad -(J-1), \quad -J,$$

and there are $(2J+1)$ numbers in this set. For example, if $J = 2$ for a certain atom, the effective moment has five possible directions, and the component μ_H in the field direction must have one of the following five values:

$$2g\mu_B, \quad g\mu_B, \quad 0, \quad -g\mu_B, \quad -2g\mu_B.$$

This is the case illustrated in Figure 3.8(c).

The maximum value of μ_H is

$$\mu_H = gJ\mu_B. \tag{3.33}$$

and the symbol μ_H, if not otherwise qualified, is assumed to stand for this maximum value [The moments given by Eqs. (3.28), (3.29), and (3.30) are μ_H values.] The relation between μ_H and μ_{eff} is shown in Fig. 3.9.

The value of J for an atom may be an integer or a half-integer, and the possible values range from $J = \frac{1}{2}$ to $J = \infty$. These extreme values have the following meanings:

1. $J = \frac{1}{2}$. This corresponds to pure spin, with no orbital contribution ($L = 0$, $J = S = \frac{1}{2}$, so that $g = 2$. Since the permissible values of M_J decrease from $+J$ to $-J$ in steps of unity, these values are simply $+\frac{1}{2}$ and $-\frac{1}{2}$ for this case. The corresponding resolved moments μ_H are then μ_B and $-\mu_B$, parallel and antiparallel to the field, as illustrated in Fig. 3.8(b).
2. $J = \infty$. Here there are an infinite number of J values, corresponding to an infinite number of moment orientations. This is equivalent to the classical distribution of Fig. 3.8(a).

To compute μ_H or μ_{eff} we must know g, as well as J, for the atom in question. The g factor is given by the Landé equation:

$$g = 1 + \frac{J(J+1) + S(S+1) - L(L+1)}{2J(J+1)}. \tag{3.34}$$

If there is no net orbital contribution to the moment, $L = 0$ and $J = S$. Then Eq. (3.34) gives $g = 2$ whatever the value of J. On the other hand, if the spins

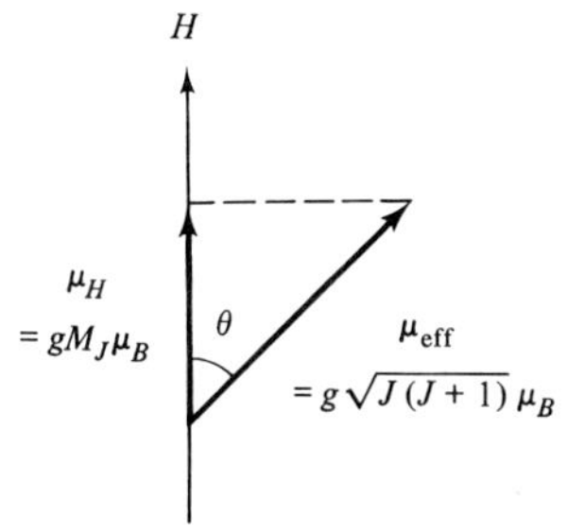

Fig. 3.9 Relation between effective moment and its component in the field direction.

cancel out, then $S = 0$, $J = L$, and $g = 1$. The g factors of most atoms lie between 1 and 2, but values outside this range are sometimes encountered.

At this point the calculation of the net magnetic moment of an atom would seem straightforward, simply by a combination of Eqs. (3.34) and (3.31). However, the values of J, L, and S are known only for *isolated atoms;* it is, in general, impossible to calculate μ for the atoms of a solid, unless certain simplifying assumptions are made. One such assumption, valid for many substances, is that there is no orbital contribution to the moment, so that $J = S$. The orbital moment is, in such cases, said to be *quenched.* This condition results from the action on the atom or ion considered of the electric field, called the *crystalline field*, produced by the surrounding atoms or ions in the solid. This field has the symmetry of the crystal involved. Thus the electron orbits in a particular isolated atom might be circular; but when that atom forms part of, for example, a cubic crystal, the orbits might become elongated along three mutually perpendicular axes because of the electric fields created by the adjoining atoms located on these axes. In any case, the orbits are in a sense bound, or "coupled," rather strongly to the crystal lattice. The spins, on the other hand, are only loosely coupled to the orbits. Thus, when a magnetic field is applied along some arbitrary direction in the crystal, the strong orbit-lattice coupling often prevents the orbits, and their associated orbital magnetic moments, from turning toward the field direction, whereas the spins are free to turn because of the relatively weak spin-orbit coupling. The result is that only the spins contribute to the magnetization process and the resultant magnetic moment of the specimen; the orbital moments act as though they were not there. Quenching may be complete or partial.

Fortunately, it is possible to measure g for the atoms of a solid, and such measurements are of great importance because they tell us what fraction of the total moment, which is also measurable, is contributed by spin and what fraction by orbital motion. Experimental g factors will be given later.

Gyromagnetic effect

Two entirely different kinds of experiments are available for the determination of g. The first involves the gyromagnetic effect, which depends on the fact that magnetic moments and angular momenta are coupled together; whatever is done to change the direction of one will change the direction of the other. From the magnitude of the observed effect a quantity g', called the *magnetomechanical factor* or *g' factor*, can be calculated. The g factor can then be found from the relation

$$\frac{1}{g} + \frac{1}{g'} = 1. \tag{3.34a}$$

If the magnetic moment is due entirely to spin, then $g = g' = 2$. If there is a small orbital contribution, g is somewhat larger than 2 and g' somewhat smaller.

Two methods of measuring the gyromagnetic effect, and thus the value of g', have been successful:

1. *Einstein-de Haas method.* A rod of the material to be investigated is suspended vertically by a fine wire and surrounded by a magnetizing solenoid. If a field is suddenly applied along the axis of the rod, the atomic moments will turn toward the axis. But this will also turn the angular momentum vectors toward the axis. Since angular momentum cannot be created except by external torques, this increase in the axial component of momentum of the atoms must be balanced by an increase, in the opposite direction, of the momentum of the bar as a whole. The result is a rotation of the bar through a very small angle, from which the value of g' can be computed. The experiment is extremely difficult with a ferromagnetic rod, even more so with a paramagnetic, because the size of the observed effect depends mainly on the magnetization that can be produced in the specimen.
2. *Barnett method.* The specimen, again in the form of a rod, is very rapidly rotated about its axis. The angular momentum vectors therefore turn slightly toward the axis of rotation and cause the magnetic moments to do the same. The rod therefore acquires a very slight magnetization along its axis, from which g' can be calculated.

The two methods may be summarily described as "rotation by magnetization" and "magnetization by rotation." Detailed accounts of both are given by Bates [G.13].

Magnetic resonance

The second kind of experiment is magnetic resonance, which measures g directly. The specimen is placed in the strong field H_z of an electromagnet, acting along the z axis. It is also subjected to a weak field H_x acting at right angles, along the x axis; H_x is a high-frequency alternating field. The atomic moments precess around H_z at a rate dependent on g and H_z. Energy is absorbed by the specimen from the alternating field H_x, and, if the intensity of H_z or the frequency v of H_x is slowly varied, a point will be found at which the energy absorption rises to a sharp maximum. In this resonant state the frequency v equals the frequency of precession, and both are proportional to the product gH_z, from which g may be calculated.

Assuming that g and J are known for the atoms involved, we can proceed to calculate the total magnetization of a specimen as a function of the field and temperature. The procedure is the same as that followed in deriving the classical (Langevin) law, except that:

1. The quantized component of magnetic moment in the field direction μ_H $(= gM_J\mu_B)$ replaces the classical term $\mu \cos\theta$.
2. A summation over discrete moment orientations replaces an integration over a continuous range of orientations.

The potential energy of each moment in the field H is

$$E_p = -gM_J\mu_B H, \tag{3.35}$$

which is the counterpart of Eq. (3.15). According to Boltzmann statistics, the probability of an atom having an energy E_p is proportional to

$$e^{-Ep/kT} = e^{gM_J\mu_B H/kT}$$

If there are n atoms per unit volume, the magnetization M is given by the product of n and the average magnetic moment resolved in the direction of the field, or

$$M = n\frac{\sum gM_J\mu_B \quad e^{gM_J\mu_B H/kT}}{\sum e^{gM_J\mu_B H/kT}}, \tag{3.36}$$

where the summations are over M_J and extend from $-J$ to $+J$. After considerable manipulation [3.6], this reduces to

$$M = ngJ\mu_B\left[\frac{2J+1}{2J}\coth\left(\frac{2J+1}{2J}\right)a' - \frac{1}{2J}\coth\frac{a'}{2J}\right], \tag{3.37}$$

where

$$a' = \frac{gJ\mu_B H}{kT} = \frac{\mu_H H}{kT}.$$

But $ngJ\mu_B = n\mu_H$, which is the product of the number of atoms per unit volume and the maximum moment of each atom in the direction of the field. Therefore $n\mu_H = M_0$, the saturation magnetization, and

$$\boxed{\frac{M}{M_0} = \frac{2J+1}{2J}\coth\left(\frac{2J+1}{2J}\right)a' - \frac{1}{2J}\coth\frac{a'}{2J}.} \tag{3.38}$$

The expression on the right is called the Brillouin function [3.7] and was first obtained in 1927. It is abbreviated $B(J,a')$. Numerical values of this function for various J and of the Langevin function have been tabulated by Smart [G.28].

When $J = \infty$, the classical distribution, the Brillouin function reduces to the Langevin function:

$$M/M_0 = \coth a' - (1/a'). \tag{3.39}$$

When $J = \frac{1}{2}$, so that the magnetic moment consists of one spin per atom, the Brillouin function reduces to

$$\boxed{M/M_0 = \tanh a'.} \tag{3.40}$$

A direct derivation of this last equation may clarify some of the physics involved. When $J = \frac{1}{2}$, the effective moment, given by Eq. (3.31), is

$$2\sqrt{(\tfrac{1}{2})(\tfrac{3}{2})}\,\mu_B \quad \text{or} \quad \sqrt{3}\,\mu_B.$$

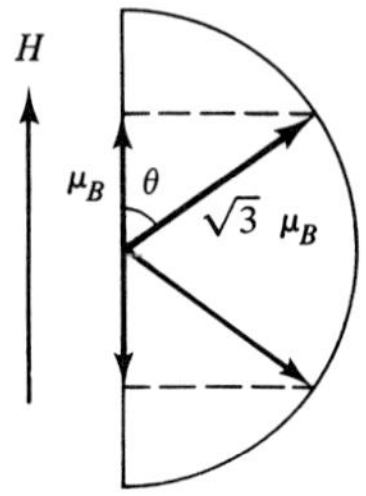

Fig. 3.10 Vector relations for $J = \frac{1}{2}$.

The component of this in the field direction is, according to Eq. (3.33), 2 ($\frac{1}{2}$) μ_B or μ_B. Note that J is always less than $\sqrt{J(J+1)}$, so that μ_H is always less than μ_{eff}. This means that the moment is never parallel to the field direction, even when J is large, and that drawings like Fig. 3.8 (b) and (c), which show such parallelism, are, strictly speaking, incorrect. The true situation is shown in Fig. 3.10, for $J = \frac{1}{2}$, where $\theta = 54.7°$. Nevertheless, a drawing like Fig. 3.8 (b) is a conventional description, and so is loose language like "spins parallel and antiparallel to the field," or, more briefly, "spins up and spins down."

Let n_+ and n_- equal the numbers of atoms with spins parallel and antiparallel to the field, respectively, per unit volume. The corresponding potential energies are $-\mu_B H$ and $+\mu_B H$. Then

$$n_+ = be^{\mu_B H/kT} \quad \text{and} \quad n_- = be^{-\mu_B H/kT}, \tag{3.41}$$

where b is a proportionality constant. Thus, at constant temperature, the number of atoms with spin up increases as the field increases, and the number of atoms with spin down decreases. We say loosely that the field has caused some of the spins to "flip over." The total number of atoms per unit volume is

$$n = n_+ + n_- = b\,(e^{\mu_B H/kT} + e^{-\mu_B H/kT}). \tag{3.42}$$

The average magnetic moment in the field direction is

$$(\mu_H)_{av} = \frac{n_+(\mu_B) + n_-(-\mu_B)}{n}. \tag{3.43}$$

Combining the previous equations yields the magnetization

$$M = n(\mu_H)_{av},$$

$$M = n\mu_B \frac{e^{\mu_B H/kT} - e^{-\mu_B H/kT}}{e^{\mu_B H/kT} + e^{-\mu_B H/kT}},$$

$$M = M_0 \tanh \frac{\mu_B H}{kT}, \tag{3.44}$$

$$M/M_0 = \tanh a'. \tag{3.40}$$

The Brillouin function, like the Langevin, is zero for a' equal to zero and tends to unity as a' becomes large. However, the course of the curve in between depends on the value of J for the atom involved. Moreover, the quantity a in the classical theory differs from the corresponding quantity a' in the quantum theory:

$$\text{(classical)} \qquad a = \mu H/kT, \tag{3.45}$$

$$\text{(quantum)} \qquad a' = \mu_H H/kT. \tag{3.46}$$

In the classical theory, μ is the net magnetic moment of the atom. The quantity which corresponds to this in quantum theory is the *effective* moment μ_{eff}, and not its component μ_H on the field direction. The parameters a and a' therefore have distinctly different physical meanings.

When H is large enough and T low enough, a paramagnetic can be saturated, and we can then compare the experimental results with the predictions of classical and quantum theory. For example, Henry [3.8] has measured the magnetization of potassium chromium alum, $KCr(SO_4)_2 \cdot 12H_2O$, at fields up to 50,000 Oe and at temperatures of 4.2°K (liquid helium) and below. The only magnetic atom, or rather ion, in this compound is the Cr^{3+} ion. The quantum numbers for this ion in the *free state* are $J = \frac{3}{2}$, $L = 3$, $S = \frac{3}{2}$, which lead to $g = \frac{2}{5}$, as the reader can verify from Eq. (3.34). The magnetic moment per molecule, or per chromium ion, is then given by Eq. (3.37):

$$M/n = gJB(J,a')\mu_B. \tag{3.47}$$

Figure 3.11 shows the experimental data as a function of H/T. The curve through the experimental points is a plot of Eq. (3.47), with $g = 2$ (not $\frac{2}{5}$) and $J = \frac{3}{2}$. The lower curve is a similar plot with $g = \frac{2}{5}$. The excellent agreement between theory and experiment for $g = 2$ shows that the magnetic moment of the chromium ion *in this compound* is entirely due to spin; in the solid, $L = 0$ rather than 3, and the orbital component has been entirely quenched. The maximum moment* of the chromium ion in the field direction is $(2)(\frac{3}{2})\,\mu_B$, or $3\mu_B$, and its effective moment is

$$\begin{aligned} \mu_{\text{eff}} &= g\sqrt{J(J+1)}\,\mu_B \\ &= 2\sqrt{(\tfrac{3}{2})(\tfrac{5}{2})}\,\mu_B \\ &= \sqrt{15}\,\mu_B = 3.87\,\mu_B. \end{aligned} \tag{3.48}$$

If this value is set equal to μ in the classical equations, (3.45) and (3.18), the upper curve of Fig. 3.11 results. It is asymptotic to 3.87 μ_B per ion, whereas the experimental and quantum curves are asymptotic to 3.00. The wide disparity between classical theory and experiment, when H/T is large, is clearly evident.

The fields required, at room and liquid-helium temperatures, to produce the values of H/T shown in Fig. 3.11, are indicated on the two H scales at the

* A method of predicting the maximum moment, when it is due only to spin, is given in Section 6.1.

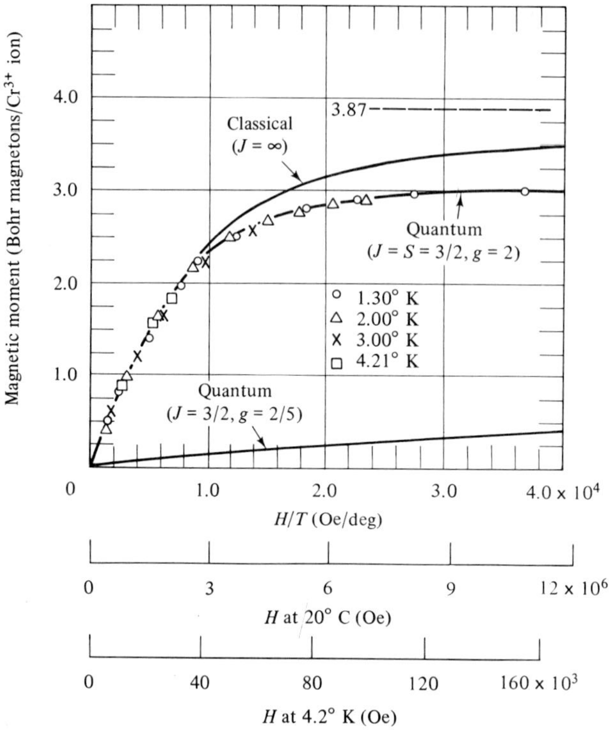

Fig. 3.11 Magnetic moment of potassium chromium alum as a function of H/T. (After Henry [3.8])

bottom. At room temperature, enormous fields, of the order of ten million oersteds, would be required to achieve saturation.

While measurements at very low temperatures and high fields are necessary in order to obtain the saturation effects which allow magnetic theories to be tested, most measurements on paramagnetics are confined to "normal" temperatures and "moderate" fields (5,000–10,000 Oe), i.e., to the region near the origin of Fig. 3.11, where the magnetization curves are linear. We therefore need a quantum-theory expression for the susceptibility. When x is small, coth x can be replaced by $(1/x + x/3)$. With this substitution, the Brillouin function, for small a', reduces to

$$B(J,a') = \frac{a'(J+1)}{3J}, \tag{3.49}$$

and

$$\begin{aligned} M &= n\,g\,J\,\mu_B\,B(J,a') \\ &= n\,g\,J\,\mu_B\left(\frac{gJ\mu_B H}{kT}\right)\left(\frac{J+1}{3J}\right) \\ &= \frac{ng^2\,J(J+1)\mu_B^2\,H}{3kT} \\ &= \frac{n\,\mu_{\text{eff}}^2\,H}{3kT}. \end{aligned} \tag{3.50}$$

The mass susceptibility is then

$$\chi = \frac{\kappa}{\rho} = \frac{M}{\rho H} = \frac{n\mu_{\text{eff}}^2}{3\rho kT} = \frac{N\mu_{\text{eff}}^2}{3AkT}. \tag{3.51}$$

This is the quantum analog of classical equation (3.23). Similarly, the Curie constant per gram is given by

$$C = \frac{N\mu_{\text{eff}}^2}{3Ak}, \tag{3.52}$$

where N is Avogadro's number and A, the atomic weight, is to be replaced by M', the molecular weight, if μ_{eff} refers to a molecule rather than an atom.

As mentioned earlier, the quantum numbers J, L, and S are normally not known for an atom or molecule in a solid. Under these circumstances it is customary to compute the magnetic moment from the susceptibility measurements, on the assumption that the moment is due only to the spin component; then $L = 0$, $J = S$, and $g = 2$. The result is called the "spin-only" moment. The value of $J(= S)$ is computed from the experimental value of C by a combination of Eqs. (3.31) and (3.52), and the moment μ_H is then given by $gJ\mu_{\text{B}} = 2J\mu_{\text{B}} = 2S\mu_{\text{B}}$. Reporting the results of susceptibility measurements in terms of a spin-only moment is merely a convention and does not imply that the orbital contribution is really absent. Other kinds of information are usually necessary to decide this point.

As an example of this kind of calculation, we can consider the data on potassium chromium alum. According to Foex [3.9], this substance follows the Curie law exactly, with a molecular Curie constant C_M of 1.85. (C_M is the Curie constant per gram molecular weight.) Therefore, Eq. (3.52) becomes

$$C_M = CM' = \frac{N\mu_{\text{eff}}^2}{3k}$$

and

$$\begin{aligned} \mu_{\text{eff}} &= \left(\frac{3kC_M}{N}\right)^{1/2} \\ &= \left[\frac{3(1.38 \times 10^{-16})(1.85)}{6.02 \times 10^{23}}\right]^{1/2} \\ &= 3.57 \times 10^{-20} \text{ erg/Oe} \\ &= \frac{3.57 \times 10^{-20}}{0.927 \times 10^{-20}}\,\mu_{\text{B}} = 3.85\,\mu_{\text{B}}. \end{aligned}$$

For pure spin, Eq. (3.31) becomes

$$2\sqrt{J(J+1)}\,\mu_{\text{B}} = \mu_{\text{eff}} = 3.85\,\mu_{\text{B}}. \tag{3.53}$$

(Some authors write S instead of J in this equation to emphasize that only spin moments are involved.) The solution of Eq. (3.53) is $J = 1.49$, which compares well with $J = \frac{3}{2}$, the value assumed in drawing the central curve of Fig. 3.11. (In this case, the good fit of the magnetization data to the curve, over the whole range of H/T, is conclusive evidence that the moment of the chromium ion in this compound is due only to spin.) We can also calculate the maximum component of the moment in the field direction, namely, $\mu_H = gJ\,\mu_B = 2(1.49) \times \mu_B = 2.98\,\mu_B$.

Writers of technical papers sometimes cause confusion by not being sufficiently specific when reporting their results. For example, the statement that "the susceptibility measurements lead to a spin-only moment of x Bohr magnetons" is ambiguous. Does he mean μ_H or μ_{eff}? Usually, μ_H is meant, but one can never be sure unless the value of the Curie constant C is also given. This should always be done, because the value of either moment can be calculated from C.

Equation (3.51) for the susceptibility applies to a substance which obeys the Curie law. A quantum relation for Curie-Weiss behavior can be obtained, just as in the classical case, by introducing a molecular field $H_m (= \gamma M)$, which adds to the applied field H. We then have, according to Eq. (3.27),

$$\chi = C/(T - \theta),$$

where C, the Curie constant per gram, is now given by Eq. (3.52). Therefore,

$$\chi = \frac{N\mu_{eff}^2}{3Ak(T - \theta)} = \frac{Ng^2J(J + 1)\mu_B^2}{3Ak(T - \theta)}, \tag{3.54}$$

where θ, as before, is a measure of the molecular field constant γ and is given by

$$\theta = \rho C\gamma = \frac{\rho N\mu_{eff}^2\gamma}{3Ak} = \frac{\rho Ng^2J(J + 1)\mu_B^2\gamma}{3Ak}. \tag{3.55}$$

Before concluding these sections on the theory of dia- and paramagnetism, it is only fair to point out the range of validity of some of the arguments advanced. On a basic level, the theory of any kind of magnetism must be an atomic theory or, more exactly, an electronic theory. But the electrons in atoms do not behave in a classical way, and to understand their behavior we must abandon the relative simplicity and "reasonableness" of classical physics for the complexity and abstractions of quantum mechanics. Classical explanations are simply not valid on the atomic level, *even when they lead to the right answer.* For example, Eq. (3.28), which states that the ratio of the magnetic moment to the angular momentum is $e/2mc$ for an electron moving in a circular orbit, is true in quantum mechanics. But the way in which this result was derived in this chapter is entirely classical. This classical treatment, and similar treatments of other basic magnetic phenomena, should be regarded more as aids to visualization than as valid analyses of the problem.

3.8 PARAMAGNETIC SUBSTANCES

These are substances composed of atoms or ions which have a net magnetic moment because of noncancellation of the spin and orbital components. Closed electron shells usually exhibit no net magnetic moment and lead to diamagnetism. Incomplete outer shells, as in metals like sodium or copper, lead to complex behavior, discussed below. However, incomplete *inner* shells, such as those of the transition metal ions and rare earth ions, can have a large net moment, and compounds of these elements are strongly paramagnetic.

Salts of the transition elements

These show the simplest behavior. The only magnetic ions in such compounds are the transition metal ions, and the magnetic moments of these are due almost entirely to spin, the orbital components being largely quenched. This is shown by the fact that the susceptibility calculated on a spin-only basis agrees well with the measured value, and we have already seen an example of that in potassium chromium alum, $KCr(SO_4)_2 \cdot 12H_2O$. Even more direct evidence is given by experimental values of the g factor, which are close to 2. In chlorides, sulfates, and carbonates of Cr, Mn, Fe, and Co, for example, the g factors are 1.95 for Cr^{3+}, 1.98 for Mn^{2+}, 1.89 for Fe^{2+}, and 1.54 for Co^{2+} [G.13, p. 270]. The transition-metal salts usually obey the Curie law, or the Curie-Weiss with a small value of θ, as shown in Fig. 3.6. (This behavior is rather surprising. The Langevin theory or its quantum-mechanical counterpart, which leads to the Curie law, was originally derived for a gas, on the assumption that there is no interaction between the individual carriers of magnetic moment, be they atoms or molecules. There is no *a priori* reason why this theory should also apply to solids, in which the atoms are packed close together. But in fact it often does apply, and we find that it applies more exactly the more "magnetically dilute" the substance is. Thus, if a compound contains a lot of water of crystallization, as $KCr(SO_4)_2 \cdot 12\,H_2O$ does, the magnetic ions, in this case Cr^{3+}, will be so far separated from each other that any interaction between them will be negligible, and the Curie law will be closely obeyed. It is thus often true that the more complicated the chemical formula of a substance, the simpler its magnetic behavior.)

Salts and oxides of the rare earths

These compounds are very strongly paramagnetic. (The effective magnetic moment varies in a regular way with atomic number and reaches a value as large as 10.6 Bohr magnetons for the trivalent ion of dysprosium.) In these substances both susceptibility and g-factor measurements show that orbital motion contributes a large part of the observed moment. In effect, the electrons in the unfilled shell responsible for the magnetic moment (the $4f$ shell in these substances) lie so deep in the ion that the outer electron shells shield them from the crystalline field of the other ions, and the orbital moments remain unquenched.

Rare-earth elements

These are also strongly paramagnetic. The magnetic moments per ion are so large that there is considerable interaction between adjacent ions, even though the moments are deep-seated, and these elements obey the Curie-Weiss law with rather large values of θ, rather than the simple Curie law. Some of the rare earths become ferromagnetic at low temperature, although gadolinium becomes ferromagnetic at just below 16° C. They are discussed further in Section 5.4.

Metals

The magnetic behavior of the metals is complex. As shown in Appendix 2, some are ferromagnetic at room temperature (Fe, Co, and Ni), one is antiferromagnetic (Cr), and the rest are para- or diamagnetic. The transition metals are ferro-, antiferro-, or paramagnetic. (All ferromagnetic substances become paramagnetic above their Curie temperatures, and the paramagnetism of Fe, Co, and Ni above their Curie points will be dealt with in the next chapter.) The susceptibility of the para- and diamagnetic metals is made up of three parts:

1. Diamagnetism of the core electrons. A metal is made up of positive ions and free electrons. These ions usually consist of closed shells which contribute a diamagnetic term to the susceptibility, just as they do in any substance.
2. Diamagnetism of the conduction electrons. When a magnetic field is present, the conduction electrons must move in curved paths. This results in an additional diamagnetic effect, for much the same reason that electron motion in an orbit causes a diamagnetic reaction when a field is applied.
3. Paramagnetism of the conduction electrons, also called *Pauli paramagnetism* or *weak spin paramagnetism*. The conduction electrons, present to the extent of one or more per atom, depending on the valence, each have a spin magnetic moment of one Bohr magneton. One would therefore expect them to make a sizable paramagnetic contribution. This does not happen, however, because the conduction electrons of a metal occupy energy levels in such a way that an applied field can reorient the spins of only a very small fraction of the total number of electrons. The resulting paramagnetism, which will be more fully explained in Sec. 4.4, is very weak and does not vary much with temperature.

The sum of these three effects, all of them small, is the observed susceptibility of the metal. If the first two are stronger, the metal is diamagnetic, like copper; if the third outweighs the other two, it is paramagnetic, like manganese or aluminum. If the net effect is paramagnetic, the resultant paramagnetism is very weak. (For example, at room temperature the susceptibility per atom of manganese in metallic manganese is less than 4 percent of the susceptibility per atom of manganese in $MnCl_2$) Moreover, the susceptibility of such a paramagnetic does not obey the Curie or Curie-Weiss law; because effects (1) and (2) are independent, and (3) almost independent, of temperature, the resulting susceptibility can decrease as the temperature increases, remain constant, or even increase.

General

When a solid solution is formed between paramagnetic or diamagnetic metals, or between a para- and a diamagnetic, the variation of the susceptibility with composition is, in general, unpredictable. One effect is clear, however. If a para- and a diamagnetic form a continuous series of solid solutions, the susceptibility must pass through zero at some intermediate composition. A substance having this composition will be completely unaffected by an applied magnetic

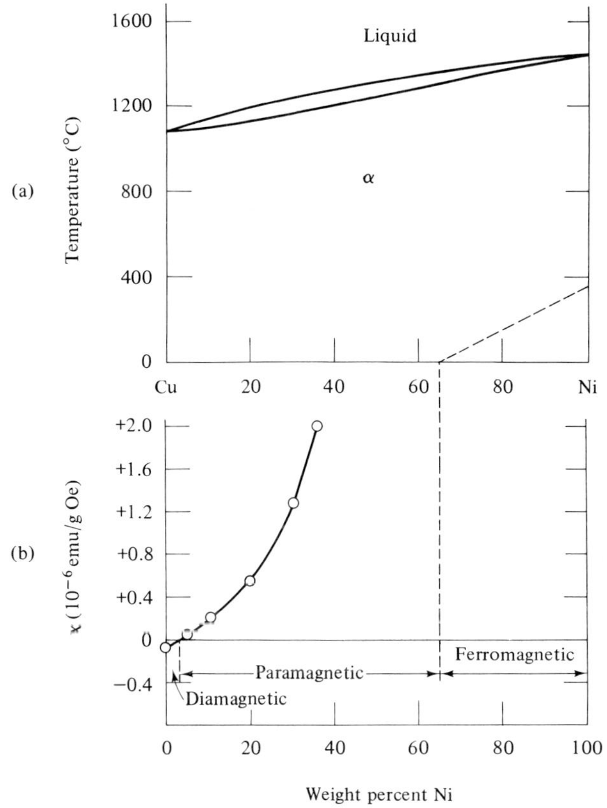

Fig. 3.12 (a) Copper-nickel phase diagram [3.11]. (b) Mass susceptibility of copper-nickel alloys at room temperature [3.9].

field and forms an exception to the general statement that all substances are magnetic. However, this zero value of the susceptibility will be retained only at one temperature, because the susceptibility of the paramagnetic constituent will generally change with temperature. An example of such a material occurs in the Cu-Ni system. The phase diagram is shown in Fig. 3.12(a), where the dashed line indicates the Curie temperatures of the ferromagnetic, Ni-rich alloys. Below about 65 percent Ni the alloys become paramagnetic at room temperature. Figure 3.12(b) shows the room temperature susceptibility of the Cu-rich alloys; it passes through zero at 3.7 weight percent Ni. Below room temperature this alloy becomes slightly diamagnetic but, at any temperature between room temperature and 2 K, its susceptibility is less than one-tenth that of pure copper [3.10]. It is therefore a suitable material for specimen holders, and other parts of equipment designed for delicate magnetic measurements, which must have a susceptibility as near zero as possible.

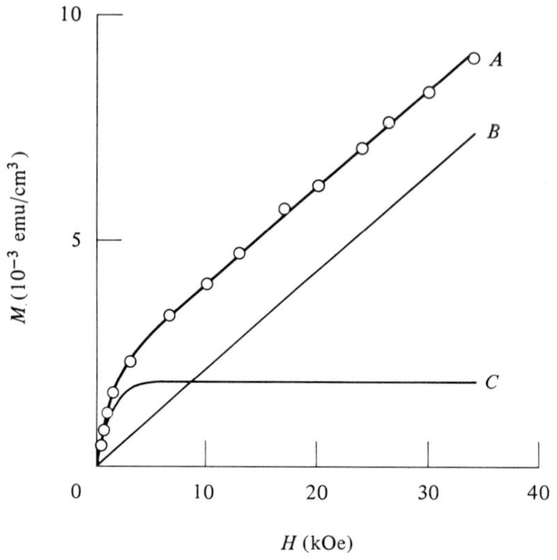

Fig. 3.13 Magnetization curve A of copper containing 0.1 % iron. (After Bitter and Kaufman [3.12])

Although the terms "weak" and "strong" paramagnetism have been used often in this section, it must be remembered that they are only relative and the susceptibility of any paramagnetic is minute in comparison with that of a ferromagnetic. This means that a small amount of a ferromagnetic impurity in a para- or diamagnetic can mask the true behavoir of the material. If the impurity is present in solid solution, the observed M,H curve will be linear from the origin, but the slope, and hence the susceptibility, will depend on the concentration of the impurity. If the impurity is present as a ferromagnetic second phase, the M,H curve will be curved initially as the second phase becomes increasingly saturated. Figure 3.13 shows this effect. A specimen of copper containing 0.1 weight percent iron gave curve A, which can be regarded as the sum of curves B and C. There is enough iron present in solid solution to change the normally diamagnetic "copper" to a paramagnetic, and B is the M,H curve of this solid solution. The rest of the iron is present as a ferromagnetic second phase, consisting of particles of an iron-rich iron-copper solid solution. C is the magnetization curve of this phase, which is saturated at a field of about 6 kOe. The susceptibility of the solid solution is given by the slope of the straight-line portion of A, which has the same slope as B. (In a substance of this kind, the applied field at which the second-phase particles saturate can depend markedly on the shape and orientation of these particles. The demagnetizing field associated with each particle will be much larger if it is spherical than if it is, for example, in the shape of a rod parallel to the applied field.)

Existing information on the magnetic state of the elements at room temperature is given in Appendix 2 in the form of a periodic table. The magnetic state of the rare earths is so dependent on temperature that data on these elements are shown separately in Appendix 3. The best source of data on dia- and paramagnetics is the compilation of Foex [3.9], which includes values reported through 1956. He gives the following data for elements, alloys, organic and inorganic compounds: atomic and molecular susceptibilities, effective number of Bohr magnetons n_{eff} for paramagnetics, and the value of θ in the Curie-Weiss law.

PROBLEMS

3.1 Assume that the electron is a solid sphere rotating about an axis through its center. Its rotational velocity is determined by the fact that the angular momentum due to spin is $sh/2\pi$, where $s = \frac{1}{2}$ and h is Planck's constant. Calculate the magnetic moment due to spin, in units of Bohr magnetons, on the assumption that the total charge on the electron is distributed uniformly (a) over its surface, and (b) throughout its volume.

3.2 For a paramagnetic which obeys the Curie-Weiss law show that:

a) The effective number of Bohr magnetons per molecule is given by

$$n_{\text{eff}} = 2.83 \sqrt{C_M},$$

where C_M is the Curie constant per molecule.

b) The molecular field is given by

$$H_m = \frac{\theta H}{T - \theta}.$$

3.3 The susceptibility of $FeCl_2$ obeys the Curie-Weiss law over the range 90° to 300°K, with $\theta = 48$°K. Its molecular susceptibility at 20°C is 1.475×10^{-2} emu/g mol Oe.

a) What is the effective magnetic moment in Bohr magnetons?

b) What are the spin-only values of J and μ_H (max)?

c) At an applied field of 10,000 Oe, what is the intensity of the molecular field at 20°C and at − 150°C?

3.4 Show that the Brillouin function $B(J,a')$ reduces to Eqs. (3.39) and (3.40) for $J = \infty$ and $\frac{1}{2}$, respectively, and to Eq. (3.49) when a' is small.

3.5 Plot the relation between M/M_0 and H, according to quantum theory, for a substance with $g = 2$ and $J = \frac{1}{2}$ for fields up to 100,000 Oe and temperatures of 20°C and 2°K.

a) What is the effective moment?
b) What is the atomic susceptibility at 20°C?
c) What percentage of the saturation magnetization is achieved at 100,000 Oe and 20°C?
d) What field is required to produce 95 percent of saturation at 2°K?

3.6 Plot the relation between M/M_0 and H, according to classical theory, for the same substance, fields, and temperatures as in the previous problem.

a) What is the atomic susceptibility at 20°C?
b) What percentage of the saturation magnetization is achieved at 100,000 Oe and 20°C?
c) How do you reconcile the fact that the susceptibilities are the same, whether calculated by the classical or quantum theory, but the saturation percentages, at 100,000 Oe and 20°C, are different?

3.7 Potassium chromium alum has $J = S = \frac{3}{2}$, $L = 0$, $g = 2$.

a) What are the possible values of μ_H? Sketch the directions of the μ_{eff} vectors which will give these values of μ_H.
b) Calculate the possible angles θ_1, θ_2,... between the effective magnetic moment of an atom and the applied field.
c) When $M/M_0 = 0.90$, calculate the fraction of the total number of atoms which has its effective moments at angles of θ_1, θ_2, ... to the field.

3.8 The susceptibility of $\alpha - \text{Mn}$ at 20°C is 9.63×10^{-6} emu/g Oe. The susceptibility of $MnCl_2$ obeys the Curie-Weiss law, with $\theta = 3.0°\text{K}$ and an effective moment of 5.7 μ_B per molecule. Calculate, for room temperature, the susceptibility per atom of manganese in $\alpha - \text{Mn}$ as a percentage of the susceptibility per atom of manganese in $MnCl_2$.

CHAPTER 4

FERROMAGNETISM

4.1 INTRODUCTION

Magnetization curves of iron, cobalt, and nickel are shown in Fig. 4.1. These curves are partly hypothetical. The experimental values of the saturation magnetization M_s are given for each metal, but no field values are shown on the abscissa. This is done to emphasize the fact that the shape of the curve from $M = 0$ to $M = M_s$ and the strength of the field at which saturation is attained are structure-sensitive properties, whereas the magnitude of M_s is not. The problems presented by the magnetization curve of a ferromagnetic are therefore rather sharply divisible into two: (1) the magnitude of the saturation value, and (2) the way in which this value is reached from the demagnetized state. We shall now consider the first problem and leave the details of the second to later chapters.

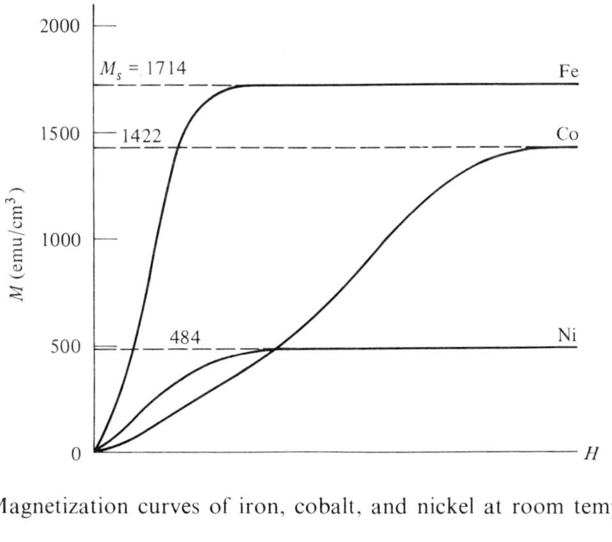

Fig. 4-1 Magnetization curves of iron, cobalt, and nickel at room temperature.

A single crystal of pure iron, properly oriented, can be brought almost to saturation with a field of about 50 Oe. Each cubic centimeter then has a magnetic moment of about 1700 emu. At the same field a typical paramagnetic will have a magnetization of about 10^{-3} emu/cm^3. Ferromagnetism therefore involves

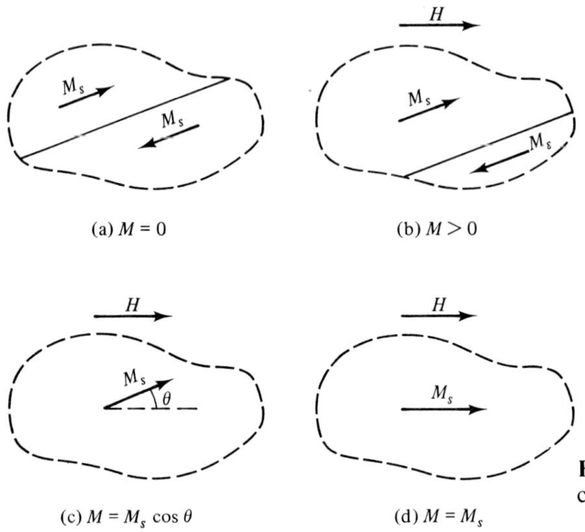

Fig. 4-2 The magnetization process.

an effect which is at least a million times as strong as any we have yet considered.

No real progress in understanding ferromagnetism was made until Pierre Weiss in 1906 advanced his hypothesis of the molecular field [4.1]. We have seen in the last chapter how this hypothesis leads to the Curie-Weiss law, $\chi = C/(T - \theta)$, which many paramagnetics obey. We saw also that θ is directly related to the molecular field H_m, because $\theta = \rho\gamma C$ and $H_m = \gamma M$, where γ is the molecular field coefficient. If θ is positive, so is γ, which means that H_m and M are in the same direction or that the molecular field aids the applied field in magnetizing the substance.

Above their Curie temperatures T_c* ferromagnetics become paramagnetic, and their susceptibilities then follow the Curie-Weiss law, with a value of θ approximately equal to T_c. The value of θ is therefore large and positive (over 1000°K for iron), and so is the molecular field coefficient. This fact led Weiss to make the bold and brilliant assumption that a molecular field acted in a ferromagnetic substance *below* its Curie temperature as well as above, and that this field was so strong that it could magnetize the substance to saturation *even in the absence of an applied field.* The substance is then self-saturating, or "spontaneously magnetized." Before we consider how this can come about, we must note at once that the theory is, at this stage, incomplete. For if iron, for example, is self-saturating, how can we explain the fact that it is quite easy to obtain a piece of iron in the unmagnetized condition?

---

* The Curie temperature (point) is a special case of a large number of "critical temperatures" in physics. T_c is thus used to stand for either "Curie temperature" or "critical temperature."

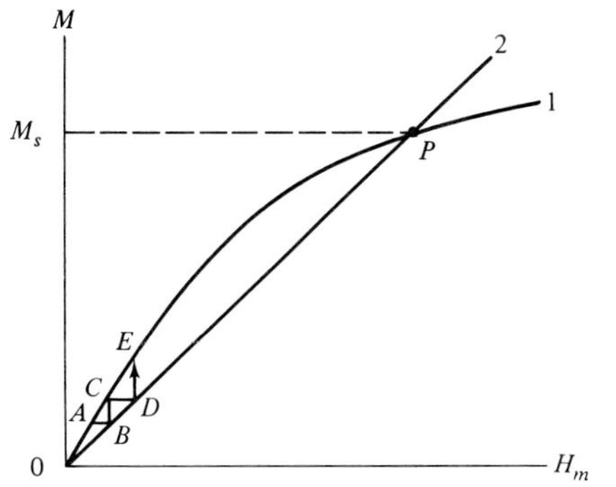

Fig. 4-3 Spontaneous magnetization by a molecular field.

Weiss answered this objection by making a second assumption: a ferromagnetic in the demagnetized state is divided into a number of small regions called *domains*. Each domain is spontaneously magnetized to the saturation value M_s, but the directions of magnetization of the various domains are such that the specimen as a whole has no net magnetization. The process of magnetization is then one of converting the specimen from a multi domain state into one in which it is a single domain magnetized in the same direction as the applied field. This process is illustrated in Fig. 4.2. The dashed line in (a) encloses a portion of a crystal in which there are parts of two domains; the boundary separating them is called a *domain wall*. The two domains are spontaneously magnetized in opposite directions, so that the magnetization of this part of the crystal is zero. In (b) a field H has been applied, causing the upper domain to grow at the expense of the lower one by downward motion of the domain wall, until in (c) the wall has moved right out of the region considered. Finally, at still higher applied fields, the magnetization rotates into parallelism with the applied field and the material is saturated, as in (d). During this entire process there has been no change in the *magnitude* of the magnetization of any region.

The Weiss theory therefore contains two essential postulates: (1) spontaneous magnetization, and (2) division into domains. Later developments have shown that both of these postulates are true. It is quite a tribute to Weiss's creative imagination that more than half a century of subsequent research has served, in a sense, only to enrich these two basic ideas.

4.2 MOLECULAR FIELD THEORY

Consider a substance in which each atom has a net magnetic moment. Assume that the magnetization of this substance increases with field, at constant temperature, according to curve 1 of Fig. 4.3, as though the substance were paramagnetic. Assume also that the only field acting on the material is a molecular field H_m proportional to the magnetization:

$$H_m = \gamma M . \tag{4.1}$$

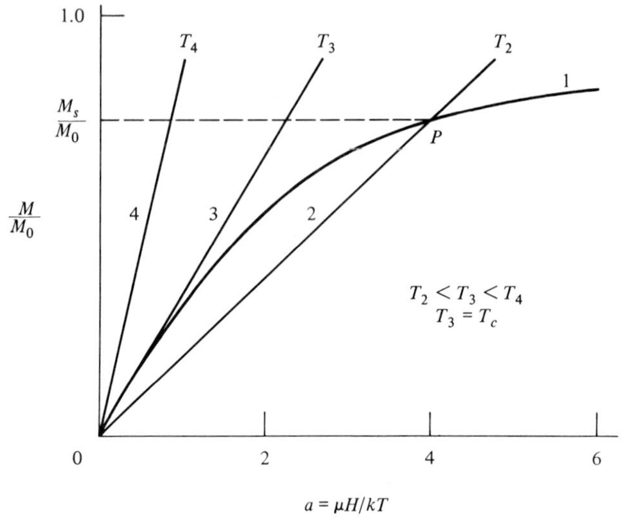

Fig. 4-4 Effect of temperature on spontaneous magnetization. Curve 1 is the Langevin function.

Line 2 in Fig. 4.3 is a plot of this equation, and the slope of the line is $1/\gamma$. The magnetization which the molecular field will produce in the material is given by the intersection of the two curves. There are actually two intersections, one at the origin and one at the point P. However, the one at the origin represents an unstable state. If M is zero anu the slightest applied field, the earth's field for example, happens to act even momentarily on the material, it will be magnetized to the point A, say. But if $M = A$, then line 2 states that H_m is B. But a field of this strength would produce a magnetization of C. Thus M would go through the values 0, A, C, E, ... and arrive at P. We know that P is a point of stability, because the same argument will show that a magnetization greater than P will spontaneously revert to P, in the absence of an applied field. The substance has therefore become spontaneously magnetized to the level of P, which is the value of M_s for the temperature in question. It is, in short, ferromagnetic. We may therefore regard a ferromagnetic as a paramagnetic subject to a very large molecular field. The size of this field will be calculated later.

We now wish to know how this behavior is affected by changes in temperature. How will M_s vary with temperature, and at what temperature will the material become paramagnetic? To answer these questions, we must replot Fig. 4.3 with a as a variable rather than H_m, where $a = \mu H/kT$ is the variable which appears in the theory of paramagnetism. Following Weiss, we will suppose that the relative magnetization is given by the Langevin function:

$$M/M_0 = L(a) = \coth(a) - (1/a). \tag{4.2}$$

(Later we will replace this with the correct quantum-mechanical relation, namely, the Brillouin function). When the applied field is zero, we have

$$a = \frac{\mu H_m}{kT} = \frac{\mu\gamma M}{kT} = \frac{\mu\gamma M M_0}{kTM_0}, \tag{4.3}$$

$$\frac{M}{M_0} = \left(\frac{kT}{\mu\gamma M_0}\right) a. \tag{4.4}$$

M/M_0 is therefore a linear function of a with a slope proportional to the absolute temperature. In Fig. 4.4, curve 1 is the Langevin function and line 2 is a plot of Eq. (4.4) for a temperature T_2. Their intersection at P gives the spontaneous fractional magnetization M_s/M_0 achieved at this temperature. An increase in temperature above T_2 has the effect of rotating line 2 counterclockwise about the origin. This rotation causes P and the corresponding magnetization to move lower and lower on the Langevin curve. The spontaneous magnetization vanishes at temperature T_3 when the line is in position 3, tangent to the Langevin curve at the origin. T_3 is therefore equal to the Curie temperature T_c. At any higher temperature, such as T_4, the substance is paramagnetic, because it is not spontaneously magnetized.

The Curie temperature can be evaluated from the fact that the slope of line 3 is the same as the slope of the Langevin curve at the origin, which is $\frac{1}{3}$. Replacing T with T_c, we have

$$\frac{kT_c}{\mu\gamma M_0} = \frac{1}{3}$$

$$T_c = \frac{\mu\gamma M_0}{3k}. \tag{4.5}$$

Therefore the slope of the straight line representing the molecular field is, at any temperature,

$$\frac{kT}{\mu\gamma M_0} = \frac{T}{3T_c}. \tag{4.6}$$

But the slope of the line determines the point of intersection P with the Langevin curve and hence the value of M_s/M_0. Therefore M_s/M_0 is determined solely by the *ratio* T/T_c. This means that all ferromagnetic materials, which naturally have different values of M_0 and T_c, have the same value of M_s/M_0 for any particular value of T/T_c. This is sometimes called the *law of corresponding states.*

[This statement of the law is very nearly, but not exactly, correct. In arriving at the Langevin law in Eq. (3.19), we considered the number n of atoms per *unit volume* and set $n\mu = M_0$. But n changes with temperature because of thermal expansion. Therefore, values of M/M_0 at different temperatures are not strictly comparable, because they refer to different numbers of atoms. When dealing with magnetization as a function of temperature, a more natural unit to use is the specific magnetization σ, which is the magnetic moment per gram, because there is then no necessity for knowing the density as a function of temperature.

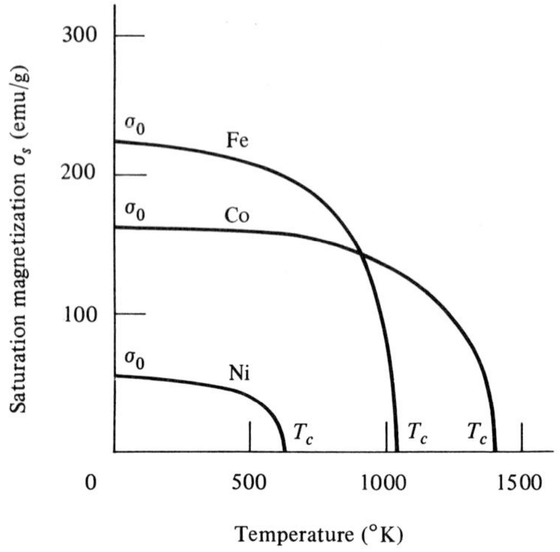

Fig. 4-5 Saturation magnetization of Fe, Co, and Ni as a function of temperature.

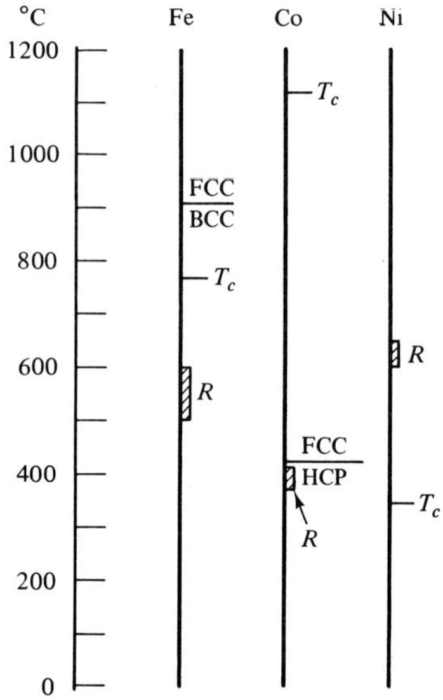

Fig. 4-6 Curie points (T_c), recrystallization temperatures (R), and phase changes in Fe, Co, and Ni. (Crystal structures: BCC, body-centered cubic; FCC, face-centered cubic; HCP, hexagonal, close-packed. Ni is FCC at all temperatures.)

If n_g is the number of atoms per gram and $\bar{\mu}$ the average component of magnetic moment in the direction of the field, then we can write Eq. (3.19) as

$$\frac{n_g\bar{\mu}}{n_g\mu} = \frac{\sigma}{\sigma_0} = \coth a - \frac{1}{a}. \tag{4.7}$$

If we then define, for a ferromagnetic material, σ_s and σ_0 as the saturation magnetizations at T° K and 0°K, respectively, an exact statement of the law of corresponding states is that all materials have the same value of σ_s/σ_0 for the same value of T/T_c. The relation between the σ and M values is

$$\frac{\sigma_s}{\sigma_0} = \frac{M_s/\rho_s}{M_0/\rho_0} = \frac{M_s\rho_0}{M_0\rho_s}, \tag{4.8}$$

where ρ_s and ρ_0 are the densities at T°K and 0°K, respectively. A change from M to σ also involves a change in the molecular field constant γ:

$$H_m = \gamma M = \gamma\rho(M/\rho) = (\gamma\rho)\sigma. \tag{4.9}$$

Thus $(\gamma\rho)$ becomes the molecular field constant, and Eqs. (4.5) and (4.6) become

$$T_c = \frac{\mu\gamma\rho\sigma_0}{3k}, \tag{4.10}$$

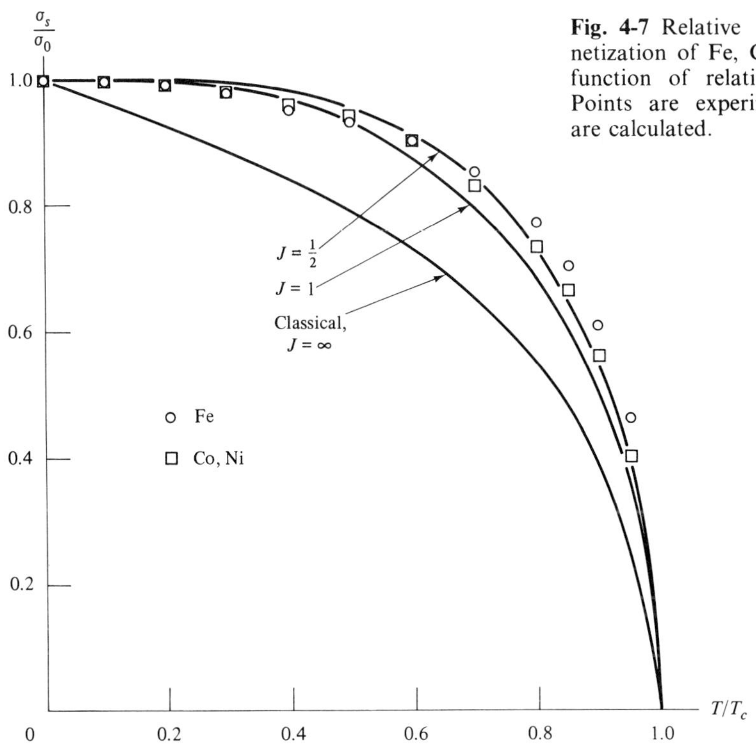

Fig. 4-7 Relative saturation magnetization of Fe, Co, and Ni as a function of relative temperature. Points are experimental. Curves are calculated.

and

$$\frac{kT}{\mu\gamma\rho\sigma_0} = \frac{T}{3T_c}. \tag{4.11}$$

Equation (4.4) therefore becomes

$$\frac{\sigma}{\sigma_0} = \left(\frac{kT}{\mu\gamma\rho\sigma_0}\right) a = \left(\frac{T}{3T_c}\right) a, \tag{4.12}$$

when the magnetization is expressed in terms of σ.]

Experimental data on the variation of the saturation magnetization σ_s of Fe. Co, and Ni with temperature are shown in Fig. 4.5. (The temperature scales shown in Fig. 4.6 give the Curie points and the temperatures of phase changes and recrystallization for the three metals. The recrystallization temperatures are the approximate minimum temperatures at which heavily cold-worked specimens will recrystallize; thus iron and cobalt can be recrystallized while still ferromagnetic, but nickel cannot.) The three curves of Fig. 4.5 have similar shapes and, when the data are replotted in the form of σ_s/σ_0 versus T/T_c as in Fig. 4.7, the points conform rather closely to a single curve. Thus Weiss's prediction of a law of corresponding states is verified. However, the *shape* of the curve of σ_s/σ_0 versus T/T_c predicted by the Weiss-Langevin theory does not agree with experiment. We can see this by finding graphically the points of intersection of the curve of Eq. (4.7) and the lines of Eq. (4.12) for various values of T/T_c. The result is shown by the curve labeled "classical, $J = \infty$" in Fig. 4.7. This disagreement is not surprising, inasmuch as we have already seen in Chapter 3 that the classical Langevin theory, which was the only theory available for Weiss to test in this manner, does not conform to experiment.

The Weiss theory may be modernized by supposing that the molecular field acts on a substance having a relative magnetization determined by a quantum-mechanical Brillouin function $B(J,a')$. In terms of specific magnetization, we have

$$\frac{\sigma}{\sigma_0} = \frac{2J+1}{2J}\coth\left(\frac{2J+1}{2J}\right)a' - \frac{1}{2J}\coth\frac{a'}{2J}, \tag{4.13}$$

where $a' = \mu_H H/kT$ from Eq. (3.46). The straight line representing the molecular field is given by

$$\frac{\sigma}{\sigma_0} = \left(\frac{kT}{\mu_H \gamma \rho \sigma_0}\right)a'. \tag{4.14}$$

The slope of the Brillouin function at the origin is $(J+1)/3J$, from Eq. (3.49). Therefore, the Curie temperature is

$$T_c = \left(\frac{\mu_H \gamma \rho \sigma_0}{k}\right)\left(\frac{J+1}{3J}\right) \tag{4.15}$$

$$= \frac{g(J+1)\mu_B \gamma \rho \sigma_0}{3k}. \tag{4.16}$$

The equation of the molecular-field line can then be written

$$\frac{\sigma}{\sigma_0} = \left(\frac{J+1}{3J}\right)\left(\frac{T}{T_c}\right)a'. \tag{4.17}$$

Values of the relative spontaneous magnetization σ_s/σ_0 as a function of T/T_c can be found graphically from the intersections of the curve of Eq. (4.13) and the lines of Eq. (4.17). A different relation will be found for each value of J. The particular value $J = \frac{1}{2}$ is of special interest. Equations (4.13) and (4.17) then become

$$\frac{\sigma}{\sigma_0} = \tanh a' \tag{4.18}$$

and

$$\frac{\sigma}{\sigma_0} = \left(\frac{T}{T_c}\right)a'. \tag{4.19}$$

These can be combined to give

$$\boxed{\frac{\sigma_s}{\sigma_0} = \tanh\frac{(\sigma_s/\sigma_0)}{(T/T_c)},} \tag{4.20}$$

Table 4.1 Values of the g factor (from Kittel [4.2])

Material	g*	g†
Fe	2.12–2.17	2.08
Co	2.22	2.18
Ni	2.2	2.09–2.19
Heusler alloy (Cu_2MnAl)	2.01	2.00
Permalloy (78 Ni, 22 Fe)	2.07–2.14	2.11
Supermalloy (79 Ni, 5 Mo, 16 Fe)	2.12–2.20	2.10

* From magnetic resonance experiments.

† Calculated, by means of Eq. (3.34a), from g' factors obtained by gyromagnetic experiments.

which can be solved for σ_s/σ_0 as a function of T/T_c by means of a table of hyperbolic tangents, since it is of the form $y = \tanh(y/x)$. The theoretical curves for $J = \frac{1}{2}$ and $J = 1$ are plotted in Fig. 4.7. Either one is in fairly good agreement with experiment, with the curve for $J = \frac{1}{2}$ perhaps giving the better fit.

If J equals $\frac{1}{2}$, the magnetic moment is due entirely to spin, the g factor is 2, and there is no orbital contribution. That this condition is closely approximated by ferromagnetic substances is also suggested by experimental values of g. Table 4.1 lists observed g factors for Fe, Co, Ni, and several ferromagnetic alloys, and they are all seen to be close to 2. We may therefore conclude that *ferromagnetism is due essentially to electron spin*, with little or no contribution from the orbital motion of the electrons. At 0°K the spins on all the atoms in any one domain are parallel and, let us say, "up". At some higher temperature, a certain fraction of the total, determined by the value of the Brillouin function at that temperature, flips over into the "down" position; the value of that fraction determines the value of σ_s.

Up to this point we have put the applied field equal to zero and considered only the effect of the molecular field. If a field H is applied, the total field acting on the substance is $(H + H_m)$, where by H we mean the true applied field, corrected for any demagnetizing effects. Therefore,

$$a' = \frac{\mu_H(H + H_m)}{kT} = \frac{\mu_H(H + \gamma\rho\sigma)}{kT}, \tag{4.21}$$

which may be written

$$\frac{\sigma}{\sigma_0} = \left(\frac{kT}{\mu_H\gamma\rho\sigma_0}\right)a' - \frac{H}{\gamma\rho\sigma_0}. \tag{4.22}$$

This is a straight line parallel to the line of Eq. (4.14) but displaced downward by an amount $H/\gamma\rho\sigma_0$, proportional to the applied field. In Fig. 4.8, lines 2 and

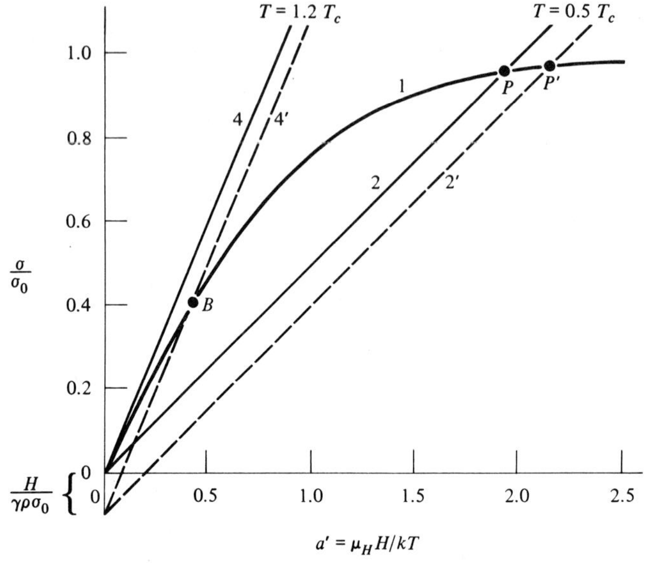

Fig. 4-8 Effects of temperature and applied field on magnetization. Curve 1 is the Brillouin function for $J = \frac{1}{2}$.

4 represent the molecular field alone, while the dashed lines 2′ and 4′ represent the molecular and applied fields.

Above the Curie point, at $T = 1.2\ T_c$, for example, the effect of the applied field is to move the point of intersection of the field line and the magnetization curve from the origin to the point B, and from this change in magnetization the susceptibility can be calculated. Inasmuch as we are interested only in the region near the origin, the Brillouin function can be approximated by the straight line

$$\frac{\sigma}{\sigma_0} = \left(\frac{J+1}{3J}\right) a' . \tag{4.23}$$

Eliminating a' from Eqs. (4.22) and (4.23), we obtain

$$\chi = \frac{\sigma}{H} = \frac{\mu_H \sigma_0 (J+1)/3kJ}{T - [\mu_H \gamma \rho \sigma_0 (J+1)/3kJ]}, \tag{4.24}$$

which has the form of the Curie-Weiss law $\chi = C/(T - \theta)$, provided that

$$C = \frac{\mu_H \sigma_0 (J+1)}{3kJ}, \tag{4.25}$$

$$\theta = \frac{\mu_H \gamma \rho \sigma_0 (J+1)}{3kJ}. \tag{4.26}$$

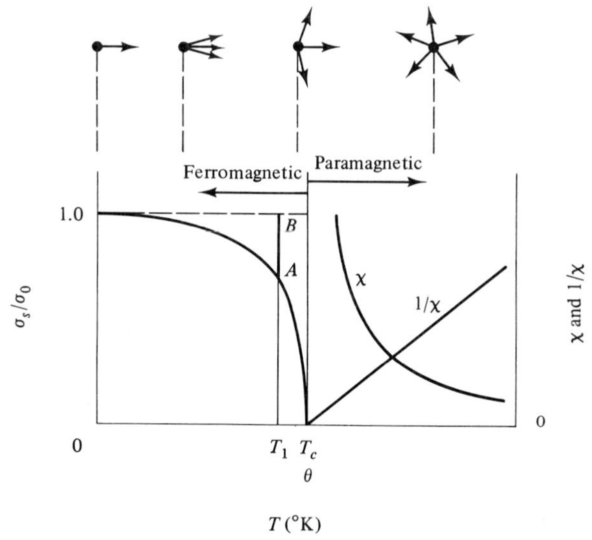

Fig. 4-9 Magnetization and susceptibility curves below and above the Curie temperature. The small sketches at the top indicate the distribution of spin direction understood in a classical sense, in zero applied field within a single domain (below T_c), and in a group of atoms (above T_c).

Since Eqs. (4.15) and (4.26) are identical, it follows from molecular-field theory that T_c, the temperature at which the spontaneous magnetization becomes zero, and θ, the temperature at which the susceptibility becomes infinite, are identical. It is left as a problem to show that the Curie constant per gram given by Eq. (4.25) is the same as that given by Eq. (3.52).

At temperatures well below the Curie temperature, for example at room temperature in most ferromagnetics, even a very strong applied field produces but a small increase in the spontaneous magnetization σ_s or M_s already produced by the molecular field. Thus, for $T = 0.5\ T_c$, line 2 in Fig. 4.8 shifts to position 2′ when a field is applied, but the increase in relative magnetization, from P to P', is very slight, because the magnetization curve is so nearly flat in this region. To increase the magnetization of a ferromagnetic specimen from zero to σ_s or M_s usually requires, at room temperature, applied fields of several hundred, or, at the most, several thousand, oersteds. (In this chapter "applied field" means the true field acting inside the specimen. It is the difference between the actual applied field and any demagnetizing field that may be present.) The magnetization of the specimen is then the same as that of each domain in the demagnetized state, and the specimen is sometimes said to be in a state of "technical saturation" or, simply, "saturation". To produce any appreciable increase in magnetization beyond this point requires fields of the order of 10^5 to 10^6 Oe or more, and such an increase is called *forced magnetization*. It represents an increase in the magne-

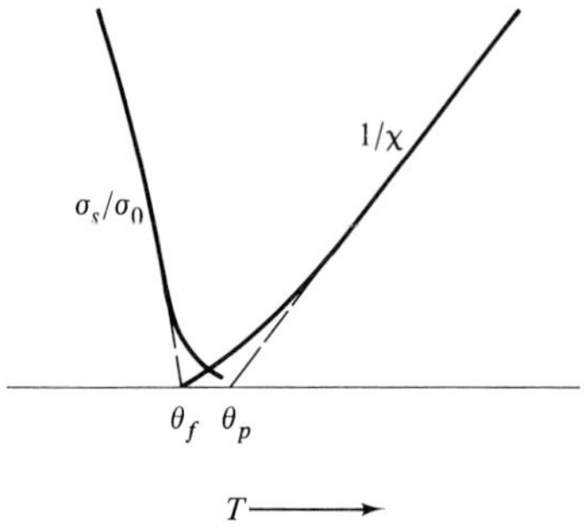

Fig. 4-10 Magnetic behavior near the Curie point.

tization of the domain itself. At any temperature above 0°K an infinite field is required to produce absolute saturation, for which $\sigma_s = \sigma_0$, or $M_s = M_0$. At 0°K this condition can be reached with fields of 10^2 to 10^3 Oe. (At 0°K the obstacles to absolute saturation no longer include thermal agitation, because σ_s equals σ_0 in each domain of the demagnetized state. The obstacles are then only the "ordinary" ones inherent in domain wall motion and domain rotation; these processes are discussed in Chapters 7 to 9.)

Forced magnetization beyond the value of the spontaneous magnetization is called the *para-process* by Russian authors, e.g., Belov [G.14]. It is represented, at temperature T_1, by points lying along the line AB of Fig. 4.9. The increase in magnetization beyond σ_s caused by a given increase in field is larger the closer T_1 is to the Curie temperature; this and other phenomena occurring near T_c have been described in detail by Belov [G.14], Bates [G.13], and Bozorth [G.4].

Figure 4.9 summarizes the results of molecular-field theory. The temperature θ at which the susceptibility χ becomes infinite, and $1/\chi$ becomes zero, is the same as the temperature T_c at which the spontaneous magnetization appears. Actually, careful measurements have shown that the situation near the Curie point is not that simple. Two deviations from the theory, illustrated in Fig. 4.10, are observed:

1. The curve of $1/\chi$ versus T is a straight line at high temperatures but becomes concave up near the Curie point. The extrapolation of the straight-line portion cuts the temperature axis at θ_p, which is called the *paramagnetic Curie point.* It is therefore equal to the θ of the Curie-Weiss law.
2. The curve of the spontaneous magnetization σ_s/σ_0 versus T does not cut the temperature axis at a large angle but bends over to form a small "tail." The temperature θ_f defined by the extrapolation of the main part of the curve is called the *ferromagnetic Curie point.*

It is usual to refer to the Curie point as T_c (or θ) except when dealing with phenomena in the immediate neighborhood of this "point"; it is then necessary to distinguish between the two Curie points θ_f and θ_p. The difference between the two is some 10° to 30°K. This behavior shows that the transition from the ferro-

Table 4.2 Atomic Moments of Iron, Cobalt, and Nickel

	Ferromagnetic		Paramagnetic		
	σ_o (emu/gm)	μ_H	μ_{eff}	Calculated μ_H	
				$J = \frac{1}{2}$	$J = 1$
Fe	221.9	2.22 μ_B	3.15 μ_B	1.82 μ_B	2.23 μ_B
Co	162.5	1.72	3.13	1.81	2.21
Ni	57.50	0.60	1.61	0.93	1.14

magnetic to the paramagnetic state is not sharp, but blurred. The fuzziness of the transition is thought to be due to *spin clusters*. These are small groups of atoms in which the spins remain parallel to one another over a small temperature range above θ_p; as such they constitute a kind of magnetic short-range order. These clusters of local spin order exist within a matrix of the spin disorder which constitutes a true paramagnetic, and they gradually disappear as the temperature is raised. Conversely, below θ_f there is a long-range order of spins even in the absence of an applied field; this is precisely what spontaneous magnetization means.

Another and perhaps more serious disagreement between molecular-field theory and experiment involves the magnitudes of the magnetic moment per atom below and above the Curie point. At absolute zero, where complete saturation is attained, the specific magnetization σ_0 is given by the maximum magnetic moment per atom in the direction of the field multiplied by the number of atoms per gram, or

$$\sigma_0 = \mu_H(N/A), \tag{4.27}$$

where N is Avogadro's number and A the atomic weight. For iron, substituting the value of σ_0 from Appendix 5, we have

$$\mu_H = \frac{(221.9)(55.85)}{6.02 \times 10^{23}} = 2.06 \times 10^{-20} \text{ erg/Oe}$$

$$= \frac{2.06 \times 10^{-20}}{0.927 \times 10^{-20}} = 2.22\ \mu_B.$$

Similar calculations for cobalt and nickel yield the values listed in the second column of Table 4.2. According to molecular-field theory, these elements should exhibit the same atomic moments above the Curie point, but this is not found to be true. From the observed Curie constants in the paramagnetic region we can calculate the effective moments per atom; these appear in the third column. Values of μ_H can then be calculated from observed values of μ_{eff} for any assumed value of J; two such values of μ_H are shown for each metal in the last two columns. [From Eqs. (3.31) and (3.33), it follows, for any value of g, that $\mu_H = \mu_{eff}/\sqrt{3}$ when $J = \frac{1}{2}$,

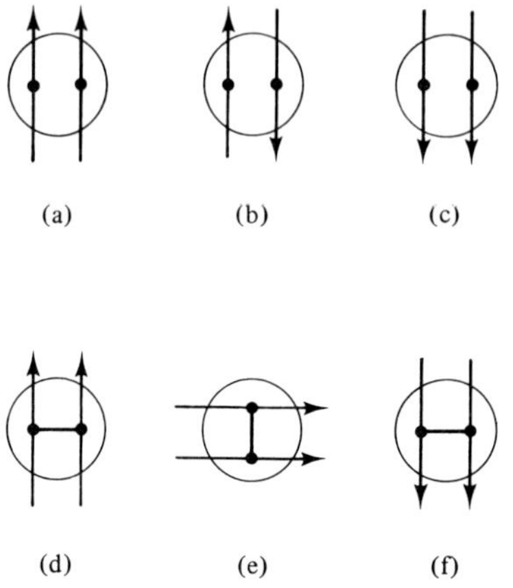

Fig. 4-11 Possible spin arrangements for uncoupled and coupled spins.

and $\mu_H = \mu_{eff}/\sqrt{2}$ when $J = 1$.] Neither for $J = \frac{1}{2}$ nor for $J = 1$ is there agreement for all three metals with the values of μ_H in the ferromagnetic region. An explanation for these differences has been given by Stoner [4.3] on the basis of the band theory of magnetism.

Certain aspects of the μ_H values in the ferromagnetic region should be noted:

1. They are nonintegral. This fact will be discussed in Section 4.4.
2. They are not what one would expect from the relation $\mu_H = gJ\mu_B$ (Eq. 3.33) and reasonable values of g and J. Thus, if $g = 2$ and $J = \frac{1}{2}$, as seems most likely, μ_H should be one Bohr magneton or one spin per atom. But the observed value, for iron, is 2.2 μ_B. Suppose we ignore the fractional part and assume exactly two spins per atom. How can this value be reconciled with $J = \frac{1}{2}$ and the tanh a' variation of σ_s/σ_0 (Eq. 4.18), which is experimentally observed? The answer is that the two spins on any one atom are not coupled together. Figure 4.11 shows atoms represented by circles and spin moments by arrows, and (a), (b), and (c) show the possible arrangements of two spins per atom on three atoms, for $J = \frac{1}{2}$. Because one spin is able to flip over, in response to a change in temperature, for example, without causing the other spin on the same atom to flip, these arrangements are equivalent to the possible arrangements of one spin per atom on six atoms, and the statistics of this situation lead to the tanh a' relation. On the other hand, if $J = 1$ and $g = 2$, the possible values of μ_H are $+2$, 0, and -2 Bohr magnetons. But here the two spins per atom are coupled; if one flips, the other must, too, as in (d), (e), and (f), and the tanh a' relation is not followed. Complete spin reversal, from (d) to (f), is a common transition when $J = 1$, but the same change, from (a) to (c), is quite unlikely when $J = \frac{1}{2}$, because it would require the simultaneous reversal of two independent spins.
3. The μ_H values in the ferromagnetic region are of the same order of magnitude as the μ_H values for paramagnetic substances generally. The huge difference

between a ferromagnetic and a paramagnetic is therefore due to the degree of alignment achieved and not to any large difference in the size of the moment per atom.

We will conclude this section on the molecular field by calculating its magnitude. From Eq. (4.26) for $J = \frac{1}{2}$, we have for the molecular field coefficient:

$$\gamma\rho = \frac{k\theta}{\mu_H \sigma_0}. \tag{4.28}$$

For iron, this becomes

$$\gamma\rho = \frac{(1.38 \times 10^{-16})(1043)}{(2.06 \times 10^{-20})(221.9)} = 3.15 \times 10^4 \text{ Oe g/emu}.$$

Therefore, the molecular field in iron at room temperature is

$$H_m = (\gamma\rho)\sigma_s = (3.15 \times 10^4)(218.0) = 6.9 \times 10^6 \text{ Oe}.$$

The corresponding values for cobalt and nickel are 11.9 and 14.7 million oersteds, respectively. These fields are very much larger than any continuous field yet produced in the laboratory, and some 10^4 times as large as the fields normally needed to achieve technical saturation. It is therefore not surprising to find that the application of even quite large fields produces only a slight increase in the spontaneous magnetization already produced by the molecular field. Finally, it should be stressed again that the molecular field is in no sense a real field, but rather a force tending to make adjacent atomic moments parallel to one another; it is called a field, and measured in field units, only because it has the same kind of effect as a real field.

4.3 EXCHANGE FORCES

The Weiss theory of the molecular field says nothing about the physical origin of this field. However, the hypothesis that H_m is proportional to the existing magnetization implies that the phenomenon involved is a cooperative one. Thus, the greater the degree of spin alignment in a particular region of a crystal, the greater is the force tending to align any one spin in that region. The cooperative nature of the phenomenon is clearly shown by the way in which σ_s decreases with increasing temperature (Fig. 4.7). Near absolute zero the decrease is slight, but, as the temperature is raised, thermal energy is able to reverse more and more spins, thus reducing the aligning force on those spins which are still aligned. The result is a more and more rapid break-down in alignment, culminating in almost complete disorder at the Curie point. (The decrease in the degree of long-range order of atomic position in certain solid solutions, such as CuZn, as the temperature is raised is represented by a curve of the same form.)

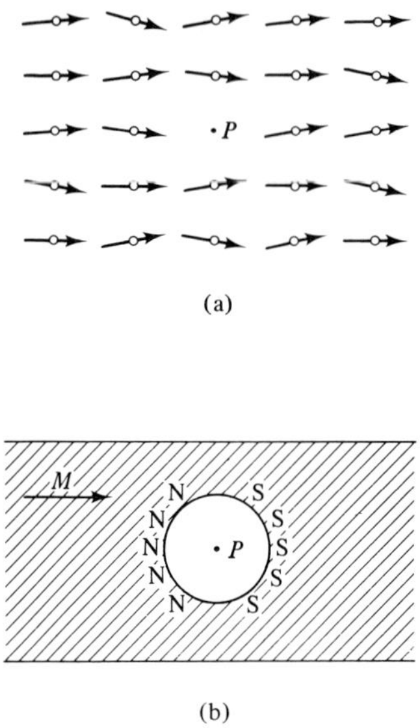

Fig. 4-12 Field at an interior point due to the surrounding magnetization.

In seeking for a physical origin of the molecular field, we might at first wonder if it could be entirely magnetic. Figure 4.12(a), for example, shows a set of atoms, represented by small circles, each having a net magnetic moment. The directions of these moments, considered as classically free to point in any direction, are indicated by the arrows. Each atom, considered as a magnetic dipole, produces an external field like that of Fig. 1.6, and the sum of the fields of all the dipoles at the point P would be a field from left to right, tending to increase the alignment of the moment on an atom placed at that point. The calculation of this field at P would involve a summation over all the dipoles in the specimen. The calculation becomes easier if we replace the set of dipoles with a continuous medium of average magnetization M, as in Fig. 4.12(b) and ask : What is the field at P in the center of a spherical hole in the material? This field, called the *Lorentz field*, is due to the north and south poles produced on the sides of the hole and is exactly equal in magnitude to the demagnetizing field of a solid magnetized sphere in empty space. We have already calculated this field in Section 2.6 and found it to be $4\pi M/3$. For saturated iron at room temperature this is equal to $(4\pi/3)$ (1714) or about 7200 Oe. Since the molecular field is about 1000 times larger, it cannot be due to purely magnetic forces.

The physical origin of the molecular field was not understood until 1928, when Heisenberg showed that it was caused by quantum-mechanical *exchange forces*. About a year earlier the new wave mechanics had been applied to the problem of the hydrogen molecule, i.e., the problem of explaining why two hydrogen atoms come together to form a stable molecule. Each of these atoms consists of a single electron moving about the simplest kind of nucleus, a single proton. For a particular pair of atoms, situated at a certain distance apart, there are certain electrostatic attractive forces (between the electrons and protons) and repulsive forces (between the two electrons and between the two protons) which can be calculated by Coulomb's law. But there is still another force, entirely non-classical, which depends on the relative orientation of the spins of the two electrons. This is the exchange force. If the spins are antiparallel, the sum of all the forces is attractive and a stable molecule is formed; the total energy of the atoms is then less for a particular distance of separation than it is for smaller or larger distances. If the spins are parallel, the two atoms repel one another. The exchange force is a consequence of the Pauli exclusion principle, applied to the two atoms as a whole. This principle states that two electrons can have the same energy only if they have opposite spins. Thus two hydrogen atoms can come so close together that their two electrons can have the same velocity and occupy very nearly the same small region of space, i.e., have the same energy, provided these electrons have opposite spin. If their spins are parallel, the two electrons will tend to stay far apart. The ordinary (Coulomb) electrostatic energy is therefore modified by the spin orientations, which means that the exchange force is fundamentally electrostatic in origin. (The term "exchange" arises in the following way. When the two atoms are adjacent, we can consider electron 1 moving about proton 1, and electron 2 moving about proton 2. But electrons are indistinguishable, and we must also consider the possibility that the two electrons exchange places, so that electron 1 moves about proton 2 and electron 2 about proton 1. This consideration introduces an additional term, the exchange energy, into the expression for the total energy of the two atoms. This interchange of electrons takes place at a very high frequency, about 10^{18} times per second in the hydrogen molecule.)

The exchange energy forms an important part of the total energy of many molecules and of the covalent bond in many solids. Heisenberg showed that it also played a decisive role in ferromagnetism. If two atoms i and j have spin angular momentum $\mathbf{S}_i h/2\pi$ and $\mathbf{S}_j h/2\pi$, respectively, then the exchange energy between them is given by

$$E_{ex} = -2J_{ex}\mathbf{S}_i \cdot \mathbf{S}_j = -2J_{ex} S_i S_j \cos\phi \tag{4.29}$$

where J_{ex} is a particular integral, called the *exchange integral*, which occurs in the calculation of the exchange effect, and ϕ is the angle between the spins. If J_{ex} is positive, E_{ex} is a minimum when the spins are parallel ($\cos\phi = 1$) and a maximum when they are anti-parallel ($\cos\phi = -1$). If J_{ex} is negative, the lowest energy state results from antiparallel spins. As we have already seen, ferromagnetism is due to the alignment of spin moments on adjacent atoms. A positive value of the exchange integral is therefore a necessary condition for ferromagnetism to occur. This is also a rare condition, because J_{ex} is commonly negative, as in the hydrogen molecule.

According to the Weiss theory, ferromagnetism is caused by a powerful "molecular field" which aligns the atomic moments. In modern language we say that "exchange forces" cause the spins to be parallel. However, it would be

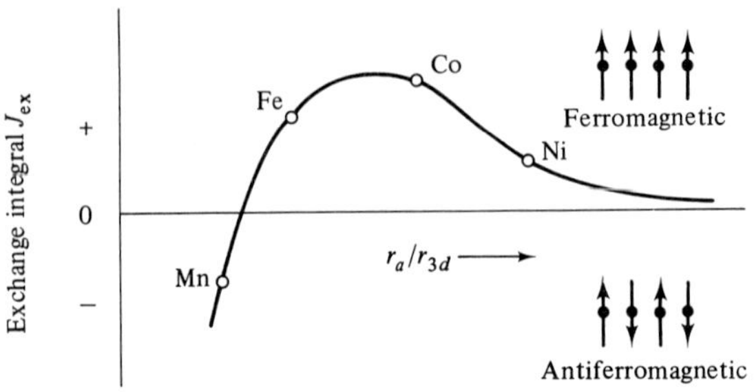

Fig. 4-13 Bethe-Slater curve (schematic).

unrealistic to conclude from this change in terminology that all the mystery has been removed from ferromagnetism. The step from a hydrogen molecule to a crystal of iron is a giant one, and the problem of calculating the exchange energy of iron is so formidable that it has not yet been solved. Expressions like Eq. (4.29), which is itself something of a simplification and which applies only to *two* atoms, have to be summed over all the atom pairs in the crystal. Exchange forces decrease rapidly with distance, so that some simplification is possible by restricting the summation to nearest-neighbor pairs. But even this added simplification does not lead to an exact solution of the problem. In the present state of knowledge, it is impossible to predict from first principles that iron is ferromagnetic, i.e., merely from the knowledge that it is element 26 in the periodic table.

Nevertheless, knowledge that exchange forces are responsible for ferromagnetism and, as we shall see later, for antiferro- and ferrimagnetism, has led to many semiquantitative conclusions of great value. For example, it allows us to rationalize the appearance of ferromagnetism in some metals and not in others. The curve of Fig. 4.13, usually called the Bethe-Slater curve, shows the postulated variation of the exchange integral with the ratio r_a/r_{3d}, where r_a is the radius of an atom and r_{3d} the radius of its $3d$ shell of electrons. (It is the spin alignment of some of the $3d$ electrons which is the immediate cause of ferromagnetism in Fe, Co, and Ni.) The atom diameter is $2r_a$ and this is also the distance apart of the atom centers, since the atoms of a solid are in contact with one another. If two atoms of the same kind are brought closer and closer together but without any change in the radius r_{3d} of their $3d$ shells, the ratio r_a/r_{3d} will decrease from large to small values. When this ratio is large, J_{ex} is small and positive. As the ratio decreases and the $3d$ electrons approach one another more closely, the positive exchange interaction, favoring parallel spins, becomes stronger and then decreases to zero. A further decrease in the interatomic distance brings the $3d$ electrons so close together that their spins must become antiparallel (negative J_{ex}). This condition is called antiferromagnetism.

The curve of Fig. 4.13 can be applied to a series of *different* elements if we compute r_a/r_{3d} from their known atom diameters and shell radii. The points so found lie on the curve as shown, and the curve correctly separates Fe, Co, and Ni from Mn and the next lighter elements in the first transition series. (Mn is antiferromagnetic below 100°K, and Cr, the next lighter element, is antiferromagnetic below 37°C; above these temperatures they are both paramagnetic.) When J_{ex} is positive, its magnitude is proportional to the Curie temperature (see below), because spins which are held parallel to each other by strong exchange forces can be disordered only by large amounts of thermal energy. The positions of Fe, Co, and Ni on the curve agree with the fact that Co has the highest, and Ni the lowest, Curie temperature of the three.

Although the theory behind the Bethe-Slater curve has received much criticism, the curve does suggest an explanation of some otherwise puzzling facts. Thus ferromagnetic alloys can be made of elements which are not in themselves ferromagnetic; examples of these are MnBi and the Heusler alloys, which have compositions approximated by the formulae Cu_2MnSn and Cu_2MnAl. Because the manganese atoms are farther apart in these alloys than in pure manganese, r_a/r_{3d} becomes large enough to make the exchange interaction positive.

Inasmuch as the molecular field and the exchange interaction are equivalent, there must be a relation between them. We can find an approximate form of this relation as follows. Let z be the coordination number of the crystal structure involved, i.e., let each atom have z nearest neighbors, and assume that the exchange forces are effective only between nearest neighbors. Then, if all atoms have the same spin S, the exchange energy between one atom and all the surrounding atoms is

$$E_{ex} = z(-2\,J_{ex}\,S^2),$$

when all the spins are parallel. But this is equivalent to the potential energy of the atom considered in the molecular field H_m. If the atom has a magnetic moment of μ_H in the direction of the field, this energy is

$$E_{pot} = -\mu_H H_m.$$

Equating these two expressions for the energy, we have

$$H_m = (\gamma\rho)\sigma_0 = \frac{2zJ_{ex}S^2}{\mu_H}. \tag{4.30}$$

But the molecular field coefficient $(\gamma\rho)$ is related to the Curie temperature θ by Eq. (4.26). When $J = S$ (pure spin), this substitution gives

$$J_{ex} = \frac{3k\theta}{2zS(S+1)}, \tag{4.31}$$

which shows that the exchange integral is proportional to the Curie temperature, as mentioned above. For a body-centered cubic structure like that of iron, for which $z = 8$, and for $S = \frac{1}{2}$, Eq. (4.31) gives $J_{ex} = 0.25\,k\theta$. A more rigorous calculation gives $J_{ex} = 0.34\,k\theta$ for this case.

Exchange forces depend mainly on interatomic distances and not on any geometrical regularity of atom position. Crystallinity is therefore not a requirement for ferromagnetism. The first discovery of an amorphous ferromagnetic was reported in 1965 by Mader and Nowick [4.4]. They made amorphous thin films of cobalt-gold alloys by co-depositing the two metals from the vapor on a substrate maintained, not at room or elevated temperatures, but at 77°K.

These alloys were both ferromagnetic and amorphous, as judged from electron-diffraction photographs, and they retained their amorphous condition when heated to room temperature.

4.4 BAND THEORY

The band (or zone) theory is a broad theory of the electronic structure of solids. It is applicable not only to metals, but also to semiconductors and insulators. It leads to conclusions about a variety of physical properties, e.g., cohesive, elastic, thermal, electrical, and magnetic. When the band theory is applied specifically to magnetic problems, it is often called the collective-electron theory. This application of band theory was first made in 1933–1936 by E. C. Stoner and N. F. Mott in England and by J. C. Slater in the United States, and the theory is still in a state of active development by solid-state physicists. Our task in this section is to apply it to Fe, Co, and Ni in an attempt to explain the μ_H values of these metals at 0°K, namely, 2.22, 1.72, and 0.60 Bohr magnetons per atom, respectively (Table 4.2). These are important numbers, and any satisfactory theory of magnetism has to account for them.

We will begin by reviewing the electronic structure of *free atoms*, i.e., atoms located at large distances from one another, as in a monatomic gas. The electrons in such atoms occupy sharply defined energy levels in accordance with the Pauli exclusion principle. This principle states that no two electrons in the atom can have the same set of four quantum numbers. Three of these numbers define the level ("shell") or sublevel involved, while the fourth defines the spin state of the electron (spin up or spin down). The Pauli principle can therefore be alternatively stated: each energy level in an atom can contain a maximum of two electrons, and they must have opposite spin. Table 4.3 lists the various energy levels and the number of electrons each can hold, in terms of x-ray notation (K, L, ...) and quantum-mechanical notation (1s, 2s, ...). The 2p subshell is actually composed of three sub-subshells of almost the same energy, each capable of holding two electrons; the 3d subshell has a similar kind of substructure, and its total capacity is ten electrons.

Table 4.3 Atomic Energy Levels

Shell	K	L		M			N	
Subshell	1s	2s	2p	3s	3p	3d	4s	...
Capacity	2	2	6	2	6	10	2	...

The filling up of levels proceeds regularly in the elements from hydrogen to argon, which has 18 electrons. At this point all levels up to and including the 3p are filled. As we go to heavier elements, however, we find irregularities in the

Table 4.4 Electron Distributions in Free Atoms

			Transition elements									
Number of electrons in shell	K	Ca	Sc	Ti	V	Cr	Mn	Fe	Co	Ni	Cu	Zn
$3d$	0	0	1	2	3	5	5	6	7	8	10	10
$4s$	1	2	2	2	2	1	2	2	2	2	1	2
$3d + 4s$	1	2	3	4	5	6	7	8	9	10	11	12

way that the $3d$ and $4s$ levels are filled, because these two have nearly the same energy and they shift their relative positions almost from atom to atom. Observations of optical spectra disclose the electron distributions listed in Table 4.4. The transition elements, those in which an incomplete $3d$ shell is being filled, are the ones of most interest to us because they include the three ferromagnetic metals. It must be emphasized that the electron distributions in Table 4.4 apply only to *free atoms*, because of the way in which these distributions were observed. (The optical spectrum of an element is obtained by placing it in an electric arc, and the temperature of the arc is high enough to convert any element into a monatomic gas.)

When atoms are brought close together to form a *solid*, the positions of the energy levels are profoundly modified. Suppose that two atoms of, say, iron approach each other from a large distance. When they are well separated, their $1s$ levels, each containing two electrons, have exactly the same energy. When they approach so closely that their electron clouds begin to overlap, the Pauli principle now applies to the two atoms as a unit and prevents them from having a single $1s$ level containing four electrons; instead, the $1s$ level must split into two levels with two electrons in each. Similarly, when N atoms come together to form a solid, each level of the free atom must split into N levels, because the Pauli principle now applies to the whole group of N atoms. However, the extent of the splitting is different for different levels, as indicated in Fig. 4.14. In the transition elements, the outermost electrons are the $3d$ and $4s$; these electron clouds are the first to overlap as the atoms are brought together, and the corresponding levels are the first to split. When the interatomic distance d has decreased to d_0, the equilibrium value for the atoms in the crystal, the $3d$ levels are spread into a band extending from B to C, and the $4s$ levels are spread into a much wider band, extending from A to D, because the $4s$ electrons are farther from the nucleus. At the same atom spacing, however, the inner core electrons ($1s$ and $2s$) are too far apart to have much effect on one another, and the corresponding energy levels show a negligible amount of splitting. If the atoms could be forced together to distances much smaller than d_0, the $1s$ and $2s$ levels would presumably broaden considerably, as indicated in the drawing. (Experimental evidence for the above

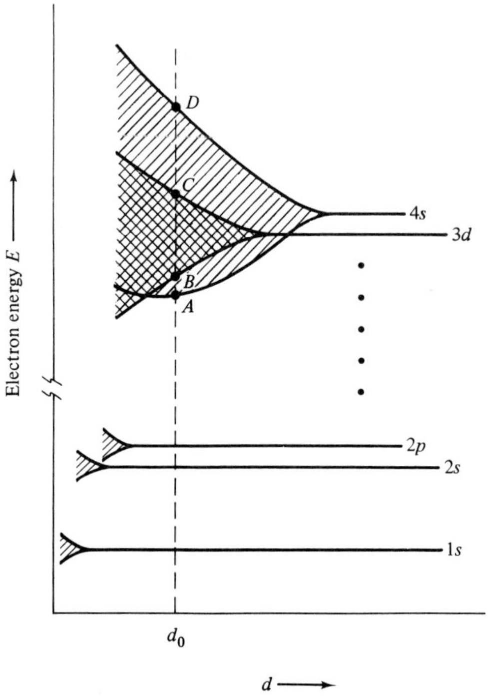

Fig. 4-14 Splitting of electron energy levels as the interatomic distance *d* decreases.

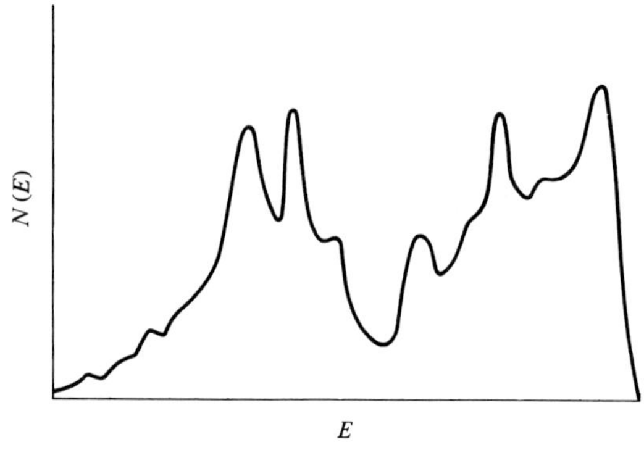

Fig. 4-15 Density of levels as a function of energy in the $3d$ band of nickel, as calculated by Koster [4.5].

statements is afforded by the x-ray emission spectra of solid metals. When electron transitions occur between two inner shells, radiation of a *single* wavelength is emitted, namely, the sharp *K*, *L*, etc. x-ray lines. The levels involved must therefore have sharply defined energies. On the other hand, when the transition is between an outer and an inner shell, the emitted radiation consists of a broad range of wavelengths.)

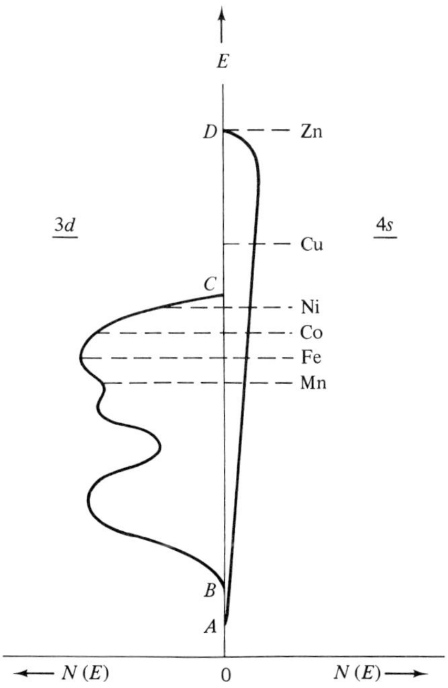

Fig. 4-16 Density of levels in the 3d and 4s bands (schematic).

There are a great many energy levels in a band, even for a small bit of crystal. For example, 55.85 grams of iron (the gram atomic weight) contain 6.02×10^{23} atoms (Avogadro's number). Thus one milligram of iron contains some 10^{19} atoms, and the Pauli principle therefore requires that each separate energy level in the free atom split into some 10^{19} levels in a one-milligram crystal. This means that the levels in a band are so closely spaced as to constitute almost a continuum of allowed energy. Nevertheless, we will still be interested in the energy difference between levels or, to put it in other terms, in the density of levels in the band. This density is often written as $N(E)$ to emphasize the fact that it is not constant but a function of the energy E itself. The product of the density $N(E)$ and any given energy range gives the number of levels in that range; thus $N(E)\,dE$ is the number of levels lying between the energies E and $E + dE$, and $1/N(E)$ is the average energy separation of adjacent levels in that range.

An important and difficult problem of the band theory is to calculate the "shape" of energy bands, i.e., the form of the $N(E)$ versus E curve for the band. The result of one such calculation, for the 3d band of nickel, is shown in Fig. 4.15. All 3d band calculations show that the variation of $N(E)$ with E is quite irregular. However, the exact shape of the band does not affect the general arguments to be advanced later.

Since the 3d and 4s bands overlap in energy (Fig. 4.14), it is convenient to draw the corresponding density curves side by side, as in Fig. 4.16. Here the density of 3d levels increases outward to the left, and that of 4s levels outward to the right;

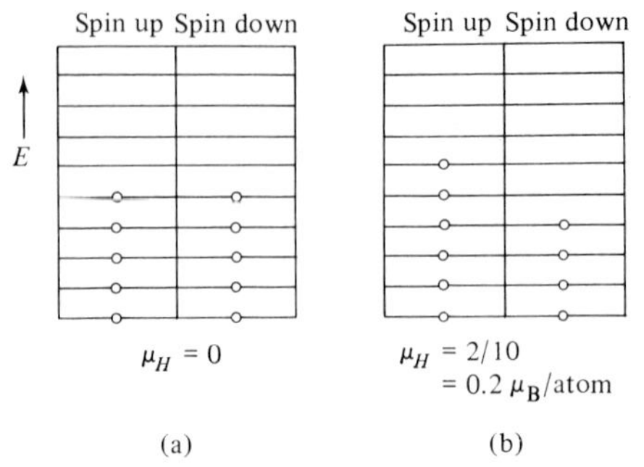

Fig. 4-17 Energy bands with (a) balanced and (b) unbalanced spins.

the letters *A*, *B*, *C*, and *D* refer to the same energies as in Fig. 4.14; and a simplified form, of schematic significance only, has been adopted for the shape of the 3*d* band. Note that the density of 3*d* levels is far greater than that of 4*s* levels, because there are five 3*d* levels per atom, with a capacity of 10 electrons, whereas there is only one 4*s* level, with a capacity of 2 electrons, as noted in Table 4.3. The area under each $N(E)$ versus E curve is equal to the total number of levels in the band.

We will now use the curves of Fig. 4.16 as a basis for discussing the electronic structure of the elements Mn through Zn; i.e., we will make the fairly reasonable assumption that the shape of these bands does not change much from one element to another in this range. (This is called the *rigid-band model.*) Note that the $N(E)$ curves show the density of *available* levels. The extent to which these levels are *occupied* by electrons depends on the number of (3*d* + 4*s*) electrons in the atom. The 3*d* band can hold a total of 10 electrons per atom, but in the transition elements it is never completely full. The extent to which it is filled in several metals is shown by dashed lines in Fig. 4.16. (The topmost filled level for any metal is called the Fermi level.) These lines also show the extent to which the 4*s* band, which can hold only 2 electrons per atom, is filled. As long as both bands are partly full, they must be filled to the same height, just as water in two interconnected tanks must reach the same level.

Nickel has a total of 10 (3*d* + 4*s*) electrons, in the solid or the free atom, and magnetic evidence, described below, indicates that 9.4 are in the 3*d* band and 0.6 in the 4*s*. (The corresponding distribution in the free atom is 8 and 2, respectively.) The Fermi level for nickel is therefore drawn just below the top of the 3*d* zone. Copper has one more electron, and its 3*d* zone is therefore completely full and its 4*s* zone half full. In zinc both zones are full.

To return to the problem of ferromagnetism we note that filled energy levels cannot contribute a magnetic moment, because the two electrons in each level have opposite spin and thus cancel each other out. This situation is depicted in Fig. 4.17(a), where a band of levels, imagined to consist of two half bands, contains an equal number of spin-up and spin-down electrons. The band shown corresponds to a highly simplified, very unreal example: suppose an atom has just one electron in a particular energy level, when the atom is free, and then suppose that

10 such atoms are brought together to form a "crystal." Then the single level in the free atom will split into 10 levels, and the lower 5 will each contain 2 electrons. If one electron reverses its spin, as in (b), then a spin imbalance of 2 is created, and the magnetic moment, or value of μ_H, is 2/10 or 0.2 μ_B per atom. The force creating this spin imbalance in a ferromagnetic is just the exchange force. (Returning to our water-in-a-tank analogy, we might say that the exchange force is like a dam holding water in one half of a tank at a higher level than in the other half.) To create a spin imbalance requires that one or more electrons be raised to higher energy levels; evidently these levels must not be too widely spaced or the exchange force will not be strong enough to effect a transfer.

The ferromagnetism of Fe, Co, and Ni is due to spin imbalance in the $3d$ band. (The $4s$ electrons are assumed to make no contribution. The density of levels in the $4s$ band is low, which means that the levels themselves are widely spaced.) Since the $3d$ band can hold 5 electrons with spin up and 5 with spin down, the maximum imbalance, i.e., the saturation magnetization, is achieved when one half-band is full of 5 electrons. Suppose we let

$$n = \text{number of } (3d + 4s) \text{ electrons per atom}$$
$$x = \text{number of } 4s \text{ electrons per atom}$$
$$n - x = \text{number of } 3d \text{ electrons per atom}$$

At saturation, five $3d$ electrons have spin up and $(n - x - 5)$ have spin down. The magnetic moment per atom is therefore

$$\mu_H = [5 - (n - x - 5)]\,\mu_B = [10 - (n - x)]\,\mu_B\,. \tag{4.32}$$

(This equation also shows that the maximum spin imbalance is equal to the number of unfilled electron states in the $3d$ band). For nickel, n is 10 and the experimental value of μ_H is 0.60 μ_B. Inserting these values in Eq. (4.32), we find $x = 0.60$. This number is proportional to the area enclosed by the lower part of the $4s$ $N(E)$ curve in Fig. 4.16 and the dashed line marked "Ni." It is therefore only a slight approximation to assume that the number of $4s$ electrons is constant at 0.60 for elements near nickel. We then have

$$\mu_H = (10.6 - n)\,\mu_B\,. \tag{4.33}$$

The magnetic moments per atom predicted by this equation are compared with experiment in Fig. 4.18 and Table 4.5. Note that theory and experiment have been made to agree for nickel and that the predicted negative moment for copper has no physical meaning, since the $3d$ band of copper is full.

Figure 4.18 shows fairly good agreement between theory and experiment for Fe, Co, and Ni, and, as we shall see in the next section, for certain alloys. However, the theory predicts that manganese and the next lighter elements would be more magnetic than iron, whereas they are, in fact, not ferromagnetic at all. In iron we have assumed 5.00 electrons with spin up and 2.40 with spin down, leading to a spin imbalance of 2.60. Since the observed spin imbalance in iron is some 20 percent less than this predicted value, and in manganese actually zero, it appears that the exchange force cannot keep one half-band full of electrons if the other half-band is less than about half full.

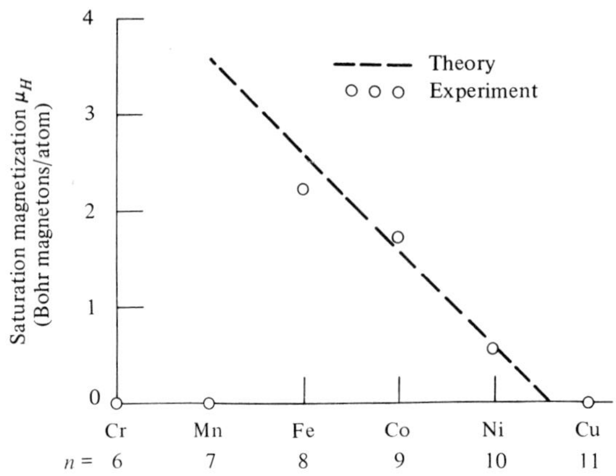

Fig. 4-18 Observed and calculated dependence of the saturation magnetization on the number n of ($3d$ + $4s$) electrons per atom.

The fact that the observed values of μ_H are nonintegral follows quite naturally from the band theory as soon as it is assumed that the $4s$ electrons contribute nothing to the magnetism. For then the division of an integral number of ($3d + 4s$) electrons between the two bands must lead, more often than not, to a nonintegral number of electrons in the $3d$ band. However, the difficulty of what value to assign to J, mentioned in Section 4.3, still remains. "Experimental" values of J can be obtained by putting the experimental value of μ_H and $g = 2$ into the relation $\mu_H = gJ\mu_B$; this leads to J values of 1.11, 0.86, and 0.30 for Fe, Co, and Ni, respectively. Inasmuch as these numbers are not integral multiples of $\frac{1}{2}$, it is not at all clear what physical significance should be attached to them.

Stoner [4.3] has also applied band theory to the important problem of predicting how the saturation magnetization σ_s of a ferromagnetic varies with temperature. This is a matter of finding the equilibrium balance between (a) the increase in electron energy caused by spin imbalance and by an increase in temperature, and (b) the strength of the exchange force. The solution is difficult and will not be described here. Suffice it to say that the relation found by Stoner agrees just about as well with experiment (Fig. 4.7) as the relation which results from the Weiss-quantum theory. Stoner's treatment also explains the observed variation of the susceptibility above the Curie point.

To summarize the results of the last two sections we can write down certain criteria for the existence of ferromagnetism in a metal:

1. The electrons responsible must lie in partially filled bands in order that there may be vacant energy levels available for electrons with unpaired spins to move into.
2. The density of levels in the band must be high, so that the increase in energy caused by spin alignment will be small.
3. The atoms must be the right distance apart so that the exchange force can cause the d-electron spins in one atom to align the spins in a neighboring atom.

Table 4.5 Saturation Magnetization

	Mn	Fe	Co	Ni	Cu
n	7	8	9	10	11
μ_H (observed) (μ_B/atom)	0	2.22	1.72	0.60	0
μ_H (calculated) (μ_B/atom)	3.60	2.60	1.60	0.60	-0.40

Requirement (1) rules out inner core electrons, and (2) rules out valence electrons, because the density of levels in the valence band is low. But the transition elements, which include the rare earth metals, have incompletely filled *inner* shells with a high density of levels, so these elements are possible candidates for ferromagnetism. However, of all the transition elements, only Fe, Co, and Ni meet requirement (3). Many of the rare earths are ferromagnetic below room temperature, as shown in Appendix 3; their spontaneous magnetization is due to spin imbalance in their $4f$ bands.

Note that all of the above criteria are the result of hindsight. They are a blend of experiment and theory, and could not have been predicted from first principles.

The band theory affords a ready explanation of the *Pauli paramagnetism* (weak spin paramagnetism) mentioned in Section 3.8. Electrons in a partially filled band of a metal occupy the available levels in accordance with the Fermi-Dirac distribution law. At 0°K this distribution is such that all levels up to the Fermi level are full and all higher levels completely empty, as shown in Fig. 4.19 (a), where the heavy lines indicate the density of occupied levels and the light lines the density of available levels. At any higher temperature, as in (b), thermal energy excites some electrons into higher levels; the density of occupied levels above the former Fermi level therefore increases from zero to some finite value, and the density of occupied levels just below the former Fermi level decreases.

To understand the effect of an applied field, imagine the band divided into two half-bands, containing electrons of opposite spin. In zero applied field, each half-band contains the same number of electrons, and the crystal as a whole has no net moment. When a field is applied at 0°K, it can reverse the spins only of those electrons which lie *at and just below the Fermi level*, and these constitute only a tiny fraction of the total number of electrons. The spins of those which lie in levels deep below E_F cannot be changed, because there are no empty levels immediately above. The result is a weak paramagnetism. (Note that here the field creates a moment on each atom, in sharp contrast to a normal paramagnetic, in which each atom has a net moment before the field is applied.) The effect of an increase in temperature is merely to excite the uppermost electrons in the band to higher levels, but this effect takes

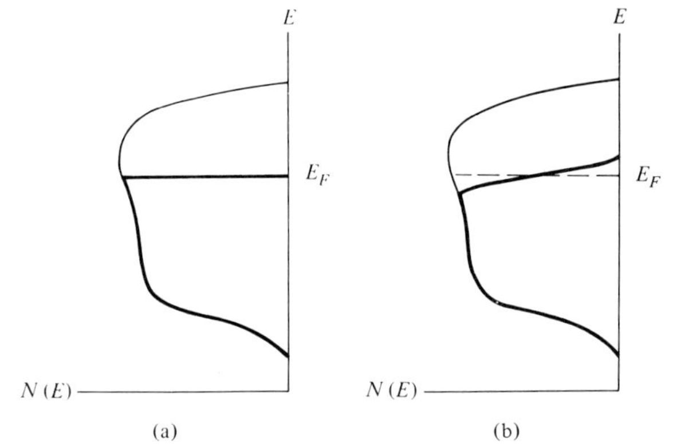

Fig. 4-19 Electron distributions (a) at 0°K, and (b) at a high temperature. E_F is the Fermi level.

place to about the same extent in both half-bands and does not change the spin imbalance created by the field. The susceptibility is therefore essentially independent of temperature.

Figure 4.19 is drawn to suggest the high, narrow 3*d* band characteristic of a transition metal like manganese, which has a high density of 3*d* levels. In a metal like potassium, however, the Pauli paramagnetism will be even weaker, because the outermost electrons are now 4*s* electrons. The density of 4*s* levels is low, which means that the spacing of levels is large. The effect of a given applied field in producing spin imbalance is therefore less than for a metal with closely spaced levels.

We can now see that the phenomena of ferromagnetism and Pauli paramagnetism are very much alike, except for one vital factor: the magnitude of the force creating spin imbalance. The band sketched in Fig. 4.19 might well indicate, in a schematic way, the 3*d* band of both manganese (paramagnetic at room temperature) and iron (ferromagnetic). The only difference would be that the Fermi level of manganese would be somewhat lower than that of iron, because manganese has one less electron. In manganese, spin imbalance can be created only by an applied field; the amount of imbalance is small; and it disappears when the field is removed. In iron, on the other hand, a very powerful molecular field (exchange force) spontaneously and permanently creates a large spin imbalance in every atom and forces the net spins of all the atoms in a single domain to be parallel to one another.

4.5 FERROMAGNETIC ALLOYS

Ferromagnetism is found in the binary and ternary alloys of Fe, Co, and Ni with one another, in alloys of Fe, Co, and Ni with other elements, and in alloys which

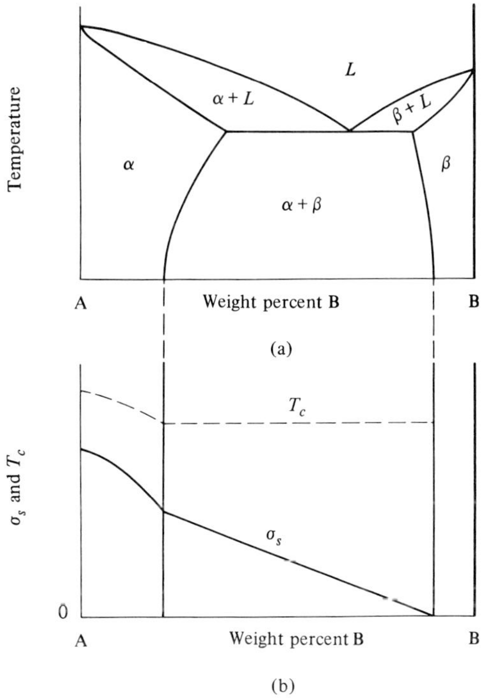

Fig. 4-20 (a) Hypothetical phase diagram of A-B alloys. (b) Variation of σ_s and T_c with composition.

do not contain any ferromagnetic elements. The ferromagnetism of alloys is therefore a very wide subject, and we can examine here only the more important trends. Bozorth [G.4] can profitably be consulted for detailed information and guidance to the literature, Crangle [4.6] for a brief review of theory, and Kouvel [4.7] for a survey of the magnetic properties of intermetallic compounds.

Certain distinctions can be made at the beginning. In terms of binary alloys, these are:

1. When two elements go into *solid solution* in each other, the variation of the saturation magnetization σ_s and the Curie temperature T_c with composition is, in general, unpredictable.

2. When an alloy consists of *two phases*, a change in total composition changes only the relative amounts of the two phases, but the composition of each phase remains constant. Therefore, if one phase is ferromagnetic, σ_s of the alloy will vary linearly with the weight percent of the added element in the alloy, and T_c will remain constant.

These two kinds of behavior are illustrated in Fig. 4.20. Element A is assumed to be ferromagnetic, and B, together with the B-rich solid solution β, is

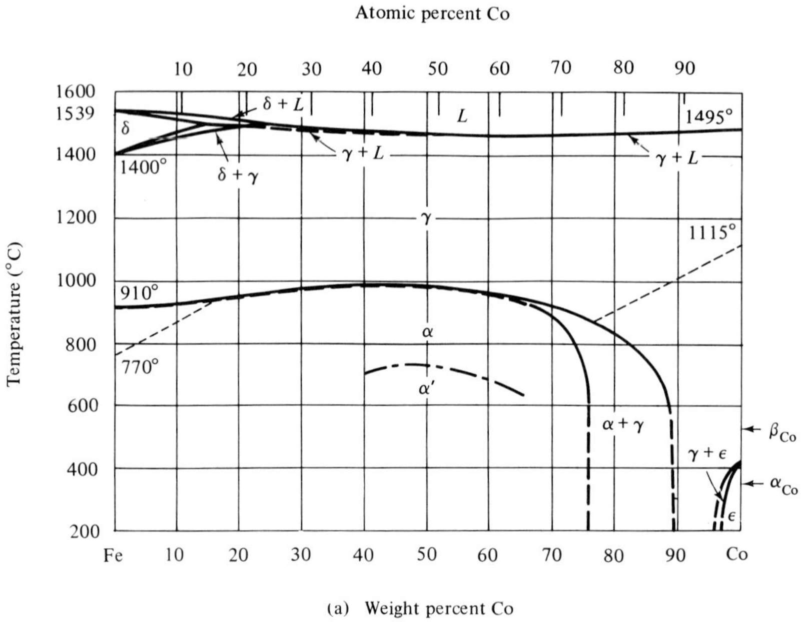

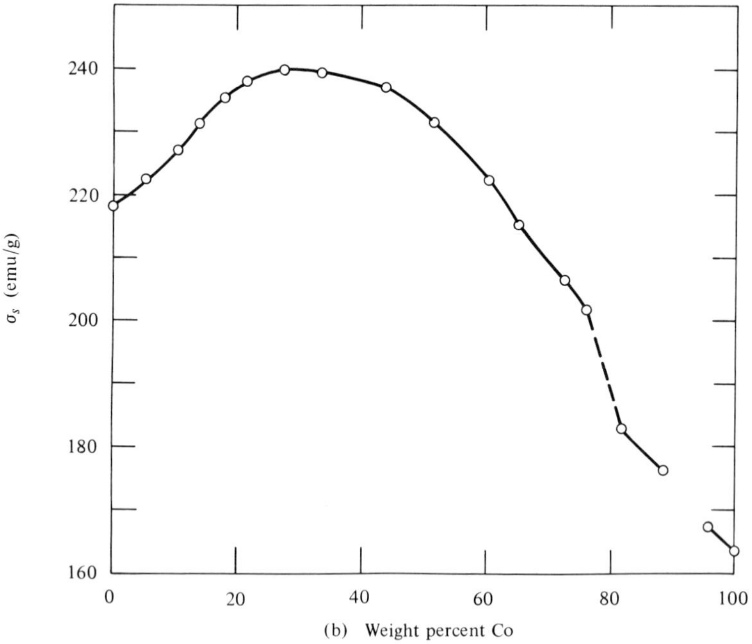

Fig. 4-21 (a) Phase diagram of iron-cobalt alloys [4.8]; **(b)** variation of σ_s at room temperature with composition (measurements of Weiss and Forrer, quoted by Hoselitz [G.6]).

assumed to be paramagnetic. When B is added to A to form the α solid solution, both σ_s and T_c are expected to decrease, but the shape of the curve of σ_s (or T_c) versus composition is impredictable. In the two-phase ($\alpha + \beta$) region, the α phase is saturated and does not change its composition, but the amount of α decreases as B is added. Therefore σ_s decreases linearly to zero at the edge of the two-phase field, and T_c remains constant.

Iron-cobalt alloys illustrate these effects. Figure 4.21 (a) shows the phase diagram; there is a very wide range of solid solubility at room temperature, extending to about 75 percent Co. The 50 percent Co alloy undergoes long-range ordering, and α' is the ordered form of α. The Curie temperature, shown by dotted lines, follows the boundaries of the ($\alpha + \gamma$) region from about 15 to 73 percent Co, a range over which these boundaries are almost coincident; T_c is then constant, but not shown on the diagram, over the two-phase region from 73 to 76 percent Co, and then rises to the value for pure Co. Figure 4.21 (b) shows the variation of σ_s with composition. The addition of cobalt, which is less magnetic than iron, *increases* the magnetization, and the 30 percent Co alloy has a higher value of σ_s at room temperature than any other known material. The cobalt-rich alloys are not spaced closely enough in composition to clearly show the expected variation of σ_s in the two-phase ($\alpha + \gamma$) and ($\gamma + \varepsilon$) regions, but the marked change in slope of the curve at about 75 percent Co, the edge of the α region, is apparent. The dashed line probably corresponds to the ($\alpha + \gamma$) field, but the phase diagram and the magnetic data are not in full accord in the Co-rich region.

In the remainder of this section we will consider only single-phase solid solutions. According to the band theory, the saturation magnetization of elements near nickel should depend only on the number n of ($3d + 4s$) electrons per atom in accordance with Eq. (4.33). By alloying we can make n take on nonintegral values and, in this way, test the theory over those ranges of n in which solid solutions exist. The results are shown in Fig. 4.22, generally called the Slater-Pauling curve. We note the following points:

1. When n is greater than about 8.3, theory and experiment are in good agreement. Both the theoretical and experimental values of μ_H go to zero in a Ni-Cu alloy containing 60 percent Cu ($n = 10.6$); this composition is such that the $3d$ band is just filled. In general, however, good agreement is obtained only when a particular value of n is arrived at by alloying *adjacent* elements.

2. For n greater than 8.3, most of the data are in marked disagreement with theory for alloys of nonadjacent elements. For example, Co containing 50 percent Ni has $n = 9.5$ and lies on the main curve, but Mn containing 83 percent Ni, for which n is also 9.5, has a value of μ_H much lower than the theoretical. There appears to be no general agreement on the reason for such deviations. (See also Section 5.5.)

3. When n is less than 8.3, there is no agreement whatever between theory and experiment, and the magnetization decreases as n decreases.

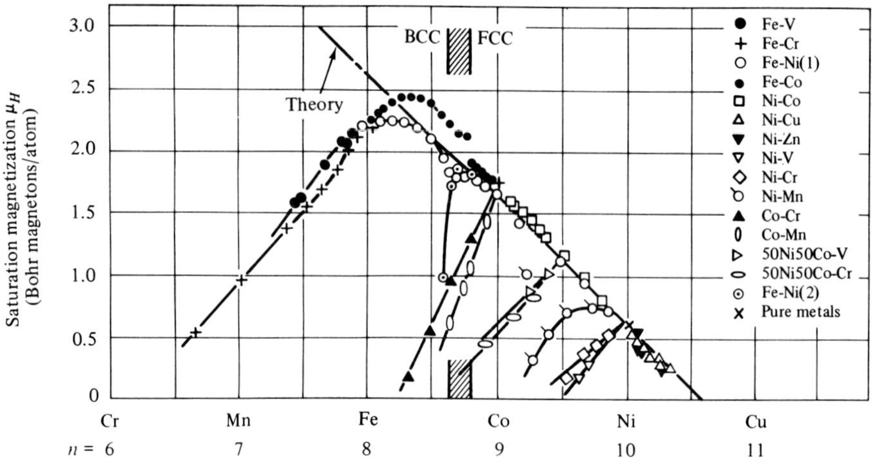

Fig. 4-22 Dependence of the saturation magnetizatism of alloys on the number n of $(3d + 4s)$ electrons per atom (Slater-Pauling curve). After Chikazumi [G-23].

When a nontransition element (e.g., Cu, Zn, Al, Si, etc.) is dissolved in Fe, Co, or Ni, the magnetization decreases, but the rate of decrease per added atom is not the same for all solutes:

1. *Ni-base alloys.* Here the magnetization decreases at a rate proportional to the valence of the solute. Thus 60 atomic percent of Cu (valence 1) is needed to reduce the magnetization to zero, but only 30 atomic percent of Zn (valence 2). The solute atom appears to contribute all its valence electrons to the $3d$ band of the alloy; the larger the valence, the more rapidly does the band fill up, and hence the more rapid is the decrease in magnetization. This behavior is in good agreement with the band theory.
2. *Fe- and Co-rich alloys.* Here the rate of decrease of magnetization, at least initially, is much the same, whatever the added atom. The solute atom appears to act as a simple diluent, i.e., the magnetization decreases as though iron atoms, for example, of moment 2.22 μ_B were being replaced by atoms of zero moment. This behavior is inexplicable in terms of the simple band theory.

Even more complex problems are presented to the theorist by binary alloys of Fe, Co, or Ni and one of the transition metals immediately below them in the periodic table (Ru, Rh, Pd, Os, Ir, and Pt). The magnetization usually decreases as one of the latter is added, but at a rate which is greater than, equal to, or less than the rate corresponding to simple dilution. And in some of the alloys the magnetization actually increases at first, before it begins to decrease.

At low temperatures, surprisingly small amounts of Fe, Co, or Ni, when alloyed with elements like Pt or Pd, can serve to make the alloy ferromagnetic. Thus ferromagnetism has been observed in Pd containing only 10 atomic percent Fe at temperatures below about 260°K. And ferromagnetism still persists when the Fe content is as low as 0.15 atomic percent, although the Curie temperature is then about 3°K. In such a dilute alloy the iron atoms are so far apart, some 8 to 10 interatomic distances, that exchange forces could not possibly keep the spins of nearest-neighbor iron atoms parallel to one another; somehow or other, the intervening palladium atoms must take part in the long-range coupling between iron atoms.

All of the above remarks on alloys apply to disordered (random) solid solutions, in which the constituent atoms, A and B, occupy the available lattice sites at random. In many alloys this disordered state is stable only at relatively high temperatures; below a certain critical temperature, long-range ordering sets in; A atoms then occupy a particular set of lattice sites and B atoms occupy another set. In the disordered state, like atoms are often adjacent to each other (AA or BB pairs), whereas ordering commonly makes all nearest neighbors unlike (AB pairs). Because the nature of the nearest neighbors in a particular alloy can influence its electronic nature, ordering usually changes the magnetic properties, sometimes dramatically.

In Fe-Co alloys, at and near the composition of FeCo, ordering takes place below a critical temperature of about 730°C, as shown in Fig. 4.21 (a). In the disordered α phase the atoms are arranged at random on the corners and at the center of a cubic unit cell. The ordered α' phase has the "CsCl structure," in which iron atoms occupy only corner sites and cobalt atoms only the cube-center sites. Ordering produces a slight increase in the saturation magnetization. The same small effect is found in $FeNi_3$. Here the disordered phase is face-centered cubic, while the ordered structure is one in which iron atoms occupy only the cube corners and nickel atoms only the centers of the cube faces. (See Fig. 10.5.) But in $MnNi_3$, which orders like $FeNi_3$, ordering has a profound effect: the disordered alloy is paramagnetic, the ordered one ferromagnetic [4.9].

Many binary alloys of a transition metal and a rare earth element are ferromagnetic. For example, alloys of the composition RCo_5, where R = Y, Ce, Pr, Nd, or Sm, are ferromagnetic and some of them are good permanent magnets. Details are given by Kouvel [4.7] and in Section 14.8.

Finally, we will consider ferromagnetism in alloys made of nonferromagnetic elements, a subject which has been reviewed by Crangle [4.10]. Most of these contain manganese or chromium. The Heusler alloys Cu_2MnSn and Cu_2MnAl, which have been known since 1898, have already been mentioned in Section 4.3. Like $MnNi_3$ they are paramagnetic when disordered and ferromagnetic when ordered. Presumably, in all three alloys, ordering makes the Mn-Mn distance large enough for the exchange interaction to become positive. MnBi and MnAl are ferromagnetic, and their coercivities are high enough to make them of interest as possible permanent-magnet materials. Some other ferromagnetic manganese alloys are Ag_5MnAl, Mn_3ZnC, Au_4Mn, MnSb, MnCrSb, and Pt_3Mn. Ferromagnetic chromium alloys or compounds include CrS, CrTe, and $CrBr_3$. One would not expect to find ferromagnetism in the Zr-Zn system, but the compound $ZrZn_2$ is ferromagnetic; however, its Curie temperature is only 35°K.

Ferromagnetism is rare in ionic compounds, in general, and even rarer in oxides.

Only two examples of the latter are known, CrO_2 and EuO. The Curie point of EuO is 77°K, but CrO_2 is ferromagnetic at room temperature ($T_c = 127°C$). The crystal structure of CrO_2 is that of the mineral rutile; it is shown in Fig. 5.12. The Cr^{4+} ions are located at the cell center and the cell corners, and the spins on these ions are all parallel, unlike the spin structure illustrated there. The properties of CrO_2 make it suitable for magnetic recording tape (see Section 14.9).

4.6 THERMAL EFFECTS

Ferromagnetic substances exhibit two unusual thermal effects. Although of little practical interest, these effects are studied by solid-state physicists because of the light they can throw on the electronic nature of a solid. They are:

1. Specific heat

The specific heat of a ferromagnetic is greater than that of a nonferromagnetic and goes through a maximum at the Curie temperature [G.4]. When heat is added to any metal, part of it increases the amplitude of thermal vibration of the ions (lattice specific heat) and the remainder increases the kinetic energy of the valence electrons (electronic specific heat). If the metal is ferromagnetic, then still additional heat is required to disorder the spins (magnetic specific heat). The number of spins disordered per degree rise in temperature increases with the temperature in accordance with the appropriate Brillouin function, and becomes very large just below the Curie temperature, where the magnetization is decreasing precipitously. The fact that the specific heat is still abnormally high just above the Curie temperature is further evidence for the spin clusters mentioned in Section 4.2.

2. Magnetocaloric effect

When heat is absorbed by a ferromagnetic, part of the heat causes a decrease in the degree of spin order. Conversely, if the spin order is increased, by the application of a large field, heat will be released. If the field is applied suddenly, the process will be essentially adiabatic and the temperature of the specimen will rise. This is the magnetocaloric effect. The increase in temperature amounts to 1° or 2°C for fields of the order of 10 to 20 kOe [4.11].

The magnetocaloric effect is not to be confused with a much smaller heating effect which occurs at room temperature during a change of magnetization from zero to the state of technical saturation, i.e., during the conversion of the specimen from the multi-domain to the single-domain state. The magnetocaloric effect is caused by the change in spin order accompanying forced magnetization, in which a high field causes an increase in magnetization of the domain itself. This increase in magnetization above σ_s, for a given applied field, is greatest at the Curie temperature and decreases as the temperature decreases below T_c. A relatively large increase in magnetization per unit of applied field can also be produced just *above* the Curie temperature because there the susceptibility of the paramagnetic state has its maximum value. As a result the temperature increase produced by this effect goes through a maximum at T_c.

It can be shown [4.12] that the molecular-field constant $\gamma\rho$ can be calculated from the temperature rise observed in the magnetocaloric effect. Such calculations show that $\gamma\rho$ increases with temperature above T_c, whereas in the Weiss molecular-field theory, $\gamma\rho$ is assumed to be independent of temperature. The change in the molecular-field "constant" indicated by the magnetocaloric measurements is a large one: a more than twofold increase in $\gamma\rho$, both for iron and nickel, in a temperature range of some 50° to 100°C above T_c. Below T_c, $\gamma\rho$ appears to be essentially constant.

4.7 THEORIES OF FERROMAGNETISM

The critical reader must by now have come to the conclusion, correctly, that the theory of ferromagnetism is in a far from satisfactory state. In this section we shall briefly examine the theory in the light of the main experimental facts. Actually, there is not a single theory, but at least two rather divergent viewpoints: (1) the localized moment theory, and (2) the band theory.

According to the *localized moment theory*, the electrons responsible for ferromagnetism are attached to the atoms and cannot move about in the crystal. These electrons contribute a certain magnetic moment to each atom and that moment is localized at each atom. This view is implicit in the molecular field theory, either in the original form given by Weiss or in the quantum-mechanical form obtained by substituting the Brillouin function for the Langevin. As we have seen, this theory accounts satisfactorily for the variation of the saturation magnetization σ_s with temperature and for the fact that a Curie-Weiss law is obeyed, at least approximately, above T_c. But it cannot explain the fact that the observed moments per atom μ_H are nonintegral for metals; since the moment is due almost entirely to spin, as shown by g factor measurements, the moment per atom, if due to localized electrons, should be an integer. Other defects of the theory are that μ_H and the molecular-field constant $\gamma\rho$ are different above and below the Curie temperature.

The Heisenberg approach is also based on the assumption of localized moments, because the expression for the exchange energy (Eq. 4.29) explicitly localizes a certain spin magnetic moment on each atom. (A substance which does behave as though its moments were localized, and there are a few, like EuO, is often called a *Heisenberg ferromagnet*.) Thus the assumption of localized moments is built into the molecular field theory, whether we call the force causing parallel spin alignment a molecular field or an exchange force.

In the *band theory*, on the other hand, all attempts at localizing the outer electrons of the atom are abandoned. As mentioned earlier, the band theory is often called the collective-electron theory, when applied to magnetic properties. Another name is the itinerant-electron theory. These alternative names emphasize the fact that the electrons responsible for ferromagnetism are considered to belong to the crystal as a whole and to be capable of motion from one atom to another, rather than localized at the positions of the atoms. This theory accounts quite naturally for the nonintegral values of the moment per atom. It also explains fairly well the relative magnitudes of μ_H in iron, cobalt, and nickel, and the value of the average magnetic moment per atom in certain alloys. These are important accomplishments of the theory. However, the band theory, at least in its simple form, cannot account for those alloys which depart from the main curve of Fig. 4.22.

The general conclusion among physicists today is that the molecular field theory, with its attendant assumption of localized moments, is simply not valid for metals. Instead, the band theory is regarded as basically correct, and the problem then becomes one of understanding the precise form of the various bands, how they are occupied by electrons, how the exchange forces operate, etc. These are problems of great difficulty and have led to rather wide divergences of opinion among experts. There is, in fact, not one band theory but several. In some, *all* the 3*d* and 4*s* electrons in a metal like iron, for example, are regarded as itinerant; in other theories, some of these electrons are regarded as bound to the atoms and the remainder as itinerant. (Theories of the latter kind therefore have some of the flavor of the localized-moment theory.) The precise action of the exchange forces is also an unsettled problem. Some believe, for example, that the conduction electrons play a vital role in producing parallel spins on adjacent atoms, by a mechanism called "indirect exchange." In this view, the basic exchange force acting is one that produces *antiparallel* spins. If an outer electron on atom A has spin up, then a nearby conduction electron will have spin down; the spin on this conduction electron will then force an adjacent electron on atom B to have its spin up. The spins on the adjacent atoms A and B are thus forced into parallelism by means of the conduction electron passing between them. These and other aspects of band theory have been reviewed in detail by Herring [4.13].

When a truly unified theory of ferromagnetism is at last achieved, it is not likely to be a simple theory, if only because the experimental facts to be explained are so diverse. Actually, the problem of ferromagnetism is only a part of the more general problem of understanding the complex electronic structure of the transition elements. The problems posed by ferromagnetism are therefore very fundamental ones, and their solution will illuminate not only ferromagnetism but many other areas of solid-state physics and metallurgy.

4.8 MAGNETIC ANALYSIS

We turn now to a very practical subject. Magnetic analysis, in the widest sense of the term, embraces any determination of chemical composition or physical structure by means of magnetic measurements (G.6, 4.14]. It therefore includes the following:

1. Measurement of the susceptibility χ of weakly magnetic substances. If the relation between χ and the chemical composition of a solid solution is known, either from the literature or from calibration experiments, a measurement of χ will give the composition. (An example was mentioned in the previous chapter, in the discussion of Fig. 3.12.) Determination of the *M*,*H* curve

will disclose the presence of a small amount of a ferromagnetic second phase, as described in the discussion of Fig. 3.13.

2. Measurement of the structure-sensitive properties, such as the initial permeability μ_0 and the coercivity H_c, of ferro- and ferrimagnetics. The interpretation of such measurements, although difficult, can yield information about the physical structure of the material, e.g., preferred crystal orientation, residual stress, and the presence of inclusions. These topics will be taken up in later portions of the book.
3. Measurement of the structure-insensitive properties of ferro- and ferrimagnetics. These are considered below.

The structure-insensitive properties are the saturation magnetization σ_s and the Curie temperature T_c. As we saw in Fig. 4.20, the value of σ_s for a two-phase alloy is simply the weighted average of the magnetizations of the two phases:

$$\sigma_s\,(\text{alloy}) = w_\alpha \sigma_{s\alpha} + w_\beta \sigma_{s\beta} = w_\alpha \sigma_{s\alpha} + (1 - w_\alpha)\,\sigma_{s\beta}, \tag{4.34}$$

where w_α and w_β are the weight fractions of the α and β phases. This equation holds whether the β phase is ferromagnetic ($\sigma_{s\beta} > 0$) or paramagnetic ($\sigma_{s\beta} = 0$). If $\sigma_{s\alpha}$ and $\sigma_{s\beta}$ are known, a measurement of σ_s(alloy) will give the amount of each phase present. At least two applications of such measurements have been made:

1. *Determination of retained austenite in hardened steel.* At a sufficiently high temperature, steel is wholly austenitic. Austenite is a paramagnetic solid solution of carbon and possibly other elements, in face-centered cubic γ-iron. When the steel is quenched in water, the austenite transforms wholly or partially to martensite, which is a ferromagnetic, supersaturated solid solution of carbon in α-iron, with a body-centered tetragonal unit cell. Any untransformed austenite is called retained austenite, and it can be present in amounts ranging from 0 to about 20 percent.
2. *Determination of martensite in stainless steel.* Many stainless steels are wholly austenitic, and paramagnetic, at room temperature; an example is the popular "18-8" variety, which contains 18 percent Cr and 8 percent Ni. However, a substantial amount of ferromagnetic martensite may be formed if the steel is severely cold-worked at room temperature or merely cooled to a temperature far below room temperature.

Although the application of Eq. (4.34) to these determinations might seem easy, there is a certain experimental difficulty in determining σ_s (alloy). The approach to magnetic saturation, as the applied field is increased, is gradual, and it is sometimes difficult to know whether or not saturation has actually been reached. (See Section 9.15.) This means in turn that greater accuracy can be achieved in determining a small amount of a ferromagnetic dispersed in a paramagnetic matrix (e.g., martensite in stainless steel) than in determining, by difference, a small amount of a paramagnetic dispersed in a ferromagnetic (e.g., retained austenite in hardened steel).

Measurements of σ_s can be helpful in determining phase diagrams. Measurement of the solid solubility of B in A is illustrated in Fig. 4.23, in which, as before,

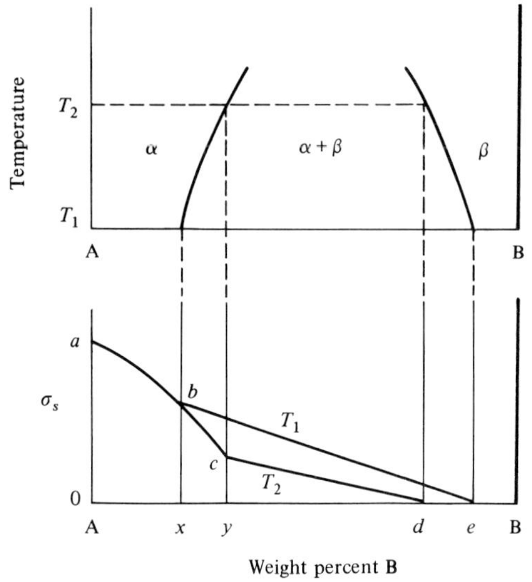

Fig. 4-23 Variation of σ_s with composition in alloys equilibrated at temperatures T_1 and T_2.

α is assumed to be ferromagnetic and β paramagnetic. If the alloys are very slowly cooled to room temperature T_1, so that they are in equilibrium at that temperature, then σ_s will vary with composition along the curve *abe*. But if the alloys are brought to equilibrium at T_2 and then quenched to room temperature, σ_s will vary along *abcd*. The intersections of the straight line portions with the curve *abc*, which shows the variation of σ_s with composition in the α region, gives the solid solubilities, namely, *x* and *y* percent B at temperatures T_1 and T_2, respectively. Although magnetic analysis can be of considerable value in the study of some alloy systems, it would be a mistake to rely on it entirely; instead, it should be supplemented with the more common techniques of microscopic examination and x-ray diffraction.

In all the above examples, the measurement of σ_s is made at room temperature. Another method, called *magnetothermal analysis*, involves measurement of σ_s as a function of temperature, leading to a curve like that of Fig. 4.24. This curve would apply to a two-phase alloy in which both phases are ferromagnetic. In principle, at least, the composition of each phase can be found from the observed Curie points, $T_{c\alpha}$ and $T_{c\beta}$, and their relative amounts from $\sigma_{s\alpha}$ and $\sigma_{s\beta}$. One difficulty with this method is that both Curie points must be well below the temperature of rapid diffusion, or alternatively, the solid-solubility limits of each phase must not change with temperature; if these conditions are not met, the mere act of heating the specimen to the measurement temperatures will change the composition of one or both phases and make quantitative interpretation of the σ_s,T curve impossible. Magnetothermal analysis can also be applied to alloys which contain only one ferromagnetic phase, and Hoselitz [G.6] has described a number of examples.

Magnetic measurements have also been used to study the precipitation process in alloys, particularly those in which a ferromagnetic phase precipitates in a paramagnetic matrix. When conditions are favorable, the size of the precipitate particles, the total amount of precipitate, and the way in which these quantities

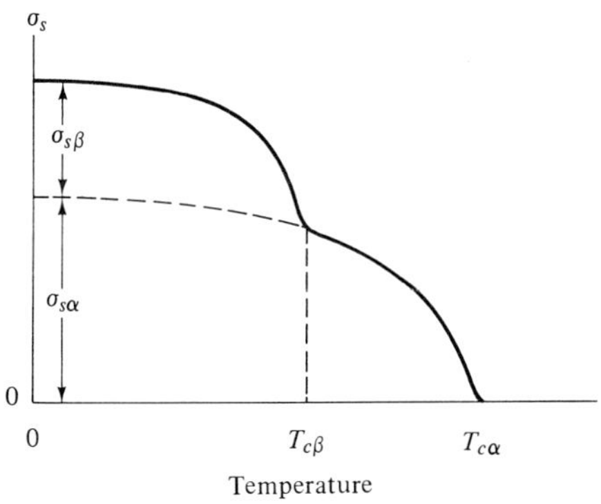

Fig. 4-24 Variation of σ_s with temperature in an alloy of two ferromagnetic phases.

change with time can be measured. The interpretation of such measurements is not always straightforward. For one thing, the size of the field required to saturate the particles depends on their shape and orientation, as mentioned near the end of Section 3.8. But more important is the fact that the magnetic properties of very fine particles are unusual. These properties are described in Chapter 11, and an example of a magnetic study of precipitation will be given there (Section 11.7).

PROBLEMS

4.1 Plot the relation between σ/σ_0 and a' for $J = 1$ and $J = \infty$. Obtain graphically from these curves the relation between σ_s/σ_0 and T/T_c. For $J = \frac{1}{2}$, obtain the same relation from a table of hyperbolic tangents. Plot the three curves as in Fig. 4.7 and the experimental values for Fe, Co, and Ni, given in Appendix 5.

4.2 Show that Eq. (4.25) is identical to Eq. (3.52).

4.3 Given that a ferromagnetic substance obeys Eq. (4.20),

a) Show that the fractional change in the relative spontaneous magnetization produced per unit of applied field H is given by

$$\left(\frac{1}{\sigma_s/\sigma_0}\right) d\left(\frac{\sigma_s}{\sigma_0}\right) = \frac{(\mu_H/kT)\,dH}{\sinh u \cosh u - (T_c/T)\tanh u},$$

where

$$u = \frac{T_c}{T}\frac{\sigma_s}{\sigma_0} + \frac{\mu_H H}{kT} = \frac{T_c}{T}\frac{\sigma_s}{\sigma_0},$$

if H is zero initially.

b) Calculate the fractional change in σ_s/σ_0 per oersted for iron at 20°C and 750°C given that $T_c = 1043$°K and $\mu_H = 2.22\ \mu_B$.

CHAPTER 5

ANTIFERROMAGNETISM

5.1 INTRODUCTION

Antiferromagnetic substances have a small positive susceptibility at all temperatures, but their susceptibilities vary in a peculiar way with temperature. At first glance, they might therefore be regarded as anomalous paramagnetics. However, closer study has shown that their underlying magnetic "structure" is so entirely different that they deserve a separate classification. The theory of antiferromagnetism was developed chiefly by Néel * in a series of papers, beginning in 1932 [5.1], in which he applied the Weiss molecular field theory to the problem. Theory and experiment have been reviewed by Lidiard [5.2], Nagamiya *et al.* [5.3], and Smart [G.28].

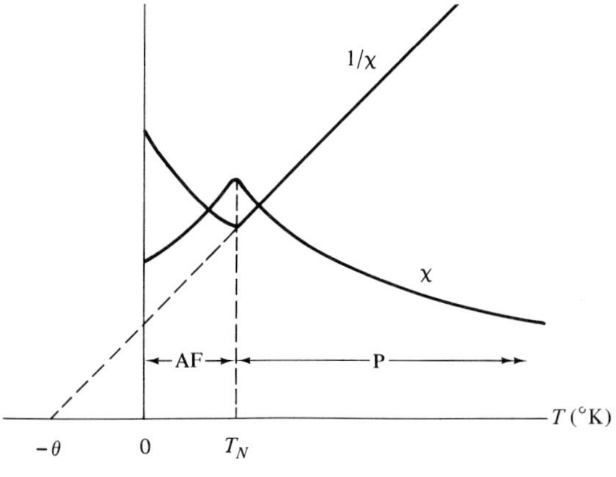

Fig. 5.1 Variation with the temperature of the susceptibility χ, and of $1/\chi$, for an antiferromagnetic (schematic). AF = antiferromagnetic, P = paramagnetic.

The way in which the susceptibility of an antiferromagnetic varies with temperature is shown in Fig. 5.1. As the temperature decreases, χ increases but

* Louis Néel (1904–), French physicist. A former student under Weiss at Strasbourg, he founded, after World War II, a flourishing center of magnetic research at the University of Grenoble. He was awarded a Nobel prize in 1970.

Table 5.1 Some Antiferromagnetic Substances*

Substance	*Metal ion arrangement*†	T_N (°K)	θ (°K)††	$\frac{\theta}{T_N}$	$\frac{\chi_p(0)}{\chi_p(T_N)}$
MnO	FCC	122	610	5.0	0.69
FeO	FCC	198	570	2.9	0.78
CoO	FCC	293	280	1.0	---
NiO	FCC	523	3000	5.7	0.67
α-MnS	FCC	154	465	3.0	0.82
β-MnS	FCC	155	982	6.3	--
α-Fe_2O_3	R	950	2000	2.1	---
Cr_2O_3	R	307	1070	3.5	0.76
$CuCl_2 \cdot 2H_2O$	R	4.3	5	1.2	---
FeS	HL	613	857	1.4	---
$FeCl_2$	HL	24	−48	−2.0	<0.2
$CoCl_2$	HL	25	−38	−1.5	~0.6
$NiCl_2$	HL	50	−68	−1.4	---
MnF_2	BCT	67	80	1.2	0.76
FeF_2	BCT	79	117	1.5	0.72
CoF_2	BCT	40	53	1.3	---
NiF_2	BCT	78	116	1.5	---
MnO_2	BCT	84	---	--	0.93
Cr	BCC	310			
α-Mn	CC	100			

* Data from Nagamiya *et al.* [5.3], Smart [G.28], and Morrish [G.27].

† FCC = face-centered cubic, R = rhombohedral, HL = hexagonal layers, BCT = body-centered tetragonal, BCC = body-centered cubic, CC = complex cubic.

†† The sign of y in this column refers to a χ, T relation in the form $\chi = C/(T - \theta)$.

finally goes through a *maximum* at a critical temperature T_N called the *Néel temperature.* The substance is paramagnetic above T_N and antiferromagnetic below it. T_N commonly lies far below room temperature, so that it is often necessary to carry susceptibility measurements down to quite low temperatures to discover if a given substance, paramagnetic at room temperature, is actually antiferromagnetic at some lower temperature. Most, but not all, antiferromagnetics are ionic compounds, namely, oxides, sulphides, chlorides, and the like. Over a hundred are now known, which makes them much more common than ferromagnetics, and the salient features of some are listed in Table 5.1. They are of considerable scientific interest, but their magnetic properties have no commercial value. How-

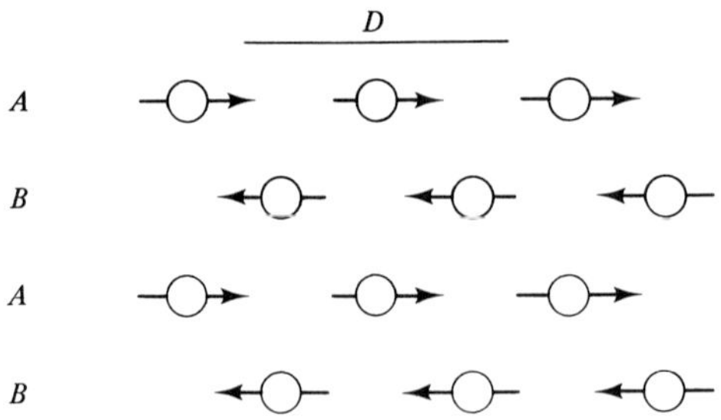

Fig. 5.2 Antiferromagnetic arrangement of A and B sublattices.

ever, the theory of these materials is worth examining in some detail because it leads naturally into the theory of ferrimagnetics, which are of great industrial importance.

Just as in the case of ferromagnetism, the clue to the behavior of an antiferromagnetic lies in the way its susceptibility varies with temperature *above* the critical temperature. Figure 5.1 shows that a plot of $1/\chi$ versus T is a straight line above T_N and that this line extrapolates to a negative temperature at $1/\chi = 0$. The equation of the line is

$$\frac{1}{\chi} = \frac{T + \theta}{C} \tag{5.1}$$

or

$$\chi = \frac{C}{T + \theta} = \frac{C}{T - (-\theta)}. \tag{5.2}$$

In other words, the material obeys a Curie-Weiss law but with a *negative* value of θ. Inasmuch as θ is proportional to the molecular field coefficient γ (see Eq. 3.27), the molecular field H_m, in the paramagnetic region, is *opposed* to the applied field H; whereas H tries to align the ionic moments, H_m acts to disalign them. If we now think of the molecular field on a very localized scale, the result is that any tendency of a particular ionic moment to point in one direction is immediately counteracted by a tendency for the moment on an adjacent ion to point in the opposite direction. In other words, the exchange force is negative.

Below the critical temperature T_N, this tendency toward an antiparallel alignment of moments is strong enough to act even in the absence of an applied field, because the randomizing effect of thermal energy is so low. The lattice of magnetic ions in the crystal then breaks up into two sublattices, designated A and B, having moments more or less opposed. The tendency toward antiparallelism becomes stronger, the lower the temperature is below T_N, until at 0°K the antiparallel arrangement is perfect, as depicted in Fig. 5.2. Only the magnetic metal ions are shown in this sketch: the other ions (oxygen, or sulphur, etc., as the case may be) are nonmagnetic and need not be considered at this point.

We now see that an antiferromagnetic at 0°K consists of two interpenetrating and identical sublattices of magnetic ions, each spontaneously magnetized to saturation in zero applied field, but in opposite directions, just as the single lattice of a ferromagnetic is spontaneously magnetized. Evidently, an antiferromagnetic has no net spontaneous moment and can acquire a moment only when a strong field is applied to it. We note also that the Néel temperature T_N plays the same role as the Curie temperature T_c; each divides the temperature scale into a magnetically ordered region below and a disordered region (paramagnetic) above. The several analogies to ferromagnetism are apparent and the name "antiferromagnetism" could not be more apt.

5.2 MOLECULAR FIELD THEORY

Before entering into the details of the molecular field theory, we should note that almost all antiferromagnetics are electrical *insulators* or, at the most, semiconductors. Their electrical resistivities are thus at least a million times larger than those of typical metals. This means that they contain essentially no free electrons and that the electrons responsible for their magnetic properties are localized to particular ions. We therefore expect greater success in applying the molecular field theory, which is a localized-moment theory, to an antiferromagnetic insulator like MnO than to a ferromagnetic conductor like iron.

We will apply the molecular field theory to the simplest possible case, namely, one for which the lattice of magnetic ions can be divided into two identical sublattices, A and B, such that any A ion has only B ions as nearest neighbors. and vice versa, as shown for two dimensions in Fig. 5.2. We assume that the only interaction is between nearest neighbors (AB) and ignore the possibility of interactions between second-nearest neighbors (AA and BB).

We now have *two* molecular fields to deal with. The molecular field H_{mA} acting on the A ions is proportional, and in the opposite direction, to the magnetization of the B sublattice:

$$H_{mA} = -\gamma M_{\mathrm{B}}, \tag{5.3}$$

where γ is the molecular field coefficient, taken as positive. Similarly,

$$H_{m\mathrm{B}} = -\gamma M_{\mathrm{A}}. \tag{5.4}$$

These two equations are valid above and below T_N. We will consider the two cases in turn.

Above T_N

In the paramagnetic region we can find an equation for the susceptibility by proceeding as in Section 3.6, according to Eqs. (3.25) to (3.27). Assuming Curie-law behavior, we have

$$\chi = \frac{M}{\rho H} = \frac{C}{T} \tag{5.5}$$

or

$$MT = \rho C H, \tag{5.6}$$

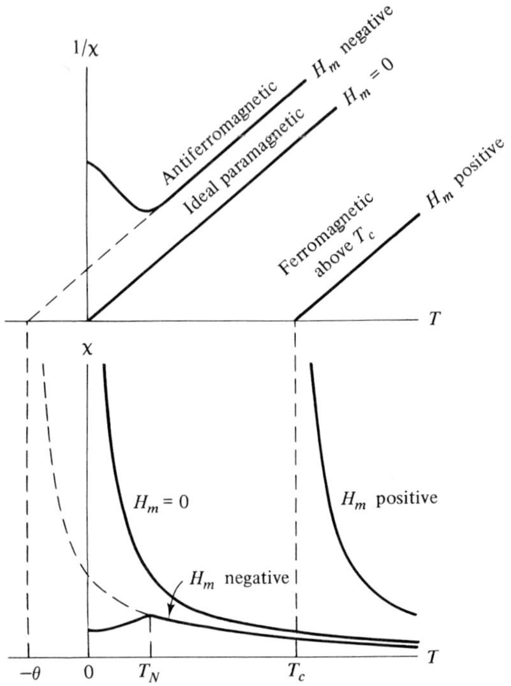

Fig. 5.3 Dependence of the susceptibility χ on the molecular field H_m, for the same value of the Curie constant C.

where H must be interpreted as the total field, applied and molecular, acting on the material. We now write Eq. (5.6) for each sublattice:

$$M_A T = \rho C'(H - \gamma M_B), \tag{5.7}$$

$$M_B T = \rho C'(H - \gamma M_A), \tag{5.8}$$

where C' is the Curie constant of each sublattice and H is the applied field. By adding these two equations we can find the total magnetization M produced by the field and hence the susceptibility:

$$\begin{aligned} (M_A + M_B)T &= 2\rho C'H - \rho C'\gamma(M_A + M_B), \\ MT &= 2\rho C'H - \rho C'\gamma M, \\ M(T + \rho C'\gamma) &= 2\rho C'H, \\ \chi &= \frac{M}{\rho H} = \frac{2C'}{T + \rho C'\gamma}. \end{aligned} \tag{5.9}$$

This relation is equivalent to Eq. (5.2), found experimentally, with

$$C = 2C' \quad \text{and} \quad \theta = \rho C'\gamma. \tag{5.10}$$

Note that, when a field is applied above T_N, each sublattice becomes magnetized in the same direction as the field, but each sublattice then sets up a molecular field, in the opposite direction to the applied field, tending to reduce both M_A and M_B. The result is that the susceptibility χ is smaller, and $1/\chi$ larger, than that of an ideal paramagnetic in which the molecular field is zero. The two are compared graphically in Fig. 5.3, which also shows how χ varies with T in a substance with a large positive molecular field, such as a ferromagnetic above its Curie point.

Below T_N

In the antiferromagnetic region, each sublattice is spontaneously magnetized, in zero applied field, by the molecular field created by the other sublattice. When H is zero,

$$M = M_A + M_B = 0,$$

and

$$M_A = -M_B, \tag{5.11}$$

at any temperature below T_N. At a temperature infinitesimally below T_N we may assume that M is still proportional to the total field, because saturation effects are unimportant near T_N, so that Eqs. (5.7) and (5.8) are still valid. At $T = T_N$ and $H = 0$, Eq. (5.7) becomes

$$M_A T_N = -\rho C' \gamma M_B. \tag{5.12}$$

Therefore,

$$\rho C' \gamma = \theta = -(M_A/M_B) T_N = T_N. \tag{5.13}$$

The Néel temperature, at which the maximum in the χ, T curve occurs, should therefore equal the θ value found from the high-temperature susceptibility measurements.

Below T_N, each sublattice is spontaneously magnetized to saturation just as a ferromagnetic is, and we can compute its magnetization in the same way. As in Section 4.2, we prefer to consider the specific magnetization σ $(= M/\rho)$ rather than M, because a range of temperature is involved. The fractional specific magnetization of the A sublattice will then be given, according to Eq. (4.13), for any temperature and field by

$$\frac{\sigma_A}{\sigma_{0A}} = B(J, a') = B\left(J, \frac{\mu_H H}{kT}\right), \tag{5.14}$$

where B is the Brillouin function. The field H which appears in Eq. (5.14) is the total field acting on the A sublattice. Since we are computing the spontaneous magnetization, the applied field is zero, and we include only the molecular field due to the B sublattice:

$$H_{mA} = -\gamma M_B = \gamma M_A = \gamma \rho \sigma_A. \tag{5.15}$$

Therefore, the fractional spontaneous magnetization of the A sublattice is given by

$$\frac{\sigma_{sA}}{\sigma_{0A}} = B\left(J, \frac{\mu_H \gamma \rho \sigma_{sA}}{kT}\right), \tag{5.16}$$

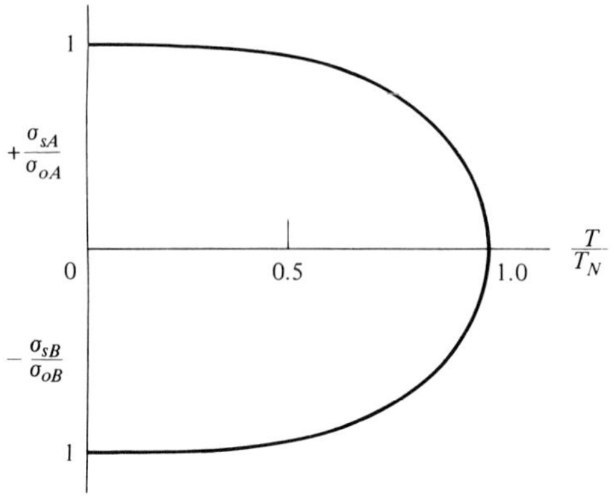

Fig. 5.4 Spontaneous magnetizations of the A and B sublattices at temperatures below T_N (schematic).

with a similar expression for the B sublattice. This equation is best solved graphically by the method of Fig. 4.4. A plot of the two sublattice magnetizations is given in Fig. 5.4.

Although the net spontaneous magnetization is zero below T_N, an applied field can produce a small magnetization. The resulting susceptibility is found to depend on the angle which the applied field makes with the axis of antiparallelism. marked D in Fig. 5.2, an axis which usually coincides with an important crystallographic direction in the crystal. For brevity we will call this the *spin axis.* (In most antiferromagnetics, the orbital contribution is almost entirely quenched, so that the net moment per magnetic ion is due essentially to spin.) We will consider two extreme cases.

Field at right angles to spin axis. The effect of the applied field H is to turn each sublattice magnetization away from the spin axis by a small angle α, as shown in Fig. 5.5 (a), where the vectors representing the magnetizations of the two sublattices are drawn from one point. This rotation immediately creates a magnetization σ in the direction of the field and sets up an unbalanced molecular field H_m in the opposite direction. The spins will rotate until H_m equals H, or

$$2(H_{mA} \sin \alpha) = H,$$

$$2\gamma\rho\sigma_{sA} \sin \alpha = H. \tag{5.17}$$

But

$$\sigma = 2\sigma_{sA} \sin \alpha.$$

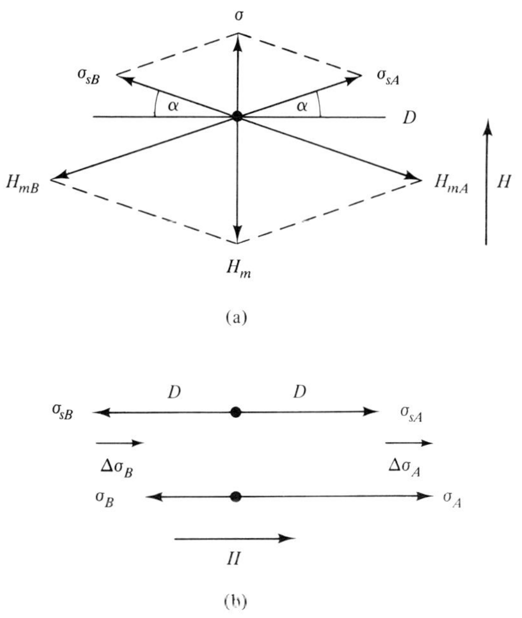

Fig. 5.5 Magnetization changes in an antiferromagnetic when the field H is applied (a) at right angles and (b) parallel to the spin axis D.

Therefore, $\gamma\rho\sigma = H$, and

$$\chi_\perp = \frac{\sigma}{H} = \frac{1}{\gamma\rho} = \frac{C}{2\theta}. \tag{5.18}$$

(We have assumed here that the sublattice magnetizations σ_{sA} and σ_{sB} change only their directions and not their magnitude when a field is applied; this is a good approximation because α is very small.) We note from Eq. (5.18) that the susceptibility at right angles to the spin axis is inversely proportional to the molecular field constant, as might be expected, and it is independent of the temperature. The high-temperature susceptibility equation (5.9) should thus give the same result as Eq. (5.18) at T_N. This can be shown by combining Eqs. (5.9), (5.10), and (5.13).

Field parallel to spin axis. Suppose the field is applied in the direction of the A sublattice magnetization. Then the effect of the field is to increase the zero-field value of the A sublattice magnetization σ_{sA} by an amount $\Delta\sigma_A$ and decrease the corresponding value σ_{sB} of the B sublattice by an amount $\Delta\sigma_B$, as shown in Fig. 5.5(b). The balance between the two sublattices is upset, and a net magnetization in the direction of the field is produced:

$$\sigma = \sigma_A - \sigma_B = |\Delta\sigma_A| + |\Delta\sigma_B|. \tag{5.19}$$

Now the magnetization of either sublattice is governed by the Brillouin function $B(J,a')$ of Eq. (4.13), as shown in Fig. 5.6. Here P represents the spontaneous magnetization σ_s, in the absence of an applied field, of either sublattice; it is de-

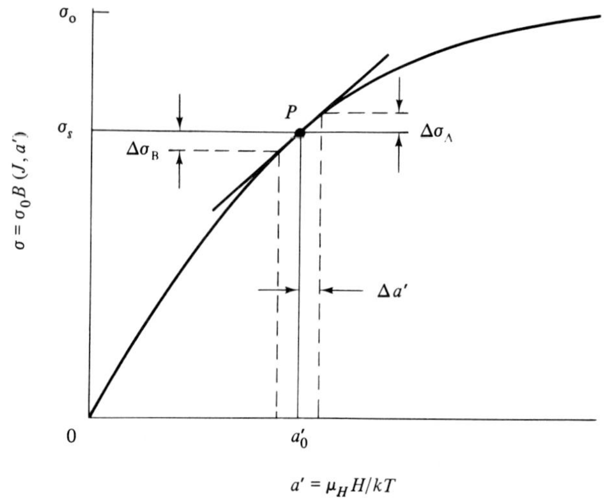

Fig. 5.6 Magnetization changes when the field is parallel to the spin axis.

termined by the particular value a'_0 of the variable a'. The effect of an applied field is to move the point P up on the curve for the A sublattice and down on the curve for the B sublattice. To simplify the calculation, we replace the Brillouin curve by its tangent at P, which amounts to assuming that the applied field produces equal changes in both sublattices:

$$\begin{aligned} |\Delta\sigma_A| &= |\Delta\sigma_B|, \\ \sigma &= 2\Delta\sigma_A. \end{aligned} \tag{5.20}$$

The value of $\Delta\sigma_A$ will be given the product of $\Delta a'$ and the slope of the magnetization curve:

$$\Delta\sigma_A = \Delta a'[\sigma_{0A} B'(J, a'_0)], \tag{5.21}$$

where $B'(J, a'_0)$ is the derivative of the Brillouin function with respect to its argument a', evaluated at a'_0. To find $\Delta a'$, we must remember that the field H in the variable $a'(= \mu_H H/kT)$ can include both an applied field, which we will now write explicitly as H_a, and a molecular field. In the present problem, the increase in a' is caused by the application of the field H_a *less* the amount the molecular field due to the B sublattice has decreased due to H_a, or

$$\Delta a' = \frac{\mu_H}{kT}\left(H_a - \gamma\rho|\Delta\sigma_B|\right) = \frac{\mu_H}{kT}\left(H_a - \gamma\rho\Delta\sigma_A\right). \tag{5.22}$$

Equation (5.21) then becomes

$$\Delta\sigma_A = \frac{n_g \mu_H^2}{2kT}(H_a - \gamma\rho\Delta\sigma_A) B'(J, a'_0), \tag{5.23}$$

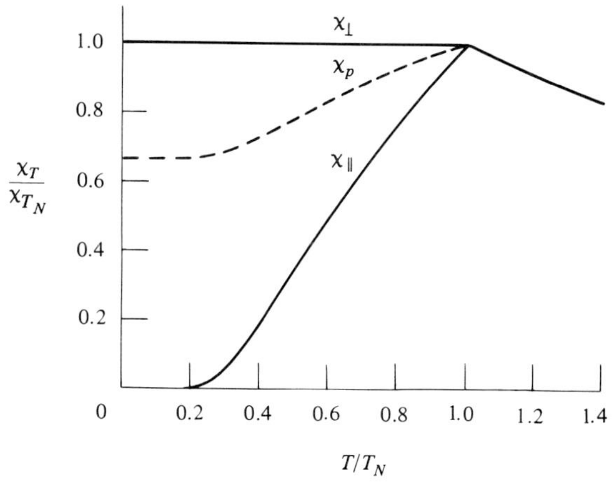

Fig. 5.7 Calculated thermal variation of the susceptibility of an antiferromagnetic near and below the Néel temperature T_N. The curve for $\chi_\parallel$ is calculated for $J = 1$.

where we have put σ_{0A}, which represents absolute saturation of the A sublattice, equal to $(n_g/2)\,\mu_H$, where n_g is the number of magnetic ions per gram. After solving Eq. (5.23) for $\Delta\sigma_A$, we arrive finally at an expression for the susceptibility parallel to the spin axis:

$$\chi_\parallel = \frac{\sigma}{H_a} = \frac{2\Delta\sigma_A}{H_a} = \frac{2n_g\mu_H^2 B'(J, a_0')}{2kT + n_g\mu_H^2\gamma\rho B'(J, a_0')}. \tag{5.24}$$

This equation may be put in another form by making the substitution

$$n_g\mu_H^2 = 3kC\left(\frac{J}{J+1}\right), \tag{5.25}$$

which follows from Eq. (3.52) and where C is the Curie constant and J the quantum number.

Equation (5.24) is quite general and holds both above and below the Néel temperature. It also holds for a ferromagnetic: the increase in magnetization caused by unit applied field, as given by Eq. (5.24), corresponds exactly to the "forced magnetization" mentioned in Section 4.2. This forced magnetization of a ferromagnetic is, however, difficult to measure because it is small and imposed on the spontaneous magnetization, which is rather large. But it is easy to measure in an antiferromagnetic, because the net magnetization, before the field is applied, is zero. It is left to the reader to demonstrate the following properties of Eq. (5.24):

1. It reduces to Eq. (5.2) at high temperatures.
2. It reduces to Eq. (5.18) at T_N.
3. It approaches zero as T approaches 0°K.

The variation of $\chi_{\parallel}$ between 0°K and T_N, relative to $\chi_{\perp}$, depends only on J and may be calculated with the aid of $B'(J, a_0')$ values tabulated by Smart [G.28]. This variation is shown in Fig. 5.7. It may be interpreted physically as follows: at 0°K, a' is infinite and the sublattice magnetization curve is perfectly flat, so that an applied field can produce no change in the magnetization of either sublattice, both of which are in a state of absolute saturation; as the temperature increases above 0°K, thermal energy decreases the spontaneous magnetization of each sublattice (as shown in Fig. 5.4), the applied field is able to reverse an increasing number of spins, and $\chi_{\parallel}$ increases.

In a powder specimen, in which there is no preferred orientation of the crystals, the spin axis D takes on all possible orientations with respect to the applied field. To find the susceptibility of a powder we must therefore average over all orientations. If the applied field H makes an angle θ with the spin axis D of a particular crystal in the powder, then the magnetizations acquired by that crystal, parallel and perpendicular to D, are

$$\begin{aligned} \sigma_{\parallel} &= \chi_{\parallel} H \cos \theta, \\ \sigma_{\perp} &= \chi_{\perp} H \sin \theta. \end{aligned} \tag{5.26}$$

The magnetization in the direction of the field is then

$$\begin{aligned} \sigma &= \sigma_{\parallel} \cos \theta + \sigma_{\perp} \sin \theta \\ &= \chi_{\parallel} H \cos^2 \theta + \chi_{\perp} H \sin^2 \theta \end{aligned}$$

or

$$\chi = \frac{\sigma}{H} = \chi_{\parallel} \cos^2 \theta + \chi_{\perp} \sin^2 \theta. \tag{5.27}$$

This susceptibility of one crystal must then be averaged over all possible values of θ to give the susceptibility of the powder:

$$\begin{aligned} \chi_p &= \chi_{\parallel} \overline{\cos^2 \theta} + \chi_{\perp} \overline{\sin^2 \theta} \\ &= \tfrac{1}{3} \chi_{\parallel} + \tfrac{2}{3} \chi_{\perp}. \end{aligned} \tag{5.28}$$

[The method of averaging $\sin^2 \theta$, for example, is the same as that involved in going from Eq. (3.8) to (3.9) in our calculations on diamagnetism.] A plot of the thermal variation of the powder susceptibility is included in Fig. 5.7.

Comparison with experiment

The molecular field theory outlined above leads to three predictions, easily compared with experiment:

1. θ/T_N should equal 1, according to Eq. (5.13).
2. The values of $\chi_{\parallel}$ and $\chi_{\perp}$ for a single crystal should vary with temperature as in Fig. 5.7.

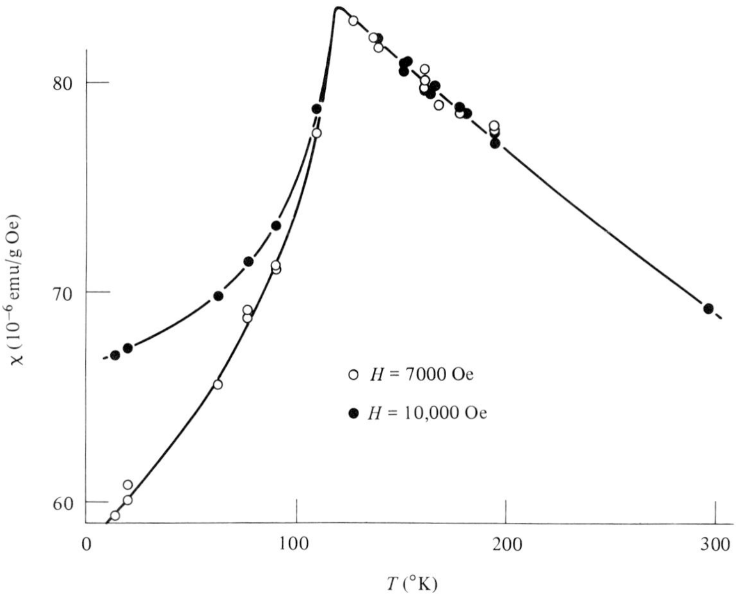

Fig. 5.8 Thermal variation of the susceptibility of MnO powder. Bizette, Squire, and Tsai [5.4].

3. The ratio of the susceptibility of a powder at 0°K to its value at T_N should equal $\frac{2}{3}$, according to Eq. (5.28), since $\chi_{\parallel} = 0$ and $\chi_{\perp} = \chi_{T_N}$ at 0°K.

Table 5-1 shows that observed values of θ/T_N range from 1 up to 5 or 6. This departure from unity does not mean that the molecular field theory has failed but that our initial assumption was too restrictive. We assumed that the only molecular field acting on the A ions was due to the B sublattice. Actually, there is no reason to exclude *a priori* the possibility that AA and BB exchange forces are also acting. Equations (5.3) and (5.4) would then be replaced by

$$H_{m\mathrm{A}} = -\gamma_{\mathrm{AB}} M_{\mathrm{B}} + \gamma_{\mathrm{AA}} M_{\mathrm{A}}, \tag{5.29}$$

$$H_{m\mathrm{B}} = -\gamma_{\mathrm{AB}} M_{\mathrm{A}} + \gamma_{\mathrm{BB}} M_{\mathrm{B}}, \tag{5.30}$$

where there are now two molecular field constants; γ_{AB} defines the strength of the AB interaction, and γ_{AA}, usually assumed equal to γ_{BB}, defines the strength of the AA interaction. The constant γ_{AA} $(= \gamma_{\mathrm{BB}})$ can be positive, negative, or zero. When γ_{AA} is not zero, the ratio θ/T_N can take on larger values than unity; the ratio $\gamma_{\mathrm{AA}}/\gamma_{\mathrm{AB}}$ can then be computed from the observed value of θ/T_N. Negative values of θ, which mean that $\chi = C/(T - \theta)$ above T_N, observed for some chlorides like $FeCl_2$, imply that γ_{AA} is negative. These extensions of the molecular field theory are described by Morrish [G.27] and Smart [G.28].

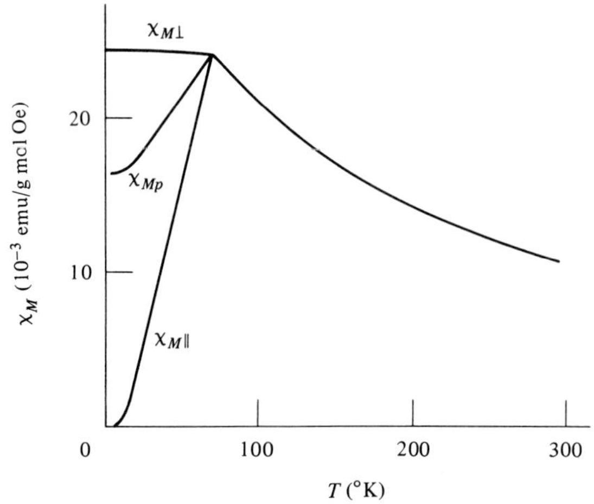

Fig. 5.9 Thermal variation of the molar susceptibility of MnF_2. The central curve below T_N is for a powder specimen, and the other two curves are for a single crystal. Bizette and Tsai [5.5].

The other predictions of the theory are reasonably well satisfied by experiment. Table 5.1 shows that $\chi_p(0)/\chi_p(T_N)$ is nearly always close to the theoretical value of 0.67. Typical χ, T curves for powders are shown in Figs. 5.8 and 5.9 and for a single crystal in Fig. 5.9; these curves are seen to agree quite closely with the theoretical curves of Fig. 5.7. (The fact that the susceptibility of MnO below T_N depends on the size of the field used to measure it is due to "crystal anisotropy," which is discussed in Section 7.9.)

The generally good agreement between theory and experiment is all the more remarkable when we realize that one of our initial assumptions is often not satisfied. This is the assumption that any A ion has only B ions as nearest neighbors. This requirement is satisfied by a body-centered cubic arrangement of metal ions, in which the cube-center ions form the A sublattice and the cube-corner ions the B sublattice, and by a body-centered tetragonal arrangement, provided the axial ratio c/a of the unit cell lies between certain limits. It is *not* satisfied for the face-centered cubic arrangement, as will become clear in the next section.

The "crystal anisotropy" of antiferromagnetics and the attendant phenomenon of *metamagnetism* are discussed in Section 7.9.

5.3 NEUTRON DIFFRACTION

The first substance to be clearly recognized as antiferromagnetic was MnO in 1938, when the results shown in Fig. 5.8 were published. Between 1938 and 1949 the evidence which had accumulated for the assumed spin arrangement in antiferromagnetics below T_N was good but rather indirect: it consisted solely in the agreement of the susceptibility data with what could be predicted from the model. In 1949 the first direct evidence was provided, when Shull and Smart [5.6] succeeded in showing, by means of neutron diffraction experiments, that the spins on the manganese ions in MnO are divided into two groups, one antiparallel to the other. Neutron diffraction has wide applicability as a research tool, and the theory, technique, and main results have been described in detail by Bacon [5.7]. Here we are interested only in its application to magnetic studies; these, incidentally, are not confined to antiferromagnetic substances, because neutron diffraction can furnish important information about ferro- and ferrimagnetics as well.

A stream of particles has many attributes of wave motion, in particular a wavelength λ, given by $\lambda = h/p$, where h is Planck's constant and p is the momentum of the particles. A stream, or beam, of neutrons can therefore be diffracted by a crystal just like a beam of x-rays, provided that the neutron wavelength is of the same order of magnitude as the interplanar spacings of the crystal. The neutrons in the core of a nuclear reactor have just the right wavelength, about one angstrom, and a beam of them can be obtained simply by cutting a narrow hole through the shielding of the reactor.

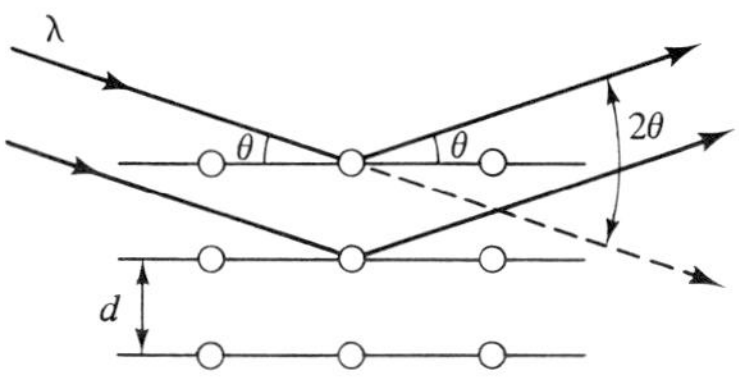

Fig. 5.10 Diffraction geometry.

The diffraction of neutrons is governed by the same Bragg law which governs the diffraction of x-rays.

$$n\lambda = 2d \sin \theta, \tag{5.31}$$

where n is an integer (0, 1, 2, ...) called the order of reflection, d is the spacing of atomic planes in the crystal, and θ is the angle between the incident beam and the atomic planes involved (Fig. 5.10). When neutrons, or x-rays, encounter an atom, they are scattered in all directions, and what we call a diffracted beam is simply a set of scattered beams which are in phase, so that they reinforce one another. The Bragg law is just the condition that rays scattered in the direction shown in the sketch, making an angle θ with the atomic planes equal to the angle of incidence, will be in phase with one another. In all other directions of space the phase relations between the scattered beams are such that they cancel one another. In experimental work, the angle 2θ, rather than θ, is usually measured; it is the angle between the diffracted beam and the transmitted beam.

Although both x-rays and neutrons obey the Bragg law, they are scattered by atoms in markedly different ways. X-rays are scattered by the electrons of the atom, because x-rays are electromagnetic radiation which can interact with the electronic charge. Neutrons are uncharged, easily penetrate the electron screen, and are scattered only by the nucleus. There is one important exception to this statement: if the scattering atom or ion has a net magnetic moment, that moment will interact with the neutron beam, because the neutron has a small magnetic moment of its own, equal to about 10^{-3} Bohr magneton. Neutron scattering from a magnetic ion therefore has two parts, one nuclear, the other magnetic; the magnetic part is due to the electrons of the ion, because it is the ion's electrons that are responsible for its magnetic moment. Neutrons can thus "see" elementary magnetic moments, whereas x-rays cannot. Furthermore, both the magnitude and direction of the magnetic moment of an atom or ion can be determined from measurements of the intensity of the magnetic scattering. It is this feature of neutron diffraction which makes it so valuable in the study of magnetic materials.

Electron diffraction is a fairly common means of investigating crystals, and the reader may wonder why it also is not effective in revealing magnetic structure. After all, each electron in the beam of electrons incident on the solid, in an electron diffraction experiment, has a magnetic moment of one Bohr magneton and therefore of the same order of magnitude as the net magnetic moment of each atom of the solid, rather than a moment of less than 10^{-3} times an atomic moment, as is typical of neutrons. The answer lies in the fact that electrons are charged. The incident electrons are scattered by atomic electrons because of the very large electrostatic (Coulomb) repulsion between them. This electrostatic interaction is so much stronger than the magnetic interaction that the latter is normally unobservable. (Extremely weak magnetic scattering of electrons has been observed under special conditions [5.12].) Neutrons, on the other hand, are uncharged, and their magnetic interaction with the scattering atoms, although much weaker than that of electrons, can easily be observed because it is not camouflaged by any electrostatic interaction.

When an antiferromagnetic is cooled below T_N, what was previously a random arrangement of spins becomes an ordered arrangement, with one set of spins antiparallel to the other. This change is very similar, especially from a diffraction point of view, to the "chemical" ordering which takes place in certain solid solutions when cooled below a critical temperature. Consider x-ray diffraction from the (100) planes of such a solid solution, consisting of elements C and D in equal atomic proportions and assumed to have a body-centered cubic structure. These planes are marked X and Z in Fig. 5.11 (a), which applies to the disordered state. If the incoming x-rays make an angle θ such that the path difference abc between scattered rays 1 and 3 equals one whole wavelength, then rays 1 and 3 will be in phase and reinforce each other. But if $abc = \lambda$, the path difference def between rays 1 and 2 is $\lambda/2$, so that these rays are exactly out of phase. Moreover, the amplitudes of 1 and 2 are exactly equal, because planes X and Y are statistically identical, inasmuch as the solution is disordered. Scattered rays 1 and 2 therefore cancel each other, as indicated in the sketch of the scattered wave form, and so do 3 and 4, 5, and 6, etc. There is no 100 reflection from the disordered solu-

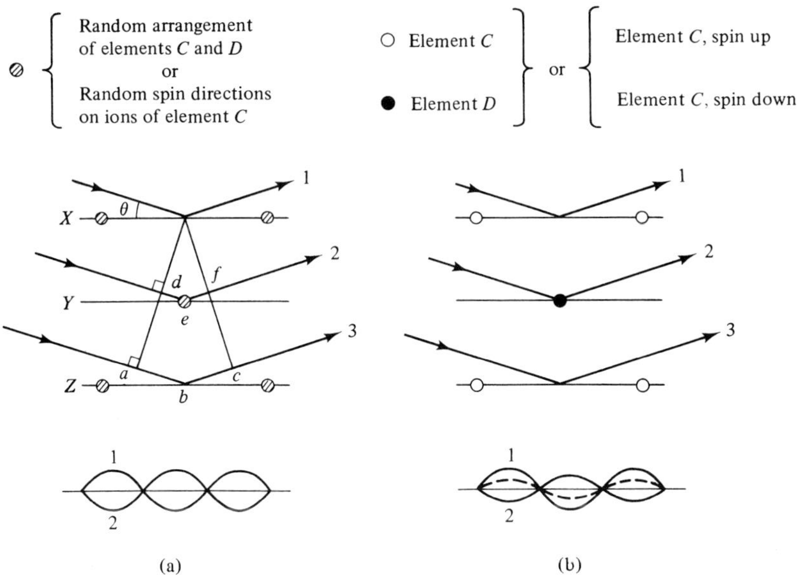

Fig. 5.11 Diffraction by (a) disordered and (b) ordered structures.

tion. [There is, however, a 200 reflection; this is obtained by increasing the angle θ until $def = \lambda$ so that rays 1 and 2, scattered from the (200) planes X and Y, are in phase.] In Fig. 5.11 (b) there is perfect order: C atoms occupy only cube corners and D atoms only cube centers. For first-order ($n = 1$) reflection from (100) planes, scattered rays 1 and 2 are again exactly out of phase. But now their amplitudes differ, because planes X and Y now contain chemically different atoms, with different numbers of electrons per atom and hence different x-ray scattering powers. Therefore, rays 1 and 2 do not cancel but combine to form the wave indicated by the dashed line in the sketch. The ordered solid solution thus produces a 100 reflection. If we examined other reflections, from planes of different Miller indices *hkl*, we would find other examples of lines which are present in the diffraction pattern of ordered solutions and absent from the pattern of disordered ones. These extra lines are called *superlattice lines*, and their presence constitutes irrefutable evidence of order.

The detection of order in magnetic systems with neutrons is exactly analogous. We now regard Fig. 5.11 (a) as representing a lattice of chemically identical ions, C ions, say, each with an identical magnetic moment randomly oriented in space. For the same reasons as in the x-ray case, there will be no 100 neutron reflection. In (b) we have magnetic order: the spins on the corner ions are "up," say, and those on the body-centered ions, "down." There will now be a 100 neutron superlattice line, because the neutron magnetic scattering is sensitive to the differing *directions* of the spin moments on adjacent planes.

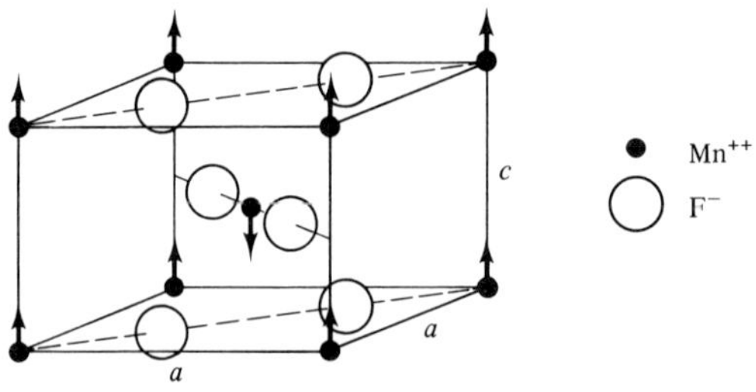

Fig. 5.12 Structure of MnF_2.

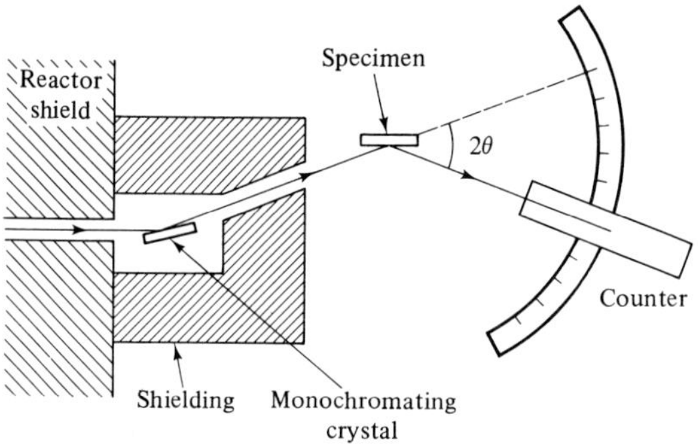

Fig. 5.13 Neutron diffractometer.

Before considering a particular example, we must qualify the remarks just made about "up" and "down" spins. No magnetic scattering at all can take place if the spin axes are *normal* to the reflecting planes, for reasons described by Bacon [5.7]. Thus, if "up" and "down" mean normal to the (100) planes, there will be no 100 superlattice reflection, not because of any cancellation effect, as in Fig. 5.11 (a), but because there is no magnetic scattering to begin with, only nuclear scattering. But a 100 superlattice reflection will occur if the axis of the antiparallel spins makes any angle other than 90° with the (100) planes.

To exemplify these general rules we will choose MnF_2, which exhibits simpler diffraction phenomena than MnO. It has the structure of the mineral rutile (TiO_2), with 2 MnF_2 per unit cell, located as follows:

2 Mn ions at 0 0 0, $\frac{1}{2}\ \frac{1}{2}\ \frac{1}{2}$.

4 F ions at $x\ x\ 0$; $\bar{x}\ \bar{x}\ 0$; $\frac{1}{2} + x, \frac{1}{2} - x, \frac{1}{2}$; $\frac{1}{2} - x, \frac{1}{2} + x, \frac{1}{2}$.

(The ionic coordinates are given as fractions of the unit-cell edges.) The value of x is 0.31. The cell is tetragonal with $a = 4.87$ Å and $c = 3.31$ Å. The unit cell is shown in Fig. 5.12.

Neutron diffraction experiments were carried out on this compound by Erickson [5.8] with an instrument called a neutron diffractometer (Fig. 5.13). The neutrons

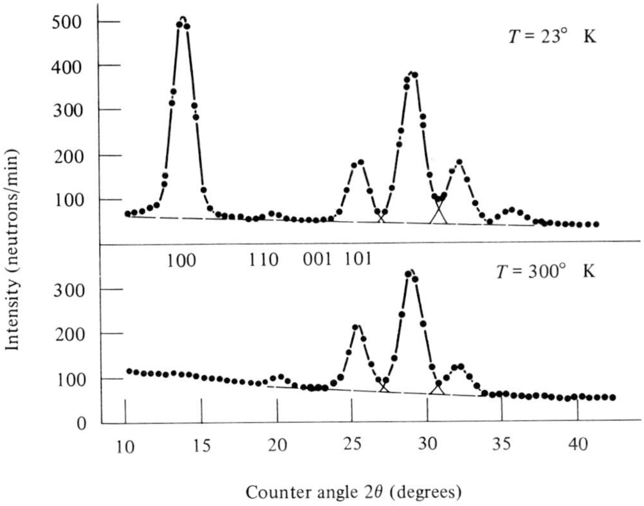

Fig. 5.14 Neutron diffraction patterns of a powder specimen of MnF_2 in the paramagnetic state (300°K) and in the antiferromagnetic state (23°K). Erickson [5.8].

which issue from a reactor have a range of wavelengths, and it is necessary to select a single wavelength from this range for the diffraction experiment. This is done by setting a single crystal, usually copper or lead, in the path of the beam at a particular angle θ of incidence, so that it will reflect, in accordance with the Bragg law, only the particular wavelength desired, usually one in the range 1.0-1.2A. The crystal "monochromator" thus reflects only one wavelength out of the many wavelengths incident on it. The reflected beam from the monochromator then encounters the specimen (which may be a single crystal or a compacted mass of powder), is diffracted by it, and enters a counter which measures its intensity. The diffraction pattern is obtained by moving the counter stepwise through various angles 2θ and measuring the intensity of the radiation diffracted by the specimen at each angle.

The result is a plot of diffracted intensity versus 2θ which is shown for MnF_2 in Fig. 5.14, for temperatures above and below the Néel temperature (67°K). The chief difference between the two patterns is the presence of the strong 100 superlattice line below T_N. This tells us immediately that the spins on the cell-corner ions are antiparallel to those on the cell-center ions. On the other hand, there is no 001 line; the spin axis is therefore normal to these planes and parallel to the c axis of the unit cell, as shown in Fig. 5.15. Detailed analysis of the intensities of the other lines in the pattern confirms this conclusion. (The inten-

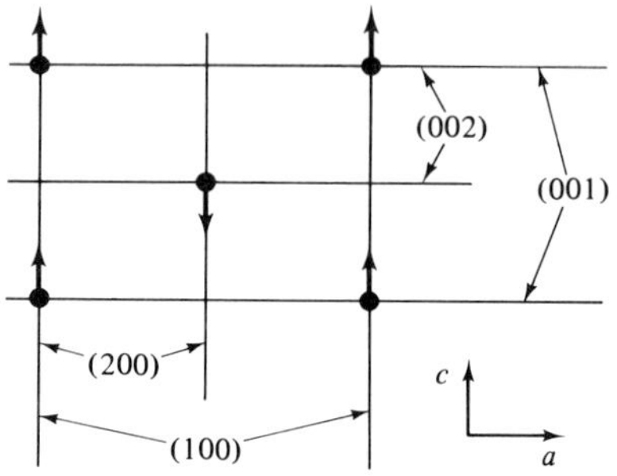

Fig. 5.15 Projection of the MnF_2 unit cell on a (010) face. Only the manganese ions are shown.

sity of the 100 line from MnF_2 depends on the degree of spin order, and it increases as the temperature decreases from T_N to 0°K. Erickson measured this intensity at a number of temperatures below T_N; from these measurements he was then able, in effect, to determine the shape of the sublattice σ, T curve, shown schematically in Fig. 5.4. Such information is unobtainable from magnetic measurements.)

It is not always possible to determine the orientation of the spin axis solely from diffraction patterns made with powder specimens. Often a single-crystal specimen is required. (In some substances the orientation of the spin axis can be found from susceptibility measurements alone, without any recourse to neutron diffraction; by trial and error two orientations of the crystal in the applied field are found for which the χ, T curves have the form of Fig. 5.7).

MnO has the face-centered cubic NaCl structure, which is like a three-dimensional checkerboard (Fig. 5.16 (a)). In (b), only the magnetic ions are shown, and the spin structure deduced by neutron diffraction. The spin axis is parallel to (111) planes and lies in the [110] direction in these planes; alternate (111) planes, shown by dashed lines in the drawing, have opposite spin. The antiferromagnetic state of MnO has one feature not found in MnF_2: the magnetic unit cell differs from the chemical, also called the nuclear, unit cell. Although a unit cell may be chosen in a great many ways, the choice must meet certain requirements. One is that the "entity" (chemical species, spin direction, etc.) at one corner

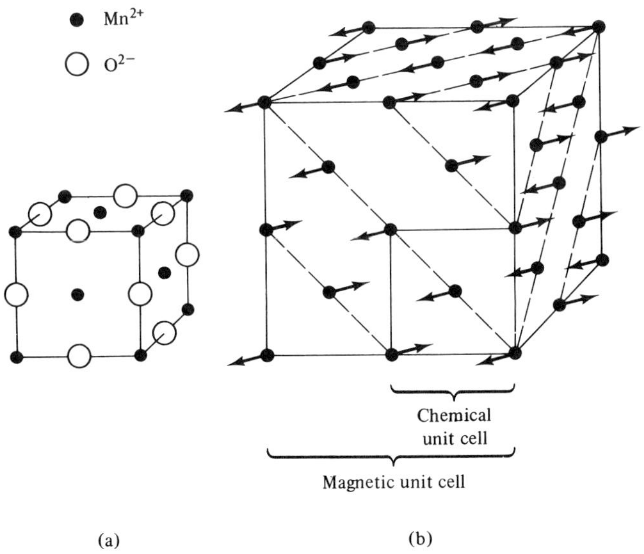

Fig. 5.16 Structure of MnO. (a) Chemical unit cell of Mn and O ions. (b) Chemical and magnetic unit cells (Mn ions only).

of the cell be the same as that at all other corners. The unit cell in Fig. 5.16 (a) is the chemical unit cell and has a manganese ion at each corner; it is also the magnetic unit cell above T_N, because the spin directions are then random and the manganese ions are, in a magnetic sense, statistically identical. But when magnetic ordering sets in, the spin direction at one corner of the chemical unit cell is opposite to that at the three nearest corners. It is then necessary to choose a magnetic unit cell twice as large along each cube edge, as shown in (b).

Neutron diffraction has disclosed spin structures in which the spins in alternate layers are not antiparallel but inclined at some angle other than 180°. $MnAu_2$ is an example and Fig. 5.17 shows its chemical unit cell. It is body-centered tetragonal, and Au atoms are arranged at a distance of about $c/3$ above and below each Mn atom, along the c axis. The spins of the Mn ions in each (002) plane are parallel to one another and to the (002) plane itself, but the spins rotate through an angle ϕ of 51° about the c axis from one (002) plane to the next. Such an arrangement of spins is called a *spiral* or *helical* structure.

Evidently we must revise our earlier definition of antiferromagnetism and make it more general, to include the possibility that the spins of the two sublattices may have any relation to each other as long as they form an ordered arrangement with no net magnetization. Even more complex spin arrangements than that of $MnAu_2$ have been found, some involving more than two sublattices, and not all of them have been solved in detail. The "easy" ones are known, but neutron

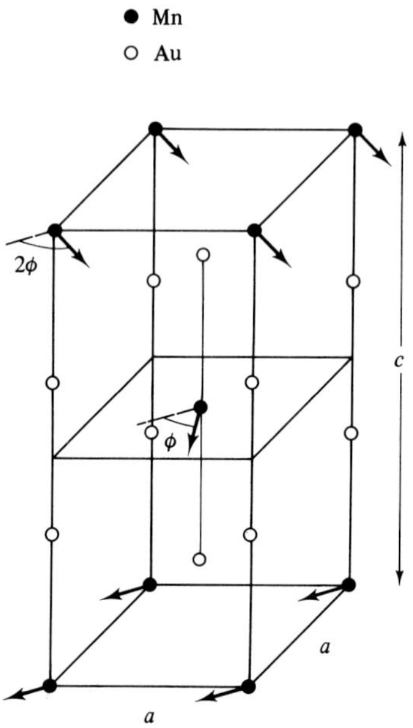

Fig. 5.17 Structure of $MnAu_2$.

crystallography, currently a very active field of research, still presents many challenges.

Slight deviations from ideal antiferromagnetism can also exist. In some substances the spins of the two sublattices are not quite antiparallel but slightly tilted or "canted" out of alignment, as indicated in Fig. 5.18. The result is a small net magnetization σ_s in one direction. From one point of view such substances are ferromagnetic; they are composed of domains, each spontaneously magnetized to a magnitude σ_s, and they show hysteresis. But they do not saturate, and in strong fields they exhibit a susceptibility χ appropriate to their basic antiferromagnetism. Such substances have a magnetization curve like that of curve A of Fig. 3.13, which can be described by

$$\sigma = \sigma_s + \chi H, \tag{5.32}$$

where the first term reaches its maximum value σ_s only in a finite field, as indicated by curve C of Fig. 3.13. In the older literature this phenomenon is called *parasitic ferromagnetism*, and some thought it was due to a ferromagnetic impurity existing as particles of a second phase. It is now recognized as having a more basic cause and is known as *canted antiferromagnetism*. Substances which show this behavior at room temperature include $\alpha - Fe_2O_3$ (hematite) and the rare-earth orthoferrites. These have the general formula $RFeO_3$, where R is yttrium or a rare earth. Their crystal structure is orthorhombic (three axes of unequal length at right angles to one another), and the spontaneous magnetization σ_s is parallel to the

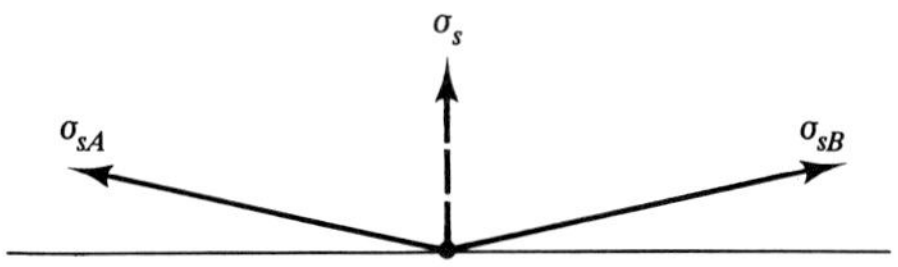

Fig. 5.18 Canted spins.

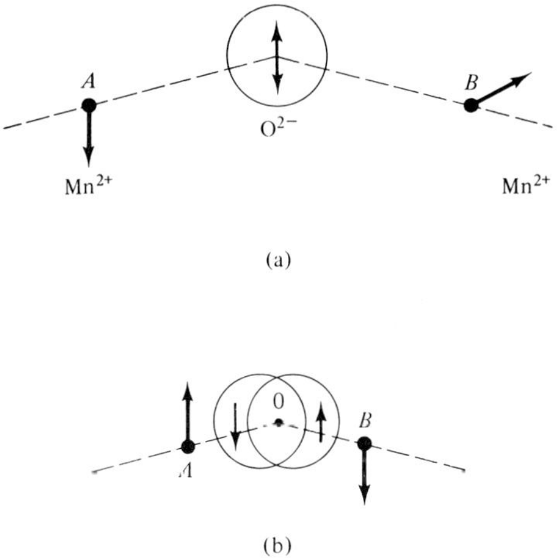

Fig. 5.19 Superexchange.

c axis (the <001> axis of the cell), except in $SmFeO_3$, where σ_s is parallel to the a axis. These orthoferrites have practical applications described in Section 13.6.

The exchange interaction in antiferromagnetic ionic solids takes place by the mechanism of indirect exchange, also called *superexchange*, already alluded to briefly near the end of Section 4.7. In these structures, the positive metal ions, which carry the magnetic moment, are too far apart for direct exchange forces to be of much consequence. Instead, they act indirectly through the neighboring anions (negative ions). Consider, for example, two Mn^{2+} ions being brought up to an O^{2-} ion from a large distance, as in Fig. 5.19 (a). The moments on these two ions are at first only randomly related. The oxygen ion has no net moment, because it has a neon-like structure of filled shells. But imagine that the outer electrons of the oxygen ion constitute two superimposed orbits, one with a net spin up, the other with a net spin down, as pictured in (a). When a manganese ion with an up spin is brought close to the oxygen ion, the up-spin part of the oxygen ion will be displaced as in (b), because parallel spins repel one another. If now another manganese ion is brought up from the right, it is forced to have a down spin when it comes close to the up-spin side of the "unbalanced" oxygen ion. The strength of the antiparallel coupling between metal ions M depends on the bond angle AOB and is generally greatest when this angle is 180° (M—O—M collinear).

To conclude this section we will consider what neutron diffraction has revealed concerning the spin structure of some transition metals:

1. *Antiferromagnetic.* Chromium is antiferromagnetic below 37° C and manganese below about 100° K. Neither has a susceptibility which varies much with temperature and neither obeys a Curie-Weiss law. (Inasmuch as they are both electrical conductors, rather than insulators, we do not expect their behavior to conform closely to a localized-moment, molecular-field theory.) But their neutron diffraction patterns both shown superlattice lines at low enough temperatures, which is sufficient evidence for antiferromagnetism. However, their spin structures have still not been solved in detail, even for such a "simple" metal as chromium, which has only two atoms per cell in a body-centered cubic arrangement. The experimental findings have been reviewed by Nathans and Pickart [5.9].

2. *Ferromagnetic.* For iron, nickel, and cobalt, neutron diffraction shows that the spins on all the atoms are parallel to one another and that the moment per atom is in accord with values deduced from measurements of saturation magnetization. (Furthermore, the diffraction experiments show that each atom has the same moment. This evidence disposes of a suggestion that had been made that a nonintegral moment, such as 0.6 μ_B/atom, was simply an average, resulting from the appropriate mixture of atoms of zero moment and atoms with a moment of one Bohr magneton.) It has even been possible to discover the way in which the magnetization is distributed around the nucleus [5.10]. In cobalt this distribution is spherically symmetrical. In iron, however, the magnetization is drawn out to some extent along the cube-edge directions of the unit cell; in nickel, it tends to bulge out in the face-diagonal and body-diagonal directions.

5.4 RARE EARTHS

The fifteen rare earth metals extend from lanthanum La (atomic number 57) to lutetium Lu (71). They are all paramagnetic at room temperature and above. At low temperatures their magnetic behavior is complex, as shown graphically in Appendix 3. Because almost all the rare earths are antiferromagnetic over at least some range of temperature, it is convenient to give their magnetic properties some brief consideration here. For details the reader is referred to various reviews [G.23] [4.10] [5.9] [5.11].

The rare earths are chemically very similar, and it is therefore difficult to separate them from one another or to obtain them in a pure state. This near identity of chemical behavior is due to the fact that the arrangement of their outer electrons is almost identical. However, the number of electrons in the inner $4f$ shell varies from 0 to 14 through the series La to Lu, and the magnetic properties are due to this inner, incomplete shell. Because the $4f$ electrons are so deep in the atom, they are shielded from the crystalline electric field of the surrounding ions; the orbital moment is therefore *not* quenched, and the total magnetic moment has both orbital and spin components. The total moment can become very large in some of the atoms and ions of the rare earths (see below).

The "light" rare earths, lanthanum (La) to europium (Eu), remain paramagnetic down to 91°K or below, and then five of the seven become antiferromagnetic. [Promethium (Pm) is not found in nature and has no stable isotope: the properties of the metallic form are therefore unknown.]

Of the eight "heavy" rare earths, six become ferromagnetic at sufficiently low temperatures, and five of these (terbium Tb through thulium Tm) pass through an intermediate antiferromagnetic state before becoming ferromagnetic. Gadolinium (Gd) just misses being ferromagnetic at room temperature; its Curie point is 16°C. All six ferromagnetic rare earths have magnetic moments per atom μ_H exceeding that of iron; if they only retained their ferromagnetism up to room temperature, they might make useful, although expensive, materials. The one with the largest moment is holmium Ho, which has $\mu_H = 10.34\ \mu_B$/atom, or almost five times that of iron (2.22 μ_B). The rare earth atoms are so heavy, however, that their saturation magnetizations σ_0 per gram at 0°K are about the same size as that of iron. For example, we may calculate, by means of Eq. (4.27) and the moment per atom given above, that σ_0 for holmium is 351 emu/g, compared to 221.9 emu/g for iron.

The rare earths and their alloys have provided a rich field for research by neutron diffraction. The spin structures of the antiferromagnetic states include helical and even more complex arrangements. Even the ferromagnetic structures are sometimes unusual. Consider, for example, gadolinium and holmium, which have the same crystal structure (hexagonal closepacked). Ferromagnetic Gd has a simple arrangement of parallel spins, like iron. Antiferromagnetic Ho has a helical spin structure like that of $MnAu_2$ in Fig. 5.17; the spins in any one hexagonal layer are all parallel, but they progressively rotate about the *c* axis from one layer to the next. In the ferromagnetic state below 20° K, this spiral spin structure is retained, but added to it is a ferromagnetic component of spins parallel to the *c* axis in every layer. (The *c* axis is normal to the hexagonal layers.) The resultant of these two components, one parallel and one at right angles to the hexagonal layers, gives ferromagnetic Ho a kind of conical spin arrangement.

5.5 ANTIFERROMAGNETIC ALLOYS

Antiferromagnetism is now known to exist in a considerable number of alloys, most of them containing Mn or Cr. It is more common in chemically ordered structures, which exist at simple atomic ratios of one element to the other, like AB or AB_2, but it has also been found, surprisingly, in some disordered solid solutions. Kouvel [4.7] and Crangle [4.10] should be consulted for details.

An example of antiferromagnetism in an *ordered* phase has already been given: $MnAu_2$ in Fig. 5.17. In the same alloy system, the phases MnAu and $MnAu_3$ are also antiferromagnetic. Some other antiferromagnetic ordered phases are CrSb, CrSe, FeRh, $FePt_3$, MnSe, MnTe, Mn_2As, and NiMn. The spin structure of the latter is interesting. The unit cell is face-centered tetragonal and the (002) planes, normal to the *c* axis, are occupied alternately by Ni and Mn atoms. Each (002) layer of atoms, whether all Ni or all Mn, is antiferromagnetic in itself, i.e., half the atoms in one layer have spins pointing in one direction and parallel to the plane of the layer, and the other half have spins pointing in the opposite direction.

(It has been suggested that antiferromagnetism may be responsible for some of the deviations from the Slater-Pauling curve of Fig. 4.22. For example, when Mn is added to Ni, the net moment per atom decreases rather than increases, as one would expect from the band theory, and appears to approach zero at a value of n equal to about 9.2. This corresponds to a Mn content of about 25 atomic percent. While this is still a long way from the 50 atomic percent in antiferromagnetic NiMn, it is conceivable that composition fluctuations in a solid solution containing, say, 15 atomic percent Mn could produce, in small regions, equal atomic proportions of the two atoms. The coupling in these regions would then be expected to be antiferromagnetic, rather than ferromagnetic; as a result, the measured net moment per atom would be less than expected. Such an explanation may indeed be true for Ni-Mn alloys, but it cannot be applied to some of the other alloys which depart markedly from the Slater-Pauling curve, such as Ni-Cr or Co-Cr, because antiferromagnetism has never been observed in these systems.)

Among *disordered* alloys antiferromagnetism has been observed in Mn-rich Mn-Cu and Mn-Au alloys. They have a face-centered tetragonal structure. All the spins in any one (002) plane are parallel to one another and to the c axis, but the spins in alternate (002) layers point "up" and "down." Disordered MnCr is also antiferromagnetic. It is body-centered cubic, with the spins on the cell-corner atoms antiparallel to those on the body-centered atoms. In none of these examples is there any chemical ordering. Each lattice site in, for example, the Mn-Cu alloys is occupied by a statistically "average" Mn-Cu atom, and each average atom appears to have a magnetic moment of the same magnitude. This behavior is understandable on the basis of the band theory, which envisages all the $3d$ and $4s$ electrons as belonging to a common pool, but not on the basis of a localized-moment theory. If the moments were localized, the various exchange interactions (molecular fields), between Mn-Mn, Mn-Cu, and Cu-Cu atoms, would have different orientations from one unit cell to the next in a disordered alloy, so that it would be difficult to understand how any long-range magnetic order could result.

Finally, it should be noted that the susceptibility-temperature curves of alloys do not usually give evidence for, or against, the existence of antiferromagnetism, because a Curie-Weiss law is not often followed. Neutron diffraction is the only sure test.

PROBLEMS

5.1 MnF_2 is antiferromagnetic and, at high temperatures, its Curie constant per mol is 4.10. Its molar susceptibility χ_M is 0.024 emu/g mol-Oe at the Néel temperature. Assume the ideal behavior described in Section 5.2, and assume that all the magnetic moment of the manganese ion is due to spin. Calculate (a) J, (b) the spontaneous magnetization σ_0 of each sublattice, and the molecular field acting on each, at O°K, and (c) the angle α in Fig. 5.5 (a) when a field of 10,000 Oe is applied at right angles to the spin axis of a single crystal at O°K.

5.2 Show that Eq. (5.24) reduces to Eq. (5.2) at high temperatures, to Eq. (5.18) at T_N, and to zero at O°K.

5.3 In a body-centered tetragonal arrangement of metal ions (Fig. 5.12), the cell-center ions form the A sublattice and the cell-corner ions the B sublattice. If an A ion is to have only B ions as nearest neighbors, between what limits must the value of c/a lie?

CHAPTER 6

FERRIMAGNETISM

6.1 INTRODUCTION

Ferrimagnetic substances exhibit a substantial spontaneous magnetization at room temperature, just like ferromagnetics, and this fact alone makes them industrially important. Again like ferromagnetics, they consist of self-saturated domains, and they exhibit the phenomena of magnetic saturation and hysteresis. Their spontaneous magnetization disappears above a certain critical temperature T_c, also called the Curie temperature, and then they become paramagnetic (Fig. 6.1). Ferrimagnetics were not recognized as forming a distinct magnetic class until 1948. It is therefore natural to find them referred to in the older literature as "ferromagnetic," because of all the similarities noted above, and this name sometimes clings to them even today. In practical importance they are second only to ferromagnetics and are superior to them in some applications.

The most important ferrimagnetic substances are certain double oxides of iron and another metal, called *ferrites*. (This mineralogical term is not to be confused with the same word applied by metallurgists to metallic solid solutions in which alpha iron is the solvent. Thus iron containing about 3 percent silicon in solution, which is also an important magnetic material, is often called silicon ferrite by metallurgists.) The ferrites were developed into commercially useful materials, chiefly during the years 1933–1945, by Snoek [G.3] and his associates at the Philips Research Laboratories in Holland. (The Philips company is one of the largest manufacturers of electrical equipment in Europe.) In a classic paper published in 1948, Néel [6.1] provided the theoretical key to an understanding of the ferrites, and the word *ferrimagnetism* is due to him. The subject has been covered at length in books by Smit and Wijn [G.10] and Standley [G.19] and in review papers by Smart [6.2], Wolf [6.3], and Gorter [6.4]. Gorter also gives a history of the development of the ferrites.

The magnetic ferrites fall mainly into two groups with different crystal structures:

1. *Cubic.* These have the general formula $MO \cdot Fe_2O_3$, where M is a divalent metal ion, like Mn, Ni, Fe, Co, Mg, etc. Cobalt ferrite $CoO \cdot Fe_2O_3$ is magnetically hard, but all the other cubic ferrites are magnetically soft. As magnetic materials these ferrites are both very old and very new, inasmuch as magnetite Fe_3O_4

(= $FeO \cdot Fe_2O_3$), which might be called iron ferrite, is the oldest magnetic material known to man, the "lodestone" of the ancients.

2. *Hexagonal.* The most important in this group is barium ferrite $BaO \cdot 6\,Fe_2O_3$, which is magnetically hard.

Ferrites are manufactured by the usual techniques of ceramics. To make nickel ferrite, for example, NiO and Fe_2O_3, in powder form, are thoroughly mixed, pressed to the desired shape, and sintered at temperatures in excess of 1200° C. (See Section 13.7 for details.) The resulting product is hard and brittle. It is also a semiconductor, which means that its electrical resistivity is at least a million times that of a metal. This very large resistivity means in turn that an applied alternating magnetic field will not induce eddy currents (Section 12.2) in a ferrite. This property makes ferrites almost ideal materials for high-frequency applications.

Many ferrites are found, usually in an impure state, as naturally occurring minerals in rocks. Knowledge of the properties of ferrites is therefore important to those geologists who are interested in rock magnetism. Studies of the magnetic properties of rocks have led to important conclusions about the strength and direction of the earth's magnetic field in past geological ages, and these conclusions form part of the evidence for the theory of continental drift.

The ferrites are ionic compounds, and their magnetic properties are due to the magnetic ions they contain. We are therefore interested in knowing what magnetic moment a particular metal ion should have. (The oxygen ion O^{2-} has no net moment.) This information is given by *Hund's rule*, which states that the spins in a partly filled shell are arranged so as to produce the maximum spin unbalance consistent with the Pauli exclusion principle. (The rule is stated here in terms of spin alone, because the orbital contribution is unimportant in ferrites.) Hund's rule was derived from a study of optical spectra, and the spin arrangements which it predicts are the result of exchange forces acting within a single atom or ion. We can apply the rule to ions of the first transition series in the following way. The outermost shell is the 3*d*, and it can contain 5 electrons with spin up and 5 with spin down. The first 5 electrons enter with spin up, say, in order to maximize the moment. The sixth electron, because of the exclusion principle, must have spin down. An ion with six 3*d* electrons, such as Fe^{2+}, must then have a spin-only moment of $5 - 1 = 4\,\mu_B$. The moments of a number of other ions are given in Table 6.1. (Note again that we are dealing with ionic compounds which are virtually insulators. In such substances, the electronic energy levels of the solid do not overlap, as they do in a metal, and therefore an integral number of electrons can be associated with each ion of the solid, just as in the free ion. This is just another way of saying that the electrons of each ion in the solid are fixed to that ion and cannot wander freely about.)

It is when we try to reconcile the ionic moments of Table 6.1 with the measured magnetization values for ferrites that we first realize the great difference between them and ferromagnetics. In nickel ferrite $NiO \cdot Fe_2O_3$, for example, there is one divalent nickel ion with a moment of $2\,\mu_B$ and two trivalent iron ions, each with a moment of $5\,\mu_B$. If positive exchange forces produced a parallel alignment of all of these moments as in a ferromagnetic, the total moment per "molecule"

Table 6.1 Spin-Only Moments of Ions of First Transition Series (after Smit and Wijn [G.10])

Ions									Number of 3d electrons	Spin-only moment in μ_B
Sc^{3+}	Ti^{4+}								0	0
	Ti^{3+}	V^{4+}							1	1
	Ti^{2+}	V^{3+}	Cr^{4+}						2	2
		V^{2+}	Cr^{3+}	Mn^{4+}					3	3
			Cr^{2+}	Mn^{3+}	Fe^{4+}				4	4
				Mn^{2+}	Fe^{3+}	Co^{4+}			5	5
					Fe^{2+}	Co^{3+}	Ni^{4+}		6	4
						Co^{2+}	Ni^{3+}		7	3
							Ni^{2+}		8	2
								Cu^{2+}	9	1
								Cu^{+} Zn^{2+}	10	0

would be $2 + 5 + 5 = 12\,\mu_B$. On the other hand, the measured saturation magnetization σ_0 at 0°K is 56 emu/g, which corresponds to 2.3 μ_B per molecule. This marked difference shows that the ionic moments cannot be aligned parallel to one another.

Close inspection of Fig. 6.1 yields further evidence that the ferrites are not ferromagnetic. The fractional magnetization σ_s/σ_0 of a typical ferrite decreases rather rapidly with increasing temperature, whereas the value of σ_s/σ_0 for iron, for example, remains large until T/T_c exceeds about 0.8. Furthermore, in the paramagnetic region, the variation of the inverse susceptibility with temperature is decidedly nonlinear, which shows that the Curie-Weiss law is not obeyed.

These several facts led Néel to the belief that the ferrites had a magnetic structure distinctly different from any previously recognized. It was known that the metal ions in a ferrite crystal occupied two crystallographically different kinds of position, called A sites and B sites. Néel made the basic assumption that the exchange force acting between an ion on an A site and an ion on a B site was *negative*, as in an antiferromagnetic. There is thus a lattice of A ions spontaneously magnetized in one direction and a lattice of B ions magnetized in the opposite direction. However, in a ferrimagnetic, the magnitudes of the A and B sublattice magnetizations are not equal. The two opposing magnetic moments do not cancel, and a net spontaneous magnetization results. (Ferrimagnetism can therefore be thought of as "imperfect antiferromagnetism.") Néel worked out all the implications of his hypothesis by means of molecular-field theory and obtained results in good agreement with experiment. Before examining this theory, we will consider the crystal structure of a cubic ferrite in some detail in order to understand the difference between A and B sites.

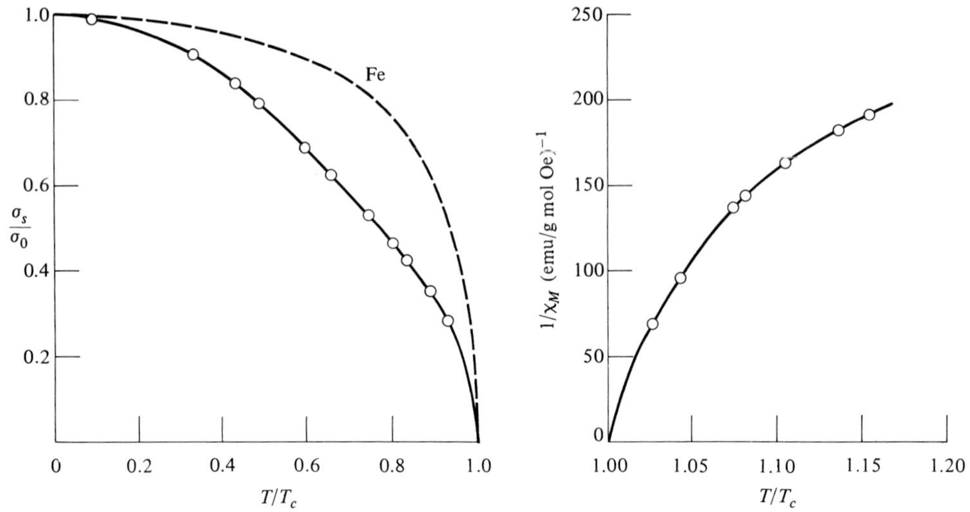

Fig. 6.1 Thermal variation of the magnetic properties of a typical ferrimagnetic ($NiO \cdot Fe_2O_3$). The fractional specific magnetization σ_s/σ_0 in the ferrimagnetic region is according to Pauthenet [6.5], and the reciprocal molecular susceptibility $1/\chi_M$ in the paramagnetic region is according to Serres [6.6]. (The dashed curve at left applies to metallic iron.)

6.2 STRUCTURE OF CUBIC FERRITES

These ferrites are said to have the spinel structure and are sometimes called *ferrospinels*, because their crystal structure is closely related to that of the mineral spinel $MgO \cdot Al_2O_3$. The structure is complex, in that there are 8 "molecules," or a total of $8 \times 7 = 56$ ions, per unit cell. The large oxygen ions (radius about 1.3 Å) are packed quite close together in a face-centered cubic arrangement, and the much smaller metal ions (radii from about 0.7 to 0.8 Å) occupy the spaces between them. These spaces are of two kinds. One is called a tetrahedral or A site, because it is located at the center of a tetrahedron whose corners are occupied by oxygen ions (Fig. 6.2 a). The other is called an octahedral or B site, because the oxygen ions around it occupy the corners of an octahedron (Fig. 6.2 b). The crystallographic environments of the A and B sites are therefore distinctly different.

The unit cell contains so many ions that a drawing of the complete cell would not be very informative. Instead we can imagine the unit cell of edge a to be divided into eight octants, each of edge $a/2$, as shown in Fig. 6.2 c. The four shaded octants have identical contents, and so do the four unshaded octants. The contents of the two lower-left octants in (c) are shown in (d). One tetrahedral site occurs at the center of the right octant of (d), and other tetrahedral sites are at certain octant corners. Four octahedral sites occur in the left octant; one is

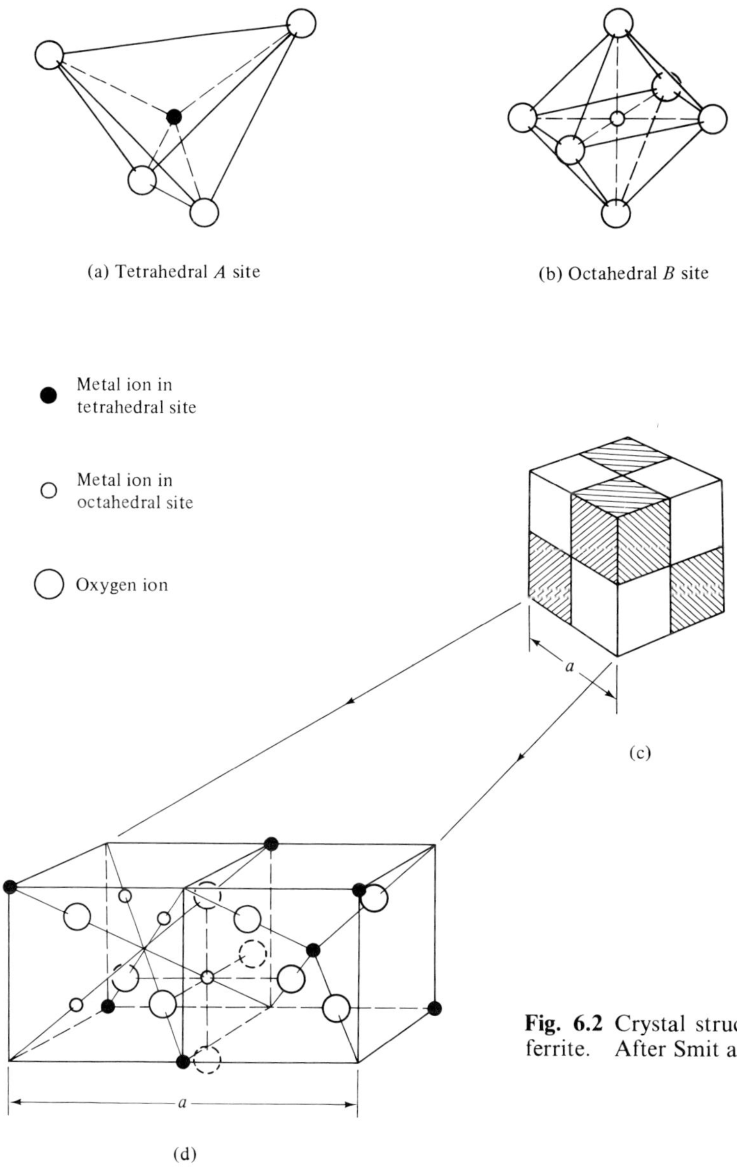

Fig. 6.2 Crystal structure of a cubic ferrite. After Smit and Wijn [G.10].

delineated by dashed lines to six oxygen ions, two of which, shown dotted, are in adjacent octants behind and below. The oxygen ions are arranged in the same way (tetrahedrally) in all octants.

By no means all of the available sites are actually occupied by metal ions. Only one-eighth of the A sites and one-half of the B sites are occupied, as shown in Table 6.2. In the mineral spinel, the Mg^{2+} ions are in A sites and the Al^{3+} ions

Table 6.2 Arrangements of Metal Ions in the Unit Cell of a Ferrite $MO \cdot Fe_2O_3$

			Occupants	
Kind of site	Number available	Number occupied	Normal spinel	Inverse spinel
Tetrahedral (A)	64	8	8 M^{2+}	8 Fe^{3+}
Octahedral (B)	32	16	16 Fe^{3+}	8 Fe^{3+} 8 M^{2+}

are in B sites. Some ferrites $MO \cdot Fe_2O_3$ have exactly this structure, with M^{2+} in A sites and Fe^{3+} in B sites. It is called the *normal spinel* structure. Both zinc and cadmium ferrite have this structure and they are both nonmagnetic, i.e., paramagnetic. Many other ferrites, however, have the *inverse spinel* structure, in which the divalent ions are on B sites, and the trivalent ions are equally divided between A and B sites. (The divalent and trivalent ions normally occupy the B sites in a random fashion, i.e., they are disordered.) Iron, cobalt, and nickel ferrites have the inverse structure, and they are all ferrimagnetic.

The normal and inverse structures are to be regarded as extreme cases, because x-ray and neutron diffraction have shown that intermediate structures can exist. Thus manganese ferrite is almost, but not perfectly, normal; instead of all the Mn^{2+} ions being on A sites, a fraction 0.8 is on A sites and 0.2 on B sites. Similarly, magnesium ferrite is not quite inverse; a fraction 0.9 of the Mg^{2+} ions is on B sites and 0.1 on A sites. In fact, the distribution of the divalent ions on A and B sites in some ferrites can be altered by heat treatment; it depends, for example, on whether the material is quenched from a high temperature or slowly cooled.

Finally, it should be noted that ferrites can be prepared containing two different kinds of divalent ion, e.g., $(Ni, Zn)O \cdot Fe_2O_3$. This is called a *mixed ferrite*, although actually it is a solid solution of $NiO \cdot Fe_2O_3$ and $ZnO \cdot Fe_2O_3$. Most of the cubic ferrites used commercially are mixed ferrites.

6.3 SATURATION MAGNETIZATION

We can calculate the saturation magnetization of a ferrite at 0° K, knowing (a) the moment on each ion, (b) the distribution of the ions between A and B sites, and (c) the fact that the exchange interaction between A and B sites is negative. (Actually, the AB, AA, and BB interactions all tend to be negative, but they cannot all be negative simultaneously. The AB interaction is usually the strongest, so that all the A moments are parallel to one another and antiparallel to the B

Table 6.3 Ion Distribution and Net Moment per Molecule of Some Typical Ferrites

Example	Substance	Structure	Tetrahedral A sites	Octahedral B sites	Net moment (μ_B/molecule)
1	$NiO \cdot Fe_2O_3$	Inverse	Fe^{3+} 5 →	Ni^{2+}, Fe^{3+} 2 ← 5 ←	2
2	$ZnO \cdot Fe_2O_3$	Normal	Zn^{2+} 0	Fe^{3+}, Fe^{3+} 5 ← 5 →	0
3	$MgO \cdot Fe_2O_3$	Mostly inverse	Mg^{2+}, Fe^{3+} 0 4.5 →	Mg^{2+}, Fe^{3+} 0 5.5 ←	1
4	$0.9\ NiO \cdot Fe_2O_3$	Inverse	Fe^{3+} 4.5 →	Ni^{2+} Fe^{3+} 1.8 ← 4.5 ←	
	$0.1\ ZnO \cdot Fe_2O_3$	Normal	Zn^{2+} 0	Fe^{3+} Fe^{3+} 0.5 ← 0.5 ←	
			4.5 →	7.3 ←	2.8

moments. The directions of these moments are of the form <111>, i.e., parallel to a body diagonal of the unit cell, in all the cubic ferrites except cobalt. In cobalt ferrite the moments are parallel to the cube edge directions <100>.)

Example 1 of Table 6.3 shows how the calculation is made for Ni ferrite. The structure is inverse, with all the Ni^{2+} ions in B sites and the Fe^{3+} ions evenly divided between A and B sites. The moments of the Fe^{3+} ions therefore cancel, and the net moment is simply that of the Ni^{2+} ion, which is 2 μ_B. Generalizing on this, we conclude that the saturation magnetization μ_H of any inverse ferrite is simply the moment on the divalent ion. This leads to the following calculated values, in μ_B per molecule, for the series of ferrites from Mn to Zn:

Ferrite	Mn	Fe	Co	Ni	Cu	Zn
Calculated μ_H	5	4	3	2	1	0
Measured μ_H	4.6	4.1	3.7	2.3	1.3	0

As stated earlier, Mn ferrite is far from being inverse, and yet its calculated net moment is still 5 μ_B per molecule. This is due to the fact that Mn^{2+} and Fe^{3+} ions each have a moment of 5 μ_B; whatever their distribution between A and B sites, the expected net moment per molecule is still 5 μ_B. The ion and spin dis-

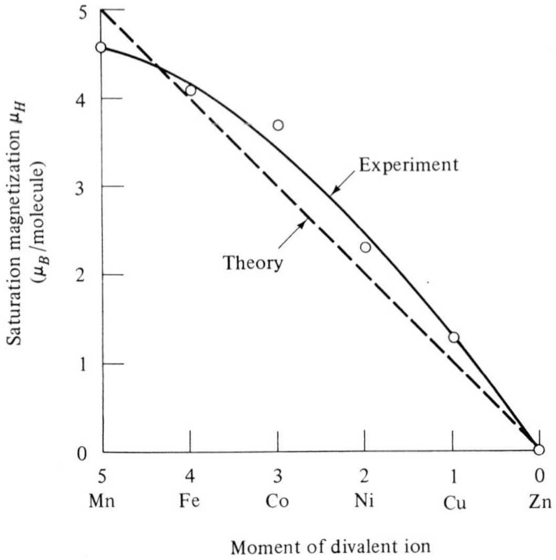

Fig. 6.3 Saturation magnetization at 0°K of some simple ferrites.

tribution in Zn ferrite are shown in Example 2 of Table 6.3. This ferrite has the normal structure, and Zn^{2+} ions of zero moment fill the A sites. There can thus be no AB interaction. The negative BB interaction then comes into play: the Fe^{3+} ions on B sites then tend to have antiparallel moments, and there is no net moment. (One would therefore expect Zn ferrite to be antiferromagnetic. It is, but only below 9°K [6.7]. It is paramagnetic down to this temperature, because the negative BB interaction is so weak that even small amounts of thermal energy can prevent the antiparallel ordering of the moments.)

Figure 6.3 offers a graphical comparison of the calculated and measured moments of the series of ferrites just discussed. The agreement is good, in general, and affords strong support to Néel's basic assumption. Even more direct support has been given by neutron diffraction, which has shown that the moments on the A and B sites are indeed antiparallel; this kind of evidence was first supplied in 1951 by Shull *et al.* [6.8], for Fe_3O_4.

The discrepancies between theory and experiment, evident in Fig. 6.3, are generally ascribed to one or both of the following:

1. Orbital moments may not be completely quenched; i.e., there may be an orbital moment, not allowed for in the theory, besides the spin moment. This is thought to be particularly true of the Co^{2+} ion.
2. The structure may not be completely inverse. And, as mentioned earlier, the degree of inversion can sometimes be changed by heat treatment. The saturation magnetization then becomes a structure-sensitive property.

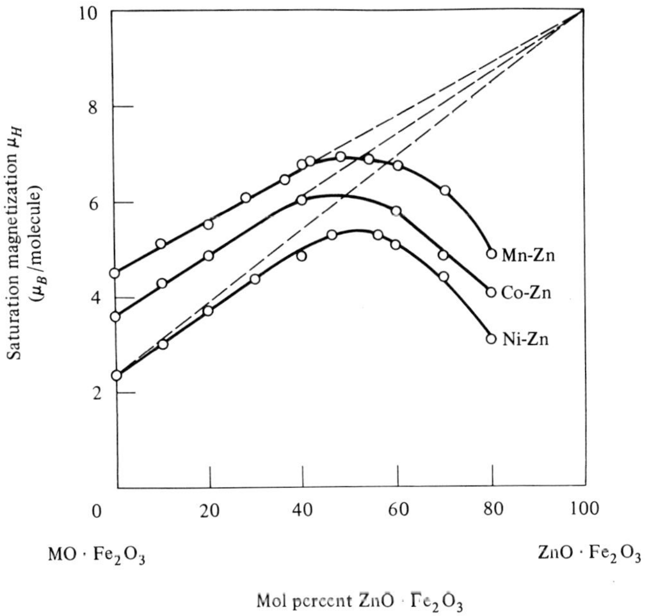

Fig. 6.4 Effect on the saturation magnetization at 0°K of adding Zn ferrite to Mn, Co, or Ni ferrite. After Guillaud [6.9].

Mg ferrite is a frequent component of mixed ferrites. If its structure were completely inverse, its net moment would be zero, because the moment of the Mg^{2+} ion is zero. But, as noted earlier, 0.1 of the Mg^{2+} ions are on A sites, displacing an equal number of Fe^{3+} ions. Then, as shown in Example 3 of Table 6.3, the A-site moment becomes 0.9 (5) = 4.5 μ_B and the B-site moment 1.1 (5) = 5.5 μ_B, giving an expected net moment of 1.0 μ_B. This agrees well with the experimental value of 1.1 μ_B.

A surprising fact about mixed ferrites containing zinc is that the addition of the nonmagnetic Zn^{2+} ion *increases* the saturation magnetization. Suppose we compute, as in Example 4 of Table 6.3, the net moment of a mixed ferrite (solid solution) containing 10 molecular percent Zn ferrite in Ni ferrite. The Zn^{2+} ions of zero moment go to the A sites as in pure Zn ferrite, thus weakening the A-site moment, and the Fe^{3+} ions from the Zn ferrite now have *parallel* moments in the B sites, because of the strong AB interaction. The expected net moment therefore increases from 2.0 μ_B, for pure Ni ferrite, to 2.8 μ_B for the mixed one. If this increase, of 0.8 μ_B per 10 molecular percent of Zn ferrite, continued with further additions, we would expect pure Zn ferrite to have a moment of 10 μ_B. This cannot occur because the A moments will soon become too weak to affect the B moments, and the net moment must sooner or later begin to decrease. However, the experimental curve does begin with a slope very close to the theoretical, as shown in Fig. 6.4, which gives data on three mixed ferrites containing zinc. (The

Table 6.4 Magnetic and Other Data for Various Ferrites and Metallic Iron (after Smit and Wijn [G.10])

Substance	Lattice parameter a(Å)	Density (g/cm³)	0° K		20° C		T_c (°C)
			σ_0 (emu/g)	M_0 (emu/cm³)	σ_s (emu/g)	M_s (emu/cm³)	
$MnO \cdot Fe_2O_3$	8.50	5.00	112	560	80	400	300
$FeO \cdot Fe_2O_3$	8.39	5.24	98	510	92	480	585
$CoO \cdot Fe_2O_3$	8.38	5.29	90	475	80	425	520
$NiO \cdot Fe_2O_3$	8.34	5.38	56	300	50	270	585
$CuO \cdot Fe_2O_3$	8.37*	5.41	30	160	25	135	455
$MgO \cdot Fe_2O_3$	8.36	4.52	31	140	27	120	440
$BaO \cdot 6Fe_2O_3$	$a = 5.88$ $c = 23.2$	5.28	100	530	72	380	450
Fe	2.87	7.87	222	1747	218	1714	770

* Cubic when quenched from above 760° C. If slowly cooled, it becomes tetragonal, with $a = 8.22$ Å and $c = 8.70$ Å.

way in which the experimental curves depart from the straight lines has been explained by Ishikawa [6.10] in terms of superparamagnetism. See Section 11.7.)

Table 6.4 summarizes magnetic and other data on various pure ferrites and compares them with similar data on metallic iron. Although the magnetic moment *per molecule* of many ferrites is rather large (several Bohr magnetons), iron has a much larger magnetization on the basis of unit mass or unit volume.

6.4 MOLECULAR FIELD THEORY

Like antiferromagnetics, most ferrimagnetics have such low electrical conductivity that their magnetic moments may be regarded as completely localized at particular ions. A molecular field (localized-moment) theory is therefore expected to be valid.

We also expect that the exchange forces between the metal ions in a ferrimagnetic will act through the oxygen ions by means of the indirect exchange (superexchange) mechanism, just as in antiferromagnetics (Fig. 5.19).

However, molecular field theory for a ferrimagnetic is inherently more complicated than for an antiferromagnetic, because the A and B sites are crystallogra-

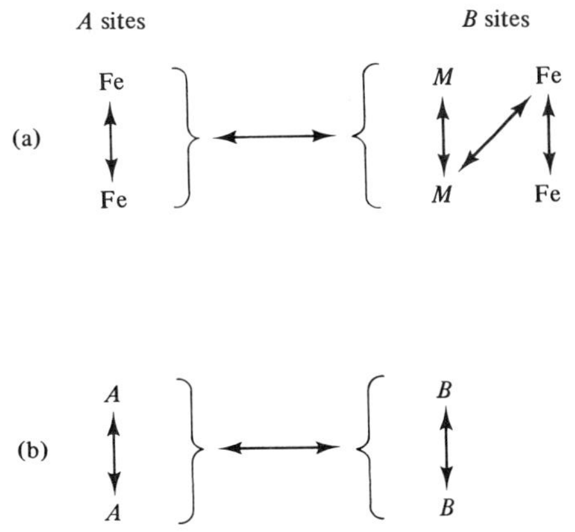

Fig. 6.5 Interactions between ions.

phically different for a ferrimagnetic, but identical for an antiferromagnetic. This means that the AA interaction in a ferrimagnetic will differ from the BB interaction, even though the ions involved are identical. The basic reason is that an ion on an A site has a different number and arrangement of neighbors than the same ion on a B site.

Figure 6.5 (a) shows the interactions (exchange forces) that would have to be considered in a rigorous treatment of an inverse ferrite $MO \cdot Fe_2O_3$. These interactions are shown by arrows, and there are 5 in all, compared to 2 (AB and AA = BB) in an antiferromagnetic. To simplify the problem, Néel [6.1] replaced the real ferrimagnetic with a model composed of *identical* magnetic ions divided unequally between the A and B sublattices. This still leaves 3 different interactions to be considered, as shown in Fig. 6.5 (b). The Néel theory is outlined below.

Let there be n identical magnetic ions per unit volume, with a fraction λ located on A sites and a fraction $\nu\,(=1-\lambda)$ on B sites. Let μ_A be the average moment of an A ion in the direction of the field at temperature T. (Even though the A and B ions are identical, μ_A is not equal to μ_B, because these ions, being on different sites, are exposed to different molecular fields.) Then the magnetization of the A sublattice is $M_A = \lambda n \mu_A$.

Put $n\mu_A = M_a$. Then $M_A = \lambda M_a$, and $M_B = \nu M_b$.

The total magnetization is

$$M = M_A + M_B = \lambda\, M_a + \nu\, M_b. \tag{6.1}$$

The molecular field acting on sublattice A is

$$H_{mA} = -\gamma_{AB} M_B + \gamma_{AA} M_A, \tag{6.2}$$

where the molecular field coefficients γ are to be regarded as positive quantities, and the signs correspond to the assumption of a negative (antiparallel) interaction between A and B ions and a positive (parallel) interaction between A ions. Similarly,

$$H_{mB} = -\gamma_{AB} M_A + \gamma_{BB} M_B. \tag{6.3}$$

The coefficients γ_{AA} and γ_{BB} are now unequal, and we express them as fractions of γ_{AB}.

$$\alpha = \frac{\gamma_{AA}}{\gamma_{AB}} \qquad \beta = \frac{\gamma_{BB}}{\gamma_{AB}}.$$

The molecular fields are then

$$H_{mA} = \gamma_{AB} (\alpha\lambda M_a - \nu M_b), \tag{6.4}$$

$$H_{mB} = \gamma_{AB} (\beta\nu M_b - \lambda M_a). \tag{6.5}$$

These two equations are valid above and below the Curie temperature.

Above T_c

In the paramagnetic region we proceed, as we did for antiferromagnetics, by assuming Curie-law behavior, namely,

$$MT = \rho C H_t$$

for each sublattice. Here ρ is the density and H_t is the total field, the sum of the applied field H and the molecular field. Then, for the two sublattices,

$$M_a T = \rho C(H + H_{mA}), \tag{6.6}$$

$$M_b T = \rho C(H + H_{mB}), \tag{6.7}$$

where C is the Curie constant per gram of the magnetic ions involved, from Eq. (3.52). By eliminating M_a, M_b, H_{mA}, and H_{mB} from Eqs. (6.1), (6.4), (6.5), (6.6), and (6.7), we find, after much tedious algebra, the following expression for the mass susceptibility χ:

$$\chi = \frac{M}{\rho H} = \frac{CT - \gamma_{AB}\,\rho C^2 \lambda\nu (2 + \alpha + \beta)}{T^2 - \gamma_{AB}\,\rho C T(\alpha\lambda + \beta\nu) + \gamma_{AB}^2\,\rho^2 C^2 \lambda\nu(\alpha\beta - 1)}. \tag{6.8}$$

This may be written in the form

$$\frac{1}{\chi} = \frac{T}{C} + \frac{1}{\chi_0} - \frac{b}{T - \theta}, \tag{6.9}$$

$$\frac{1}{\chi} = \frac{T + (C/\chi_0)}{C} - \frac{b}{T - \theta}, \tag{6.10}$$

where

$$\frac{1}{\chi_0} = \gamma_{AB}\,\rho(2\lambda\nu - \alpha\lambda^2 - \beta\nu^2),$$

$$b = \gamma_{AB}^2\,\rho^2 C\lambda\nu\,[\lambda(1 + \alpha) - \nu(1 + \beta)]^2,$$

$$\theta = \gamma_{AB}\,\rho C\lambda\nu\,(2 + \alpha + \beta).$$

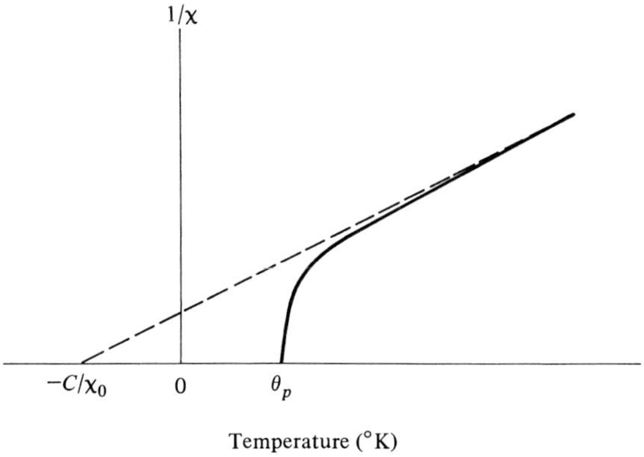

Fig. 6.6 Theoretical variation of the reciprocal susceptibility with temperature for a ferrimagnetic above the Curie point.

Equation (6.10) represents a hyperbola, and the physically meaningful part of it is plotted in Fig. 6.6. It cuts the temperature axis at θ_p, called the paramagnetic Curie point. At high temperatures the last term of Eq. (6.10) becomes negligible, and the equation reduces to a Curie-Weiss law:

$$\chi = \frac{C}{T + (C/\chi_0)}.$$

This is the equation of a straight line, shown dashed in Fig. 6.6, to which the $(1/\chi)$, T curve becomes asymptotic at high temperatures.

Equation (6.10) is in good agreement with experiment, except near the Curie point. Figure 6.7 shows the data for Mg ferrite. The temperature θ_f (or T_c) at which the susceptibility becomes infinite and spontaneous magnetization appears is called the ferrimagnetic Curie point; in the example shown, it was determined from measurements made in the ferrimagnetic region. This disagreement between theory and experiment in the region of the Curie "point" recalls the similar disagreement in ferromagnetism (Fig. 4.10) and is presumably due to the same cause: short-range spin order (spin clusters) at temperatures above θ_f.

By fitting Eq. (6.10) to the experimental points at temperatures sufficiently above θ_f, the constants χ_0, b and θ can be evaluated. For the curve of Fig. 6.7, for example, they have the values $1/\chi_0 = 296.7$, $b = 14{,}700$, and $\theta = 601.8$, with C equal to 4.38 for the Fe^{3+} ion, which is assumed to be the only magnetic ion present. Values of γ_{AB}, α, β, and λ can then be calculated from χ_0, b, and θ, by a method given by Néel [6.1], provided that the saturation magnetization at 0°K is also known. Néel analyzed the data on several ferrites in this way and found γ_{AB} to be large and positive, as expected, but α and β small and *negative*, which means that γ_{AA} and γ_{BB} are small and negative. Recalling the assumptions behind the signs of Eqs. (6.2) and (6.3), we conclude that the AA and BB interactions are weakly antiparallel.

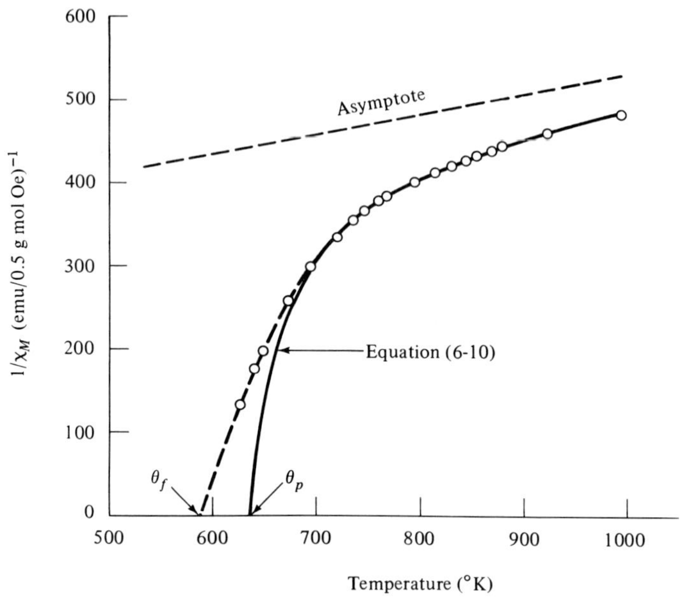

Fig. 6.7 Reciprocal susceptibility of Mg ferrite. (Here χ_M refers to a half molecule of the ferrite, i.e. to one mol of Fe^{3+}.) Experimental data by Serres [6.6]; curve from constants given by Néel [6.1].

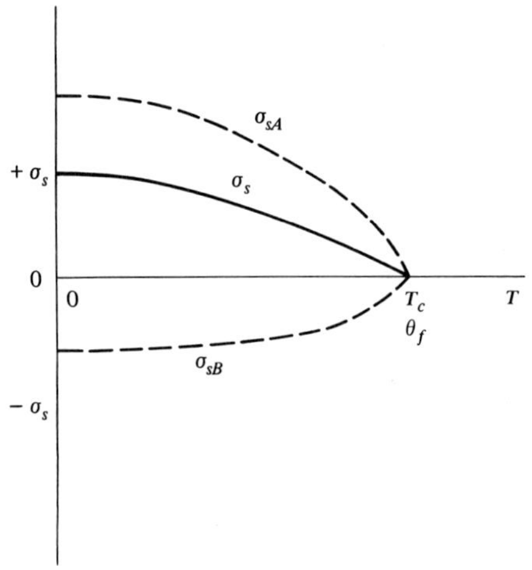

Fig. 6.8 Spontaneous magnetizations of the A and B sublattices, and the resultant σ_s (schematic).

Below T_c

In the ferrimagnetic region each sublattice is spontaneously magnetized by the molecular field acting on it, but the two sublattice magnetizations are opposed to each other. The net (observable) magnetization is then

$$|M| = |M_A| - |M_B|.$$

Each sublattice magnetization is governed by the same relation as a ferromagnetic, namely, Eq. (4.13). In terms of the magnetization per gram σ $(= M/\rho)$, the fractional specific magnetization of the A sublattice is given by

$$\frac{\sigma_A}{\sigma_{0A}} = B(J, a') = B\left(J, \frac{\mu_H H}{kT}\right), \tag{6.11}$$

where B is the Brillouin function. The field H here is to be put equal to the molecular field H_{mA} acting on the A lattice, because we are computing the spontaneous magnetization in the absence of an applied field. In terms of σ rather than M, Eq. (6.4) becomes

$$H_{mA} = \gamma_{AB}\, \rho\, (\alpha\lambda\sigma_a - \nu\sigma_b).$$

The two sublattice fractional spontaneous magnetizations are then given by

$$\frac{\sigma_{sA}}{\sigma_{0A}} = B\left(J, \frac{\mu_H\, \gamma_{AB}\, \rho(\alpha\lambda\sigma_a - \nu\sigma_b)}{kT}\right), \tag{6.12}$$

$$\frac{\sigma_{sB}}{\sigma_{0B}} = B\left(J, \frac{\mu_H\, \gamma_{AB}\, \rho(\beta\nu\sigma_b - \lambda\sigma_a)}{kT}\right). \tag{6.13}$$

These two equations cannot be solved separately by the simple graphical method of Fig. 4.4, because they are not independent. The extent to which the A lattice is magnetized depends on the extent to which the B lattice is magnetized, and vice versa. Instead, the equations must be solved simultaneously, and Néel has given a complex graphical method of doing so. The solutions in a typical case might appear like Fig. 6.8, where the dashed lines show the sublattice magnetizations and the full line is the resultant. (Note that the two sublattices must have the same Curie point. If not, then, at some temperature between the two Curie points, one lattice would have zero moment and so could not align the moments on the other.)

If the values of the constants γ_{AB}, α, β, and λ have been calculated for a particular substance from an analysis of its paramagnetic behavior, then the σ_s, T curve of that substance below T_c can be calculated, and the result is in fairly good agreement with experiment. However, it is necessary to follow a rather arbitrary procedure. In an inverse ferrite $MO \cdot Fe_2O_3$, the B sites are actually occupied by M^{2+} and Fe^{3+} ions with different moments. In the calculation it is necessary, to conform to the assumptions of the molecular-field theory, to replace these two kinds of ions on B sites with a single fictitious kind of ion having a moment intermediate between that of M^{2+} and Fe^{3+}. Pauthenet [6.5] should be consulted for details.

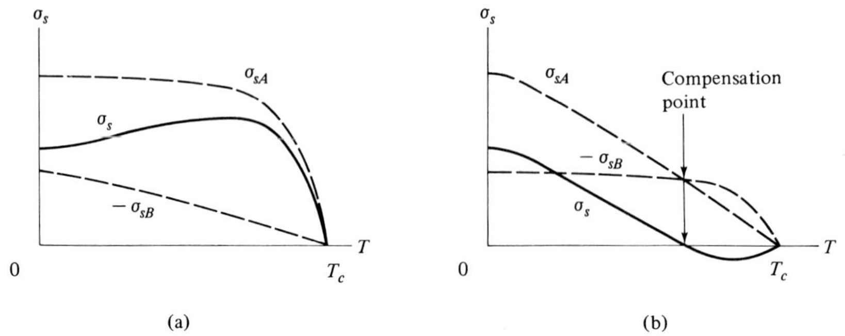

Fig. 6.9 Unusual σ_s, T curves.

The sublattice magnetizations given by Eqs. (6.12) and (6.13) depend on the molecular field constants γ_{AB}, α, and β and on the magnetic ion distribution parameter λ. The shapes of the sublattice σ_s, T curves thus depend on the values of these constants, and the shape of the curve for lattice A will generally differ from that of lattice B. Since the observed, resultant curve is the difference between these two, it follows that slight changes in the shapes of the sublattice curves can yield resultant curves of quite unusual shape. Néel has determined the various forms the resultant curve can assume, as a function of γ_{AB}, α, β, and λ. Two quite unexpected forms, shown in Fig. 6.9, were predicted, and both have since been observed.

In Fig. 6.9, the sublattice magnetization curves are both plotted on the positive side of the temperature axis. In (a) the resultant magnetization *increases* with temperature and goes through a maximum before finally falling to zero, because $|\sigma_{sA}|$ decreases less rapidly with increasing temperature than $|\sigma_{sB}|$. The chromite $NiO \cdot Cr_2O_3$, which has the spinel structure, behaves like this. In (b) we see the opposite behavior: the resultant magnetization decreases to zero below T_c and then becomes "negative." The temperature at which the resultant becomes zero is that at which the opposing sublattice magnetizations are exactly balanced; it is called a *compensation point*. $Li_{0.5}Fe_{1.25}Cr_{1.25}O_4$, which also has the spinel structure, shows this peculiar behavior.

[Actually, it is not accurate to say that σ_s becomes negative above the compensation point, because that would imply diamagnetism. If a rod of the material is placed parallel to a saturating field directed from left to right, then, at any temperature below T_c, the magnetization σ_s will be directed from left to right. The σ_s, T curve should thus be plotted as in Fig. 6.10 (a) rather than as in Fig. 6.9 (b). It is the *remanent* magnetization σ_r which changes sign with change in temperature. Let the rod be saturated at temperature T_1 and the field then removed; its remanent magnetization at this temperature $\sigma_r(T_1)$ is then represented by point a in Fig. 6.10 (b). (The effect of the demagnetizing field of the rod has been neglected in this illustration.) If the rod is now heated, still in zero field, from T_1 to T_2, its remanent magnetization will decrease, become zero, and then reverse

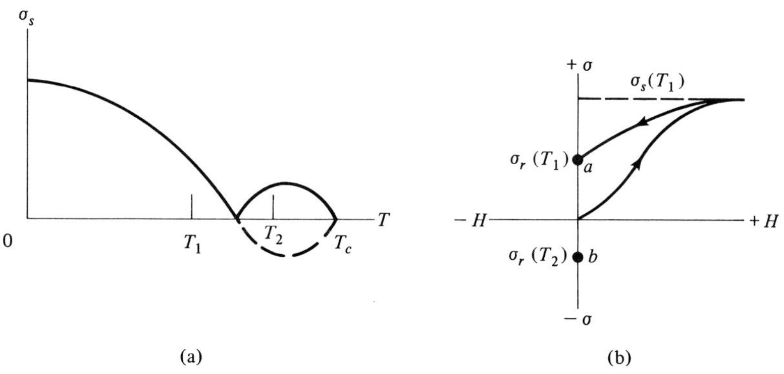

Fig. 6.10 Behavior of a ferrimagnetic with a compensation point.

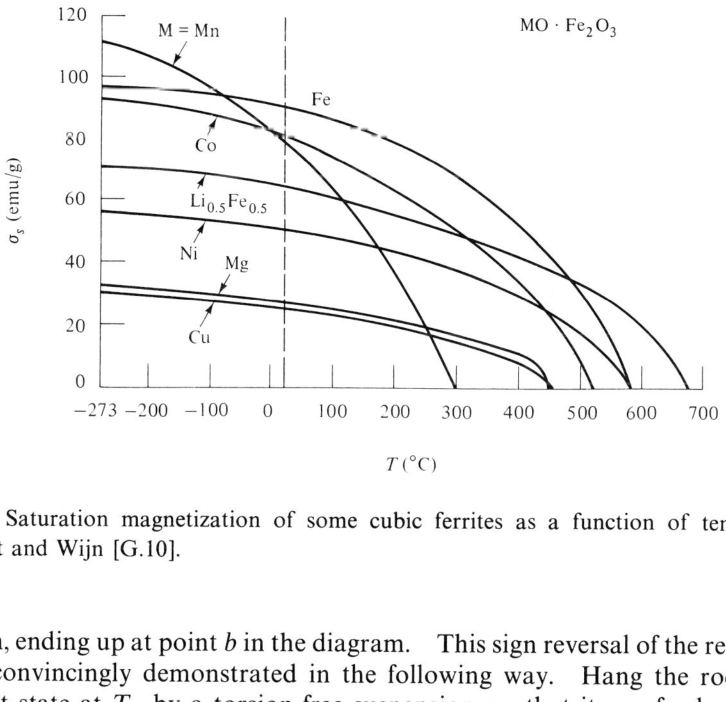

Fig. 6.11 Saturation magnetization of some cubic ferrites as a function of temperature, after Smit and Wijn [G.10].

direction, ending up at point b in the diagram. This sign reversal of the remanence can be convincingly demonstrated in the following way. Hang the rod, in the remanent state at T_1, by a torsion-free suspension, so that it can freely rotate in a horizontal plane, in a weak field. The field should be too weak to alter the magnetic state of the rod appreciably but strong enough to align it. When the rod is then heated through the compensation point, it will rotate through 180°. Only a ferrimagnetic of this peculiar kind will behave in this way; this experiment is thus, in a sense, a crucial test of the theory of ferrimagnetism.]

The two examples just described are indeed unusual. The saturation magnetization of most ferrimagnetics decreases continously, but more rapidly than that of a ferromagnetic, to zero at T_c. Typical examples are shown in Fig. 6.11.

General conclusions

We have seen that the Néel molecular field theory successfully accounts for a whole new class of magnetic materials and is in generally good agreement with experiment. In particular, it offers satisfactory explanations for (a) the marked curvature of the $(1/\chi),T$ plot, which has been called the most characteristic single property of ferrimagnetics, and (b) the unusually shaped σ_s,T curves shown in Fig. 6.9.

The success of the theory may seem surprising in view of the simplified magnetic structure of the model adopted, a structure which rarely corresponds to that of a real ferrimagnetic. However, according to Smart [G.28], "it can be shown that a generalization of the model to include the possibility [of more than one type of magnetic ion] merely introduces more adjustable parameters into the theory and does not change the general characteristics of the susceptibility and magnetization curves already predicted."

6.5 HEXAGONAL FERRITES

There are many hexagonal ferrimagnetic oxides, but the only ones of commercial importance are barium ferrite $BaO \cdot 6\,Fe_2O_3$ ($= BaFe_{12}O_{19}$) and strontium ferrite with the same formula. Barium ferrite has the same crystal structure as magnetoplumbite, which is a mineral with the approximate composition $PbFe_{7.5}Mn_{3.5}Al_{0.5}Ti_{0.5}O_{19}$. (Thus the Fe ions in barium ferrite occupy the same positions as the mixture of Fe, Mn, Al, and Ti ions in magnetoplumbite.)

The hexagonal unit cell of barium ferrite contains 2 "molecules," or a total of $2 \times 32 = 64$ atoms. It is very long in the c direction, with $c = 23.2$ Å and $a = 5.88$ Å. The Ba^{2+} and O^{2-} ions are both large, about the same size, and nonmagnetic; they are arranged in a close-packed fashion. The smaller Fe^{3+} ions are located in the interstices.

They key to an understanding of this large complex cell lies in the relation between the hexagonal-close-packed and the face-centered-cubic structures. Both are built up by stacking identical layers of atoms one on top of another in a particular sequence. Within each layer the atoms are located at the corners of a network of adjoining equilateral triangles, as shown at the top of Fig. 6.12. If the layers are stacked in the sequence ABABAB..., i.e., with the third layer directly over the first, the resulting structure is hexagonal close-packed. If the stacking sequence is ABCABC..., so that the sequence does not repeat until the fourth layer, the result is face-centered cubic. The cubic ferrites with the spinel structure may be thought of in this way, i.e., as being composed of layers of oxygen ions stacked in the ABCABC sequence, with the M^{2+} and Fe^{3+} ions in the interstices. The moments of the magnetic ions are normal to the plane of the oxygen layers, in a direction of the form $\langle 111 \rangle$.

In the barium ferrite unit cell, shown schematically in Fig. 6.12, there are 10 layers of large ions (Ba^{2+} or O^{2-}), with 4 ions per layer. Eight of these layers are wholly oxygen, while two contain one barium ion each, as indicated. The whole block of 10 layers can be regarded as made up of four blocks, two cubic and two hexagonal. In the cubic blocks the arrangement of oxygen ions, occupied

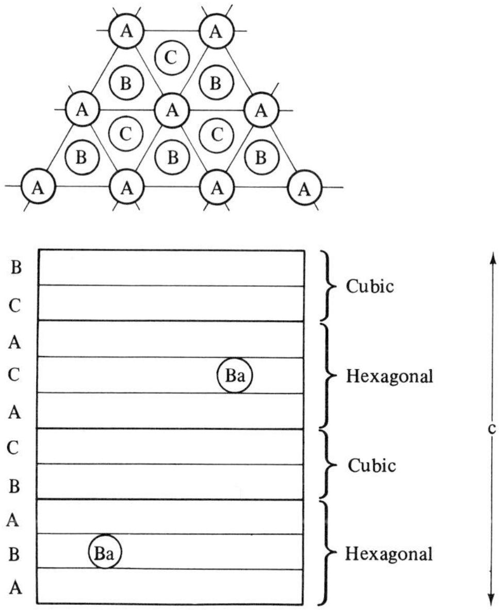

Fig. 6.12 Schematic representation of the barium ferrite structure.

tetrahedral sites, and occupied octahedral sites is exactly the same as in the cubic spinels. In each hexagonal block a barium ion substitutes for an oxygen ion in the central of the three layers, and the layers are stacked in the hexagonal sequence. A study of the stacking sequence indicated in the drawing shows that the cubic and hexagonal sections overlap; thus the four layers between those containing barium have cubic packing, and the five layers centered on a barium ion have hexagonal packing. The unit cell as a whole has hexagonal symmetry. Smit and Wijn [G.10] should be consulted for a detailed description and drawings of this structure.

The only magnetic ions in barium ferrite are the Fe^{3+} ions, each with a moment of 5 μ_B. These are located in *three* crystallographically different kinds of sites: tetrahedral, octahedral, and hexahedral. (The last named is a site surrounded by five equidistant oxygen ions, arranged at the corners of a bipyramid with a triangular base. One of these sites occurs in each barium-containing layer.) The Fe^{3+} ions have their moments normal to the plane of the oxygen layers, and thus parallel or antiparallel to the $+c$ axis of the hexagonal cell, which is a $\langle 0001 \rangle$ direction. Of the 24 Fe^{3+} ions per cell, 4 are in tetrahedral sites, 18 in octahedral, and 2 in hexahedral. By starting with the known spin directions of the Fe^{3+} ions in the cubic sections of the cell and by applying the known principles governing the superexchange force, one can proceed from ion to ion throughout the cell and predict the direction of its spin moment, i.e., whether it is [0001] or [000$\bar{1}$]. In this way one arrives at a predicted value, per cell, of 16

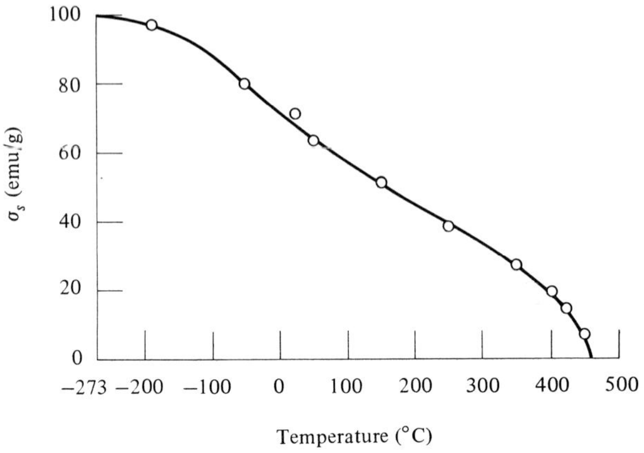

Fig. 6.13 Temperature variation of the saturation magnetization σ_s of barium ferrite $BaO \cdot 6\,Fe_2O_3$, after Smit and Wijn [G.10].

ions with spins in one direction and 8 with spins in the other. The predicted magnetic moment per cell is therefore $(16 - 8)5 = 40\ \mu_B$ per cell or $20\ \mu_B$ per molecule of $BaO \cdot 6\,Fe_2O_3$. This quantity corresponds to 100 emu/g and agrees exactly with the measured value of the saturation magnetization at 0°K. The variation of σ_s with temperature is shown in Fig. 6.13. At 20°C it has fallen to 72 emu/g ($M_s = 380$ emu/cm^3) and the Curie temperature is 450°C.

Other ferrimagnetic oxides with a hexagonal structure include the following:

$$BaO \cdot 2\,MO \cdot 8\,Fe_2O_3 \qquad W$$

$$2(BaO \cdot MO \cdot 3\,Fe_2O_3) \qquad Y$$

$$3BaO \cdot 2\,MO \cdot 12\,Fe_2O_3 \qquad Z$$

Here M is a divalent ion as before, and the letter symbols at the right serve as abbreviations. Thus, Co_2Z stands for $3\,BaO \cdot 2\,CoO \cdot 12\,Fe_2O_3$. The structures and magnetic properties of these compounds are described by Smit and Wijn [G.10].

6.6 OTHER FERRIMAGNETIC SUBSTANCES

Besides the ferrites already described, there are a number of other ferrimagnetics of considerable interest. We will briefly survey them now.

γ-Fe_2O_3

This compound, called maghemite, has a cubic structure and is made by oxidizing magnetite:

$$2\,Fe_3O_4 + \tfrac{1}{2}\,O_2 \rightarrow 3\,Fe_2O_3.$$

It is unstable and transforms to α-Fe_2O_3 (hematite) on heating above 400°C. (Hematite is rhombohedral and a canted antiferromagnetic, with a Néel temperature of 950°K.) γ-Fe_2O_3 is currently the most popular material for magnetic recording tapes.

All the magnetic ions in γ-Fe_2O_3 are identical, namely Fe^{3+}, and ferrimagnetism arises from an unequal distribution of these ions in A and B sites. (This substance therefore corresponds exactly to the model adopted by Néel for his theory of ferrimagnetism.) The higher O/Fe ratio of γ-Fe_2O_3, compared to that of Fe_3O_4, is achieved, not by adding oxygen, but by removing iron. The unit cell of γ-Fe_2O_3 is tetragonal, with $c/a = 3$. It is made by piling up three of the cubic spinel cells of Fe_3O_4 and then removing 8 Fe ions from octahedral (B) sites. The unit cell of Fe_3O_4 contains 8 molecules, so that the conversion to γ-Fe_2O_3 can be written, in terms of the tetragonal cell:

$$(3)(8)(Fe_3O_4) - 8\ Fe = Fe_{64}O_{96} = 32(Fe_2O_3).$$

From the information in Table 6.2 we can then conclude that the Fe^{3+} ions in a unit cell of γ-Fe_2O_3 are distributed as follows:

$$\text{A sites: } 8 \times 3 = 24,$$

$$\text{B sites: } (16 \times 3) - 8 = 40.$$

Because each ion has a moment of $5\mu_B$, the net moment is $(40 - 24)(5) = 80\mu_B$ per unit cell or $80/32 = 2.50\ \mu_B$ per molecule of Fe_2O_3. This is in good agreement with the experimental value of σ_0 of 2.39 μ_B/molecule or 83.5 emu/g. At 20°C σ_s is 76.0 emu/g.

Garnets

The semiprecious stone garnet is actually a group of isomorphous minerals with a complex cubic structure. A typical composition is $3\ MnO \cdot Al_2O_3 \cdot 3\ SiO_2$, but certain other divalent ions can be substituted for Mn^{2+} and certain other trivalent ions for Al^{3+}. By substituting certain trivalent ions for the mixture of divalent (Mn^{2+}) and tetravalent (Si^{4+}) ions in natural garnet, it is possible to make silicon-free garnets with the composition $3\,M_2O_3 \cdot 5\,Fe_2O_3$. The most magnetically interesting of these synthetic garnets are those in which M is yttrium (Y) or one of the rare earths from gadolinium (Gd) to lutetium (Lu), inclusive. These are all ferrimagnetic, but rather weakly so; σ_s in the neighborhood of room temperature is less than 10 emu/g. Yttrium-iron garnet, commonly known as YIG, has a normal σ_s,T curve, but most of the rare-earth garnets exhibit a compensation point (Fig. 6.10). YIG has important applications at very high frequency, in the microwave region.

The cubic unit cell is large, with a lattice parameter of more than 12 A, and it contains 160 atoms. Three crystallographically different kinds of sites exist, occupied as follows: $16\,Fe^{3+}$ in A, $24\,Fe^{3+}$ in B, and $24\,M^{3+}$ in C. The interaction between the Fe^{3+} ions in A and B sites is strongly antiparallel. In the rare-earth garnets, the moment on the rare-earth ions in C sites is antiparallel to the resultant moment of the Fe^{3+} ions. The Y^{3+} ion in YIG has no moment, so that the net moment of YIG is solely due to an unequal distribution of the same kind of ions (Fe^{3+}) in A and B sites, as in γ-Fe_2O_3.

Alloys

Ferrimagnetic intermediate phases occur in several alloy systems. Perhaps the best known is Mn_2Sb. Its tetragonal cell contains two atoms of manganese in different kinds of sites. The moment of a Mn atom in an A site is antiparallel and unequal to the moment of a Mn atom in a B site, leading to a net moment of 0.94 μ_B per Mn atom. (This magnetic structure for Mn_2Sb was proposed by Guillaud in 1943, five years before the publication of Néel's general theory of ferrimagnetism, and was later confirmed by neutron diffraction. Mn_2Sb was the first strongly magnetic substance to be recognized as ferrimagnetic rather than ferromagnetic.)

Other ferrimagnetic metallic phases include Mn_2Sn, Mn_3Ga, Mn_3Ge_2, Mn_3In, $FeGe_2$, FeSe, Cr_3As_2, and $CrPt_3$.

Many binary alloys of a rare-earth element, especially a heavy rare earth, and a transition metal are ferrimagnetic. Typical examples are RCo_5 alloys, where R = Gd, Tb, Dy, Ho, Em, or Tm, all of which, incidentally, show a compensation point. In these alloys the cobalt moments are antiparallel to those of the rare earths. On the other hand, when R is a light rare earth, all moments are parallel, resulting in ferromagnetism, as mentioned in Section 4.5. The conditions governing parallel or antiparallel alignment are explained by Kouvel [4.7].

6.7 SUMMARY: KINDS OF MAGNETISM

We have surveyed, in Chapters 3 through 6, the five main kinds of magnetism exhibited by matter. A graphical summary of this material is shown in Fig. 6.14. Here a circle represents an atom or ion, and an arrow through that circle represents its net magnetic moment. Open and solid circles represent atoms or ions of different valence or chemical species. The magnetic structures depicted are those which exist in zero applied field.

The five kinds of magnetism can be divided into two broad categories:

1. Diamagnetism and ideal (Curie law) paramagnetism, characterized by noncooperative (statistical) behavior of the individual magnetic moments.
2. Nonideal (Curie-Weiss law) paramagnetism, ferromagnetism, antiferromagnetism, and ferrimagnetism, which are all examples of cooperative phenomena.

PROBLEMS

6.1 The approximate values σ_s/σ_o for magnetite Fe_3O_4 as a function of T/T_c, as measured by Weiss [3.3], are as follows:

σ_s/σ_o	0.92	0.88	0.83	0.77	0.68	0.58	0.43	0.32	0.22	0.03
T/T_c	0.23	0.33	0.43	0.54	0.66	0.78	0.89	0.94	0.95	0.98

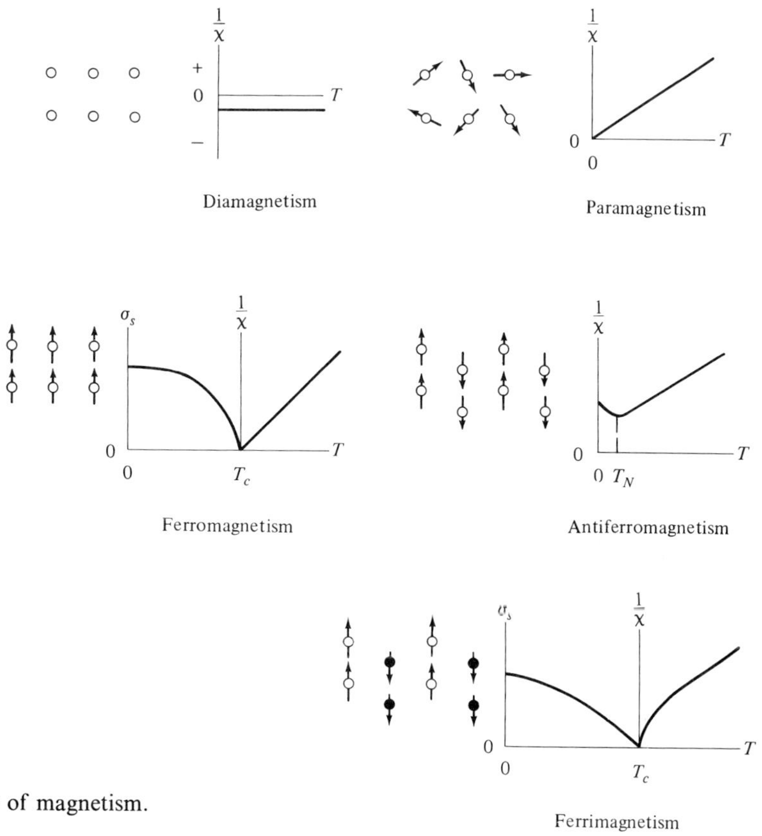

Fig. 6.14 Kinds of magnetism.

Plot these values together with the curve derived from the Weiss-Langevin theory of ferromagnetism. (This curve was calculated in Problem 4.1 and appears as the $J = \infty$ curve of Fig. 4.7.) (In 1907, when these results were published, Weiss, and everyone else, regarded magnetite as ferromagnetic and essentially similar to iron. The excellent agreement between his theory and the experimental data on magnetite gave Weiss great confidence in the basic soundness of his molecular-field theory, although he was well aware of the marked disagreement between theory and experiment for iron, cobalt, and nickel. This kind of situation – agreement between incorrect theory and experiment – arises not infrequently in the history of science. Here the data on magnetite, a ferrimagnetic in which both positive and negative molecular fields are acting, quite accidentally happen to agree with a theory designed to apply to a ferromagnetic in which only a positive molecular field acts. Hindsight is so easy.)

6.2 The measured saturation magnetization σ_0 of $NiO \cdot Fe_2O_3$ is 56 emu/g. Calculate the magnetic moment per molecule in Bohr magnetons.

6.3 A mixed Co-Zn ferrite contains cobalt and zinc in the ratio of 4 to 1 by weight. What is its theoretical saturation magnetization μ_H in Bohr magnetons per molecule? Assume pure Co ferrite to have the theoretical, spin-only moment of a completely inverse ferrite.

MAGNETIC PHENOMENA

CHAPTER 7

MAGNETIC ANISOTROPY

7.1 INTRODUCTION

The remainder of this book will be devoted, almost without exception, to the strongly magnetic substances, namely, ferro- and ferrimagnetics. Chapters 7 through 12 deal mainly with structure-sensitive properties, those which depend on the prior history (thermal, mechanical, etc.) of the specimen. In these chapters we shall be concerned chiefly with the *shape* of the magnetization curve, i.e., with the way in which the magnetization changes from zero to the saturation value M_s. The value of M_s itself will be regarded simply as a constant of the material. If we understand the several factors which affect the shape of the M,H curve, we will then understand why some materials are magnetically soft and others magnetically hard.

One factor which may strongly affect the shape of the M,H (or B,H) curve, or the shape of the hysteresis loop, is magnetic anisotropy. This term simply means that the magnetic properties depend on the direction in which they are measured. This general subject is of considerable practical interest, because anisotropy is exploited in the design of most magnetic materials of commercial importance. A thorough knowledge of anisotropy is thus the key to an understanding of these materials.

There are several kinds of anisotropy:

1. *Crystal anisotropy*, also called *magnetocrystalline anisotropy*.
2. *Shape anisotropy*.
3. *Stress anisotropy* (Section 8.5).
4. Anisotropy induced by
   a) Magnetic annealing (Chapter 10).
   b) Plastic deformation (Chapter 10).
   c) Irradiation (Chapter 10).
5. *Exchange anisotropy* (Section 11.8).

Of these, only crystal anisotropy is intrinsic to the material. Strictly, then, all the others are extrinsic or "induced." However, it is customary to limit the term "induced" to the anisotropies listed under (4) above. All the anisotropies from (1) through (4 b) are important, and any one may become predominant in special circumstances. The last two, (4 c) and (5), are rather uncommon. In this chapter we will consider only crystal and shape anisotropy.

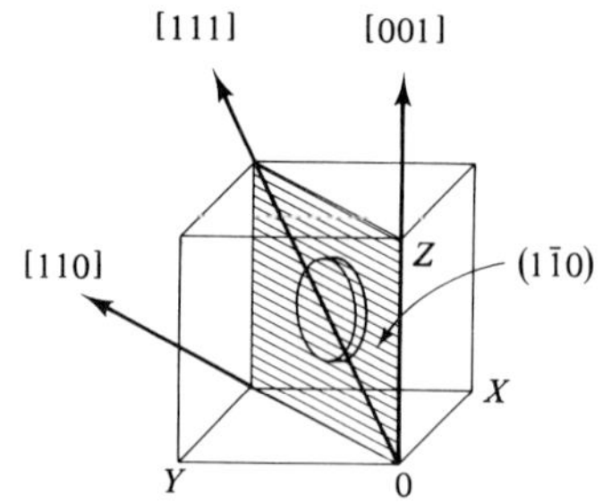

Figure 7.1

7.2 ANISOTROPY IN CUBIC CRYSTALS

Suppose a single crystal is cut in the form of a disk parallel to a plane* of the form {110}. This specimen will then have directions of the form ⟨100⟩, ⟨110⟩, and ⟨111⟩ as diameters, as shown in Fig. 7.1 for the plane (1$\bar{1}$0). Measurements of magnetization curves along these diameters, in the plane of the disk, will then give information about three important crystallographic directions. The results for iron, which has a body-centered cubic structure, are shown in Fig. 7.2(a), and those for nickel (face-centered cubic), in Fig. 7.2(b).

For iron these measurements show that saturation can be achieved with quite low fields, of the order of a few tens of oersteds at most, in the ⟨100⟩ direction. which is accordingly called the "easy direction" of magnetization. This tells us something about domains in iron in the demagnetized state. As will become clear later, a domain wall separating two domains in a crystal can be moved by a small applied field. If we tentatively assume that domains in demagnetized iron are spontaneously magnetized to saturation in directions of the form ⟨100⟩, then a possible domain structure for a demagnetized crystal disk cut parallel to (001) would be that shown in Fig. 7.3(a). It has four kinds of domains, magnetized parallel to four of the six easy directions, namely, [010], [100], [0$\bar{1}$0], and [$\bar{1}$00]. (Actually, an iron crystal disk of diameter, say, 1 cm, would contain many hundreds of domains, rather than the four shown. However, it would still be true that these hundreds of domains would be of only four *kinds*, namely those with M_s vectors in the [010], [100], [0$\bar{1}$0], and [$\bar{1}$00] directions. Domain sketches like those of Fig. 7.2 are therefore highly simplified, but they do contain all or most of the necessary information, depending on the problem involved.) If a field H

* The following convention is followed for the indices of planes and directions. Planes of a form are planes related by symmetry, such as the six faces of a cube: (100), (010), (001), ($\bar{1}$00), (0$\bar{1}$0), and (00$\bar{1}$). The indices of any one, enclosed in braces {100}, stand for the whole set.

The indices of particular directions are enclosed in square brackets, such as the six cube-edge directions: [100], [010], [001], [$\bar{1}$00], [0$\bar{1}$0], and [00$\bar{1}$]. These are directions of a form, and the whole set is designated by the indices of any one, enclosed in angular brackets ⟨100⟩.

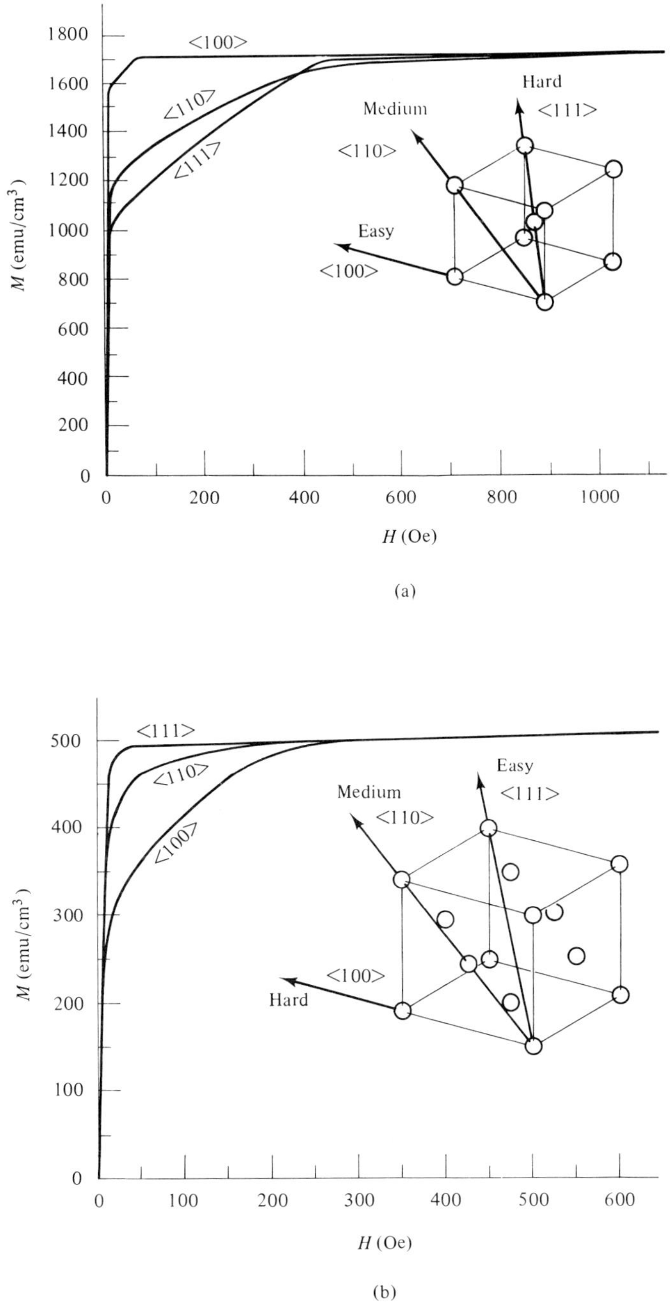

Fig. 7.2 Magnetization curves for single crystals of (a) iron (by Honda and Kaya [7.1]), and (b) nickel (by Kaya [7.2]).

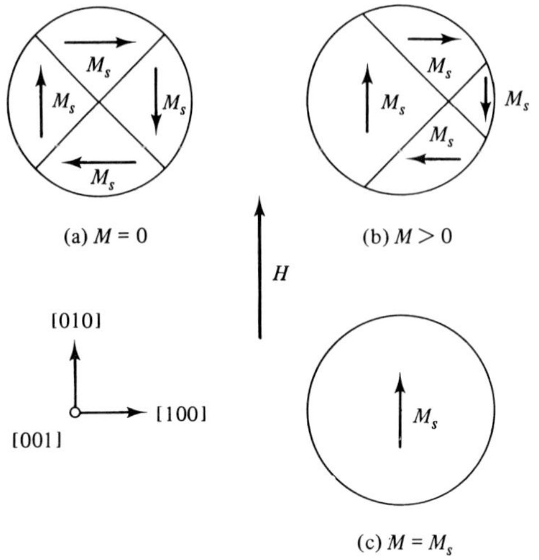

Fig. 7.3 Changes in the domain structure of a crystal of iron (schematic). H is in direction [010].

is now applied in, say, the [010] direction, the [010] domain will grow in volume by the mechanism of domain-wall motion, as indicated in Fig. 7.3(b). It does so because the magnetic potential energy of the crystal is thereby lowered; Eq. (1.5) shows that the energy of a [010] domain in the field is $-M_sH$ per unit volume, that of a $[0\bar{1}0]$ domain is $+M_sH$, and that of a [100] or $[\bar{1}00]$ domain is zero. Continued application of the field eliminates all but the favored domain, and the crystal is now saturated (Fig. 7.3 c). This has been accomplished simply by applying the low field required for domain wall motion. Since experiment shows that only a low field is needed to saturate iron in a $\langle 100 \rangle$ direction, we conclude that our postulated domain structure is correct and, more generally, that *the direction of easy magnetization of a crystal is the direction of spontaneous domain magnetization in the demagnetized state.* In nickel, Fig. 7.2(b) shows that the direction of easy magnetization is of the form $\langle 111 \rangle$, the body diagonal of the unit cell. The direction $\langle 111 \rangle$ is also the direction of easy magnetization in all the cubic ferrites, except cobalt ferrite or mixed ferrites containing a large amount of cobalt. The latter have $\langle 100 \rangle$ as an easy direction.

[Note that, on the scale of a few domains, as in Fig. 7.3(b), a partially magnetized crystal is never uniformly magnetized, in the sense that **M** is everywhere equal in magnitude and direction, whether or not the crystal is ellipsoidal in shape. The notion of uniform magnetization predates the domain hypothesis. It has meaning, for a crystal containing domains, only when applied either (a) to a volume less than that of one domain, or (b) to a volume so large that it contains many domains and has a net magnetization **M** equal to that of the whole crystal.]

Figure 7.2(a) shows that fairly high fields, of the order of several hundred oersteds, are needed to saturate iron in a $\langle 110 \rangle$ direction. For this orientation of the field, the domain structure changes as in Fig. 7.4. Domain wall motion,

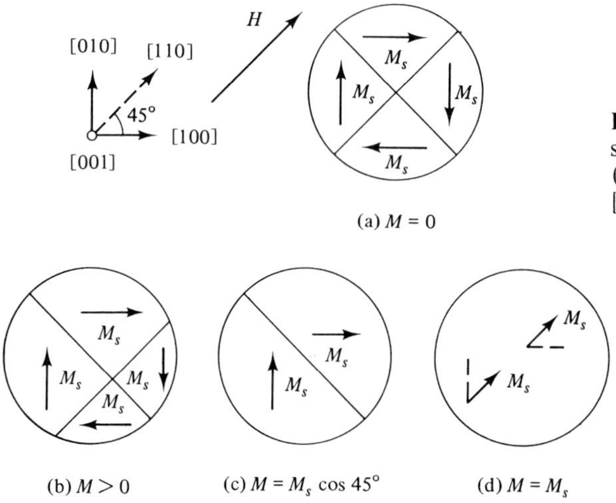

Fig. 7.4 Changes in the domain structure of a crystal of iron (schematic). H is in direction [110].

in a low field, occurs until there are only two domains left (Fig. 7.4 c), each with the same potential energy. The only way in which the magnetization can increase further is by rotation of the M_s vector of each domain until it is parallel with the applied field. This process is called *domain rotation.* (The domain itself, which is a group of atoms, does not rotate. It is the net magnetic moment of each atom which rotates.) Domain rotation occurs only in fairly high fields, because the field is then acting against the force of crystal anisotropy, which is usually fairly strong. Crystal anisotropy may therefore be regarded as a force which tends to bind the magnetization to directions of a certain form in the crystal. When the rotation process is complete (Fig. 7.4 d), the domain wall in (c) disappears, and the crystal is saturated.

Because the applied field must do work against the anisotropy force to turn the magnetization vector away from an easy direction, there must be energy stored in any crystal in which M_s points in a noneasy direction. This is called the *crystal anisotropy energy E.* The Russian physicist Akulov showed in 1929 that E can be expressed in terms of a series expansion of the direction cosines of M_s relative to the crystal axes. In a cubic crystal, let M_s make angles a, b, c with the crystal axes, and let $\alpha_1, \alpha_2, \alpha_3$ be the cosines of these angles, Then

$$E = K_0 + K_1(\alpha_1^2\alpha_2^2 + \alpha_2^2\alpha_3^2 + \alpha_3^2\alpha_1^2) + K_2(\alpha_1^2\alpha_2^2\alpha_3^2) + \qquad (7.1)$$

where $K_0, K_1, K_2,$... are constants for a particular material and are expressed in ergs/cm^3. Higher powers are generally not needed, and sometimes K_2 is so small that the term involving it can be neglected. The first term, which is simply K_0, is independent of angle and is usually ignored, because normally we are interested only in the change in the energy E when the M_s vector rotates from

Table 7.1 Crystal Anisotropy Energies for Various Directions in a Cubic Crystal

$[uvw]$	a	b	c	α_1	α_2	α_3	E
[100]	0	90°	90°	1	0	0	K_0
[110]	45°	45°	90°	$1/\sqrt{2}$	$1/\sqrt{2}$	0	$K_0 + K_1/4$
[111]	54.7°	54.7°	54.7°	$1/\sqrt{3}$	$1/\sqrt{3}$	$1/\sqrt{3}$	$K_0 + K_1/3 + K_2/27$

Table 7.2 Directions of Easy, Medium, and Hard Magnetization in a Cubic Crystal (from Bozorth [G.4])

K_1	+	+	+	−	−	−
K_2	$+\infty$ to $-9K_1/4$	$-9K_1/4$ to $-9K_1$	$-9K_1$ to $-\infty$	$-\infty$ to $9\lvert K_1\rvert/4$	$9\lvert K_1\rvert/4$ to $9\lvert K_1\rvert$	$9\lvert K_1\rvert$ to $+\infty$
Easy	⟨100⟩	⟨100⟩	⟨111⟩	⟨111⟩	⟨110⟩	⟨110⟩
Medium	⟨110⟩	⟨111⟩	⟨100⟩	⟨110⟩	⟨111⟩	⟨100⟩
Hard	⟨111⟩	⟨110⟩	⟨110⟩	⟨100⟩	⟨100⟩	⟨111⟩

one direction to another. Table 7.1 gives the value of E when the M_s vector lies in a particular direction $[u\,v\,w]$.

When K_2 is zero, the direction of easy magnetization is determined by the sign of K_1. If K_1 is positive, then $E_{100} < E_{110} < E_{111}$, and ⟨100⟩ is the easy direction, because E is a minimum when M_s is in that direction. Thus iron and the cubic ferrites containing cobalt have positive values of K_1. If K_1 is negative, $E_{111} < E_{110} < E_{100}$, and ⟨111⟩ is the easy direction. K_1 is negative for nickel and all the cubic ferrites that contain no cobalt.

When K_2 is not zero, the easy direction is determined by the values of both K_1 and K_2. The way in which the values of these two constants determine the directions of easy, medium, and hard magnetization is shown in Table 7.2.

7.3 ANISOTROPY IN HEXAGONAL CRYSTALS

Magnetization curves of cobalt, which has a hexagonal close-packed structure, are shown in Fig. 7.5. The hexagonal c axis is the direction of easy magnetization, and, within the accuracy of the measurements, any direction in the basal plane is found to be equally hard. Under these circumstances the anisotropy energy E depends on only a single angle, the angle θ between the M_s vector and the c axis. Therefore,

$$E = K_0' + K_1' \cos^2\theta + K_2' \cos^4\theta + \cdots \tag{7.2}$$

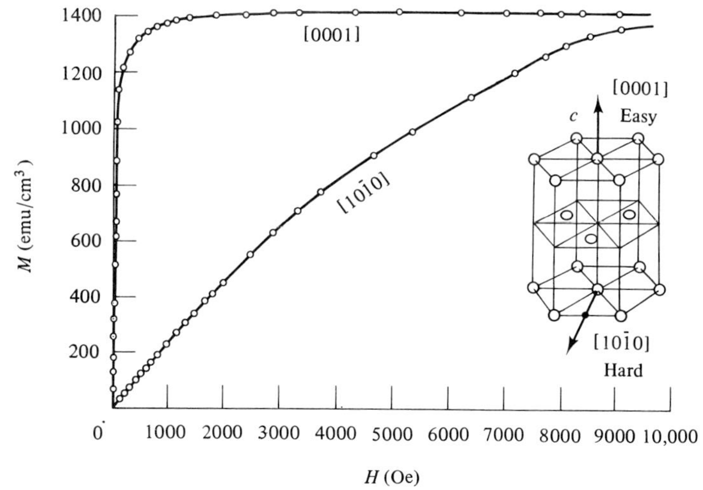

Fig. 7.5 Magnetization curves for a single crystal of cobalt (by Kaya [7.3]).

However, it is customary to write the equation for E in hexagonal crystals in powers of $\sin\theta$. Putting $\cos^2\theta = 1 - \sin^2\theta$ into Eq. (7.2), we have

$$E = K_0 + K_1 \sin^2\theta + K_2 \sin^4\theta + \cdots \tag{7.3}$$

When K_1 is positive and $K_2 > -K_1$, the energy E is a minimum for $\theta = 0$, and the c axis is one of easy magnetization. These conditions are met for cobalt and hexagonal barium ferrite $BaO \cdot 6Fe_2O_3$. A crystal with a single easy axis, along which the magnetization can point either up or down, is referred to as a *uniaxial crystal.* Its domain structure in the demagnetized state is particularly simple (Fig. 7.6).

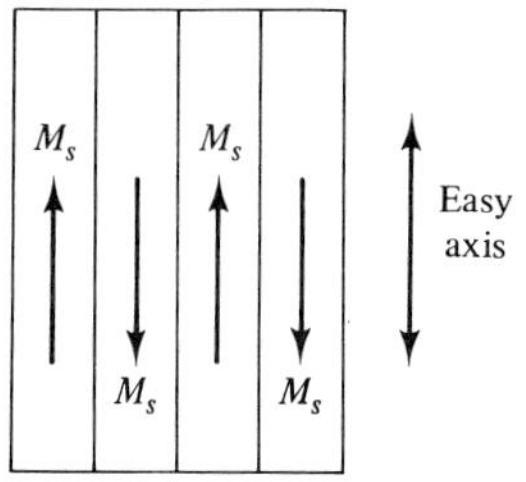

Fig. 7.6 Domain structure of a uniaxial crystal.

The energy E will have a minimum for $\theta = 90°$ if K_1 is negative and $K_2 < |K_1|/2$, or if K_1 is positive and $K_2 < -K_1$. The basal plane is then an *easy plane* of magnetization. A stable intermediate value of θ can also exist: if K_1 is negative and $K_2 > |K_1|/2$, then E is a minimum for $\theta = \sin^{-1}(-K_1/2K_2)^{1/2}$. The M_s vector may then lie anywhere on the surface of a cone with a semivertex angle of θ. Both of these states (a minimum in E at $\theta = 90°$ or a particular value between 0 and 90°) are unusual but have been observed in cobalt at high temperatures.

7.4 PHYSICAL ORIGIN OF CRYSTAL ANISOTROPY

Crystal anisotropy is due mainly to *spin-orbit coupling*. By coupling is meant a kind of interaction. Thus we could speak of the exchange interaction between two neighboring spins as a spin-spin coupling. This coupling is very strong and keeps neighboring spins parallel, or antiparallel, to one another. But the associated exchange energy is isotropic; it depends only on the angle between adjacent spins, as stated by Eq. (4.29), and not at all on the direction of the spin axis relative to the crystal lattice. The spin-spin coupling therefore cannot contribute to the crystal anisotropy.

The orbit-lattice coupling is also strong. This follows from the fact that orbital magnetic moments are almost entirely quenched, as discussed in Section 3.7. This means, in effect, that the orientations of the orbits are fixed very strongly to the lattice, because even large fields cannot change them.

There is also a coupling between the spin and the orbital motion of each electron. When an external field tries to reorient the spin of an electron, the orbit of that electron also tends to be reoriented. But the orbit is strongly coupled to the lattice and therefore resists the attempt to rotate the spin axis. The energy required to rotate the spin system of a domain away from the easy direction, which we call the anisotropy energy, is just the energy required to overcome the spin-orbit coupling. This coupling is relatively weak, because fields of a few hundred oersteds are usually strong enough to rotate the spins. Inasmuch as the "lattice" is really constituted by a number of atomic nuclei arranged in space, each with its surrounding cloud of orbital electrons, we can also speak of a spin-lattice coupling and conclude that it too is weak. These several relationships are summarized in Fig. 7.7.

The strength of the anisotropy in any particular crystal is measured by the magnitude of the anisotropy constants K_1, K_2, etc. Although there is no doubt that crystal anisotropy is due to spin-orbit coupling, the details are not clear, and it is not yet possible to calculate the value of the anisotropy constants from first principles.

Nor is there any simple relationship between the easy, or hard, direction of magnetization and the way atoms are arranged in the crystal structure. Thus in iron, which is body-centered cubic, the direction of greatest atomic density, i.e., the direction in which the atoms are most closely packed, is $\langle 111 \rangle$, and this

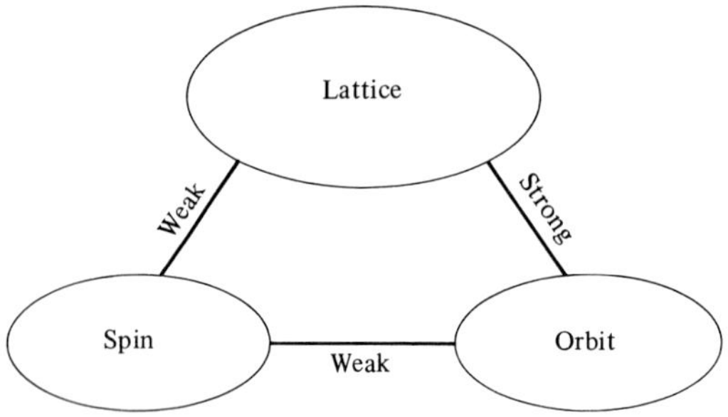

Fig. 7.7 Spin-lattice-orbit interactions.

is the hard axis. But in nickel (face-centered cubic) the direction of greatest atomic density is ⟨110⟩, which is an axis of medium hard magnetization. And when iron is added to nickel to form a series of face-centered cubic solid solutions, the easy axis changes from ⟨111⟩ to ⟨100⟩ at about 25 percent iron, although there has been no change in crystal structure.

7.5 ANISOTROPY MEASUREMENT (BY TORQUE CURVES)

The anisotropy constants of a crystal may be measured by the following methods:

1. Torque curves.
2. Torsion pendulum.
3. Magnetization curves (Section 7.6).
4. Magnetic resonance (Section 12.7).

The first method is by far the most popular and will be described, along with the closely related torsion-pendulum method, in this section. The other methods are left to later sections.

Torque curves

These depict the torque required to rotate the magnetization away from an easy direction as a function of the angle of rotation. We will first consider a *uniaxial* crystal, such as a hexagonal crystal with an easy axis parallel to the c axis. Let it be cut in the form of a thin disk with the c axis in the plane of the disk. It is then suspended by a torsion wire at its center, so that the plane of the disk is horizontal and parallel to the strong field of an electromagnet (Fig. 7.8). If the upper end of the torsion wire is turned clockwise, the c axis will rotate away from the field direction by the angle θ, as shown. If the field is large enough to saturate the crystal, then M_s is parallel to H and the angle between c and M_s will also be θ.

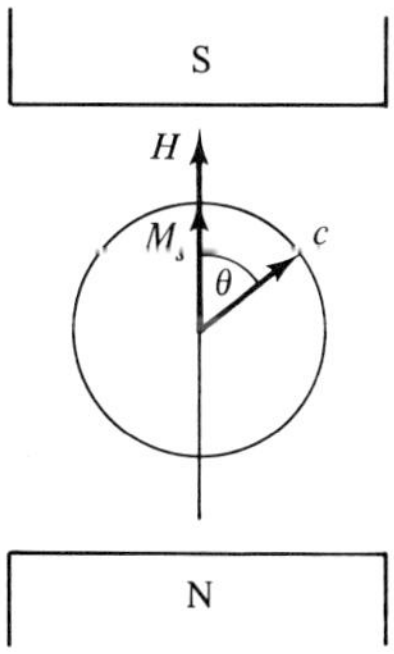

Fig. 7.8 Horizontal section through a disk specimen in the field of an electromagnet. c = easy axis.

We now need an expression for the torque. From Eq. (7.3) the θ-dependent part of the anisotropy energy, if K_2 is negligible, is given by

$$E = K_1 \sin^2 \theta. \tag{7.4}$$

When the energy of a system depends on an angle, the derivative of the energy with respect to the angle is a torque. Thus $dE/d\theta$ is the torque exerted by the crystal on M_s, and $-dE/d\theta$ is the torque exerted *on* the crystal *by* M_s. (Clockwise torques are taken as positive, and the positive direction of θ is measured from M_s to c.) Then the torque on the crystal per unit volume is

$$L = -dE/d\theta, \tag{7.5}$$

$$L = -2K_1 \sin\theta \cos\theta = -K_1 \sin 2\theta. \tag{7.6}$$

The torque L is in dyne-cm/cm^3 if E is in ergs/cm^3. Figure 7.9 shows how E and L vary with angle. For positive K_1, the 0 and 180° positions are energy minima, and $\theta = 90°$, which is a direction of difficult magnetization, is a position of instability. The slope of the torque curve at $L = 0$ is negative for the positions of stability ($\theta = 0$ and 180) and positive for the unstable position ($\theta = 90°$). Experimentally, the value of K_1 is found simply from the amplitude of the torque curve.

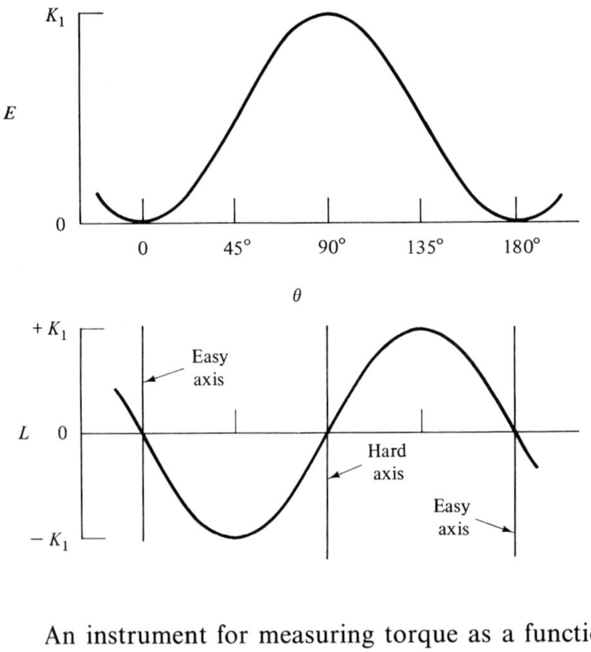

Fig. 7.9 Variation with θ of the anisotropy energy E and the torque L ($= -dE/d\theta$) for a uniaxial crystal. θ is the angle between M_s and the easy axis.

An instrument for measuring torque as a function of angle is called a *torque magnetometer*. These devices are all homemade, and their design varies almost from one laboratory to the next. Figure 7.10 shows a simple version. The disk-shaped specimen is fixed to a rigid rod which is suspended by a torsion wire. A weight is hung below the specimen to counteract any sidewise force that may

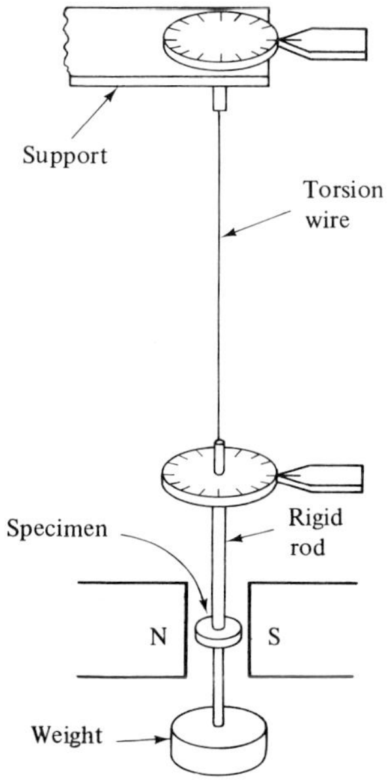

Fig. 7.10 A torque magnetometer.

act on the specimen. [If the field is not perfectly uniform, there will be a sidewise force according to Eq. (2.52). A force will also act, even in a uniform field, if the specimen is not exactly centered in the gap of the electromagnet; this is the "image force", caused by the attraction of the specimen to the closer of its magnetic images (Fig. 2.35) in the magnet pole pieces.] If the anisotropy to be measured is so large that some frictional force can be tolerated, the rod can be held in place by bearings and the lower weight dispensed with. The position of a reference line on the specimen, such as the c axis in the example considered above, is known with respect to the lower graduated dial; the reading on this dial then gives the angle θ between the reference line and the field. The position of the specimen in the field is changed by rotating the torsion head (upper dial). The angle of twist ϕ of the torsion wire is found from the difference in readings between the two dials, and the torque L is obtained from ϕ. In some instruments the entire electromagnet, instead of the torsion head, is rotated.

The ideal specimen shape is that of an ellipsoid of revolution (oblate spheroid), which is relatively easy to saturate. However, an ellipsoidal specimen is difficult to make, and most investigators settle for a disk with a diameter/thickness ratio of the order of 10/1. But as saturation is approached, high demagnetizing fields develop near the sharp edges of the disk, so that these edges are hard to saturate.

The length and diameter of the torsion wire must be chosen so that the torsional stiffness of the wire is of the right order of magnitude relative to the anisotropy being measured. If the wire is too stiff, the twist angle ϕ will be too small to be measured accurately. If the wire is not stiff enough, the anisotropy of the specimen will tend to keep the easy axis parallel or nearly parallel to the field direction, even when ϕ is very large. (Another difficulty with a wire of insufficient stiffness occurs near points of instability, as at $\theta = 90$ in Fig. 7.9. The specimen will then suddenly rotate through a large angle, and it is difficult to obtain enough points to delineate the torque curve.) If C is the torsion constant of the wire, then

$$C = \frac{L_w}{\phi} = \frac{\pi G r^4}{2l} \text{ dyne-cm/radian,} \tag{7.7}$$

where L_w = torque on wire (dyne-cm), ϕ = angle of twist (radians), G = shear modulus of the wire material (dynes/cm^2), r = wire radius (cm), and l = wire length (cm). It is difficult to give typical values of C, because the anisotropies to be measured and specimen sizes vary between such large limits; however, the data of Problem 7.2 illustrate a particular case. Once the required value of C is approximately known, Eq. (7.7) will suggest appropriate values of r and l. When the wire material (tungsten and phosphor bronze are popular) and wire dimensions are decided on, C should be measured, because an accurate value of C cannot be obtained from Eq. (7.7) and a handbook value of G. Two methods of measuring the torsion constant C follow.

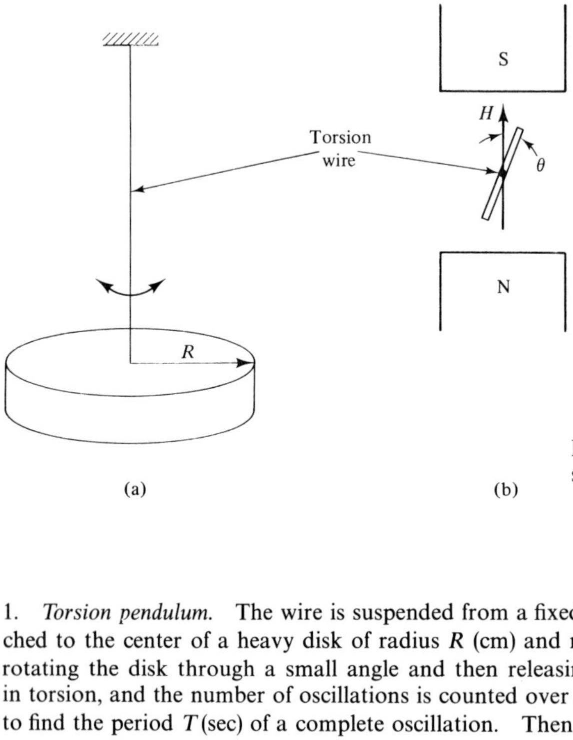

Fig. 7.11 Measurement of the torsion constant of a wire.

1. *Torsion pendulum.* The wire is suspended from a fixed support and its lower end is attached to the center of a heavy disk of radius R (cm) and mass M (g), as in Fig. 7.11 (a). By rotating the disk through a small angle and then releasing it, the wire is made to oscillate in torsion, and the number of oscillations is counted over a measured length of time in order to find the period T(sec) of a complete oscillation. Then

$$T = 2\pi \sqrt{I/C}, \tag{7.8}$$

where I is the moment of inertia, equal to $MR^2/2$ for a disk. Therefore

$$C = \frac{2\pi^2 M R^2}{T^2}. \tag{7.9}$$

2. *Measurement of a known anisotropy.* A thin ferromagnetic rod placed in a magnetic field tends to remain parallel to the field, and a finite torque must be exerted on the rod to turn it away from the field direction, because of the shape anisotropy (Section 7.10) of the rod. This torque can be calculated. We therefore replace the usual disk specimen in the torque magnetometer with a thin rod specimen, usually of iron or nickel, suspended at its center as in Fig. 7.11 (b), which depicts a horizontal section through the specimen. We then determine the twist ϕ of the torsion wire required to rotate the rod through a small angle θ from the field direction. The torque required is given by

$$L = C\phi = \frac{2\pi M_s^2 H V \theta}{H + 2\pi M_s} \text{ dyne-cm}, \tag{7.10}$$

from which C can be determined. Here M_s = saturation magnetization of the rod (emu/cm^3), H = field strength (Oe), and V = volume of rod (cm^3). This equation is derived by Luborsky and Morelock [7.4] and is valid if θ is small and the length/diameter ratio of the rod is greater than 10.

In the determination of crystal anisotropy, the total torque on a crystal of volume V is LV. But this is equal to the torque on the torsion wire, which is L_w or $C\phi$. Therefore, the torque on unit volume of the crystal is $L = C\,\phi/V$.

It is important that the component parts of a torque magnetometer be properly aligned. The following axes should coincide with the center of the magnet air gap: axis of specimen, axis of rotation of torsion head (upper dial of Fig. 7.10), and axis of instrument (line from upper support of torsion wire through the center of gravity of all suspended parts). Improper alignment will introduce spurious torques which distort the experimental curve. The alignment can be checked by determining the torque curve of a specimen having a known, and simple, anisotropy, such as uniaxial, and comparing this curve with that theoretically expected.

Much more elaborate instruments than that shown in Fig. 7.10 have been built, and the trend is to those which will automatically record the torque curve. Figure 7.12 shows an example. A rigid rod carries the specimen S, coil C, and mirror M; the rod itself is supported at top and bottom by taut torsion wires T, T. When the rod is in its zero position, the light beam reflected by the mirror is split equally by the prism P and falls on a pair of photocells G. When a torque is exerted on the specimen, the rod rotates slightly, causing unequal amounts of light to fall on the photocells, and therefore unequal conduction currents result. This current imbalance, which is proportional to the torque on the specimen, is amplified and fed to the coil C in such a way that the torque on the coil, due to its location in the magnetic field of the permanent magnet D, returns the suspension to its zero position. A voltage proportional to the coil current is applied to the Y axis of an X,Y recorder. At the same time the electromagnet E is slowly rotated. Through a friction drive, the magnet turntable rotates the rotor of a potentiometer, which therefore supplies a voltage proportional to the angular position of the magnet. This voltage is applied to the X axis of the recorder. The result is a curve of torque on the specimen versus the angle between the applied field and a reference line on the specimen.

While torque magnetometers of the highest sensitivity will probably always be designed around a torsion wire and an optical lever, others have been made in which torsional strain in a rod supporting the specimen is translated into tensile

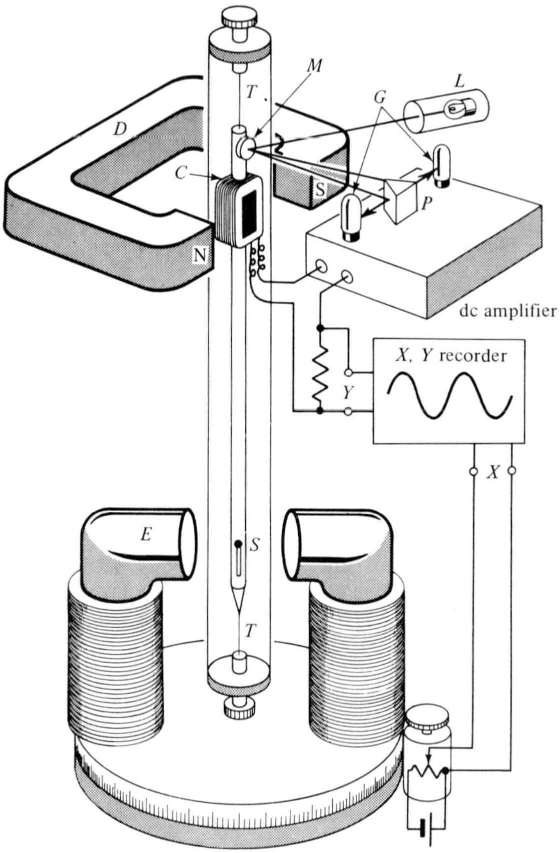

Fig. 7.12 Automatic recording torque magnetometer. After Chikazumi [G.23], based on the designs of Penoyer [7.5] and Pearson [7.6].

or compressive strain of an electrical-resistance strain gage. The resistance of this gage then becomes a measure of the torque. Such types lend themselves easily to automatic recording.

Up to now we have assumed that the applied field H is strong enough to keep M_s parallel to H during the torque measurement. This is not always true, and M_s may instead lie between H and c in Fig. 7.8. Under these circumstances, the peaks on the torque curves are distorted (skewed) and of lesser height than the peaks observed at saturation (M_s parallel to H.) The value K' of an anisotropy constant measured at a field H is then less than the true value K corresponding to infinite field. The relation between these quantities, based on measurements at fields up to about 5 kOe, has been found to be

$$K' = K(1 - a/H), \tag{7.11}$$

where a is a constant. It has therefore been customary to plot K' against $1/H$ and extrapolate to $1/H = 0$ to obtain K. But Kouvel and Graham [7.7] showed, by measurements at fields up to 10 kOe, that extrapolation against $1/\sqrt{H}$ gave better straight lines. However, it is not clear that this conclusion is valid for all

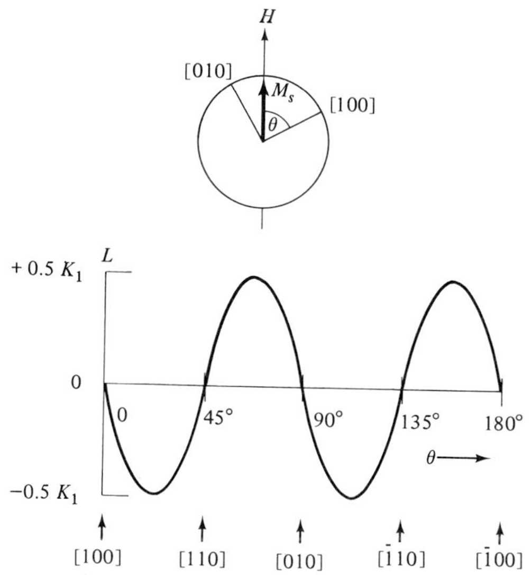

Fig. 7.13 Variation of torque L with angle θ in the (001) plane of a cubic crystal.

kinds and shapes of specimens. Their paper should be consulted for a discussion of the problems involved in torque measurements on unsaturated specimens.

Having examined the methods of measuring torque curves, we will now return to the curves themselves. They are usually more complex than the one of Fig. 7.9, which applies to a uniaxial crystal. Consider, for example, a disk cut parallel to the (001) plane of a *cubic crystal*, in which ⟨100⟩ are the easy directions. This disk will have *biaxial anisotropy*, because it has two easy directions in its own plane. The top of Fig. 7.13 shows the orientation after the [100] axis has been rotated by an angle θ away from M_s and the field H, which is assumed to be very strong. The direction cosines of M_s are then $\alpha_1 = \cos\theta$, $\alpha_2 = \cos(90^\circ - \theta) = \sin\theta$, and $\alpha_3 = 0$. Putting these values into Eq. (7.1), we find the crystal anisotropy energy

$$E = K_0 + K_1 \sin^2\theta \cos^2\theta, \tag{7.12}$$

which is independent of K_2. It can be written

$$E = K_0 + (K_1/4)\sin^2 2\theta. \tag{7.13}$$

The torque on the crystal is then

$$L = -\frac{dE}{d\theta} = -K_1 \sin 2\theta \cos 2\theta,$$

$$L = -\frac{K_1}{2}\sin 4\theta. \tag{7.14}$$

This equation is plotted in Fig. 7.13; the torque goes through a full cycle in a 90° rotation of the disc. The amplitude of the curve equals $K_1/2$, but K_2 cannot be determined. The polar diagram of Fig. 7.14 clearly shows the minima in anisotropy energy in ⟨100⟩ directions and the maxima in ⟨110⟩.

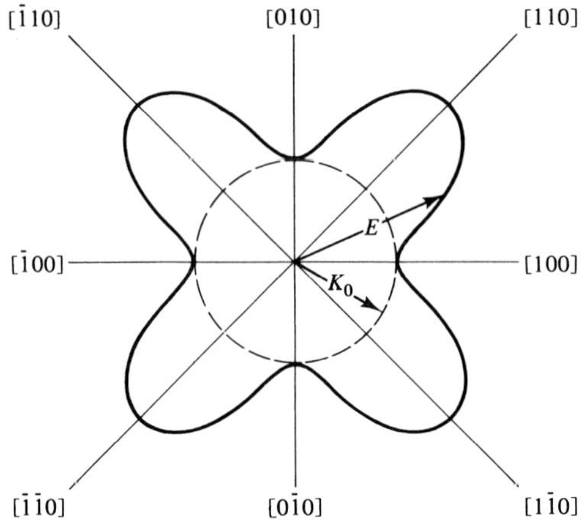

Fig. 7.14 Polar plot of the crystal anisotropy energy E as a function of direction in the (001) plane. K_1 is positive and equal to 5 K_0.

If a disk is cut parallel to {110}, as in Fig. 7.1, it will have three important crystal directions in its plane, and both K_1 and K_2 will affect the torque curve. If M_s is in the ($1\bar{1}0$) plane of Fig. 7.1 and at an angle θ to [001], the direction cosines of M_s are $\alpha_1 = \alpha_2 = (\sin\theta)/\sqrt{2}$ and $\alpha_3 = \cos\theta$. Equation (7.1) then becomes

$$E = K_0 + (K_1/4)(\sin^4\theta + \sin^2 2\theta) + (K_2/4)(\sin^4\theta\cos^2\theta). \tag{7.15}$$

When this equation is differentiated to find the torque, the result is an equation in powers of $\sin\theta$ and $\cos\theta$. This may be transformed into an equation in the sines of multiple angles:

$$L = -dE/d\theta = -(K_1/4 + K_2/64)\sin 2\theta - (3K_1/8 + K_2/16)\sin 4\theta + (3K_2/64)\sin 6\theta. \tag{7.16}$$

This form of the equation shows immediately the various components of the torque: the term in $\sin 2\theta$ is the uniaxial component (like Eq. 7.6), the term in $\sin 4\theta$ is the biaxial component (like Eq. 7.14), etc. Figure 7.15 shows the form of the torque curve obtained by Williams [7.8] on a {110} disk cut from a crystal of 3.85 percent silicon iron (iron containing 3.85 percent silicon in solid solution). The points are experimental, and the curve is a plot of Eq. (7.16) with values of the constants chosen to give the best fit. (The reader will encounter several references to research studies on single-crystal alloys of iron and 3 to 4 percent silicon. The main reason for selecting such an alloy is that it is "ironlike" – it has ⟨100⟩ easy directions – and its behavior thus sheds light on the behavior of iron, and yet it is far easier to prepare in single-crystal form, for reasons given in Section 13.4. In addition, these alloys have wide applications for the magnetic cores of electrical machines.)

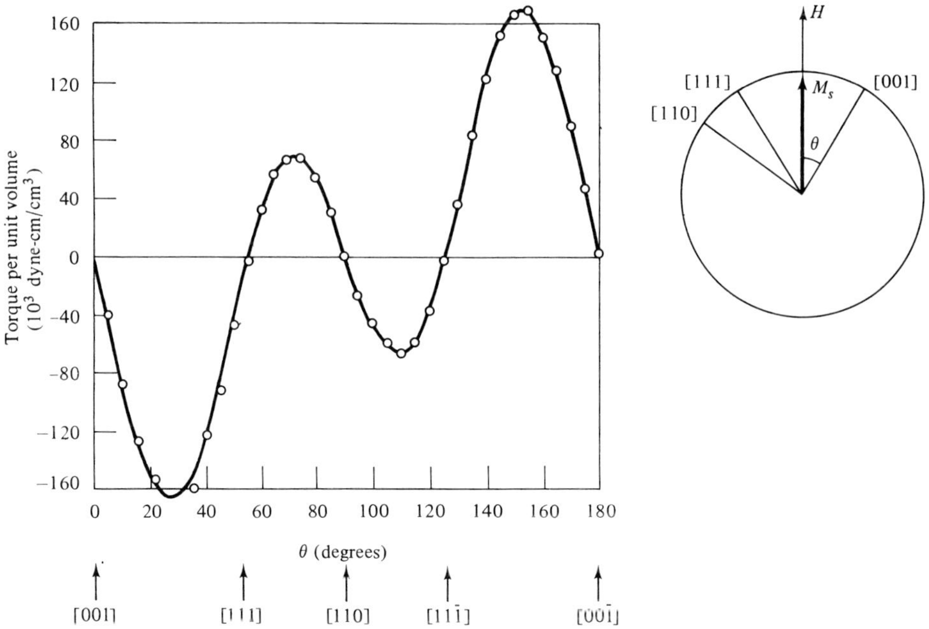

Fig. 7.15 Torque measurements on a {110} disk of an Fe-Si crystal. The curve is drawn for $K_1 = 2.87 \times 10^5$ ergs/cm^3 and $K_2 = 1.00 \times 10^5$ ergs/cm^3. After Bozorth [G.4], from measurements by Williams [7.8].

If a disk is cut parallel to {111} and M_s makes an angle θ with a ⟨110⟩ direction, the crystal anisotropy energy may be written

$$E = K_0 + K_1/4 + (K_2/108)(1 - \cos 6\theta). \tag{7.17}$$

The torque is then

$$L = -dE/d\theta = -(K_2/18)\sin 6\theta. \tag{7.18}$$

The torque curve is a simple sine curve, repeating itself every 60°, with an amplitude of $K_2/18$. In principle, a disk cut parallel to {111} is thus a better specimen for the determination of K_2 than one cut parallel to {110}, because the torque on a {111} specimen is determined only by K_2, whereas the torque on a {110} specimen is a function of both K_1 and K_2. In practice, however, {111} disks often yield distorted sin 6θ curves, presumably because slight misorientation of the specimen gives a relatively large contribution from K_1. In general, accurate K_2 values are not easy to obtain from torque measurements, and the values reported in the literature are often unreliable.

Fourier analysis of experimental torque curves offers a means of sorting out the various contributions to the observed torque. The torque is expressed as a Fourier series:

$$L = A_1 \cos y + A_2 \cos 2\theta + \cdots + B_1 \sin \theta + B_2 \sin 2\theta + \cdots \tag{7.19}$$

Fourier analysis of the experimental curve then yields the values of the various coefficients A_n and B_n, which in turn describe the kinds of anisotropy present. For example, suppose a torque curve is obtained from a disk cut parallel to $\{100\}$ in a cubic crystal. If conditions are perfect, Fourier analysis of the experimental curve would show that all Fourier coefficients are zero except B_4, because Eq. (7.14) shows that the torque varies simply as sin 4θ. But suppose that the specimen or torque magnetometer, or both, is slightly misaligned and that this misalignment introduces a spurious uniaxial component into the torque curve. This will be reflected in a nonzero value for B_2, because Eq. (7.6) shows that uniaxial anisotropy causes a sin 2θ variation of torque. The nonzero value of B_2, in this particular example, discloses the misalignment, while the value of B_4 yields the quantity desired, namely, $K_1 = -2B_4$.

Slight experimental imperfections do not distort a torque curve so much that its basic character is unrecognizable. Thus, in the example just described, the experimental curve would be somewhat distorted but still recognized as basically similar to the curve of Fig. 7.13, which describes pure biaxial anisotropy. The function of Fourier analysis is then to separate out the spurious torques and leave only the torque due to the crystal itself. But specimens are also encountered in which two, or even three, causes of anisotropy are simultaneously present, and with more or less the same "strength." Fourier analysis of the torque curve then becomes, not merely a means of refining slightly imperfect experimental data, but an absolutely necessary method for disentangling the various components of the anisotropy.

Torsion-pendulum method

This method of measuring anisotropy is described by Rathenau and Snoek [7.9] and in greater detail by Zijlstra [7.10, G.31]. The specimen is a disk, suspended by a torsion wire in the air gap of an electromagnet, just as in a torque magnetometer. Suppose the crystal is uniaxial, with the easy axis c in the plane of the disk. The initial, minimum-energy position of the specimen is one with c parallel to the field H and no twist in the torsion wire. The specimen is then rotated away from H to the position shown in Fig. 7.8, released, and allowed to oscillate back and forth about the field direction at the natural frequency of the suspended system. The experimenter then measures this frequency. When the specimen is in the deflected position, two restoring torques act on it: (1) the torque in the wire, and (2) the crystal anisotropy torque which tries to rotate c back into parallelism with M_s and H. The period T of the oscillatory motion is still given by an equation of the general form of Eq. (7.8) but modified to include the anisotropy torque:

$$\frac{1}{f} = T = 2\pi\sqrt{\frac{I}{C_w + C_s}} \tag{7.20}$$

where f is frequency, I moment of inertia of suspended system, C_w torsion constant of the wire, and C_s the torsional stiffness of the specimen. The quantity C_s is the rate of change of torque with angle and is therefore given by $dL/d\theta$. But $|L| = dE/d\theta$, so that $C_s = d^2E/d\theta^2$, where E is the anisotropy energy. From the measured frequency, C_s may be calculated, because I and C_w are known, and the anisotropy constant may then be determined from C_s. Zijlstra should be consulted for details, particularly for the effect of field strength on the measurement.

Anisotropy can be measured by the torsion-pendulum method in a few seconds. Zijlstra's instrument had a frequency of 10 to 100 cycles/sec, depending on specimen anisotropy, and the frequency was measured by an electrical method. He was able to measure rapid changes in the anisotropy of a permanent-magnet alloy occurring during heat treatment; some 16 anisotropy determinations were made within 5 minutes. Although very fast, the torsion-pendulum method is seldom used, possibly because of the fairly elaborate instrumentation needed for the measurement of the frequency.

7.6 ANISOTROPY MEASUREMENT (BY MAGNETIZATION CURVES)

Anisotropy constants may determined from the magnetization curves of single crystals in two ways:

1. By fitting a calculated magnetization curve to the observed one.
2. By measuring, on an M,H graph, the area included between the magnetization curves for two different crystal directions.

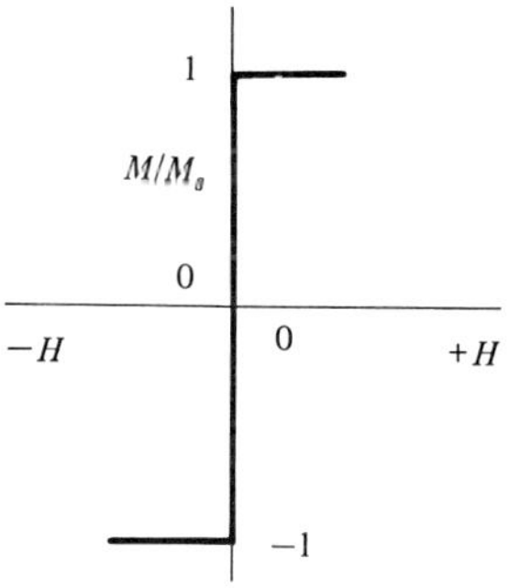

Fig. 7.16 Magnetization curves of a crystal parallel to an easy axis.

First method

This is not very common, but it is instructive to find out how magnetization curves are calculated, because the calculation will tell us something about the magnetization process. In such calculations we ignore everything but the crystal anisotropy forces; that is, we assume that domain walls can be moved with negligibly small fields, but that M_s can be rotated out of the easy direction only by fields strong enough to overcome the anisotropy forces.

The simplest case is that of a crystal magnetized in one of its easy directions, e.g., an iron crystal magnetized in one of the directions ⟨100⟩, as illustrated in Fig. 7.3, or a uniaxial crystal (Fig. 7.6), magnetized parallel to its easy axis. Here the whole process, from the demagnetized state to saturation, occurs by wall motion only, at an (assumed) negligibly small field. The magnetization curve, shown in Fig. 7.16, is simply a vertical line, and the hysteresis "loop" encloses zero area.

When a field is applied to an iron crystal in the [110] direction, wall motion occurs until there are only two kinds of domains left, namely those with M_s vectors in the [010] and [100] directions, the two easy directions closest to the field (Fig. 7.4). The magnetization is then $M = M_s \cos 45° = M_s/\sqrt{2} = 0.71\ M_s$. Fur-

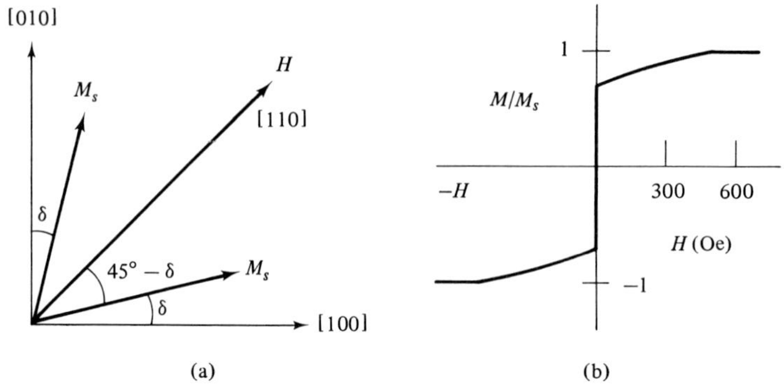

Fig. 7.17 Magnetization of an iron crystal in a [110] direction.

ther increase in field rotates the M_s vectors away from the easy directions by an angle δ in the (001) plane (Fig. 7.17 a). The direction cosines of M_s relative to the crystal axes are then $\alpha_1 = \cos\delta$, $\alpha_2 = \cos(90^\circ - \delta)$, and $\alpha_3 = 0$, for the [100] domains. (These comprise half the volume of the crystal. It is enough to base the following calculation on them alone, because the behavior of the [010] domains is exactly similar.) The anisotropy energy is then, from Eq. (7.13),

$$E_a = K_0 + (K_1/4)\sin^2 2\,\delta.$$

The magnetic potential energy is, from Eq. (1.5),

$$E_p = -M_s H \cos(45^\circ - \delta).$$

The larger the angle δ, the larger is the anisotropy energy and the smaller the potential energy. The angle δ will therefore be such as to minimize the total energy E_t.

$$E_t = K_0 + (K_1/4)\sin^2 2\,\delta - M_s H \cos(45^\circ - \delta).$$

To minimize E_t we put

$$dE_t/d\delta = [K_1 \sin 2\,\delta \cos 2\,\delta] - [M_s H \sin(45^\circ - \delta)] = 0. \tag{7.21}$$

(This problem may be thought of in terms of torques, rather than energies. The first term in the last equation is the torque exerted on M_s by the crystal, the second term is the torque on M_s by the field, and the equation states that these torques are equal and opposite.) The component of M_s in the field direction is the measured magnetization:

$$M = M_s \cos(45^\circ - \delta). \tag{7.22}$$

Eliminating δ from Eqs. (7.21) and (7.22), we find

$$H = \frac{4K_1}{M_s}\,\frac{M}{M_s}\left[\left(\frac{M}{M_s}\right)^2 - \frac{1}{2}\right], \tag{7.23}$$

which gives the field required to reach any given level of magnetization. This field is directly proportional to K_1 and independent of K_2. The field required to saturate in the [110] direction is

$$H = \frac{2K_1}{M_s}. \tag{7.24}$$

Figure 7.17 (b) shows the magnetization curve for iron at room temperature, calculated for $M_s = 1714$ emu/cm^3 and $K_1 = 4.5 \times 10^5$ ergs/cm^3.

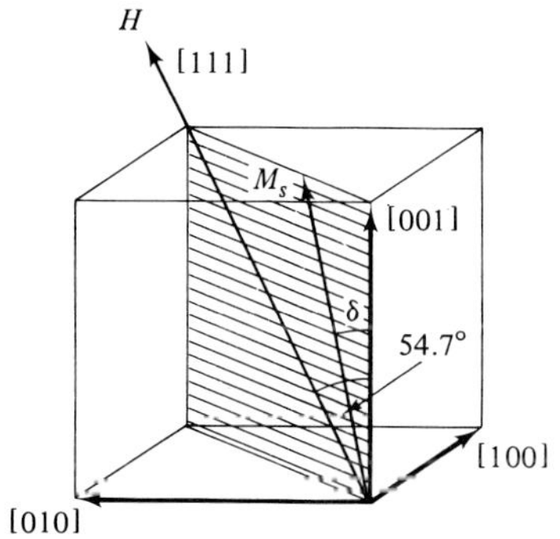

Fig. 7.18 Magnetization of an iron crystal in a [111] direction.

The magnetization curve of an iron crystal, or any crystal having $\langle 100 \rangle$ easy directions, in the [111] direction is calculated in similar fashion. Wall motion in low fields will eliminate all but three kinds of domains — [100], [010], and [001] — and M_s in each will be equally inclined, at 54.7°, to the [111] field direction. The magnetization will then be $M = M_s \cos 54.7° = M_s/\sqrt{3} = 0.58 M_s$. Further increase in field will rotate the M_s vectors in $\{110\}$ planes, as shown for one of these in Fig. 7.18. The equation for the magnetization curve, as given by Brailsford [G.30], is complex:

$$\begin{aligned} HM_s = {} & (K_1/3)\left[\sqrt{2 - 2m^2}\,(4m^2 - 1) + m(7m^2 - 3)\right] \\ & - (K_2/18)\left[\sqrt{2 - 2m^2}\,(10m^4 - 9m^2 + 1) - m(23m^4 - 16m^2 + 1)\right], \end{aligned} \tag{7.25}$$

where $m = M/M_s$. The field required to saturate in the [111] direction is

$$H = \frac{4(3K_1 + K_2)}{9M_s}. \tag{7.26}$$

For a crystal like iron, this is less than the field required for saturation in the [110] direction, in agreement with the experimental results shown in Fig. 7.2 (a).

It is also of interest to calculate the magnetization of a uniaxial crystal like cobalt, when the field is applied in a direction at right angles to the easy axis. When the field is strong enough to rotate M_s away from the easy axis by an angle θ, the anisotropy energy is, from Eq. (7.3),

$$E_a = K_0 + K_1 \sin^2\theta + K_2 \sin^4\theta.$$

The magnetic potential energy is

$$E_p = -M_s H \cos(90^\circ - \theta).$$

The condition for minimum total energy is

$$2K_1 \sin\theta \cos\theta + 4K_2 \sin^3\theta \cos\theta - M_s H \cos\theta = 0.$$

Also,

$$M = M_s \cos(90^\circ - \theta).$$

Elimination of θ from these two equations gives

$$H = \frac{2K_1}{M_s}\left(\frac{M}{M_s}\right) + \frac{4K_2}{M_s}\left(\frac{M}{M_s}\right)^3, \tag{7.27}$$

with saturation being attained a field of

$$H = \frac{2K_1 + 4K_2}{M_s}. \tag{7.28}$$

If K_2 is zero, the magnetization curve becomes a straight line,

$$H = \frac{2K_1 M}{M_s^2}, \tag{7.29}$$

and the saturating field becomes

$$H = \frac{2K_1}{M_s}, \tag{7.30}$$

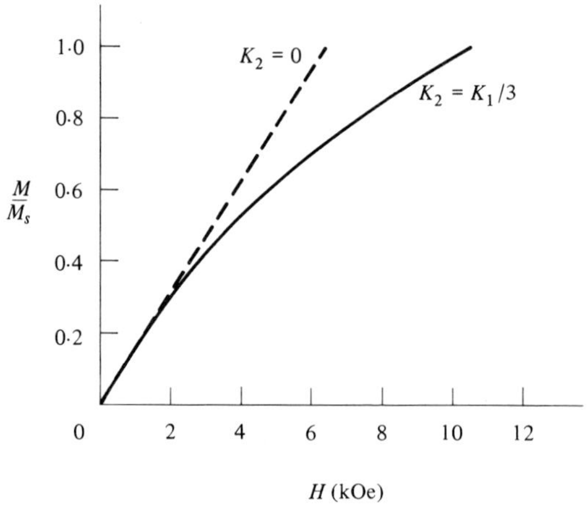

Fig. 7.19 Calculated magnetization curve for cobalt, with field at right angles to easy axis (full line).

Figure 7.19 shows the room-temperature magnetization curve for cobalt, calculated for $M_s = 1422$ emu/cm^3, $K_1 = 4.5 \times 10^6$ ergs/cm^3, and $K_2 = 1.5 \times 10^6$ ergs/cm^3. Fields in excess of 10,000 Oe are needed for saturation. The dashed line shows the magnetization behavior if K_2 were zero.

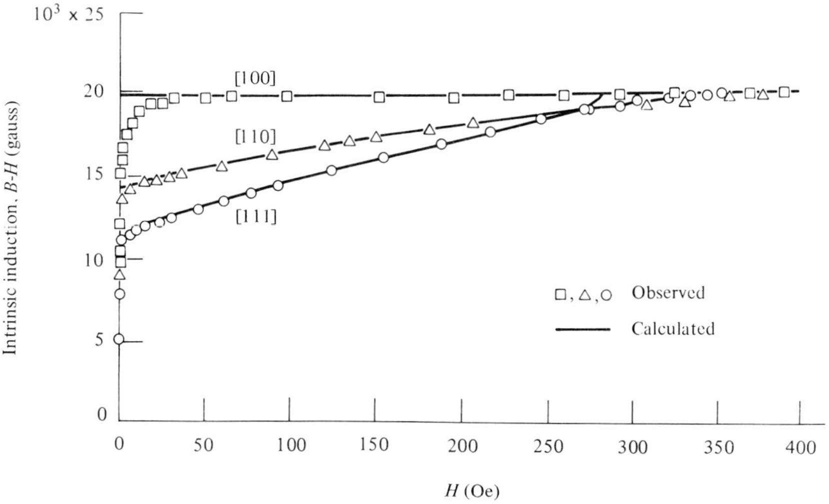

Fig. 7.20 Calculated and observed magnetization curves for Fe-Si crystals. After Bozorth [G.4], from measurements by Williams [7.8].

For iron, nickel, and cobalt crystals the general features of the experimental magnetization curves are well reproduced by the calculated ones. Figure 7.20 shows this kind of a comparison for crystals of 3.85 percent silicon iron. The that finite fields are required for saturation in the easy direction, evident in Figs. 7.2, 7.5, and 7.20, shows that domain walls encounter obstacles to their easy motion; this topic will be pursued in Chapter 9.

Up to this point we have examined the magnetization curve of a single crystal only when the field was applied parallel to an axis of symmetry in the crystal. If the field is not so directed, the magnetization process becomes more complicated. The demagnetizing field H_d must then be considered and a distinction made between the applied field H_a and the true field H, equal to the vector sum of H_a and H_d, inside the specimen. H and M are no longer always parallel, and neither is always parallel to H_a. This more complex kind of magnetization has been analyzed by Lawton and Stewart [7.11] for an iron crystal and by Barnier, Pauthenet, and Rimet [7.12] for a cobalt crystal.

Second method

This method of determining anisotropy constants from magnetization curves is based directly on the definition of the anisotropy energy E, namely, the energy stored in a crystal when it is magnetized to saturation in a non-easy direction. If we can determine W, the work done on the crystal to bring it to saturation, we can equate E and W and so determine the anisotropy constants.

One way of finding an expression for W is to calculate the electrical work done in magnetizing a rod specimen by means of a current in a solenoidal coil wound on the rod. Assume that the rod is so long that the demagnetizing field can be neglected. Let the rod be of length l and cross-sectional area A, wound with n

turns. When the current increases by an amount di, the induction increases by dB and the flux by $d\phi = AdB$. This change in flux causes a back emf e in the coil, and work must be done to overcome this emf. (We neglect the work done per second in producing heat in the coil, equal to i^2R, where R is the resistance. This work contributes nothing to the magnetization of the rod.) The total work done in time dt is

$$VdW = ei\,dt \text{ joules,} \tag{7.31}$$

where V is the volume of the rod and W the work per unit volume. From Eq. (2.6) we have

$$e = 10^{-8}\,n\frac{d\phi}{dt} = 10^{-8}\,n\,A\frac{dB}{dt} \text{ volts.} \tag{7.32}$$

The field produced by the current is, from Eq. (1.12),

$$H = \frac{4\pi\,ni}{10\,l} \text{ Oe.} \tag{7.33}$$

Combining the last three equations and noting that $V = Al$, we obtain

$$dW = \frac{10^{-7}}{4\pi} HdB \text{ joules/cm}^3 = \frac{HdB}{4\pi} \text{ ergs/cm}^3. \tag{7.34}$$

Then the work done per unit volume in changing the induction from 0 to B is

$$\boxed{W = \frac{1}{4\pi}\int_0^B HdB} \text{ ergs/cm}^3, \tag{7.35}$$

where H is in oersteds and B in gauss. Because $B = H + 4\pi M$, at the same field $dB = 4\pi\,dM$, and

$$\boxed{W = \int_0^M HdM} \text{ ergs/cm}^3, \tag{7.36}$$

where M is in emu/cm^3. The work done in magnetization is simply the area between the M,H curve and the M axis, shown shaded in Fig. 7.21 (a).

It is convenient to mention here the energy loss due to hysteresis, although hysteresis has nothing to do with the measurement of anisotropy from magnetization curves. In fact, complete reversibility (no hysteresis) is usually assumed in such measurements. But if hysteresis is present, then removal of the magnetizing field will return energy equal to the area shaded in Fig. 7.21 (b) to the magnetizing circuit. The energy stored in the material at its remanence point M_r is the area shaded in (c).

When a specimen is made to go through one complete cycle, the total work done on the specimen is the *hysteresis loss* W_h, which is equal to $1/4\pi$ times the area enclosed by the B,H loop, shown in Fig. 7.21 (d). This work appears as heat in the specimen.

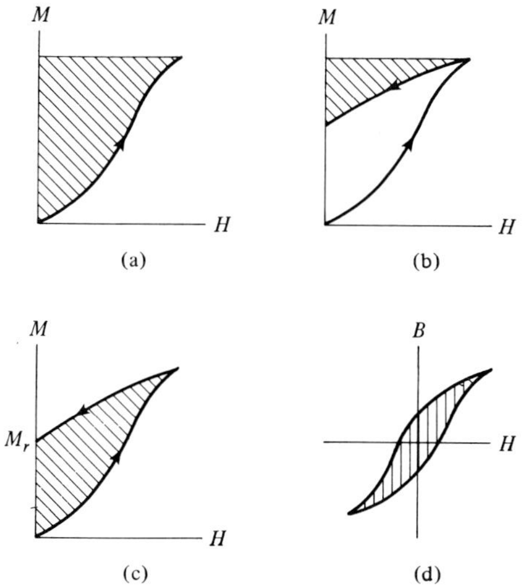

Fig. 7.21 Work done in magnetization.

[For a substance of constant permeability μ, such as a dia-, para-, or antiferromagnetic, $B = \mu H$, $dB = \mu\, dH$, and Eq. (7.35) becomes

$$W = \frac{\mu H^2}{8\pi}. \tag{7.37}$$

If the substance is air, then $\mu \approx 1$, and

$$W = \frac{H^2}{8\pi}. \tag{7.38}$$

This is the energy stored in a magnetic field in air, per unit volume.]

Another, more direct, way of arriving at Eq. (7.36) is given by Trevena [7.13]. The magnetization of a specimen can increase either by domain rotation or domain wall motion, or both. Actually, both mechanisms have the same result, because the spins in the region swept out by a moving wall have their orientation rotated through a definite angle.

Consider a small volume of the specimen with a magnetic moment m oriented at an angle θ to the magnetizing field H. This moment has a component $m \cos \theta$ parallel to the field. Summing over unit volume of the specimen, we have

$$\sum m \cos \theta = M. \tag{7.39}$$

When the field increases from H to $H + dH$, the moment of the small volume considered will rotate from orientation θ to $\theta - d\,\theta$. The work done by the field is (couple)(angle) = $(mH \sin \theta)(-d\,\theta)$. (See Section 1.3). Summed over unit volume, the work done is

$$dW = -\sum mH \sin \theta\, d\theta. \tag{7.40}$$

From Eq. (7.39),

$$dM = d(\sum m \cos \theta) = -\sum m \sin \theta\, d\theta. \tag{7.41}$$

Combination of Eqs. (7.40) and (7.41) gives

$$dW = H\, dM.$$

Table 7.3 Anisotropy Constants of Fe +3.85 percent Si

Method	K_1	K_2
	(10^5 ergs/cm^3)	
Torque curves	2.87	1.00
Fitting magnetization curves	2.80	1.00
Areas between magnetization curves	2.72	1.50

If W is the area between a particular M,H curve and the M axis, then W equals the anisotropy energy E stored in a crystal magnetized in that particular direction. We have already worked out these energies for cubic crystals, and they appear in Table 7.1. Therefore,

$$\left.\begin{aligned} W_{100} &= E_{100} = K_0, \\ W_{110} &= E_{110} = K_0 + K_1/4, \\ W_{111} &= E_{111} = K_0 + K_1/3 + K_2/27. \end{aligned}\right\} \qquad (7.42)$$

These equations may be solved for the anisotropy constants:

$$\left.\begin{aligned} K_0 &= W_{100}, \\ K_1 &= 4(W_{110} - W_{100}), \\ K_2 &= 27(W_{111} - W_{100}) - 36(W_{110} - W_{100}). \end{aligned}\right\} \qquad (7.43)$$

Here an expression like $(W_{110} - W_{100})$ is to be understood as the area included between the M,H curves for the [110] and [100] directions. As mentioned earlier, experimental M,H curves in the easy direction usually show a finite area between the curve and the M axis, indicating that the field has had to overcome hindrances to domain wall motion. These hindrances can be considered roughly constant for any direction of the applied field relative to the crystal axes. Therefore, equations like (7.43), which are based on the area *between* certain curves, yield anisotropy constants which should be essentially free of any disturbing effects due to difficult wall motion.

It is rare to find more than one experimental method used in a single investigation, and the literature therefore contains few comparisons of rival techniques. Williams, however, measured the anisotropy constants of his silicon iron crystals by three methods, with the results shown in Table 7.3 [7.8]. Differences between the three results reflect, not only experimental error, but also differences in what is actually being measured. The torque measurement is probably the most fundamental, in that a high-field torque measurement involves only the rotation of M_s relative to the axes of (ideally) a single-domain crystal; no wall motion is included.

In concluding these two sections on measurements, it is worth noting that the crystal anisotropy forces which bind the spontaneous magnetization M_s of any domain to an easy direction can also be expressed in an indirect but sometimes useful way that does not explicitly involve anisotropy constants. This way is in terms of a fictitious *anisotropy field* H_K, which is in turn related to the usual constants. The anisotropy field is parallel to the easy direction and of such a magnitude that it exerts the same torque on M_s, for small angular deviations θ of M_s from the easy direction, as the crystal anisotropy itself. The torque due to the anisotropy field is $H_K M_s \sin\theta$ or $H_K M_s \theta$, for small values of θ. The torque due to crystal anisotropy depends on the crystal structure and easy axis direction. For example, in a cubic crystal with $\langle 100 \rangle$ easy directions, the torque exerted on M_s by the crystal is, from Eq. (7.14), $+ (K_1/2) \sin 4\theta$, or $2K_1\theta$ for small θ. Equating these torques, we have

$$H_K M_s \theta = 2K_1 \theta,$$

$$H_K = \frac{2K_1}{M_s} . \tag{7.44}$$

If $\langle 111 \rangle$ is the easy direction, the reader can show that

$$H_K = \frac{-4(3K_1 + K_2)}{9M_s}. \tag{7.45}$$

The last two equations are valid whatever the plane of rotation of M_s away from the easy direction. For a uniaxial crystal we find, through Eq. (7.4), that

$$H_K = \frac{2K_1}{M_s}. \tag{7.46}$$

From Eq. (7.30) this is also the value of the field, applied at 90° to the easy axis, that is required to saturate when K_2 is zero. Thus for a uniaxial crystal the anisotropy field has the added physical significance of the field, applied at 90° to the easy axis, which can completely overcome the anisotropy forces by rotating M_s through 90°.

7.7 ANISOTROPY CONSTANTS

Table 7.4 shows the values of the room temperature anisotropy constant for Fe, Co, Ni, and some of the more common ferrites. Uniaxial substances, such as those with a hexagonal crystal structure, commonly have larger anisotropy constants than cubic materials. The extremely large anisotropy of YCo_5 is typical of that of other RCo_5 phases, where R is Y (yttrium) or a rare-earth element.

Table 7.4 Anisotropy Constants

Structure	Substance	K_1	K_2
		(10^5 ergs/cm^3)	
Cubic	Fe	4.8	± 0.5
	Ni	-0.5	-0.2
	$FeO \cdot Fe_2O_3$	-1.1	
	$MnO \cdot Fe_2O_3$	-0.3	
	$NiO \cdot Fe_2O_3$	-0.62	
	$MgO \cdot Fe_2O_3$	-0.25	
	$CoO \cdot Fe_2O_3$	20.	
Hexagonal	Co	45.	15.
	$BaO \cdot 6Fe_2O_3$	33.	
	YCo_5	550.	
	MnBi	89.	27.

The reader will also note that the values of these constants are known only to two significant figures, or even less, and that the accuracy of K_2 is less than that of K_1. Furthermore, the results of different investigators are often in poor agreement. This may appear surprising, in that it is natural to think of crystal anisotropy as a structure-insensive property (and therefore reproducible from one specimen to another), inasmuch as it is due to spin-orbit coupling. However, one is forced to conclude that, although crystal anisotropy may be fundamentally structure-insensitive, various crystal imperfections (residual strain, lattice vacancies, inclusions, porosity, etc.) interfere with its accurate measurement in such a way that crystal anisotropy becomes effectively structure-sensitive.

Anisotropy constants almost always decrease as the temperature increases and become essentially zero even before the Curie temperature is reached. There is then no preferred crystallographic direction for the magnetization of a domain. Figure 7.22 shows the behavior of iron and nickel and Fig. 7.23 that of cobalt. (The data for iron, which are quite old, are in poor agreement with the room-temperature values given in Table 7.4.) The thermal variation of K_1 and K_2 for cobalt is such that the easy axis is (a) parallel to the c axis up to 245°C, (b) inclined to the c axis, but having any azimuthal position about it, at an angle θ which increases from 0 to 90° as the temperature increases from 245° C to 325° C, and (c) in any direction in the basal plane above 325° C.

The anisotropy constants of alloys vary markedly with composition. These values will be given later when particular alloy systems are discussed.

7.8 POLYCRYSTALLINE MATERIALS

If the constituent crystals (grains) of a polycrystalline body are oriented randomly in space, a rather rare circumstance, then the anisotropy of the individual grains

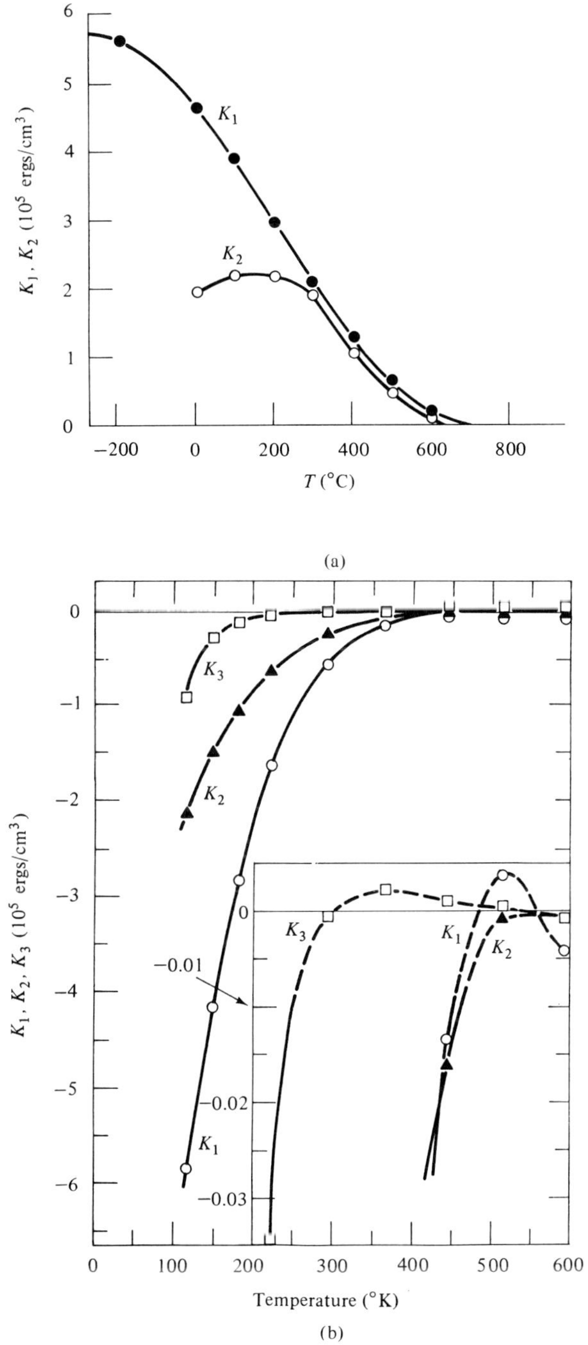

Fig. 7.22 Variation of anisotropy constants with temperature; (a) iron, after Bozorth [7.14] from measurements by Honda *et al.* [7.15], and (b) nickel, after Aubert [7.16].

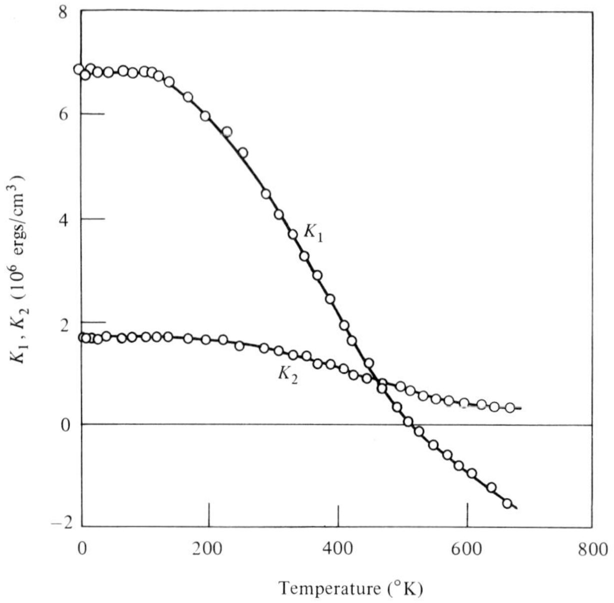

Fig. 7.23 Variation of anisotropy constants of cobalt with temperature, after Barnier *et al.* [7.12].

will average out, and the body on the whole will exhibit no crystal anisotropy. If, on the other hand, the crystals have a *preferred orientation*, also called a *texture*, then the polycrystalline aggregate itself will have an anisotropy dictated by that of the individual crystals.

The kind of texture possessed by a body depends on its shape and how it was formed [4.9], [7.24]. Thus a round wire, rod, or bar has a *fiber texture:* each grain has a certain crystallographic direction $\langle u\, v\, w \rangle$ parallel, or nearly parallel, to a single direction called the *fiber axis*, which, in the case of a wire, coincides with the wire axis. However, the grains can have any rotational position about this axis. Iron wire, for example, has a $\langle 110 \rangle$ fiber texture. (Double textures are also possible. Wires of some face-centered-cubic metals have a $\langle 111 \rangle$ + $\langle 100 \rangle$ texture; some grains have $\langle 111 \rangle$ directions parallel to the wire axis, others have $\langle 100 \rangle$.) Electrodeposited or evaporated layers, deposited on a flat surface, also have a fiber texture, but here the fiber axis is normal to the surface of deposition.

Sheet made by rolling has a texture in which, in each grain, a certain plane $\{h\, k\, l\}$ is parallel to the sheet surface and a certain direction $\langle u\, v\, w \rangle$ in that plane is parallel to the direction in which the sheet was rolled. Such *sheet textures* are described by the symbolism: {rolling plane} $\langle$rolling direction$\rangle$. Thus the so-called "cube texture," found in some metals and alloys, is $\{100\}$ $\langle 001 \rangle$; a cube plane $\{100\}$ is parallel to the sheet surface, and a cube-edge direction $\langle 001 \rangle$ is parallel to the rolling direction.

Textures are also distinguished by whether they are formed during deformation (deformation textures) or during a recrystallization heat treatment (recrystallization textures).

Crystal anisotropy is often exploited in the manufacture of magnetic materials by inducing a texture such that the easy directions of magnetization in all grains are parallel. The polycrystalline body as a whole then has an easy direction. Some control of the degree and, to a lesser extent, of the kind of preferred orientation is possible in metals and alloys formed by the usual processes of casting and working by rolling or wire drawing. But the metallurgist has not yet learned how to produce a particular required texture at will. Thus both the deformation and recrystallization textures of iron wire are $\langle 110 \rangle$, and it stubbornly resists any attempts to rotate the $\langle 100 \rangle$ easy directions into parallelism with the wire axis. In fact, efforts to make marked changes of any kind in the $\langle 110 \rangle$ texture, including recrystallization in a magnetic field [7.17], have so far been unsuccessful.

On the other hand, control of easy-axis orientation is relatively easy when the manufacturing operation is one of pressing and sintering a powder, metallic or nonmetallic. It is simply a matter of applying a strong magnetic field in the required direction during the pressing operation. When the powder particles are still a "loose" assemblage, the field automatically lines them up with their easy axes parallel to one another and the field; the compacting die then locks this preferred orientation together as the powder is compressed. An important limitation is that each particle of the powder must be a single crystal; if not, no alignment will occur.

(The aligning effect of a magnetic field is also brought into play in a method for *determining* the easy axis of a material, when a single-crystal specimen is not available. A powder is prepared by grinding or filing, annealed, and mixed with a solution of a binder so that it forms a powder suspension. A few drops are then placed on a glass slide, or other flat plate, and allowed to dry in the presence of a magnetic field applied normal to the slide surface. The dried powder specimen on the slide then has a marked preferred orientation, with easy axes $\langle u\,v\,w \rangle$ normal to the slide surface. It is then examined in an x-ray diffractometer in the usual way, i.e., with the incident and diffracted beams making equal angles with the slide surface. Under these circumstances, only those grains which have their $\{h\,k\,l\}$ planes parallel to the slide surface can contribute to a particular $h\,k\,l$ reflection. For the field-oriented specimen, certain x-ray reflections will be abnormally strong, namely, those from planes at right angles to the $\langle u\,v\,w \rangle$ easy axis. The direction $\langle u\,v\,w \rangle$ can then be determined from a knowledge of the crystal structure of the specimen. For example, a field-oriented powder specimen of hexagonal barium ferrite ($BaO \cdot 6\,Fe_2O_3$) produces abnormally strong basal-plane $\{0002\}$ reflections. This means that the c axis $\langle 0002 \rangle$ is the easy direction. Smit and Wijn [G.10, p. 190] give details.)

If a polycrystalline specimen in the form of a sheet, for example, has preferred orientation, then a disk cut from the sheet will normally show magnetic anisotropy when examined in a torque magnetometer. If it is a cubic material, with $\langle 100 \rangle$ easy axes, and has the cube texture, then its torque curve will resemble Fig. 7.13, except for possibly smaller amplitudes. (Textures are never perfectly sharp, and

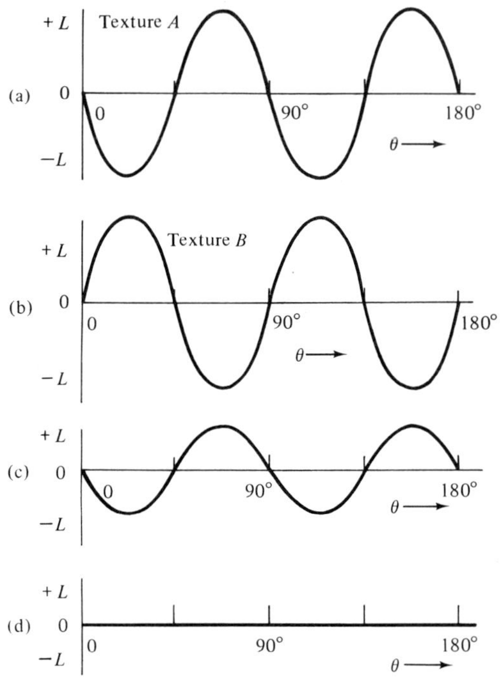

Fig. 7.24 Effect of mixed textures on torque curves, after Becker [7.18].

some scatter of the crystal orientations about the ideal orientation {100} ⟨001⟩ introduces a degree of randomness and therefore decreases the amplitude of the torque-curve peaks.)

While it is thus possible to predict the torque curve of a polycrystalline disk from a knowledge of its texture, the texture cannot be unambiguously deduced from the torque curve, as pointed out by Becker [7.18]. For example, consider again a cubic material with ⟨100⟩ easy axes. If one sheet had the cube texture {100} ⟨001⟩, called texture *A* for short, and another sheet had texture *B*, {100} ⟨011⟩, the corresponding torque curves would resemble (a) and (b) of Fig. 7.24. If a third sheet had a double texture, composed of 75 percent *A* and 25 percent *B*, its torque curve would look like (c), the weighted sum of the (a) and (b) curves. But curve (c) would also be produced by a mixture of 50 percent *A* and 50 percent random orientations. Furthermore, an equal mixture of textures *A* and *B* would produce curve (d), which exhibits no anisotropy, even though the specimen has preferred orientation. (See also Section 7.11.)

Dunn and Walter [7.19] have also shown how misleading a torque curve can be, when an attempt is made to interpret it in terms of preferred orientation. They examined a specimen of silicon steel sheet in a torque magnetometer, and obtained a torque curve like that of a single crystal in the {110} ⟨001⟩ orientation, i.e., like the curve of Fig. 7.15, except that the torque maxima were not so high. This result suggested that some of the grains in the sheet had the {110} ⟨001⟩ orientation and the balance were randomly oriented. They then determined the orientation of all the grains by means of x-ray diffraction. The true texture of the sheet was found to have more than 10 components, *none of which was* {110} ⟨001⟩.

The only sure way of determining the kind and degree of preferred orientation is by x-ray diffraction. On the other hand, even though torque curves are ambiguous, in that more than one kind of texture corresponds to a given curve, torque curves are nevertheless very useful in any work aimed at the development of magnetically useful textures. A torque curve requires much less time than an x-ray texture determination, and it does after all indicate the magnetically easy direction in the sheet. When torque and x-ray measurements are carried out together, the effect of processing variables on texture, and thus on magnetic properties, can be evaluated more quickly than with either technique alone.

7.9 ANISOTROPY IN ANTIFERROMAGNETICS

We digress now from the main subject of strongly magnetic substances to consider unusual effects which occur in some antiferromagnetics. These effects are due to their crystal anisotropy, which is about as strong as in ferro- or ferrimagnetics. In Section 5.2 we noted that the spins of the two sublattices are parallel, in zero applied field, to an important crystallographic axis, labelled D in Fig. 5.2. When a field was applied at right angles to the D axis, we saw that the sublattice magnetizations rotated away from D, as shown in Fig. 5.5 (a), until the reverse molecular field equalled the applied field. Actually, there is another force tending to resist the rotation of the spins, and that is the crystal anisotropy which tends to bind the spin directions to the D axis. Inclusion of an anisotropy term does not alter the main conclusions reached in Section 5.2, but anisotropy forces *are* responsible for the following effects:

1. *Field-dependent susceptibility of powders.* When the specimen is a powder composed of randomly oriented crystals, the angle between the applied field and the D axis takes on all values between 0 and 90°. Under these circumstances the susceptibility increases as the field strength increases, as shown for MnO powder in Fig. 5.8.

2. *Spin flopping.* When a substance of mass susceptibility χ and density ρ is magnetized by a field H, its magnetization M is $\chi\rho H$ and its potential energy in the field is $(-\chi\rho H^2)$, from Eq. (1.5). In an antiferromagnetic below the Néel temperature, $\chi_\perp$ is greater than $\chi_\parallel$, which means that the state with spins at right angles to H is of lower energy than that in which spins are parallel and antiparallel to H. Thus when H is parallel to the spin directions and the D axis, as in Fig. 7.25 (a), there is a tendency for the spin directions to rotate into orientation (b). Counteracting this is the binding of the spin directions to the D axis by the crystal anisotropy forces. As the field increases, a critical value will be reached when these forces are overcome; the spins then "flop over" from orientation (a) to (b), causing a sudden increase in magnetization; further increase in field then rotates the spins slightly from the perpendicular orientation, and M increases with H at a rate governed by the value of $\chi_\perp$. Figure 7.26(a) shows an example of this behavior in single crystals of MnF_2 at 4.2°K. The angles marked on the curves are those between H and the spin direction, which is the c axis in these crystals. (See Fig. 5.12 for the spin structure of MnF_2.) A quantitative treatment of spin flopping and of the effect described under (1) above is given by Morrish [G.27, p. 470].

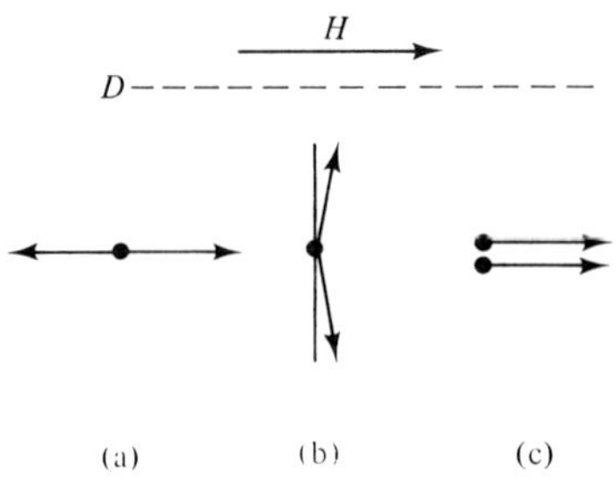

Fig. 7.25 Spin flopping and metamagnetism.

3. *Metamagnetism.* If the anisotropy forces are *very* strong and the field is applied parallel to *D*, the magnetization will first increase with *H* according to the value of $\chi_{\parallel}$. Then, at a high, critical value of *H*, the spins antiparallel to the field will flip over into parallelism, and the substance will saturate. Thus an abrupt transition is made from (a) to (c) of Fig. 7.25; because of the very strong binding of the spin directions to the *D* axis, the intermediate state (b) is not stable. This behavior is known as *metamagnetism*, and Fig. 7.26 (b) shows an example. The specimen of $FeCl_2$ was in the form of a powder with a very high degree of preferred orientation, and the field direction was such that it was almost parallel to the *D* axis in every powder particle. Note that metamagnetic behavior represents a change in *magnetic state* from antiferromagnetic to ferromagnetic, because the final state of parallel spins is, by definition, ferromagnetic. This transition is brought about solely by an increase in field at constant temperature, and it thus differs from the antiferromagnetic-to-ferromagnetic transition which occurs spontaneously in some rare earths on cooling through a critical temperature (see Appendix 3). The conditions governing metamagnetism are discussed by Williams [7.22].

7.10 SHAPE ANISOTROPY

Consider a polycrystalline specimen having no preferred orientation of its grains; it therefore has no crystal anisotropy. If it is spherical in shape, the same applied field will magnetize it to the same extent in any direction. But if it is nonspherical, it will be easier to magnetize it along a long axis than along a short axis. The reason for this is contained in Section 2.6, where we saw that the demagnetizing field along a short axis is stronger than along a long axis. The applied field along a short axis then has to be stronger to produce the same true field inside the specimen. Thus shape alone can be a source of magnetic anisotropy.

In order to treat shape anisotropy quantitatively, we need an expression for the magnetostatic energy E_{ms} of a permanently magnetized body in zero applied field. If a body is magnetized by an applied field to some level *A* (Fig. 7.27) and the applied field is then removed, the magnetization will decrease to *C* under the action of the demagnetizing field H_d. Here *O C* is the demagnetizing-field line, with a slope of $-1/N_d$, where N_d is the demagnetizing coefficient. The specimen then contains stored energy E_{ms} equal to the area of the shaded triangle *OCD*, accord-

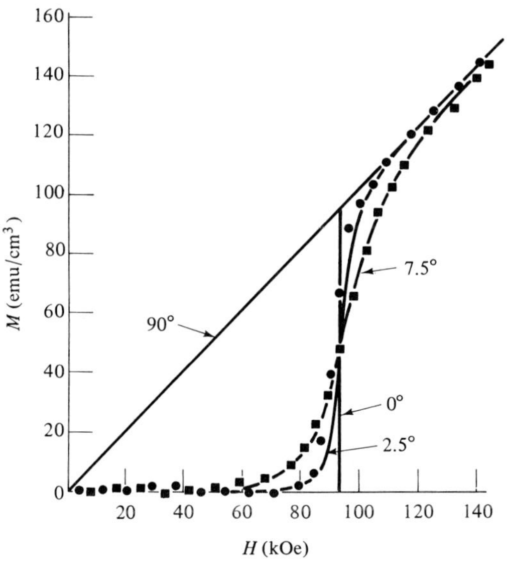

(a)

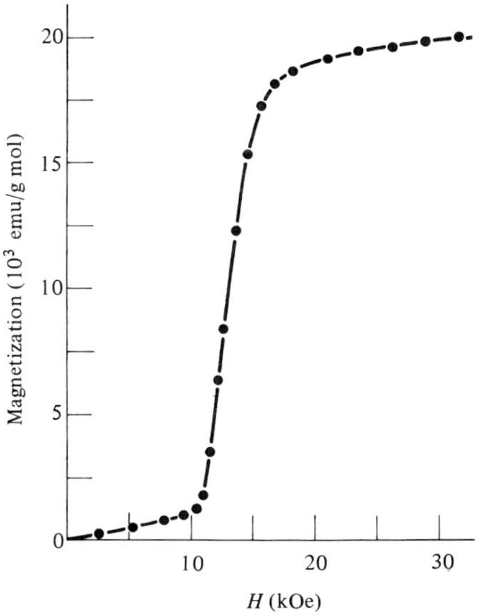

(b)

Fig. 7.26 (a) Spin flopping in MnF_2 at 4.2 K, after Jacobs [7.20]. (b) Metamagnetism in $FeCl_2$ at 13.9 K, after Starr, Bitter, and Kaufmann [7.21].

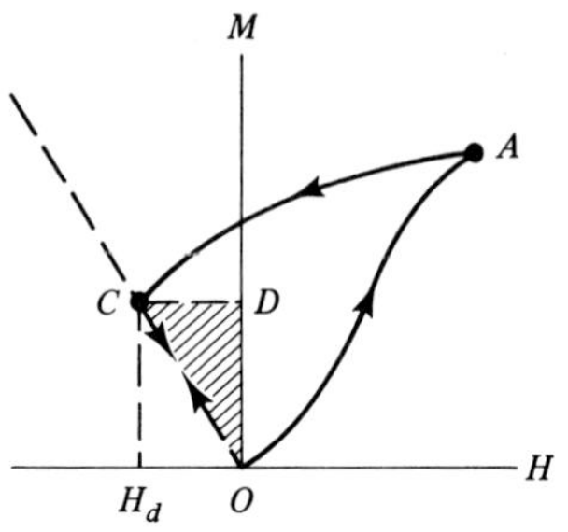

Fig. 7.27 Magnetostatic energy of a magnetized body in zero applied field.

ing to Eq. (7.36). This energy is that associated with the demagnetizing field of the specimen, and is variously called the magnetostatic energy, the self-energy, or the energy of a magnet in its own field. From Eq. (7.38), this energy is

$$E_{ms} = \frac{1}{8\pi} \int H_d^2 \, dv \text{ ergs,} \tag{7.47}$$

where dv is an element of volume and the integration extends over all space. The distribution of H_d in space is seldom known accurately and, even when it is, the evaluation of this integral would be difficult. It is easier to compute the area of the triangle OCD in Fig. 7.27:

$$E_{ms} = \tfrac{1}{2} H_d M \text{ ergs/cm}^3, \tag{7.48}$$

where M is the level of magnetization at point C. This energy can be written in vector form as

$$E_{ms} = -\tfrac{1}{2} \mathbf{H}_d \cdot \mathbf{M}, \tag{7.49}$$

because $\mathbf{H}_d$ is antiparallel to $\mathbf{M}$.

On the other hand, the potential energy per unit volume of a magnet in an *applied* field H_a is, from Eq. (1.6),

$$E_p = -\mathbf{H}_a \cdot \mathbf{M}. \tag{7.50}$$

The expressions for the energy of a magnet in its own field and in an applied field are therefore similar in form, except for the factor $\frac{1}{2}$. Equation (7.48) can be written in terms of N_d, by the substitution $H_d = N_d M$:

$$\boxed{E_{ms} = \tfrac{1}{2} N_d M^2} \text{ ergs/cm}^3. \tag{7.51}$$

(The reasoning leading to Eq. (7.48) is more physically meaningful when one realizes that the point representing the state of the specimen, or part of it, in Fig. 7.27 can be made to move back and forth along the line OC. If C represents the room-temperature state, then the point moves from C to O when the specimen is heated from room temperature to the Curie point. If one now imagines a small volume,

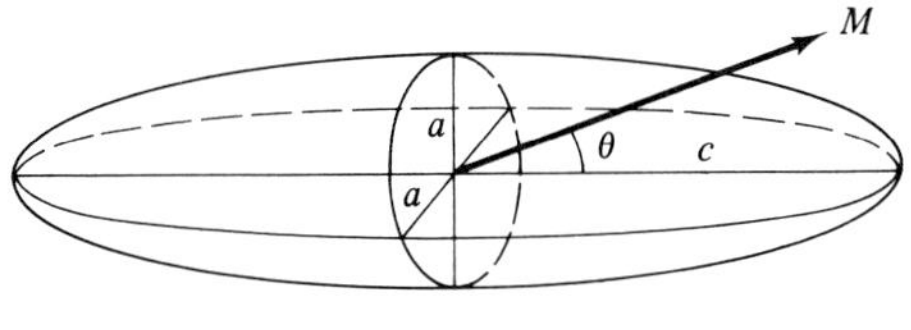

Figure 7.28

smaller that the normal volume of a domain, at a temperature above T_c, then, on cooling through T_c, this small volume becomes spontaneously magnetized to a level M_s which increases as the temperature decreases. Simultaneously, a demagnetizing field H_d is set up. The point representing this small volume therefore moves from O toward C. At room temperature the magnetization of the whole specimen, which is the sum of the M_s vectors in all the domains, is at point C.)

To return to shape anisotropy, we now consider a specimen in the shape of a prolate spheroid (rod) with semi-major axis c and semi-minor axes of equal length a (Fig. 7.28). Let it be magnetized to a level M at an angle θ to c. Then, taking components of M parallel and at right angles to c, we have

$$E_{ms} = \tfrac{1}{2}\,[(M\cos\theta)^2 N_c + (M\sin\theta)^2 N_a], \tag{7.52}$$

where N_c and N_a are demagnetizing coefficients along c and a, respectively, as given by Eqs. (2.20) and (2.21). Substituting $\cos^2\theta = 1 - \sin^2\theta$, we find

$$E_{ms} = \tfrac{1}{2}M^2N_c + \tfrac{1}{2}(N_a - N_c)M^2\sin^2\theta. \tag{7.53}$$

This expression for the magnetostatic energy has an angle-dependent term of exactly the same form as uniaxial crystal anisotropy energy (Eq. 7.4). The long axis of the specimen plays the same role as the easy axis of the crystal, and the shape-anisotropy constant K_s is given by

$$K_s = \tfrac{1}{2}(N_a - N_c)M^2. \tag{7.54}$$

Magnetization is easy along the c axis and equally hard along any axis normal to c. If c shrinks until it equals a, the specimen becomes spherical, $N_a = N_c$, $K_s = 0$, and shape anisotropy disappears. (The long axis of the spheroid is labelled $2c$ in Fig. 7.28 to conform with Fig. 2.27. It is not to be confused with the c axis of a hexagonal crystal.)

Magnetization of an oblate spheroid (disk) (Fig. 2.27) is difficult along the short a axis and equally easy along any axis normal to a, i.e., in the plane of the disk, which is why specimens for crystal-anisotropy measurements are made in the form of disks.

As Eq. (7.54) shows, the "strength" of shape anisotropy depends both on the axial ratio c/a of the specimen, which determines $(N_a - N_c)$, and on the extent

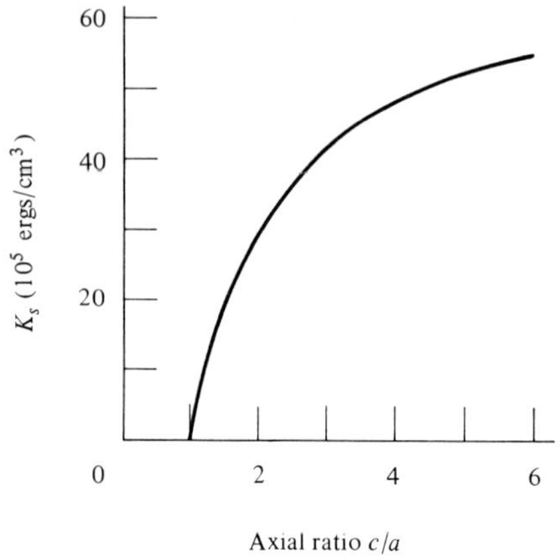

Fig. 7.29 Shape-anisotropy constant K_s for saturated cobalt in the form of a prolate spheroid.

of the magnetization M. To illustrate the sort of numbers involved, we will put $M = 1422$ emu/cm^3, which is the room-temperature saturation magnetization of cobalt, a uniaxial crystal, and calculate the value of the shape-anisotropy constant K_s as a function of c/a for a prolate spheroid of polycrystalline cobalt with no preferred orientation. [Stoner and Wohlfarth [7.23] give values of $(N_a - N_c)$ as a function of c/a for both prolate and oblate spheroids.] Figure 7.29 shows the results. At an axial ratio of about 3.5, K_s is about 45×10^5 ergs/cm^3, which is equal to the value of the first crystal-anisotropy constant K_1 of cobalt. In other words, and neglecting K_2, we can say that a prolate spheroid of saturated cobalt, with $c/a = 3.5$ and without any crystal anisotropy, would show the same uniaxial anisotropy as a spherical cobalt crystal with its usual crystal anisotropy.

7.11 MIXED ANISOTROPIES

The calculation of the last paragraph suggests that we consider a more realistic situation, namely, one in which two anisotropies are present together. The discussion will be limited to uniaxial anisotropies. (We have already touched on the problem of mixed anisotropies in Section 7.8, where the effect of double textures on the anisotropy of polycrystalline sheet was discussed. Here we are interested in the combined effect of two anisotropies of different physical origin, such as crystal and shape anisotropy, on the resultant anisotropy of a single crystal.)

We might have, for example, a rod-shaped crystal of a uniaxial substance like cobalt, with its easy crystal axis at right angles to the rod axis. Will it be easier to magnetize along the rod axis, as dictated by shape anisotropy, or at right angles to the rod axis, as dictated by crystal anisotropy? Both anisotropy energies are given, except for constant terms, by expressions of the form

$$\text{energy} = (\text{constant}) \sin^2 (\text{angle between } M \text{ and easy axis}).$$

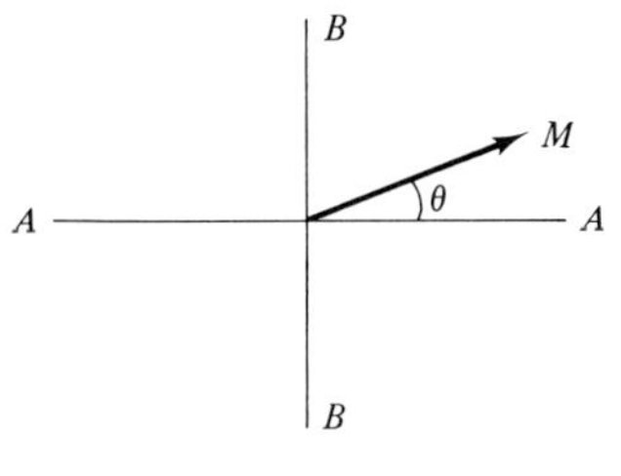

Figure 7.30

The problem is generalized in Fig. 7.30, where AA represents one easy axis and BB the other. The separate anisotropy energies, distinguished by subscripts, are

$$E_A = K_A \sin^2 \theta,$$
$$E_B = K_B \sin^2 (90^\circ - \theta).$$

The total energy is

$$E = K_A \sin^2 \theta + K_B \cos^2 \theta$$
$$= K_B + (K_A - K_B) \sin^2 \theta. \tag{7.55}$$

If the two anisotropies are of equal strength ($K_A = K_B$), then E is independent of angle and there is *no* anisotropy. (Thus two equal uniaxial anisotropies at right angles are not equivalent to biaxial anisotropy.) If they are not equal, we want the value of θ for which E is a minimum:

$$\frac{DE}{d\theta} = (K_A - K_B) \sin 2\theta = 0. \tag{7.56}$$

Solutions are $\theta = 0$ and $\theta = 90^\circ$. To find whether these are minima or maxima, we take the second derivative,

$$\frac{d^2E}{d\theta^2} = 2(K_A - K_B) \cos 2\theta,$$

which must be positive for a minimum. Therefore, $\theta = 0$ is a minimum-energy position if $K_A > K_B$, and $\theta = 90^\circ$ if $K_A < K_B$. The direction of easiest magnetization is not, as might be expected, along some axis lying between AA and BB. The easy direction is along AA if the A anisotropy is stronger, and along BB if B is stronger. The basic reason for this behavior is that a uniaxial anisotropy exerts no torque on M when M is at 90° to the uniaxial axis.

If the two easy axes of Fig. 7.30 are at some angle α to each other, rather than at right angles, the reader can show that they are together equivalent to a new uniaxial axis CC, which either

a) lies midway between AA and BB, if $K_A = K_B$, and has a strength of $K_C = K_A = K_B$, or

b) lies closer to AA, if $K_A > K_B$, and has a strength $K_C > K_A$.

Nesbitt *et al.* [7.25] have examined the problem of mixed anisotropies experimentally. They determined the torque curves of synthetic specimens composed of two, or three, mutually perpendicular wires embedded in plastic in order to understand the torque curves of an alloy containing precipitate particles with shape anisotropy.

PROBLEMS

7.1 Prove the statements made in Section 7.3 regarding the equilibrium values which θ assumes, in a hexagonal crystal, for various relative values of K_1 and K_2.

7.2 a) The torsion wire of a torque magnetometer is made of tungsten, 4 in. long and 0.01 in. in diameter. It is calibrated by the torsion-pendulum method, being affixed to a metal disk weighing 1610 g and having a diameter of 8 in. The time required for 10 oscillations is 242 sec. What is the torsion constant C of the wire, in dyne-cm/degree?

b) The wire is used to measure a weak uniaxial anisotropy, for which the energy is given by $E = K_1 \sin^2 \theta$, with $K_1 = 2000$ ergs/cm^3. The specimen is in the form of a disk, 0.600 in. in diameter and 0.043 in. thick. What is the twist angle ϕ of the wire (in degrees) at the point of maximum amplitude of the torque curve?

7.3 Derive Eq. (7.16).

7.4 Derive Eq. (7.18).

7.5 a) Find relations between M and H for the magnetization in the $\langle 100 \rangle$ and $\langle 110 \rangle$ directions of a cubic crystal like nickel, which has $\langle 111 \rangle$ easy directions. Assume $K_2 = 0$.

b) Compute and plot the magnetization curves of a nickel crystal in the $\langle 100 \rangle$, $\langle 110 \rangle$, and $\langle 111 \rangle$ directions. Take $K_1 = -0.5 \times 10^5$ ergs/cm^3, $K_2 = 0$, and $M_s = 484$ emu/cm^3.

c) What are the fields required to saturate in the $\langle 100 \rangle$ and $\langle 110 \rangle$ directions?

7.6 Calculate the anisotropy energy stored in a cubic crystal saturated in a $\langle 110 \rangle$ direction by finding the area between the M,H curve and the M axis. The easy direction is $\langle 100 \rangle$.

7.7 Derive Eq. (7.45).

7.8 A recrystallized sheet has a double texture, with components A and B present in equal amounts. Component A is (110) $[1\bar{1}0]$ and B is (110) [001]. K_1 is positive and K_2 is zero.

a) What is the nature of the resulting anisotropy in the plane of the sheet?

b) Find an equation for the torque in terms of the angle α between M_s and the rolling direction.

7.9 If the ratios of their long to short axes are equal, which has the greater shape anisotropy, a prolate spheroid (rod) or an oblate spheroid (disk)?

7.10 A cobalt single crystal has the form of an obblate spheroid with a polar (minor) axis, normal to the disk, of $2a$ and an equatorial (major) axis, in the plane of the disk, of $2c$, and

$c/a = 2$. The $\langle 0001 \rangle$ axis of the crystal is normal to the plane of the disk. Take $K_1 = 45 \times 10^5$ ergs/cm^3, $K_2 = 0$, and $M_s = 1422$ emu/cm^3. Determine the direction (normal or parallel to plane of disk) in which it is easier to saturate this crystal by comparing

a) shape and crystal anisotropy constants, and
b) *applied* fields required to saturate.

7.11 In terms of M_s, K_1, and the demagnetizing coefficients N_a and N_c, show that criteria (a) and (b) of the preceding problem lead to the same condition governing the relative ease of magnetization in the two directions. What is this condition?

7.12 At what value of c/a will the disk of Problem 7.10 become as easy to magnetize in one direction as in any other?

CHAPTER 8

MAGNETOSTRICTION AND THE EFFECTS OF STRESS

8.1 INTRODUCTION

When a substance is exposed to a magnetic field, its dimensions change. This effect is called *magnetostriction.* It was discovered as long ago as 1842 by Joule, who showed that an iron rod increased in length when it was magnetized lengthwise by a weak field. The fractional change in length $\Delta l/l$ is simply a strain, and, to distinguish it from the strain ε caused by an applied stress, we give the magnetically induced strain a special symbol λ:

$$\lambda = \frac{\Delta l}{l}. \tag{8.1}$$

The value of λ measured at magnetic saturation is called the saturation magnetostriction λ_s, and, when the word "magnetostriction" is used without qualification, λ_s is usually meant.

The longitudinal, sometimes called Joule, magnetostriction just described is not the only magnetostrictive effect. Others include the magnetically induced torsion or bending of a rod. These effects, which are really only special cases of the longitudinal effect, will not be described here.

Magnetostriction occurs in all pure substances. However, even in strongly magnetic substances, the effect is small: λ_s is typically of the order of 10^{-5}. The smallness of this strain may perhaps be better appreciated if it is translated into terms of stress. If Young's modulus is 30×10^6 lb/in^2, a strain of 10^{-5} would be produced by an applied stress of only $(10^{-5})(30 \times 10^6) = 300$ lb/in^2 or 0.2 kg/mm^2. In weakly magnetic substances the effect is even smaller, by about two orders of magnitude, and can be observed only in very strong fields. We will not be concerned with magnetostriction in such materials.

Although the direct magnetostrictive effect is small there exists an inverse effect (Section 8.5) which causes such properties as permeability and the size of the hysteresis loop to be highly dependent on stress in many materials. Magnetostriction therefore has many practical consequences, and a great deal of research has accordingly been devoted to it.

The value of the saturation longitudinal magnetostriction λ_s can be positive, negative, or, in some alloys, zero. The value of λ depends on the extent of magnetization and hence on the applied field, and Fig. 8.1 shows how λ typically varies with H for a substance with positive magnetostriction. As mentioned in the preceding chapter, the process of magnetization occurs by two mechanisms, domain-wall motion and domain rotation; most of the magnetostrictive change in length usually occurs during domain rotation.

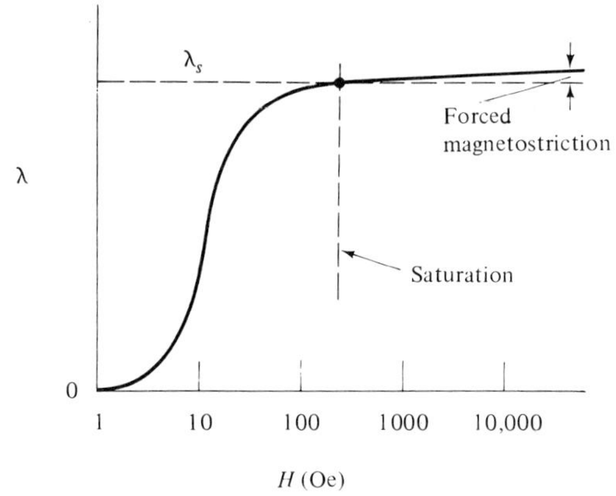

Fig. 8.1 Variation of magnetostriction λ with field H (schematic). Note logarithmic scale of H. [8.17].

Between the demagnetized state and saturation the volume of a specimen remains constant, to a quite good approximation. This means that there will be a transverse magnetostriction λ_t very nearly equal to one-half the longitudinal and opposite in sign, or

$$\lambda_t = -\frac{\lambda}{2}. \tag{8.2}$$

When technical saturation is reached at any given temperature, in the sense that the specimen has been converted into a single domain magnetized in the direction of the field, further increase in field causes forced magnetization (Section 4.2). This causes a slow change in λ with H called *forced magnetostriction*, and the logarithmic scale of H in Fig. 8.1 roughly indicates the fields required for this effect to become appreciable. It is caused by an increase in the degree of spin order which very high fields can produce.

The longitudinal, forced-magnetostriction strain λ shown in Fig. 8.1 is a consequence of a small volume change, of the order of $\Delta V/V = 10^{-10}$ per oersted, occurring at fields beyond saturation and called *volume magnetostriction*. It causes an equal expansion or contraction in all directions.

Forced magnetostriction is a very small effect and has no bearing on the behavior of practical magnetic materials in ordinary fields.

The *measurement* of longitudinal magnetostriction is straightforward. While early investigators had perforce to use mechanical and optical levers to magnify the magnetostrictive strain to an observable magnitude, today this measurement is almost always made with an electrical-resistance strain gage cemented to the specimen. Zijlstra [G.31] gives details. (Less commonly, capacitance or inductance gages are preferred. These have the advantage, for some measurements, that nothing has to be cemented to the specimen. No time is lost in waiting for the cement to dry, and difficulties in bonding strain gage to specimen at high or low temperatures are avoided.) If the applied field is produced by a solenoid, heat from the solenoid can cause thermal expansion of the specimen and introduce serious error. Inasmuch as the expansion coefficient of most metals is of the order of 10^{-5} per °C, a temperature increase of only 1°C can cause a thermal strain of 10^{-5}, which is about equal to the effect being measured. To avoid this error, the specimen temperature must be strictly controlled, or the magnetostriction must be measured so quickly that no appreciable heat flow into the specimen occurs during the time of the measurement. Other, more subtle sources of error, due to uncertainties about the nature of the demagnetized state, are mentioned later.

The subject of magnetostriction has been reviewed in detail by Lee [8.1], Birss [8.2], and Carr [8.3]. Bozorth [G.4] gives a wealth of data on particular metals and alloys. The data on magnetostriction of ferrites are less plentiful; they are summarized by Smit and Wijn [G.10] and Chikazumi [G.23].

8.2 MAGNETOSTRICTION OF SINGLE CRYSTALS

When an iron single crystal is magnetized to saturation in a [100] direction, the length of the crystal in the [100] direction is found to increase. From this we infer that the unit cell of ferromagnetic iron is not exactly cubic, but slightly *tetragonal*. (A tetragonal cell has three axes at right angles; two are equal to each other, and the third is longer or shorter than the other two.) This conclusion follows from what we already know about the changes occurring in an iron crystal during magnetization in a [100] direction. These changes consist entirely in domain wall motion and were described with reference to Fig. 7.3. If the saturated crystal is longer in the direction of its magnetization than the demagnetized crystal, then the single domain which comprises the saturated crystal must be made up of unit cells which are slightly elongated in the direction of the magnetization vector.

The same is true of each separate domain in the demagnetized state. Figure 8.2(a) depicts this state in terms of four sets of domains, [100], [$\bar{1}$00], [010], and [0$\bar{1}$0]. Unit cells are shown by dashed lines; their tetragonality is enormously exaggerated and so is their size relative to the domain size. But the main point to notice is that these cells are all longer in the direction of the local M_s vector than they are in directions at right angles to this vector. Thus, when a region originally occupied by, say, a [0$\bar{1}$0] domain is replaced by a [100] domain, by the mechanism of wall motion, that region must expand in the [100] direction and contract in directions at right angles. The length of the whole crystal therefore changes from l to $l + \Delta l$, where $\Delta l/l = \lambda_s$ = the saturation magnetostriction in the [100] direction.

The unit cell of iron is exactly cubic only when the iron is above the Curie temperature, i.e., only when it is paramagnetic, and subject to no applied field.

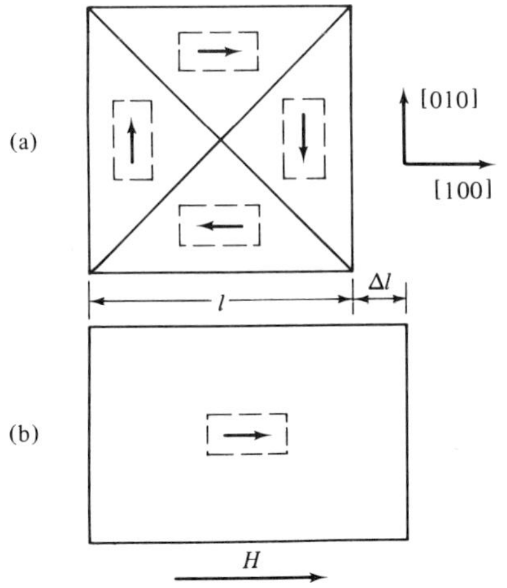

Fig. 8.2 Magnetostriction of an iron crystal in the [100] direction [8.17].

As soon as it cools below T_c, spontaneous magnetization occurs and each domain becomes spontaneously strained, so that it is then made up of unit cells which are slightly tetragonal.

There are therefore two basic kinds of magnetostriction: (1) spontaneous magnetostriction, which occurs in each domain when a specimen is cooled below the Curie point, and (2) forced magnetostriction, which occurs when a saturated specimen is exposed to fields large enough to increase the magnetization of the domain above its spontaneous value. Both kinds are due to an increase in the degree of spin order. The spontaneous magnetostriction is difficult to observe directly, but it *is* evidenced by a local maximum at T_c in the variation of the thermal expansion coefficient with temperature. (This is the reason why Invar, an Fe-Ni alloy containing 36 percent Ni, has such a low expansion coefficient near room temperature. Its Curie point is near room temperature, and its spontaneous magnetostriction varies with temperature in such a way that it almost compensates the normal thermal expansion.) The "ordinary," field-induced magnetostriction which concerns us, and in which λ changes from 0 to λ_s (Fig. 8.1), is caused by the conversion of a demagnetized specimen, made up of domains spontaneously strained in various directions, into a saturated, single-domain specimen spontaneously strained in *one* direction. Figure 8.2 shows one special case of such a conversion, in which the only mechanism of magnetization change is domain wall motion.

Domain walls are described by the angle between the M_s vectors in the two domains on either side of the wall. Two types exist: 180° walls and non-180° walls. The uniaxial crystal of Fig. 7.6 has only 180° walls, while the iron crystal of Fig. 8.2(a) has only 90° walls. Nickel, which has ⟨111⟩ easy directions, can have 180°, 110°, or 71° walls. Non-180° walls are often called 90° walls for brevity, whether the actual angle is 90°, 110°, or 71°. The domain structure of real single crystals is normally such that 180° and non-180° walls both exist.

The magnetostrictive effect of the motion of the two kinds of walls is quite different. Because the spontaneous strain is independent of the *sense* of the magnetization, the dimensions of a domain do not change when the direction of its spontaneous magnetization is reversed. Since passage of a 180° wall through a certain region reverses the magnetization of that region, we conclude that *180° wall motion does not produce any magnetostrictive change in dimensions.* Thus, when the uniaxial crystal of Fig. 7.6 is saturated in the axial direction by an applied field, only 180° wall motion is involved, and the length of the crystal does not change in the process. On the other hand, magnetization of the iron crystal of Fig. 8.2 is accomplished by 90° wall motion, and a change in the length of the crystal does occur.

Rotation of the M_s vector of a domain always produces a dimensional change, because the spontaneous magnetostriction depends on the direction of the M_s vector relative to the crystal axes. Thus, in the general case of a crystal being magnetized in a noneasy direction, the magnetization process will involve 180° and 90° wall motion and domain rotation. The last two of these three processes will be accompanied by magnetostriction.

We now need expressions for the strain which a crystal undergoes in a certain direction when it is magnetized either in the same direction or another one.

Cubic crystals

The saturation magnetostriction λ_{si} undergone by a cubic crystal in a direction defined by the cosines $\beta_1, \beta_2, \beta_3$ relative to the crystal axes, when it changes from the demagnetized state to saturation in a direction defined by the cosines $\alpha_1, \alpha_2, \alpha_3$, is given by

$$\boxed{\begin{aligned}\lambda_{si} &= \tfrac{3}{2}\lambda_{100}\left(\alpha_1^2\beta_1^2 + \alpha_2^2\beta_2^2 + \alpha_3^2\beta_3^2 - \tfrac{1}{3}\right)\\ &\quad + 3\lambda_{111}(\alpha_1\alpha_2\beta_1\beta_2 + \alpha_2\alpha_3\beta_2\beta_3 + \alpha_3\alpha_1\beta_3\beta_1),\end{aligned}} \tag{8.3}$$

where λ_{100} and λ_{111} are the saturation magnetostrictions when the crystal is magnetized, and the strain is measured, in the directions $\langle 100 \rangle$ and $\langle 111 \rangle$, respectively. This equation is valid for crystals having either $\langle 100 \rangle$ or $\langle 111 \rangle$ as easy directions. It is sometimes written in terms of the constants h_1 and h_2, where $h_1 = 3\lambda_{100}/2$ and $h_2 = 3\lambda_{111}/2$. We often wish to know the strain in the same direction as the magnetization; then $\beta_1, \beta_2, \beta_3 = \alpha_1, \alpha_2, \alpha_3$, and Eq. (8.3) becomes

$$\lambda_{si} = \tfrac{3}{2}\lambda_{100}\left(\alpha_1^4 + \alpha_2^4 + \alpha_3^4 - \tfrac{1}{3}\right) + 3\lambda_{111}\left(\alpha_1^2\alpha_2^2 + \alpha_2^2\alpha_3^2 + \alpha_3^2\alpha_1^2\right). \tag{8.4}$$

This can be further reduced, by means of the relation

$$\left(\alpha_1^2 + \alpha_2^2 + \alpha_3^2\right)^2 = \left(\alpha_1^4 + \alpha_2^4 + \alpha_3^4\right) + 2\left(\alpha_1^2\alpha_2^2 + \alpha_2^2\alpha_3^2 + \alpha_3^2\alpha_1^2\right) = 1, \tag{8.5}$$

to the expression

$$\lambda_{si} = \lambda_{100} + 3(\lambda_{111} - \lambda_{100})\left(\alpha_1^2\alpha_2^2 + \alpha_2^2\alpha_3^2 + \alpha_3^2\alpha_1^2\right). \tag{8.6}$$

Equation (8.3) is called the "two-constant" equation for magnetostriction. It is partly empirical, partly theoretical, and its derivation is given by Lee [8.1] and Kittel [8.4], among others. Like the rather similar equation for crystal anisotropy energy (Eq. 7.1), it can be expanded to higher powers of the direction cosines. The next approximation involves five constants; its use is rarely justified by the accuracy of presently available magnetostriction data.

Equation (8.3) and others derived from it require a word of caution. They give the field-induced strain when the crystal is brought from the demagnetized to the saturated state. The saturated state is, by definition, one in which the whole specimen consists of a single domain with its M_s vector parallel to the applied field. The demagnetized state, on the other hand, is by no means well defined. All that is required is that all the domain magnetizations, each properly weighted by its volume, add up vectorially to zero. However, an infinite number of domain arrangements and relative volumes, each resulting in zero net magnetization of the whole specimen, is possible. Equation (8.3) is based on an arbitrary definition of the demagnetized state, namely, one in which *all possible types of domains have equal volumes.* For example, in a cubic crystal like iron, with $\langle 100 \rangle$ easy directions, this state is one in which the total volume of the crystal is divided equally among the six kinds of domains: $[100]$, $[\bar{1}00]$, $[010]$, $[0\bar{1}0]$, $[001]$, and $[00\bar{1}]$. If this "ideal" demagnetized state is not achieved, Eq. (8.3) is invalid, and the measured magnetostriction will be larger or smaller than the calculated value. Differences in the magnetostriction values observed by different investigators for the same metal are usually due to differences in the demagnetized states of their specimens.

The special symbol λ_{si} is attached to Eq. (8.3) and similar ones, where the subscript i refers to the ideal demagnetized state. The reader should note that, to the author's knowledge, this symbol is not used by any other writer on this subject. It is introduced in this book in an attempt to inject greater clarity into the discussion of magnetostrictive strains. The single symbol λ_s is ambiguous, because it is used in the literature to refer both to ideal and nonideal demagnetized states. We therefore have two kinds of saturation magnetostriction:

1. λ_{si}, measured from the ideal demagnetized state. This value is a *constant of the material.* Because it is defined, through Eq. (8.3), in terms of λ_{100} and λ_{111}, the two latter are also λ_{si} values, measured in these particular crystal directions.

2. λ_s, measured on a particular specimen having a particular, nonideal demagnetized state. This value is a property only of that specimen. The quantity λ_s is highly structure sensitive, in that it depends on the mechanical, thermal, and magnetic history of the specimen. If the demagnetized state is nonideal, i.e., if all possible domains are not present in equal volumes, it is said to be a state of *preferred domain orientation.* Thus a preferred *domain* orientation can exist in the demagnetized state of a single crystal, or in the individual grains of a polycrystal, in addition to the preferred *grain* orientation that may exist in a polycrystal.

The domain arrangement shown for the demagnetized crystal in Fig. 8.2(a) is an example of preferred domain orientation, because it contains only four kinds

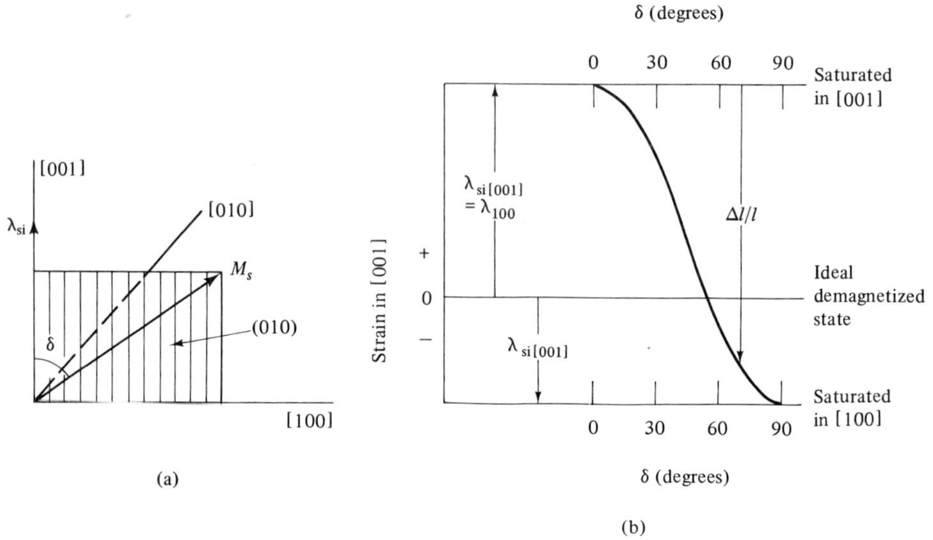

Fig. 8.3 Magnetostrictive strains in a cubic crystal for which λ_{100} is positive.

of domains. If the missing domains [001] and [00$\bar{1}$] were present, the demagnetized crystal would be shorter in both the [100] and [010] directions than the crystal shown. This means, for magnetization and strain measurement both in the direction [100], that the magnetostriction for the ideal state would be *greater* than that for the crystal shown.

Equation (8.3) is of greater utility than may first appear. Properly manipulated, it allows us (a) to calculate the dimensional change of a *single domain* due to a rotation of its M_s vector out of the easy direction, and (b) to circumvent, in magnetostriction measurements, the uncertainty about the demagnetized state. The first application arises from the fact that a saturated single crystal is a single domain. If we compute, by means of Eq. (8.3), the values of λ_{si} for two different orientations of M_s in the saturated state, then the difference between these two values is the strain undergone by the saturated, single-domain crystal when M_s rotates from one orientation to the other. For example, suppose we wish to know the length of a cube-edge direction $\langle 100 \rangle$ in a single domain changes as the M_s vector rotates away from it. In Fig. 8.3(a) let M_s rotate away from [001] by an angle δ in the plane (010). The direction cosines of M_s are $\alpha_1 = \cos(90° - \delta) = \sin\delta$, $\alpha_2 = 0$, $\alpha_3 = \cos\delta$. We wish to know the strain along the [001] direction; therefore, $\beta_1 = \beta_2 = 0$ and $\beta_3 = 1$. Substituting these values into Eq. (8.3), we find

$$\lambda_{si}(\delta = \delta) = \tfrac{3}{2}\lambda_{100}(\cos^2\delta - \tfrac{1}{3}). \tag{8.7}$$

This is the strain along [001] which occurs when an ideally demagnetized crystal

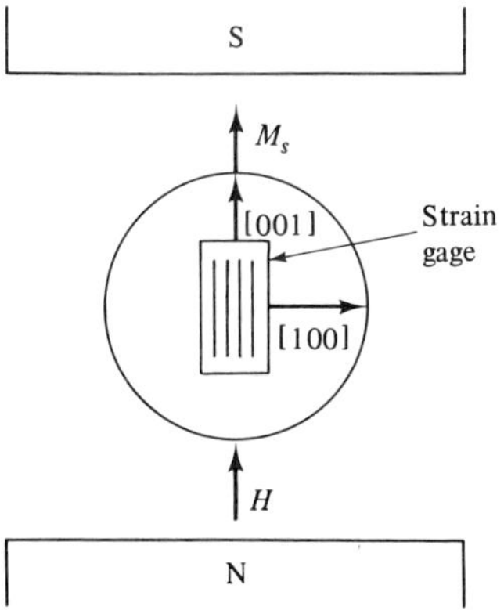

Fig. 8.4 Magnetostriction measurements on a single crystal.

is saturated in the direction δ. When $\delta = 0$, this expression reduces to

$$\lambda_{si}(\delta = 0) = \lambda_{100}, \tag{8.8}$$

as it should. If we take the state of saturation along [001] as the initial state, then the strain along [001] in a single domain when M_s rotates by an angle δ away from [001] is

$$\begin{aligned}\frac{\Delta l}{l} &= \lambda_{si}(\delta = \delta) - \lambda_{si}(\delta = 0)\\ &= \tfrac{3}{2}\lambda_{100}(\cos^2\delta - \tfrac{1}{3}) - \lambda_{100}\\ &= -\tfrac{3}{2}\lambda_{100}\sin^2\delta.\end{aligned} \tag{8.9}$$

In iron, λ_{100} is positive and $\langle 100\rangle$ is an easy direction. Therefore, when M_s rotates through an angle of 90° out of an easy direction, the domain contracts fractionally in that direction by an amount $3\lambda_{100}/2$. The M_s vector may rotate away from [001] in any plane, not only (010), and Eq. (8.9) will still apply, because a change in the plane of rotation changes only α_1 and α_2. Inasmuch as these appear only in terms involving β_1 or β_2, both zero, they do not affect the final result. These several changes in the length of the crystal along [001] are illustrated in Fig. 8.3(b). If the demagnetized state is nonideal, the zero of strain in this diagram will be shifted up or down, and the λ_{si} values shown will become λ_s values. However, the strain $\Delta l/l$, resulting from a change from one saturated state to another, will remain the same.

These results show that magnetostriction constants can be determined, without any uncertainty regarding the demagnetized state, by making strain measurements as the M_s vector rotates from one orientation to another in a saturated crystal. For example, λ_{100} can be determined by cutting a disk from a crystal parallel to the plane (010). A strain gage is cemented to the disk with its axis parallel to the chosen direction of measurement, namely, [001], as in Fig. 8.4. The disk

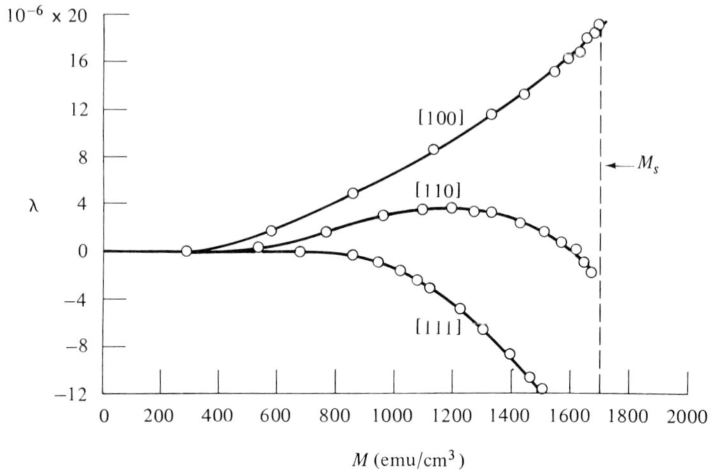

Fig. 8.5 Magnetostriction as a function of magnetization of iron single crystals in the form of rods cut parallel to the principal crystal directions. Webster [8.14].

is then placed in the strong field of an electromagnet. When this is done, the disk magnetostrictively strains, of course, but this strain is ignored. With the disk in the position shown, the strain gage reading is noted. The disk is then rotated by 90° in its own plane to make [100] parallel to M_s, which is parallel to the applied field, and the gage reading is again noted. The difference between these two readings multiplied by $-\frac{2}{3}$ gives λ_{100}, according to Eq. (8.9). Actually, it is better to cut the disk parallel to $\{110\}$, because both λ_{100} and λ_{111} can then be determined from measurements on a single specimen. Details are given in Problem 8.5. By this technique, i.e., without any reference to or knowledge of the demagnetized state, we can determine the value of λ_{100}, for example, even though λ_{100} is defined as the strain in $\langle 100 \rangle$ occurring in a crystal when it passes from the ideal demagnetized state to saturation in $\langle 100 \rangle$.

Figure 8.5 shows experimental curves for magnetostriction in various directions in an iron crystal. The behavior is complex. When the field is parallel to [100], the strain in that direction is a simple expansion, as noted earlier. When the field is parallel to [111], 180° wall motion occurs until the crystal contains only three sets of domains – [100], [010], and [001] – with M_s in each set equally inclined at 55° to the field; during this process the dimensions of the crystal do not change. Further increase of field causes M_s vectors to rotate toward [111], and this rotation causes a contraction along [111].

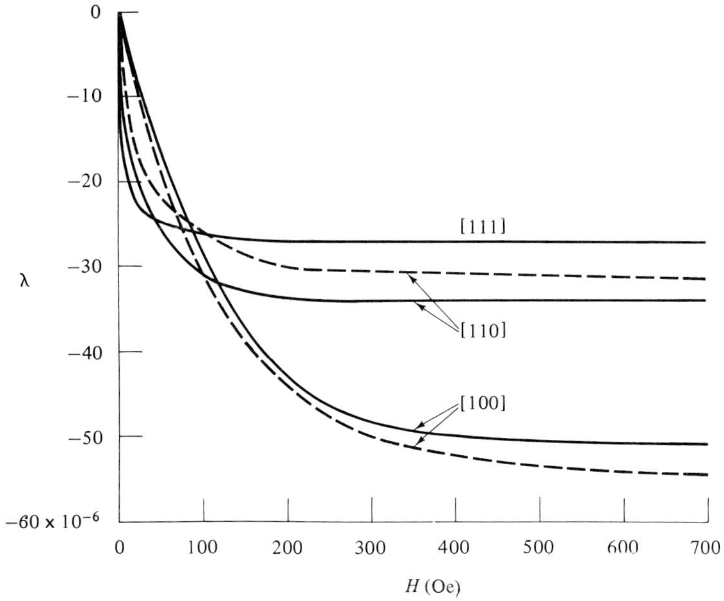

Fig. 8.6 Magnetostriction as a function of field and crystal direction for nickel single crystals in the form of oblate spheroids (disks) cut parallel to the (110) plane (solid curves) and the (100) plane (dashed curves). Masiyama [8.15].

When H is parallel to [110] in iron, the crystal first expands in that direction and then contracts. These changes can be understood by reference to Fig. 7.4, if one imagines the addition of [001] and $[00\bar{1}]$ domains to the initial state depicted there. In response to the applied field, 90° and 180° wall motion will take place until the crystal contains only two sets of domains, those corresponding to the two easy axes nearest the applied field (Fig. 7.4 c). During this process, [001] and $[00\bar{1}]$ domains have disappeared. Inasmuch as these domains are spontaneously contracted in a direction parallel to the field direction [110], their removal causes an expansion in the [110] direction, as observed. With further increase in field, the magnetization in the remaining [100] and [010] domains rotates into the [110] direction, causing an additional strain of $3\lambda_{111}/4$ along [110] (Problem 8.2). Because λ_{111} is negative, this strain is a contraction, and it is large enough to make the crystal shorter at saturation than it was initially.

A nickel crystal contracts in all three principal directions when magnetized, as shown in Fig. 8.6. From the observed contraction in the [111] direction and the fact that the easy directions in nickel are ⟨111⟩, it follows that the unit cell of ferromagnetic nickel is slightly distorted from cubic to rhombohedral, with one cell diagonal, the one parallel to the local direction of magnetization, slightly shorter than the other three. It follows that, when the M_s vector in a domain is rotated away from a ⟨111⟩ easy axis, that axis becomes longer (Problem 8.3).

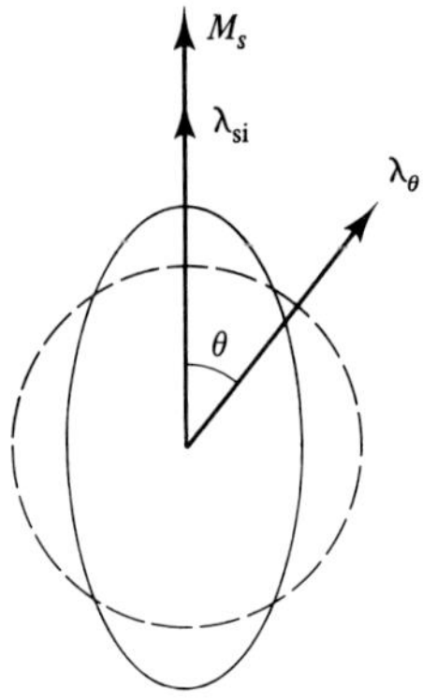

Fig. 8.7 Isotropic magnetostriction [8.17].

One can then understand, by arguments similar to those given above for iron, why λ_{100} and λ_{110} are both negative in nickel.

These qualitative descriptions of the variation of λ with H (or M) below saturation can be made quantitative without much difficulty, although certain rather arbitrary assumptions have to be made about the sequence of 180° and 90° wall motion. Lee [8.1] gives a quantitative treatment.

If the magnetostriction of a particular material is isotropic, we can put $\lambda_{100} = \lambda_{111} = \lambda_{si}$. Then Eq. (8.3) becomes, with the introduction of a new symbol,

$$\lambda_\theta = \tfrac{3}{2}\lambda_{si}[(\alpha_1^2\beta_1^2 + \alpha_2^2\beta_2^2 + \alpha_3^2\beta_3^2 - \tfrac{1}{3}) + 2(\alpha_1\alpha_2\beta_1\beta_2 + \alpha_2\alpha_3\beta_2\beta_3 + \alpha_3\alpha_1\beta_3\beta_1)],$$

$$\lambda_\theta = \tfrac{3}{2}\lambda_{si}[(\alpha_1\beta_1 + \alpha_2\beta_2 + \alpha_3\beta_3)^2 - \tfrac{1}{3}],$$

$$\boxed{\lambda_\theta = \tfrac{3}{2}\lambda_{si}(\cos^2\theta - \tfrac{1}{3})} \tag{8.10}$$

Table 8.1 Magnetostriction Constants of Cubic Substance (Units of 10^{-6})

	λ_{100}	λ_{111}	λ_p^*
Fe	+ 21	− 21	− 7
Ni	− 46	− 24	− 34
$FeO \cdot Fe_2O_3$	− 20	+ 78	+ 40
$Co_{0.8}Fe_{0.2}O \cdot Fe_2O_3$	− 590	+ 120	
$CoO \cdot Fe_2O_3$			− 110
$Ni_{0.8}Fe_{0.2}O \cdot Fe_2O_3$	− 36	− 4	
$NiO \cdot Fe_2O_3$			− 26
$MnO \cdot Fe_2O_3$			− 5
$MgO \cdot Fe_2O_3$			− 6

* Experimental values for polycrystalline specimens.

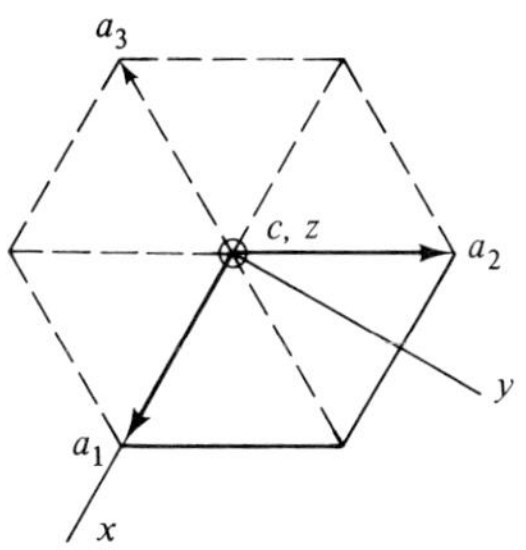

Fig. 8.8 Relation of orthogonal to hexagonal axes.

where λ_θ is the saturation magnetostriction at an angle θ to the direction of magnetization, measured from the ideal demagnetized state. (If θ is the angle between two directions defined by cosines $\alpha_1, \alpha_2, \alpha_3$ and $\beta_1, \beta_2, \beta_3$, then $\cos\theta = \alpha_1\beta_1 + \alpha_2\beta_2 + \alpha_3\beta_3$.) Because of isotropy, no reference to the crystal axes appears in Eq. (8.10), and the magnetostrictive effect can be illustrated quite simply by Fig. 8.7, which shows a demagnetized sphere distorted into an ellipsoid of revolution when saturated, for a positive value of λ_{si}. Figure 8.6 shows that the magnetostrictive behavior of nickel is approximately isotropic, and Eq. (8.10) is often applied to nickel.

Table 8.1 lists λ_{si} values for some cubic metals and ferrites. (The variation with composition of λ_{si} for alloys will be described later.) In general, the magnetostriction of the ferrites is of about the same order of magnitude as that of the metals, with the notable exception of cobalt ferrite. Here the spontaneous distortion of the crystal unit cell, from cubic to tetragonal, is so large that it can be detected by x-ray diffraction. This ferrite also has an unusually large value of crystal anisotropy (Table 7.4).

Hexagonal crystals

The magnetostriction of a hexagonal crystal [8.5] is given by the following equation, which corresponds to Eq. (8.3) for a cubic crystal:

$$\begin{aligned}\lambda_{si} &= \lambda_A\left[(\alpha_1\beta_1 + \alpha_2\beta_2)^2 - (\alpha_1\beta_1 + \alpha_2\beta_2)\alpha_3\beta_3\right]\\ &\quad + \lambda_B\left[(1-\alpha_3^2)(1-\beta_3^2) - (\alpha_1\beta_1 + \alpha_2\beta_2)^2\right]\\ &\quad + \lambda_C\left[(1-\alpha_3^2)\beta_3^2 - (\alpha_1\beta_1 + \alpha_2\beta_2)\alpha_3\beta_3\right]\\ &\quad + 4\lambda_D(\alpha_1\beta_1 + \alpha_2\beta_2)\alpha_3\beta_3. \qquad (8.11)\end{aligned}$$

[Although this expression has four constants, it is the first approximation, like Eq. (8.3). The next approximation, involving higher powers of the direction cosines, has nine constants.] It is important to note that the direction cosines in Eq. (8.11) relate, not to hexagonal axes, but to orthogonal axes x, y, z. Figure 8.8

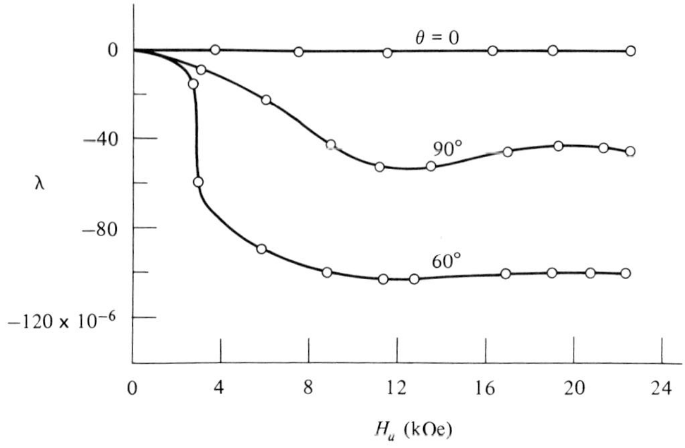

Fig. 8.9 Magnetostriction of a cobalt crystal as a function of applied field. λ and H_a are parallel, and θ is the angle between them and the hexagonal c axis. After Bozorth [8.5].

shows the relation between the two. The usual hexagonal axes are a_1, a_2, a_3, and c. The orthogonal axes are chosen so that x is parallel to a_1, a_2, or a_3, and z is parallel to c. The base of the hexagonal unit cell is outlined. The c and z axes are normal to the plane of the drawing. Equation (8.11) is valid only for crystals in which the c axis is the easy direction.

When the magnetostriction is measured in the same direction as the magnetization, then $\beta_1, \beta_2, \beta_3 = \alpha_1, \alpha_2, \alpha_3$, and Eq. (8.11) reduces to the two-constant expression

$$\lambda_{si} = \lambda_A [(1 - \alpha_3^2)^2 - (1 - \alpha_3^2)\alpha_3^2] + 4\lambda_D(1 - \alpha_3^2)\alpha_3^2, \tag{8.12}$$

because $\alpha_1^2 + \alpha_2^2 + \alpha_3^2 = 1$. Inasmuch as only α_3 appears in Eq. (8.12), the value of λ_{si} in, for example, the basal plane, is the same in any direction. Equations (8.11) and (8.12) therefore express cylindrical, rather than hexagonal, symmetry. Hexagonal symmetry appears only in the next approximation. Bozorth [8.5] finds that the behavior of cobalt is adequately described by the following constants:

$$\lambda_A = -45 \times 10^{-6}, \qquad \lambda_B = -95 \times 10^{-6},$$
$$\lambda_C = +110 \times 10^{-6}, \qquad \lambda_D = -100 \times 10^{-6}.$$

Magnetostriction as a function of field strength is shown in Fig. 8.9. As expected, λ_{si} parallel to the c axis is zero, because only 180° wall motion is involved. The contraction observed at 60° to the c axis is much larger than in the basal plane ($\theta = 90°$) and is, in fact, the maximum contraction for any value of θ.

General

Magnetostriction constants usually decrease in absolute magnitude as the temperature increases, and they approach zero at the Curie point. Figure 8.10 shows the behavior of iron.

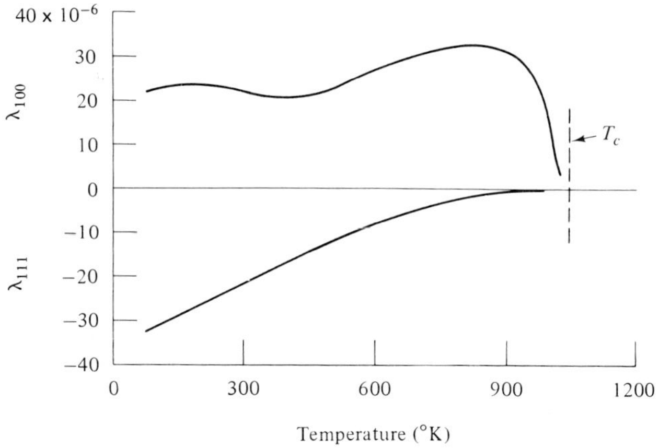

Fig. 8.10 Temperature dependence of the magnetostriction constants of iron. Tatsumoto and Okamoto [8.16].

Before leaving the topic of single crystals it is well to realize that any demagnetized crystal that contains 90° walls is never completely stress free at room temperature. The various domains simply do not fit together exactly. Figure 8.11(a) depicts a single crystal of iron, for example, at a temperature above the Curie point; the dashed lines indicate where domain walls will form below T_c. As the crystal cools below T_c, it spontaneously magnetizes in four different directions in various parts of the crystal, thus forming domains. At the same time, each domain strains spontaneously. If the domains were free to deform, they would separate at the boundaries, as shown in (b), because each domain lengthens in the direction of M_s and contracts at right angles. But the strains involved are much too small to cause separation of the domains. The result is a deformed state, something like (c), in which each domain exerts stress on its neighbor. Saturation removes the 90° walls, the cause of the misfit, and an elongated, single-domain, stress-free crystal results.

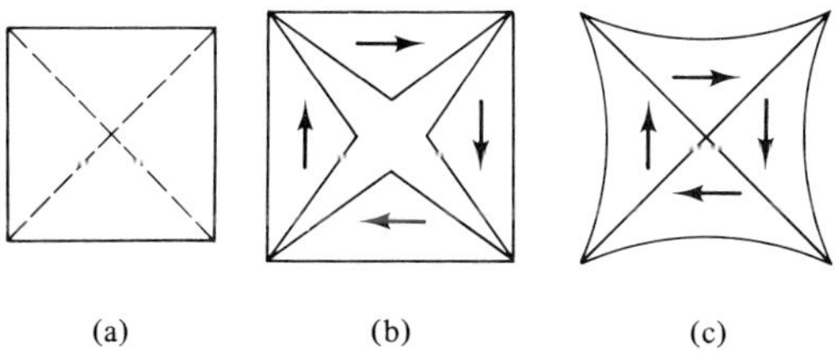

Fig. 8.11 Strains in a demagnetized single crystal.

8.3 MAGNETOSTRICTION OF POLYCRYSTALS

The saturation magnetostriction of a polycrystalline specimen, parallel to the magnetization, is characterized by a single constant λ_p. Its value depends on the magnetostrictive properties of the individual crystals and on the way in which they are arranged, i.e., on the presence or absence of preferred domain or grain orientation.

If the grain orientations are completely random, the saturation magnetostriction of the polycrystal should be given by some sort of average over these orientations. Just how this averaging should be carried out, however, is not entirely clear. When a polycrystal is saturated by an applied field, each grain tries to strain magnetostrictively, in the direction of the field, by a different amount than its neighbors, because of its different orientation. One of two extreme assumptions may then be made: (1) stress in uniform throughout, but strain varies from grain to grain, or (2) strain is uniform, and stress varies.

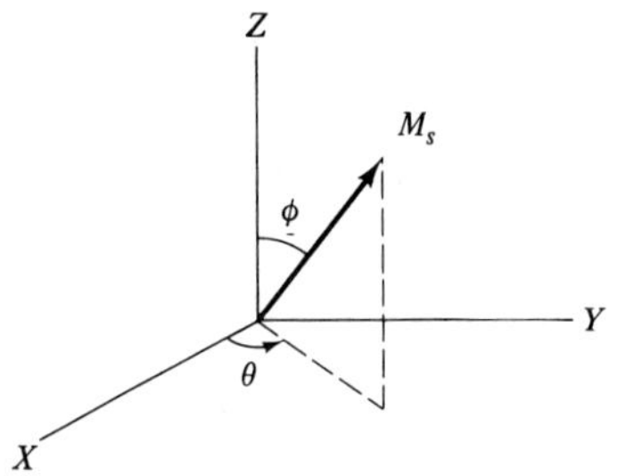

Figure 8.12

The condition of uniform strain is usually considered to be physically more realistic. It is then a question of averaging the magnetostriction in the field direction over all crystal orientations, or, what amounts to the same thing, averaging Eq. (8.6), for cubic crystals, over all orientations of the M_s vector with respect to a set of fixed crystal axes. We first express $\alpha_1, \alpha_2, \alpha_3$ in terms of the angles θ and ϕ of Fig. 8.12. The relations are

$$\alpha_1 = \sin\phi\cos\theta, \qquad \alpha_2 = \sin\phi\sin\theta, \qquad \alpha_3 = \cos\phi.$$

On the surface of a sphere of unit radius centered on the origin, the element of area is $dA = d\phi\,(\sin\phi)\,d\theta$. Averaging over the upper hemisphere, we find the average value of λ_{si} to be

$$\overline{\lambda_{si}} = \frac{1}{2\pi}\int_{\theta=0}^{\pi/2}\int_{\phi=0}^{\pi/2} \lambda_{si}\sin\phi\,d\phi\,d\theta. \tag{8.13}$$

Substitution of Eq. (8.6) for λ_{si} and integration lead to

$$\overline{\lambda_{si}} = \frac{2\lambda_{100} + 3\lambda_{111}}{5}. \tag{8.14}$$

On the other hand, Callen and Goldberg [8.6] claim that an assumption intermediate between those of uniform stress and uniform strain leads to an equation which better represents the data available for a number of polycrystalline metals, alloys, and ferrites. This assumption leads to the following equation

$$\overline{\lambda_{si}} = \lambda_{111} + \left(\frac{2}{5} - \frac{\ln c}{8}\right)(\lambda_{100} - \lambda_{111}), \tag{8.15}$$

where $c = 2c_{44}/(c_{11} - c_{12})$, and c_{44}, c_{11}, and c_{12} are single-crystal elastic constants. A crystal is elastically isotropic if $c = 1$; in that case, Eqs. (8.14) and (8.15) become identical.

When the single-crystal data for iron are substituted into these relations, $\overline{\lambda_{si}}$ is found to be -4×10^{-6} from Eq. (8.14) and -9×10^{-6} from Eq. (8.15). The usually accepted experimental value of λ_p is about -7×10^{-6}. However, it should be noted that the magnetostriction of polycrystalline materials is usually measured on rod specimens, and iron rods almost invariably have a more or less pronounced $\langle 110 \rangle$ fiber texture, as mentioned in Section 7.8. Inasmuch as the value of λ_{110} is -10×10^{-6}, the presence of a $\langle 110 \rangle$ component would tend to make the value of λ_p measured on a rod specimen more negative than the value to be expected for a polycrystal with randomly oriented grains. This suggests that Eq. (8.14) is more accurate for iron.

The usually accepted experimental value of λ_p for nickel, -34×10^{-6}, compares well with -33×10^{-6} given by Eq. (8.14), and with -30×10^{-6} given by Eq. (8.15). However, the values of λ_p reported by individual investigators cover a surprisingly wide range, from -25×10^{-6} to -47×10^{-6}. Part of this spread may be due to differences in preferred grain orientation. While magnetostriction in nickel is commonly regarded as approximately isotropic, the magnitude of λ_{100} is in fact almost double that of λ_{111}. Like other face-centered-cubic metals, nickel in rod form can be expected to have a double $\langle 100 \rangle + \langle 111 \rangle$ fiber texture, as mentioned in Section 7.8. Variations in the relative amounts of these two components could cause large changes in λ_p. So could insufficient annealing of the specimens. The magnetic measurements might then have been made on specimens containing residual stress left over from the cold-worked state. This stress could introduce large errors, because the magnetic properties of nickel are very stress sensitive, as we shall see in Sections 8.5 and 8.6.

If we wish to know the magnetostriction at an angle θ to the magnetization, we can use Eq. (8.10), namely,

$$\lambda_\theta = \tfrac{3}{2}\lambda_p \left(\cos^2\theta - \tfrac{1}{3}\right), \tag{8.16}$$

where λ_p has been substituted for λ_{si}. Because this equation was derived for an isotropic specimen, its application to a polycrystal requires that the specimen

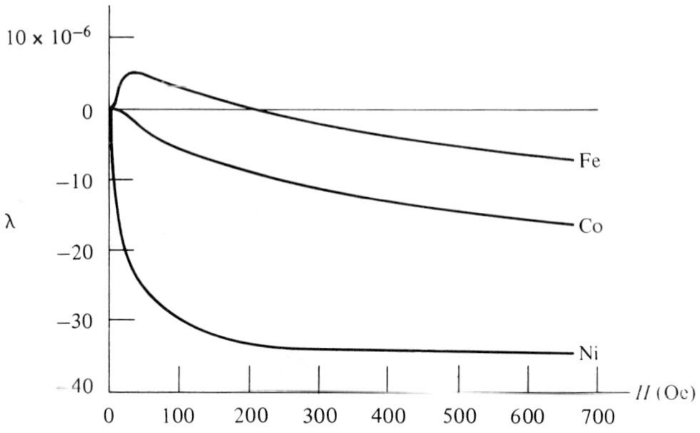

Fig. 8.13 Magnetostriction of polycrystalline iron, cobalt, and nickel. After Lee [8.1].

have no preferred orientation or that it be composed of grains which are themselves magnetically isotropic.

Figure 8.13 shows typical λ,H curves for polycrystalline iron, cobalt and nickel. The shapes of such curves, as reported by different investigators, can vary widely, usually because of differences in preferred orientation. Unfortunately, however, the measurement of magnetostriction is hardly ever accompanied by a determination of the kind and degree of preferred orientation.

The shapes of λ,H (or λ,M) curves below saturation are discussed by Lee [8.7] from a theoretical standpoint. The problem of rationalizing these shapes is difficult, chiefly because of uncertainty about the contribution which each magnetization process (180° wall motion, 90° wall motion, and domain rotation) is making to the increase in magnetization at any particular fraction of saturation.

8.4 PHYSICAL ORIGIN OF MAGNETOSTRICTION

Magnetostriction is due mainly to spin-orbit coupling. This coupling, as we saw in Section 7.4, is also responsible for crystal anisotropy. It is relatively weak, because applied fields of a few hundred oersteds usually suffice to rotate the spins away from the easy direction.

The relation between magnetostriction and spin-orbit coupling can be crudely pictured in terms of Fig. 8.14, which is a section through a row of atoms in a crystal. The black dots represent atomic nuclei, the arrows show the net magnetic moment per atom, and the oval lines enclose the electrons belonging to, and distributed nonspherically about, each nucleus. The upper row of atoms depicts the paramagnetic state above T_c. If, for the moment, we assume that the spin-orbit coupling is *very strong*, then the effect of the spontaneous magnetization occurring

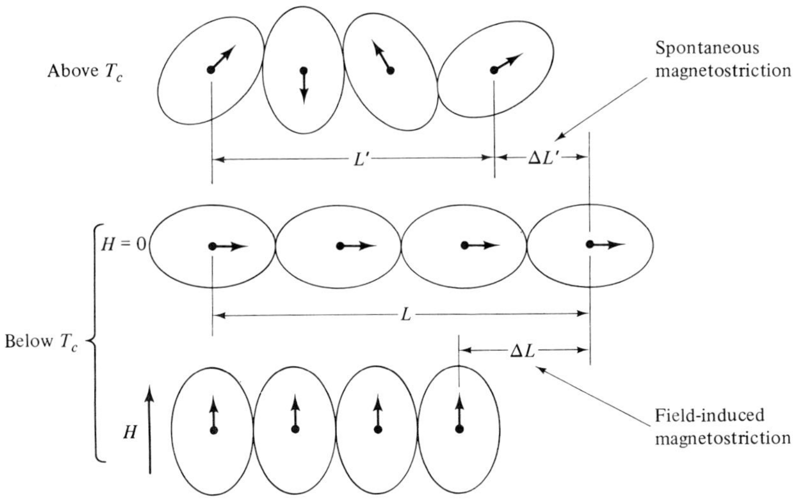

Fig. 8.14 Mechanism of magnetostriction [8.17].

below T_c would be to rotate the spins *and* the electron clouds into some particular orientation determined by the crystal anisotropy, left to right, say. The nuclei would be forced further apart, and the spontaneous magnetostriction would be $\Delta L'/L'$. If we then apply a strong field vertically, the spins and the electron clouds would rotate through 90°, and the domain of which these atoms are a part would magnetostrictively strain by an amount $\Delta L/L$.

The strains pictured are enormous, of the order of 0.3. Actually, we know that the magnetostrictive strain produced in a domain or a crystal, when its direction of magnetization is changed, is usually very small indeed, of the order of 10^{-5}. This means that the reorientation of electron clouds takes place only to a very small extent. This conclusion is in turn supported by the fact that orbital magnetic moments are almost entirely quenched, i.e., not susceptible to rotation by an applied field, in most materials, as shown by measurements of the g or g' factors (Section 3.7 and Table 4.1.)

(That no large reorientation of the electron cloud occurs in materials like iron and nickel when the magnetization is rotated was also demonstrated by certain x-ray experiments performed about 1920. X-rays are scattered by the electrons of an atom, and the intensity of a diffracted x-ray beam depends on the spatial distribution of the electrons about the nucleus. When specimens were subjected simultaneously to x-rays and a magnetic field, variations in field magnitude and orientation had no detectible effect on the intensity of diffracted beams.)

The rare-earth metals are exceptions to the above statements. Many of them are ferromagnetic, at temperatures usually far below room temperature, and their orbital moments are not quenched, i.e., the spin-orbit coupling is strong. Moreover, the electron cloud about each nucleus is decidedly nonspherical. Therefore, when an applied field rotates the spins, the orbits rotate too and considerable distortion results. At 22°K the saturation magnetostriction of dysprosium is about 4.5×10^{-3} in the basal plane, or some 100 times that of "normal" metals and alloys.

Inasmuch as magnetostriction and crystal anisotropy are both due to spin-orbit coupling, we would expect some correlation between the two. In fact, a large (numerical) value of the anisotropy constant K_1 is usually accompanied by a large (numerical) value of λ_{si}. For example, hexagonal substances tend to have larger values of both $|K_1|$ and $|\lambda_{si}|$ than cubics. And in binary alloys, the addition of a second element in solid solution often decreases both $|K_1|$ and $|\lambda_{si}|$. These are just general trends, however, and there are many exceptions.

The close physical connection between crystal anisotropy and magnetostriction has been emphasized by Kittel [8.4]. He extends the usual definition of anisotropy energy by regarding it as a function, not only of the orientation of M_s relative to the crystal axes, but also of the interatomic distance, i.e., the strain. Thus a unit cell of, for example, cubic iron will spontaneously distort to tetragonal below the Curie point if the anisotropy energy is thereby reduced by an amount greater than the elastic energy is increased. Furthermore, the amount of this distortion will depend on the direction of M_s relative to the crystal axes. Kittel develops this approach in detail and shows that it leads naturally to the general equation for magnetostrictive strain, namely, Eq. (8.3).

When a specimen is magnetized, a strain $(\Delta l/l)_f$ occurs which has a physical origin entirely different from that of magnetostriction but which can be erroneously ascribed to magnetostriction, if the investigator is not aware of it. Because this strain depends on the shape of the specimen, it is said to be due to the *form effect*.

This effect occurs because of the tendency of a magnetized body to minimize its magnetostatic energy. Suppose a specimen is magnetized to the saturation level M_s in a direction along which its demagnetizing factor is N_d. Then, according to Eq. (7.51), its magnetostatic energy is $(N_d M_s^2/2)$ ergs/cm^3. The specimen can decrease this energy by lengthening a fractional amount $(\Delta l/l)_f$ in the direction of M_s, because a longer specimen has a smaller value of N_d. Superimposed on the form-effect strain $(\Delta l/l)_f$ is the strain λ_{si} due to magnetostriction alone, and the two together make up the observed strain. For an iron sphere, $(\Delta l/l)_f$ is about 4×10^{-6}, so the form effect is by no means negligible with short specimens.

Because it is difficult to calculate $(\Delta l/l)_f$ with any accuracy, it is better to avoid the form effect entirely by choosing a specimen shape with small N_d. The magnetostatic energy will then be so small that $(\Delta l/l)_f$ will become negligible. The condition for this, according to Zijlstra [G.31], is that the length/diameter ratio of a rod specimen be greater than 5, or that the length/diameter ratio of a rod specimen be greater than 5, or that the diameter/thickness ratio of a disk specimen be greater than 10.

8.5 EFFECT OF STRESS ON MAGNETIZATION

"No part of our subject is more interesting than that which deals with the effects of mechanical stress in altering the susceptibility, the retentiveness, and other qualities of the three magnetic metals." So wrote Ewing [G.1] in 1900, and his

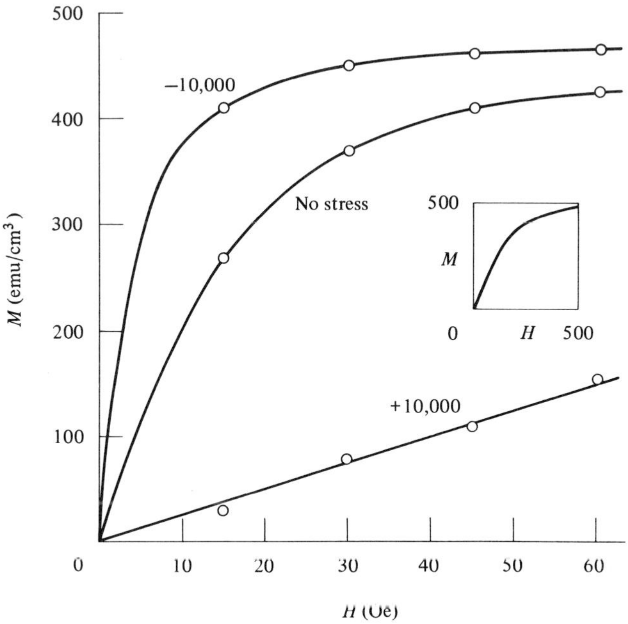

Fig. 8.15 Effects of applied tensile (+) and compressive (−) stress on the magnetization of nickel. Numbers on curves are stresses in lb/in^2. Insert: behavior of nickel under 10,000 lb/in^2 tension at higher fields. After Bagchi [8.8].

remark remains no less true today. Figure 8.15 shows the marked effects of applied stress on the magnetization behavior of polycrystalline nickel. At a field H of 10 Oe, a compressive stress of 10,000 lb/in^2 almost doubles the permeability μ, while the same amount of tensile stress reduces μ to about one-tenth of the zero-stress value and makes the M,H curve practically linear. Nickel is not unique in this respect. Materials are known in which the low-field permeability is changed by a factor of 100 by an applied stress of the order of 10,000 lb/in^2.

The magnetostriction of nickel is negative. For a material with positive magnetostriction, like 68 Permalloy, the effect of stress is just the opposite. (The word Permalloy refers to a family of Ni-Fe alloys; sometimes they also contain small additions of other elements. The number before the alloy name gives the nickel content. Thus "68 Permalloy" means an alloy containing 68 percent Ni and 32 percent Fe.) Applied tensile stress *increases* the permeability of this alloy, as Fig. 8.16 shows.

The magnetostriction of polycrystalline iron is positive at low fields, then zero, then negative at higher fields, as shown in Fig. 8.13. As a result, the magnetic behavior under stress is complicated. At low fields tension raises the B,H curve and at higher fields lowers it; the crossover of the two curves at a particular field strength, which depends on the stress and on preferred orientation, is called the *Villari reversal.* In the measurements shown in Fig. 8.17, tension has no appreciable

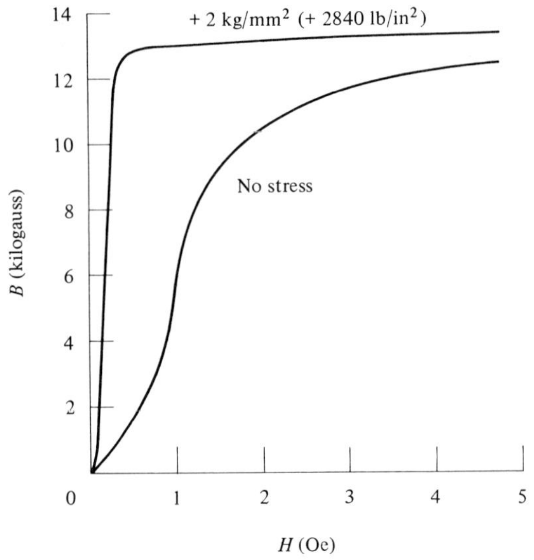

Fig. 8.16 Effect of applied tensile stress on the magnetization of 68 Permalloy. After Bozorth [G.4].

effect until B exceeds about 10 kilogauss; the Villari reversal occurs at about 20 Oe. (Bozorth [G.4, p. 602] shows tensile-stress and zero-stress B,H curves deviating from one another right from the origin.) Compression has a reverse, and larger, effect, lowering B at low fields and raising it at large fields.

The experimental results of Figs. 8.15 to 8.17 show that there is a close connection between the magnetostriction λ of a material and its magnetic behavior under stress. (As a result, the effect of stress on magnetization is sometimes called the *inverse magnetostrictive effect*, but more commonly is referred to simply as a *magnetomechanical effect*. It has nothing to do with the magnetomechanical factor of Section 3.7.) In fact, these results could have been anticipated by a general argument based on Le Chatelier's principle. If, for example, a material has positive λ, it will elongate when magnetized; applied tensile stress, which tends to elongate it, will therefore increase the magnetization, and applied compressive stress will decrease it. These conclusions are valid whether or not a field is acting, as long as M is not zero. Thus, in Fig. 8.18, if a field H_1 produces a magnetization of A at zero stress, then application of a stress $+\sigma_1$ will raise the magnetization to B at constant field. The magnetization in the remanent state at zero stress is C, and the same stress $+\sigma_1$ will increase this to D. But a stress applied to a demagnetized specimen will not produce any magnetization, as shown by the intersection of the full-line and dashed curves at the origin.

So far we have tacitly assumed that H, M (or B), and σ were all parallel, but, in general, M and σ may not be parallel. Now we know from Eq. (8.3) that the

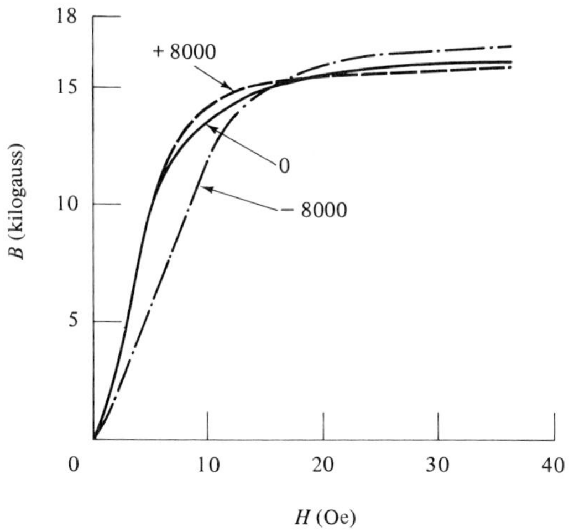

Fig. 8.17 Effects of applied tensile (+) and compressive (—) stress on the magnetization of iron. Numbers on curves are stresses in lb/in². Kuruzar [8.9].

amount of magnetostrictive strain exhibited by a crystal in a particular direction depends on the direction of the magnetization. If we now impose an additional strain by applying a stress, we expect that the direction of the magnetization will change. We therefore need a general relation between the direction of M_s within a domain and the direction and magnitude of σ. But we know that, in the absence of stress, the direction of M_s is controlled by crystal anisotropy, as characterized by the first anisotropy constant K_1. Therefore, when a stress is acting, the di-

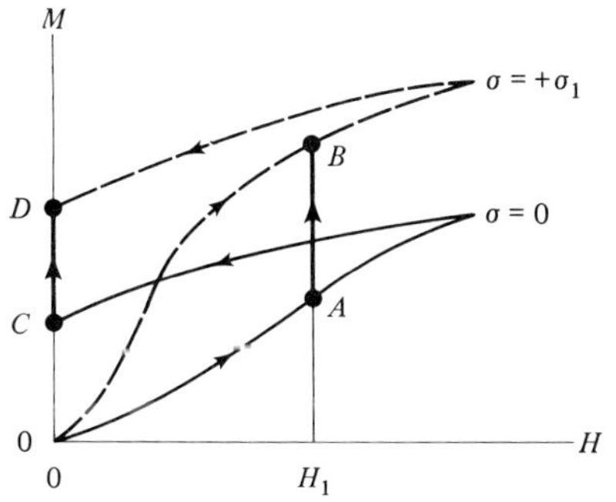

Fig. 8.18 Effect of tension on the magnetization of a material with positive magnetostriction (schematic).

rection of M_s is controlled by both σ and K_1. These two quantities are therefore involved in the expression for that part of the energy which depends on the direction of M_s, which is, for a cubic crystal,

$$E = K_1 (\alpha_1^2 \alpha_2^2 + \alpha_2^2 \alpha_3^2 + \alpha_3^2 \alpha_1^2)$$

$$- \tfrac{3}{2} \lambda_{100} \sigma(\alpha_1^2 \gamma_1^2 + \alpha_2^2 \gamma_2^2 + \alpha_3^2 \gamma_3^2)$$

$$- 3 \lambda_{111} \sigma(\alpha_1\alpha_2\gamma_1\gamma_2 + \alpha_2\alpha_3\gamma_2\gamma_3 + \alpha_3\alpha_1\gamma_3\gamma_1) \qquad (8.17)$$

where α_1, α_2, α_3, are the direction cosines of M_s, as before, and γ_1, γ_2, γ_3 are the direction cosines of σ. The units of E are ergs/cm^3 if σ is expressed in dynes/cm^2. (1 dyne/cm^2 = 1.02×10^{-8} kg/mm^2 = 1.45×10^{-5} lb/in^2.) This relation is derived by Lee [8.1] and Kittel [8.4].

The first term of Eq. (8.17) is the crystal anisotropy energy, taken from Eq. (7.1) in its abbreviated form. The next two terms, which involve the magnetostrictive strains and the stress, comprise what is usually called the *magnetoelastic energy* E_{me}. The equilibrium direction of M_s is that which makes E a minimum, and this direction is seen to be a complicated function of K_1, λ_{100}, λ_{111}, and σ, for any given stress direction γ_1, γ_2, γ_3. But we can note, qualitatively, that the direction of M_s will be determined largely by crystal anisotropy when K_1 is much larger than $\lambda_{100}\,\sigma$ and $\lambda_{111}\,\sigma$; when this inequality is reversed, the stress will control the M_s direction.

When the magnetostriction is isotropic, so that $\lambda_{100} = \lambda_{111} = \lambda_{si}$, the last two terms of Eq. (8.17) reduce to a very simple form for the magnetoelastic energy:

$$\boxed{E_{me} = -\tfrac{3}{2} \lambda_{si}\, \sigma \cos^2 \theta,} \qquad (8.18)$$

where θ is the angle between M_s and σ. Alternatively, we can substitute $(1 - \sin^2 \theta)$ for $\cos^2 \theta$, drop a constant term, and write the energy as

$$E_{me} = \tfrac{3}{2} \lambda_{si}\, \sigma \sin^2 \theta. \qquad (8.19)$$

The two forms are equivalent with respect to the angular dependence of E_{me} and differ only in what is taken as the zero of energy. In one form or the other these relations are often used to determine the effect of stress on magnetic behavior. We note that the way in which a material responds to stress depends only on the sign of the *product* of λ_{si} and σ; a material with positive λ_{si} under tension behaves like one with negative λ_{si} under compression.

A direct derivation of Eq. (8.19) brings out its physical meaning. Suppose a tensile stress σ is applied to the unit cube of Fig. 8.19 and that M_s is initially parallel to the stress. Let M_s then rotate through an angle θ. As it does so, the material will contract along the stress axis because λ_θ is less than λ_{si}, as shown by Eq. (8.10), when λ_{si} is positive. This contraction, in the presence of a tensile stress, means that work is done on the material. This work is

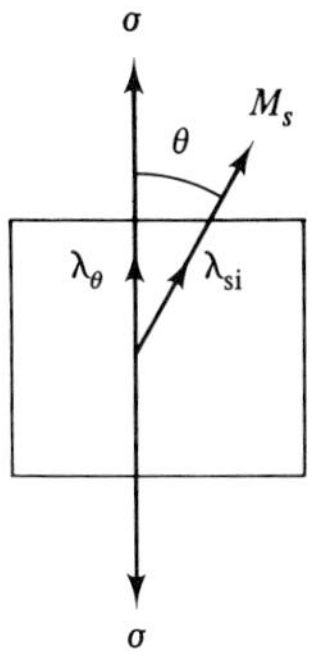

Figure 8.19

stored as magnetoelastic energy in the material and is given by $dE_{me} = -\sigma d\lambda$ for an infinitesimal rotation of M_s. Therefore,

$$\int_0^{E_{me}} dE_{me} = -\sigma \int_{\lambda_{si}}^{\lambda_\theta} d\lambda, \tag{8.20}$$

$$E_{me} = -\sigma\left[\tfrac{3}{2}\lambda_{si}(\cos^2\theta - \tfrac{1}{3}) - \lambda_{si}\right], \tag{8.21}$$

$$E_{me} = \tfrac{3}{2}\lambda_{si}\,\sigma\sin^2\theta. \tag{8.22}$$

This form of the equation states that the magnetoelastic energy is zero when M_s and σ are parallel and that it increases to a maximum of $\frac{3}{2}\lambda_{si}\sigma$ when they are at right angles, provided that $\lambda_{si}\sigma$ is positive. If this quantity is negative, the minimum of energy occurs when M_s and σ are at right angles.

We are now in a position to understand why it is so easy to magnetize a material with positive λ, such as 68 Permalloy, when it is stressed in tension (Fig. 8.16). If the crystal anisotropy is weak, as it is in this alloy, the direction of M_s in the absence of a field will be controlled largely by stress, and, in a polycrystalline specimen with no preferred orientation, Eq. (8.18) or (8.19) will apply. Let Fig. 8.20(a) represent a small portion of the specimen, comprising four domains. The application of a small tensile stress to the demagnetized specimen, as in (b), will cause domain walls to move in such a way as to decrease the volume of domains magnetized at right angles to the stress axis, because such domains have a high magnetoelastic energy. These domains are completely eliminated by some higher value of the stress, as in (c), and E_{me} is now a minimum. The domain structure is now identical with that of a uniaxial crystal, shown in Fig. 7.6. Only a small applied field is now required to saturate the specimen, because the transition from Fig. 8.20(c) to (d) can be accomplished solely by the relatively easy process of 180° wall motion. If the crystal anisotropy is zero, and if no impediments to wall motion, of the kind to be discussed in Chapter 9, exist, then an infinitesimal stress and an infinitesimal field should suffice for saturation. These

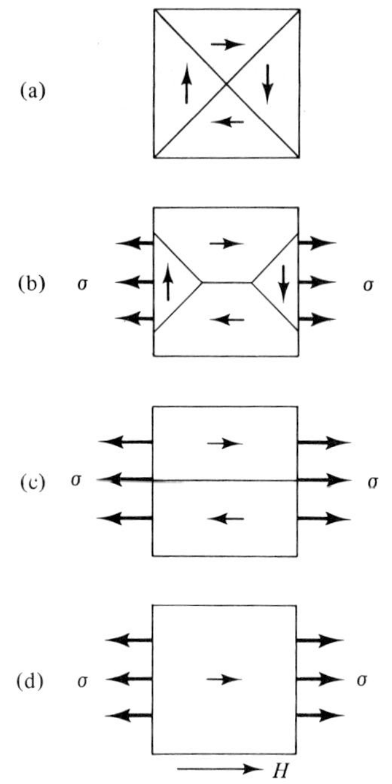

Fig. 8.20 Magnetization of a material with positive magnetostriction under tensile stress (schematic). Positive $\lambda_{si}\sigma$.

conditions are never met, and we find in practice that a particular stress and a particular field are required. For 68 Permalloy, Fig. 8.16 shows that + 2840 lb/in^2 and 0.5 Oe are enough to raise the magnetization almost to saturation.

In this example, $\lambda_{si}\,\sigma$ is a positive quantity. The mechanism of Fig. 8.20 will therefore also apply to nickel under compression, because $\lambda_{si}\,\sigma$ is again positive, and we see in Fig. 8.15 that compression of nickel does indeed make magnetization easier.

Two points emerge from this examination of magnetization under stress:

Table 8.2 Summary of Some Uniaxial Anisotropies

Kind of anisotropy	Energy responsible	Governing relation $E = K_u \sin^2\theta$	Equation
Crystal	Crystal anisotropy	$K_u = K_1$	(7.4)
Shape	Magnetostatic	$K_u = K_s = \frac{1}{2}(N_a - N_c)M^2$	(7.53)
Stress	Magnetoelastic	$K_u = K_\sigma = \frac{3}{2}\lambda_{si}\sigma$	(8.23)

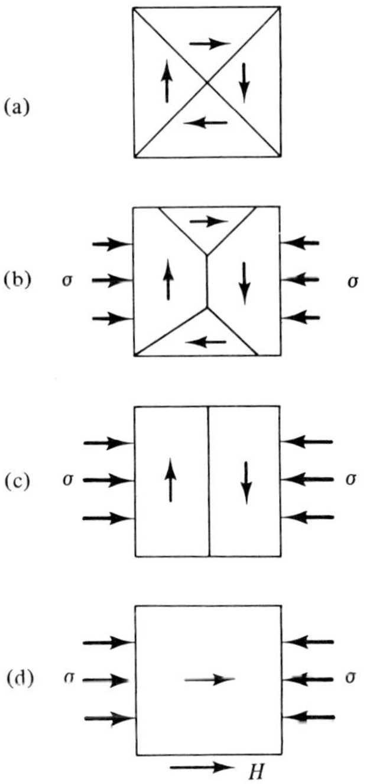

Fig. 8.21 Magnetization of a material with positive magnetostriction under compressive stress (schematic). Negative $\lambda_{si}\sigma$.

1. In the demagnetized state, *stress alone can cause domain wall motion.* This motion must be such as to ensure zero net magnetization for the whole specimen. This condition is not difficult to meet, however, because there is an infinite number of domain arrangements which make M equal to zero.

2. Stress alone can create an easy axis of magnetization. Therefore, when stress is present, *stress anisotropy* must be considered, along with any other anisotropies that may exist. It is a uniaxial anisotropy, and the relation which governs it, namely, Eq. (8.19) is of exactly the same form as Eq. (7.4) for uniaxial crystal anisotropy or Eq. (7.53) for shape anisotropy. We therefore write for the stress anisotropy energy, which is a magnetoelastic energy,

$$E_{me} = K_\sigma \sin^2 \theta, \tag{8.23}$$

where the stress anisotropy constant K_σ is given by $\frac{3}{2}\lambda_{si}\sigma$, from Eq. (8.19). The axis of stress is an easy axis if $\lambda_{si}\sigma$ is positive. If this quantity is negative, the stress axis is a hard axis and the plane normal to the stress axis is an *easy plane* of magnetization. The three anisotropies we have met so far are summarized in Table 8.2, in terms of a general uniaxial anisotropy constant K_u.

When $\lambda_{si}\sigma$ is negative, as it is for nickel under tension, the stress axis becomes a hard axis, because the field now has to supply energy equal to the magnetoelastic energy in order to rotate the M_s vector of each domain by 90° into the field direction (Fig. 8.21 c). When this rotation is complete, the domain wall simply disappears, and the saturated state of Fig. 8.21 (d) results. The magnetization curve is then expected to be a straight line, just like the M,H curve of a uniaxial crystal such as cobalt when H is at right angles to the easy axis. This latter example was discussed in Section 7.6, and the relation between M and H was given by Eq. (7.29), in terms of a single anisotropy constant. Translating this equation into terms of stress anisotropy, we have

$$M = \frac{M_s^2 H}{2K_\sigma} = \frac{M_s^2 H}{3\lambda_{si}\sigma}. \qquad (8.24)$$

The lower curve of Fig. 8.15 shows that the M,H behavior of nickel under tension is indeed linear, as required by Eq. (8.24), at least up to a field of 60 Oe. However, over the whole range of about 500 Oe required to saturate, the M,H relation is decidedly nonlinear, as shown by the insert to Fig. 8.15.

This disagreement suggests that we examine the validity of Eq. (8.24). It is derived on the basis that only stress anisotropy is present, i.e., that any other anisotropy which may exist, such as crystal anisotropy, is negligible in comparison. Now crystal anisotropy, for example, is a constant of the material, but stress anisotropy, for a given material, depends on the stress. We might therefore ask: at what stress does the stress anisotropy become equal to the crystal anisotropy? This stress can be found approximately by equating the stress anisotropy energy $\frac{3}{2}\lambda_{si}\sigma$ to the crystal anisotropy energy K_1. For nickel, this critical stress will be

$$\sigma = \frac{2K_1}{3\lambda_{si}} = \frac{2(0.5 \times 10^5)}{3(34 \times 10^{-6})} \qquad (8.25)$$

$$= 10^9 \text{ dynes/cm}^2 = 15{,}000 \text{ lb/in}^2.$$

(The sign of λ_{si} is irrelevant in this calculation.) Therefore, at a stress of 10,000 lb in^2, at which the data of Fig. 8.15 were obtained, stress anisotropy is actually weaker than crystal anisotropy. The stress would presumably have to be many times 15,000 lb/in^2 before Eq. (8.24) would strictly apply, all the way to saturation.

(The calculation just made is only approximate, because it ignores the complex details of domain rotation. In a single domain of nickel, for example, there are four easy axes determined by the crystal anisotropy and one easy axis determined by the stress. As the M_s vector of this domain rotates from its initial position, in response to an increasing applied field, its rotation is sometimes helped and sometime hindered by the crystal anisotropy, depending on the relative orientation at that time of M_s and the easy crystal axes.)

Of the three cases illustrated in Figs. 8.15 to 8.17 (nickel, 68 Permalloy, and iron), the effect of stress on magnetization is smallest for iron. This is because iron has a relatively large value of K_1 and a small value of λ_{si}. This combination of properties means that crystal anisotropy is more important than stress anisotropy; the critical stress at which the two become equal is found from Eq. (8.25) to have the very high value of 660,000 lb/in^2. This stress is some ten times the fracture stress of iron and is therefore unattainable.

This section has been devoted mainly to the effect of stress on domain rotation. It can also affect domain wall motion, and this topic will be discussed in Section 9.11.

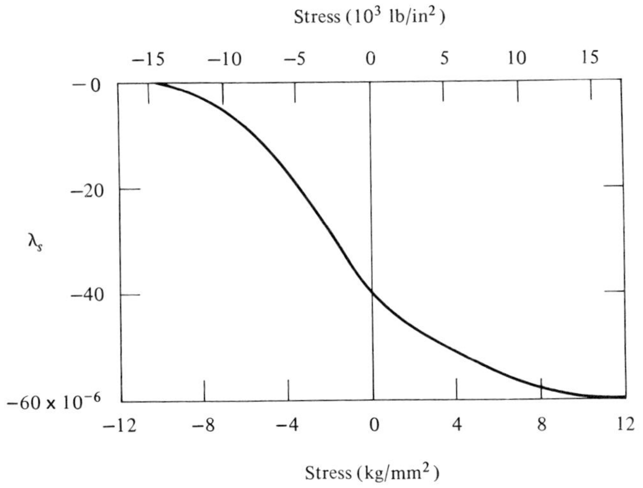

Fig. 8.22 Saturation magnestostriction λ_s of nickel under tension and compression. After Kirchner [8.10].

8.6 EFFECT OF STRESS ON MAGNETOSTRICTION

Stress not only alters the character of a magnetization curve, but it can produce large changes in the observed magnetostriction λ_s. This effect is nicely illustrated by the results of Kirchner [8.10], shown in Fig. 8.22. He subjected nickel rods to axial tension and compression and measured the saturation magnetostriction λ_s in the axial direction. Compression reduced the magnitude of λ_s and, at a stress of about 12 kg/mm^2 (17,000 lb/in^2), the magnetostriction disappeared. Tension, on the other hand, increased the magnitude of λ_s and, at a stress of about 12 kg/mm^2, λ_s had become $\frac{3}{2}$ of the zero-stress value, which Kirchner found to be -40×10^{-6}.

This behavior can be understood in terms of the *preferred domain orientation* set up in the demagnetized state by the applied stress *before the magnetostriction measurement is begun.* Thus, for nickel under compression, $\lambda_{si}\sigma$ is positive, and sufficient compression will produce the domain arrangement shown in Fig. 8.20 (c). This specimen can be brought to saturation entirely by 180° wall motion, which produces no magnetostrictive change in length. Similarly, a high tensile stress will result in the domain arrangement of Fig. 8.21 (c); saturation now requires 90° domain rotation over the total volume of the specimen; the result is a numerically larger magnetostriction than that observed when the structure of Fig. 8.21 (a) is saturated, because, in (a), domain rotation is required in only a part of the total volume.

These arguments can be made quantitative by means of Eq. (8.10):

$$\lambda_\theta = \frac{3}{2}\lambda_{si}\left(\cos^2\theta - \frac{1}{3}\right). \tag{8.10}$$

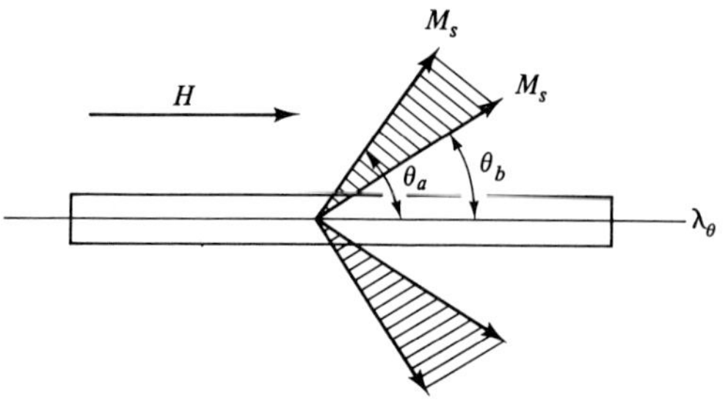

Figure 8.23

Here λ_θ is the saturation magnetostriction, measured at an angle θ to the magnetization, when the initial state is the *ideal demagnetized state.* Suppose now that the initial state is the ideal demagnetized state and that the final state is, not saturation, but the state of partial magnetization indicated in Fig. 8.23, where the domain vectors all lie within a range of angles θ_a to θ_b with the specimen axis. Then the observed magnetostriction parallel to the specimen axis (λ_θ) would be

$$\lambda = \tfrac{3}{2}\lambda_{si}(\langle\cos^2\theta\rangle - \tfrac{1}{3}), \tag{8.26}$$

where the angular brackets indicate an average of $\cos^2\theta$ over all orientations of the M_s vectors in the final state, namely, over the range of angles θ_a to θ_b. If now the state of a specimen changes from state 1 to state 2, each defined by average values $\langle\cos^2\theta\rangle_1$ and $\langle\cos^2\theta\rangle_2$, respectively, then the observed magnetostriction during this change will be

$$\begin{aligned}\lambda &= \tfrac{3}{2}\lambda_{si}[(\langle\cos^2\theta\rangle_2 - \tfrac{1}{3}) - (\langle\cos^2\theta\rangle_1 - \tfrac{1}{3})] \\ &= \tfrac{3}{2}\lambda_{si}[\langle\cos^2\theta\rangle_2 - \langle\cos^2\theta\rangle_1]. \end{aligned} \tag{8.27}$$

If the final state 2 is one of saturation, as it usually is, then $\theta = 0$ for all domains, $\langle\cos^2\theta\rangle_2 = 1$, and

$$\lambda_s = \tfrac{3}{2}\lambda_{si}[1 - \langle\cos^2\theta\rangle_1]. \tag{8.28}$$

Equation (8.28) is very useful, because it permits a calculation of the saturation magnetostriction for *any* initial state, providing we know the distribution of domain vectors in that state. For example, the ideal demagnetized state of a polycrystal has domains randomly oriented in space, and the average value of $\cos^2\theta$ is then 1/3. If this is substituted into Eq. (8.28), λ_s becomes equal to λ_{si}, as it should. On the other hand, demagnetized nickel under high compressive stress has all domain vectors parallel to the specimen axis. The value of $\langle\cos^2\theta\rangle_1$ is 1, and the observed λ_s is zero, in accordance with Fig. 8.22. Under high tensile stress all vectors are at right angles to the axis, $\langle\cos^2\theta\rangle_1 = 0$, and the observed λ_s is 3/2 of the normal, stress-free value (λ_{si}), again in accordance with experiment.

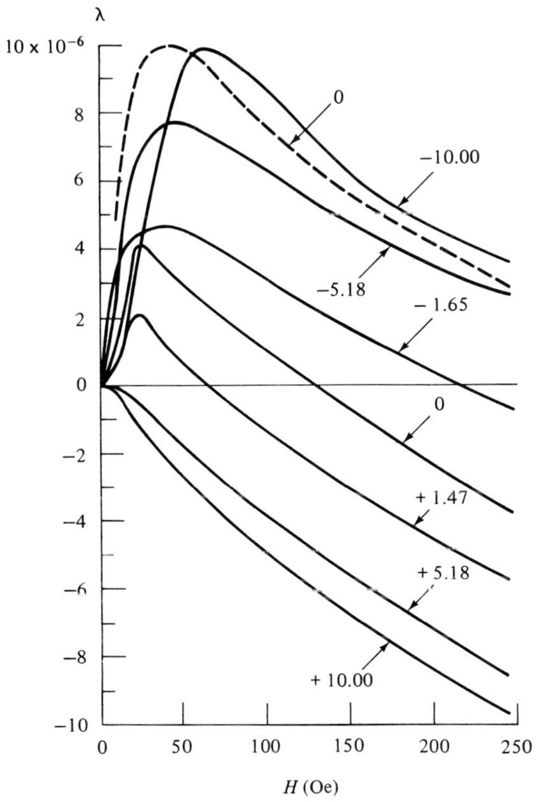

Fig. 8.24 Magnetostriction λ as a function of field H for iron under applied tensile and compressive stress. The number on each curve is the stress in units of 1000 lb/in². The dashed curve is for a pre-stretched specimen. After Kuruzar *et al.* [8.11] [8.17].

(One aspect of Kirchner's results is puzzling, however. If we repeat the kind of calculation made near the end of Section 8.5, but with Kirchner's value of -40×10^6 for λ_{si}, we find that the stress at which the stress and crystal anisotropies become equal is 8.5 kg/mm^2 or 12,000 lb/in^2. One would therefore think that a stress many times 8.5 kg/mm^2 would be necessary before crystal anisotropy could be neglected. Yet Kirchner's data show that a stress of about 12 kg/mm^2 is enough to produce the domain rotations expected in the absence of crystal anisotropy.)

Figure 8.24 shows the effect of stress on the magnetostriction of iron. The trend is the same as in nickel. Increasing tension makes λ more negative and finally suppresses the positive λ observed at low fields and zero stress. Increasing compression shifts λ in the positive direction. (The solid curves in Fig. 8.24 apply to an annealed specimen subjected to applied stresses within the elastic

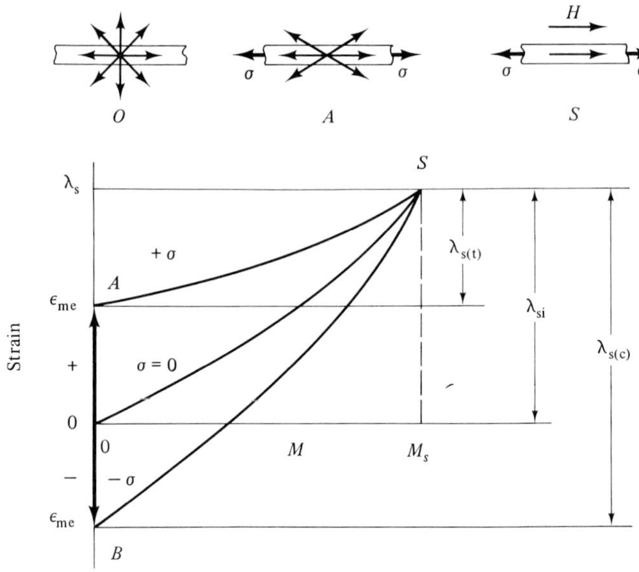

Fig. 8.25 Effects of stress and changes in magnetization on strains of magnetic origin. Sketches at the top indicate domain vector orientations in the three states O, A, and S. Positive λ_{si} [8.17].

limit. The dashed curve refers to a specimen which had been prestrained by 3.2 percent plastic elongation; it was then unloaded and the λ,H curve measured without any applied stress; the interpretation of this curve is given in Section 9.10.)

The effect of stress on λ_s is summarized in Fig. 8.25, which shows the strains of magnetic origin which occur when a material of positive λ_{si} is first stressed and then magnetized to saturation. The point O at the origin represents the ideal demagnetized state of the specimen. If it is then magnetized to saturation at zero stress, it goes from O to the saturation point S along the curve marked $\sigma = O$. However, if it is first loaded in tension, domain vectors rotate toward the stress axis. (Actually, the process is one of growth of axially magnetized domains at the expense of those transversely magnetized, but it is convenient to think of the change as domain rotation.) The result is a magnetoelastic strain $\varepsilon_{me} = OA$. [This strain is given by Eq. (8.27), but it is difficult to calculate because of the difficulty of determining $\langle\cos^2\theta\rangle_2$ except in the limit of very high stress.] If, still under tension, the specimen is magnetized to saturation, it goes from A to S along the curve marked $+\sigma$. The observed saturation magnetostriction $\lambda_{s(t)}$ is then smaller than λ_{si}. Similarly, under compression, the specimen passes from O to B to S, and $\lambda_{s(c)}$ is larger than λ_{si}. (The purely elastic strain ε_{el} due to the stress is neglected in this discussion, but its inclusion would not alter the results. This strain, in a loaded specimen, is present in both the demagnetized and saturated states and therefore does not affect the observed value of λ_s.)

We have seen that Eq. (8.28) can be used to predict the observed value of λ_s when the initial domain orientations are known. It can also be applied in reverse: the measured value of λ_s is substituted on the left-hand side and the equation is solved for $\langle \cos^2 \theta \rangle_1$. Any departure of $\langle \cos^2 \theta \rangle_1$ from the value 1/3 is evidence of a nonideal, preferred domain orientation in the demagnetized state. Thus an important property of a magnetostriction measurement is that it can *disclose preferred domain orientations.* This property is often exploited in research. Other variables besides stress can cause preferred domain orientations, and an investigator can show by a magnetostriction measurement that such variables are operating. Bozorth [G4, p. 637] describes two examples of preferred domain orientations disclosed by magnetostriction measurements, one due to heat treatment in a magnetic field (magnetic annealing) and the other due to preferred grain orientation in polycrystalline sheet. (See Fig. 10.3.)

In connection with the experimental problem of measuring λ_s, it is of interest to enquire into the shape of λ,M curves. First, consider an extreme example, nickel under sufficient tension so that all M_s vectors are initially at right angles to the axis. Then Eq. (8.27) becomes

$$\lambda = \tfrac{3}{2}\lambda_{si} \cos^2 \theta, \tag{8.29}$$

where θ is the angle between M_s vectors and the axis at any particular field strength. But

$$M = M_s \cos \theta. \tag{8.30}$$

Therefore,

$$\lambda = \tfrac{3}{2}\lambda_{si} \left(\frac{M}{M_s}\right)^2. \tag{8.31}$$

The linear relation between λ and M^2 predicted by this equation has been verified by experiment. Now, in measurements on "ordinary" polycrystalline specimens, it is sometimes impossible for the investigator to reach magnetic saturation because his field source is not strong enough. He then often finds that a plot of λ versus M^2 is reasonably straight in the high-field region, namely, for values of M greater than about 0.7 M_s or 0.8 M_s. He can then extrapolate this line to saturation to find λ_s. The reason for this behavior is that the magnetization of most specimens changes almost entirely by rotation in the high-field region, just as the magnetization of the stressed nickel specimen changes by rotation over the whole range of M, from 0 to M_s.

8.7 APPLICATIONS OF MAGNETOSTRICTION

When a material is subjected to an alternating magnetic field, the variation of B (or M) with H traces out a hysteresis loop. At the same time, the variation of λ with H traces out another loop. Actually, the latter is a double loop, as illustrated for nickel in Fig. 8.26, because the magnetostrictive strain is independent

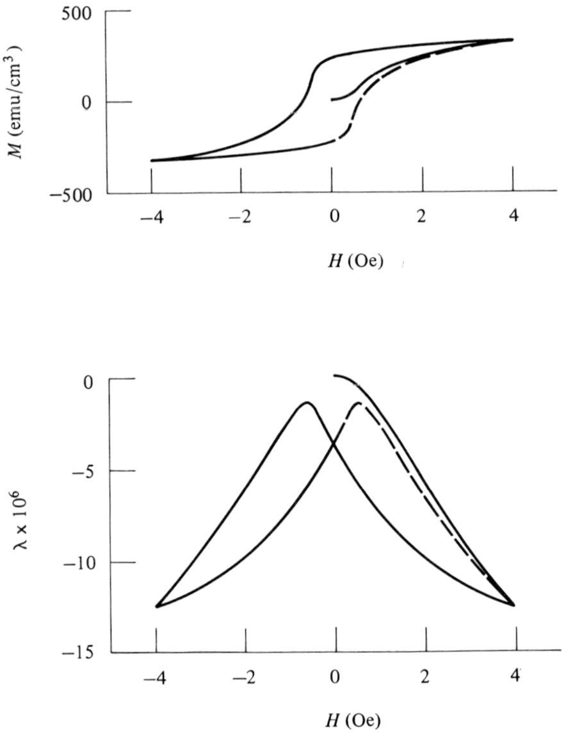

Fig. 8.26 Hysteresis in the magnetostriction of nickel, as measured by McKeehan and Cioffi [8.12].

of the sense of the magnetization. The material therefore vibrates at twice the frequency of the field to which it is exposed. This magnetostrictive vibration is the major source of the humming sound emitted by transformers. (A transformer contains a "core" of magnetic material subjected to the alternating magnetic field generated by the alternating current in the primary winding.)

Conversely, if a partially magnetized body is mechanically vibrated, its magnetization will vary in magnitude about some mean value because of the inverse magnetostrictive effect, and this alternating magnetization will induce an alternating emf in a coil wound around the body.

These two effects are exploited in the *magnetostrictive transducer*. It is one of a large group of electromechanical transducers which can convert electrical energy into mechanical energy, and vice versa. The shape and size of a magnetostrictive transducer depends on the nature of the application; one simple form is a ring carrying a toroidal winding, as in Fig. 1.10. Magnetostrictive transducers are best suited for operation at frequencies of about 20 to 100 kHz. They have two main applications:

1. *Underwater sound.* The detection of an underwater object, such as a submarine, is accomplished by a *sonar* (*so*und *na*vigation and *r*anging) system. An "active" sonar generates a sound signal with a transmitting transducer and listens for the sound reflected from the submerged object with a receiving transducer, called a *hydrophone.* A "passive" sonar listens for sound, such as engine noise, generated by the submerged object. An *echo sounder* is an active sonar designed to measure the depth of the ocean bottom. While magnetostrictive transducers still have a role to play in underwater sound applications, they have been largely replaced by *piezoelectric transducers.* (A piezoelectric material, like barium titanate or quartz, is one which becomes electrically polarized when it is mechanically strained; i.e., electric charges of opposite sign are formed on its ends. Conversely, it becomes strained when subjected to an electric field. The analogy between these electrical effects and certain magnetostrictive effects have caused magnetostrictive transducers to be sometimes called *piezomagnetic* transducers. The analogy is not exact, however, and the word *piezomagnetic* should be reserved for those few antiferromagnetic single crystals which develop a small magnetic moment when mechanically strained in the absence of a field; CoF_2 and MnF_2 are examples [G.32].) The reader should note that considerable confusion is possible when one moves from the literature of magnetism to the literature of transducers (acoustics), or vice versa, because the same symbol, and even the same name, are applied to quite different quantities in the two areas. Thus, in acoustics the quantity dP/dB at constant strain, where P is stress and B is induction, is called the "magnetostriction constant" and is designated λ, whereas the field-induced magnetostrictive strain is called S.

2. *Ultrasonic sound generators.* These are also transducers, but here the emphasis is not on sound as a signal but on sound as a mechanical disturbance, usually in some liquid medium, capable of bringing about some desired effect. Probably the most important application of this kind is ultrasonic cleaning of metal and other parts during manufacturing operations, such as before electroplating or final assembly. The parts to be cleaned are immersed in a solvent which is agitated by an ultrasonic generator. Removal of dirt and grease is much faster by this method than by simple immersion.

Magnetostriction finds another application in the *acoustic delay line.* A delay line is a device designed to delay the passage of an electrical signal for a predetermined length of time. A purely electrical line, such as a coaxial cable, will produce delays of the order of microseconds, due chiefly to the capacitance of the cable. For longer delays, of the order of milliseconds, acoustic delay lines are needed; these involve the conversion of an electrical pulse into an acoustic pulse. Since the velocity of an electrical signal along a conducting wire is about equal to the velocity of light, 3×10^{10} cm/sec, and the velocity of sound in a metal is about 5×10^5 cm/sec, an acoustic pulse travels with a velocity some 10^{-5} times that of an electrical pulse.

A magnetostrictive delay line (Fig. 8.27a) consists of a thin rod or wire, usually of nickel or a nickel alloy, with a transmitting coil T wound on one end and a receiving coil R wound on the other. When a pulse of current is fed into the

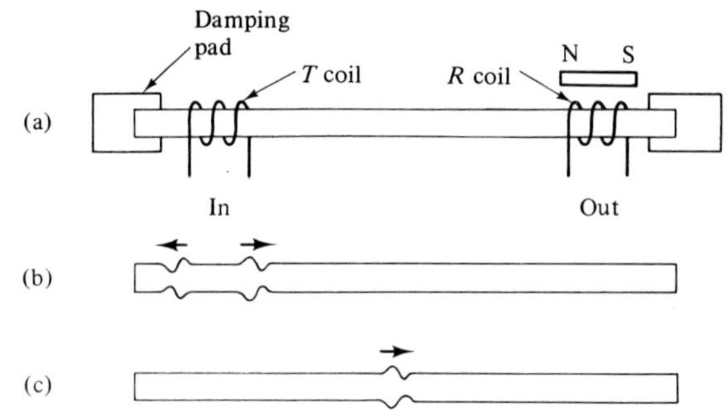

Fig. 8.27 Magnetostrictive delay line.

T coil, a momentary magnetic field is produced within the coil, parallel to the axis of the nickel rod. The rod therefore contracts axially (negative magnetostriction) in this region and expands transversely, sending out an acoustic pulse in each direction. [These pulses have approximately the shape sketched in (b). When the field pulse begins, the local axial contraction sends a tensile wave down the rod, contracting the diameter. This is immediately followed, when the field pulse stops, by a compression wave, expanding the diameter.] The acoustic pulse (strain wave) travels down the rod, as in (c), and reaches the *R* coil. Near this is a small permanent magnet, whose external field keeps the nickel rod within the *R* coil in a state of partial magnetization. When the strain wave reaches this part of the rod, it changes the extent of magnetization through the inverse magnetostrictive effect. The external field of the rod therefore changes and a momentary emf (output pulse) is induced in the *R* coil. Damping pads at each end of the rod act as acoustic absorbers to prevent reflection of pulses from the ends. The travel time of the acoustic pulse is about 5 microseconds per inch of line. When the length required for a particular delay time exceeds a few inches, the rod is coiled into a flat spiral to save space. Such delay lines are used in radar circuits and other applications.

If the damping pads are omitted and the rod properly supported, an alternating emf applied to the *T* coil will cause the whole rod to vibrate at its resonant frequency, thus generating an alternating emf in the *R* coil. This process is efficient only when the input frequency is equal, or nearly equal, to the resonant frequency of the rod, which is fixed by its dimensions. The device then acts as an *electromechanical (magnetostrictive) filter*, passing only a narrow range of frequencies.

The delay line of Fig. 8.27 (a) can also serve as a *storage device*, or *memory*. Suppose information in digital form is represented by a series of closely spaced electrical pulses fed into coil *T*. A series of acoustic pulses will then travel along the rod and produce a series of output pulses in coil *R*. These pulses are fed into an amplifier, to make up for losses, and the output pulses from the amplifier are applied to coil *T*. The series of pulses then circulates indefinitely in the closed electrical-mechanical circuit until such time as it is led off, through appropriate switches, into some other circuit. Such storage devices are used in automatic telephone exchanges and in some special-purpose computers. Most computers, however, have memory units made of the ferrite cores described in Section 13.6.

8.8 ΔE EFFECT

Another consequence of magnetostriction is a dependence of Young's modulus E of a material on its state of magnetization. When an originally demagnetized specimen is saturated, its modulus *increases* by an amount ΔE. The value of $\Delta E/E$ is about 6 percent for nickel and less than 1 percent for iron.

When a stress is applied to a demagnetized specimen, two kinds of strain are produced:

1. Elastic ε_{el}, such as occurs in any substance, magnetic or not.
2. Magnetoelastic ε_{me}, due to the reorientation of domain vectors by the applied stress. This strain is zero in the saturated state, because no domain reorientation can occur. For an applied tensile stress, ε_{me} is always positive, whatever the sign of λ_{si}. (If a rod of positive λ_{si} is stressed in tension, M_s vectors will rotate toward the axis, and the rod will lengthen. If λ_{si} is negative, M_s vectors will rotate away from the axis, and the rod will lengthen.)

As a result of these two kinds of strain, the modulus in the demagnetized state is

$$E_d = \frac{\sigma}{\varepsilon_{el} + \varepsilon_{me}}, \tag{8.32}$$

and the modulus in the saturated state is

$$E_s = \frac{\sigma}{\varepsilon_{el}}. \tag{8.33}$$

These two relations lead to

$$\frac{\Delta E}{E} = \frac{E_s - E_d}{E_d} = \frac{\varepsilon_{me}}{\varepsilon_{el}}. \tag{8.34}$$

We have seen in Section 8.6 that ε_{me} depends on the magnitude of the applied stress and on the strength of whatever other anisotropy, such as crystal anisotropy, is present. Figure 8.28 shows three kinds of stress-strain curves, with the differences between them much exaggerated. One is for a saturated specimen, and the other two are for demagnetized specimens: (a) one with some strong other anisotropy, and (b) one with weak anisotropy. In (b) the maximum magnetoelastic strain is soon developed, and the curve rapidly becomes parallel to the curve for the saturated specimen. In (a) this occurs only at a higher stress, because of the stronger opposing anisotropy. We see therefore that the ΔE effect is actually rather complicated, in that E depends, not only on the degree of magnetization, but also on the stress (or strain) and the strength of the other anisotropy present. Conventionally, ΔE is taken as the modulus difference for a vanishingly small stress; i.e., it is the difference in the initial slopes of the stress-strain curves of a demagnetized and a saturated specimen.

The ΔE effect is a special case of what is more generally called the *modulus defect*. Whenever any mechanism is present which can contribute an extra strain (inelastic strain) over and above the elastic strain, the modulus will be smaller than normal. Magnetoelastic strain

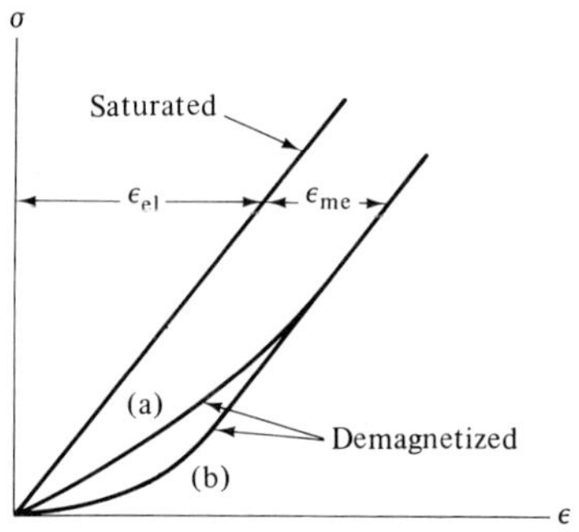

Fig. 8.28 Elastic and magnetoelastic components of tensile strain.

is just one example of such an extra strain; other examples are the strain contributed by dislocation motion and the strain due to carbon atoms in iron moving into preferred positions in the lattice.

8.9 MAGNETORESISTANCE

The *magnetoresistance effect* is a change in the electrical resistance R of a substance when it is subjected to a magnetic field. The value of $\Delta R/R$ is extremely small for most substances, even at high fields, but is relative large (a few percent) for strongly magnetic substances. The resistance of nickel increases about 2 percent, and that of iron about 0.3 percent, when it passes from the demagnetized to the saturated state. The magnetoresistance effect is mentioned here because of its close similarity to magnetostriction, and because a magnetoresistance measurement, like a magnetostriction measurement, can disclose the presence or absence of preferred domain orientations in the demagnetized state.

As in magnetostriction, we will be interested only in the change in R between the demagnetized and saturated states (the region of domain wall motion and domain rotation), and we will ignore the small change in R that occurs during forced magnetization beyond saturation. A magnetoresistance measurement is usually made on a rod or wire specimen with the measuring current i and the applied field H both parallel to the rod axis. Then, in most ferromagnetics, the observed effect is an *increase* in R of any domain as the angle θ between the current i and the M_s vector of that domain *decreases*. The physical origin of the magnetoresistance effect lies in spin-orbit coupling: as M_s rotates, the electron cloud about each nucleus deforms slightly, as shown by the existence of magnetostriction, and this deformation changes the amount of scattering undergone by the conduction electrons in their passage through the lattice.

The relation between the saturation magnetoresistance change $(\Delta R/R)_s$ and the initial domain arrangement is exactly analogous to Eq. (8.28) for magnetostriction, namely,

$$\left(\frac{\Delta R}{R}\right)_s = \frac{3}{2}\left(\frac{\Delta R}{R}\right)_{si}\left[1 - \langle\cos^2\theta\rangle_1\right], \tag{8.35}$$

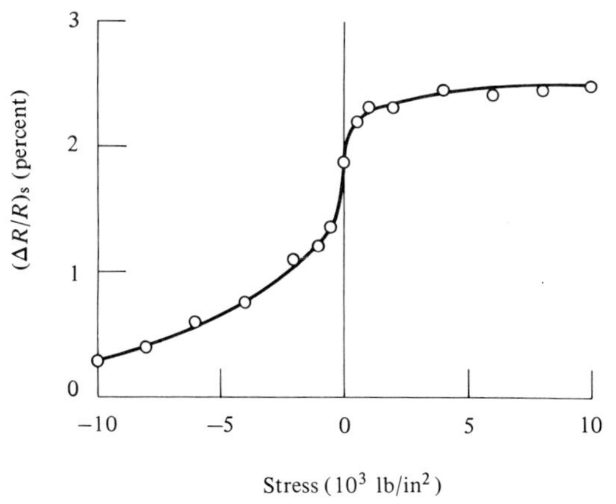

Fig. 8.29 Effect of applied stress on the saturation magnetoresistance of nickel. Bagchi *et al.* [8.13].

where $(\Delta R/R)_{si}$ is the change observed when a specimen is brought from the ideal demagnetized state to saturation.

Stress alone can change the resistance, an effect called *elastoresistance*, because stress alters the orientations of the M_s vectors. Stress can therefore change the observed magnetoresistance, just as it can change magnetostriction. The nature of these changes can be shown by a diagram exactly analogous to Fig. 8.25, with $(\Delta R/R)$ as the ordinate instead of strain, for a material with positive λ_{si}. Tensile stress applied along the specimen axis causes M_s vectors to rotate toward the axis, resulting in an increase of resistance from O to A. This is the elastoresistance change, denoted $(\Delta R/R)_\sigma$. When a field is then applied, the resistance changes from A to S along the curve marked $+\sigma$, and the result is that the saturation magnetoresistance measured under tension $(\Delta R/R)_{s(t)}$ is less than the normal value $(\Delta R/R)_{si}$. Compression has the opposite effect, because a negative elastoresistance change, from O to B, is involved.

These effects are reversed for a material with negative λ_{si} like nickel. Tension will then increase $(\Delta R/R)_s$ and compression will decrease it, as shown in Fig. 8.29. These results are very similar to the corresponding data on magnetostriction in Fig. 8.22, especially if it is noted that tension acts to increase the magnitude of both effects.

We see then that a preferred domain arrangement can be disclosed by either a magnetostriction or a magnetoresistance measurement. Which one is chosen depends only on experimental convenience.

PROBLEMS

8.1 With reference to Fig. 8.3, find an expression for the strain in [100] in a single domain of a cubic substance as M_s rotates away from [001] by an angle δ toward [100].

8.2 Find an expression for the strain in [110] in a single domain of a cubic substance as M_s rotates away from [100] toward [110] by an angle γ in the (001) plane.

8.3 Find an expression for the strain in [111] in a single domain of a cubic substance as M_s rotates away from [111] toward [001] by an angle θ in the $(1\bar{1}0)$ plane.

8.4 Find the saturation magnetostriction λ_{si} of a cubic crystal in the direction $\langle 110 \rangle$ in terms of λ_{100} and λ_{111}.

8.5 In order to determine the constants λ_{100} and λ_{111} of a cubic substance, a single-crystal disk is cut parallel to the plane $(1\bar{1}0)$. A strain gage is cemented to one side of the disk parallel to [001], and another gage is cemented to the other side parallel to [110]. The disk is placed in a strong field, with [001] initially parallel to the field. The disk is then rotated in its own plane in the direction which increases the angle between [001] and the field and decreases the angle between [110] and the field. Derive an expression for the strain indicated by each gage, measured from the initial position of the disk, as a function of the rotation angle θ from this position.

8.6 Compute λ_{si} for a cobalt crystal for angles of 60° and 90° between λ_{si} (and M_s) and the c axis. Compare with Fig. 8.9.

8.7 The M_s vector of a single domain of a cubic substance points in the [010] direction. A small tensile stress σ is then applied along the [100] axis. How does the direction of M_s change as σ is increased from zero to a large value? λ_{100}, λ_{111}, and K_1 are positive, and K_2 is zero.

8.8 Show that the average of $\cos^2 \theta$ is 1/3 for domains oriented randomly in space.

8.9 A nickel rod is subjected to a tensile stress of 17,000 lb/in^2 and is then magnetized to saturation. Compute the elastic and magnetoelastic strains.

a) Construct a diagram like Fig. 8.25 except (i) include both ε_{el} and ε_{me}, and (ii) draw the strain axis to scale. Take the modulus of demagnetized nickel as 30×10^{-6} lb/in^2, and take the values of λ_{si} and of λ_s under stress from Kirchner's data in Fig. 8.22.

b) What is $\Delta E/E$ in percent, evaluated at 17,000 lb/in^2?

CHAPTER 9

DOMAINS AND THE MAGNETIZATION PROCESS

9.1 INTRODUCTION

The two previous chapters, on anisotropy and magnetostriction, were in a sense merely prologue to the present one. There we were concerned only incidentally with the processes of domain wall motion and domain rotation. Now we must examine these processes in detail in order to better understand how they contribute to the magnetization process. To do this requires study of the domain itself, particularly with respect to the structure and orientation of the walls which surround it.

The magnetic domain, in theory and experiment, has had a curious history. Of the two great concepts introduced by Weiss in 1906, namely, the domain hypothesis and that of spontaneous magnetization by a molecular field, Weiss himself stressed only the latter. Later investigators added very little. For a period of 43 years they made virtually no application of the domain idea to the problems of explaining the shape of a magnetization curve or the mechanism of magnetic hysteresis. It is true that during this period some isolated, though important, theoretical work was done, to be referred to later, but the domain hypothesis was not brought into the mainstream of research on magnetic materials. Not until 1949 was there any direct experimental evidence for, and clear understanding of, the domain structure of a real material; in that year Williams, Bozorth, and Shockley [9.1] published their work, performed at the Bell Telephone Laboratories, on domains in silicon-iron single crystals. Since that time, domain theory has become absolutely central to any discussion of magnetization.

Domain theory and observation have been treated in review articles by Kittel [8.4], Kittel and Galt [9.2], Craik and Tebble [9.3], and Dillon [9.4], in the general book by Chikazumi [G.23], and in specialized books by Craik and Tebble [G.26] and Carey and Isaac [G.29].

9.2 DOMAIN WALL STRUCTURE

Domain walls are interfaces between regions in which the spontaneous magnetization has different directions. At or within the wall the magnetization must change direction.

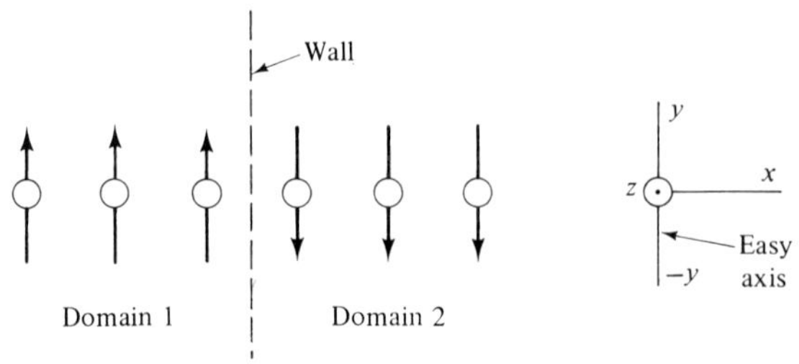

Fig. 9.1 Hypothetical, infinitely thin 180° wall.

We might at first imagine this change to be abrupt, occurring from one atom to the next as in Fig. 9.1. A row of atoms is shown, parallel to x, with the 180° domain wall lying in the yz plane; the easy axis is $\pm y$. But the exchange energy in a ferromagnetic is a minimum only when adjacent spins are parallel. Therefore, the wall of Fig. 9.1 would have a large exchange energy associated with it, because the spins adjacent to the wall are antiparallel. This exchange energy can be decreased if we allow the 180° change in spin direction to take place gradually over N atoms, so that the angle ϕ between adjacent spins, equal to π/N, has some small value, as in Fig. 9.2, which is drawn for $\phi = 30°$. The total exchange energy is then reduced because, as shown below, it varies as ϕ^2 rather than as ϕ.

On the other hand, the spins within the wall of Fig. 9.2 are pointing in noneasy directions, so that the crystal anisotropy energy within the wall is higher than it is in the adjoining domains. While the exchange energy tries to make the wall as wide as possible, in order to make the angle ϕ between adjacent spins as small as possible, the anisotropy energy tries to make the wall thin, in order to reduce the number of spins pointing in noneasy directions. (The hypothetical wall of Fig. 9.1 has no extra anisotropy energy.) As a result of this competition, the wall has a certain finite width and a certain structure. Also, like any other interface (such as a grain, twin, or phase boundary), the wall has a certain energy per unit area of its surface, because the spins in it are not quite parallel to one another and not parallel to an easy axis. The first theoretical examination of the structure of a domain wall was made by Bloch in 1932, and domain walls are accordingly often called *Bloch walls*.

We can make an approximate calculation of domain wall width and energy as follows. According to Eq. (4.29), the exchange energy for a pair of atoms of the same spin S is

$$E_{ex} = -2JS^2 \cos\phi, \tag{9.1}$$

where the subscript on the exchange integral J has been dropped. The series expansion of $\cos\phi$ is

$$\cos\phi = 1 - \frac{\phi^2}{2} + \frac{\phi^4}{24} - \cdots. \tag{9.2}$$

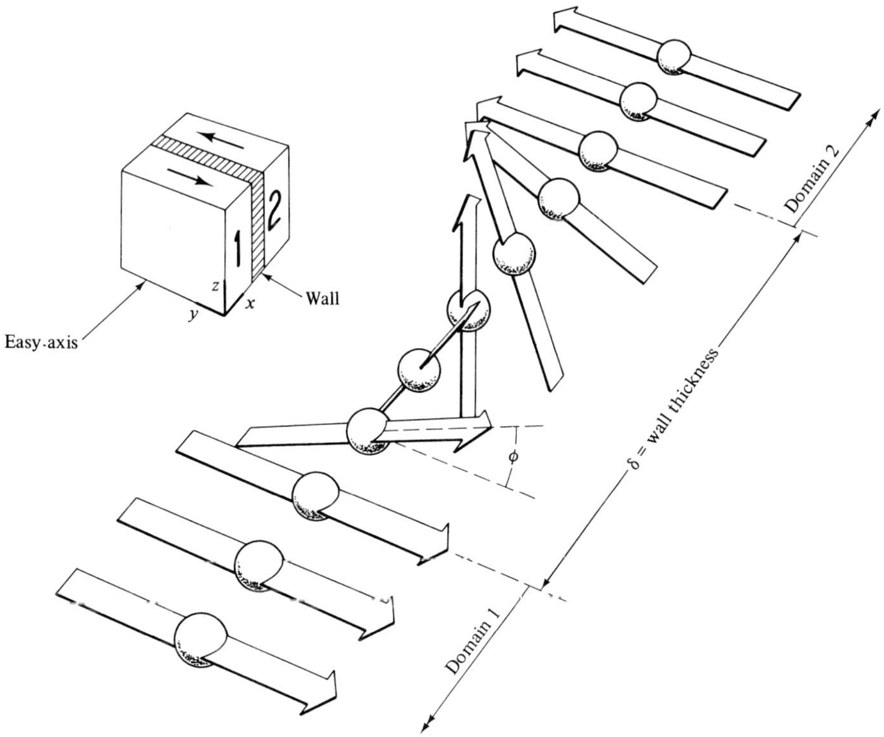

Fig. 9.2 Structure of a 180° wall.

Dropping the term in ϕ^4 and higher powers, because ϕ is small, and substituting into Eq. (9.1), we have

$$E_{ex} = JS^2 \phi^2 - 2JS^2. \tag{9.3}$$

The second term is independent of angle and has the same value within a domain as within the wall, and it can therefore be dropped. The *extra* exchange energy existing within the wall is given by the first term, $JS^2 \phi^2$, per spin pair.

To find the exchange energy per unit area of domain wall, we must make an assumption about the crystal structure. For simplicity we assume simple cubic, with an atom at each corner of a cell of edge a, and the plane of the wall parallel to a cube face $\{100\}$. The wall is N atoms thick, and, per unit area of wall, there will be $1/a^2$ rows of N atoms. Therefore, the extra exchange energy per unit area of wall is

$$\gamma_{ex} = (JS^2 \phi^2)(N)(1/a^2). \tag{9.4}$$

Putting $\phi = \pi/N$ for a 180° wall, we have

$$\gamma_{ex} = \frac{JS^2 \pi^2}{Na^2}. \tag{9.5}$$

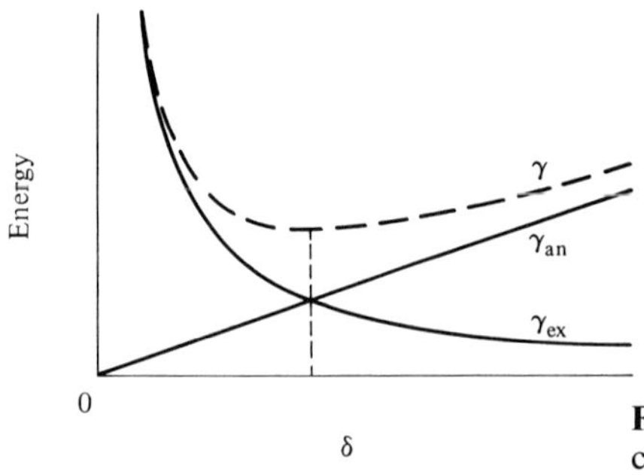

Fig. 9.3 Dependence of total wall energy γ (dashed curve) on wall thickness δ.

The anisotropy energy is of the order of the anisotropy constant K times the volume of the wall. Therefore, per unit area of wall,

$$\gamma_{an} = KNa. \tag{9.6}$$

The total wall energy per unit area, for a wall of thickness $\delta = Na$, is

$$\gamma = \gamma_{ex} + \gamma_{an} = \frac{JS^2 \pi^2}{\delta a} + K\delta. \tag{9.7}$$

This energy has a minimum for a particular value of δ, as shown by the dashed curve of Fig. 9.3, which also shows how the exchange and anisotropy energies vary with δ. This minimum is given by

$$\frac{d\gamma}{d\delta} = -\frac{JS^2 \pi^2}{\delta^2 a} + K = 0, \tag{9.8}$$

$$\delta = \sqrt{\frac{JS^2 \pi^2}{Ka}}. \tag{9.9}$$

According to Eq. (4.31), the exchange integral J is proportional to the Curie temperature T_c. Therefore,

$$\delta \sim \sqrt{T_c/K}.$$

The smaller the anisotropy constant, the thicker the wall; therefore, wall thickness increases with temperature, because K almost always decreases with rising temperature. Putting Eq. (9.9) into (9.7), we find

$$\gamma = \sqrt{\frac{JS^2 \pi^2 K}{a}} + \sqrt{\frac{JS^2 \pi^2 K}{a}} = 2K\delta. \tag{9.10}$$

Therefore, the minimum in the total energy occurs when the exchange and anisotropy energies are equal, as shown in Fig. 9.3.

We will now calculate δ and γ for iron and nickel. Neither has a simple cubic structure, but we ignore this and put a equal to the distance of closest approach (distance between centers of nearest-neighbor atoms) in each metal. The value of J can be taken as roughly equal to 0.3 kT_c, in accordance with the discussion near the end of Section 4.3. With $S = 1/2$, Eq. (9.9) becomes

$$\delta = \sqrt{\frac{0.3\, kT_c\, \pi^2}{4Ka}}. \tag{9.11}$$

With $T_c = 1043°\text{K}$, $K = 4.8 \times 10^5$ ergs/cm³, and $a = 2.48$ Å, we have for iron

$$\begin{aligned}\delta &= \sqrt{\frac{0.3\,(1.38 \times 10^{-16})(1043)\,\pi^2}{4(4.8 \times 10^5)(2.48 \times 10^{-8})}}\\ &= 3.0 \times 10^{-6}\ \text{cm} = 300\ \text{Å}\\ &= \frac{300}{2.48} = 120\ \text{atoms}.\end{aligned}$$

The angle ϕ between adjacent spins is therefore 180°/120 or 1.5°, much smaller than the value of 30° shown in Fig. 9.2. The expression for the wall energy becomes

$$\begin{aligned}\gamma &= 2\sqrt{\frac{JS^2\,\pi^2 K}{a}} = \sqrt{\frac{0.3\, kT_c\, \pi^2 K}{a}}\\ &= \sqrt{\frac{0.3(1.38 \times 10^{-16})(1043)(\pi^2)(4.8 \times 10^5)}{2.48 \times 10^{-8}}}\\ &= 2.9\ \text{ergs/cm}^2.\end{aligned} \tag{9.12}$$

More exact calculations give values of γ ranging from about 1 to 2 ergs/cm² and show that spin rotation through the wall is not exactly uniform, as assumed here. Instead, the angle ϕ between adjacent spins is somewhat smaller near either edge of the wall than in the center.

Similar calculations for nickel, with $K = 0.5 \times 10^5$ ergs/cm³, $T_c = 631°\text{K}$, and $a = 2.49$ Å, lead to

$$\delta = 720\ \text{Å} = 290\ \text{atoms},$$

$$\gamma = 0.7\ \text{erg/cm}^2.$$

Calculations for other materials yield similar results, leading to the general conclusion that domain walls are several hundred angstroms thick and have energies of a few ergs per square centimeter. [While the energy per unit area is rather small, as interface energies go, the energy *density* within a wall is large because the wall is so thin. For example, with the values obtained above for iron, the energy density in a 180° wall is $(2.9)/(1)(3 \times 10^{-6})$ or 10^6 ergs/cm³. This is about five times the density of crystal anisotropy energy stored in an iron crystal when it is saturated in a hard ⟨111⟩ direction.]

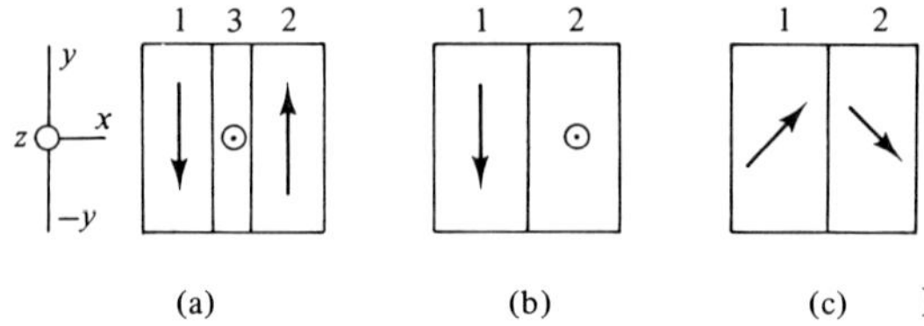

Fig. 9.4 (a) Hypothetical splitting of 180° wall. (b) and (c) Two types of 90° wall.

Various methods have been applied to the *measurement* of the wall energy γ. The results are not yet particularly reliable, as partly evidenced by the fact that wall energies are never stated to more than two significant figures. Better methods of directly measuring γ are needed, not only because of interest in the wall energy itself, but also because such measurements would lead, through Eq. (9.10), to better values of the fundamental constant J, the exchange integral.

What is the structure of domain walls in ferrimagnetics? A drawing of such a wall would necessarily be more complicated than Fig. 9.2, because of the presence of chemically different atoms with oppositely directed spins. However, one can still think of the spin axis as slowly rotating through the wall from one domain to the other.

In our discussion of the wall of Fig. 9.2, we have tacitly assumed that we were dealing with a uniaxial crystal. However, suppose the material is iron, with $\langle 100 \rangle$ easy axes. Then the $+z$ direction is an easy one, as well as $\pm y$. One might then expect the spins in the middle of the wall to "lock in" to the easy $+z$ direction over a considerable distance; the 180° wall would then split into two 90° walls, and a new domain 3 would form between domains 1 and 2 as in Fig. 9.4(a). However, domain 3 would not fit in properly. It would magnetostrictively lengthen in the $+z$ direction and contract in $\pm y$, whereas domains 1 and 2 are lengthened along $\pm y$. If domain 3 were constrained to fit, the magnetoelastic energy of the system would go up. As a result, the expected splitting does not occur, and the 180° wall remains intact.

Two types of 90° wall are possible, and they have different structures. One kind is shown in Fig. 9.4(b); the spins in the adjoining domains are parallel to the wall, which therefore has a structure identical with half of a 180° wall. In the other kind, represented in Fig. 9.4(c), the spins in the domains are at 45° to the wall. The structure of this kind is harder to visualize: the spins rotate through the wall in such a way that they make a constant angle of 45° both with the wall normal and the wall surface.

The energy of a domain wall also depends on the orientation of the wall in the crystal. This point is discussed in Section 9.5.

9.3 DOMAIN WALL OBSERVATION

Domains are normally so small that one must use a microscope to see them. What one sees then depends on the technique involved. The most popular techniques fall into two groups:

1. Those which disclose domain *walls* (Bitter method, electron microscope). The individual domains, whatever their direction of magnetization, look more or less the same, but the domain walls are delineated.

2. Those which disclose *domains* (optical methods involving the Kerr or Faraday effects). Here domains magnetized in different directions appear as areas of different color, and the domain wall separating them appears merely as a line of demarcation where one color changes to the other.

Methods of the first type are described in this section.

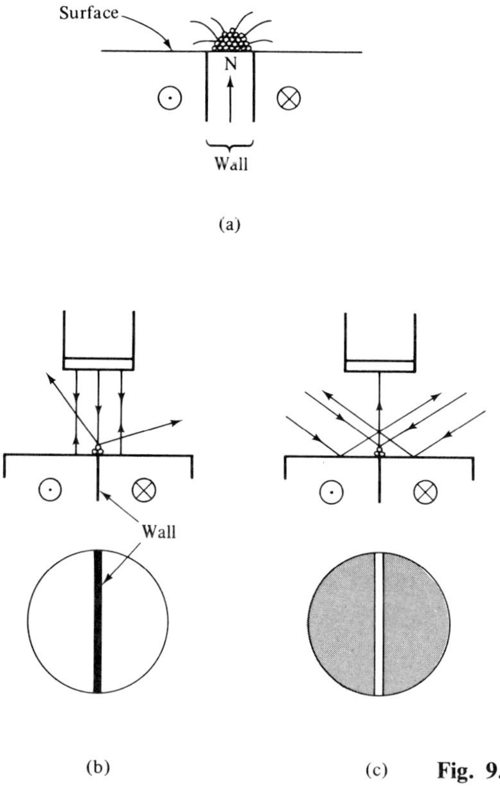

Fig. 9.5 Bitter method of observing domain walls.

Bitter method

The Bitter or powder method is the most popular of all because of its greater convenience. Devised by Bitter [9.5] in 1931, it involves the application of an aqueous suspension of extremely fine (colloidal) particles of magnetite Fe_3O_4 to the polished surface of the specimen. Imagine a 180° wall intersecting the surface, as in Fig. 9.5 (a), where the spins in the wall are represented simply by the one in the center, normal to the surface. A north pole is therefore formed as shown,

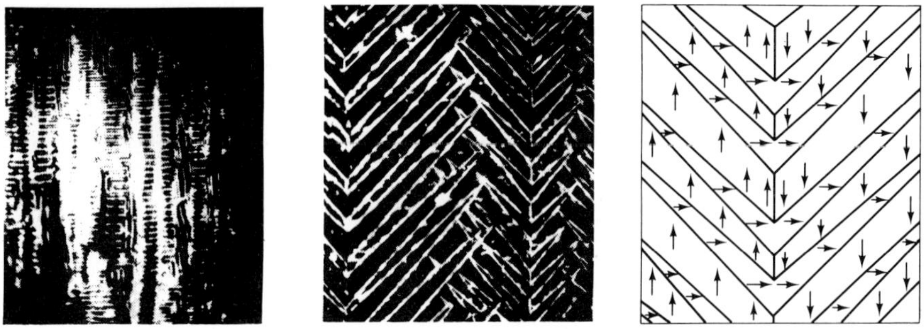

(a) Powder pattern on a mechanically polished surface

(b) Pattern on the same surface, electrolytically polished

(c) Interpretation of pattern in (b); arrows show directions of domain magnetization

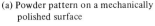

Crystal axes +

0.1 mm

Fig. 9.6 Powder patterns on a single crystal of iron + 3.8 percent silicon, dark-field illumination. Williams *et al.* [9.1], Bozorth [9.6].

and this is the origin of an *H* field which spreads out above the surface as indicated. The fine particles of magnetite are attracted to this region of nonuniform field, depositing as a band along the edge of the domain wall, normal to the plane of the drawing. If the surface is then examined with a metallurgical (reflecting) microscope under the usual conditions of "bright-field" illumination, as in Fig. 9.5(b), the domain wall will show up as a dark line on a light background; the domains on either side of the wall reflect the vertically incident light back into the microscope and so appear light, while the particles on top of the wall scatter the light to the side and appear dark. Under "dark-field" illumination (c), the incident light strikes the specimen at an angle; the domain wall appears as a light line on a dark background, because only light reflected from the particles along the wall is scattered into the microscope. In either case, the arrangement of lines (domain walls) seen under the microscope is commonly called a *powder pattern* or *Bitter pattern.*

Chikazumi [G.23] gives the following method of preparing the colloidal magnetic suspension that is used in the Bitter method. Dissolve 2 g $FeCl_2 \cdot 2H_2O$ and 5.4 g $Fe_3Cl_3 \cdot 6H_2O$ in 300 cc of water and maintain the solution at 30° to 40°C. To precipitate Fe_3O_4, slowly add a solution of NaOH (5 g in 50 cc of water) with vigorous stirring. Filter out the precipitate and wash it several times with water. The final suspension is made by adding 1 cc of precipitate to about 30 cc of a 0.3 percent soap solution. A drop of the suspension is then placed on the surface to be examined, and covered with a thin microscope cover glass to spread out the suspension into a uniform film.

Careful specimen preparation is extremely important. The surface of a metallic specimen is first mechanically polished and then electrolytically polished to remove the strained layer produced by the mechanical polishing. (Electropolishing is accomplished by making the specimen the anode in an electrolytic cell and passing a fairly heavy current.) This second step is essential. Early workers did not realize this and obtained "maze patterns" similar to Fig. 9.6(a). This pattern

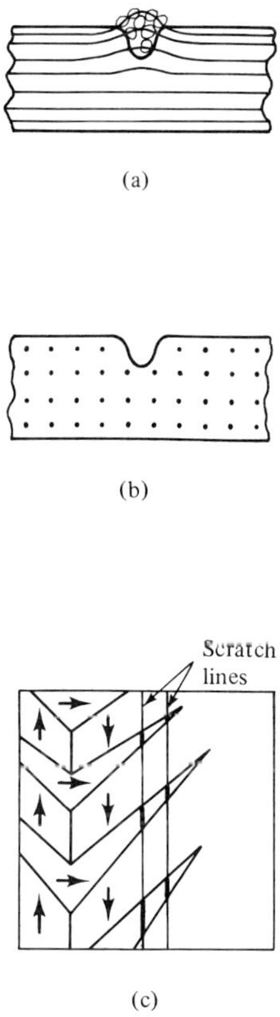

Fig. 9.7 Effect of scratches on colloid deposition.

is determined entirely by the strains left in the surface after mechanical polishing and reveals nothing of the true domain structure shown in (b). (The interpretation of this rather complex structure will be given later.) Ferrites, being nonconductors, cannot be electropolished, but strain-free surfaces can be prepared by fine mechanical polishing with diamond dust, followed by heating to about 1200°C, according to Craik and Tebble [G.26].

The essential features of most domain structures can be seen clearly at magnifications of a few hundred diameters.

The Bitter method can detect moving, as well as stationary, domain walls. When a wall moves in response to an applied field or stress, the line of colloid particles follows the intersection of the wall with the surface, as long as the wall is not moving too fast. Observation of how walls move, causing one domain to grow at the expense of another, can be very fruitful. For this reason it is usual to have some means, such as Helmholtz coils or a small electromagnet, of applying a field to the specimen while it is under the microscope.

When greater detail is desired, static Bitter patterns may be examined at the much higher magnification of the electron microscope by a replica technique. Some water-soluble plastic is added to the magnetite suspension, which is then spread on the specimen and allowed to dry. The powder particles go to the domain walls, as usual, and are trapped in these positions as the suspension dries. The result is a thin film, containing the powder particles, which can be peeled away from the specimen surface and examined in transmission in the electron microscope.

The first step in understanding an observed domain structure is to determine the directions in which the various domains are magnetized. This can be done by observing the behavior of the colloid particles at accidental or deliberate scratches, or other irregularities, on the specimen surface. Figure 9.7 shows the principle involved. When M_s in a particular domain is parallel to the surface and at right angles to a scratch, as in (a), the flux lines tend to bow out into the air at the scratch, and this nonuniformity of field attracts the powder particles. This effect does not occur when M_s is parallel to the scratch, as in (b), and few powder particles are attracted. Thus scratches crossing a domain structure like that of (c) will appear dark when they are at right angles to M_s and light when they are parallel. It remains then to determine the *sense* of the M_s vector in each domain; this is done by applying a field parallel to the M_s axis of a certain domain, and noting whether that domain grows in volume or shrinks. If it grows, then M_s must be parallel, rather than antiparallel, to the applied field.

(The tendency of powder particles to collect at surface flaws which are transverse to the magnetization is exploited, on a much grosser scale, in the Magnaflux method of detecting cracks in steel objects. The object to be inspected is magnetized by a strong field and immersed in a suspension of magnetic particles. When withdrawn, previously invisible cracks are made visible to the naked eye by the powder particles attracted to them.)

The Bitter method has two limitations: (1) If the anisotropy constant K of the material becomes less than about 10^3 ergs/cm^3, the domain walls become so broad that powder particles are only weakly attracted to them, and (2) the method can be applied only over a rather restricted temperature range.

Transmission electron microscope

This instrument can disclose domain walls in specimens thin enough, about 1000 A or less, to transmit electrons. Since a moving electric charge is acted on by a force when it is in a magnetic field, the electrons passing through the specimen will be deviated by an amount and in a direction determined by the magnitude and direction of the local M_s vector. In a domain wall this vector has different orientations at different positions within the wall; the result is that the wall shows up as a line, either dark or light, on the image of the specimen. However, the microscope must be slightly under- or overfocused in order to make the wall visible. [This technique is often called *Lorentz microscopy*, because the force F on the electron is known as the Lorentz force. This force, on an electron of charge $-e$, is given by $-e/c$ ($\mathbf{v} \times \mathbf{B}$), where $\mathbf{v}$ is the electron velocity, $\mathbf{B}$ the induction, and c the velocity of light. The greater part of $\mathbf{B}$ at any point in the specimen is made up of the local value of $4\pi \mathbf{M}_s$, the balance being due to the vector sum of any applied or demagnetizing $\mathbf{H}$ fields present.]

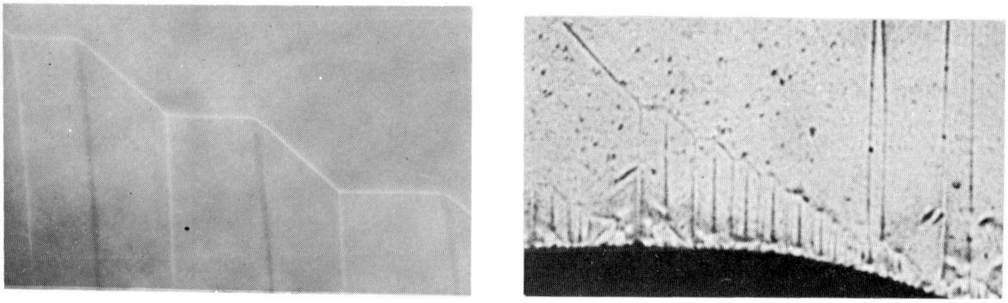

(a) (b)

Fig. 9.8 Domain photographs of iron by (a) transmission electron microscopy (5600 ×), and (b) Bitter technique (600 ×). Both photographs are of the same foil but not necessarily of the same area. Michalak and Glenn [9.7].

Two kinds of specimen are of interest:

1. Those which are thin already. These will normally have been made by evaporation or electrodeposition and are called "films." Their magnetic properties are of great interest because of digital computer applications. Description of these materials is postponed to Chapter 11.

2. Bulk specimens which have been thinned down by grinding and etching. These are usually called "foils." Figure 9.8 (a) shows an unusually clear example of domains in a foil of high-purity iron. The domain walls are the light and dark lines. Normally, crystal imperfections such as dislocations and stacking faults are also visible.

Lorentz microscopy has the advantage of high resolution, which allows the examination of the fine detail of domain structure. It also permits the direct observation of interactions between domain walls and crystal imperfections and grain boundaries, a field of study still in its infancy.

9.4 DOMAIN OBSERVATION

Two magneto-optic effects can distinguish one domain from another, either as a difference in color or in the degree of light and dark.

Kerr effect

This effect is a rotation of the plane of polarization of a light beam during *reflection* from a magnetized specimen. The amount of rotation is small, much less than one degree, and depends on the direction and magnitude of the magnetization relative to the plane of incidence of the light beam. Light from a source passes

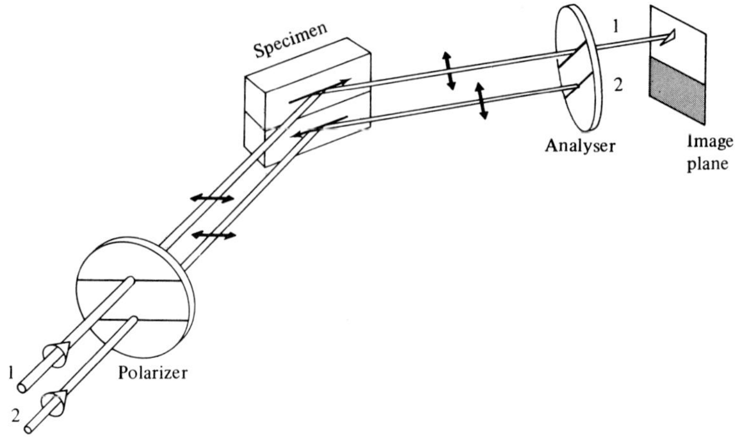

Fig. 9.9 Domain observation by the Kerr effect. After Prutton [G.24].

through a polarizer which transmits only plane polarized light (Fig. 9.9). This is then incident on the specimen, supposed to contain only two domains, magnetized antiparallel to each other as indicated by the arrows. During reflection the plane of polarization of beam 1 is rotated one way and that of beam 2 the other way, because they have encountered oppositely magnetized domains. The light then passes through an analyser and into a telescope or low-power microscope. The analyser is now rotated until it is "crossed" with respect to reflected beam 2; this beam is therefore extinguished and the lower domain appears dark. However, the analyser in this position is not crossed with respect to beam 1, because the plane of polarization of beam 1 has been rotated with respect to that of beam 2. Therefore beam 1 is not extinguished, and the upper domain appears light. Figure 11.36(c) shows domains in a thin film revealed in this way; the light and dark bands are domains magnetized in opposite directions.

The Kerr method is not easy to apply. Because of the small angle of rotation of the plane of polarization, the contrast between adjoining domains tends to be low, unless all optical parts are of high quality and well adjusted. (Further increase in contrast can be obtained by evaporating a very thin layer of zinc sulphide on to the specimen surface and inserting a quarter-wave plate in the reflected beam [G.29]. On the credit side, however, the Kerr method is ideal for studies of domain walls in motion and has largely supplanted the Bitter method for such studies. It has no limitations with respect to specimen temperature beyond the usual ones of thermal insulation and protection against oxidation. It can be applied both to bulk specimens and thin films.

(Note that the term "Kerr effect" is also applied to an *electro-optic* effect. If certain organic liquids are placed in a glass container, called a Kerr cell, and subjected to an electric field, plane polarized light passing through the cell will be rotated by an amount depending on the applied voltage.)

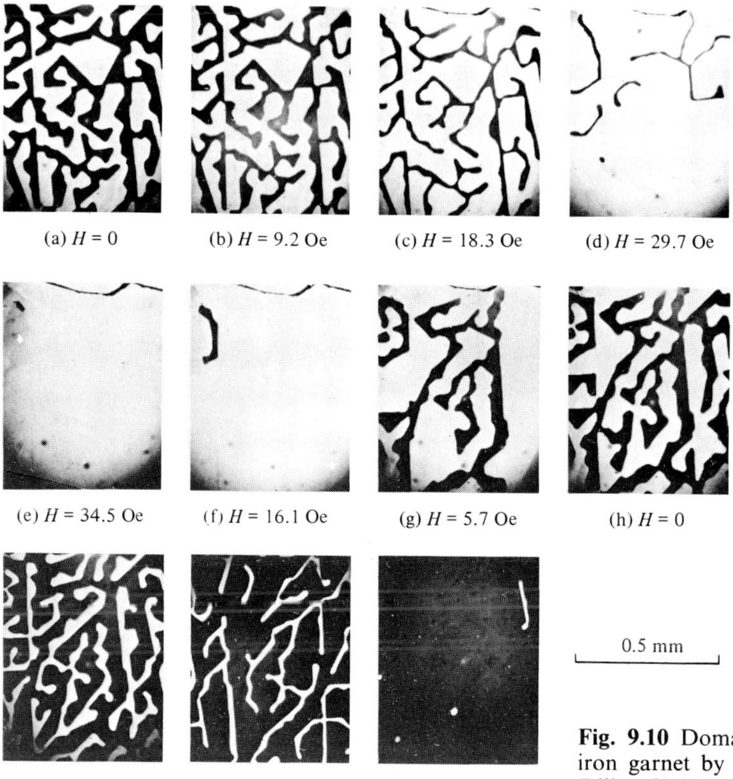

(a) $H = 0$ (b) $H = 9.2$ Oe (c) $H = 18.3$ Oe (d) $H = 29.7$ Oe

(e) $H = 34.5$ Oe (f) $H = 16.1$ Oe (g) $H = 5.7$ Oe (h) $H = 0$

(i) $H = -11.4$ Oe (j) $H = -20.1$ Oe (k) $H = -27.4$ Oe

Fig. 9.10 Domains in gadolinium iron garnet by the Faraday effect. Dillon [9.4a].

Faraday effect

This effect is a rotation of the plane of polarization of a light beam as it is *transmitted* through a magnetized specimen. The optical system is the same as the Kerr-effect system, except that source, polarizer, specimen, analyser, and microscope are all in line. The method is, of course, limited to specimens thin enough to transmit light; it is applied most often to thin sections of ferrimagnetic oxides, up to about 0.1 mm in thickness, although metallic films less than 400 Å thick have also been examined.

For thin sections of oxides, the amount of the Faraday rotation is of the order of several degrees. This results in high contrast between adjoining domains and yields photographs of remarkable clarity, as shown in Fig. 9.10. These are pictures of a thin section of a gadolinium iron garnet crystal, with a field of varying strength applied normal to the plane of the section. Strain in the crystal has given it an easy axis normal to its plane, and the black and white areas are domains with M_s pointing in or out of the section. An applied normal field therefore causes one kind of domain to grow at the expense of the other. The crystal is demagnetized in (a). Increasing field then causes the white domains to grow

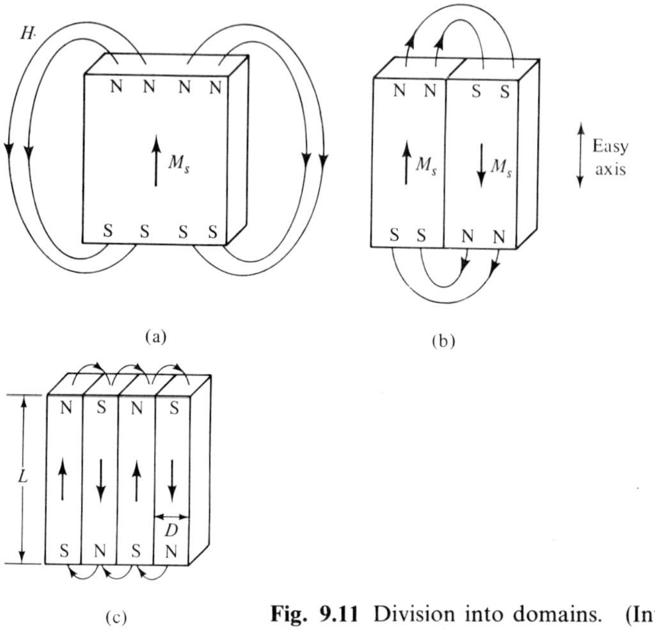

Fig. 9.11 Division into domains. (Internal H fields not shown.)

at the expense of the black, until almost complete saturation is achieved in (e). The field is then reduced to zero, in (f), (g), and (h). Hysteresis is apparent when we compare (c) with (f), which are roughly at the same field strength, because (f) on the descending branch of the hysteresis loop is more nearly saturated (higher M) than (c) on the ascending branch. Finally, when the field is increased in the opposite direction, in (i), (j), and (k), near saturation is achieved in the opposite direction.

In general, the clearest domain pictures of all are obtained by the Faraday method when it is applied to the kind of specimens for which it is suited. Like the Kerr method, it is unrestricted as to temperature and is excellent for wall motion studies.

9.5 MAGNETOSTATIC ENERGY AND DOMAIN STRUCTURE

We turn now from the observation of domains to an examination of the reasons for their formation and their relative arrangement in any given specimen. We will find that *magnetostatic energy* plays a primary role.

Uniaxial crystals

Consider a large single crystal of a uniaxial substance. Suppose it is entirely one domain, spontaneously magnetized parallel to the easy axis, as in Fig. 9.11 (a). Then free poles form on the ends, and these poles are the source of a large H field. The magnetostatic energy of this crystal is $(1/8\pi)\int H^2 dv$, evaluated over all space where H is appreciable. This considerable energy can be approximately halved, if the crystal splits into two domains magnetized in opposite directions as in (b), because this brings north and south poles closer to one another, thus decreasing the spatial extent of the H field. If the crystal splits into four domains as in (c), the magnetostatic energy again decreases, to about one-fourth of its original value, and so on. But this division into smaller and smaller domains cannot continue indefinitely, because each wall formed in the crystal *adds* energy. Eventually an equilibrium domain size will be reached.

The magnetostatic energy of the single-domain crystal is, from Eq. (7.51),

$$E_{ms} = N_d M_s^2/2 \tag{9.13}$$

per unit volume, where N_d is the demagnetizing factor. The value of N_d for a flat plate, in a direction normal to the plate, is 4π. If we take this value as applying approximately to the crystal of Fig. 9.11 (a), certainly within a factor of 2, the magnetostatic energy of the crystal per unit area of its top surface is

$$E_{ms} = 2\pi M_s^2 L, \tag{9.14}$$

where L is the thickness.

The calculation, which is not easy, of the magnetostatic energy of the multidomain crystal of Fig. 9.11 (c) has been given by Chikazumi [G.23]. This energy, per unit area of the top surface, is

$$E_{ms} = 1.7\, M_s^2 D, \tag{9.15}$$

where D is the thickness of the slab-like domains, provided that D is small compared to L. The total energy is the sum of the magnetostatic and wall energies:

$$E = E_{ms} + E_{\text{wall}},$$

$$E = 1.7\, M_s^2 D + \gamma L/D, \tag{9.16}$$

where γ is the domain wall energy per unit area of wall and L/D is the wall area per unit area of the top surface of the crystal. [This equation, incidentally, is of exactly the same functional form as Eq. (9.7).] The minimum energy occurs when

$$\frac{dE}{dD} = 1.7 M_s^2 - \frac{\gamma L}{D^2} = 0,$$

$$D = \sqrt{\frac{\gamma L}{1.7\, M_s^2}}. \tag{9.17}$$

Putting this in Eq. (9.16) gives

$$E = 2\sqrt{1.7\, M_s^2 \gamma L}. \tag{9.18}$$

For cobalt, taking $\gamma = 7.6$ ergs/cm^2 and $L = 1$ cm, we find

$$D = \sqrt{\frac{(7.6)(1)}{(1.7)(1422)^2}} = 1.5 \times 10^{-3} \text{ cm},$$

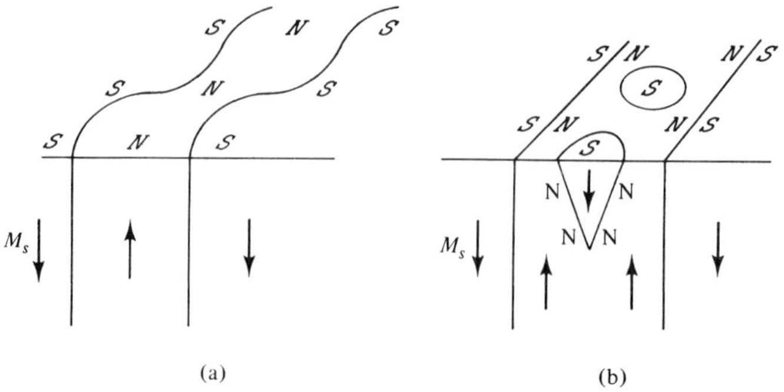

Fig. 9.12 Curved walls and surface spike domains.

which means $1/1.5 \times 10^{-3}$ or some 700 domains in a crystal 1 cm cube in size. The ratio of total energy before and after division into domains is

$$\frac{E\,(\text{single-domain crystal})}{E\,(\text{multi-domain crystal})} = \frac{2\pi M_s^2 L}{2\sqrt{1.7 M_s^2 \gamma L}}$$

$$= 2.4\sqrt{\frac{M_s^2 L}{\gamma}} = 2.4\sqrt{\frac{(1422)^2(1)}{7.6}} = 1200, \qquad (9.19)$$

where the values appropriate to cobalt have been inserted. Thus a thousand-fold reduction in energy has been effected by the creation of domains.

A still larger reduction in magnetostatic energy will result if the unlike poles on each end of the crystal are "mixed" more intimately. This can be done if the domain walls become curved rather than flat, although still parallel to the easy axis, as in Fig. 9.12(a). A section of such a crystal parallel to the easy axis will show straight lines separating the domains, and a section normal to the easy axis will show curved lines, as in Fig. 9.10. Curvature of the walls increases the wall area, and this type of domain structure is therefore found mainly in very thin crystals. In thick crystals, wall curvature involves too much extra wall energy, and another method of reducing magnetostatic energy is favored (Fig. 9.12 b). Here spike-shaped domains of reversed magnetization are formed at the surface. This has the desired effect of producing a fine mixture of opposite poles on the end surfaces without adding too much wall energy, because the spike domains are short. However, there is a discontinuity in the normal component of M_s on the walls of the spike domains, and free north poles must form there in accordance with Eq. (2.29). These interior poles are the source of an H field and therefore contribute to the magnetostatic energy. The number and size of the spike domains will be such as to balance the reduction in main magnetostatic energy due to the surface poles against the increase in wall energy and in magnetostatic energy due to interior poles.

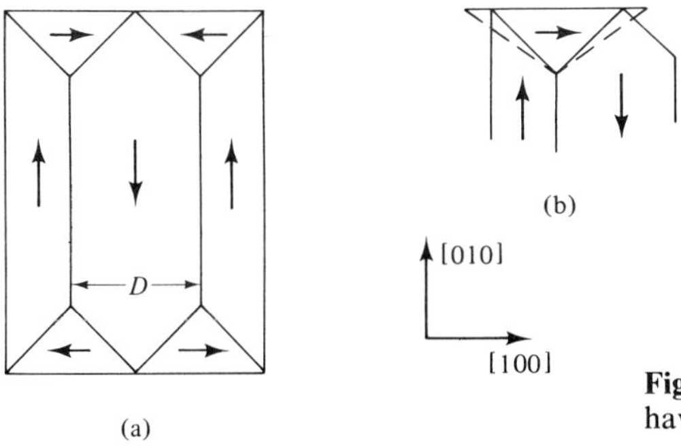

Fig. 9.13 Closure domains in a cubic crystal having ⟨100⟩ easy axes.

Cubic crystals

The domain structure of cubic crystals tends to be more complicated, because there are now 3 or 4 easy axes, depending on the sign of the anisotropy constant K_1. On the other hand, it is now possible for the flux to follow a closed path *within* the specimen; no surface or interior poles are formed, and the magnetostatic energy is therefore reduced to zero. Figure 9.13 (a) shows how this is done. Triangular domains are formed at the ends and, because they are paths by which the flux can close on itself, they are called *closure domains*. At first glance, one might think that the domains in such a crystal could be very large, since the only obvious source of energy is wall energy. However, there is also magnetoelastic energy. If λ_{100} is positive, as it is in iron, then the [100] closure domain would strain magnetostrictively to the dotted lines shown in (b), if not restrained by the main [010] and [0$\bar{1}$0] domains on either side. The closure domains are therefore strained, and the magnetoelastic energy stored in them is proportional to their volume. The total closure-domain volume can be reduced by decreasing the width D of the main domains. The crystal will therefore split into more and more [010] and [0$\bar{1}$0] domains, until the sum of the magnetoelastic and domain wall energies becomes a minimum. Closure domains of the type shown in Fig. 9.13 have often been seen at the edge of cubic crystals. (They were postulated in 1935 by Landau and Lifshitz [9.8] on theoretical grounds and were first observed by Williams in 1949 in an iron-silicon crystal [9.6].) Note that a closure-domain structure should have M_s parallel to the surface at all free surfaces.

Closure domains are also thought to occur in some uniaxial crystals. If the crystal anisotropy constant K is not too large, the magnetization near the crystal surface will turn away from the easy axis in order to close the flux path within the crystal and avoid free pole formation.

The avoidance of free poles is also the guiding principle controlling the *orientation* of domain walls. For example, a 180° wall must be parallel to the M_s vectors in the adjacent domains; if not, as in the spike domains of Fig. 9.12 (b), free poles will form on the wall, creating magnetostatic energy. For the same reason, a 90° wall, such as those bounding the closure domains of Fig. 9.13, must

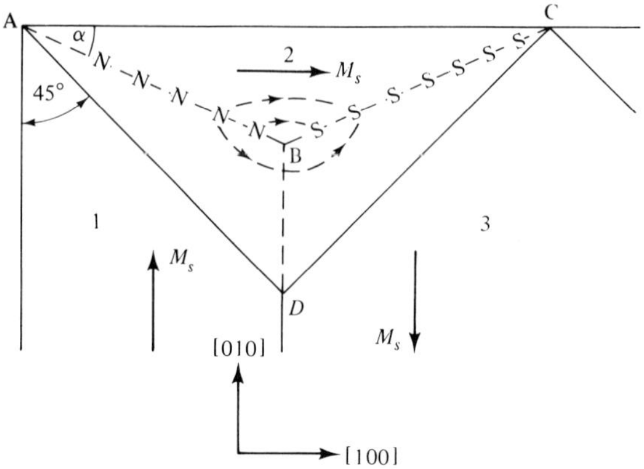

Fig. 9.14 Fields controlling wall orientation. After Stewart [G.7].

lie at 45° to the adjoining M_s vectors. Figure 9.14 affords a closer look at this effect. Suppose the closure domain walls, normally at AD and DC, should move up to the positions AB and BC, where AB makes an angle α with the [100] direction. Then the component of magnetization normal to the wall AB is $M_s \cos \alpha$ in domain 1 and $M_s \sin \alpha$ in domain 2. The discontinuity across the wall is M_s $(\cos \alpha - \sin \alpha)$; this quantity is positive and therefore north poles will form on the wall, in accordance with Eq. (2.29). Similarly, south poles will form on the wall BC. These poles set up an H field, shown by the dashed lines, which is predominantly directed from left to right in the neighborhood of B. This field therefore favors magnetization in the [100] direction; i.e., it favors the growth of domain 2 at the expense of domains 1 and 3. The walls therefore move down until α equals 45° and the free poles disappear.

The principle of free-pole avoidance does not entirely fix the orientation of a domain wall. For example, suppose a 180° wall, in a material with ⟨100⟩ easy directions, separates two domains magnetized along [100] and [$\bar{1}$00]. Then the wall could tilt about [100] as an axis through an infinity of orientations and still remain parallel to the M_s vectors on either side of it, thus remaining free of poles. Calculations [G.7] [G.23] [G.26] then show that the energy γ per unit area of wall varies with the tilt angle and goes through a maximum for some orientations and a minimum for others. Similar conclusions apply to 90° walls. But the ratio of maximum to minimum values of γ is less than 2, and this effect is small compared to the creation of magnetostatic energy by walls so oriented that they have free poles.

Keeping the principle of free-pole avoidance firmly in mind will prevent one from drawing hypothetical domain structures like that sketched in Fig. 9.15, where the heavily outlined walls and exterior surfaces have large discontinuities of magnetization across them. They are accordingly the sites of free poles and the source of magnetostatic energy.

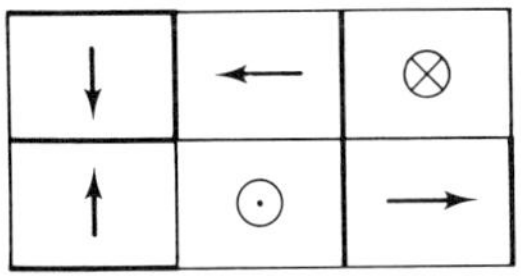

Fig. 9.15 A highly unlikely domain structure.

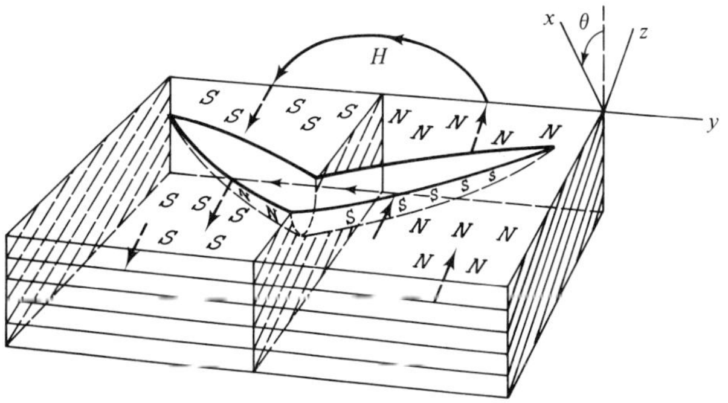

Fig. 9.16 Interpretation of the "tree" pattern. Lines on side surfaces are traces of (100) planes. Bozorth [G.4].

Simple domain arrangements, giving rise to a set of parallel 180 walls, as in the central portion of the crystal of Fig. 9.13, are seen only when the surface of the crystal is accurately parallel to a plane like (100). If the surface deviates only a few degrees from (100), the complex "tree" pattern of Fig. 9.6 (b) is formed. It has "branches" jutting out from the main 180° walls, as shown in detail in Fig. 9.16. Here the (100) planes make an angle θ with the crystal surface, and so do the M_s vectors in the two main domains; north and south poles are therefore formed on the top surfaces of these domains. To decrease the resulting magnetostatic energy the branch domains form. These carry flux *parallel* to the crystal surface, in the easy y or [010] direction, and therefore have no poles on their top surface. These branch domains are shallow and are bounded on the bottom by curved, nearly-90° walls which have some free poles distributed on them. Nevertheless, the total energy is still reduced by the formation of the branch domains.

The complex tree pattern is instructive, when one considers that well below the surface the domain structure of this crystal could not be more simple. This observation is quite generally true. Many of the complex domain arrangements seen on some crystal or grain surfaces would not exist if that surface had not been exposed by cutting; they are closure domains which form when the cut is made, in order to reduce magnetostatic energy. Surface domain structures can therefore be very deceptive guides to the real domain structure of the interior. (On the other hand, domains in very thin specimens, such as films and foils, normally extend completely through the specimen thickness. The domain structure re-

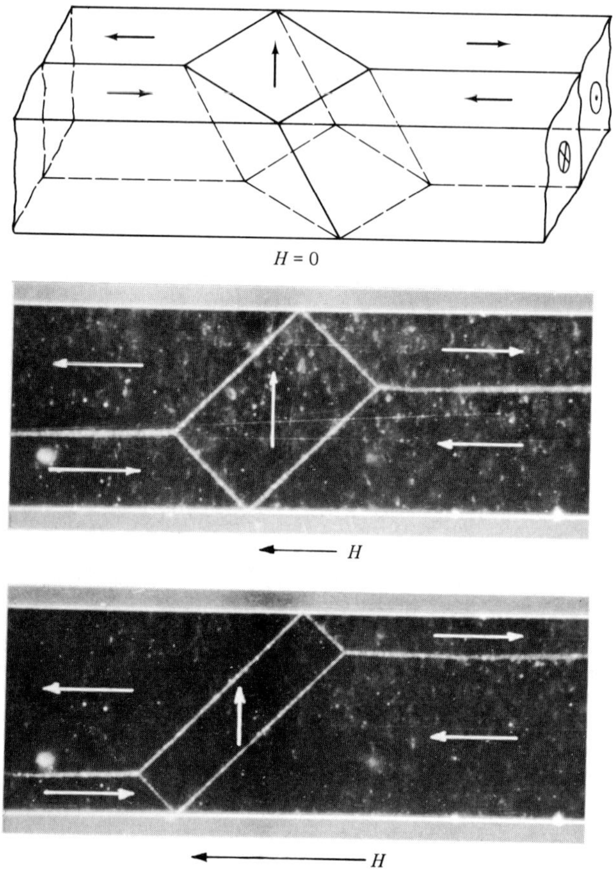

Fig. 9.17 Reversible wall motion in an iron whisker. Bitter patterns under dark-field illumination. Courtesy of Deblois and Graham, [9.9].

vealed by surface examination, by the Bitter or Kerr techniques, is therefore the same as the interior structure.)

The simplest kinds of domain structure are seen in properly prepared single crystals, and the most complex in polycrystals. (The great accomplishment of Williams, Bozorth, and Shockley [9.1] lies precisely in the fact that they established this conclusion. They succeeded in showing that the complex pattern of Fig. 9.6 (b), for example, was nothing but a surface artifact, caused by the fact that the specimen surface was not accurately parallel to a $\{100\}$ plane.) The most nearly perfect single crystals available are *whiskers*. These are fine filaments of the order of several millimeters in length and some tens of microns thick. [1 micron (μ) = 10^{-6} m = 10^4 A.] They are usually grown by the reduction of metal bromide at 800°C by hydrogen. Whiskers first attracted scientific interest in 1952

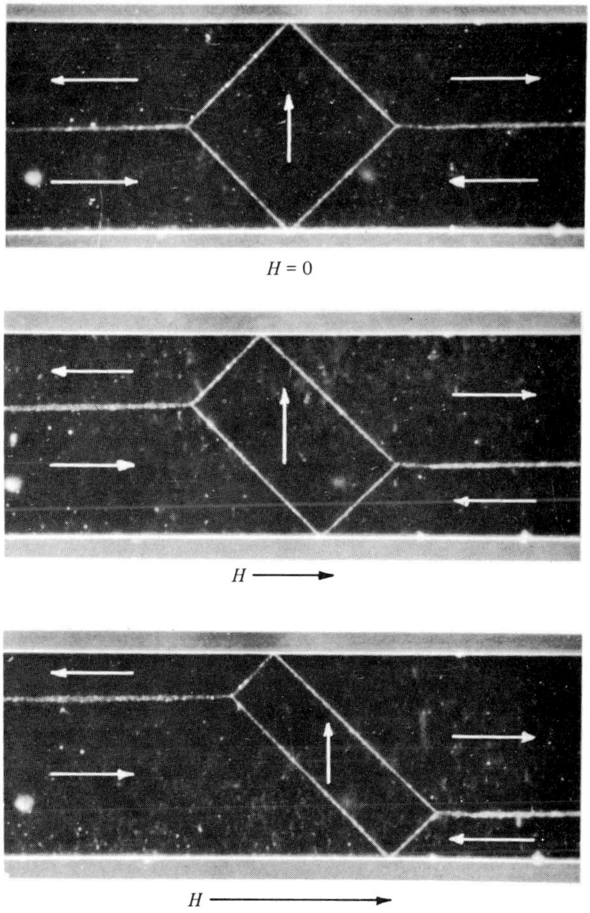

Figure 9.17 *(continued)*

because of their extremely high mechanical strength. Later it was realized that they offered excellent opportunities for magnetic domain studies; many iron whiskers, for example, grow with a ⟨100⟩ axis and with sides parallel to {100} planes. Moreover, these sides are optically flat, which means that no specimen preparation is needed before the magnetic suspension is added to form a Bitter pattern. Figure 9.17 shows the domain structure observed on a {100}⟨100⟩ iron whisker and the wall motion that occurs in an applied field. The sketch at the upper left shows the domain structure through the volume of the whisker, and the Bitter patterns are of the top face. The arrows below each pattern indicate the direction and magnitude of the applied field, which had a maximum value of about 10 Oe. The changes shown are entirely reversible, i.e., the configuration at top right could be regained at any stage of the magnetization process by reducing the field to zero.

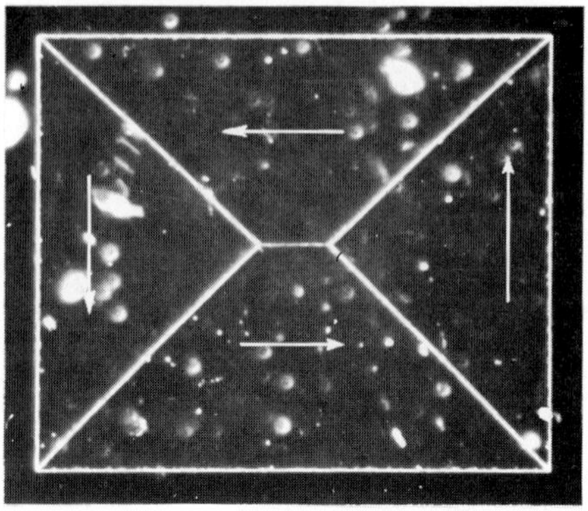

(a)

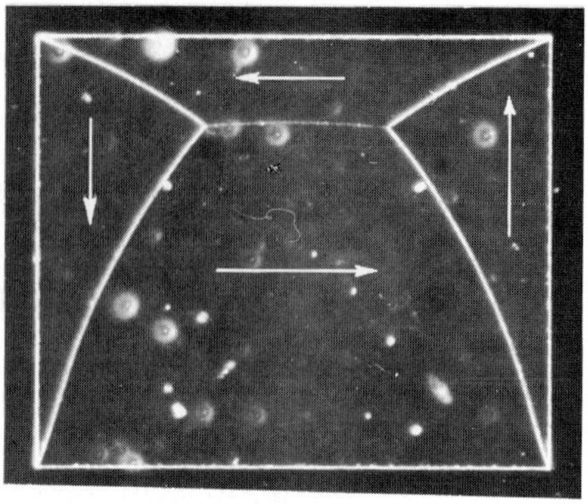

(b)

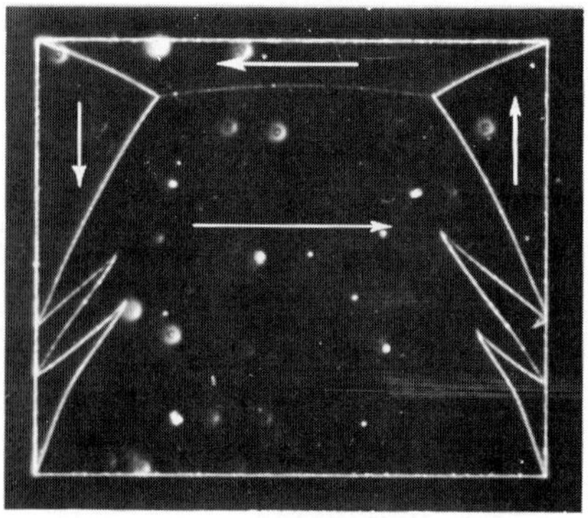

(c)

Fig. 9.18 Wall motion in an Fe-Co alloy platelet, 165 μ wide. Bitter patterns, dark-field illumination. Courtesy of R. W. De-Blois [9.10].

Closely analogous to whiskers are metal "platelets," which are grown by the same methods and are also single crystals. Here the growth habit is edgewise rather than axial; the result is a platelet some hundreds of microns in its lateral dimensions and from less than 0.1 micron (1000 A) to over 10 microns in thickness. The platelet surfaces and edges are crystal planes and directions of low indices, like {100} and ⟨100⟩. These platelets are structurally more nearly perfect than the usual thin films formed by evaporation of a metal in vacuum onto a substrate. Figure 9.18 (a) shows an exceptionally simple domain structure in a demagnetized Ni-Co alloy platelet. The crystal structure is face-centered cubic, with ⟨100⟩ easy directions (positive K_1). When a field of 3.6 Oe is applied to the right, the domain walls move reversibly to the positions shown in (b). Here the corners of the platelet have pinned the ends of the 90° walls, forcing the latter to bend. (The normal component of M_s is no longer continuous across these walls, so they are now the sites of free poles.) With further increase in field, to 4.1 Oe, the domain structure snaps irreversibly to (c). The reason that the 180° walls are so thin, or invisible, in these photographs is that the crystal anisotropy is low in this alloy;

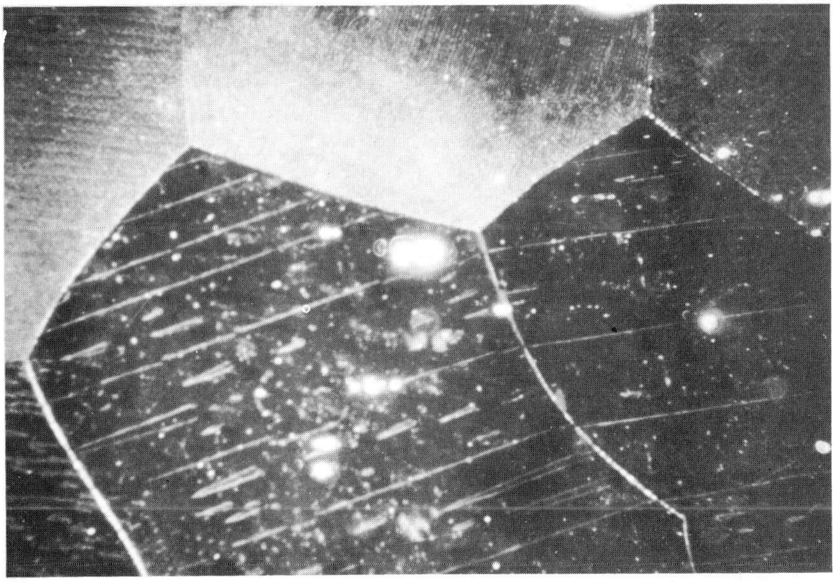

Fig. 9.19 Domains in a polycrystalline specimen of Fe + 3 percent Si. Photo by C. D. Graham, Jr. [9.11].

the walls are accordingly very broad, produce only a weak field gradient above the surface, and attract little or no colloid. The 90° walls are thinner and attract more colloid.

The domain structure observed on the surface of a polycrystalline specimen is usually quite complex, because (a) the grain surfaces are rarely even approximately parallel to crystal planes of low indices like {100} or {111}, except in certain specimens having a high degree of preferred orientation, and (b) the grain boundaries tend to interrupt the continuity of magnetization from grain to grain. Because the easy directions of magnetization have different orientations in two adjoining grains, the grain boundary between them is also a domain boundary. The normal component of M_s is rarely continuous across a grain boundary, which therefore becomes the seat of free poles and magnetostatic energy. Although domains are not continuous across a grain boundary, domain walls often are, and Fig. 9.19 shows an example of this. Spike domains often form at grain boundaries to reduce the free pole density there, just as they do at the surface of some single crystals (Fig. 9.12 b).

9.6 SINGLE-DOMAIN PARTICLES

We saw in the previous section that a crystal will spontaneously break up into a number of domains in order to reduce the large magnetostatic energy it would have if it were a single domain, and this process was illustrated in Fig. 9.11. We also

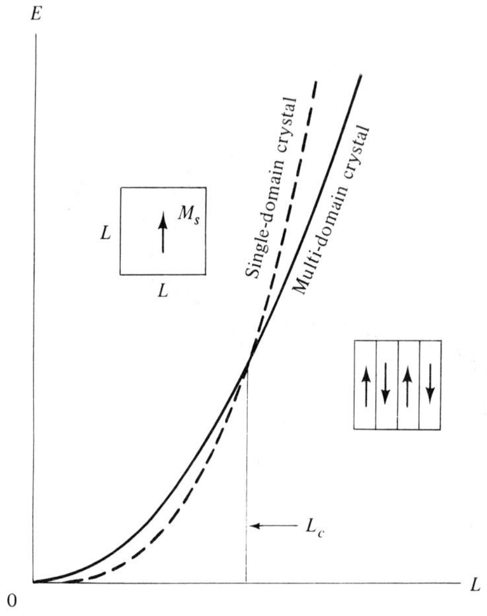

Fig. 9.20 Relation between the energy E of a crystal and its linear dimensions L for two kinds of magnetic state (schematic).

found, in Eq. (9.19), that the ratio of the energies before and after division into domains varied as $\sqrt{L}$, where L is the crystal thickness. Thus, as L becomes smaller, the reduction in energy becomes smaller; this suggests that at very small L the crystal might prefer to remain in the single-domain state.

Another way of looking at this is to note that the total magnetostatic energy of a single-domain crystal, in the form of a cube of edge L, varies as L^3, from Eq. (9.14). In the multi-domain state, (a) if the crystal is cubic with closure domains, the magnetostatic energy is zero and the total energy is wall energy, which varies as wall area or as L^2, or (b) if the crystal is uniaxial, the total energy will be the sum of magnetostatic and wall energy, and this sum will vary as $L^{5/2}$, from Eq. (9.18). In either case, the relation between energy and L will resemble the curves of Fig. 9.20. Because the term in L^3 becomes so very large when L is large, and so very small when L is small, there will be a critical size L_c below which the single-domain crystal will have the lower energy.

How can the value of L_c be calculated? For a crystal in the form of a cube of edge L, we might approach this problem by equating its energy as a single domain, given by Eq. (9.14), to its energy when divided into domains, given by Equation (9.18), and solve the resulting equation for L_c. The result is

$$L_c = \frac{1.7\,\gamma}{\pi^2 M_s^2}, \tag{9.20}$$

where the numerical coefficient may be uncertain by a factor of about 2. When values of γ and M_s for particular substances are substituted into this relation, we find that two cases must be distinguished:

1. *Moderate crystal anisotropy.* For materials like iron and nickel, Eq. (9.20) gives values of L_c of less than one-tenth of the normal domain wall thickness δ. Even cobalt has a computed value of L_c less than δ, as shown by Problem 9.1. For such materials Eq. (9.20) is invalid, because its derivation involves the energy of walls of normal thickness. However, other more complex calculations [9.2] lead to the result that L_c is of the order of 100–500 Å, or about equal to the normal wall thickness. This result is reasonable. Suppose a single-domain particle has a diameter about equal to the normal wall thickness. If we attempt to reduce its magnetostatic energy by dividing it into two domains by a single wall, the wall will have to be so thin that the exchange energy created in the wall will exceed the decrease in the magnetostatic energy of the particle. Hence the wall does not form.

2. *High crystal anisotropy.* When the crystal anisotropy constant K is large, then the wall energy γ is large and the wall thickness δ is small. Equation (9.20) is now valid because it gives values of L_c of the order of 500–1000 Å, which is several times the value of δ. Because $\gamma = 2K\delta$, the ratio $L_c/\delta = 0.4K/M_s^2$.

This discussion has ignored the value of M_s. Even though Eq. (9.20) is not always valid, it does show the correct functional dependence of L_c on γ and M_s. If γ is large, then it will cost a good deal of energy to put a wall in an originally single-domain particle, so L_c tends to be large. If M_s is small, then a single-domain particle can become larger without much increase in magnetostatic energy. It is for this reason that barium ferrite, which has a smaller value of K, and hence of γ, than cobalt, but a much smaller value of M_s, has a critical size for single-domain behavior several times that of cobalt. See Problem 9.2.

One point not yet considered is the *shape* of the single-domain particle. Compared to a cubical or spherical particle, an elongated particle will have a lower demagnetizing factor parallel to its long axis and therefore lower magnetostatic energy per unit volume. The elongated particle can then have a larger volume, and even a larger width, than a spherical particle before breaking up into domains. In fact, calculations [7.23] show that a rod-shaped particle with a length/diameter ratio of 10/1 has a critical diameter, for single-domain behavior, several times that of a sphere. The critical diameter of such a particle for a material like iron, is thus of the order of 1000 Å or 0.1 μ, and its length is about 1 μ.

Experiment has shown that single-domain particles actually exist, and they are of great theoretical and practical interest. Much more will be said about them in later portions of this book. Such particles are unique, in that they cannot be demagnetized, and, having no domain walls to be moved by a field, their magnetization can be reversed only by rotation. This rotation may be difficult, because it is resisted by whatever anisotropy forces (crystal, shape, stress, etc.) happen to be present.

Single-domain particles were first postulated in 1930 by Frenkel and Dorfman [9.12]. Subsequent calculations of their critical size and behavior were made by Kittel [9.13], Néel [9.14], and Stoner and Wohlfarth [7.23].

9.7 MICROMAGNETICS

What we have done in the two preceding sections was, in effect, to determine the direction of the local magnetization vector $\mathbf{M}_s$ in a specimen subjected to no external field. We reached two conclusions:

1. If the specimen exceeded a certain critical size, it would divide itself into *domains*, in each of which $\mathbf{M}_s$ was everywhere parallel, separated by *domain walls*, in which the direction of $\mathbf{M}_s$ varied with position.
2. If the specimen size was less than a critical value, $\mathbf{M}_s$ was everywhere parallel.

As far as large specimens are concerned, the problem is one of finding the domain arrangement of lowest total energy. For example, in considering the cubic crystal of Fig. 9.13, we realized that the crystal would not consist solely of slab-like domains magnetized parallel and antiparallel to [010], because the magnetostatic energy of such an arrangement could be eliminated by putting in closure domains at the ends. But these domains contribute magnetoelastic energy. It was then necessary to make the closure domains smaller without, however, adding too much wall energy. Continuing in this manner, we finally arrive at an equilibrium width D of the main domains, which in turn defines the sizes of the other domains and the total amount of wall area. This whole process of devising a domain configuration of minimum energy has been criticized by Brown [G.15] as being nonrigorous. He points out that "the particular configuration devised is dependent on the ingenuity of the theorist who devised it; conceivably a more ingenious theorist could devise one with even lower free energy."

Brown instead advocates a rigorous approach, called *micromagnetics.* Here we forget about domains and domain walls. But we allow the $\mathbf{M}_s$ vector, of constant magnitude, to have a direction which is a continuous function of its position x, y, z in the crystal. We then express the various energies (exchange, anisotropy, magnetostatic, etc.) in terms of these directions throughout the crystal. The resulting equations are then to be solved for the spin directions at all points. If the crystal is a large one, then the existence of domains (regions of parallel spin) and the positions of domain walls (regions of rapid change of spin direction) should come quite naturally out of the solution. If the crystal is very small, then the solution should indicate that the spins are everywhere parallel, making it a single domain. In each case we must begin with a complete physical description of the crystal; this would include, in the main, such things as its crystal anisotropy, magnetostrictive behavior, size, shape, and the presence or absence of imperfections.

While there is no doubt that micromagnetics is more rigorous and intellectually satisfying than domain theory, the mathematics involved in the micromagnetics approach is of formidable complexity. As a result, only certain rather limited kinds of problems have been attempted. One such problem is that of calculating the magnitude of the applied field required to reverse the magnetization of a crystal that contains no domains. Micromagnetics calculations yield a result which agrees with experiment when shape anisotropy predominates, but not when crystal anisotropy is the controlling factor. This difficulty has not yet been resolved. (See Section 11.4 and 11.5.)

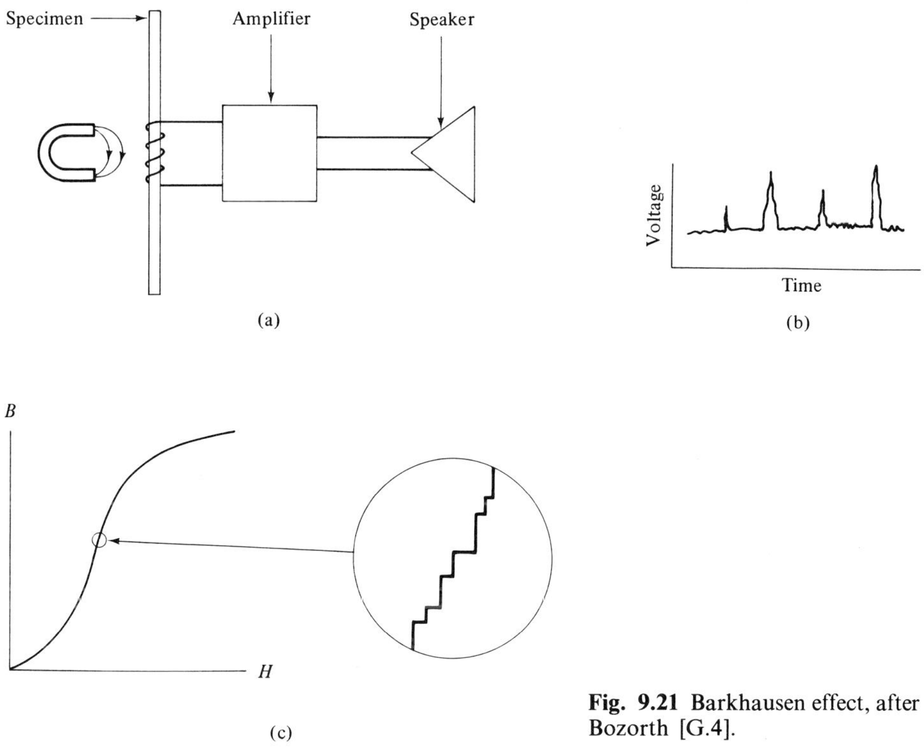

Fig. 9.21 Barkhausen effect, after Bozorth [G.4].

(In passing, we might note that the problem of determining the spin structure of a domain wall, outlined in Section 9.2, is a problem in micromagnetics, because there we allowed the spin direction to be a function of position, subject to certain constraints.)

The reader interested in pursuing this subject further should consult the book by Brown [G.15] and a review paper by Shtrikman and Treves [9.15].

9.8 DOMAIN WALL MOTION

Up to this point our attention has been focused on static domain structures. We must now consider how domain walls move in response to an applied field. This motion is often observed to be jerky and discontinuous, rather than smooth.

This effect, known as the *Barkhausen effect*, was discovered in 1919 and can be demonstrated with the apparatus shown in Fig. 9.21 (a). A search coil is wound on a specimen and connected through an amplifier to a loudspeaker or earphones. The specimen is then subjected to a smoothly increasing field, as by approaching it with a hand-held magnet. No matter how smoothly and continuously the field is increased, a crackling noise is heard from the speaker. If the search coil is connected to an oscilloscope, instead of a speaker, irregular spikes will be observed on the voltage-time curve, as in (b). These voltage spikes are known as *Barkhausen noise*. [The emf induced in the search coil is, by Eq. (2.6), proportional to the rate of change of flux through it, or to dB/dt. But even when dH/dt is constant, and even on those portions of the B,H curve which are practically linear, the induced voltage is not constant with time but shows many discontinuous changes.] The effect is strongest on the steepest part of the magnetization curve and is evidence for sudden, discontinuous changes in magnetization. This is indicated in Fig. 9.21 (c), where the magnification factor applied to one portion of the curve is of the order of 10^9.

The Barkhausen effect was originally thought to be due to sudden rotations of the M_s vector from one orientation to another in various small volumes of the specimen. It is now known to be due mainly to domain walls making sudden jumps from one position to another. (Whether interpreted in one way or the other, the Barkhausen effect is evidence for the existence of domains, and it was the first evidence in support of Weiss' hypothesis of thirteen years earlier.)

In a classic paper published in 1949, Williams and Shockley [9.16] reported direct visual evidence of jerky domain wall motion. They made a single crystal of Fe + 3.8 percent Si in the form of a hollow rectangle ("picture frame") with each side parallel to an easy direction of the form $\langle 100 \rangle$ and all faces parallel to $\{100\}$. The overall dimensions were 19×13 mm, and each side was 1 mm wide and 0.7 mm thick. They examined the whole surface of the polished crystal by the Bitter technique and found the particularly simple domain structure shown in Fig. 9.22 (a). This shows the demagnetized state, and the crystal contains only eight domains. The domain walls, shown as dashed lines, were found, by observation of the other side, to pass straight through the crystal normal to the plane of the drawing. They then wound H and B coils on opposite legs and connected the B coil to a recording fluxmeter. When the applied field H was clockwise, the domain wall in each leg moved outward until it reached the edge of the specimen at saturation; each leg of the crystal was then a single domain. While the wall AB, for example, was moving, they could watch it with a microscope focused on the top leg, as indicated in (a). They observed that the wall motion was generally fairly smooth, but now and then jerky when the wall encountered an inclusion. The nature of this interaction will be described in the next section.

However, the Williams-Shockley experiment has an importance much more fundamental than its clarification of the Barkhausen effect. When the applied field was changed from its maximum clockwise value to its maximum counterclockwise value and back again, the crystal went through its hysteresis cycle, and the recorded loop is shown in (b). At the same time photographs of the wall position in one leg were made through the microscope. It was found (c) that there was a direct linear relation between wall position measured on the film and the magnetization M of the specimen, as would be expected from the previous observation that the walls went straight through the crystal. This was the first experimental demonstration of a relation between domain wall motion and change in magnetization. It showed that observation of a domain wall at its intersection with a surface could be correlated with a real volume effect, as measured with the B coil.

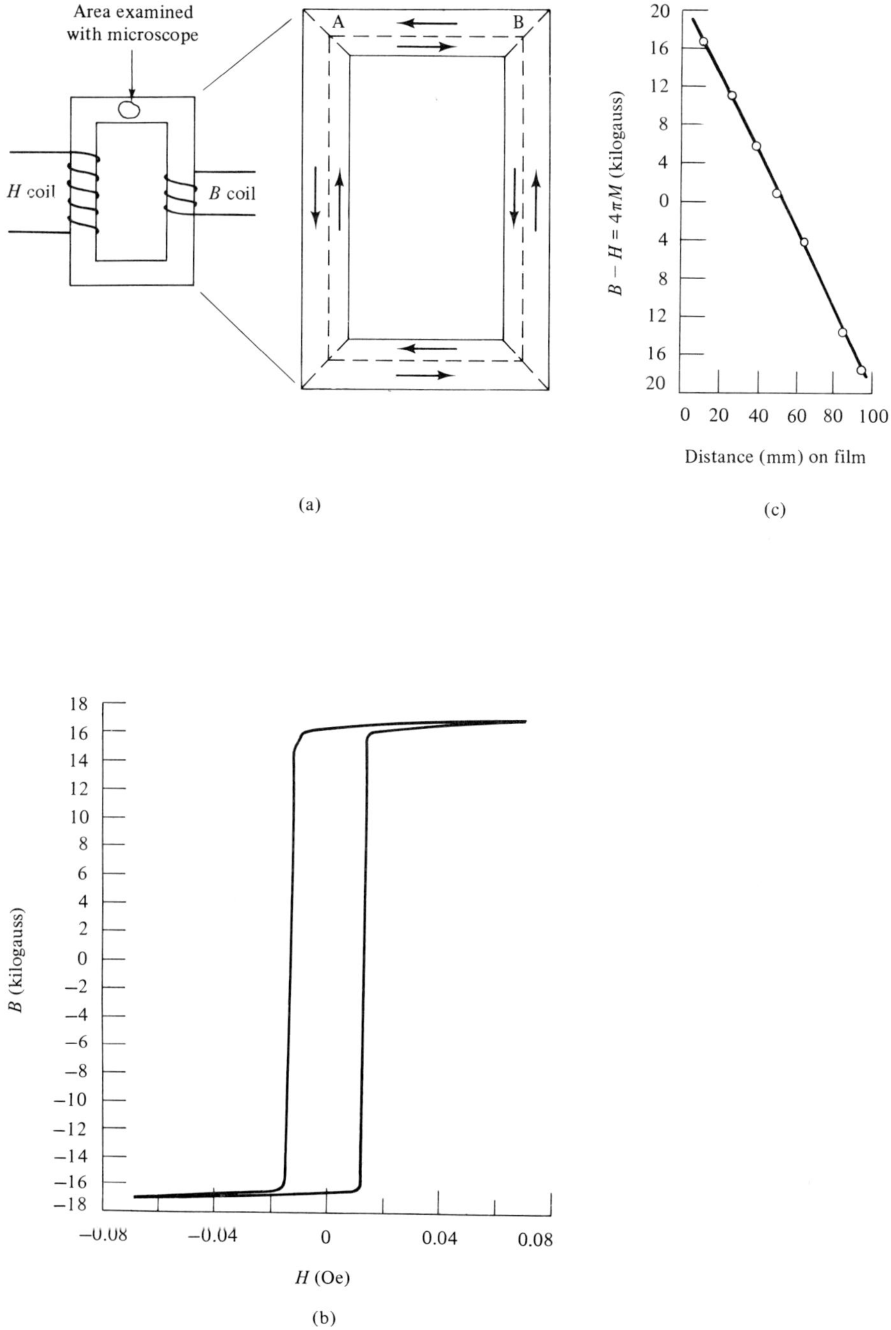

Fig. 9.22 The picture-frame experiment. Williams and Shockley [9.16].

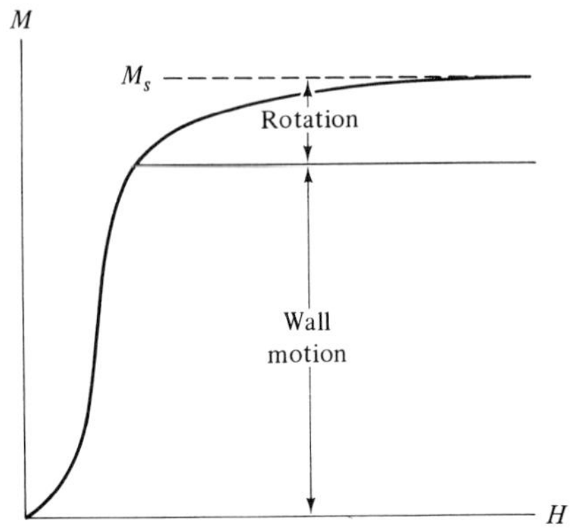

Fig. 9.23 Magnetization processes.

Before leaving this beautiful experiment, we might note two other points of interest:

1. The hysteresis loop is "square," where this word refers to the relative squareness of the loop corners rather than to the loop shape. The coercive force of less than 0.02 Oe is a measure of the extreme magnetic softness of this well-prepared specimen. And when the magnetization begins to reverse, a field change of less than 0.005 Oe is enough to effect essentially complete reversal. Materials with square hysteresis loops are important in many applications, and we shall return to this subject in later pages.

2. When a magnetized rod-shaped crystal is heated above the Curie point and cooled again, it becomes demagnetized, in order to reduce its demagnetizing field and the associated magnetostatic energy. But when this picture-frame crystal was cooled from the Curie point, it was observed to be *saturated*, clockwise or counterclockwise, each leg a single domain. Because it forms a closed magnetic circuit, this crystal can have no demagnetizing field. Therefore, its state of minimum energy is one of minimum domain wall area; this is the saturated state, with only four short domain walls at the corners. (Since the crystal anisotropy energy is zero, the only other source of energy is some magnetoelastic energy due to domain misfit at the corners.)

In previous chapters we noted that magnetization can change as a result both of domain wall motion and domain rotation. The question then arises: in a typical polycrystalline specimen, what proportion of the total change in M is due to wall motion and what rotation? This question has no precise answer, but a rough division is indicated in Fig. 9.23. Wall motion is the main process up to about the "knee" of the magnetization curve. From there to saturation, rotation predominates; in this region work must be done against the anisotropy forces, and a rather large increase in H is required to produce a relatively small increase in M.

This division of the magnetization curve is rather arbitrary, because wall motion and rotation are not sharply divisible processes. In fact, at any one

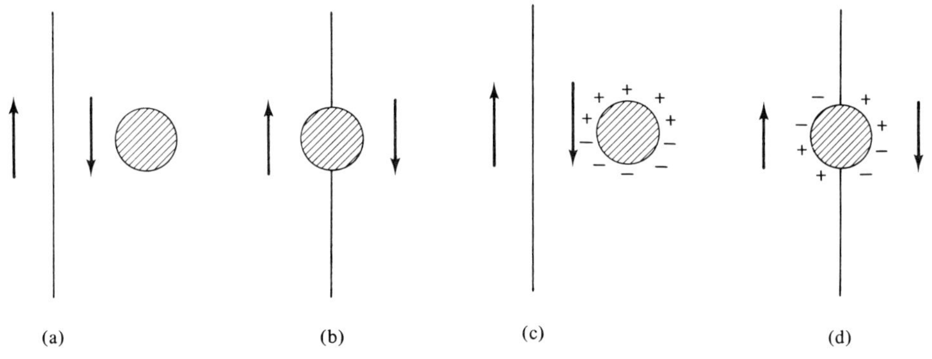

Fig. 9.24 Interactions of domain walls with inclusions.

level of M, wall motion may be occurring in one portion of a specimen and rotation in another. And in certain orientations of a single-crystal specimen relative to the applied field, wall motion and rotation can occur simultaneously in the same part of the specimen.

9.9 HINDRANCES TO WALL MOTION (INCLUSIONS)

Even in the very special case of the picture-frame crystal, a nonzero applied field was required to move domain walls through the material. In other specimens, substantial wall motion may require fields of tens or hundreds of oersteds. Evidently real materials contain crystal imperfections of one sort or another which hinder the easy motion of walls. These hindrances are of two kinds: *inclusions* and *residual microstress.*

Inclusions may take many forms. They may be particles of a second phase in a binary alloy, present because the solubility limit has been exceeded. They may be oxide or sulphide particles and the like, existing as impurities in a metal or alloy. They may be simply holes or cracks. From a magnetic point of view, an "inclusion" in a domain is a region which has a different spontaneous magnetization from the surrounding material, or none at all. We will regard an inclusion simply as a nonmagnetic region.

One reason that an inclusion might impede wall motion is that the wall might tend to cling to the inclusion in order to decrease the area, and hence the energy, of the wall. When a wall passes from outside an inclusion to a position bisecting the inclusion, as from (a) to (b) of Fig. 9.24, the wall area decreases by πr^2, for a spherical inclusion of radius r, and the wall energy decreases by $\pi r^2 \gamma$. But in 1944 Néel [9.17] pointed out that free poles on an inclusion would form a far greater source of energy. An inclusion entirely within a domain, as in Fig. 9.24 (c), would have free poles on it and an associated magnetostatic energy of $(N_d M_s^2/2)$ (volume) $= (4\pi/3)(M_s^2/2)(4\pi r^3/3) = (8\pi^2 M_s^2 r^3/9)$. When the wall moves to (d), bisecting the inclusion, the free poles are redistributed as shown, and the mag-

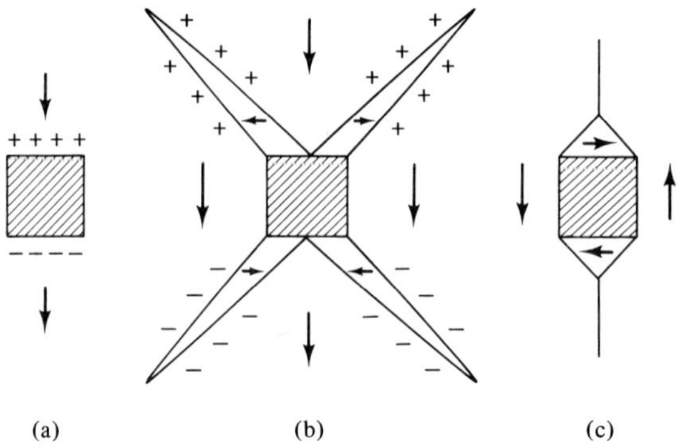

Fig. 9.25 Spike and closure domains on inclusions. Néel [9.17].

netostatic energy is approximately halved, just as it is when a single-domain crystal is divided into two oppositely-magnetized domains (Fig. 9.11 b). Therefore, when a wall moves from a position away from an inclusion to a position bisecting it, the ratio of the energy reduction by the free-pole effect to the energy reduction by the wall-area effect is, for a 1 μ diameter inclusion in iron,

$$\frac{(4\pi^2 M_s^2 r^3/9)}{\pi r^2 \gamma} = \frac{4\pi M_s^2 r}{9\gamma} = \frac{(4\pi)(1714^2)(0.5\times 10^{-4})}{(9)(1.5)} = \frac{140}{1}, \tag{9.21}$$

when γ is taken as 1.5 ergs/cm^2. This ratio, being proportional to r, becomes even larger for larger inclusions, showing that the wall-area effect is negligible.

However, Néel also pointed out that the magnetostatic energy of an inclusion isolated within a domain could be decreased if subsidiary spike domains formed on the inclusion, and reduced to zero by closure domains when a wall bisected the inclusion. Thus an inclusion, taken as cubical for simplicity and wholly within a domain, might have spike domains attached to it as in Fig. 9.25 (b). The total free pole strength in (a) is spread over a larger surface in (b), and such "dilution" always decreases magnetostatic energy. (See Problem 9.3 for a particular example of this effect.) If the walls bounding the spike domains were all at exactly 45° to the magnetization of the surrounding domain, there would be no discontinuity in the normal component of M_s and hence no free poles; however, such walls would extend to infinity and add an infinite amount of wall energy to the system. Spikes of finite length and having walls at nearly, but not exactly, 45° to the adjacent M_s vectors represent a compromise structure. Spike domains on inclusions were first seen in 1947 by Williams [9.18], in an iron-silicon crystal. They differed from those predicted by Néel three years earlier only in having two spikes, rather than four, on each inclusion; examples are shown in Fig. 9.26.

If a wall bisects an inclusion, magnetostatic energy can be reduced to zero, at the cost of a little wall energy, if closure domains form as shown in Fig. 9.25 (c).

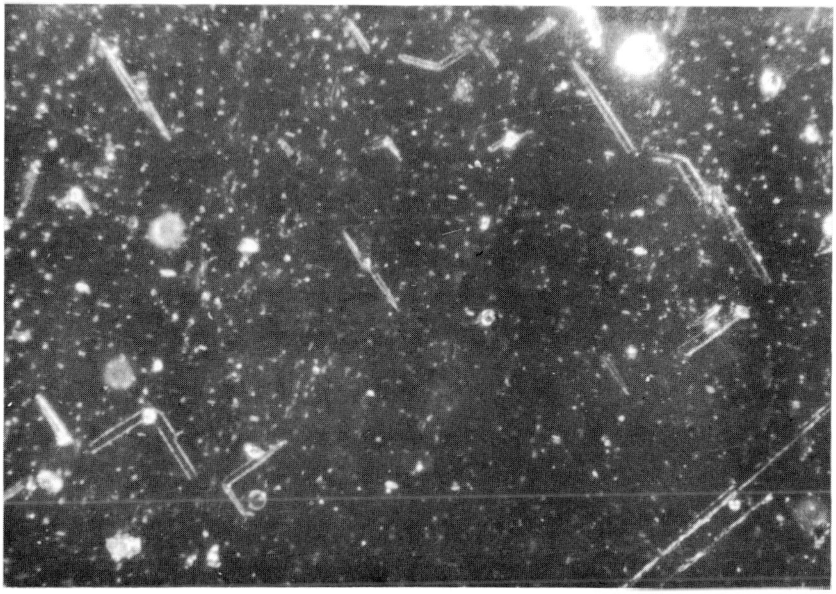

Fig. 9.26 Spike domains on inclusions in an iron-silicon crystal. C. D. Graham, Jr. [9.11].

Observation of moving domain walls in crystals has shown that wall motion is impeded by interaction of the moving wall with the spike domains normally attached to inclusions rather than by interaction with the inclusions themselves. A typical sequence is shown in Fig. 9.27. In response to an upward applied field, the wall in (a) moves to the right, as in (b), dragging out the closure domains into the form of tubes and creating a new domain just to the right of the inclusion.

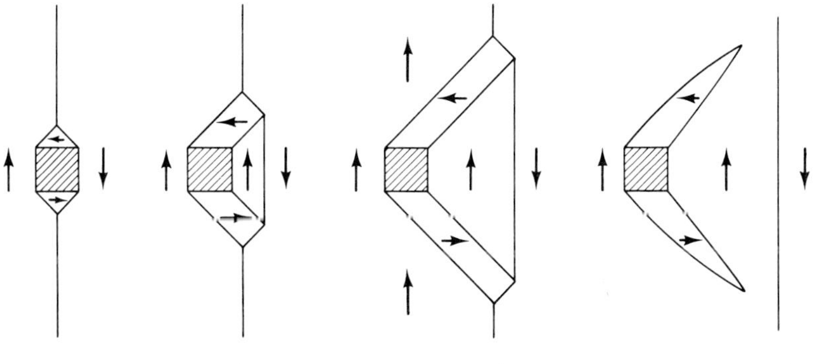

Fig. 9.27 Passage of a domain wall through an inclusion.

Further motion of the main wall lengthens the tubular domains, as in (c). The change from (a) to (b) to (c) is *reversible*, and the domain arrangement of (a) can be regained if the field is reduced. But if the field is continually increased, the tubular domains do not continue to lengthen indefinitely, because their increasing surface area adds too much wall energy to the system. A point is reached when the main wall suddenly snaps off the tubular domains *irreversibly* and jumps a distance to the right, leaving two spike domains attached to the inclusion, as in (d). This is a Barkhausen jump. If the field is now reduced to zero and reversed, the reader can visualize these changes occurring in reverse; if the reversed field is strong enough to drive the wall well to the left of the inclusion, the inclusion will be left with spike domains pointing to the left.

The magnetostatic energy associated with the "naked" inclusion of Fig. 9.25 (a) is proportional to its volume. The results of Problem 9.3 suggest that, in a material like iron, inclusions with a diameter of about 1μ or larger will have spike domains attached to then in order to reduce this energy. Smaller inclusions will remain bare, because their magnetostatic energy is small. When the inclusion size is of the order of 0.01μ (= 100 A), it is smaller than the usual domain wall thickness. When such an inclusion is within a wall, it reduces the wall energy and hence tends to anchor the wall. Inclusions are most effective, per unit volume of inclusion, when the inclusion diameter is about equal to the wall thickness.

Thus both large and small inclusions tend to impede wall motion, large ones because they have subsidiary domains which tend to stick to walls, small ones because they reduce the energy of any walls that contain them.

9.10 RESIDUAL STRESS

The second kind of hindrance to domain wall motion is residual microstress. Before examining the magnetic effects of such a stress, we will digress in this section to consider residual stress in general, in order to get a clearer notion of what is meant by microstress.

Stress may be divided into applied stress (the stress in a body due to external forces) and residual stress (the stress existing in a body after all external forces have been removed). Residual stress is often mistakenly called "internal stress," which is a completely uninformative term. All stress, whether applied or residual, is internal.

Residual stress may in turn be divided into *macro* and *micro*, depending on scale. Residual macrostress is reasonably constant in magnitude over distances many times the normal grain diameter in a metal, while residual microstress varies rapidly in magnitude and sometimes in sign over distances about equal to, or smaller than, the normal grain diameter. Residual macrostress, which is the kind of most concern to the engineer because of its effect on such phenomena as fatigue and fracture, is due to nonuniform plastic flow that has occurred at some time in the previous history of the material; its creation by various processes and removal by annealing are well described by Baldwin [9.19] and Richards [9.20]. Residual microstress is caused by crystal imperfections of various kinds, par-

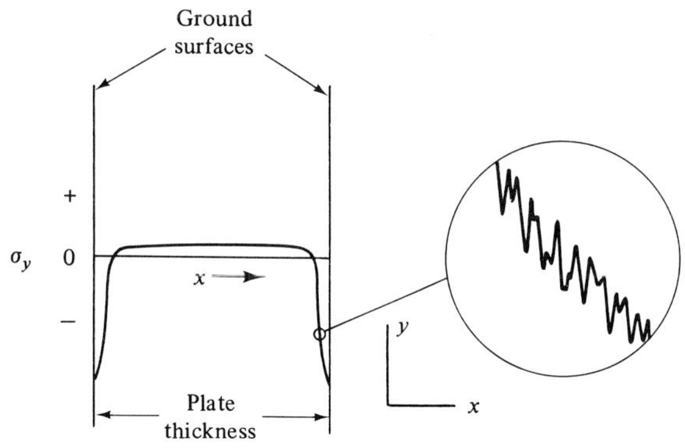

Fig. 9.28 Residual stress distribution due to grinding.

ticularly dislocations. Although it may be reduced to quite low levels by annealing, it is never entirely absent.

(The distinction between residual macrostress and microstress in terms of scale is necessarily rather vague. A more useful distinction can be made on the basis of the effect of the stress on the x-ray diffraction pattern of the material. The depth to which reflected x-rays penetrate a metal surface is typically of the order of 20–30 μ, while the grain size of most metals lies in the range 10–100 μ. Thus a macrostress will be essentially constant over the depth examined by x-rays, while a microstress will vary within this depth. As a result, residual macrostress causes an x-ray *line shift*, a change in direction of the diffracted beam, because the lattice-plane spacing d in Eq. (5.31) is essentially constant, but different from the stress-free value, in the region examined. On the other hand, residual microstress causes *line broadening*, because of the variation of plane spacing within this region [9.21].)

Residual macro- and microstress are quite often found together. For example, the residual macrostress caused by grinding the opposite faces of a metal strip or plate with an abrasive wheel is shown at the left of Fig. 9.28. Each ground surface is in residual compression in a direction parallel to the surface. This stress rapidly decreases to zero with depth and changes to tension in most of the interior. (Residual stresses must of necessity form a balanced force system. Regions in residual compression must be accompanied by regions in tension.) X-ray examination of the material at the surface, or just below (accomplished by etching part of the material away), shows the x-ray lines to be both shifted and broadened. We can then imagine the microstress distribution in one small region to resemble that sketched at the right of Fig. 9.28.

There is, however, an ambiguity in the interpretation of line broadening, because *two* effects can cause line broadening: (1) microstress, and (2) small crystal size, less than about 0.1 μ (1000 Å). Thus line broadening cannot be interpreted solely in terms of either cause unless the other can be shown to be absent. (Plastic deformation of a metal commonly has both effects. It introduces microstress

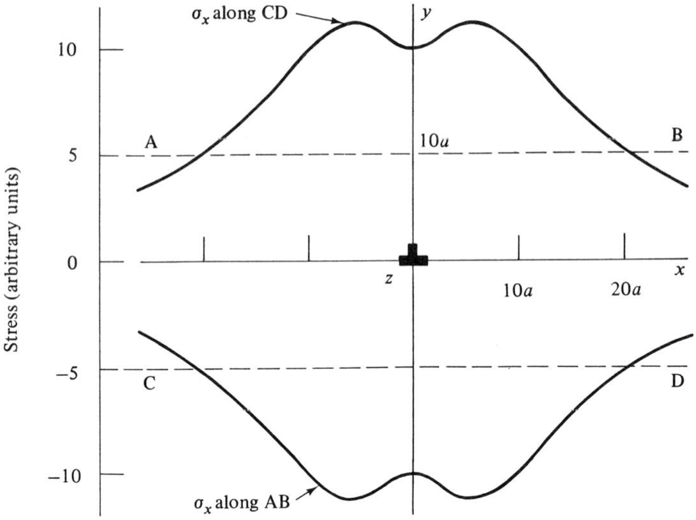

Fig. 9.29. Stresses near a dislocation; a = lattice parameter.

and breaks up the grains into smaller fragments, and the relative importance of each effect varies from metal to metal.) However, Warren [9.22, 9.23] has shown that the two causes of broadening can be separated if the entire shape of the broadened line is measured, and not merely its breadth. The result of analyzing the line shape, if microstress is present, is a value for the root-mean-square strain averaged over any chosen distance. Thus, while x-rays can tell us something about the distribution of microstress, they cannot reveal the details of this distribution. On the other hand, they comprise the only direct method available of obtaining quantitative information about the variation of stress on a micro scale.

As mentioned earlier, residual microstress is never entirely absent, even in a well-annealed specimen. Consider these effects:

1. *Dislocations* [9.24]. All specimens, except a few of the most carefully grown single crystals, contain a goodly number of dislocations and each has a stress field associated with it, because the dislocation distorts the surrounding material. The nature of the stress field depends on whether the dislocation is of the edge or screw type. Figure 9.29 shows one example. A positive edge dislocation, symbolized by the inverted T, is at the origin, parallel to the z axis. The stress σ_x in the x direction is compressive above the dislocation and tensile below it. The curves show how σ_x varies along two lines (y = constant) ten lattice constants above and below the dislocation, for a simple cubic lattice. These stresses are quite high near the dislocation and decrease as the reciprocal of the distance away from it. In a real material, dislocation lines run in many different directions, forming a complex network and a very irregular distribution of microstress.

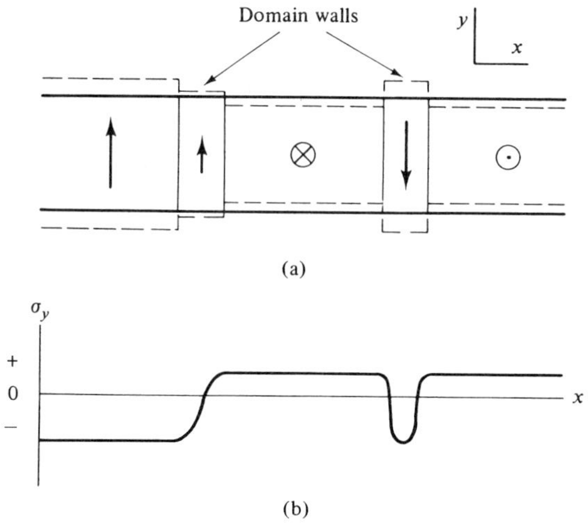

Fig. 9.30 Microstresses due to magnetostriction.

2. *Magnetostriction.* When a ferro- or ferrimagnetic is cooled below the Curie point, spontaneous magnetostriction tries to distort different domains in different directions. Because the domains are not free to deform independently, microstresses are set up. The same argument shows that the stress inside a domain wall differs from the stress in the adjoining domains. Figure 9.30 shows some examples for a material, like iron, in which the magnetostriction is positive in the direction of the spontaneous magnetization. Three domains are shown in (a), separated by one 90° and one 180° wall. Dashed lines indicate the dimensions of the various portions if they were free to deform in the y direction; strains in the x direction are ignored here. The corresponding variation of σ_y with x is shown in (b), where the zero-stress level refers to the paramagnetic material above the Curie point. Stress of magnetostrictive origin are rather small, being of the order of $\lambda_{si}E$, where E is Young's modulus (see Section 8.1). They are large enough, however, to cause interactions between domains, or domain walls, and crystal imperfections. (In fact, if one takes the point of view that anything which sets up microstress is a crystal imperfection, then domains and domain walls *are* crystal imperfections.) The localized distortion within domain walls has been revealed by special methods of x-ray diffraction [9.25] [9.25 a].

Plastic deformation (cold work) has marked effects on many magnetic properties, and it has often been studied. In this context, two kinds of plastic deformation should be distinguished:

1. *Nonuniform*, such as the rolling of sheet, the drawing or swaging of wire and rod through a die, bending, twisting, etc. These processes invariably produce both macro and micro residual stress. The resulting macrostress distribution

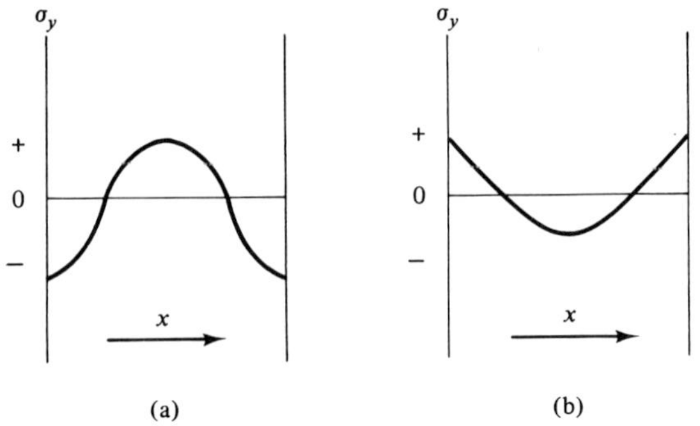

Fig. 9.31 Distribution of the longitudinal residual stresses σ_y across the diameter x of cold-drawn rods.

depends on the kind and amount of working, and its form is not always predictable in advance. Figure 9.31, for example, shows two stress distributions that can be produced by cold drawing, depending on the shape of the die; in (a) the rod is composed of an outer "case" in compression and an inner "core" in tension; in (b) this distribution is reversed. Nonuniform modes of deformation are quite unsuitable for fundamental studies of the effect of cold work, because the inside and outside of the same specimen can have grossly different magnetic properties.

2. *Uniform*, such as uniaxial tension and compression. These processes produce only micro residual stress, as indicated in Fig. 9.32 for a stretched rod. However, the microstress distribution is irregular, as shown in the sketch at the right: most of the lattice, estimated at some 90 percent, is in residual compression, balanced by small regions under high tensile stress. Although the details of this stress distribution are hypothetical, as is true of any microstress distribution in the present state of our knowledge, its general form is supported by both x-ray and magnetic evidence [9.26, 9.27]. (Part of the magnetic evidence is given in Fig. 8.24. The dashed curve there refers to a specimen which had been plastically elongated by 3.2 percent. Its λ,H behavior, under zero applied stress, is much like that of an annealed specimen under 10,000 lb/in^2 compressive stress, indicating that most of the volume of the pre-stretched specimen is in a state of residual compression. For evidence of another kind, see Section 10.6.) While tensile deformation has the advantage over swaging, for example, of not introducing macrostress, the amount of uniform tensile deformation that can be achieved is limited by the onset of "necking" (local contraction); this limits the amount of uniform deformation, as measured by the reduction in area of cross section to some 30–35 percent for most materials. Reductions of double this amount can usually be achieved by swaging or wire drawing before cracking begins.

The reader should note one simplification running through all the above discussion of residual stress: only the stress acting in one direction has been con-

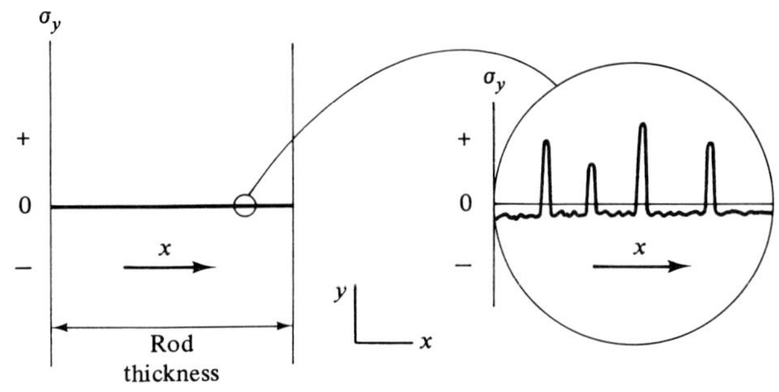

Fig. 9.32 Distribution of longitudinal residual stress σ_y across the diameter x of a rod after plastic elongation in the y direction.

sidered. The real situation is much more complex. At the surface of a body the residual stress can be, and usually is, biaxial (two stresses at right angles) and, in the interior, triaxial (three stresses at right angles). Baldwin [9.19] has given clear depictions of these complex stress states.

9.11 HINDRANCES TO WALL MOTION (MICROSTRESS)

Residual microstress hinders domain wall motion because of magnetostriction. The first detailed calculations of this effect were made in 1937 by Kondorski [9.28]. The behavior of 90° walls (non-180° walls) and 180° walls is quite different, and in the following discussion only reversible wall motion is considered.

90° walls

Suppose the 90° wall of Fig. 9.33 (a) is moved to the right by a field applied in the $+x$ direction. Consider unit area of material normal to the direction of motion, between the two thin lines, and let the wall move a distance dx between positions 1 and 2. Let the residual stress σ_y in this region vary with x as shown in (b), where the variation is assumed to be linear over a short distance:

$$\sigma_y = g\,x, \tag{9.22}$$

where $g = d\,\sigma_y/dx$ is the stress gradient. Assume also that the material, like iron, has positive magnetostriction in the easy direction ⟨100⟩. Then, as the wall moves from 1 to 2, it sweeps out a volume dx. Before this motion the magneto-elastic energy of this volume was zero, by Eq. (8.19); after the motion, it is $\frac{3}{2}\lambda_{100}\,\sigma_y dx$. The increase is

$$dE_{me} = \tfrac{3}{2}\,\lambda_{100}\,\sigma_y\,dx. \tag{9.23}$$

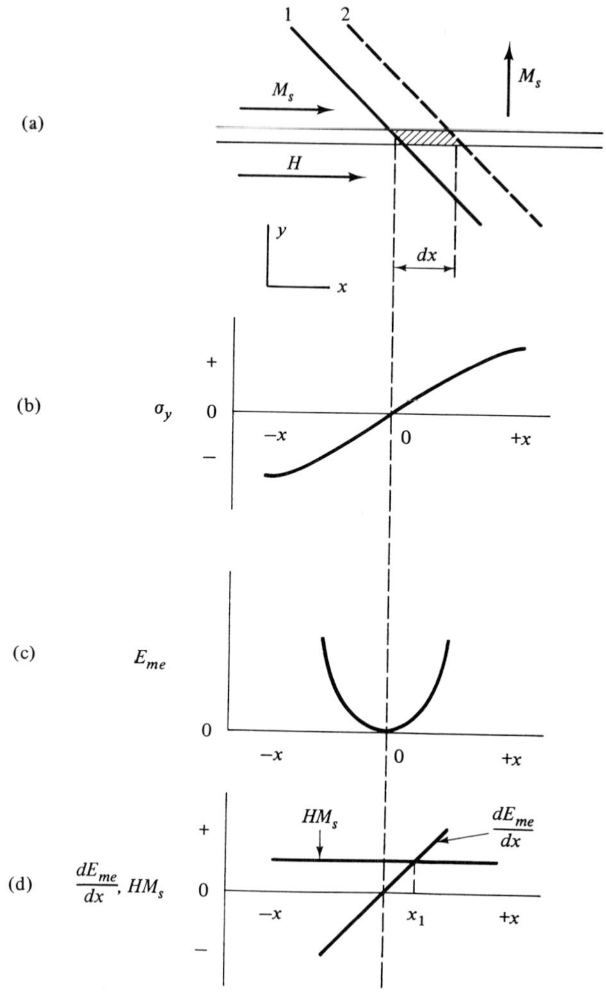

Fig. 9.33 Motion of a 90° wall in a stress field.

(It is here assumed that the crystal anisotropy energy K_1 is large compared to the magnetoelastic energy $\frac{3}{2}\lambda_{100}\,\sigma_y$, so that M_s in the left-hand domain is not turned out of the x direction by the tensile stress in the y direction.) The potential energy of the same volume in the field H is zero before and $-HM_s\,dx$ after. Therefore,

$$dE_p = -HM_s\,dx. \tag{9.24}$$

Combining these three equations, we find for the change in total energy:

$$dE = dE_p + dE_{me} = -HM_s\,dx + \tfrac{3}{2}\lambda_{100}gx\,dx. \tag{9.25}$$

The equilibrium position of the wall is given by

$$dE/dx = -HM_s + \tfrac{3}{2}\lambda_{100}g\,x = 0. \tag{9.26}$$

Now dE/dx is a force and, since we are dealing with unit area normal to x, the first term HM_s can be regarded as the force per unit area, or *pressure*, exerted by the field on the wall. At equilibrium, this is balanced by a pressure in the other direction given by the second term. This resistance is due to the fact that the region swept out by the moving wall tends to contract in the y direction, because M_s and its magnetostrictive elongation are now parallel to x, and it must do this against a tensile stress. Integration of Eq. (9.25) from 0 to x gives

$$E = E_p + E_{me} = -HM_s x + \tfrac{3}{4}\lambda_{100}g\,x^2. \tag{9.27}$$

The variation of E_{me} and its gradient is sketched in Fig. 9.33 (c, d). When H is zero, we note from (c) that the equilibrium position of the wall is at $x = 0$, where the stress is zero. When H is positive, we can find the wall position from (d), which is the graphical equivalent of Eq. (9.26); we draw a horizontal line HM_s at a height above zero proportional to H; its intersection with the line dE_{me}/dx then gives the position x_1 of the wall for this value of the field.

(The conclusion that the stable position of the wall of Fig. 9.33 at zero field is a point of zero stress is not always true. If the stress gradient is negative at the zero-stress point, the magnetoelastic energy is a maximum rather than a minimum at that point, and the wall will move spontaneously away from it. And when we consider the other kind of 90° wall, where the domain to the left of the wall has M_s in the y rather than the x direction, these conclusions are reversed. Table 9.1 summarizes the situation for both positive and negative magnetostriction in the easy direction. There are also situations in which stress has *no* effect on wall position. For example, suppose the stress is parallel to the wall and both lie in the yz plane, with the stress making any angle with the y axis. Let M_s be in the [110] direction in the domain to the left of the wall and in [1$\bar{1}$0] in the domain to the right. Then the wall has no preferred position. The total magnetoelastic energy of the two domains is constant, whatever the sign of the stress and whatever the wall position.)

We can also calculate the initial susceptibility κ_0 associated with motion of the wall of Fig. 9.33. When the wall moves a distance x, the magnetization

Table 9.1 Stability Conditions for a 90° Domain Wall in Zero Field

Kind of wall	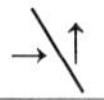		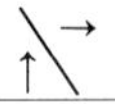	
Stress gradient at $\sigma_y = 0$	$+\lambda$	$-\lambda$	$+\lambda$	$-\lambda$
$+g$	S	U	U	S
$-g$	U	S	S	U

S = stable, U = unstable.

of a length l of material in the direction of the field increases from zero to $M_s x/l$, and x is given by Eq. (9.26). Therefore,

$$\kappa_0 = \frac{M}{H} = \frac{M_s x}{Hl} = \frac{2M_s^2}{3l\lambda_{100} g}. \tag{9.28}$$

(The quantity $1/l$ equals $1\sqrt{2}$ times the wall area per unit volume.)

When a material is free from inclusions, it should therefore have high initial susceptibility if its magnetostriction and the residual stress gradients within it are small, as far as 90° wall motion is concerned.

180° walls

The magnetostrictive strain in the domains on either side of a 180° wall is the same, because it is independent of the sense of the magnetization. Therefore, when the wall moves, there can be no change in the magnetoelastic energy of the domains, whatever the stress distribution. The only effect of stress is to change the energy of the wall itself.

The energy γ of a domain wall is made up of exchange and anisotropy energy. In deriving Eq. (9.10),

$$\gamma = 2\,K\,\delta, \tag{9.10}$$

where δ is the wall thickness, we considered only crystal anisotropy energy, represented by the constant K. But if the crystal is under stress, parallel to the wall and to the easy axis, then we should add another term representing the stress anisotropy. Then,

$$\gamma = 2\,\delta\,(K_1 + K_\sigma), \tag{9.29}$$

where K_1 represents crystal anisotropy and K_σ stress anisotropy. [Here we ignore the fact that the crystal and stress anisotropies do not necessarily vary with M_s orientation in the same way. This is unimportant, however, in view of the fact that Eq. (9.10) is itself approximate.] Then, from Eq. (8.23),

$$\gamma = 2\,\delta\,[K_1 + \tfrac{3}{2}\,\lambda_{100}\,\sigma_y], \tag{9.30}$$

for a stress in the y direction.

We might now assume that the wall is in position 1 of Fig. 9.34 (a), where, as before, the stress σ_y is zero and, near the point $x = a$, varies linearly with x:

$$\sigma_y = g(x - a). \tag{9.31}$$

Putting this value into the expression for γ, we have

$$\gamma = 2\,\delta\,[K_1 + \tfrac{3}{2}\,\lambda_{100} g(x - a)]. \tag{9.32}$$

Since γ increases linearly with x, the wall would move spontaneously to the left of $x = a$ in zero field. The point of zero stress is not a stable one.

However, we can show that the wall will be stable in position 1′, where the stress is assumed to go through a minimum and which is taken as the origin. As-

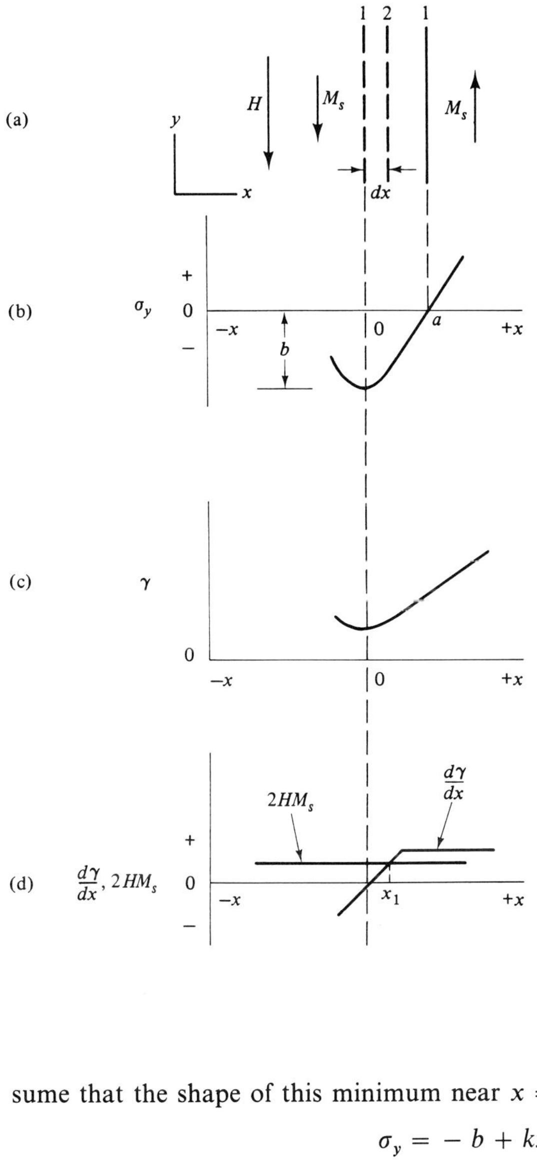

Fig. 9.34 Motion of a 180° wall in a stress field.

sume that the shape of this minimum near $x = 0$ is parabolic:

$$\sigma_y = -b + kx^2, \tag{9.33}$$

where b and k are constants. Then,

$$\gamma = 2\delta[K_1 - \tfrac{3}{2}\lambda_{100}b + \tfrac{3}{2}\lambda_{100}kx^2], \tag{9.34}$$

and γ goes through a minimum at $x = 0$, as shown in Fig. 9.34(c). (The wall is stable at a stress minimum, in zero field, whether the stress is tensile or compressive at the minimum.) If a field in the direction indicated moves the wall by an amount dx from 1′ to 2, the change in potential energy per unit area of wall is

$$dE_p = -2HM_s\,dx. \tag{9.35}$$

The equilibrium position of the wall is given by

$$\frac{dE}{dx} = \frac{dE_p}{dx} + \frac{d\gamma}{dx} = -2HM_s + 6\,\delta\,\lambda_{100}kx = 0. \tag{9.36}$$

This balance of field pressure and increasing wall energy is illustrated in (d). As before, we can calculate an initial susceptibility:

$$\kappa_0 = \frac{2M_s^2}{3l\delta\,\lambda_{100}k}, \tag{9.37}$$

where $1/l$ is the wall area per unit volume. [As the field is increased, for the arbitrary stress distribution shown in (b), the wall would move to the right from position 1′ until it reaches the position where the stress gradient becomes constant; it would then move continuously to the right without any further increase of field.] The smaller the constant k, the more "shallow" is the stress minimum and the lower the stress gradient at points near the minimum. Therefore, the conditions for high susceptibility with respect to 180° wall notion are much like those for 90° wall motion, namely, low magnetostriction and low stress gradients.

In the above treatment of 180° wall motion, it is tacitly assumed that the wall is thin with respect to the scale of the stress variation, i.e., that the stress is essentially constant across the width of the wall. But the stress may vary rapidly with distance and go through several maxima and minima even within the wall thickness. Although the above treatment is then invalid, the susceptibility is expected to be a minimum when the average distance between stress maxima is about equal to the wall thickness. These matters are discussed by Chikazumi [G.23]. Actually, when the stress varies on the scale of the wall thickness, the concepts of domains and domain walls lose their significance and a micromagnetics approach is preferable, at least when $\lambda\sigma$ is large compared to K.

Remarks

The calculations in this section are subject to several qualifications:

1. The energy of a 90° wall must also depend on stress, but we have ignored it. However, the result of Problem 9.5 shows that the change in wall energy is small compared to the change in the magnetoelastic energy of the adjoining domains.

2. The calculated susceptibilities refer only to the motion of *one* wall in the particular kind of stress field assumed, and not to the motion of a large number of walls in the complex stress fields of real specimens.

3. Only a single stress σ_y has been considered, and it has been assumed to depend only on x. Actually, various parts of a 180° wall, for example, can be expected to be subjected to quite different stresses. Suppose that σ_y is a function of both x and y. Then the wall energy γ will vary not only with x, as in Fig. 9.34 (c), but also with y. The variation of γ with x and y might appear as in Fig. 9.35, which shows energy "hills" in the xy plane. The full line shows the initial position of the wall, lying in a valley where the energy is a minimum, and the dashed line

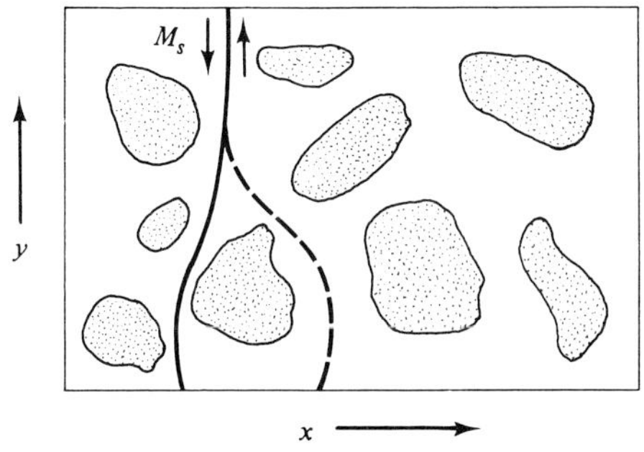

Fig. 9.35 Motion of a 180° wall in a stress field which has stress variations in two directions. After Bozorth [G.4].

shows the wall's position after a portion of it has been moved over a hill by a field. If the wall is curved as shown, to fit in energy valleys, then free poles must form on the wall and a new form of energy (magnetostatic) is introduced. On the other hand, if the wall remains straight in order to avoid free poles, parts of it must have high energy because they are located in regions where the stress is not a minimum. The wall will therefore assume a shape that minimizes the sum of the magnetostatic and wall energies.

Because of these restrictions, the results of this section can have only qualitative value. They lead to the conclusion that in the absence of inclusions, wall motion is easiest when the magnetostriction and stress gradients are low. Not much more can be said until our knowledge of real microstress distributions is improved. (Some writers have assumed regular, periodic stress variations, such as sinusoidal, and then have made fairly elaborate calculations on the basis of such a model. The results of such calculations are no better than the assumed stress distribution, which is quite unreal.)

As mentioned earlier, one source of microstress that is always present, to a greater or lesser degree, is dislocations. These can interact with domain walls because of the magnetostrictive stress associated with the walls, with the result that walls and dislocations can exert forces on one another. The subject of wall-dislocation interactions is complex, because there is a variety of kinds of walls and kinds of dislocations, and this subject has received relatively little attention, theoretical or experimental. As a result, only a few simple generalizations are possible:

1. If a dislocation intersects a 180° wall, there is no interaction, in the sense that the wall can move with respect to the dislocation without any change in the energy of the system.

2. When a 180° wall is parallel to a dislocation, there is an interaction. (Domain walls are usually flat and dislocations curved, so that a dislocation line might pass through one part of a domain wall and be parallel to another.)

3. A dislocation lying within a domain is the source of extra energy, mainly magnetoelastic, the amount depending on the relative orientation of the dislocation and the M_s vector of the domain. If the stress field of the dislocation

rotates nearby spins out of the easy direction by an appreciable amount, extra exchange and magnetostatic energy will be created, the latter because of the divergence of the magnetization. Thus an array of dislocations, as may be produced in a single crystal by appropriate deformation, may determine the M_s directions and hence the domain structure.

Vicena [9.29], Seeger *et al.* [9.30] and Scherpereel *et al.* [9.30 a] have described wall-dislocation interactions in terms of both theory and experiment.

The theoretical work performed so far on this problem has been confined to the interaction of one wall and one dislocation. Closer to reality, and much more complex, is the interaction of a wall and one of the various kinds of *arrays* of dislocations.

9.12 HINDRANCES TO WALL MOTION (GENERAL)

At zero applied field, a domain wall will be in a position which minimizes the energy of the system, where the "system" means the wall itself and/or the adjoining domains. Thus, if it is a 180° wall, it will tend to bisect an inclusion and be located at the point where the microstress goes through a minimum.

If a small field is then applied, the wall will move, but it moves against a force tending to restore it to its original position. This restoring force, if caused by an inclusion, is due mainly to the increase in wall energy and magnetostatic energy resulting from the wall motion. If caused by microstress, the restoring force is due to an increase in wall and magnetoelastic energy.

Hindered wall motion of either kind may be discussed in terms of the variation of a single energy E with wall position x, where E stands for any or all of the various energies mentioned above. We may suppose that E varies with x as in Fig. 9.36 (a). The gradient of the energy is shown in (b), along with the line cH representing the pressure of the field on the wall. (The value of the constant c depends on the kind of wall and the orientation of the field; for the case illustrated in Fig. 9.34, $c = 2M_s$.) At $H = 0$, the wall is at position 1, in an energy minimum. As H is increased from zero, the wall moves *reversibly* to 2; if the field were removed in this range, the wall would return to 1. But point 2 is a point of maximum energy gradient (maximum restoring force); if the field is sufficient to move the wall to 2, it is sufficient to make it take an *irreversible* jump to 3, which is the only point ahead of the wall with an equally strong restoring force. This is a Barkhausen jump. If the field is then reduced to zero, the wall will go back, not to point 1, but to 4, which is the nearest energy minimum, thus exhibiting the phenomena of hysteresis and remanence. A reverse field will then drive the wall reversibly from 4 to 5 and by another Barkhausen jump from 5 to 6. If the diagram in (b) is rotated 90°, it takes the form of the elemental hysteresis loop shown in (c), because wall motion x is equivalent to change in magnetization M, and dE/dx is proportional to H.

This hysteresis loop refers to only one small region of a specimen and to a restricted range of H. The hysteresis loop of a real specimen is the sum of a great

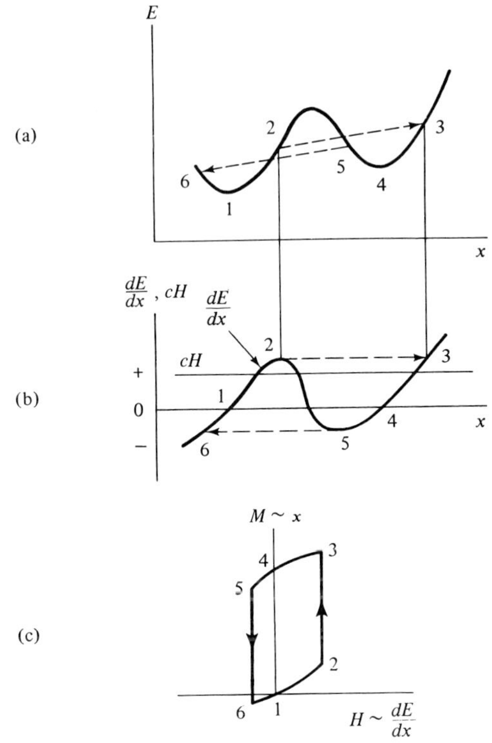

Fig. 9.36 Reversible and irreversible wall motion.

number of these elemental loops, of various shapes and sizes, summed over the whole volume of the specimen. Although we are ignorant of the exact form of the E,x curve, diagrams like Fig. 9.36 have proved to be valuable aids to thought in many magnetic problems. And, as will become clear later, Fig. 9.36 can be generalized to include magnetization changes by rotation as well as by wall motion.

One form of energy change not accurately described by Fig. 9.36 (a) is the sudden motion of a wall as it jumps away from the tubular domains connecting it to an inclusion (Fig. 9.27 d). Further comment on this inherently irreversible process will be made in Section 13.3.

9.13 MAGNETIZATION BY ROTATION

We have already made some calculations, in Chapter 7, of magnetization change by rotation in single crystals of particular orientations. We now wish to obtain more general results. As it was convenient, in the preceding sections on wall motion, to consider only one wall, it will now be convenient to consider the rotation process in isolation. We can do this by treating only single-domain particles, from which domain walls are necessarily absent. This problem was examined in great detail by Stoner and Wohlfarth [7.23] in a classic paper published in 1948. Their calculations have an important bearing on the theory of permanent-magnet materials, because some of these materials are thought to consist of single domains.

When an applied field rotates the M_s vector of a single domain out of the easy direction, the rotation takes place against the restoring force of some anisotropy, usually the anisotropy of shape, stress, or crystal. We will treat the problem in terms of shape anisotropy, in which most of the other forms can be included. By letting the particle have the shape of an ellipsoid of revolution, we include all the particle shapes of physical interest: rod (prolate spheroid), sphere, and disk (oblate spheroid).

Prolate spheroid

Let c be the semi-major axis, the axis of revolution, and a the semi-minor axis. Then c is the easy axis of magnetization, and the anisotropy energy is given by

$$E_a = K_u \sin^2 \theta, \tag{9.38}$$

where θ is the angle between M_s and c, and the uniaxial anisotropy constant K_u can be written in terms of the demagnetizing coefficients along a and c by means of Eq. (7.54). For the present, though, we leave Eq. (9.38) in its simple form, because it can also represent the anisotropy energy of a spherical crystal subjected to a stress or that of a spherical uniaxial crystal. The values of K_u for these three forms of anisotropy have been given in Table 8.2.

Let the applied field H make an angle α with the easy axis, as in Fig. 9.37(a). Then the potential energy is

$$E_p = -HM_s \cos(\alpha - \theta), \tag{9.39}$$

and the total energy is

$$E = E_a + E_p = K_u \sin^2 \theta - HM_s \cos(\alpha - \theta). \tag{9.40}$$

The equilibrium position of M_s is given by

$$\frac{dE}{d\theta} = 2K_u \sin\theta \cos\theta - HM_s \sin(\alpha - \theta) = 0, \tag{9.41}$$

and the magnetization resolved in the field direction is given by

$$M = M_s \cos(\alpha - \theta). \tag{9.42}$$

Suppose the field is normal to the easy axis, so that α is 90°. Then,

$$2K_u \sin\theta \cos\theta = HM_s \cos\theta,$$

and

$$M = M_s \sin\theta.$$

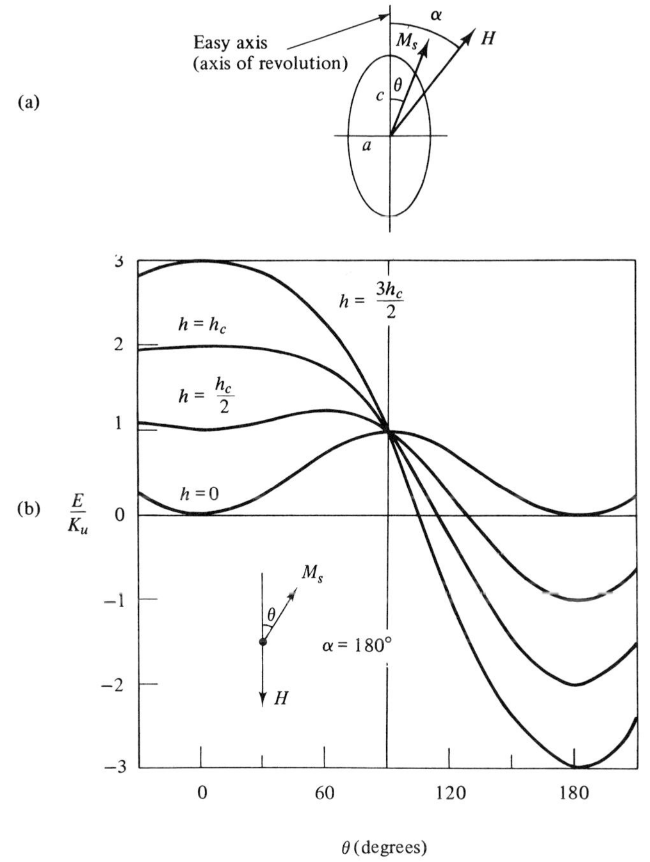

Fig. 9.37 Rotation of magnetization in an ellipsoid. h = reduced field.

Therefore,

$$2K_u(M/M_s) = HM_s.$$

Put $M/M_s = m$ = reduced magnetization. Then,

$$m = H(M_s/2K_u). \tag{9.43}$$

This shows that the magnetization is a linear function of H, with no hysteresis. Saturation is achieved when $H = H_K = 2K_u/M_s$ = anisotropy field, as we saw in Eq. (7.30) for a similar problem. If we put h = reduced field $= H/H_K = HM_s/2K_u$, then $m = h$ when α is 90 .

For the general case, Eqs. (9.41) and (9.42) may now be written

$$\sin\theta\cos\theta - h\sin(\alpha - \theta) = 0, \tag{9.44}$$

$$m = \cos(\alpha - \theta). \tag{9.45}$$

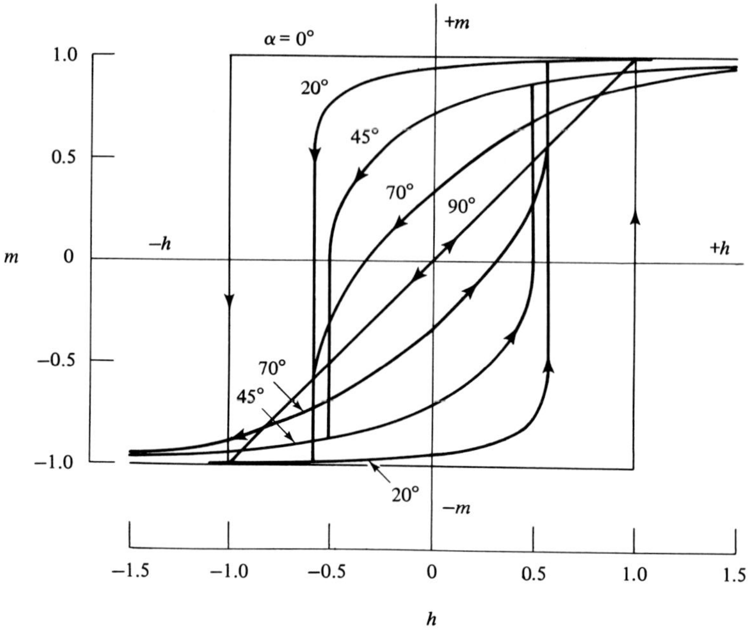

Fig. 9.38 Hysteresis loops for uniaxial anisotropy. (α = angle between field and easy axis.)

Suppose now that the field is along the easy axis ($\alpha = 0$), and that H and M_s both point along the positive direction of this axis. Then let H be reduced to zero and then increased in the negative direction ($\alpha = 180°$). Although H and M_s are now antiparallel and the field exerts no torque on M_s, the magnetization will become unstable at $\theta = 0$ and will flip over to $\theta = 180°$ (parallelism with H) when H reaches a sufficiently high value in the negative direction. To find this critical value we note that a solution to Eq. (9.44) does not necessarily correspond to a minimum in the total energy E, a point of stable equilibrium. It might also correspond to an energy maximum (unstable equilibrium), depending on the sign of the second derivative. If $d^2E/d\theta^2$ is positive, the equilibrium is stable; if it is negative, the equilibrium is unstable; if it is zero, a condition of stability is just changing to one of instability. Thus the critical field is found by setting

$$\frac{d^2E}{d\theta^2} = \cos^2\theta - \sin^2\theta + h\cos(\alpha - \theta) = 0. \tag{9.46}$$

Simultaneous solution of Eqs. (9.44) and (9.46) leads to the following equations, from which the critical field h_c and the critical angle θ_c, at which the magnetization will flip, may be calculated:

$$\tan^3\theta_c = -\tan\alpha, \tag{9.47}$$

$$h_c^2 = 1 - \tfrac{3}{4}\sin^2 2\theta_c. \tag{9.48}$$

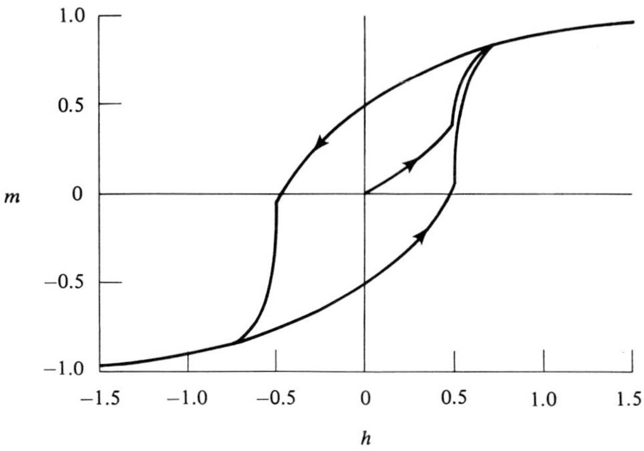

Fig. 9.39 Hysteresis loop of an assembly of single-domain, uniaxial particles having their easy axes randomly oriented. Stoner and Wohlfarth [7.23], Rhodes [9.31].

When $\alpha = 180°$, $\theta_c = 0$ and $h_c = 1$, or $H = H_k$. The hysteresis loop is then rectangular, as shown in Fig. 9.38.

The way in which the total energy E varies with the angular position θ of the M_s vector for $\alpha = 180°$ is shown in Fig. 9.37 (b) for various field strengths. It is apparent there how the original energy minimum at $\theta = 0$ changes into a maximum when $h = h_c$. These curves are plotted from Eq. (9.40). They are the analogues, for the rotational process, of the curve of Fig. 9.36 (a), which showed how energy varies with domain wall position.

The reduced magnetization m as a function of reduced field h for an intermediate angle, say $\alpha = 20°$, is calculated as follows. For positive values of h, the angle θ will vary between 0 and 20^0. For selected values of θ in this range, corresponding values of h and m are found from Eqs. (9.44) and (9.45). When h is negative, $\alpha = 180° - 20° = 160°$, and Eq. (9.47) gives the critical value θ_c at which the magnetization will flip. Values of h and m are again found from Eqs. (9.44) and (9.45), with α equal to 160°, for selected values of θ in the range 0 to θ_c.

Figure 9.38 shows hysteresis loops calculated for various values of α. In general, these loops consist of *reversible* and *irreversible* portions, the latter constituting Barkhausen jumps. (We see, therefore, that reversible and irreversible changes in magnetization can occur by domain rotation as well as by domain wall motion.) The portion of the total change in m due to irreversible jumps varies from a maximum at $\alpha = 0$ to zero at $\alpha = 90°$. The critical value of reduced field h_c, at which the M_s vector flips from one orientation to another, decreases from 1 at $\alpha = 0$ to a minimum of 0.5 at $\alpha = 45°$ and then increases to 1 again as α approaches 90°; these critical values are equal for any two values of α, such as 20° and 70°, symmetrically located about $\alpha = 45°$. On the other hand, the reduced intrinsic coercivity h_{ci} (the value of h which reduces m to zero) decreases from 1 at $\alpha = 0$ to zero at $\alpha = 90°$. For values of α between 0 and 90°, cyclic variation of H in a fixed direction has a curious result: M_s makes one complete revolution per cycle, although it does not rotate continously in the same direction.

Stoner and Wohlfarth [7.23] and Rhodes [9.31] also calculated the hysteresis loop of an assembly of *noninteracting* particles, with their easy axes randomly oriented in space so that the assembly as a whole is magnetically isotropic (Fig. 9.39). This hysteresis loop is characterized by a retentivity m_r of 0.5 and a coercivity h_{ci} of 0.48. (The stipulation that the particles are not interacting means, in effect, that the external field of each particle, which is due to its own magnetization and which acts on surrounding particles in addition to the field applied to the assembly as a whole, is neglected. This is a serious restriction, which is discussed in Section 11.3.)

We will now translate the results of the above calculations from the rather abstract reduced field h into the more concrete actual field H for three kinds of uniaxial anisotropy (shape, stress, and crystal). By definition, $H = h(2K_u/M_s)$. Combining this with the values of K_u from Table 8.2, we have

$$\text{(shape)} \qquad H = h(N_a - N_c)M_s, \tag{9.49}$$

$$\text{(stress)} \qquad H = h\left(\frac{3\lambda_{si}\sigma}{M_s}\right), \tag{9.50}$$

$$\text{(crystal)} \qquad H = h\left(\frac{2K_1}{M_s}\right), \tag{9.51}$$

where N_a and N_c are the demagnetizing coefficients parallel to the a and c axes, λ_{si} the saturation magnetostriction (assumed isotropic), σ the stress, and K_1 the crystal anisotropy constant. If we wish to calculate the coercivity, for example, under similar conditions (same inclination α of easy axes to field direction), then h is constant in these three equations; this means then that the intrinsic coercivity H_{ci} varies directly as M_s for shape anisotropy but inversely with M_s for stress and crystal anisotropy. Thus, to maximize H_{ci}, attention must be paid not only to the anisotropy itself but also to the saturation magnetization of the material.

As an indication of the coercivities that result from shape anisotropy alone, Table 9.2 gives values of $(N_a - N_c)$ for various values of c/a, along with calculated H_{ci} values for iron particles (M_s = 1714 emu/cm^3). (Crystal anisotropy is assumed absent, so that H_{ci} for spherical particles is assumed to be zero.) We note that the particle shape has to depart only slightly from spherical in order to produce a coercivity of several hundred oersteds, and that an increase in c/a from 1.1 to 1.5 quadruples the coercivity. Increasing c/a beyond about 5 produces only moderate increases.

It is of interest to calculate the stresses which would have to be applied to a spherical particle in order to attain the same coercivities. (These are uniaxial

Table 9.2 Calculated Coercivities of Single-Domain Particles (Easy Axes Aligned with Field)*

Shape anisotropy			Stress anisotropy		Crystal anisotropy
c/a	$N_a - N_c$	H_{ci}(Oe)	σ (lb/in^2)	σ (kg/mm^2)	K_1 (10^5 ergs/cm^3)
1.0	0	0	0	0	0
1.1	0.472	810	340,000	240	7
1.5	1.892	3,240	1,350,000	950	28
5.	5.231	8,950	3,700,000	2,600	77
10.	5.901	10,100	4,200,000	3,000	87
20.	6.156	10,500	4,400,000	3,100	90
∞	$6.283 = 2\pi$	10,800	4,500,000	3,200	93

* $M_s = 1714$ emu/cm^3, $\lambda_{si} = 20 \times 10^{-6}$.

stresses, tensile for positive λ_{si} and compressive for negative λ_{si}.) The same value of M_s is assumed, but λ_{si} is arbitrarily taken as 20×10^{-6}, which is intermediate between the values for iron and nickel. The results are shown in the central portion of Table 9.2. The required stresses are seen to be very large. For example, a stress of 340,000 lb/in^2, which produces the same coercivity as an unstressed nonspherical particle with $c/a = 1.1$, is above the yield strength of the best heat-treated steels. This result does not necessarily mean that this stress could not be supported by particles a few hundred angstroms in diameter, because mechanical strength is known to increase markedly as the "specimen" size decreases. However, it does raise questions as to how such a stress is to be applied to the particles. The conclusion is that high coercivity is much easier to attain by shape, rather than stress, anisotropy.

Finally, in the third part of Table 9.2, values of the uniaxial crystal anisotropy constant K_1 are shown, which would produce in spherical particles the coercivities listed in the first part. The same value of M_s is assumed. Here the required values of K_1 are of the same order of magnitude as those available in existing materials. Thus, barium ferrite has K_1 equal to 33×10^5 ergs/cm^3, and cobalt has K_1 equal to 45×10^5 ergs/cm^3, leading to calculated coercivities of several thousand oersteds. (The fact that M_s for these substances is less than the value of 1714 emu/cm^3 assumed in the calculations would make the computed values even higher. The calculated value for barium ferrite, with $M_s = 380$ emu/cm^3, is 17,000 Oe and for cobalt, with $M_s = 1422$ emu/cm^3, it is 6,300 Oe.)

Although the calculations leading to the stress and crystal anisotropy figures in Table 9.2 are based on very arbitrarily selected values of M_s and λ_{si}, they do serve to fix the order of magnitude of the requirements.

All of the above values apply to an assembly of particles with their easy axes parallel to one another and to the applied field. If the particles are randomly oriented, the calculated coercivities are to be multiplied by 0.48.

If the crystal structure is *cubic*, then the rotational processes in an unstressed, spherical particle are much more difficult to calculate. There is now not one easy axis, but three (K_1 positive) or four (K_1 negative), and the anisotropy constants

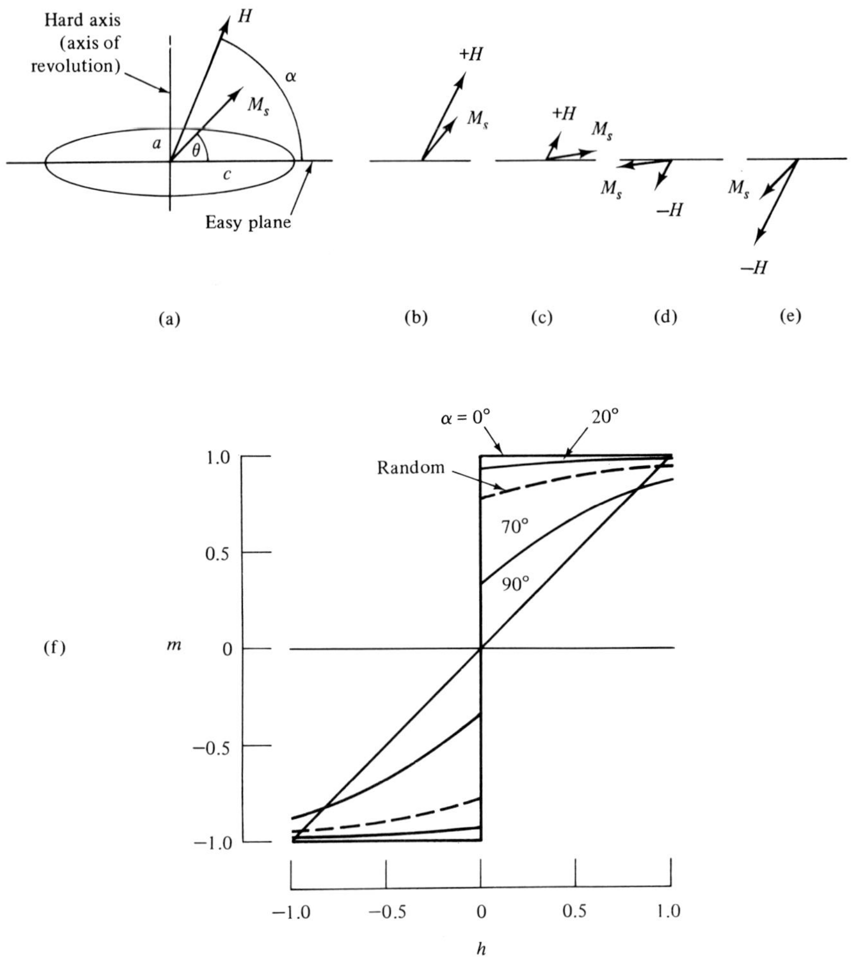

Fig. 9.40 Hysteresis "loops" for oblate spheroids.

are generally low. Stoner and Wohlfarth estimate that if the easy axes of an assembly of particles are aligned with the field, the maximum value of H_{ci} will be about 400 Oe for iron and about 200 Oe for nickel.

Oblate spheroid

Let a be the semi-minor axis, the axis of revolution, and c the semi-major axis, as in Fig. 9.40 (a). The a axis is now a hard axis and cc an *easy plane* of magnetization. We define α as the angle between H and the easy plane and θ as the angle between M_s and the easy plane. Then Eqs. (9.38), (9.44), (9.45), and (9.49) are again valid.

However, a new kind of rotational behavior now occurs: whatever the value of α, there is no hysteresis and the coercivity is zero. This is shown by the sequence of sketches in Fig. 9.40, (b) to (e). As H is reduced from a large positive value (b) to a small positive value (c), M_s rotates reversibly toward the easy plane. Then, as H changes to a small negative value (d), M_s rotates by 180° in the easy plane, changing m abruptly from a positive to a negative value at $h = 0$. Further increase in H in the negative direction rotates M_s away from the easy plane. Calculated hysteresis "loops" for several angles α are shown in (f). (Note that the curve for $\alpha = 0$ is identical with that for a multidomain single crystal magnetized along its easy axis, as shown by Fig. 7.16. There the magnetization change was assumed to occur entirely by unhindered domain wall motion.) The dashed lines show the behavior of an assembly of oblate spheroids randomly oriented in space, as calculated by Stoner and Wohlfarth [7.23].

Remarks

The hysteresis loops of Figs. 9.38 to 9.40 exhibit a remarkable variety of size and shape. They range from square loops to straight lines to curved-and-straight lines without hysteresis. They offer a challenge to the materials designer, who seemingly has at his fingertips the ability to produce a magnetic material tailored to almost any application. (This vista is not quite perfect, however. For particles with a single easy axis, Fig. 9.38 shows that retentivity and coercivity are not independently variable; if one increases, so does the other.)

The great difficulty, of course, is to make particles small enough to be single domains, and then imbed them in some controlled manner in a suitable matrix without sacrificing too much of the potential inherent in a single isolated particle.

Some considerable success in this direction has already been achieved, as we shall see in Chapter 14 on permanent-magnet materials. These materials must have a high coercivity, and magnets with coercivities roughly in the range of 500–2000 Oe are currently in production. Thus the even larger values shown in Table 9.2 as theoretically attainable in small particles are of great practical interest, and the processes of magnetization rotation in small particles have been the subject of much experimental and theoretical research.

Throughout this section the tacit assumption has been made that the rotational process in a single-domain particle is one in which all spins in the particle remain parallel to one another during the rotation. This is not always true, as will be shown in Section 11.4. We will find there that the shape-anisotropy coercivities listed in Table 9.2 cannot be obtained in some particles, even in principle.

9.14 MAGNETIZATION IN LOW FIELDS

The magnetization curves and hysteresis loops of real materials have quite variable shapes. Only in three circumstances, however, do we have algebraic expressions to fit the observed curves:

1. High-field magnetization curves of single crystals, like those of Fig. 7.20.

2. High-field magnetization curves of polycrystalline specimens, to be described in the next section. In both (1) and (2) the magnetization change is by domain rotation
3. Low-field magnetization curves and hysteresis loops of polycrystalline specimens. These are discussed in this section.

By low fields are meant fields from zero to about one oersted. (Since the earth's field amounts to a few tenths of an oersted, precautions have to be taken to ensure that this field does not interfere with the measurements.) This range of magnetization was first investigated in 1887 by Lord Rayleigh [2.17] and is accordingly known as the *Rayleigh region.* In this region, magnetization is believed to change entirely by domain wall motion, except, of course, in specimens composed of single-domain particles.

By means of a suspended-magnet magnetometer, Rayleigh measured the low-field behavior of iron and steel wire. In extremely low fields, 4×10^{-5} to 4×10^{-2} Oe, he found that the permeability μ was *constant*, independent of H, which means that B (or M) varies linearly and reversibly with H. At higher fields, in the range 0.08 to 1.2 Oe, hysteresis appeared, and μ was no longer constant but increased linearly with H:

$$\mu = \mu_0 + \nu H, \tag{9.52}$$

where μ_0, the initial permeability, and ν are called the Rayleigh constants. This relation is called the *Rayleigh law.* (In very low fields, the term νH becomes negligibly small with respect to μ_0, which explains Rayleigh's finding of a constant μ. Alternatively stated, the observation of linear, reversible magnetization depends on the sensitivity of the measuring apparatus in detecting, or not detecting, the term νH. Rayleigh's experimental method was extremely sensitive.) Equation (9.52) forms the basis of the standard procedure for finding the initial permeability: experimental values of μ are plotted against H and extrapolated to zero field.

If Eq. (9.52) is multiplied by H, we have

$$B = \mu_0 H + \nu H^2, \tag{9.53}$$

which is the equation of the normal induction curve in the Rayleigh region. The term $\mu_0 H$ represents the reversible part, and νH^2 the irreversible part, of the total change in induction. By substituting $(H + 4\pi M)$ for B, we can obtain the equivalent expression for M:

$$M = \kappa_0 H + (\nu/4\pi) H^2, \tag{9.54}$$

where κ_0 is the initial susceptibility.

Rayleigh also showed that the hysteresis loop was composed of two parabolas:

$$B = (\mu_0 + \nu H_m) H \pm (\nu/2)(H_m^2 - H^2), \tag{9.55}$$

where H_m is the maximum field applied, and where the plus and minus signs apply to the descending and ascending portions of the loop, respectively. Figure 9.41

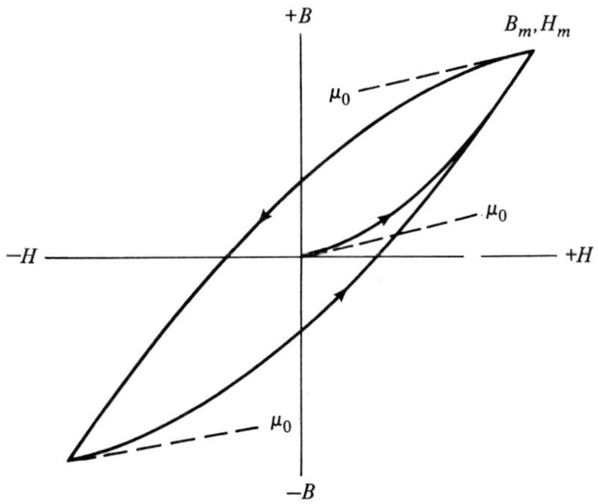

Fig. 9.41 Normal induction curve and hysteresis loop in Rayleigh region. Dashed lines indicate where slopes are μ_0.

shows such a loop, together with the initial curve. The slope of the curve leaving each tip is the same as the initial slope of the normal induction curve, namely μ_0. The hysteresis loss is, from Eq. (7.35),

$$\text{hysteresis loss} = (1/4\pi)(\text{area of } B,H \text{ loop})$$
$$= \frac{\nu H_m^3}{3\pi} \text{ erg/cm}^3/\text{cycle}. \tag{9.56}$$

The loss increases rapidly with H_m but is independent of μ_0. (Hysteresis loss is caused only by irreversible changes in magnetization and therefore depends on ν; the value of μ_0 affects the inclination of the loop but not its area.) The coercive force, at which $B = 0$, is given by

$$H_c = \left(\frac{\mu_0}{\nu} + H_m\right) - \sqrt{\left(\frac{\mu_0}{\nu} + H_m\right)^2 + H_m^2}, \tag{9.57}$$

and the remanence, at which $H = 0$, by

$$B_r = \frac{\nu}{2} H_m^2. \tag{9.58}$$

The Rayleigh relations have been confirmed for many materials. The constants μ_0 and ν vary over a wide range: 30 to 100,000 for μ_0 and 0.5 to 12,000,000 for ν, according to Bozorth [G.4]. Any change in the physical condition of a given material, such as a change in temperature or degree of cold work, usually causes μ_0 and ν to change in the same direction; sometimes there is a simple linear relation between the two constants. The maximum value of H, beyond which the Rayleigh relations no longer hold, can only be found by experiment.

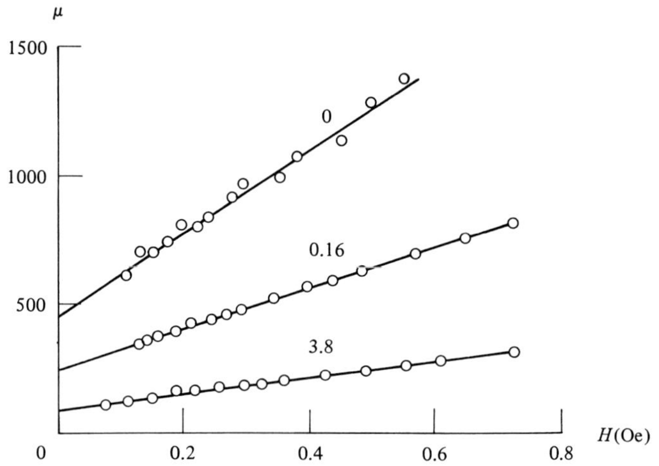

Fig. 9.42 Permeability μ *versus* field strength H for iron in the Rayleigh region as a function of prestrain. Numbers on curves are percent plastic elongation. Rusnak [9.32].

The initial permeability μ_0 almost invariably increases with rise in temperature, goes through a maximum just below the Curie point T_c, and decreases abruptly to unity at T_c. This behavior is related to the fact that the crystal anisotropy constant K and the magnetostriction λ_{si} both decrease to zero at or below T_c. Now domain wall energy is proportional to $\sqrt{K}$, and wall energy is the main contribution to the hindrance offered to wall motion by inclusions. The hindrance due to microstress depends only on the product $\lambda_{si}\sigma$. Therefore, whether the hindrances to wall motion are inclusions or microstress or both, they are expected to become less effective as the temperature increases, leading to an increased permeability.

Cold work decreases both Rayleigh constants. Figure 9.42 shows three μ,H curves, linear in accordance with Eq. (9.52); the material is Armco ingot iron, annealed in hydrogen to reduce the interstitial content (carbon and nitrogen in interstitial solid solution) to 23 ppm C and 20 ppm N. (1 ppm = 1 part per million = 10^{-4} percent.) The upper curve is for an annealed specimen and the other two for specimens prestrained in tension. These curves show, incidentally, the great sensitivity of the low-field magnetic behavior to a small amount of cold work. (For some materials the μ,H curve bends downward as H approaches zero. Measured values of μ at very low fields are then somewhat less than the value of μ_0 found by extrapolating the main linear portion of the curve.)

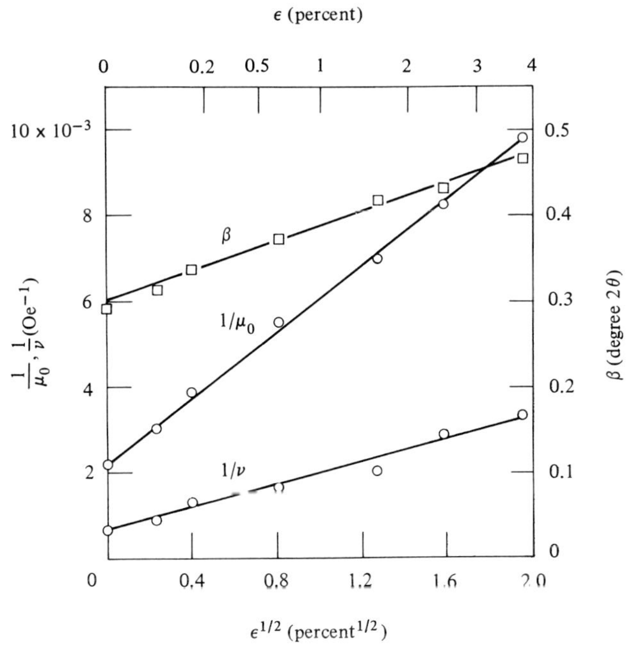

Fig. 9.43 Relations between the Rayleigh constants μ_0 and v and the x-ray line width β and the strain ε for cold-worked iron. Rusnak *et al.* [9.33].

For these iron specimens both Rayleigh constants are rather simply related to the strain ε; as Fig. 9.43 shows, the reciprocal of either constant varies linearly with $\sqrt{\varepsilon}$. The x-ray diffraction line width β was taken as a measure of the crystal imperfections produced by cold work; β also varies linearly with $\sqrt{\varepsilon}$. From these observations Rusnak *et al.* [9.33] concluded that μ_0 was proportional to the dislocation spacing, which was derived from the β values, and that dislocations, which are generated by cold work, provided the main hindrances to wall motion. It was also found that μ_0/v was constant, independent of strain in the range 0 to 3.8 percent. This means that the relative contribution of reversible (μ_0) and irreversible (vH) wall motion to the total permeability is the same for cold-worked and annealed specimens at the same value of H. This leads to the conclusion that plastic deformation produces the same percent reduction in reversible and irreversible wall motion in this material.

There have been several formal theories of the Rayleigh region. Most of them take as point of departure the general view of wall motion expressed in Fig. 9.36 and then attempt to make it particular by assuming some probable form, or distribution of forms, for the variation of energy with wall position. If the proper kind of assumptions are made, the result is the correct relation, the Rayleigh relation, between μ and H. The defect of such theories is that they fail to connect the Rayleigh constants with any microstructural features of the material. What is really needed is more experimental work designed to relate the Rayleigh constants to such variables as inclusion content, microstress, dislocation density and arrangement, domain size, etc.

The just-mentioned domain size is important because it determines the domain wall area per unit volume. If that is known, then the average distance a wall moves can be calculated from the observed change in magnetization. And this distance is important *vis-à-vis* the scale of the structural irregularities in the specimen. As an example of such a calculation we will take the annealed iron specimen whose μ,H curve appears in Fig. 9.42. At $H = 0.1$ Oe, $\mu = 617$, $B = 62$ gauss, and $M = 5.0$ emu/cm^3. When unit area of a single 180° wall is moved a distance x by a field parallel to the M_s vector in one of the adjacent domains, the magnetization of that region changes in the direction of the field by an amount $2M_s x$. If the spacing of the walls is d, then the number of walls per unit length is $1/d$, and

$$M = (2M_s)(x)(1/d). \tag{9.59}$$

If the field is almost normal, instead of parallel, to the M_s vector, then the factor $2M_s$ in this expression, resolved parallel to the field, is almost zero. When we consider the motion of both 180° and 90° walls, variously oriented with respect to the applied field and assumed to be present in equal numbers, the factor $2M_s$ in Eq. (9.59) is reduced to a value of roughly $(3M_s/4)$. Then,

$$x = \frac{4Md}{3M_s}. \tag{9.60}$$

For the annealed iron specimen referred to, d was estimated as 3μ from Bitter patterns. Therefore, $x = (4)(5.0)(3 \times 10^{-4})/(3)(1714) = 1.2 \times 10^{-6}$ cm $= 120$ Å. Thus the average distance a domain wall moves in a field of 0.1 Oe is *less than half* of the domain wall thickness, estimated at 300–400 Å in iron.

Such calculations suggest that the conventional view of wall motion, as in Fig. 9.36, wherein a wall moves reversibly from an energy minimum to a position of maximum energy gradient and then jumps to a new position, is grossly out of scale in the low-field region. Instead, it appears that in the demagnetized specimen many walls are already poised at metastable positions, ready to move irreversibly when even slight fields are applied, because v *is* finite in this region. The above calculation assumes that all walls in the specimen move when a field is applied. An alternative view is that many of the walls are pinned so strongly by imperfections that they do not move at all in these weak fields; the remainder would then have to move several times the distance calculated above in order to produce the observed magnetization. (It is interesting to translate into wall motion the results of Elwood [9.34], whose measurements of magnetization are possibly the most sensitive ever made. With a special galvanometer operated in a special way, he was able to detect an induction B of 2×10^{-4} gauss in a specimen of compressed iron powder. If the domain wall spacing in his specimen was 3μ, as above, and all the walls moved, then their average distance of motion was 4×10^{-4} Å! This distance is not only smaller than the wall thickness but only some 10^{-4} times the atom spacing. Another alternative is that only a fraction 10^{-4} of the walls were unpinned, and these moved an average distance of 4 A. In either case, the notion of wall "motion" becomes rather strained, and one is left instead only with the image of the spins in the walls being rotated through minute angles by the field.)

9.15 MAGNETIZATION IN HIGH FIELDS

Between the low-field Raylegh region and the high-field region near saturation there exists a large section of the magnetization curve, comprising most of the change of magnetization between zero and saturation. The main processes occurring here are large Barkhausen jumps, and the shape of this portion of the magnetization curve varies widely from one kind of specimen to another. It is not possible to express M as a simple function of H in this intermediate region.

In the high-field region, on the other hand, domain rotation is the predominant effect, and this phenomenon obeys fairly simple rules. The relation between M and H in this region is called the "law of approach" to saturation and is usually written as

$$M = M_s\left(1 - \frac{a}{H} - \frac{b}{H^2}\right) + \kappa H. \tag{9.61}$$

The term κH represents the field-induced increase in the spontaneous magnetization of the domains, or forced magnetization (see Problem 4.3); this term is very small at temperature well below the Curie point and may be neglected. There remain the terms a/H and b/H^2. Considerable theoretical interest attaches to the constants a and b; a is generally interpreted as due to inclusions and/or microstress, and b as due to crystal anisotropy. In fields over a few thousand oersteds, the term a/H is usually dominant, so that the above expression reduces to

$$M = M_s\left(1 - \frac{a}{H}\right). \tag{9.62}$$

High-field measurements of M can therefore be plotted against $1/H$ and extrapolated to $1/H = 0$ in order to find the value of M_s.

The law of approach has been discussed by Bozorth [G.4], Stoner [9.35], and Chikazumi [G.23].

9.16 SHAPES OF HYSTERESIS LOOPS

So far we have examined the shapes of major hysteresis loops of assemblies of single-domain particles and the minor loops, in the Rayleigh region, of polycrystalline specimens. Now we will consider the shapes of major loops of polycrystalline specimens.

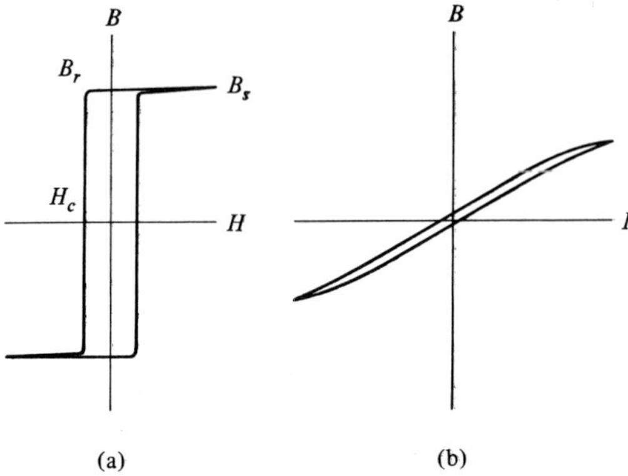

Fig. 9.44 Hysteresis loops of uniaxial materials: (a) field parallel to easy axis, (b) field normal to easy axis.

Two extreme kinds are sketched in Fig. 9.44. Both apply to a material with a strongly developed uniaxial anisotropy, due, for example, to stress, uniaxial crystal anisotropy together with preferred orientation, or other causes. In (a) the applied field is parallel to the easy axis. The domain walls are predominantly 180° walls parallel to the easy axis, as idealized, for example, in Fig. 8.20 (c). Magnetization reversal, from saturation in one direction to saturation in the other, occurs almost entirely by motion of these walls, in the form of large Barkhausen jumps occurring at a field equal to the coercivity H_c. The magnitude of H_c depends on the extent to which crystal imperfections impede the wall motion or on the mechanism discussed in the next paragraph. The result is a "square loop," with vertical or almost vertical sides and retentivity B_r almost equal to the saturation induction B_s.

In many square-loop materials, the loop shape is not a real characteristic of the material but rather an artifact, due to the usual method of measuring the loop. Normally the saturated specimen is exposed to a constantly increasing reverse field. Essentially no change in B is observed until the point A in Fig. 9.45 is reached; then, when the coercivity H_{c1} is exceeded, a large change in B, practically equal to $2B_s$, is observed as the induction changes along the dashed line. However, special electrical circuits (feedback circuits) can be assembled which will sense the beginning of this large change in B and quickly reduce the magnitude of the reverse field. Ideally, such a circuit will provide only enough reverse field, at any level of B, to cause B to slowly decrease toward zero. In such a case, the irregular line crossing the H axis at the "true" coercivity H_{c2} would be traced out, the irregularities reflecting irregular impediments to wall motion. Re-entrant loops of this kind have been observed in an Fe-Si alloy by Stewart [12.3] and in an Fe-Ni-Co alloy by Williams and Goertz [13.31]. Two interpretations of what occurs at point A are possible:

1. The specimen is truly saturated, in the sense of consisting of one domain. If the magnetization is to reverse by wall motion, the reversal can be initiated only

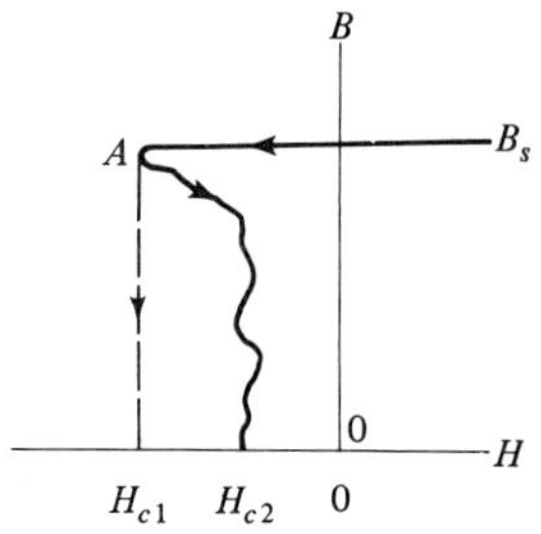

Fig. 9.45 Re-entrant hysteresis loop.

if one or more small reverse domains are nucleated. This is a relatively difficult process, requiring the field H_{c1}.

2. The specimen is not truly saturated. Some small reverse domains persist, but their walls are so strongly pinned that the field H_{c1} is required to free them.

In either case, mobile walls are suddenly created at point A; these walls can then be made to keep on moving by a field much weaker than the field required to initiate their motion.

The other extreme kind of loop is shown in Fig. 9.44 (b). Here the applied field is at right angles to the easy axis. The change of B with H is nearly linear over most of its range, which is an advantage for some applications but obtained at the cost of decreased permeability. The retentivity and coercivity both approach zero.

Between these extremes of square loops and almost linear loops lie those for specimens with more or less randomly oriented easy axes. Certain parameters of such loops can sometimes be calculated. Consider, for example, a uniaxial material like cobalt in which each grain has a single easy axis and in which the grains are randomly oriented. Then Fig. 9.46 illustrates several states of magnetization. The arrangements of M_s vectors in space is represented there by a set of vectors drawn from a common origin, each vector representing a group of domains. The ideal demagnetized state is shown at point O. When a positive field is applied, domains magnetized in the minus direction are eliminated first, by 180° wall motion, leading to the distribution shown at point B. Further increase in field rotates vectors into the state of saturation shown at C. When the field is now removed, the domain vectors fall back to the easy direction in each grain nearest to the $+H$ direction. Because the easy axes are assumed to be randomly distributed, the domain vectors are then uniform spread over one half of a sphere, as indicated at D. If M_s in any one domain makes an angle θ with the $+H$ direction, the magnetization of that domain is $M_s \cos\theta$, and the reten-

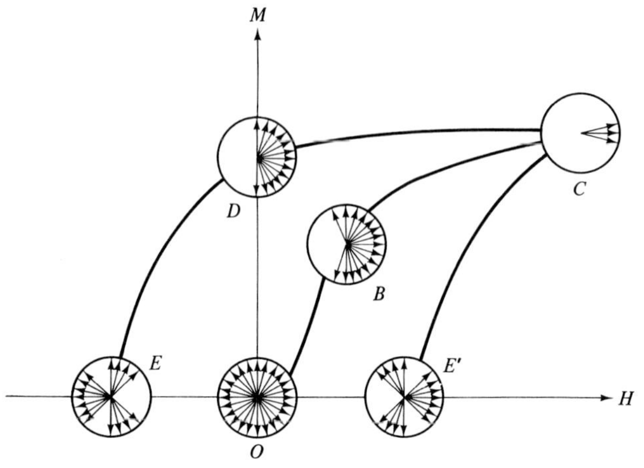

Fig. 9.46 Domain arrangements for various states of magnetization. Chikazumi [G.23].

tivity M_r of the specimen as a whole is given by the average of $M_s \cos\theta$ over all domains. This average is found by the method of Fig. 3.2 (b) to be

$$M_r = \int_0^{\pi/2} M_s \cos\theta \sin\theta \, d\theta = M_s/2. \tag{9.63}$$

Thus M_r/M_s, which might be termed the retentivity ratio, is 0.5. Chikazumi [G.23] has shown how this ratio can be calculated for other kinds of materials, with the following results, where random grain orientation is assumed throughout:

1. Cubic crystal anisotropy, K_1 positive (three $\langle 100 \rangle$ easy axes). $M_r/M_s = 0.83$.
2. Cubic crystal anisotropy, K_1 negative (four $\langle 111 \rangle$ easy axes). $M_r/M_s = 0.87$.

However, Chikazumi points out that such calculations ignore the free poles that form on most grain boundaries. These free poles set up demagnetizing fields which can cause the actual value of M_r/M_s to be substantially less than that calculated.

Returning to Fig. 9.46, we note that the effect of applying a negative field to the remanent state is to first reverse the domain magnetizations pointing in the $+H$ direction, leading to state E at the coercivity point. We note also that states E, O, and E' all have $M = 0$ but quite different domain distributions; of these, E and E' are unstable in the sense that an applied field of one sign or the other is necessary to maintain these distributions.

Although not a hysteresis loop, it is convenient to mention here the *ideal* or *anhysteretic* (= without hysteresis) magnetization curve. A point on this curve is obtained by subjecting the specimen to a constant unidirectional field H_1 together with an alternating field of amplitude large enough to saturate the specimen. The amplitude of the alternating field is then slowly reduced to zero. The magnetization M_1, or induction B_1, resulting from this treatment is then measured.

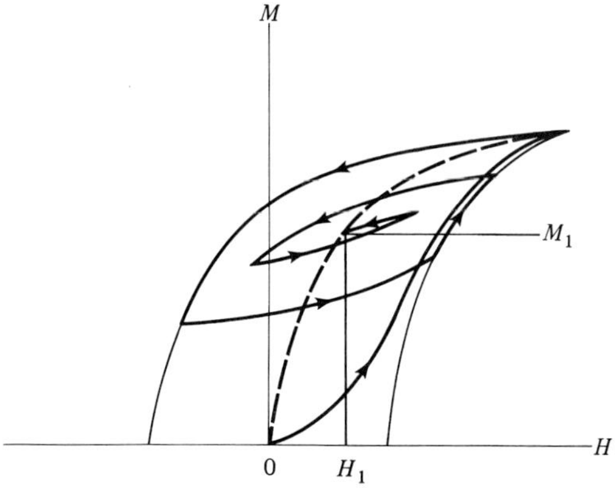

Fig. 9.47 Anhysteretic magnetization curve (dashed line).

Figure 9.47 shows how the final state is arrived at. This process is repeated for several values of H_1, and the resulting anhysteretic curve, shown by the dashed line, is then plotted.

(The constant and alternating fields may be generated by currents in the same or separate coils. The magnetization M_1 attained after the alternating field is reduced to zero may be measured by the extraction, vibrating-sample, or magnetometer methods, as described in Section 2.7.)

The anhysteretic curve has no points of inflection, lies above the normal magnetization curve, passes approximately through the midpoints of horizontal chords of the major loop, and is independent of the previous magnetic history of the specimen. It has an important bearing on the behavior of magnetic tapes for sound recording (Section 14.9).

9.17 EFFECT OF COLD WORK

Cold work (plastic deformation) makes magnetization more difficult. The M,H (or B,H) curve is lower than that for the fully annealed condition. The hysteresis loop rotates clockwise, becomes wider (larger coercivity), and has a bigger area (larger hysteresis loss) for the same maximum value of M (or B). The mechanical strength and hardness also increase.

These several changes are caused by the increased numbers of dislocations and other lattice defects. In polycrystalline materials, severe cold work multiplies the dislocation density by a factor of about 10^4. The resulting microstress impedes both domain wall motion and domain rotation, increasing the magnetic

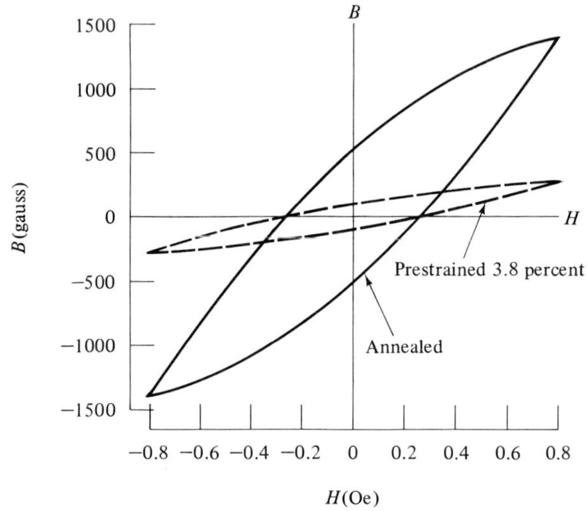

Fig. 9.48 Hysteresis loops for iron, calculated from measured Rayleigh constants.

hardness. It also impedes the motion of the dislocations themselves and the generation of new dislocations, thus increasing the mechanical hardness.

In the low-field region the effect of cold work is illustrated by Fig. 9.48, which applies to the hydrogen-annealed ingot iron mentioned earlier. The hysteresis curves shown were calculated by Eq. (9.55) from values of μ_0 and v derived from the μ,H curves of Fig. 9.42. Note that the coercive force H_c is the same for both the annealed and prestrained specimens, which is unusual. This agrees with the fact that μ_0/v was found to be independent of strain in this material. Equation (9.57) states that H_c, for a given value of H_m, depends only on the ratio of μ_0 to v and not on the individual values of these constants.

It is instructive to consider the changes in magnetic and other properties caused by still larger amounts of cold work, and how these properties change during subsequent annealing [9.36]. Figure 9.49 (a) shows how the hardness of commercially pure nickel increases with tensile deformation, up to 35 percent elongation (26 percent reduction in area). The x-ray line width β increases similarly, because of increased microstress and grain fragmentation. Conversely, the maximum permeability μ_m (see Fig. 1.14) precipitously decreases with even a small amount of cold work.

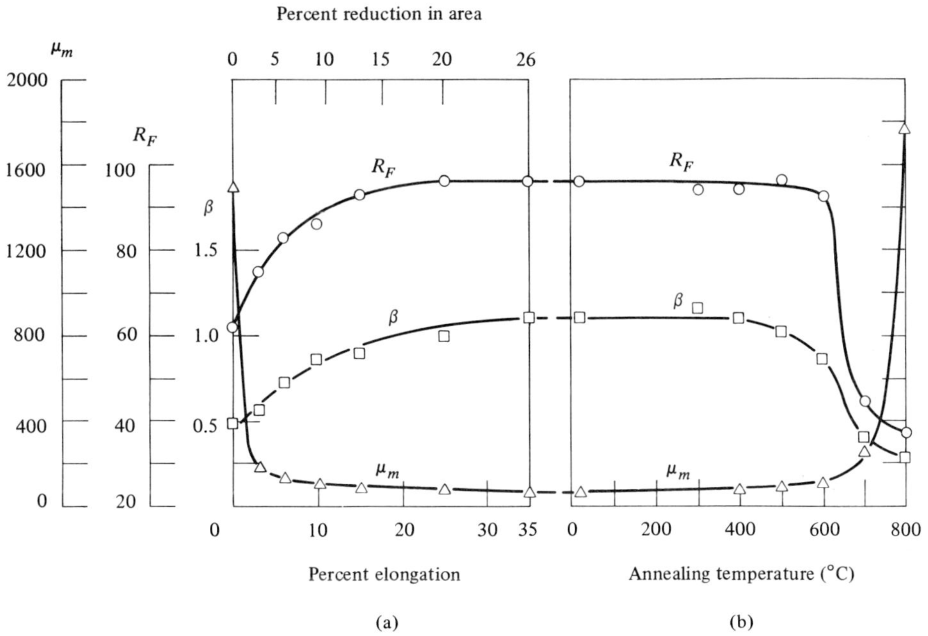

Fig. 9.49 Changes in physical properties of nickel during tensile deformation and subsequent annealing. R_F = hardness on the Rockwell F scale, β = width in degrees 2θ of the 420 x-ray line (Cu$K\alpha_1$ radiation), μ_m = maximum permeability. Chou *et al.* [9.36].

Specimens elongated 35 percent were then annealed for one hour at various temperatures up to 800°C, with the results shown in Fig. 9.49(b). Three phenomena occur, depending on the temperature:

1. *Recovery*, up to about 600°C. The main process taking place here is dislocation rearrangement, leading to partial relief of residual stress (macro and micro). As a result, at the higher temperatures in this range, β begins to decrease and μ_m to increase. However, the dislocation density remains high and so does the hardness.

2. *Recrystallization*, in the range 600° to 700°C. The cold-worked material is replaced by an entirely new, almost stress-free, grain structure, and many properties change abruptly. The hardness and dislocation density drop to low values, and β continues to decrease. There is a moderate increase in μ_m.

3. *Grain growth*, above about 700°C. The new grains grow in size at the expense of one another, causing a further, but small, decrease in hardness, due to the reduction in the amount of grain boundary material per unit volume. Residual stress, as evidenced by β, decreases to its minimum level. The increase in grain size, permitting more extensive motion of domain walls, and the decrease in stress

result in an abrupt increase in μ_m, and magnetic softness is restored. [The fact that the hardness, for example, of the material annealed at 800°C is less than that of the unworked material, shown at the extreme left of Fig. 9.49 (a), shows that this material had been incompletely annealed before the cold-working experiments were begun.]

It is clear from these measurements that a cold-worked material can be restored to a condition of maximum magnetic softness only by annealing above the recrystallization temperature. This temperature depends on the material, and, for a given material, it is lower the greater the purity and the larger the amount of previous cold work. Figure 4.6 shows the approximate recrystallization temperatures of iron, cobalt, and nickel.

Plastic deformation, in addition to lowering the permeability, can also produce magnetic anisotropy, particularly in certain alloys. This effect will be described in Sections 10.5 and 10.6.

The close connection between mechanical and magnetic hardness has been exploited in a device for continuously monitoring the hardness of low-carbon sheet steel as it issues from an annealing furnace [9.37]. The steel is in strip form and is moving so rapidly that measurements of mechanical hardness, to see if it has been properly annealed, are impractical. Instead a magnetic device continuously measures the magnetic hardness of the moving strip. It consists of three units mounted about a foot away from one another and 3/8 inch from the strip. The first is a demagnetizing coil to remove any residual magnetism the steel may possess. The strip then passes under a magnetizing head which produces a pulsed periodic field; this prints a succession of magnetized spots on the strip. When these spots pass below the third unit, a pick-up coil, an emf is induced in the coil proportional to the remanent induction B_r of the spots. The greater the mechanical hardness of the steel, the greater is its magnetic hardness and the smaller B_r, and, for steel of a given composition, the magnetic hardness measured in this way varies approximately linearly with mechanical hardness. The output of the monitor is continuously recorded, so that the operator has a record of hardness variations from one end of the strip to the other.

The results in Fig. 9.49 and similar studies are evidence of the close connection between magnetic and mechanical hardness, in that both increase when a metal is cold worked. And when carbon is added to iron to form steel, both the magnetic and mechanical hardness increase with the carbon content. It would seem that anything done to increase mechanical hardness will also increase magnetic hardness. While this is a fairly good rule, it does have important exceptions:

1. *Any* element that goes into solid solution in a metal will increase the mechanical hardness, because the solute atoms interfere with dislocation motion. However, the effect on magnetic behavior is not predictable. The addition of silicon to iron, for example, makes the material magnetically *softer*, because the silicon addition decreases the crystal anisotropy constant K_1 and the magnetostriction λ (Section 13.4).

2. If elongated single-domain particles of, for example, pure iron, itself rather

soft mechanically, are dispersed in a matrix of lead, the resulting composite material is magnetically hard and mechanically soft (Section 14.7).

PROBLEMS

9.1

a) Calculate the energy and width of a domain wall in cobalt. Ignore the fact that cobalt is not cubic but hexagonal close-packed, and assume that Eq. (9.9) is valid. Take a = distance of closest approach = 2.51 A, and $K = 45 \times 10^5$ ergs/cm^3. Obtain the exchange integral J from Eq. (4.31), with S = spin = $\frac{1}{2}$, and z = number of nearest neighbors = 12.

b) Find the value of L_c from Eq. (9.20) and compare with δ.

9.2 Find γ,δ, and L_c for barium ferrite, for which $K = 33 \times 10^5$ ergs/cm^3, $T_c = 450°$C, and $M_s = 380$ emu/cm^3. Find δ from Eq. (9.11) with a put equal to 3 Å. [This calculation is very approximate, because Eq. (9.11) was derived for a simple cubic structure and not for a complex structure like that of $BaO \cdot 6\,Fe_2O_3$. There is the additional uncertainty of what value should be assigned to a.]

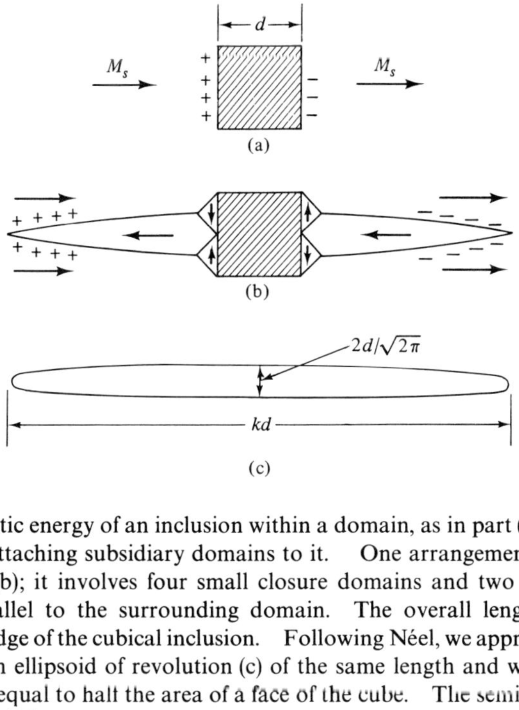

9.3 The magnetostatic energy of an inclusion within a domain, as in part (a) of the illustration, can be reduced by attaching subsidiary domains to it. One arrangement envisaged by Néel [9.17] is shown in (b); it involves four small closure domains and two long spike domains magnetized antiparallel to the surrounding domain. The overall length of this structure is kd, where d is the edge of the cubical inclusion. Following Néel, we approximate this domain arrangement with an ellipsoid of revolution (c) of the same length and with a cross-sectional area at its midpoint equal to half the area of a face of the cube. The semi-axes of the ellipsoid are thus $d/\sqrt{2\pi}$, $d/\sqrt{2\pi}$, $kd/2$. If S is the surface area and V the volume of the ellipsoid and N its axial demagnetizing coefficient, show that

$$S = (\pi^3/8)^{1/2} kd^2,$$
$$N = (8/k^2)(\ln k\sqrt{2\pi} - 1),$$
$$V = kd^3/3,$$

where the expressions for S and N are more nearly exact, the larger the value of k. The total energy is

$$E_t = S\gamma + 2NM_s^2V.$$

(If the ellipsoid of magnetization M_s were isolated in a nonmagnetic medium, its magnetostatic energy would be $NM_s^2V/2$, but here it is imbedded in a medium of magnetization $-M_s$, which doubles the surface pole strength and quadruples the energy.) Show that the value of k which minimizes the total energy is given by

$$\frac{3}{32}\sqrt{\frac{\pi^3}{2}}\frac{\gamma}{dM_s^2} = \frac{1}{k^2}(\ln k\sqrt{2\pi} - 2).$$

Verify the values tabulated below for iron ($\gamma = 1.5$ ergs/cm^2, $M_s = 1714$ emu/cm^3). E_0 is the magnetostatic energy of the isolated cube shown in (a), taken as equal to the magnetostatic energy of a sphere of the same volume.

k	$d(\mu)$	E_t(erg)	E_t/E_0
10	0.15	1.9×10^{-8}	0.88
50	1.7	9.5×10^{-6}	0.34
100	5.4	1.9×10^{-4}	0.21

(These values show that there is little difference in energy for inclusions a few tenths of a micron in size. However, inclusions several microns in size can reduce their energy considerably by the formation of spike domains, and the equilibrium spike length increases rapidly as the inclusion size increases.)

9.4 In terms of the fields created by free poles, explain why the domain with M_s upward should form just to the right of the inclusion in Fig. 9.27 (b).

9.5 Consider a 90° wall moving in the linear stress field of Fig. 9.33 (b). Let ΔE_{me} be the increase in magnetoelastic energy, given by Eq. (9.27) with no consideration given to wall energy, when the wall moves a distance $x = a\delta$ from the position of zero stress, where a is a constant and δ the wall thickness. Let $\Delta\gamma$ be the increase in wall energy during the same motion. Assume that the energy of a 90° wall is half that of a 180° wall. Show that $\Delta\gamma/\Delta E_{me} = 2/a$.

9.6 Consider magnetization by rotation in a single-domain particle with uniaxial anisotropy, as illustrated by the hysteresis loops of Fig. 9.38.

a) Compute and plot the loop when the angle α between easy axis and field is 20 . What are the critical values h_c and θ_c?
b) Show that the reduced intrinsic coercivity (the field h that makes m zero) is equal to h_c for α less than 45° and equal to $(-\sin 2\alpha)/2$ for α greater than 45°.

9.7 Derive Eq. (9.56).

9.8 Derive Eq. (9.57).

CHAPTER 10

INDUCED MAGNETIC ANISOTROPY

10.1 INTRODUCTION

So far in this book we have encountered three kinds of magnetic anisotropy: crystal, shape, and stress. Various other kinds may be induced in certain materials, chiefly solid solutions, by appropriate treatments. These induced anisotropies are of considerable interest both to the physicist, for the light they throw on basic magnetic phenomena, and to the technologist, who may wish to exploit them in the design of magnetic materials for specific applications.

The following treatments can induce magnetic anisotropy:

1. *Magnetic annealing.* This means heat treatment in a magnetic field, sometimes called a *thermomagnetic* treatment. This treatment can induce anisotropy in certain alloys. (Here the term "alloys" includes not only metallic alloys but also mixed ferrites.) The results depend on the kind of alloy:

a) Two-phase alloys. Here the cause of anisotropy is the shape anisotropy of one of the phases and is therefore not basically new. However, it is industrially important because it affects the behavior of some of the Alnico permanent-magnet alloys. It will be described in Chapter 14.

b) Single-phase solid-solution alloys. Here it will be convenient to discuss substitutional and interstitial alloys in separate sections.

2. *Stress annealing.* This means heat treatment of a material that is simultaneously subjected to an applied stress.

3. *Plastic deformation.* This can cause anisotropy both in solid solutions and in pure metals, but by quite different mechanisms.

4. *Magnetic irradiation.* This means irradiation with high-energy particles in a magnetic field.

10.2 MAGNETIC ANNEALING (SUBSTITUTIONAL SOLID SOLUTIONS)

When certain alloys are heat treated in a magnetic field and then cooled to room temperature, they develop a permanent uniaxial anisotropy with the easy axis parallel to the direction of the field during heat treatment. They are then magnetically softer along this axis than they were before treatment. The heat treat-

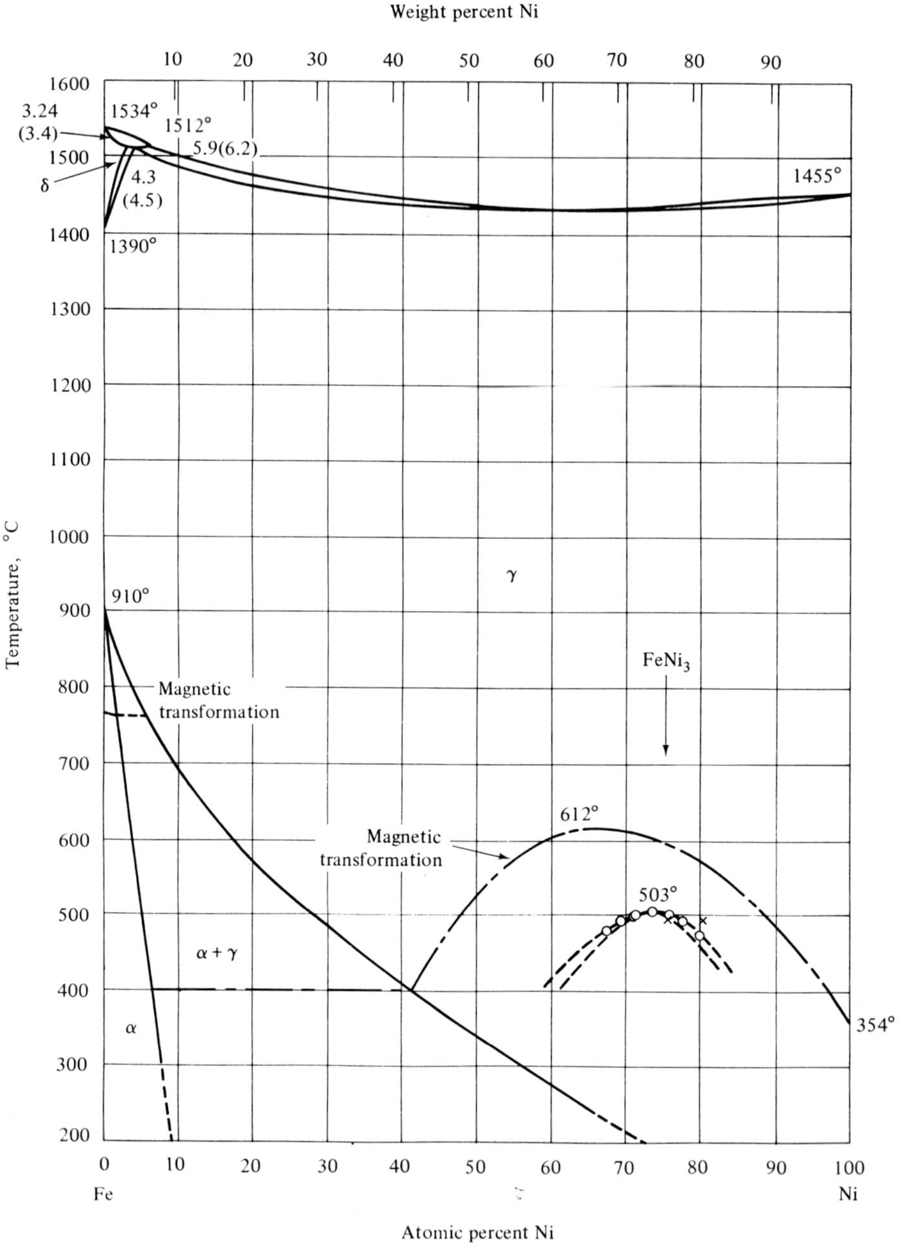

Fig. 10.1 Equilibrium diagram of Fe-Ni alloys. Hansen and Anderko [10.6].

ment may consist only of cooling through a certain temperature range in a field, rather than prolonged annealing; the cooling range or annealing temperature must be below the Curie point of the material and yet high enough, usually above

400 C, so that substantial atomic diffusion can occur. An alternating or unidirectional field is equally effective; all the field does is determine an easy axis, rather than direction, of easy magnetization. The field must be large enough to saturate the specimen during the magnetic anneal, if the resulting anisotropy is to develop to its maximum extent. Usually a field of some 10 Oe or less is sufficient; the material is magnetically soft to begin with, and its permeability at the magnetic-annealing temperature is higher than at room temperature. The term "magnetic annealing" is applied both to the treatment itself and to the phenomenon which occurs during the treatment; i.e., an alloy is often said to magnetically anneal if it develops a magnetic anisotropy during such an anneal. The subject of magnetic annealing has been reviewed by Graham [10.1], Slonczewski [10.2], and Chikazumi and Graham [10.3]; Graham's review contains a large bibliography classified by material composition.

The phenomenon of magnetic annealing was first discovered in 1913 by Pender and Jones [10.4] in an alloy of Fe + 3.5 percent Si. They found that cooling the alloy from about 800° C to room temperature in an alternating field, of about 20 Oe maximum value, caused a substantial increase in maximum permeability. Many years later Goertz [10.5] made measurements on a picture-frame single crystal, with ⟨100⟩ sides, of an alloy of Fe + 6.5 percent Si; heat treatment in a field increased its maximum permeability from 50,000 to 3.8×10^6, the highest value yet reported for any material.

However, most of the research on magnetic annealing has been devoted to the binary and ternary alloys of Fe, Co, and Ni. Compositions which respond well to magnetic annealing are Fe + 65–85 percent Ni, Co + 30–85 percent Ni, Fe + 45–60 percent Co, and the ternary alloys containing 20–60 percent Ni, 15–35 percent Fe, balance Co. Magnetic annealing has been studied most often in binary Fe-Ni alloys, for which the equilibrium diagram is shown in Fig. 10.1. Both the α (body-centered cubic) and the γ (face-centered cubic) phases are ferromagnetic. There is a large thermal hysteresis in the $\alpha \rightarrow \gamma$ and $\gamma \rightarrow \alpha$ transformations because of low diffusion rates below about 500 C, and the equilibrium shown in Fig. 10.1 is very difficult to achieve. For example, the $\gamma \rightarrow \alpha$ transformation on cooling is so sluggish that it is easy to obtain 100 percent γ at room temperature in alloys containing more than about 35 percent Ni by air cooling γ from an elevated temperature. Hansen and Anderko [10.6] should be consulted for further details.

Typical of the magnetic-annealing results obtained on Fe-Ni alloys are those shown for 65 Permalloy in Fig. 10.2. Comparison of the hysteresis loop of (c) with (a) or (b) shows the dramatic effect of field annealing: the sides of the loop become essentially vertical, as expected for a material with a single easy axis. Conversely, if the loop is measured parallel to the hard axis, i.e., at right angles to the annealing field, the sheared-over, almost linear loop shown in (d) is obtained, where the change in the H scale should be noted. [Specimens for magnetic annealing studies are sometimes in the form of rods, either straight or made into a hollow rectangle in order to have a closed magnetic circuit. In any case it is not usually practical to apply a field transverse to the rod axis because of the very large demagnetizing factor, equal to 2π, in that direction. Instead, a direct cur-

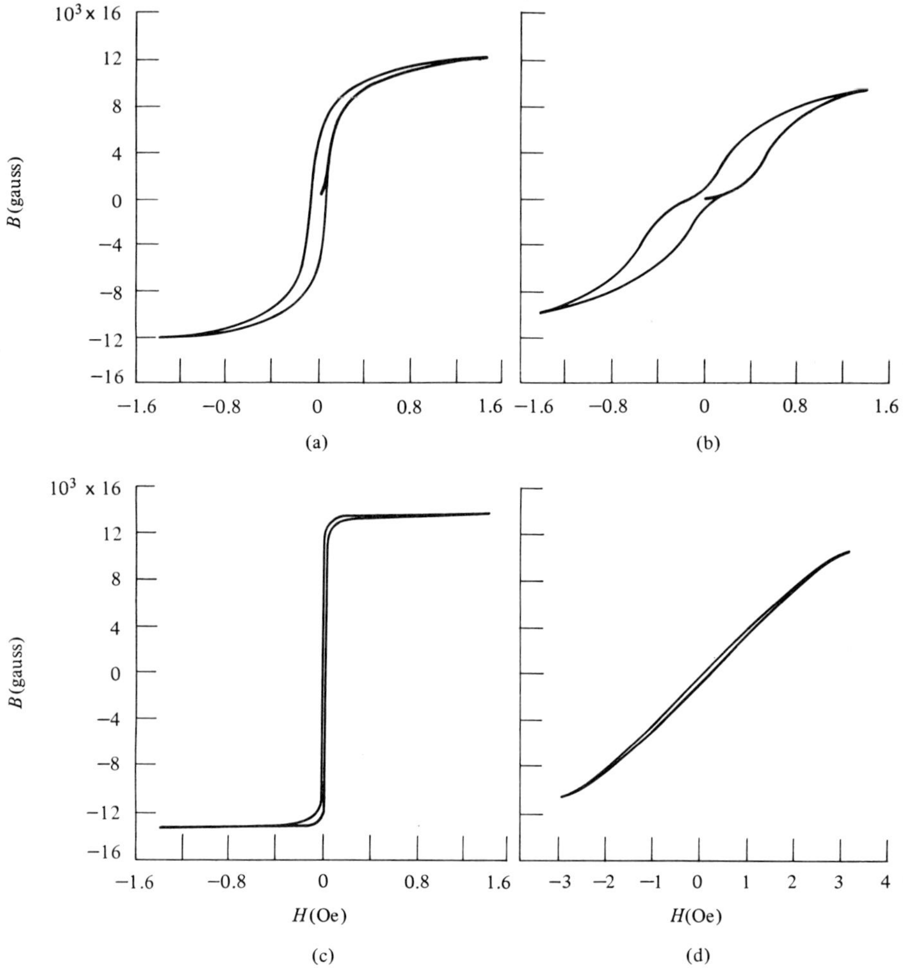

Fig. 10.2 Hysteresis loops of a 65 Ni-35 Fe alloy after various heat treatments: (a) annealed at 1000° C and cooled quickly, (b) annealed at 425° C or cooled slowly from 1000° C, (c) annealed at 1000° C and cooled in a longitudinal field, (d) same as (c) but with a transverse field. Bozorth [G.4].

rent is passed along the rod axis during the anneal, producing a circular field around the axis (Section 1.6). This field can easily be made strong enough to saturate the specimen circumferentially, except for a relatively small volume near the axis. If a magnetic measurement is subsequently made parallel to the axis in the usual way, the measurement direction is then at right angles to that of the annealing field. A longitudinal annealing field is achieved simply by wrapping the rod with a helical magnetizing winding, suitably protected by an insulator that will withstand the annealing temperature. If the specimen is in the form

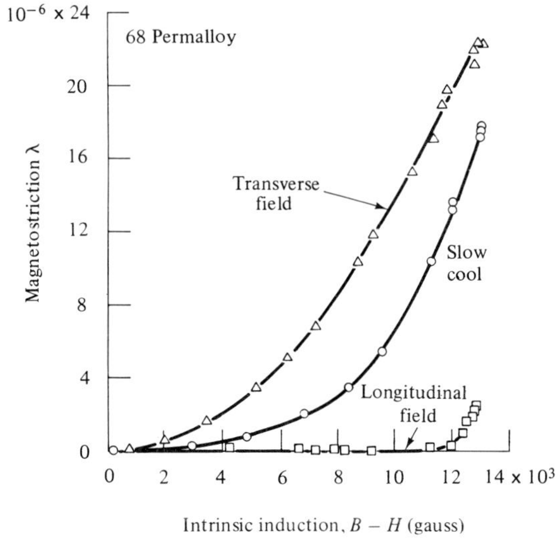

Fig. 10.3 Magnetostriction of a 68 Ni-32 Fe alloy after various treatments. Williams *et al.* [10.7]

of a disk, there is no problem in applying the annealing field in any chosen direction in the plane of the disk]. Alloys which show the magnetic-annealing effect commonly have the peculiar, "constricted" loop shown in (b) when they are slowly cooled in the absence of a field; this kind of loop will be discussed later.

Magnetic annealing evidently creates a preferred domain orientation in the demagnetized state, with M_s vectors parallel to the annealing field. Subsequent magnetization in this direction can then take place by 180° wall motion. As a result, magnetostriction and magnetoresistance in this direction are essentially zero. Figure 10.3 illustrates this point for an alloy of slightly different composition. The central curve applies to a specimen cooled slowly from 1000°C in the absence of a field; slow cooling from 600°C in a longitudinal field (lower curve) decreases the longitudinal magnetostriction and the same treatment in a transverse field increases it. Furthermore, when the data for the upper curve are replotted in the form of λ versus $(B - H)^2$, the result is a straight line. This behavior agrees with Eq. (8.31) because $(B - H)$ is proportional to M. The magnetization process for this specimen (annealing field transverse to axis) is therefore one of pure rotation of the domain vectors through 90° from their initial positions transverse to the axis.

The anisotropy created by magnetic annealing is due to *directional order* in the solid solution, an idea originated by Chikazumi [10.8]. The theory of this effect was proposed independently by Néel [10.9], Taniguchi and Yamamoto [10.10], Taniguchi [10.11], and Chikazumi and Oomura [10.12]. By directional order is meant a preferred orientation of the axes of like-atom pairs. An example,

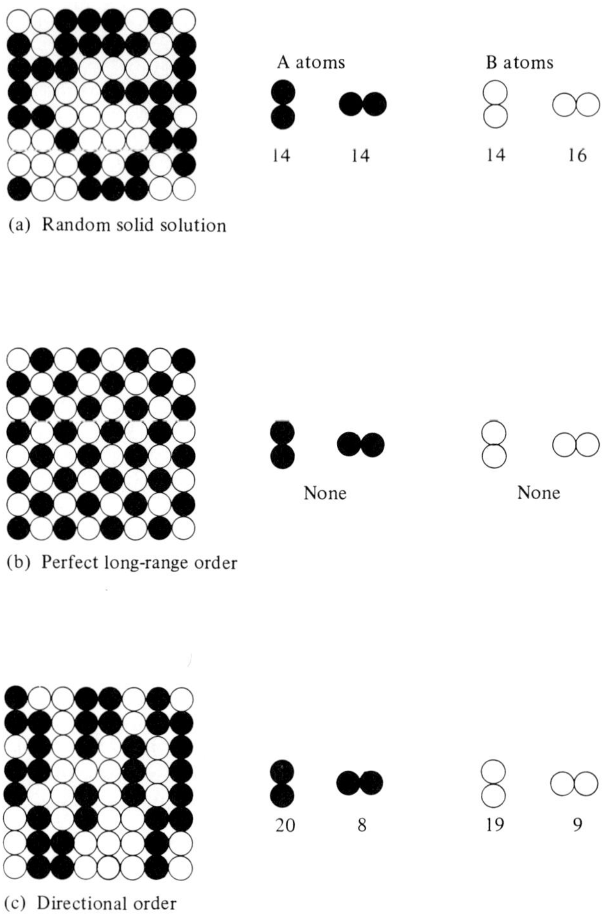

Fig. 10.4 Possible atom arrangements in a square lattice. After Graham [10.1].

for atoms arranged on a two-dimensional square lattice, is shown in Fig. 10.4. The arrangement in (a) approximates a random solid solution of a 50-50 "alloy" of A and B; the atom positions were determined by drawing black and white balls from a box. If the solution were truly random, there would be 56 black-white (AB) nearest-neighbor pairs, 28 AA pairs, and 28 BB pairs; the arrangement in (a) comes close to this, since it has 54 AB, 28 AA, and 30 BB pairs. Perfect ordering is shown in (b); all nearest-neighbor pairs are AB, and no AA or BB pairs exist. The arrangement in (c) has directional order in the vertical direction. It has the same 28 AA pairs as (a), but in (c) 20 of these have vertical axes and only 8 have horizontal axes. The same kind of preferred orientation is shown by BB pairs; 19 are vertical and 9 horizontal.

Note that directional order can be achieved in a solid solution which is perfectly random in the usual crystallographic sense. Thus, in terms of nearest-neighbors only, a 50-50 solution is random if the neighbors of any given A atom, for example, are on the average half A and half B. This solution could deviate from randomness in either of two ways: (1) *short-range order*, in which more than half of the neighbors of an A atom would be B atoms, and (2) *clustering*, in which more than half of the neighbors would be A atoms. The solution of Fig. 10.4 (c) shows neither a tendency to short-range order (a preponderance of AB pairs) nor to clustering (a preponderance of AA and BB pairs); it has 56 unlike-atom pairs and 56 like-atom pairs. The like-atom pairs are, however, preferentially oriented. The word "order" in the term "directional order" should therefore not be misconstrued. Whether short-range order or clustering exists depends on the number and kind of atoms surrounding a given atom but not on their relative positions.

What does directional order have to do with magnetic anisotropy? The basic hypothesis of the theory is that there is a magnetic interaction between the axis of like-atom pairs and the direction of the local magnetization such that the two tend to be parallel. (The exact physical nature of this interaction is not specified; presumably it is related, like crystal anisotropy, to spin-orbit coupling.) Thus, if a saturating field is applied at a high temperature, the magnetization will be everywhere in one direction and diffusion will occur until there is a preferred orientation of like-atom pairs parallel to the magnetization and the field. On cooling to room temperature, this directional order will be frozen in; the domains that form when the annealing field is removed will then have their M_s vectors bound to the axis of the directional order. The anisotropy energy, for random polycrystals, is then of the same form as for other uniaxial anisotropies:

$$E = K_u \sin^2 \theta, \tag{10.1}$$

where K_u is the anisotropy constant and θ is the angle between M_s and the direction of the annealing field.

The value of K_u for polycrystals may be determined by measuring the area between magnetization curves made in the easy and hard directions or by the analysis of torque curves, if due allowance is made for the possible contribution of preferred orientation to the observed anisotropy. For single crystals K_u may be found from torque curves, if these are Fourier analysed to separate the field-induced anisotropy from the crystal anisotropy. The values of K_u so found are not particularly large; they are typically of the order of a few thousand ergs/cm^3, or some 10^{-1} to 10^{-2} times crystal anisotropy constants. The variation of K_u with composition and heat treatment is qualitatively what would be expected from the theory:

1. K_u decreases as the alloy composition changes toward the pure metal. (In a pure metal all pairs are like-atom pairs and there can be no directional order.) Figure 10.5 shows the results for the Fe-Ni system.

2. K_u is zero if long-range order is present, because all pairs are then unlike-atom pairs and no directional order is possible. This can occur at 75 atom percent

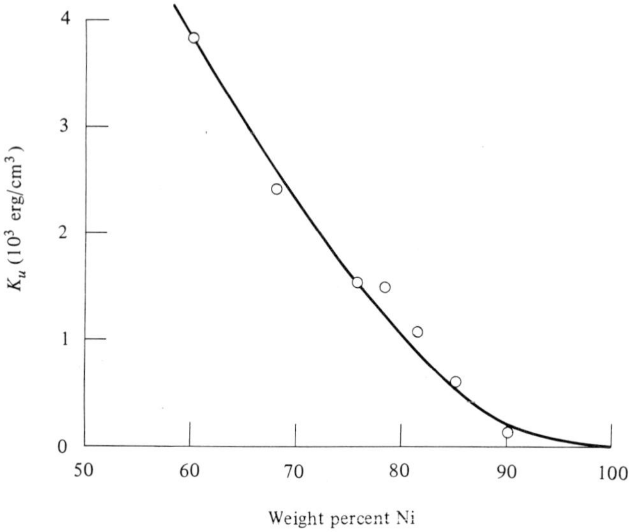

Fig. 10.5 Uniaxial anisotropy constant K_u of polycrystalline Ni-Fe alloys, cooled in a field from 600° to 80° C at 10° C/hour. Chikazumi and Oomura [10.12].

Ni in the Fe-Ni system (Fig. 10.1), where the $FeNi_3$ superlattice shown in Fig. 10.6 can form below a critical ordering temperature of 503°C; the nearest neighbors in this structure are the corner atoms and those at the centers of the faces. (The reader may then wonder how directional order can be produced at this composition. The reason is that crystallographic long-range ordering of this alloy is very sluggish, requiring some 160 hours for completion; field annealing of a random solid solution can produce directional order before much long-range order is established. However, in general, the fact that many alloys which show magnetic annealing are also capable of long-range ordering is a complication that must be kept in mind in the experimental study of magnetic annealing.)

3. When magnetic annealing is carried out at constant temperature rather than by continuous cooling, K_u is found to decrease as the field-annealing temperature is increased. The higher the temperature, the greater is the randomizing effect of thermal energy and the smaller the degree of directional order.

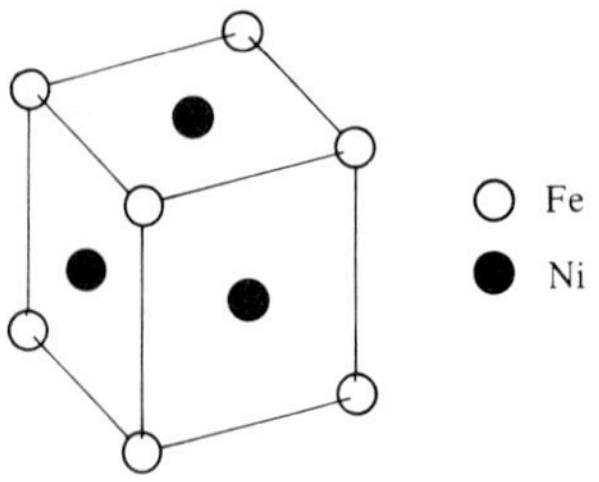

Fig. 10.6 The ordered form of $FeNi_3$.

As mentioned above, the anisotropy induced by field annealing is not particularly strong. This suggest that the directional order is not very strong, certainly not as pronounced as that shown in Fig. 10.4(c). If it were, such properties as lattice parameter and elastic constants should be detectibly different in directions parallel and perpendicular to the directional-order axis in a single crystal; such differences have not been observed. Nor has x-ray or neutron diffraction ever furnished any direct evidence for the existence of directional order. According to Slonczewski [10.2], the preference of like-atom pairs for the local magnetization direction, or the preference of solute atoms for certain interstitial sites (see next section), is never more than about one percent.

The behavior of field-annealed single crystals is more complex than that of polycrystals, and Eq. (10.1) is not adequate to describe all the phenomena observed. Whereas the annealing field may be made parallel to any crystallographic direction that the experimenter chooses, the axis of directional order is determined by the structure of the crystal. If directional order is due to an interaction between the M_s vector and the axis of *nearest-neighbor* pairs, as is usually assumed, then the order axis must be an axis on which nearest neighbors are located, i.e., the direction of closest packing for the structure involved. This direction is the face diagonal $\langle 110 \rangle$ for the face-centered cubic structure and the cell diagonal $\langle 111 \rangle$ for the body-centered cubic. If M_s is not parallel to this direction, directional order can still occur and create an easy axis, but this axis may or may not be parallel to M_s. For example, suppose the annealing field is parallel to [001] in a FCC crystal. Then directional order is equally favored along four of the six $\langle 110 \rangle$ directions, namely, [011], [$\bar{1}$01], [0$\bar{1}$1], and [101], which are all equally inclined at 45° to M_s, but not along the [110] and [$\bar{1}$10] directions, which are at right angles to M_s; as a result an easy axis is created parallel to M_s. In one particular case theory predicts that K_u will be zero. If the annealing field is along [001] in a BCC crystal, then all four $\langle 111 \rangle$ directions are equally inclined to [001] and no anisotropy should result, because none of the four close-packed directions is favored over the others.

The experimental results on single crystals are as follows. When the annealing field is parallel to a simple crystallographic direction like $\langle 100 \rangle$, $\langle 110 \rangle$, or $\langle 111 \rangle$ in a cubic crystal, the resulting easy axis is parallel to the annealing field direction, but K_u is no longer a true constant; its value is different for each of these directions. (When the annealing field is parallel to $\langle 100 \rangle$ in a BCC crystal, K_u is observed to be small, but not zero as predicted by theory. Presumably, directional order of second-nearest neighbors, which lie on $\langle 100 \rangle$ axes, is taking place.) When the annealing field is not parallel to one of these simple directions, an easy axis results, but it is not parallel to the annealing field direction; in fact, it can deviate from it by as much as 20°. For a fuller account of single-crystal behavior and a summary of directional-order theory, the reader should consult Chikazumi [G.23] and Chikazumi and Graham [10.3].

Magnetic annealing has also been observed in solid solutions of cobalt ferrite and iron ferrite (called cobalt-substituted magnetite) and in some other mixed ferrites containing cobalt. The effect depends on, but is not caused by, the presence of metal-ion vacancies in the lattice. (Such vacancies can be created by heating the ferrite in oxygen.) The Co^{2+} ion in a cubic spinel occupies an octahedral site (Table 6.2, Fig. 6.2) lying on a ⟨111⟩ axis of the crystal, and, for reasons described by Chikazumi [G.23] and Slonczewski [10.2], this ion creates a local uniaxial anisotropy parallel to that ⟨111⟩ axis on which it resides. In a non-field-annealed crystal all four ⟨111⟩ axes will be equally populated by Co^{2+} ions and the local anisotropies will cancel out, leaving only the cubic crystal anisotropy. In the presence of an annealing field, however, the Co^{2+} ions will tend to occupy the ⟨111⟩ axis closest to the field, producing an observable and rather large uniaxial anisotropy. Directional order of Co^{2+}-Co^{2+} pairs is thought to be a contributing factor. The role of vacancies is simply to insure enough diffusion of the metal ions for substantial ordering to take place.

It is important to realize that directional order, whether of like-atom pairs in an alloy or Co^{2+} ions in a cobalt-containing ferrite, is not due to the field applied during annealing, but to the local magnetization. All that the applied field does is to saturate the specimen, thereby making the direction of spontaneous magnetization uniform throughout. It follows that a magnetically annealable material will undergo *self-magnetic-annealing* if it is heated, in the demagnetized state and in the absence of a field, to a temperature where substantial diffusion is possible. Directional ordering will then take place in each domain, parallel to the M_s vector of that domain, and in each domain wall, parallel to the local spin direction at any place inside the wall. (This statement is not entirely true, because the pair-ordering axis in a domain, a direction of closest packing, does not necessarily coincide with M_s, which is the easy direction of magnetization determined by crystal anisotropy. Furthermore, within the domain wall, directional order actually occurs along the close-packed direction which is nearest to the M_s direction at that position in the wall. However, for the sake of brevity we can simply say that directional order takes place "parallel to M_s," with the above complications being understood.) This directional order, differing in orientation from one domain to another, is then frozen in when the specimen is cooled to room temperature. The result is that domain walls tend to be stabilized in the positions they occupied during the anneal.

The effect differs in kind for 90° and 180° walls. When a 90° wall is moved a distance x by an applied field, it sweeps out a certain volume. In the volume swept out, the M_s vector is now at 90° to the axis of directional order. The potential energy of the system, which is just anisotropy energy, must therefore increase from 0 to K_u per unit volume swept out, according to Eq. (10.1). The potential energy then varies with x as shown by the dashed line of Fig. 10.7 (a). The slope of this curve is the force on the wall which the field must exert to move the wall. This force soon becomes constant, as shown in (b), and so does the restoring force tending to move the wall back to its original position. A 180° wall, on the other hand, separates domains in which the directional order is identical. But the directional order axes within the wall itself differ from the axis in the adjoining domains. The energy of the system therefore increases as the wall is moved from its original position, but soon becomes constant when the wall is wholly within one of the original domains. The force to move the wall, and the balancing restoring force, increases to a maximum and then returns to zero.

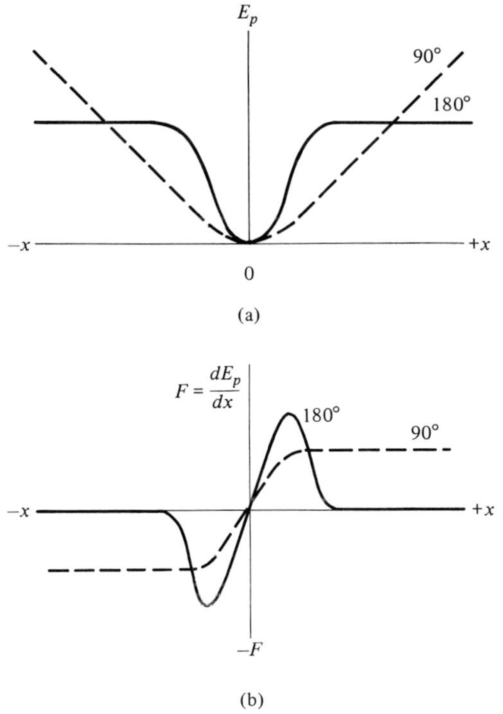

Fig. 10.7 (a) Potential energy E_p as a function of the displacement x of 90° and 180° walls from a stabilized position. (b) Force F required to move the wall. (Schematic.)

Each kind of wall is stabilized by the potential well built up around it by directional order, but the shape of the well depends on the type of wall.

These considerations explain the constricted or "wasp-waisted" hysteresis loops found in self-magnetically-annealed alloys and ferrites. Such a loop was first observed in Perminvar and is often called a "Perminvar loop." This alloy usually contains 25 percent Co, 45 percent Ni, and 30 percent Fe, but the composition may vary over wide limits. It has the remarkable and useful property of constant permeability (hence its name) and zero hysteresis loss at low field strengths, as shown in Fig. 10.8 (a). As the field is cycled between increasingly larger limits, the B,H line progressively opens up into a loop but with essentially zero remanence and coercive force, as in (c). The "Perminvar loop" is shown in (d), and the loop at near saturation in (e). If the alloy is annealed at 600°C, a temperature too high for directional order to be established, and then quickly cooled, minor hysteresis loops of normal shape are observed at all field strengths.

The magnetic behavior shown in Fig. 10.8 can be qualitatively understood in terms of the potential wells of Fig. 10.7. There is the added complication

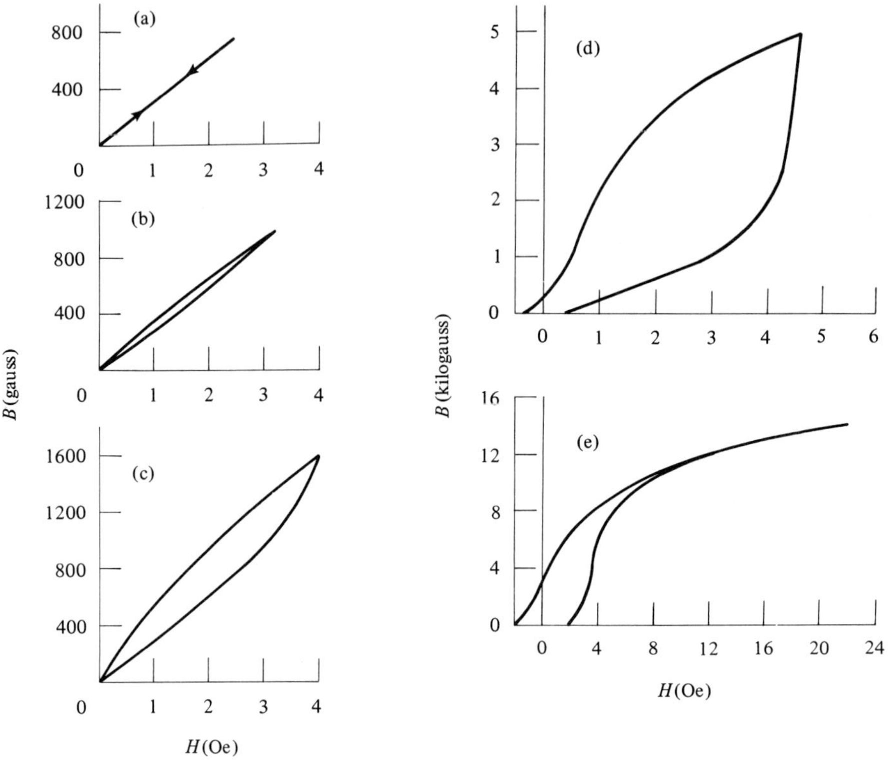

Fig. 10.8 Hysteresis loops of Perminvar annealed at 425° C. Bozorth [G.4].

that in a real material 90° and 180° walls form a complex network, so that one kind of wall cannot move independently of the other. But at sufficiently low fields, neither kind of wall moves out of its potential well, and the reversible behavior of Fig. 10.8 (a) results. Higher fields cause larger wall displacements, some of which are irreversible. However, when the field is reduced to zero, most of the walls return to their original positions, leading to the abnormally low remanence characteristic of the Perminvar loop in (d). (Presumably the directional-order potential wells are deeper than inclusion wells and the microstress wells that are always present because of dislocations.) It will be noted in (d) that the permeability is small up to a field of about 3 Oe and then more or less abruptly becomes larger; this is the critical field (often called the *stabilization field*) required to move 180° walls out of their potential wells; after they are out, subsequent motion becomes much easier and the permeability increases. Finally, if the material is driven to near saturation, as in (e), the original domain structure is completely destroyed; when the field is then reduced to zero, a new structure forms with walls in positions quite unrelated to the existing potential wells. The hysteresis loop is then of more or less normal shape.

10.3 MAGNETIC ANNEALING (INTERSTITIAL SOLID SOLUTIONS)

The title of this section suggests a greater generality than really exists. Actually, the effects to be described here have been observed only in one solid solution, that of carbon and/or nitrogen in body-centered-cubic α-iron. The C and N atoms are so small that they can fit into the holes between the iron atoms, either at the centers of the edges or at the centers of the faces of the unit cell (Fig. 10.9), without producing too much distortion. We can distinguish *x*, *y*, and *z* sites for the interstitial atoms; an *x* site, for example, is one on a [100] edge. A face-centered site such as *f* is entirely equivalent to an edge site. The one shown is on a [100] line joining the body-centered iron atom to a similar atom in the next cell (not shown); this *f* site is therefore equivalent to an *x* site.

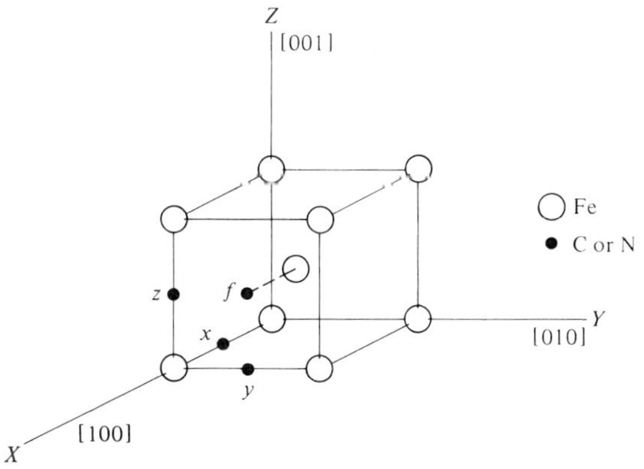

Fig. 10.9 Positions of carbon or nitrogen atoms in the unit cell of iron.

The solubility of C or N in iron is so low that not more than one of these available sites is occupied in any one unit cell. The equilibrium solubility of C, for example, ranges from essentially zero at room temperature to a maximum of about 0.025 weight percent at 723°C. The latter figure corresponds to only one C atom in about 500 unit cells, about the number of cells contained in a cubical block of 8 cells on each edge. The solubility of N is larger, but still small.

[For experimental purposes one wishes to control the amount of C and N in solution. This is done by heat treatment in an appropriate gaseous atmosphere. Both C and N can be removed from the iron by annealing for several hours at 800°C in hydrogen containing a small amount of water vapor. These interstitials can then be added in controlled amounts by heating the iron at 725°C in a mixture of hydrogen and a carbon-bearing gas like carbon monoxide or methane (to add C) or in a mixture of hydrogen and ammonia (to add N). The amount introduced at high temperature can be retained in solution at room temperature by quenching.]

In paramagnetic α-iron, the *x*, *y*, and *z* sites are crystallographically equivalent and equally populated by any C or N in solution. (For brevity, only C will be

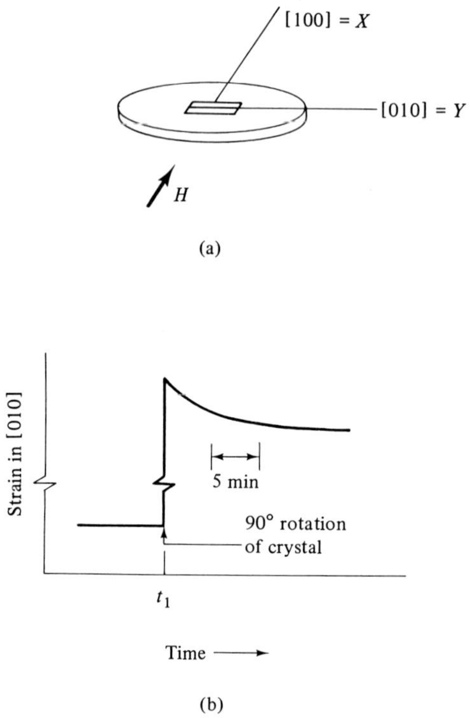

Fig. 10.10 Experiment to determine the preferred interstitial sites in iron. After DeVries *et al.* [10.16].

mentioned from now on. Both interstitials have the same effect.) Below the Curie point, however, in the presence or absence of an applied field, some sites will be preferred over others, depending on the direction of the local magnetization. This preference was first suggested by Néel [10.13, 10.14], who was following up an earlier idea due to Snoek [10.15]. Just what sites are preferred was demonstrated by DeVries *et al.* [10.16] by means of an ingenious experiment which is described here in a slightly simplified form. They took a single-crystal disk of iron cut parallel to (001) and containing 0.008 weight percent C (80 ppm C) in solution, and cemented a strain gage on it to measure strain in the [010] direction, as indicated in Fig. 10.10 (a). The strain gage was connected to a recorder which plotted strain versus time. The disk was placed in a saturating field parallel to [100] and maintained at a temperature of $-23\,^{\circ}$C. After the C atoms had reached equilibrium positions with H parallel to [100], the disk was quickly rotated 90° at time t_1 to make [010], and the strain gage, parallel to the field. What happened is shown in (b). Because the magnetostriction λ_{100} is positive in iron, the gage was contracted before the rotation. When rotated, the gage showed an instantaneous elongation because of the positive magnetostrictive strain along its length, followed immediately by a slow *contraction*, amounting to about 4 percent of λ_{100}. Now the C atoms in iron are somewhat too big for the holes they occupy, and they cause a slight expansion of the lattice. The observed contraction shows that C atoms originally in y sites began to diffuse away from these sites as soon as the Y axis became parallel to the magnetization. Evidently, before the rotation when H was parallel to X, there were more C atoms in y and z sites than in x sites. The C atoms therefore preferentially occupy sites in a plane *normal* to M_s.

This result was unexpected. Because each domain in iron is spontaneously elongated parallel to M_s, and contracted at right angles, it was felt that there would be more room for a C atom in an x site than in a y or z site in a domain which was magnetized in the [100] or X direction. The contrary is true. This result can be understood by assuming that the electron cloud around an iron atom is elongated parallel to the spontaneous magnetization, as indicated in Fig. 10.11. This assumption explains both the positive value of λ_{100} and the greater amount of room for C atoms in y or z sites when M_s is parallel to X.

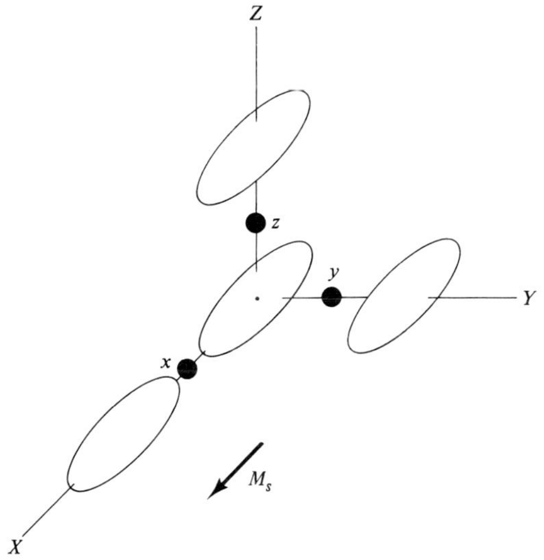

Fig. 10.11 Relation of interstitial sites x, y, z to iron atoms elongated in the magnetization direction.

(As suggested in the previous section, there is no direct evidence to support the assumption that like-atom pairs in a substitutional solid solution have their axes parallel to the local direction of M_s. The other possibility is that these pair axes lie in a plane normal to M_s, which would make the axes of unlike-atom pairs preferentially parallel to M_s. An experiment on substitutional alloys to decide between these possibilities, analogous to the experiment of De Vries *et al.*, has not yet been done.)

We see then that the interstitial sites in iron are preferentially occupied by C atoms in a manner governed by the local magnetization, namely, in a plane normal to the M_s direction. This direction becomes an easy axis, and the resulting anisotropy is described by Eq. (10.1). The value of K_u is quite small, being 166 ergs/cm^3 for C at -34 C and 71 ergs/cm^3 for N at -47 C [10.2]. The cause of this anisotropy, like that due to directional order of solute-atom pairs, is assumed to be entirely magnetic, i.e., if the magnetization M_s is in, say, the X direction, then the energy of a C atom in an x site is larger than it is in a y or z site.

Although the induced anisotropy in, for example, an Fe-Ni alloy and the induced anisotropy in iron containing C are formally the same (both are due to an anisotropic distribution of solute atoms), there is a vast difference in the kinetics of the two, because interstitial atoms diffuse much faster than substitutional atoms. For example, the frequency at which C atoms jump from one interstitial site to a neighboring one in iron at room temperature is about once a second. This means that it is impossible to set up a preferred distribution of C atoms by magnetic annealing at an elevated temperature and preserve it by fast cooling to room temperature with or without a field; soon after the annealing field is removed, diffusion destroys the anisotropy. It also means that iron containing C will self-magnetically-anneal at room temperature and that domain-wall stabilization will occur by virtue of the potential wells of Fig. 10.7. (In fact, the form of these wells follows from calculations made by Néel [10.13, 10.14] for the case of interstitials in iron. Although the potential wells due to directional order of solute-atom pairs are of slightly different shape, we have assumed that wall stabilization due to either cause can be at least qualitatively discussed in terms of Fig. 10.7.)

As the reader may have noticed, all quantitative measurements mentioned in this section were made at subzero temperatures in order to slow down the diffusion of carbon to a point where the measurement could be made in a convenient time. Even then the measurement was dynamic, rather than static, in character. For example, in the experiment of Fig. 10.10, the disturbed distribution of C atoms immediately began to change to a new distribution and the change was complete in some 20 minutes at -23°C.

This does not mean that domain-wall stabilization does not lead to observable magnetic effects at room temperature. It does, but the experiments required are of a different nature than those we have so far encountered, and the effects are called "time effects." We shall therefore postpone their discussion to Chapter 12.

10.4 STRESS ANNEALING

When a uniaxial stress is applied to a solid solution, magnetic or not, interstitial or substitutional, the distribution of solute atoms will become anisotropic if the temperature is high enough for rapid diffusion. In an interstitial solution, for example, x sites will be preferred if a tensile stress is applied along the X axis, simply because of the elongation along that axis. In a substitutional solution, the axes of like-atom pairs may be oriented parallel or perpendicular to the axis of tension, depending on the alloy. In either case, the resulting anisotropic distribution of solute can be frozen in by cooling to a temperature where diffusion is negligible.

If the material is magnetic, the effect becomes more complicated. Because of the magnetoelastic interaction, the stress σ will change the domain structure by reorienting the domain vectors. If $\lambda_{si}\sigma$ is large enough and positive, M_s vectors will be parallel to the stress axis; if it is negative, they will be at right angles to the stress. If the alloy is magnetically annealable, then the solute distribution will change in response to the local direction of M_s. Two independent forces, stress and magnetization, are thus acting to change the solute distribution. These may favor the same or different distributions, depending on the alloy, and the result is not predictable. The resulting magnetic anisotropy, if any, is found to be uniaxial.

Stress annealing has been studied only in a few substitutional alloys. In 50 Ni-50 Co the easy axis is parallel to the tension axis; in 45 Ni-55 Fe it is at right angles; K_u is less than 10^3 ergs/cm^3 for a stress of 2.5 kg/mm^2 (3600 lb/in^2). In Fe + 22 atomic percent Al, cooling from 550°C under a tensile stress of 14 kg/mm^2 (20,000 lb/in^2) produces an easy axis parallel to the stress, with K_u equal to about 17×10^3 ergs/cm^3; this is almost double the value of K_u obtained by cooling in a field [10.17].

10.5 PLASTIC DEFORMATION (ALLOYS)

A uniaxial anisotropy can be induced in certain substitutional alloys simply by plastic deformation at room temperature. Ni-Fe alloys have received the most attention. The usual method of deformation has been by rolling, and the effect is often called *roll magnetic anisotropy*. Equation (10.1) still applies, where θ, measured in the plane of the sheet, is now the angle between M_s and the rolling direction (R.D.). The magnitude of K_u can be considerable, more than 2×10^5 ergs/cm^3; this is larger than the crystal anisotropy constant of nickel, of any high-Ni alloy in the Ni-Fe system, and of most cubic ferrites, but smaller than that of iron (Table 7.4). Depending on the crystallographic orientation of the specimen before rolling, the easy axis can be parallel to the R.D. (positive K_u) or at right angles to the R.D. and parallel to the transverse direction (T.D.) (negative K_u). The main observations on rolled Ni-Fe alloys can be grouped as follows:

1. *Polycrystals with randomly oriented grains.* When this material is rolled, the easy axis is parallel to the R.D. and K_u varies with Ni content according to curve (a) of Fig. 10.12.

2. *Polycrystals with* $\{100\}$ $\langle 001 \rangle$ *preferred orientation.* By heavy cold rolling followed by recrystallization, the cube texture $\{100\}$ $\langle 001 \rangle$ can be obtained [4.9]. When this material is again rolled, an easy axis at right angles to the R.D. develops. K_u is therefore negative; its magnitude is shown by curve (b) in Fig. 10.12. (A 50 Ni-50 Fe alloy so treated is called Isoperm, because its permeability in the R.D. is constant right up to saturation; in the demagnetized state, its M_s vectors lie at 90° to the R.D., and magnetization in the R.D. is accomplished entirely by domain rotation.)

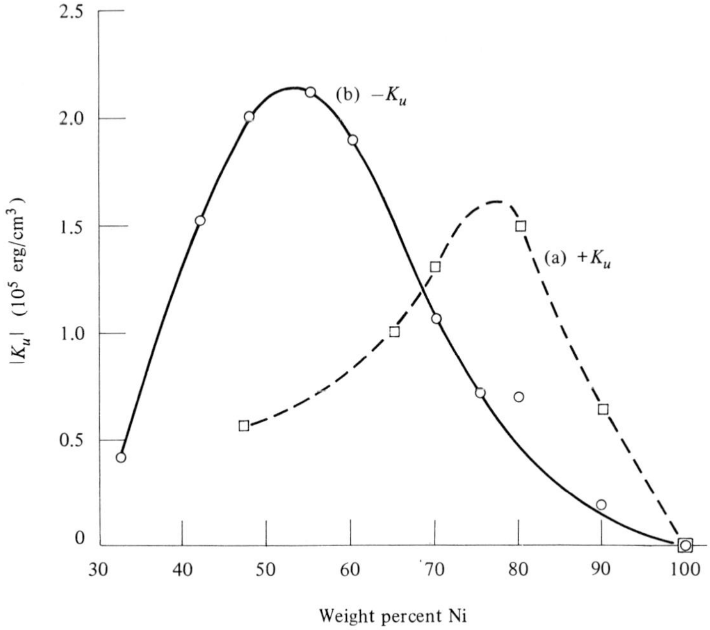

Fig. 10.12 Uniaxial anisotropy constant K_u of rolled polycrystalline Ni-Fe alloys. (a) Randomly oriented grains rolled to 33 percent reduction in thickness. (b) Cube-texture material rolled to 55 percent reduction. After Rathenau and Snoek [7.9].

3. *Single crystals.* The easy axis of a rolled single crystal may be either parallel or at right angles to the R.D., depending on the crystal orientation. Examples are given later.

The anisotropy induced by plastic deformation has been explained by Chikazumi, Suzuki, and Iwata in terms of directional order; the theory is summarized by Chikazumi [G.23] and theory and experiment by Chin [10.23]. This directional order is created not by diffusion, as in magnetic annealing, but by slip.

How this can occur is shown in Fig. 10.13 for a perfectly ordered superlattice like $FeNi_3$. Slip by unit distance on the (111) plane in the $[01\bar{1}]$ direction, resulting from the passage of a single dislocation, has created like-atom pairs across the slip plane where none existed before; the double lines indicate BB pairs, with axes in the [011] direction. This direction therefore becomes a local easy axis. If a second dislocation passes, the crystal becomes perfectly ordered again, with no like-atom pairs. Thus directional order is created only if an odd number of dislocations pass along the slip plane. When a face-centered cubic crystal is deformed by rolling, at least two slip systems (combination of a slip plane of the form {111} and a slip direction of the form ⟨110⟩) must operate to accomplish the observed change in shape, which is a lengthening in the rolling direction, a reduction in thickness, and little or no increase in width. There are thus at

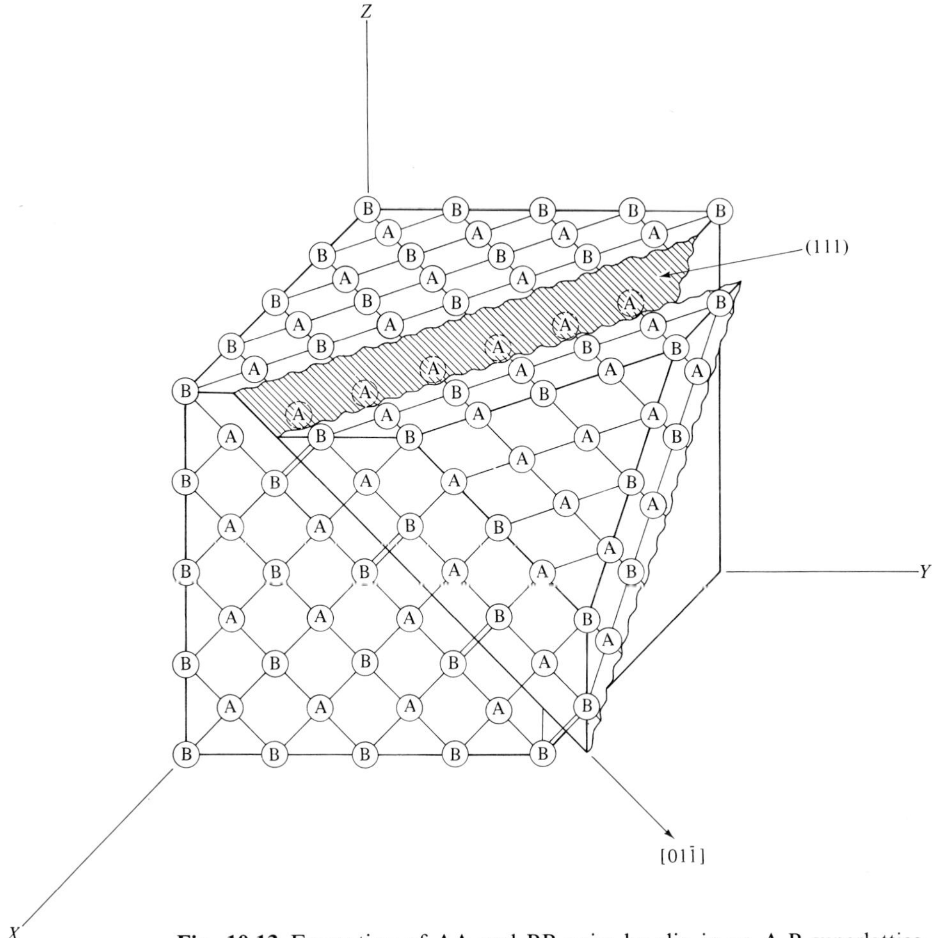

Fig. 10.13 Formation of AA and BB pairs by slip in an A_3B superlattice. Chikazumi [G.23].

least two orientations of the local easy axes, corresponding to the two orientations of active slip planes.

A further complication is that an ordered crystal is never completely ordered from one side to the other, but is made up of a number of "antiphase" regions. (In order-disorder language, these are called domains, but the word "region" will be substituted here to avoid confusion with magnetic domains.) The crystal structure is out of step at the boundaries of these regions, and only within them is the order perfect. Now when the slip distance is large enough to bring a substantial portion of one antiphase region into contact with another across the slip plane, new orientations of like-atom pairs become possible. In this way slip on the (111) plane, for example, can create **BB** pairs not only in the [011] direction shown in Fig. 10.13 but also in [110] and [101]. The same result is expected when only short-range order exists. The various orientations of like-atom-pair axes in the deformed crystal then combine to give a single easy axis for the whole specimen.

The theory of Chikazumi *et al.* takes into account the orientations of the active slip systems when a crystal of a particular orientation is rolled, together with the orientations of like-atom pairs created by slip on these systems, and it yields a prediction of the easy axis of the rolled crystal. Chikazumi and his co-workers tested such predictions by rolling single crystals of the $FeNi_3$ composition and measuring the resultant anisotropy; the agreement with theory was very good.

If we leave aside the complex details of the theory, we can still understand, in a qualitative way, why rolling an alloy single crystal should induce in it a magnetic anisotropy. Given the facts (1) that directional order of like-atom pairs creates an easy axis in some alloys, and (2) that slip can create like-atom pairs where none existed before, and (3) that the axes of such pairs must be oriented in particular ways with respect to the slip planes, and, finally, (4) that the slip planes which operate have particular orientations in the crystal, it follows that the local easy axes due to like-atom pairs must have particular orientations throughout the crystal. These local easy axes then combine to give a net overall anisotropy to the crystal as a whole. This general argument cannot, of course, predict the orientation of the resultant easy axis, or, for that matter, even show that the resultant anisotropy should be uniaxial rather than, say, biaxial.

When a Ni-Fe polycrystal with randomly oriented grains is rolled, the R.D. becomes an easy direction (Fig. 10.12). This result is not understood. Bitter-pattern examination of the rolled material shows that the direction of the easy axis, as judged by the orientation of 180° walls, varies from one grain to another; the observed anisotropy is thus an average over all the grains in the specimen. Theoretical analysis would be almost forbiddingly complex. When a polycrystal is deformed, as many as five slip systems can operate within a single grain, because the change in shape of that grain must conform to the changes in shape of all the neighboring grains, a requirement not imposed on a rolled single crystal.

There is little doubt that slip-induced anisotropy is due to directional order, because the magnitude of K_u varies with composition, and thus with the likelihood of having like-atom pairs, in much the same way after rolling (Fig. 10.12) as after magnetic annealing (Fig. 10.5). (However, it is not clear why the two curves in Fig. 10.12 should peak at different compositions.) The magnitude of K_u after rolling is some 50 times larger than after magnetic annealing, which shows that slip can produce a far larger concentration of like-atom pairs.

Roll magnetic anisotropy has also been observed in Fe-Al, Ni-Co, and Ni-Mn alloys. Chin *et al.* [10.18, 10.19] have extended the theory and experiments to deformation by wire drawing, roll flattening (rolling of wire), flat drawing (drawing through a rectangular hole), and plane strain compression. The last of these simulates rolling, but the compression is carried out in a die which prevents the lateral spreading which occurs when single crystals of certain orientations are rolled. Some of these deformation methods are of industrial interest because of the application of thin, narrow Permalloy tapes as magnetic memory elements (Section 13.5).

10.6 PLASTIC DEFORMATION (PURE METALS)

Krause *et al.* [9.27] cut disk specimens from a flat strip of annealed, polycrystalline nickel and measured the torque on them with a torque magnetometer in an applied field of 9 kOe. The result is shown in Fig. 10.14, where θ is the angle between the applied field and the long axis of the strip, which is the direction in which the strip was rolled before being annealed. There is a very weak, biaxial anisotropy, like that described by Eq. (7.12), due to a small amount of preferred orientation; the easy axes are at 0 and 90° to the strip axis.

If the strip is now stretched a few percent, the torque curve is radically changed. The maxima are now some ten times higher and occur at different values of θ. A fairly strong uniaxial anisotropy has been created, with the easy axis parallel to the direction of previous elongation. Compression has the reverse effect; the anisotropy is again uniaxial, but the easy axis is at 90° to the direction of previous compression. This anisotropy is also governed by Eq. (10.1), where θ is the angle between M_s and the deformation axis (tensile or compressive). Fourier analysis of the torque curves for deformed specimens permits the separation of the uniaxial

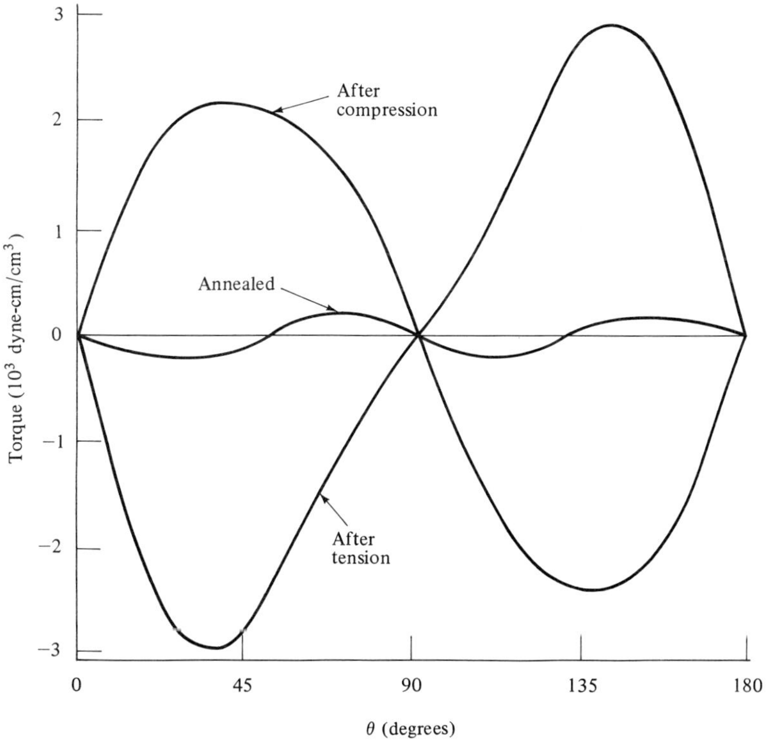

Fig. 10.14 Torque curves for annealed nickel and for nickel after plastic tension and compression of 2.5 percent. After Krause *et al.* [9.27].

anisotropy caused by deformation from the weak biaxial anisotropy caused by texture. The value of K_u was found to increase rapidly with tensile strain up to a maximum of 2700 ergs/cm^3 at 2.5 percent strain; it then decreased slowly with further deformation. After 2.5 percent compression, K_u was $-$ 2400 ergs/cm^3.

Examination of the deformed specimens by x-ray diffraction revealed the presence of residual stress, because the x-ray lines were both shifted and broadened. The line shift, as mentioned in Section 9.10, indicates a stress which is more or less constant over most of the specimen volume, and the broadening reflects stress variations about the mean in this volume. The x-ray observations are consistent with the assumed stress distribution of Fig. 9.32, for a specimen previously deformed in tension. The material can then be imagined as consisting of regions in longitudinal compression (C regions) and regions in tension (T regions). The C regions comprise most of the specimen volume and are responsible for the x-ray effects, while the smaller T regions add nothing observable to the x-ray pattern. The C regions have been tentatively identified with the subgrains which form within each plastically deformed grain, and the T regions with the subgrain boundaries. After plastic compression, the stress distribution of Fig. 9.32 is inverted; most of the specimen is then in a state of tensile stress in the direction of previous compression. These conditions are not peculiar to nickel but have been observed in a number of metals and alloys. (Actually, the stress state at the surface of the deformed nickel was found to be one of biaxial compression after plastic tension and biaxial tension after plastic compression. This complication does not invalidate the argument given below and will be ignored in what follows.)

The observed magnetic anisotropy after plastic deformation is attributed to the residual stress described above. Because residual stresses must form a balanced-force system, one might at first think that they could not cause any net anisotropy. To see this, we take Eq. (8.23), which governs stress anisotropy, and write the magnetoelastic energy as

$$E_{me} = \tfrac{3}{2}\lambda_{si}\sigma \sin^2\theta. \tag{10.2}$$

Suppose that a residual tensile stress σ_T exists, in certain portions of the specimen, parallel to a particular axis from which θ is measured, and a residual compressive stress σ_C exists in other portions, parallel to the same axis. Let f_T and f_C be the volume fractions of the specimen in residual tension and compression, respectively. Then the magnetoelastic energy per unit volume of such a specimen becomes

$$E_{me} = \tfrac{3}{2}\lambda_{si}(f_T\sigma_T + f_C\sigma_C)\sin^2\theta. \tag{10.3}$$

If relative cross-sectional areas of the regions under tension and compression are assumed equal to their relative volumes, a balance of forces requires that

$$f_T\sigma_T + f_C\sigma_C = 0, \tag{10.4}$$

where σ_T is taken as positive and σ_C as negative. The magnetoelastic energy is therefore zero, provided that the specimen is *completely saturated* at all values of θ. Then M_s in the tensile regions, M_s in the compressive regions, and the applied field are all parallel to one another, and it is then justifiable to write Eq. (10.3) in terms of a single angle θ between M_s and the stress axis. Then, whatever the nature of the residual stress state, no magnetic anisotropy will result. (Physically, for any value of θ, the torque on M_s due to the regions in tension is exactly balanced by an opposing torque on M_s due to the regions in compression.)

Some portions of the specimen, however, are not fully saturated at all angles θ, and this condition causes the observed anisotropy. After plastic tension, for example, most of the specimen volume is in compression parallel to the deformation axis; these regions (the C regions) are therefore easy to magnetize along this axis, because λ_{si} is negative for nickel. The average stress in the T regions is necessarily larger than that in the C regions, because the cross section of the T regions is smaller. Therefore, the M_s vectors in the T regions, initially at right angles to the deformation axis, strongly resist rotation by a field applied along that axis. The T regions are thus difficult to saturate, but they comprise such a small fraction of the total volume, estimated at about one-tenth, that the specimen as a whole acts as though it had an easy axis parallel to the deformation axis. The same argument in reverse will account for the easy axis at right angles to the axis of previous compression.

There is no reason to believe that plastic tension or compression would not produce the same kind of anisotropy in other polycrystalline ferromagnetic metals, or in alloys. In alloys this anisotropy would be superimposed on any anisotropy that might result from deformation-induced directional order.

10.7 MAGNETIC IRRADIATION

The physical and mechanical properties of any material are usually changed when it is bombarded with neutrons, ions, electrons, or gamma rays. These property changes are due to the atomic rearrangements, called *radiation damage*, brought about by the radiation, and the kind of rearrangement depends on the kind of radiation. For example, neutrons, being uncharged particles, are highly penetrating. When incident on a solid, they can travel relatively large distances between collisions with atoms of the solid. Each collision, however, causes displacement of one or more atoms into interstitial positions, leaving vacancies behind. (Interstitials and vacancies are collectively called *point defects* to distinguish them from a line imperfection like a dislocation.) Neutron bombardment is therefore expected to create point defects distributed rather widely through the irradiated material.

A curious aspect of radiation damage is the fact that it can produce a substantial increase in the mechanical strength and hardness of many metals and alloys, just as cold work can, but without a corresponding increase in magnetic hardness. (In fact, irradiation can make some materials magnetically softer, as described below.) Radiation-induced defects are apparently more effective in impeding dislocations than in impeding domain walls.

The effect of radiation damage on magnetic properties has been studied mainly in Ni-Fe alloys subjected to fast neutron bombardment.

Schindler *et al.* [10.20, 10.21] investigated polycrystalline Ni-Fe alloys containing from 50 to 80 percent Ni. When these alloys, which already had fairly square hysteresis loops, were irradiated with neutrons at room temperature and without a magnetic field, the coercive force H_c increased, the remanence B_r decreased, and the hysteresis loop became constricted at top and bottom. These results were ascribed to radiation-induced directional order in each domain; this can occur at room temperature because the vacancies created by irradiation speed up diffusion. [Why the hysteresis loop becomes constricted at top and bottom, rather than in the center like the Perminvar loop of Fig. 10.8(d), is not clear. Possibly a central constriction would have been observed in a loop measured at a lower maximum field.] When the irradiation was carried out in the presence of a saturating magnetic field, H_c was somewhat larger that that of an unirradiated specimen, but B_r became considerably larger, i.e., the loop became even more square. These results are consistent with the assumption that directional order was created parallel to the field applied during irradiation, again because of the diffusion-enhancing effect of radiation.

Néel *et al.* [10.22] found that a very large uniaxial anisotropy could be created in a single crystal of 50 Ni-50 Fe (atomic percent) by subjecting it simultaneously to (1) heat (a 295°C anneal), (2) fast neutron irradiation, and (3) a magnetic field. When the field is parallel to [100], that direction becomes an easy axis with an anisotropy energy given by

$$E = K_1 \sin^2 \theta + K_2 \sin^4 \theta, \tag{10.5}$$

where K_1 and K_2 are equal to 3.2×10^6 and 2.3×10^6 ergs/cm^3, respectively, and θ is the angle between M_s and [100]. X-ray diffraction showed that the face-centered-cubic random solid solution had undergone long-range ordering by changing to the "AuCu structure" shown in Fig. 10.15. This structure consists of alternate layers of iron and nickel atoms, parallel to (001). The unit cell is very slightly tetragonal, with the c axis normal to the layers. Moreover, the FeNi "crystal" was found to have changed into three sets of crystallites, with their c axes distributed among the three cube axes of the original crystal. This distribution is not uniform, however; 60 percent of the crystallites had their c axes parallel to [100], the direction of the field during treatment, 20 percent along [010], and 20 percent along [001], where these directions refer to the original cubic crystal. The function of the field is to promote the formation of one of the three kinds of crystallites and thus create a uniaxial anisotropy, while the function of the thermal energy and neutron kinetic energy is to cause long-range ordering to take place rapidly. The anisotropy here is not due to directional order in

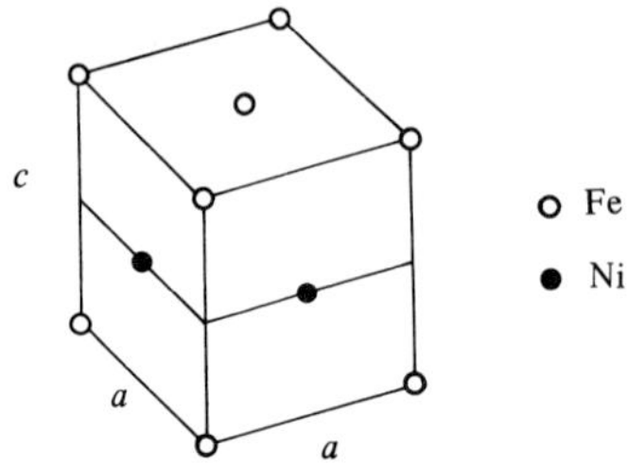

Fig. 10.15 Ordered FeNi, which has the same structure as ordered AuCu.

a solid solution, but to three superimposed crystal anisotropies, with one stronger, per unit volume of specimen, than the other two. These results of Néel *et al.* are remarkable, not only for the large anisotropy produced, but also for the fact that long-range ordering had never before been observed in this system at the composition FeNi.

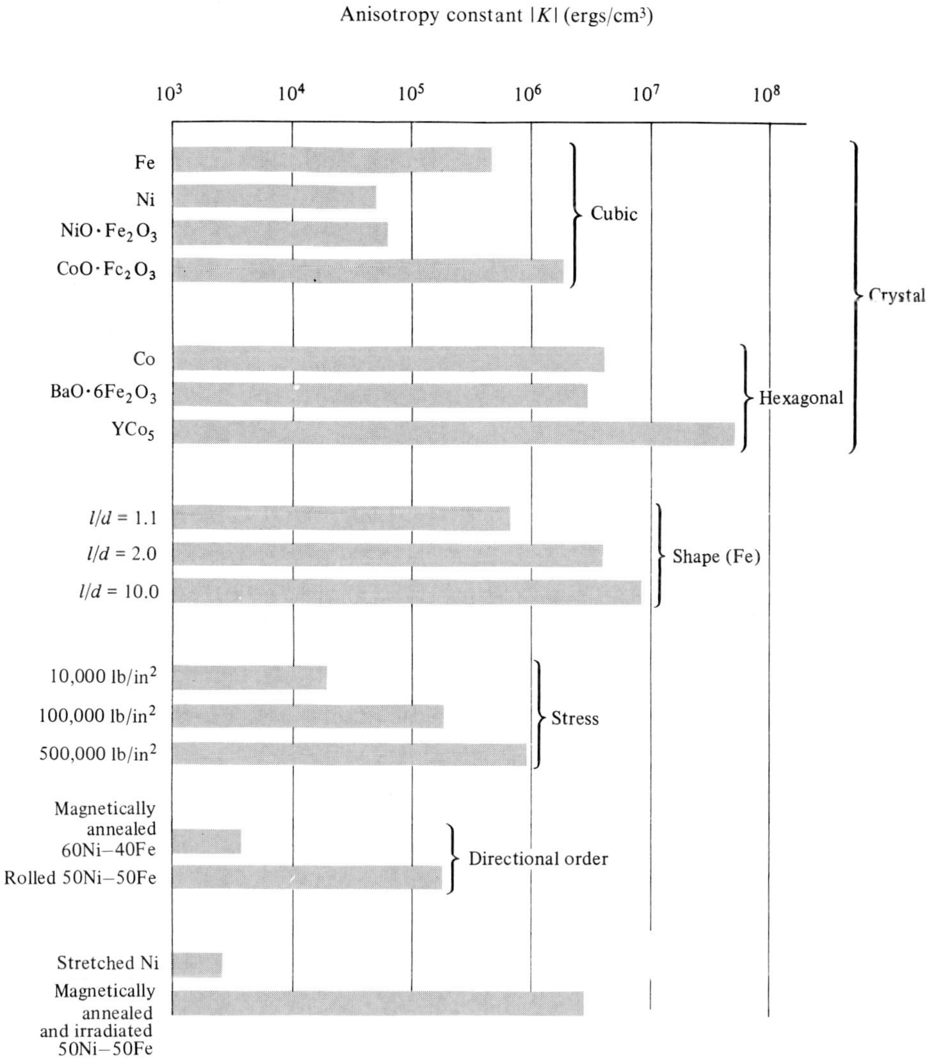

Fig. 10.16 Comparison of anisotropy strengths. See next page for details.

10.8 SUMMARY OF ANISOTROPIES

The magnitudes $|K|$ of the anisotropy constants at room temperature are compared in Fig. 10.16 for the various anisotropies we have met:

1. Grouped in the upper part of the figure are the crystal anisotropies of Fe, Ni, and Co, of some ferrites, and of YCo_5. The last is typical of the huge anisotropy found in the RCo_5 compounds, where R is yttrium or certain of the rare earths.

2. Next are some anisotropies due to shape, for prolate ellipsoids. Shape anisotropy not only depends on the length/diameter ratio l/d, but also is proportional to M_s^2. The values shown were calculated for iron ($M_s = 1714$ emu/cm^3).

3. The stress anisotropy constant is proportional to the product of stress and the magnetostriction λ_{si}. The anisotropies shown were calculated for $\lambda_{si} = 20 \times 10^{-6}$, taken as typical of many materials.

4. The two examples of anisotropy due to directional order are (a) polycrystalline, magnetically annealed 60 Ni-40 Fe (Fig. 10.5), and (b) polycrystalline 50 Ni-50 Fe with a cube texture, rolled 55 percent (Fig. 10.12).

5. "Stretched Ni" refers to the uniaxial anisotropy found in polycrystalline nickel plastically elongated 2.5 percent (Section 10.6).

6. The final entry in the figure shows the anisotropy developed in a single crystal of 50 Ni-50 Fe by the treatment described in Section 10.7.

CHAPTER 11

FINE PARTICLES AND THIN FILMS

11.1 INTRODUCTION

When the measured value of some physical or mechanical property is found to depend on the size of the specimen, that property is said to exhibit a *size effect.* Thus the yield stress in tension of an iron whisker 50 microns in diameter is more than a thousand times that of the same iron in the form of a single-crystal rod 1 cm in diameter. On the other hand, both specimens have the same density. In general, only structure-sensitive properties show a size effect.

Among magnetic properties, the saturation magnetization M_s, for example, is independent of specimen size, but the coercivity H_c shows a marked size effect. The coercivity of elongated iron particles 150 Å in diameter is some 10^4 times that of iron in bulk. In this chapter we will examine some of the magnetic properties of specimens which have been made very small in one dimension (thin films), two dimensions (fine wires), or three dimensions (small particles). Of these, thin films are important in computer applications, and fine particles in permanent magnets. Fine wires are of less importance and will be treated in the last section.

This chapter begins with a discussion of the fundamental properties of fine particles; the more practical aspect of their application in magnets is reserved for Chapter 14. Similarly, the material on thin films is divided mainly between this chapter and Chapter 12, where the mechanism of flux reversal (switching) is described; computer applications are postponed to Chapter 13.

There is an immense literature, both theoretical and experimental, on fine particles and thin films. Fine particles have been reviewed by Paine [11.1], Jacobs and Bean [11.2], and Wohlfarth [11.3, 11.4]. Thin films have been treated in review articles by Goodenough and Smith [11.5], Jacobs and Bean [11.2], Smith [11.6], and Pugh [11.7], and in books by Prutton [G.24] and Soohoo [G.25].

11.2 SINGLE-DOMAIN VERSUS MULTI-DOMAIN BEHAVIOR

We have already examined, in Section 9.6, the theoretical reasons for believing that a single crystal will become a single domain when its size is reduced below a critical value L_c of a few hundred angstroms. What is the experimental evidence for the existence of single-domain particles? There is today a great deal of evidence, but perhaps the most clear-cut is that provided by a simple experiment performed in 1950 by Kittel, Galt, and Campbell [11.8]. It was known at that time that an assembly of very fine particles was magnetically hard, i.e., that it had a large coercivity. Some argued that this effect was due to the fact that the particles were single domains and that magnetization reversal could take place only by rotation of M_s vectors against strong anisotropy forces (see Table 9.2); others believed that the particles were multidomain, and so strained that large microstress gradients prevented easy wall motion. To distinguish between these two views, Kittel *et al.* made two specimens, each composed of a very dilute suspension of spherical nickel particles in paraffin wax; in one specimen the particles were smaller than L_c, which they estimated as 600 Å, and in the other they were larger. (The suspension was made dilute to avoid magnetic interactions between particles, and the particles were spherical to eliminate shape anisotropy.) They then measured, not the coercivity, but the field required to *saturate* the specimen. The results were as follows:

1. *Specimen A.* Particle diameter, 200 Å. Field required to saturate, 550 Oe. While this is somewhat larger than expected for single-domain particles, it is not unreasonably larger. To overcome the crystal anisotropy of nickel, fields of the order of 200–300 Oe are needed, as shown by Fig. 7.2. The presence of some unavoidable shape or stress anisotropy can account for the higher field actually required.

2. *Specimen B.* Particle diameter, 80,000 Å (8 μ). Field required to saturate, 2100 Oe. Inasmuch as this field is almost four times the field required to saturate Specimen *A*, the magnetization mechanism must be entirely different. It is, in fact, the mechanism appropriate to a multidomain specimen, namely, wall motion and rotation. To saturate, the applied field must at least overcome the demagnetizing field of the specimen, which is $4\pi M_s/3$ at saturation, where $4\pi/3$ is the demagnetizing factor of the spherical particles. For nickel, $4\pi M_s/3$ is $4\pi\,(484)/3$ or 2020 Oe, in good agreement with experiment.

These results prove that the 200 Å particles were single domains. Otherwise, a field of at least 2020 Oe would be needed to saturate them. An originally multidomain particle can be kept in a saturated state only by a field larger than the demagnetizing field; once this field is removed, the magnetostatic energy associated with the saturated state breaks the particle up into domains and reduces M, and the demagnetizing field, to some lower value. But a single-domain particle is by definition always saturated (in the sense of being spontaneously magnetized in one direction throughout its volume, but not in the sense of having its M_s vector necessarily parallel to an applied field). An applied field does not have to overcome the demagnetizing field in order to rotate M_s; in the limit of zero anisotropy, M_s can be rotated by an infinitesimally small field. (A well-pivoted compass needle is an exact analogue to the M_s vector of a low-anisotropy single-domain particle. The needle is a permanent magnet with a demagnetizing field of more than 50 Oe, but the earth's field, of a few tenths of an oersted, can easily rotate it.)

Another way of viewing this experiment is to focus on the energies involved. A saturated sphere, whatever its size, must always have a magnetostatic energy

E_{ms} of $\frac{1}{2}(4\pi/3)(M_s^2)$ per unit volume, according to Eq. (7.51). If the sphere is originally multidomain, this energy must be supplied by the applied field; if it is a single domain, this energy is always present, at all field strengths including zero.

(The above statement that a large particle will spontaneously revert from the saturated state to the multidomain state when the field is removed is true of all "normal" materials. It is not true of a few substances of very high crystal anisotropy in which domains of reversed magnetization are nucleated only with great difficulty. See Section 11.5. However, the existence of such substances does not invalidate the main argument above, which is based on the large difference between the fields required to *saturate* large and small particles.)

Morrish [G.27] describes a different experiment that proves the existence of single-domain particles. The M,H loop of a single particle was measured and shown to have the rectangular shape expected for a uniaxial, single domain.

11.3 COERCIVITY OF FINE PARTICLES

In magnetic studies on fine particles the single property of most interest is the coercivity, for two reasons: (1) it must be high, at least exceeding a few hundred oersteds, to be of any value for permanent-magnet applications, and (2) it is a quantity which comes quite naturally out of theoretical calculations of the hysteresis loop.

The coercivity of fine particles has a striking dependence on their size. As the particle size is reduced, it is typically found that the coercivity increases, goes through a maximum, and then tends toward zero. This is clearly shown for three different materials in Fig. 11.1; for the other three the maximum in coercivity has not yet been reached. The very large range of the variables should be noted; the coercivities vary over three orders of magnitude and the particle sizes over five; the smallest particles are less than ten unit cells thick, while the largest have a 0.1 mm diameter and would be retained on a 170-mesh screen.

An understanding of the shape of the curves of Fig. 11.1 has come slowly and only after much experimental and theoretical research. It can now be claimed that the magnetic behavior of fine particles is broadly, but not precisely, understood. The main conclusion is that the mechanism by which the magnetization of a particle changes differs from one part of the size range to another, and much of this chapter is devoted to an examination of these different mechanisms. In anticipation of the results of later sections, Fig. 11.2 shows very schematically how the size range is divided, in relation to the variation of coercivity with particle diameter D. Beginning at large sizes, we can distinguish the following regions:

1. *Multidomain.* Magnetization changes by domain wall motion (Section 11.5). For most, but not all, materials the size dependence of the coercivity is experimentally found to be given approximately by

$$H_{ci} = a + \frac{b}{D}, \tag{11.1}$$

where a and b are constants. This relation is only partially understood.

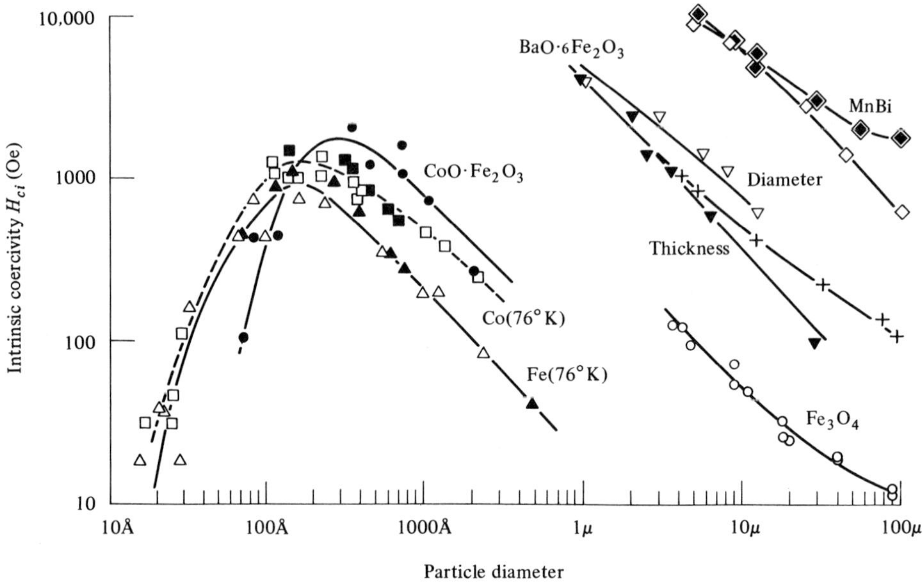

Fig. 11.1 Variation of coercivity with particle size for particles deriving their coercivity principally from crystal anisotropy. Luborsky [11.9]

2. *Single-domain.* Below a critical diameter D_s, which is not well defined, the particles become single domains, and in this size range the coercivity reaches a maximum. (The quantity D_s is equivalent to L_c of Chapter 9.) Particles of size D_s and smaller change their magnetization by spin rotation, but more than one mechanism of rotation can be involved (Section 11.4).

a) As the particle size decreases below D_s the coercivity decreases, because of thermal effects, according to

$$H_{ci} = g - \frac{h}{D^{3/2}}, \tag{11.2}$$

where g and h are constants. This relation is well understood (Section 11.6).

b) Below a critical diameter D_p the coercivity is zero, again because of thermal effects, which are now strong enough to spontaneously demagnetize a previously saturated assembly of particles. Such particles are called *superparamagnetic* (Section 11.6).

The magnetic hardness of most fine particles is due to the forces of shape and/or crystal anisotropy. In order to study the effect of either of these alone, the experimenter can try to make essentially spherical particles, to eliminate shape anisotropy, or make elongated particles of a material having low or zero crystal anisotropy. Then, in order to form a practical magnet or a specimen for research

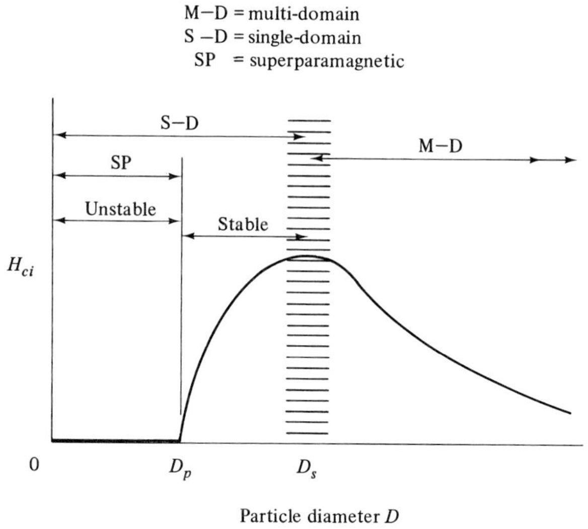

Fig. 11.2 Variation of intrinsic coercivity H_{ci} with particle diameter D (schematic).

studies, the fine magnetic particles must be compacted, with or without a nonmagnetic binder, into a rigid assembly. Then an important variable is the *packing fraction p*, defined as the volume fraction of magnetic particles in the assembly. The variation of coercivity with p depends on the kind of anisotropy present.

When *shape anisotropy* prevails, the coercivity H_{ci} decreases as p increases, because of particle interactions. The following relation has been proposed on theoretical grounds,

$$H_{ci}(p) = H_{ci}(0)(1 - p), \tag{11.3}$$

where $H_{ci}(0)$ is the coercivity of an isolated particle ($p = 0$), but the underlying theory is not considered sound [11.4]. Some materials obey this relation but many do not. Figure 11.3 shows the behavior of elongated Fe-Co alloy particles having an axial ratio greater than 10 and a diameter of 305 A.

The nature of the particle-interaction problem is indicated in Fig. 11.4, for elongated single-domain particles. In (a), part of the external field of particle A is sketched, and this field is seen to act in the $+ z$ direction on particle B below it, but in the $- z$ direction on particle C beside it. Thus the "interaction field" depends not only on the separation of the two particles but also on their positions relative to the magnetization direction of the particle considered as the source of the field. Suppose the M_s vectors of these particles had all been turned upward by a strong field in the $+ z$ direction. This field is then reduced to zero and increased

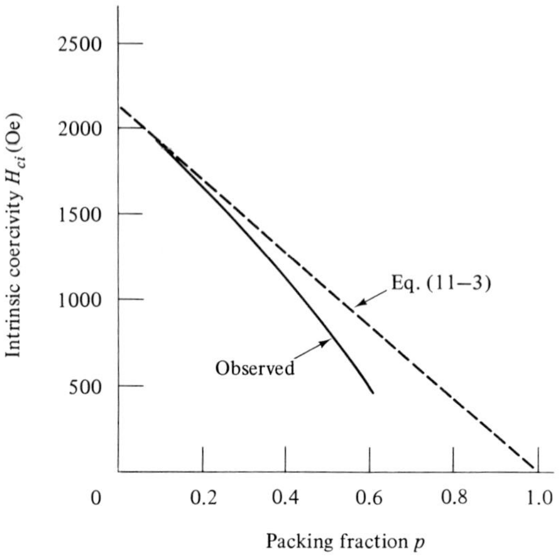

Fig. 11.3 Variation of H_{ci} with p for elongated Fe-Co particles. After Luborsky [11.9]

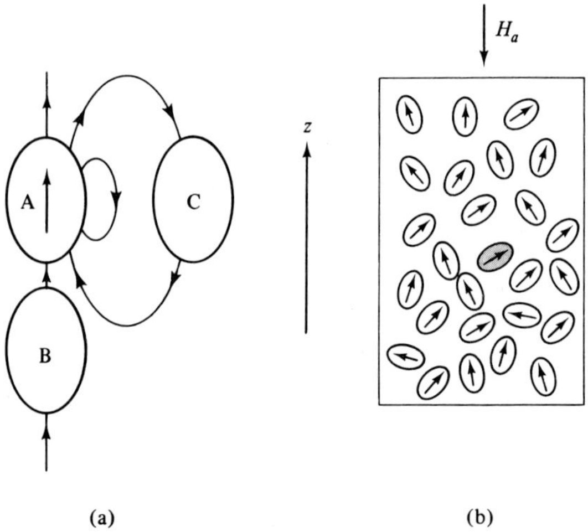

Fig. 11.4 Particle interactions.

in the $-z$ direction. The field of A at C now aids the applied field, and C would reverse its magnetization at a lower applied field than if A were absent; the coercivity would therefore be lowered. The opposite conclusion would be reached if we considered only the pair of A and B particles. In either case we have a relatively easy two-body problem to deal with. A quantitative study of the interactions of three particles is very difficult, and the exact solution of the many-body problem represented by Fig. 11.4 (b), a cylindrical compact of a large number of particles, is virtually impossible. If a reversing field H_a is applied in the $-z$

direction, what is the true field acting on the shaded particle in the interior of the compact? Note that it is not enough to correct H_a for the demagnetizing field of the whole compact. In order to compute the coercivity we would have to know, in addition, the field at the interior particle due to all the other particles, both at the start of and *during* the reversal. As p increases, the particles come closer together, the interactions become stronger, and the coercivity continues to decrease. Finally, at $p = 1$, all particles are everywhere in contact, shape anisotropy is lost, and the coercivity becomes zero, if other forms of anisotropy are absent.

On the other hand, when *crystal anisotropy* prevails, the coercivity is expected to be independent of p, and this view is supported by experiment. This kind of anisotropy is due to forces (spin-orbit coupling) which are "internal" to the particle and not, like shape anisotropy, to magnetostatic fields external to the particle. At any value of p, including unity, the crystal anisotropy forces remain constant.

11.4 MAGNETIZATION REVERSAL BY SPIN ROTATION

In Section 9.13 we examined the hysteresis loops of uniaxial single-domain particles reversing their magnetization by rotation, as calculated by Stoner and Wohlfarth [7.23]. The tacit assumption made there was that the spins of all the atoms in the particle remained parallel to one another during the rotation. This mode of reversal is called *coherent rotation, rotation in unison*, or the *Stoner-Wohlfarth mode.*

For spherical single-domain particles of iron, with their easy axes aligned with the field, the intrinsic coercivity H_{ci} due to crystal anisotropy is equal to $2K/M_s = 2(4.8 \times 10^5)/1714 = 560$ Oe, from Eq. (9.51). Aligned elongated particles, reversing by coherent rotation, have a coercivity due to shape anisotropy given by Eq. (9.49), namely,

$$H_{ci} = (N_a - N_c)M_s, \tag{11.4}$$

where N_a is the demagnetizing factor along the short axis and N_c along the long one. The maximum value of $(N_a - N_c)$ for infinite elongation is 2π, so that the maximum attainable coercivity from shape anisotropy is $2\pi\ M_s$. For aligned iron particles we calculated, in Table 9.2, that H_{ci} should be 10,100 Oe for an axial ratio c/a of 10 and 10,800 Oe for infinite elongation. About 1955 a method was developed for making very thin, elongated, iron particles with axial ratios from about 1 to well over 10 by electrodeposition on a mercury cathode. (Details of the method are given in Section 14.7.) Measurements on aligned, dilute compacts of these particles showed that H_{ci} increased with c/a but did not exceed about 1800 Oe for c/a larger than 10 [11.9]. These results showed that the observed coercivity certainly could not be explained by crystal anisotropy, but neither was it as large as expected for shape anisotropy. In fact, it was only some 18 percent of the theoretical value. This discrepancy forced theoreticians to question the assumption on which Eq. (11.4) is based, namely, coherent rotation of the spins. They therefore examined possible modes of *incoherent* rotation, in which all spins do

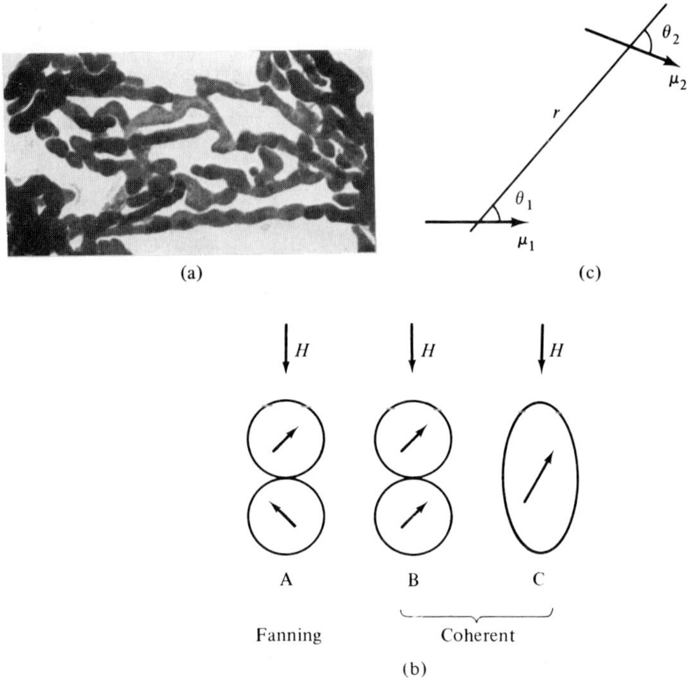

Fig. 11.5 (a) Electron micrograph of electrodeposited iron particles [11.9]. (b) Models of particle shape and reversal modes. (c) Dipole-dipole interactions

not remain parallel, to see if such modes would give coercivities in better accord with experiment. The two most important of these incoherent modes are magnetization *fanning* and *curling*.

Fanning

This mode was suggested by the shape of the electrodeposited iron particles observed with the electron microscope. Figure 11.5 (a) shows that these particles have a kind of "peanut shape," characterized by periodic bulges rather than smooth sides. This suggests that they can be approximated by the "chain of spheres" model shown at A and B of (b) for two-sphere chains. Jacobs and Bean [11.10] considered two possible reversal mechanisms: (1) symmetric fanning (A), in which the M_s vectors of successive spheres in the chain fan out in a plane by rotating in alternate directions in alternate spheres, and (2) coherent rotation (B) in which the M_s vectors of all the spheres are always parallel. The coercivities calculated for these mechanisms were then compared with that calculated for coherent rotation in a prolate spheroid (C) of the same axial ratio as the chain of spheres.

It is assumed that each sphere is a single domain with no anisotropy of its own and that the spins in each one reverse coherently. Each sphere is treated as a dipole of magnetic moment μ and diameter a. Now the mutual potential energy of two dipoles, which is basically magnetostatic energy, is shown by Eq. (7) of Appendix 4 to be

$$E_{ms} = \frac{\mu_1 \mu_2}{r^3}\left[\cos(\theta_1 - \theta_2) - 3\cos\theta_1 \cos\theta_2\right], \tag{11.5}$$

where r is their distance apart and θ_1 and θ_2 are defined in Fig. 11.5 (c). For a two-sphere chain reversing by fanning, $\mu_1 = \mu_2 = \mu, r = a, \theta_1 = \theta, \theta_2 = -\theta$, and Eq. (11.5) reduces to

$$\text{(fanning, A)} \quad E_{ms} = -\frac{\mu^2}{a^3}(1 + \cos^2\theta). \tag{11.6}$$

This energy depends on θ in exactly the same way as the various forms of uniaxial anisotropy energy we have previously encountered. The magnetostatic coupling between two dipoles therefore causes the pair to have a uniaxial anisotropy with an easy axis along the line joining the dipoles. This has been called *interaction anisotropy*.

(In the fanning mode, the spins in one sphere are not parallel to those in the adjacent sphere at the point of contact. Some exchange energy is therefore introduced. But exchange energy is essentially short range, which means that the spins contributing to this energy in the fanning mode form only a small fraction of the total. Thus the total exchange energy is considered to be small, and it can be made still smaller by imagining the spheres to be slightly separated. It is neglected in these calculations.)

To find the coercivity of the two-sphere chain in a field H parallel to the chain axis, we note that the potential energy E_p in the field is $2\mu H \cos\theta$, when H is antiparallel to μ. The total energy is then

$$E = E_{ms} + E_p = -\frac{\mu^2}{a^3}(1 + \cos^2\theta) + 2\,\mu H \cos\theta. \tag{11.7}$$

This equation is of the same form as Eq. (9.40) for a uniaxial particle reversing coherently. Therefore, a fanning reversal is also characterized by a rectangular hysteresis loop, and the coercivity is the field at which the moments will flip from $\theta = 0$ to $\theta = 180°$. To find it we set $d^2E/d\theta^2$ equal to zero and proceed in exactly the same way as in Section 9.13. The result, for the intrinsic coercivity, is

$$\text{(fanning, A)} \quad H_{ci} = \frac{\mu}{a^3} = \frac{\pi M_s}{6}, \tag{11.8}$$

because $\mu = (M_s)(4\pi/3)(a/2)^3$.

For a two-sphere chain rotating coherently (B), $\theta_1 = \theta_2 = \theta$, with the result that

$$\text{(coherent, B)} \quad E_{ms} = \frac{\mu^2}{a^3}(1 - 3\cos^2\theta). \tag{11.9}$$

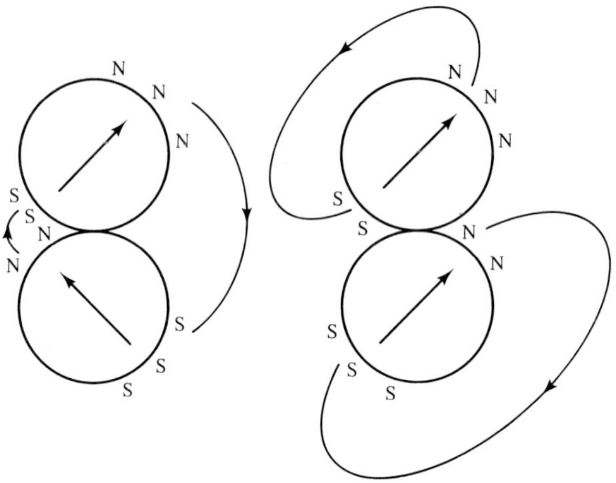

Fig. 11.6 External fields of spheres reversing by fanning and by coherent rotation.

The coercivity is then, for H parallel to the chain axis,

$$\text{(coherent, B)} \quad H_{ci} = \frac{3\mu}{a^3} = \frac{\pi M_s}{2} \tag{11.10}$$

or three times as large as H_{ci} for fanning. The easier reversal for the fanning mode is due to the fact that the field must overcome an energy barrier for reversal only one-third as high for fanning as for coherent rotation (Problem 11.4). The physical reason is suggested in Fig. 11.6. Fanning brings north and south poles closer together, thus reducing the spatial extent of the external fields of the spheres and hence the total magnetostatic energy.

Jacons and Bean [11.10] also computed the coercivity for linear chains longer than $n = 2$, where n is the number of spheres in the chain, both for chains aligned with the field and for chains randomly oriented. Similar calculations were made for reversal mechanism C, where n is now the axial ratio of the prolate spheroid, by means of Eq. (11.4). Their results, for chains aligned with the field, are shown in Fig. 11.7 both in terms of the intrinsic coercivity H_{ci} of iron particles and in terms of the reduced intrinsic coercivity

$$h_{ci} = \frac{H_{ci}}{2\pi M_s}, \tag{11.11}$$

which gives H_{ci} as a fraction of the maximum attainable coercivity in a prolate spheroid undergoing coherent rotation. The predictions of the fanning theory [11.10] are seen to be in rather close accord with the experimental results on

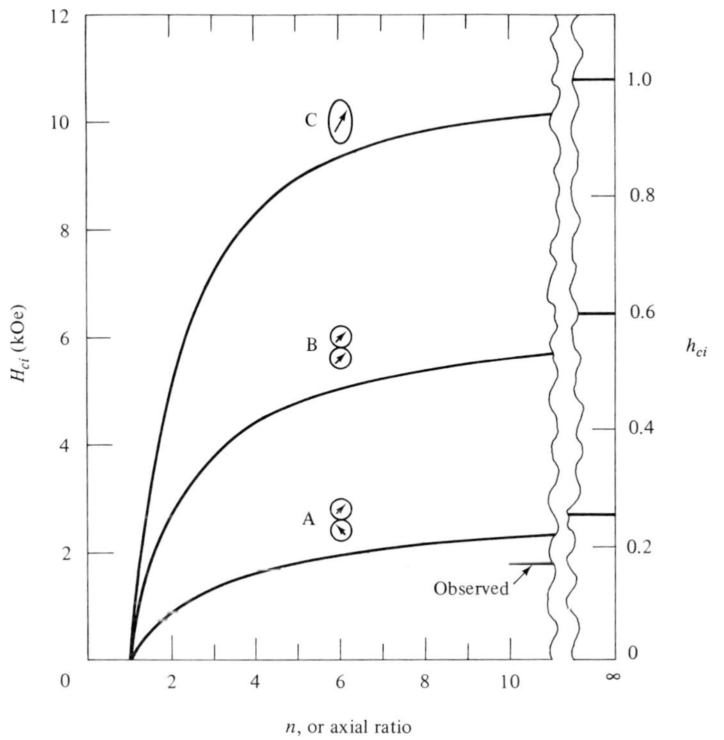

Fig. 11.7 Calculated coercivities for aligned n-sphere chains (A and B) and prolate spheroids of axial ratio n (C). The observed values are for elongated iron particles [11.9]

highly elongated iron particles [11.9], pointing to incoherent rotation as the operative reversal mechanism. In a sense this good agreement appears rather surprising. As mentioned earlier, fanning mechanism A implies negligible exchange interaction between adjacent spheres because of their "point" contact. But if the contact area between spheres is large, the exchange forces would favor coherent-rotation mechanism B. The actual shape of these particles, seen in Fig. 11.5 (a), is that of a cylinder with periodic bulges, which is closer to a chain of squashed-together spheres than to a chain of spheres in point contact. Nevertheless, there must be something in the structure of these particles that forces them to reverse incoherently rather than coherently.

Curling

This mode of incoherent reversal was investigated theoretically by Frei, Shtrikman, and Treves [11.11, 9.15] by the methods of micromagnetics. Their calculations are too intricate to reproduce here, and only the main results will be given. Consider a single-domain particle in the shape of a prolate spheroid, initially magne-

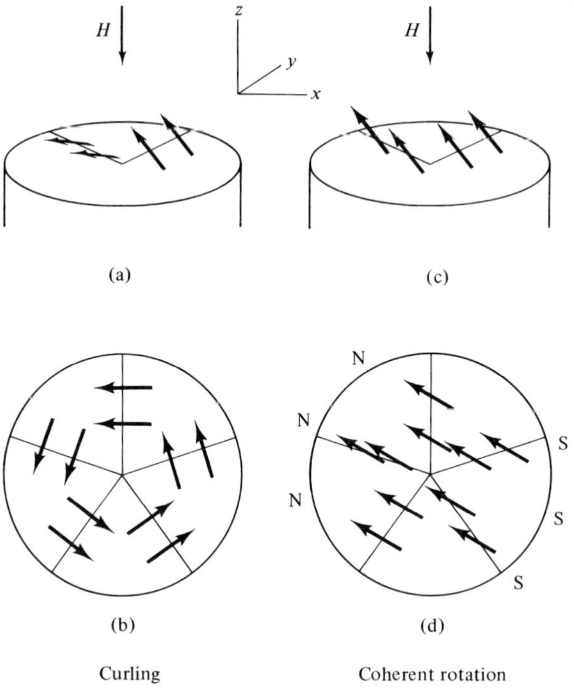

Fig. 11.8 Curling and coherent-rotation modes. (b) and (d) are cross sections normal to the z axis after a 90° rotation from the $+z$ direction

tized in the $+z$ direction parallel to its long axis. Suppose a field is then applied in the $-z$ direction, causing each spin to rotate about the radius, parallel to the xy plane, on which it is located. This curling mode is illustrated in perspective in Fig. 11.8 (a), where the reversal is shown as about one-fourth completed. When the reversal has gone half way, the spins are all parallel to the xy plane and form closed circles of flux in all cross sections, as shown in (b). If the axial ratio of the spheroid approaches infinity, so that it approximates an infinite cylinder, the spins are always parallel to the surface during a curling reversal, so that no free poles are formed and no magnetostatic energy is involved. The energy barrier to a curling reversal is then entirely exchange energy, because the spins are not all parallel to one another during the reversal. In contrast, the coherent rotation shown in (c) and (d) produces free poles on the surface, and therefore magnetostatic energy, but no exchange energy. We can also conclude that, if highly elongated particles in an assembly reverse by curling, the coercivity should be independent of the packing fraction, because no magnetostatic energy is involved, again in sharp contrast to coherent rotation.

As the axial ratio c/a of the spheroid (c = semi-major axis, a = semi-minor axis) changes from infinity (for the cylinder) to unity (for the sphere), some magnetostatic energy will be generated during a curling reversal, because the spins are no longer always parallel to the particle surface. The energy barrier then includes exchange and magnetostatic energy, and the importance of the latter increases as c/a decreases. For finite values of c/a the coercivity will depend on the packing fraction.

The calculations show that the coercivity in the curling mode is markedly size-dependent. A convenient and fundamental unit of length with which to measure size is

$$D_0 = \frac{2\,A^{1/2}}{M_s}, \tag{11.12}$$

where A is the *exchange constant.* It is a measure of the force tending to keep adjacent spins parallel to one another, i.e., of the torsional stiffness of the spin-spin coupling; it is related to the *exchange integral* J_{ex} of Eq. (4.31) by

$$A = \frac{2J_{ex}\,S^2}{a} \quad \text{for a BCC structure,} \tag{11.13}$$

$$A = \frac{4J_{ex}\,S^2}{a} \quad \text{for a FCC structure,} \tag{11.14}$$

where S is the spin and a the lattice parameter.

The micromagnetics calculation of shape anisotropy begins with the assumption that the particle is a single domain with all spins initially parallel to the $+z$ direction in zero field. Crystal anisotropy is ignored and so are thermal effects. The coercivity is then found by calculating the field in the $-z$ direction that is just enough to supply the energy required for a reversal by curling. The result is found to depend both on particle size and shape, as follows:

1. *Infinite cylinder.* If the diameter of the cylinder is D, then the reduced coercivity is

$$h_{ci} = \frac{H_{ci}}{2\pi M_s} = \frac{1.08}{(D/D_0)^2}. \tag{11.15}$$

2. *Prolate spheroid.* If N_c is the demagnetizing factor along the long axis c, and $D = 2a$, where a is the semi-minor axis, then

$$h_{ci} \geqslant \frac{N_c}{2\pi} - \frac{k}{(D/D_0)^2}, \tag{11.16}$$

where k depends on the axial ratio c/a and varies from 1.08 for the infinite cylinder to 1.39 for the sphere. (The dependence of k on c/a is given in graphical form by Aharoni [11.12].)

3. *Sphere.*

$$h_{ci} \geqslant \frac{2}{3} - \frac{1.39}{(D/D_0)^2}. \tag{11.17}$$

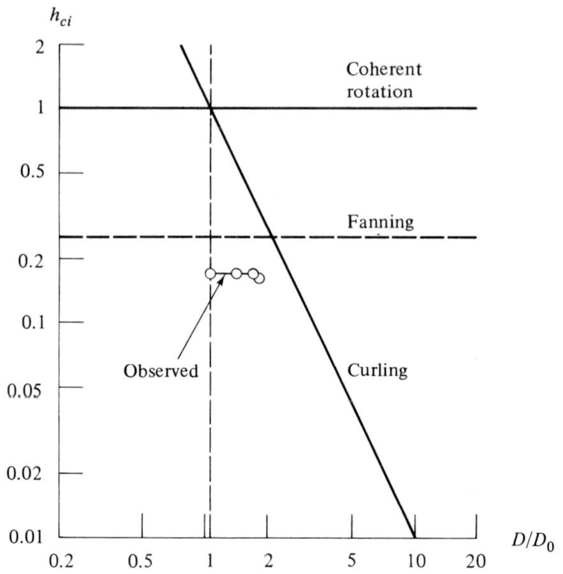

Fig. 11.9 Calculated reduced coercivities for coherent and curling reversals in an infinite cylinder, and for fanning reversal in an infinite chain of spheres, as a function of the reduced diameter D/D_0. Crystal anisotropy is neglected. The observed values are for elongated iron particles [11.9]

This equation is valid only for D/D_0 values greater than 1.44. [The inequality signs in Eqs. (11.16) and (11.17) mean that the coercivities are expected to be somewhat larger than the values given by these expressions but definitely smaller than the coercivity of the infinite culinder.]

Figure 11.9 shows, for the infinite cylinder, the size dependence of h_{ci} for curling and the size independence for coherent rotation. The particle will reverse by that mechanism which has the lower coercivity. Thus below a critical diameter $D = (1.08)^{1/2}\ D_0 = 1.04\ D_0$, coherent rotation is favored, while larger particles will reverse by curling. [If Fig. 11.9 is replotted for a finite value of c/a (prolate spheroid), both curves will move to lower h_{ci} values and their intersection to larger D/D_0 values. For a sphere the critical diameter separating the two modes of reversal is 1.44 D_0.] In Fig. 11.10 the theoretical curves for the infinite cylinder are plotted in a different way; the scales are now linear, rather than logarithmic, and the contribution of crystal anisotropy, which is the same for either reversal mode, is included. The easy crystal axis is assumed to be parallel to the cylinder axis. (Another possible incoherent reversal mode, called *magnetization buckling*, was also investigated by Frei *et al.* [11.11]. It was found to give a coercivity lower than that by coherent rotation or curling only over a very narrow range of D/D_0 values near unity. For that reason details of this mechanism are ignored here.)

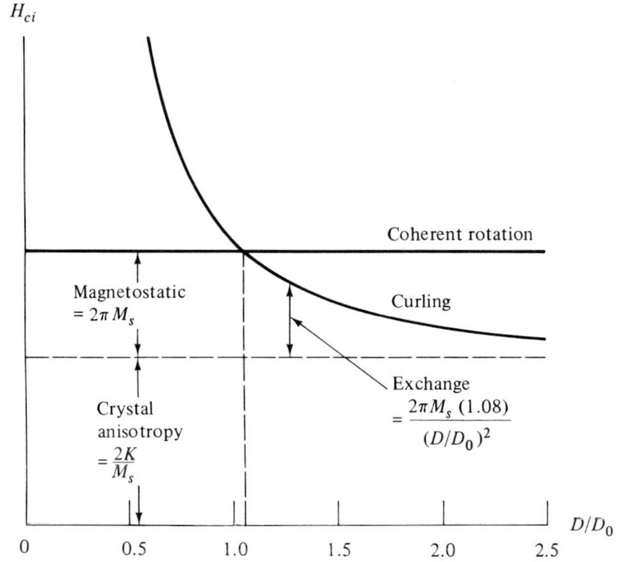

Fig. 11.10 Calculated coercivities, with the effect of crystal anisotropy included, for coherent and curling reversals in an infinite cylinder as a function of the reduced diameter D/D_0. After Brown [G.15]

The main conclusion of the theory, then, is that small particles will reverse coherently and large particles by curling. The main reason is the rapid increase of magnetostatic energy with particle size for coherent rotation, because this energy is directly proportional to particle volume. But so is the potential energy available from the reversing field, and the result is that the coercivity is independent of particle size. In a curling reversal, on the other hand, the mean angle between adjacent spins, averaged over the cross section of the particle (Fig. 11.8 b), is less, and the exchange energy per unit volume is smaller, the larger the diameter of the particle. The total exchange energy during reversal therefore increases with particle volume at a less than linear rate, with the result that the coercivity decreases as particle diameter increases. Large particles can therefore reverse by curling, which involves exchange energy but complete flux closure and no magnetostatic energy during reversal (infinite cylinder) or partial flux closure and low magnetostatic energy (prolate spheroid or sphere). In very small particles, the magnetostatic energy required for coherent reversal is less than the exchange energy required for curling, so that coherent reversal is preferred.

To what actual sizes do the critical diameter ratios D/D_0 calculated above correspond? This question is difficult to answer precisely because of uncertainty in the value of the exchange constant A. For iron, for example, estimates of A range from 0.3×10^{-6} to 2×10^{-6} erg/cm. If we take A as 1×10^{-6} erg/cm, then D_0 for iron is 120 Å. Then the critical diameters vary from (1.04)(120) = 120 Å for the infinite cylinder to (1.44)(120) = 170 Å for the sphere.

Workers in the realm of micromagnetics have taken the position that a particle is truly a single domain only when its spins are always parallel, both before and during a reversal. Thus two definitions of critical size now exist:

1. *Static.* A particle is a single domain when all its spins are parallel in zero field. If it spontaneously breaks up into domains in zero field, then the critical size D_s has been exceeded. This is the sense of the term "single domain" appearing up to now in this book. With this criterion we estimated, in Section 9.6, that the critical size for single-domain particles was several hundred angstroms and that the critical size would be larger, the larger the crystal anisotropy K.

2. *Dynamic.* A particle is a single domain only when its spins are all parallel, both in zero field and during reversal in an applied field. Thus, according to this micromagnetics definition, only particles which reverse coherently are single domain, and the critical size for single-domain behavior is to be found from the intersection of curves like those shown in Figs. 11.9 and 11.10. For spherical iron particles this critical size was estimated above as 170 A. Now this is certainly smaller than the size estimated according to static definition (1) above. But note that micromagnetics *assumes* that a particle larger than the critical size according to dynamic definition (2), a particle which would reverse by curling, is a single domain *in zero field*, no matter how greatly its size exceeds the micromagnetics critical size. This assumption is certainly valid for the infinite cylinder, which has a zero demagnetizing factor along its axis, hence no magnetostatic energy when uniformly magnetized along its axis, hence no tendency to break up into domains. However, this assumption is invalid for the prolate spheroid and sphere, which have finite demagnetizing fields. But, given the severe restraints of this exacting discipline, micromagnetics must make this assumption because it has not yet found any rigorous way of calculating the largest size a spherical particle, say, can attain in zero field before breaking up into domains; i.e., it cannot calculate the critical size D_s according to static definition (1). When it does, incidentally, micromagnetics will have to invent some descriptive term for a particle with a size in between the critical sizes given by definitions (1) and (2). For example, a spherical particle of iron 250 Å in diameter is a single domain in zero field but not a single domain, according to micromagnetics, during reversal.

What experimental evidence is there for reversal by curling? The electrodeposited, elongated iron particles examined by Luborsky [11.9], and mentioned earlier, had c/a values larger than 10 and diameters $D\,(=2a)$ ranging from 125 to 215 Å, corresponding to D/D_0 values of 1.04 to 1.79. Their coercivities H_{ci} were all about 1800 Oe ($h_{ci} = 0.17$), as shown in Fig. 11.9. This size independence of the coercivity means that these particles reverse by fanning rather than curling. This is not unexpected, because the peanutlike shape of these particles departs rather far from that of a cylinder.

In order to deal with particles more nearly resembling the idealized models of micromagnetics, Luborsky and Morelock [7.4] examined the properties of iron and iron-cobalt alloy *whiskers.* These have straight, smooth sides and are structurally the most nearly perfect of all crystals. The coercivities were found to be very size dependent over the large range of whisker diameters investigated, from about 200 Å to 10^6 Å (100 μ), and Luborsky and Morelock concluded from the nature of this size dependence that whiskers thinner than 1000 Å reversed by curling. The behavior of thicker whiskers deviated from the predictions of curling theory, so that some other reversal mechanism must be acting instead of, or in addition to, curling.

The discussion of incoherent reversal mechanisms in this section hardly does justice to the large amount of theoretical and experimental work done, by the authors cited and others, to examine and test these hypotheses. In particular, two lines of investigation in addition to those mentioned above have been pursued:

1. *Angular variation of the coercivity.* The way which h_{ci} varies with the angle α between the applied field and the axis of the specimen depends in a predictable way on the mode of reversal. Therefore, a measurement of this variation offers a means of distinguishing between various possible modes. (The discussions of this section were limited to the case of $\alpha = 0$.)

2. *Rotational hysteresis.* (See also Section 13.3.) Suppose a specimen is rotated 360° in an applied field H of constant magnitude. Then, if any irreversible processes (of spin rotation or domain wall motion) occur during this rotation, there will be a rotational hysteresis loss W_r. The value of W_r, which is a function of H, can be found from a torque curve; W_r is equal to the average ordinate of the curve multiplied by 2π. The *rotational hysteresis integral* R is defined by

$$R = \int_0^\infty \frac{W_r}{M_s}\, d\left(\frac{1}{H}\right). \tag{11.18}$$

The quantity R is dimensionless and, for single-domain particles, the value of R depends in a predictable way on the mode of reversal. The measurement of R thus affords still another means of distinguishing one mode from another. Details are given by Jacobs and Luborsky [11.13].

11.5 MAGNETIZATION REVERSAL BY WALL MOTION

Relatively large particles can have quite substantial coercivities, if their crystal anisotropy constant K is large. For example, the curves for barium ferrite ($K = 3.3 \times 10^6$ erg/cm^3) in Fig. 11.1 show that a coercivity H_{ci} of about 3000 Oe can be obtained in particles 1 μ in diameter. Yet we estimated, in Problem 9.2, that the critical size D_s for single domains in barium ferrite was 730 Å. Therefore, 1μ (= 10,000 Å) particles must surely be multidomain. Two problems then arise:

1. Why are the coercivities so high, if domain-wall motion is a relatively easy process, as is usually assumed?
2. Why does the coercivity decrease as the particle size increases?

We shall look for answers to these questions in this section. Note that the discussion is restricted to particles owing their coercivity to crystal, rather than shape, anisotropy. The properties of such particles have been surveyed for a number of materials by Becker [11.14].

The two questions posed above are made even more pointed by some micromagnetics calculations made by Brown. Consider a *perfect* crystal in the form of a spheroid, with the spheroid axis and the easy crystal axis parallel to the z axis. It is saturated in the $+z$ direction and is therefore a single domain. The saturating field is then removed, and the crystal is assumed to remain a single domain. A reversing field is then applied in the $-z$ direction. Brown showed that the field required for reversal is given by

$$H_{ci} = \frac{2K}{M_s}, \tag{11.19}$$

for a crystal *of any size*, whether larger or smaller than D_s. (Here H_{ci} is a particular value of the true field H acting on the crystal, where H is the sum of the applied field H_a and the demagnetizing field H_d. Both of these are acting in the $-z$ direction, and negative signs are understood.) This value of H_{ci} is just the value obtained in Section 9.13 for coherent rotation. For barium ferrite $2K/M_s = 2(3.3 \times 10^6)/380 = 17{,}000$ Oe. It would not be surprising to find this value approached in single-domain particles smaller than D_s, for example, in 100 Å particles. In fact, coercivities as large as 11,000 Oe have been observed in barium ferrite [11.14]. But in particles larger than D_s the coercivities are much less than expected; in 1μ particles H_{ci} is only 3000 Oe, and it decreases still further as the particle size increases (Fig. 11.1). This large discrepancy between theory and experiment for large particles is known as "Brown's paradox" [11.4, 9.15].

What is the mechanism of magnetization reversal in particles larger than D_s, i.e., in particles large enough to contain domain walls? It cannot be coherent rotation, because the observed coercivities are much less than $2K/M_s$. The only other alternative is that one or more domain walls are *nucleated* in the particle; the wall then moves through the particle and reverses its magnetization. Suppose we consider this wall nucleation process in some detail, with reference to the perfect spheroidal crystal mentioned earlier, now assumed to be larger than D_s. When saturated in the $+z$ direction, it is a single domain with all spins parallel. When a reversing field H acts in the $-z$ direction, wall formation presumably begins at the surface, at the point B, say, of Fig. 11.11 (a). The spins along a line AB intersecting the surface are shown in (b), with the spin on the surface atom rotated 45° out of the xz plane by the field. This rotation tends to rotate the spin on the next atom, because of exchange coupling, and (c), (d), and (e) show successive steps in the rotation. In (e) a complete wall, assumed for simplicity to be only four atoms thick, has been formed. In (f) the wall has moved inward, producing a new domain 2 magnetized antiparallel to the original domain 1. Because the crystal has been assumed perfect, there are no hindrances to the motion of the newly formed wall; under pressure of the nucleating field, it flashes across the crystal in one huge Barkhausen jump, and the magnetization reversal is completed. (The just-nucleated wall will move entirely across the crystal only if the *applied* field exceeds the demagnetizing field. See the discussion

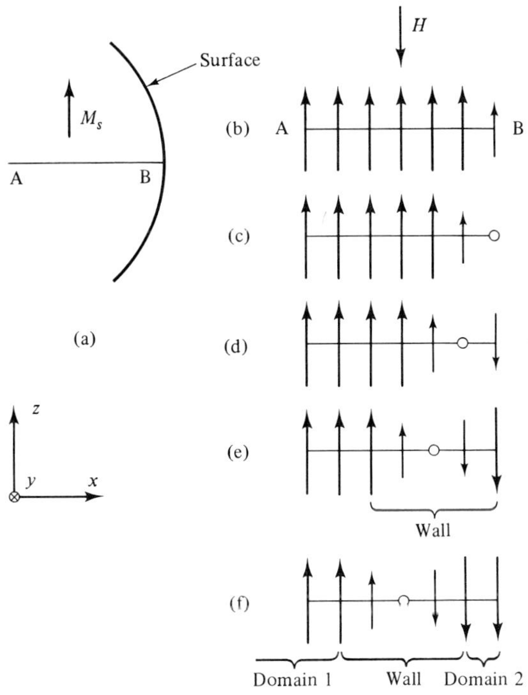

Fig. 11.11 Domain wall nucleation at a crystal surface. Only the projections of the spins on the *xz* plane are shown.

of Fig. 11.16.) The hysteresis loop is rectangular, just at it would be if all the spins reversed coherently. But this whole process of wall nucleation and wall motion has its beginnings in spin rotation, on a few atoms at the surface, against the crystal anisotropy forces, and this requires a field equal to $2K/M_s$. The essential point to note is that the process of nucleating a domain wall in a *perfect* crystal is just as difficult as coherent rotation; either process requires spin rotation against the same anisotropy forces. (A close analogy exists in the mechanical behavior of crystals. If one part of a crystal moves as a unit relative to another part, in response to an applied stress, the process is called block slip; it is the mechanical analogue of coherent spin rotation. Calculations show that the stress required to cause block slip in a *perfect* crystal is very high. It is, in fact, just as high as the stress required to nucleate a dislocation at the surface of the crystal; once nucleated, the dislocation moves easily across the crystal, and the net result is the same as that of block slip, namely, motion of one part of the crystal through unit distance relative to the other part. Slip by dislocation motion takes place by an atom-by-atom motion, just as magnetization reversal by wall motion takes place by spin-by-spin rotation. Both processes are hard to imitate in a perfect crystal. Real crystals already contain dislocations and other imperfections, and the stress required to cause slip in them is very low.)

We might digress to examine a tacit assumption made in earlier parts of this book, namely, that a saturated crystal larger than D_s will break up into domains when the saturating field is reduced to zero, in order to lower its magnetostatic energy. In a *perfect* crystal this may not be true, because the difficulty of wall nucleation is an energy barrier between the single-domain state and the lower-energy, multidomain state. The crystal will spontaneously divide into domains only if

$$H_d = N_d M_s > \frac{2K}{M_s}, \tag{11.20}$$

$$M_s > \left(\frac{2K}{N_d}\right)^{1/2}, \tag{11.21}$$

where N_d is the demagnetizing factor along the axis of magnetization z. For a given material in the form of a spheroid, the process of subdivision into domains is easiest when N_d has a maximum value of 4π, in a thin oblate spheroid (disk) with the plane of the disk normal to z. For such a disk of barium ferrite, $(2K/N_d)^{1/2} = [(2)(3.3 \times 10^6)/4\pi]^{1/2} = 720$. Because this is larger than M_s (= 380), a perfect crystal of barium ferrite in the form of a spheroid cannot spontaneously break up into domains, whatever its size or shape. (The same is not true of a high-M_s, low-K material like iron, as shown in Problem 11.6. A perfect spherical iron crystal will form domains, but one in the form of a prolate spheroid of sufficiently high axial ratio will not.) The demagnetizing field of the thin disk of barium ferrite ($= 4\pi M_s = 4800$ Oe) amounts to a substantial fraction of the theoretical coercivity of 17,000 Oe. Thus the applied field required to reverse the magnetization is 17,000 − 4,800 = 12,000 Oe. Inasmuch as actual crystals of barium ferrite tend to be plate-shaped, with easy axis normal to the plate, the figure of 12,000 Oe is a more realistic value of the theoretical upper limit of coercivity for reversal by wall motion than 17,000 Oe, which applies to a crystal so shaped as to have a negligible demagnetizing field.

Brown's paradox is based on calculations relative to a perfect crystal in the form of a perfect spheroid. Real crystals contain imperfections and are irregularly shaped. Interior imperfections include such defects as dislocations, solute atoms, interstitials, and vacancies; not only may the shape depart from spheroidal in various ways, but the surface is usually marred by pits, bumps, steps, cracks, and scratches. Brown's paradox can only be resolved by invoking the aid of some of these imperfections in lowering the field required to nucleate a domain wall. The condition for wall nucleation in a single particle is that

$$H_a + H_d > \frac{2K}{M_s}. \tag{11.22}$$

The applied field required may be much less than expected if (a) H_d or M_s is larger than normal, or (b) K is smaller than normal. The maximum value of H_d is $4\pi M_s$ only in a perfect spheroid; near sharp corners, for example, H_d can approach infinity. Because the value of M_s is determined by the magnetic moment per atom and the exchange coupling between adjacent atoms, it may change locally, up or down, in the vicinity of vacancies, interstitials, or the core of dislocations where the strains are very large. Similarly, the local value of K, which is due to spin-orbit coupling, may be changed by imperfections or small-scale inhomogeneities in the chemical composition of the particle.

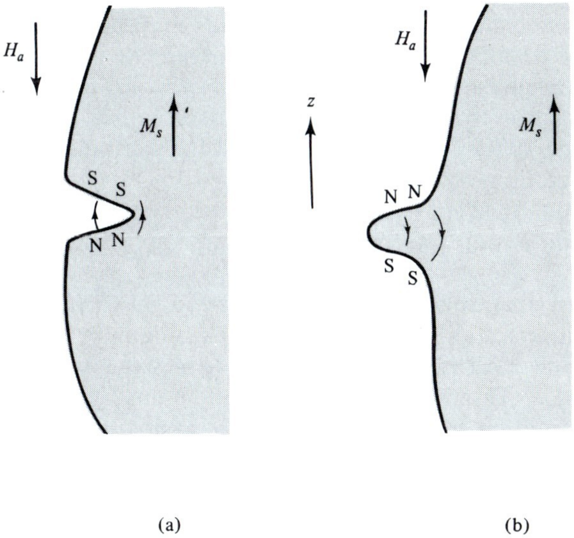

Fig. 11.12 Local fields near pits and bumps

Of these several possibilities, local variations in the demagnetizing field H_d are usually assumed to be the most likely nucleators of domain walls. Among surface imperfections, pits are no help at all. As shown in Fig. 11.12 (a), the field at the base of the pit, due to the poles on the pit walls, acts to increase the magnetization at the base of the pit and to counteract the field H_a which is trying to reverse the magnetization. (Mechanical analogy: a pit or notch acts as a stress raiser, increasing the stress at the base of the notch above the average stress applied to the specimen.) On the other hand, a surface bump, shown in (b), produces local fields which act to demagnetize the particle and aid H_a, as pointed out by Kittel and Galt [9.2].

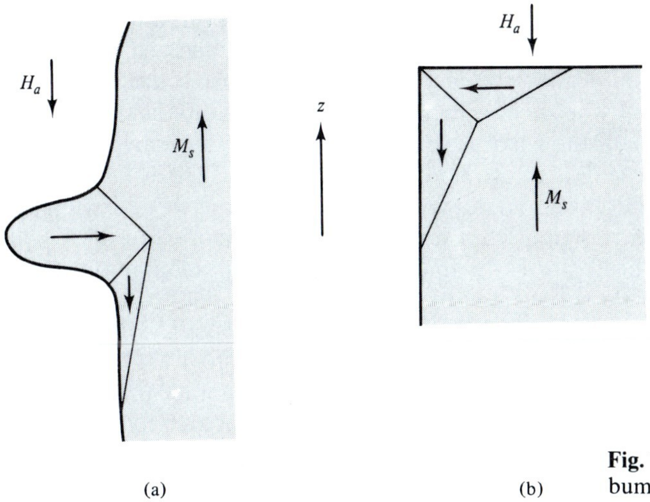

Fig. 11.13 Closure domains near bumps and sharp corners.

The magnetostatic energy associated with these local fields may be so large that the system is more likely to form closure domains instead, as suggested by Fig. 11.13 (a); the domain arrangement indicated would decrease the density of, but not eliminate, the free poles on the surface of the bump. Similarly, closure domains may be formed at sharp corners, as in (b), suggested by Shtrikman and Treves [11.15]. In fact, closure domains have been observed by Fowler *et al.* [11.16] at the square tips of iron whiskers held in a "saturated" state by a field applied along their axis, and these domains persisted even at applied fields of several thousand oersteds.

These considerations suggest that small domains of reversed magnetization may always be present near imperfections, even in specimens nominally "saturated." It is difficult to determine experimentally the difference between $M = M_s$ and, for example, $M = 0.995\ M_s$. This means that reverse domains occupying a volume fraction of the order of a few tenths of a percent can easily remain undetected in a measurement of magnetization. If so, then magnetization reversal becomes a question, not of nucleating walls, but of freeing or "unpinning" walls that already exist.

The role of imperfections in nucleating magnetization reversals was nicely demonstrated by experiments of DeBlois and Bean [11.17, 11.18] on single whiskers of iron. These whiskers were about 1 cm long, from 1 to 20μ thick, and had $\langle 100 \rangle$ axes. Each whisker was enclosed for protection in a quartz capillary which was then inserted in a tiny solenoid, scarcely longer than 1 cm, as shown in Fig. 11.14 (a). An isolated whisker appears just above the solenoid, and the bits of graph paper with 1-mm squares give the scale. An even shorter reversing coil, 13 turns and 0.8 mm long, was wound on the center of the solenoid. (The very small scale of this experiment must surely qualify it as "experimental micromagnetics" or, at the very least, minimagnetics.) The whisker is initially saturated in one direction by applying a 50-Oe pulse with the solenoid; it remains saturated in that direction after the field has returned to zero. A large pulse field is then applied in the opposite direction with the short central coil. If the region of the whisker inside this coil reverses its magnetization, an emf will be induced in the solenoid and can be detected with an oscilloscope. If no reversal is observed, the magnitude of the reverse pulse field is increased until one is observed. The capillary and whisker is then moved a short distance in the solenoid to expose another portion of the whisker to the central coil. An example of such measurements along the length of an 8 mm whisker, which varies in thickness from 19μ to 1μ along its length, is shown in (b). The coercivity is extremely variable along the length, ranging from less than 50 to a maximum of 504 Oe at a point where the whisker is 12.4μ thick. Microscopic examination of the whisker showed that visible surface defects corresponded to the positions of the local minima in the nucleating field. The converse was not always true.

The theoretical coercivity of iron is $2K/M_s = 2(4.8 \times 10^5)/1714 = 560$ Oe. The maximum observed value of 504 Oe is 0.90 of the theoretical, the largest fraction so far attained in any material. (Incidentally, this approach to the theoretical coercivity for crystal anisotropy is much closer than has been attained in materials owing their coercivity to shape anisotropy, where the largest observed

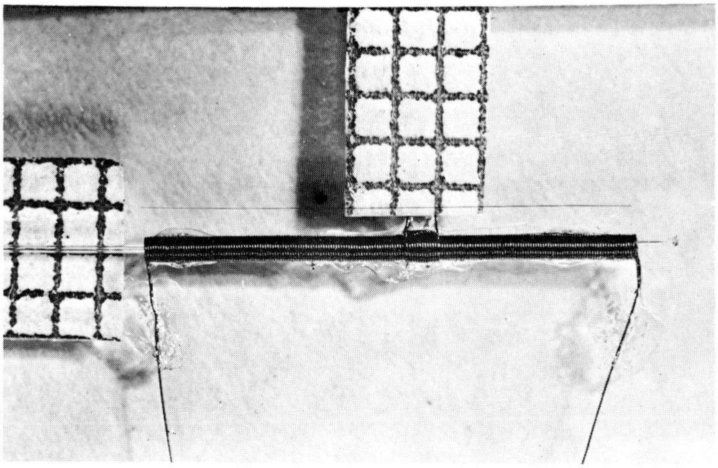

(a)

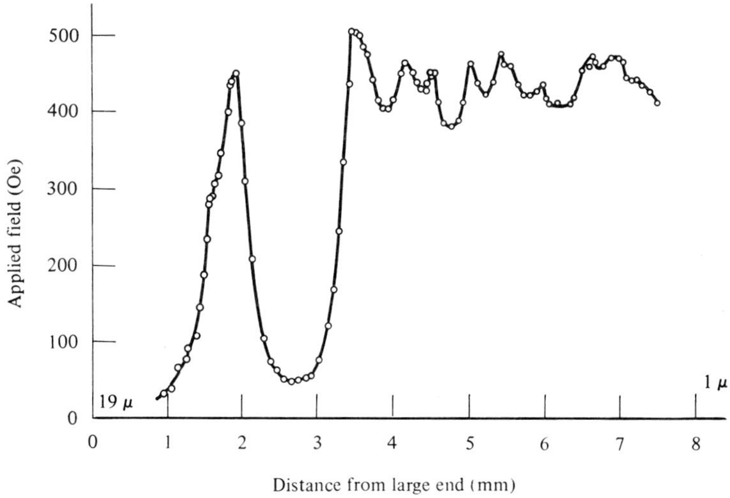

(b)

Fig. 11.14 Measurement of magnetization reversal in iron whiskers. (a) Apparatus. (b) Applied field required for reversal at various points along the length of the whisker. DeBlois and Bean [11.17] value of H_{ci} is only 18 percent of $2\pi M_s$.) These experiments therefore support the view that a perfect crystal does have a coercivity of $2K/M_s$ and that imperfections can drastically lower this value.

(The experiments just described relate to the difficulty of nucleating, or unpinning, a domain wall and give no information on the field required to move a wall, once it is free, or on its velocity. In other work, however, DeBlois [11.18] observed that a field as small as 0.008 Oe could move a domain wall in an iron whisker; this field is about 10^{-5} times that required to nucleate a wall in the more nearly perfect portions of a whisker. This result is not too surprising, because whiskers are the most nearly perfect crystals known; they therefore contain a minimum amount of the usual hindrances to wall motion, namely, microstress and inclusions. DeBlois also measured domain wall velocity and found it very high. This velocity depends on the applied field; at 10 Oe it was about 1 km/sec or 2200 mph.)

Further evidence of the role of surface imperfections as nucleating points comes from measurements of coercivity before and after polishing. The coercivity of iron whiskers can be increased by several hundred oersteds by electropolishing [11.19] and that of YCo_5 particles by several thousand oersteds by chemical polishing [11.20]. This means that there is not a single curve relating coercivity to particle size for a given material, as in Fig. 11.1, but a whole family of curves, each for a different degree of surface roughness.

A most unusual effect was uncovered by Becker, who showed that the coercivity of particles increased, not only with increasing surface smoothness, but also with increasing magnitude of the field used to magnetize the particles [11.21]. Now a dependence of the *coercive force*, as defined in Section 1.8, on the maximum magnetizing field H_m is quite normal, because the coercive force is the reverse field required to reduce M or B to zero after the specimen has been partially magnetized in the forward direction. Thus in Fig. 11.15 (a), an increase in H_m from 1 to 2 increases the intrinsic coercive force H_{ci} from 1 to 2. But that the *coercivity*, measured after saturation in the forward direction, should depend on H_m is quite unexpected; whether H_m has the value 3 or 4, the coercivity should have the value 3. Actually, the coercivity is found to increase from 3 to 4 when H_m is increased from 3 to 4. The experimental data are shown in (b) for assemblies of particles of $SmCo_5$ in various size ranges. The easy axes of these particles (the c axes of their hexagonal unit cells) were aligned with the field, and saturation is essentially complete for fields of 10 k Oe and above. Yet substantial increases in H_{ci} are observed, even for changes in H_m beyond 20 kOe. (One might also take the position that coercivity is the limiting value of the coercive force as H_m is increased. Then the quantity plotted in Fig. 11.15 (b) is coercive force, and the fields H_m are not really saturating the material.) The retentivity M_r also increases with H_m but the effect is small. Thus, when H_{ci} doubles in size, the increase in M_r is only about 10 percent.

If we regard the high coercivity of particles such as these as due to the difficulty of freeing domain walls that are pinned or trapped at imperfections, then the dependence of coercivity on H_m must mean that the "strength" of a pinning site must depend on how hard the main wall is driven into it by the magnetizing field in the forward direction. The mechanism of this behavior is not understood.

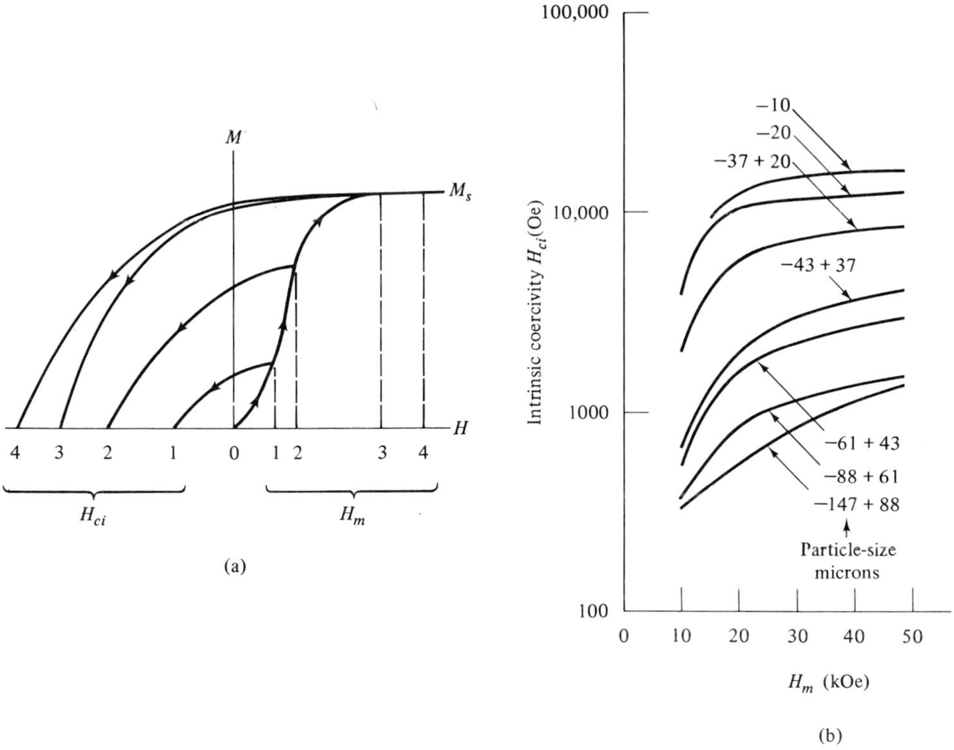

Fig. 11.15 Dependence of coercivity on maximum value H_m of magnetizing field. (a) Schematic. (b) Experimental results on aligned particles of $SmCo_5$. Becker [11.21]

(In terms of Fig. 9.36, where wall motion was discussed in terms of energy changes in the system, pinning sites must correspond to potential energy wells with almost vertical sides, and the sides must become more nearly vertical as H_m increases.) If a pinning site is an arrangement of closure domains near an imperfection, as suggested in Fig. 11.13, it is conceivable that the particular configuration of the closure domains, and therefore the strength of the pinning site, could depend on the magnitude H_m of the field that originally drove the main wall into the vicinity of the imperfection.

Becker has suggested [11.20, 11.21] that a single particle can have a number of pinning sites, each characterized by the field H_n necessary to nucleate, or unpin, a domain wall, and that the particular site that operates depends on the value of the previously applied field H_m. The predicted behavior of a single particle is shown in Fig. 11.16 (a) where M is plotted against the applied field H_a. If walls are present and free to move, the hysteresis loop will have a low coercivity H_w, the wall-motion coercivity. If the interior defect concentration is so low

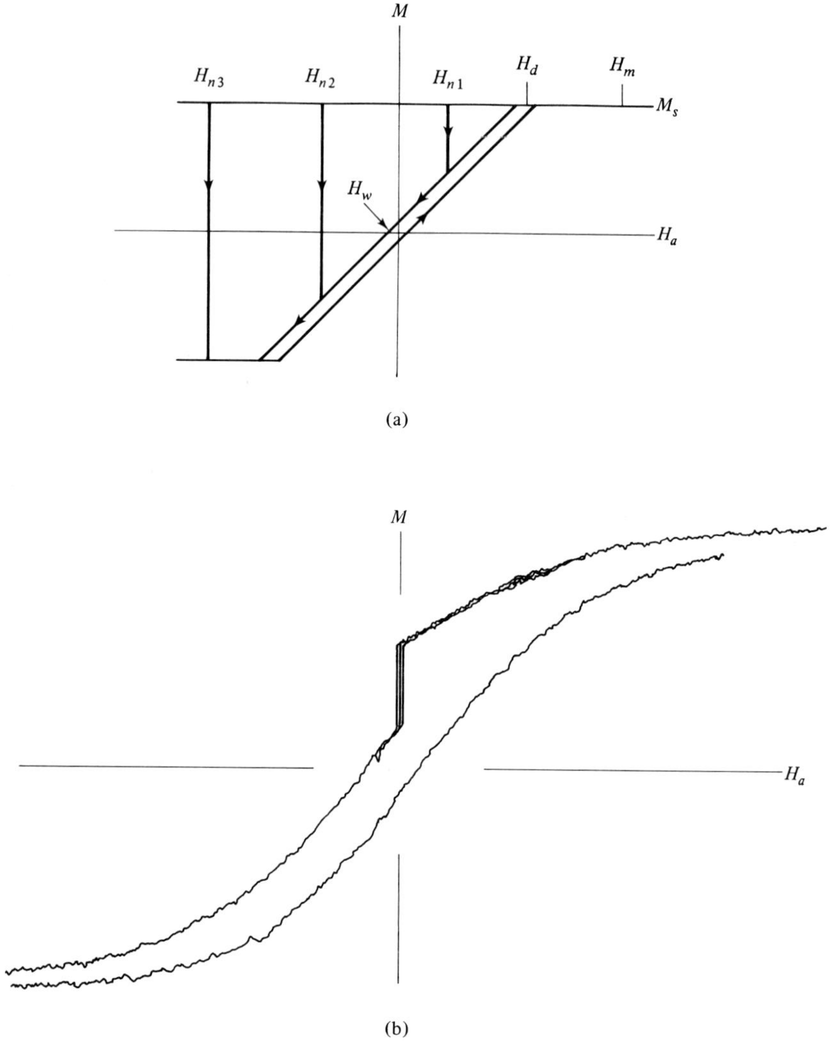

Fig. 11.16 Hysteresis loops of a single particle. (a) Theoretical. (b) Observed for a particle of $SmCo_5$. Becker [11.20]

that there is negligible resistance to wall motion, then the loop will have a negligible width and a slope of $1/N_d$, where N_d is the demagnetizing factor of the particle. Saturation is achieved at $H_a = H_d$. If the field is increased to H_m, walls are pinned so strongly that nucleating fields H_n, which depend on H_m, are needed to cause reverse wall motion. If a wall nucleates at H_{n1}, it would immediately move until M drops to the upper branch of the wall-motion hysteresis loop, which would then describe the magnetization as H_a was further decreased. If a wall did not nucleate until H_{n2}, this field would also be the coercivity. If the wall nucleated at H_{n3}, the loop would be square.

With a vibrating-sample magnetometer, Becker was able to measure the hysteresis loop of a single particle of $SmCo_5$, about 0.2 by 0.5 mm in size, and the result is shown in Fig. 11.16 (b). A field of 18 kOe was applied before drawing the curves shown. The narrow portion of the loop is due to wall motion, and it is curved because of the irregular shape of the sample. The four closely spaced vertical lines are four magnetization jumps that took place after four successive magnetizations to +18 kOe. If H_m was less than 13 kOe, no such jumps were observed. When H_m was between 17 and 28 kOe, jumps occurred at H_a equal to about zero, as shown. When H_m was increased to 30 kOe, the jump did not take place until H_a was -2000 Oe. (These values refer to the upper branch of the loop; the values for the lower loop were similar but not exactly the same; in fact, for the loop shown, no jumps are evident for $H_m = -18$ kOe.) These observations demonstrate that the strong dependence of H_{ci} on H_m for assemblies of particles (Fig. 11.15 b) is an inherent property of the individual particles, caused by a variation in the strength of pinning sites with H_m. (The fact that a crystal can have more than one wall-nucleating field, for a constant value of H_m, has been demonstrated for barium ferrite by Kooy and Enz [11.22].)

As mentioned in Section 11.3, the coercivity of an assembly of multidomain particles varies inversely with the diameter D of the particles, in accordance with Eq. (11.1). This variation can be understood, at least qualitatively, with the following assumptions:

1. The particles are uniaxial, with their easy axes all parallel to the reversing field.

2. When a free 180° wall is formed by a reverse field H_n, either by nucleation or unpinning, it moves completely through the particle and reverses its magnetization, which means that H_n must exceed H_d.

3. Walls are nucleated only at surface defects.

4. The "strength" of these defects varies in such a way that all values of H_n from zero up to an upper limit of $2K/M_s$ are equally probable. (A defect is strong if a small field can nucleate or unpin a wall at the defect.)

5. The probability of a defect existing, per unit surface area of particle, is constant.

If the particles are large, the probability that a particular particle has a strong defect is high, because of the large surface area per particle. That particle will therefore reverse at a low H_n. (The particle considered can be expected to have strong defects as well as weak ones, but the strongest defect present determines the reversing field H_n.) Moreover, the reversal of this particle will cause a relatively large change in the magnetization M of the assembly because of the large volume per particle. For example, if there are only eight particles, the reversal of one of them would cause M to decrease from M_s to $3\,M_s/4$, as in Fig. 11.17 (a). The coercivity H_{ci} of the assembly (the field required to reverse half the particles) is expected to be low because of the high probability that many particles will contain strong defects. Small particles, on the other hand, each have a small

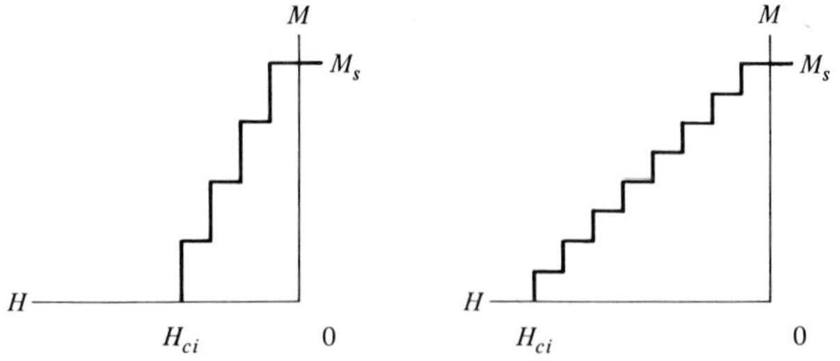

(a) Large particles

(b) Small particles

Fig. 11.17 Demagnetization curves of assemblies of particles.

surface area and therefore fewer defects. The range of probable H_n values per particle is thus more restricted; some particles will contain only weak defects and some only strong ones. The demagnetization curve is therefore expected to resemble (b), with a large coercivity.

11.6 SUPERPARAMAGNETISM IN FINE PARTICLES

Consider an assembly of uniaxial, single-domain particles, each with an anisotropy energy density $E = K \sin^2 \theta$, where K is the anisotropy constant (in erg/cm^3) and θ the angle between M_s and the easy axis. If the volume of each particle is V, then the energy barrier ΔE that must be overcome before a particle can reverse its magnetization is KV ergs. Now in any material, fluctuations of thermal energy are continually occurring on a microscopic scale. In 1949 Néel [11.23] pointed out that, if single-domain particles became small enough, KV would become so small that energy fluctuations could overcome the anisotropy forces and spontaneously reverse the magnetization of a particle from one easy direction to the other, even in the absence of an applied field. Each particle has a magnetic moment $\mu = M_s V$ and, if a field is applied, the field will tend to align the moments of the particles, whereas thermal energy will tend to disalign them. This is just like the behavior of a normal paramagnetic, with one notable exception. The magnetic moment per atom or ion in a normal paramagnetic is only a few Bohr magnetons. But a spherical particle of iron, say, 50 Å in diameter, contains 5560 atoms and has the relatively enormous moment of $(5560)(2.2) = 12{,}000\ \mu_B$. As a result, Bean has coined the very apt term *superparamagnetism* to describe the magnetic behavior of such particles. This subject has been reviewed by Bean and Livingston [11.24] and Jacobs and Bean [11.2].

If $K = 0$, so that each particle in the assembly has no anisotropy, then the moment of each particle can point in any direction, and the classical theory of paramagnetism will apply. Then the magnetization curve of the assembly, consisting of magnetic particles in a nonmagnetic matrix, will be given by Eq. (3.19), or

$$M = n\mu\, L(a). \tag{11.23}$$

Here M is the magnetization of the assembly, n the number of particles per unit volume of the assembly, $\mu\,(= M_s V)$ the magnetic moment per particle, $a = \mu H/kT$, and $n\mu = M_{sa}$ = saturation magnetization of the assembly. (As a consequence of the large value of μ, the variable $a = \mu H/kT$ can assume large values at ordinary fields and temperatures. Thus the full magnetization curve, up to saturation, of superparamagnetic particles can be easily observed, whereas very high fields and low temperatures are required for ordinary paramagnetic materials, as we saw in Chapter 3.) At the other extreme, if K is finite and the particles are aligned with their easy axes parallel to one another and to the field, then the moment directions are severely quantized, either parallel or antiparallel to the field, and quantum theory will apply. Then, in accordance with Eq. (3.40),

$$M = n\mu \tanh a, \tag{11.24}$$

where the hyperbolic tangent is just a special case of the Brillouin function. In the intermediate case of nonaligned particles of finite K, neither Eq. (11.23) nor (11.24) will apply. Nor will these equations be obeyed if, as is usually true, all particles in the assembly are not of the same size, because then the moment per particle is not constant. Nevertheless, two aspects of superparamagnetic behavior are always true:

1. Magnetization curves measured at different temperatures superimpose when M is plotted as a function of H/T.
2. There is no hysteresis; i.e., both the retentivity and coercivity are zero. We are therefore dealing with particles having diameters smaller than the critical value D_p of Fig. 11.2.

Both of these features are illustrated by the measurements on fine iron particles dispersed in solid mercury shown in Fig. 11.18. The curves for 77° and 200°K in (a) show typical superparamagnetic behavior, and they superimpose when plotted as a function of H/T, as shown in (b). But at 4.2°K the particles do not have enough thermal energy to come to complete thermal equilibrium with the applied field during the time required for the measurement, and hysteresis appears. (The 4.2°K curve is only half of the complete hysteresis loop.)

Hysteresis will appear and superparamagnetism disappear, when particles of a certain size are cooled to a particular temperature, or when the particle size, at constant temperature, increases beyond a particular diameter D_p. To determine these critical values of temperature or size, we must consider the rate at which thermal equilibrium is approached, and we will do this first for the case of zero applied field. Suppose an assembly of uniaxial particles has been brought to some initial state of magnetization M_i by an applied field, and the field is then turned off at time $t = 0$. Some particles in the assembly will immediately reverse their magnetization, because their thermal energy is larger than the average,

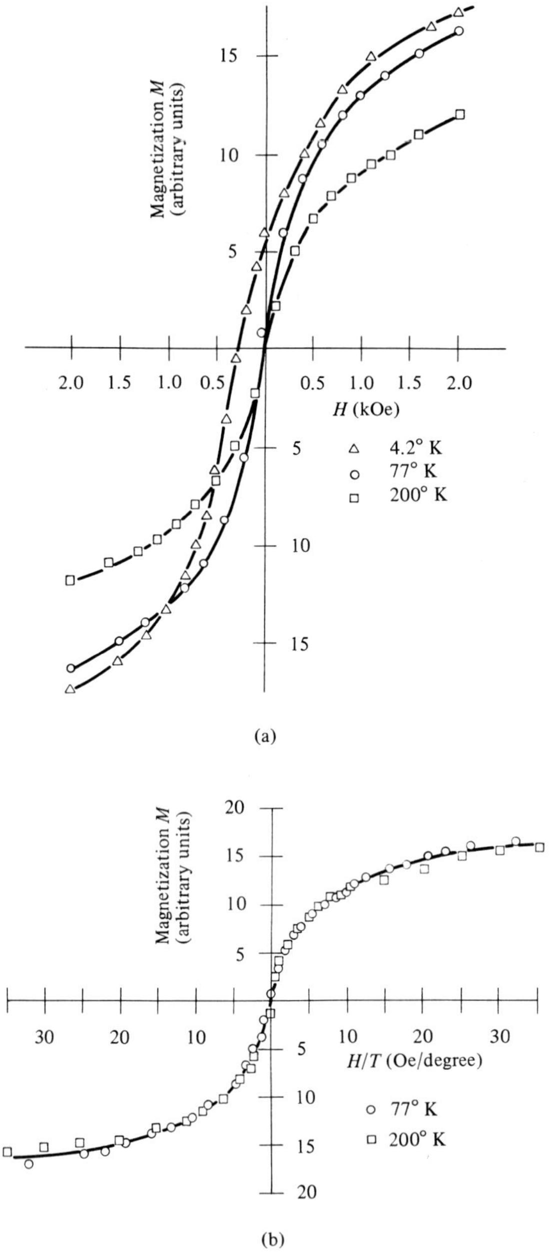

Fig. 11.18 Magnetization curves of iron particles 44 A in diameter. Jacobs and Bean [11.2].

and the magnetization of the assembly will begin to decrease. The rate of decrease at any time will be proportional to the magnetization existing at that time and to the Boltzmann factor $e^{-KV/kT}$, because this exponential gives the probability that a particle has enough thermal energy to surmount the energy gap $\Delta E = KV$ required for reversal. Therefore,

$$-\frac{dM}{dt} = f_0 M e^{-KV/kT} = \frac{M}{\tau}. \tag{11.25}$$

Here the proportionality constant f_0 is called the frequency factor and has a value of about 10^9 sec^{-1}; this value is slightly field dependent, but this dependence will be ignored here. The constant τ is called the relaxation time. (For spherical particles of a cubic substance the energy barrier is not KV but $KV/4$ if K is positive, with $\langle 100\rangle$ easy directions, or $KV/12$ if K is negative and $\langle 111\rangle$ are easy directions.) To find how the remanence M_r decreases with time we rearrange the terms of Eq. (11.25) and integrate:

$$\int_{M_i}^{M_r} \frac{dM}{M} = -\int_0^t \frac{dt}{\tau}, \tag{11.26}$$

$$\ln \frac{M_r}{M_i} = -\frac{t}{\tau}, \tag{11.27}$$

$$M_r = M_i e^{-t/\tau}. \tag{11.28}$$

The meaning of τ is now apparent; it is the time for M_r to decrease to $1/e$ or 37 percent of its initial value. (If the initial state is the saturated state, then $M_i = M_s$.) From Eq. (11.25) we have

$$\frac{1}{\tau} = f_0 e^{-KV/kT}. \tag{11.29}$$

Because the particle volume V and the temperature T are both in the exponent, the value of τ is strongly dependent on these quantities. For example, a spherical particle of cobalt which is 68 Å in diameter has a relaxation time τ at room temperature, given by Eq. (11.29), of only 10^{-1} sec. An assembly of such particles would therefore reach thermal equilibrium ($M_r = 0$) almost instantaneously; M_r would appear to be zero in any normal measurement and the assembly would be superparamagnetic. On the other hand, if the particle diameter is increased to only 90 A, the value of τ jumps to 3.2×10^9 sec or 100 years. An assembly of such particles would be very stable, with M_r essentially fixed at its initial value.

Because τ varies so very rapidly with V, it follows that relatively small changes in τ do not produce much change in the corresponding value of V. Thus it becomes possible to define, rather closely, an upper limit V_p for superparamagnetic behavior by rather arbitrarily letting a value of τ equal to 100 sec mark the transition to stable behavior. (This is roughly the time required to measure the remanence of a specimen. If this time were increased to, say, 1000 sec, the corresponding value of V_p would be only slightly increased.) With this value of τ, Eq. (11.29) becomes

$$10^{-2} = 10^9 e^{-KV_p/kT}, \tag{11.30}$$

whence KV_p/kT equals 25. Therefore, the transition to stable behavior occurs when the energy barrier becomes equal to $25kT$. For uniaxial particles,

$$V_p = \frac{25kT}{K}, \tag{11.31}$$

and the corresponding diameter D_p can be calculated for any given particle shape. It is this value which in Fig. 11.2 marks the upper limit of superparamagnetism, and the terms "stable" and "unstable" in that diagram refer to particles which have relaxation times longer and shorter, respectively, than 100 sec. The value of D_p for spherical cobalt particles is 76 Å at room temperature.

For particles of constant size there will be a temperature T_B, called the *blocking temperature*, below which the magnetization will be stable. For uniaxial particles and the same criterion of stability,

$$T_B = \frac{KV}{25k}. \tag{11.32}$$

The iron particles of Fig. 11.18 must have a blocking temperature between 77° and 4.2 K, because they ceased to behave superparamagnetically on cooling through this range. The results of Problem 11.9 show that these particles must have been slightly elongated, rather than spherical.

Figure 11.19 summarizes some of these relations for spherical cobalt particles. The variation of D_p with temperature shows that 20°C is the blocking temperature for particles 76 Å in diameter; above this temperature particles of this size have enough thermal energy to be superparamagnetic, while below this temperature they are stable and show hysteresis. The other curve shows how the relaxation time of 76 Å particles varies with temperature; at 20°C τ is 100 sec.

The next point to consider is the effect of an applied field on the approach to equilibrium. Assume an assembly of uniaxial particles with their easy axes parallel to the z axis. Let the assembly be initially saturated in the $+z$ direction. A field H is then applied in the $-z$ direction, so that M_s in each particle makes an angle θ with $+z$. The total energy per particle is then

$$E = V(K \sin^2 \theta + HM_s \cos \theta), \tag{11.33}$$

which is the same as Eq. (9.40) with $\alpha = 180°$. The energy barrier for reversal is the difference between the maximum and minimum values of E, and it is left to the reader to show that this barrier is

$$\Delta E = KV\left(1 - \frac{HM_s}{2K}\right)^2. \tag{11.34}$$

The barrier is therefore reduced by the field, as shown by the curves of Fig. 9.37. Particles larger than D_p are stable in zero field and will not thermally reverse in 100 sec. But when a field is applied, the energy barrier can be reduced to $25\,kT$, which will permit thermally activated reversal in 100 sec. This field will be the coercivity H_{ci}, given by

$$\Delta E = KV\left(1 - \frac{H_{ci}M_s}{2K}\right)^2 = 25\,kT. \tag{11.35}$$

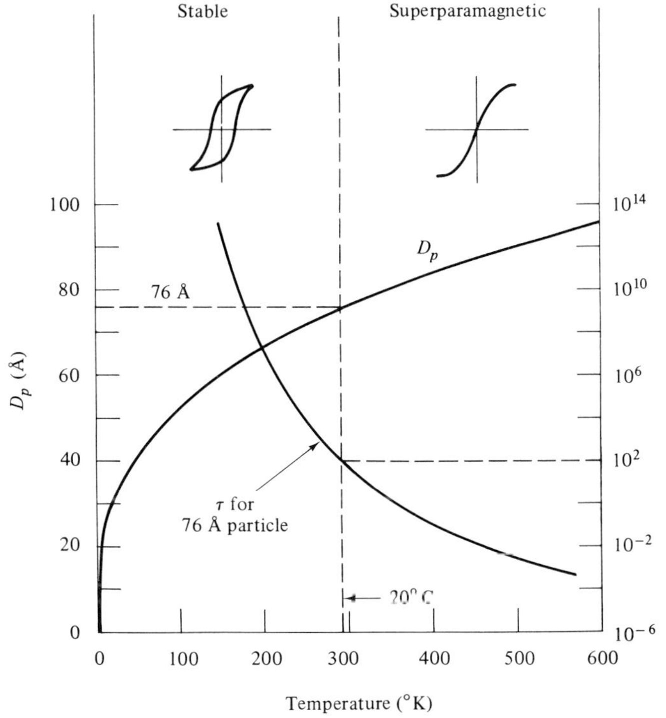

Fig. 11.19 Temperature dependence of the relaxation time τ for spherical cobalt particles 76 Å in diameter and of the critical diameter D_p of spherical cobalt particles.

The solution of this is

$$H_{ci} = \frac{2K}{M_s}\left[1 - \left(\frac{25kT}{KV}\right)^{1/2}\right]. \tag{11.36}$$

When V becomes very large or T approaches zero, H_{ci} approaches $2K/M_s$ as it should, because $2K/M_s$ is the coercivity when the field is unaided by thermal energy. If we put this limiting value equal to $H_{ci,0}$ and substitute Eq. (11.31) into (11.36), we obtain the reduced coercivity

$$h_{ci} = \frac{H_{ci}}{H_{ci,0}} = 1 - \left(\frac{V_p}{V}\right)^{1/2} = 1 - \left(\frac{D_p}{D}\right)^{3/2}. \tag{11.37}$$

The coercivity therefore increases as the particle diameter D increases beyond D_p, as indicated qualitatively in Fig. 11.2. A quantitative comparison of Eq. (11.37) with experiment appears in the lower part of Fig. 11.20 for slightly elongated, randomly oriented particles of a 60 Co-40 Fe alloy dispersed in mercury. The

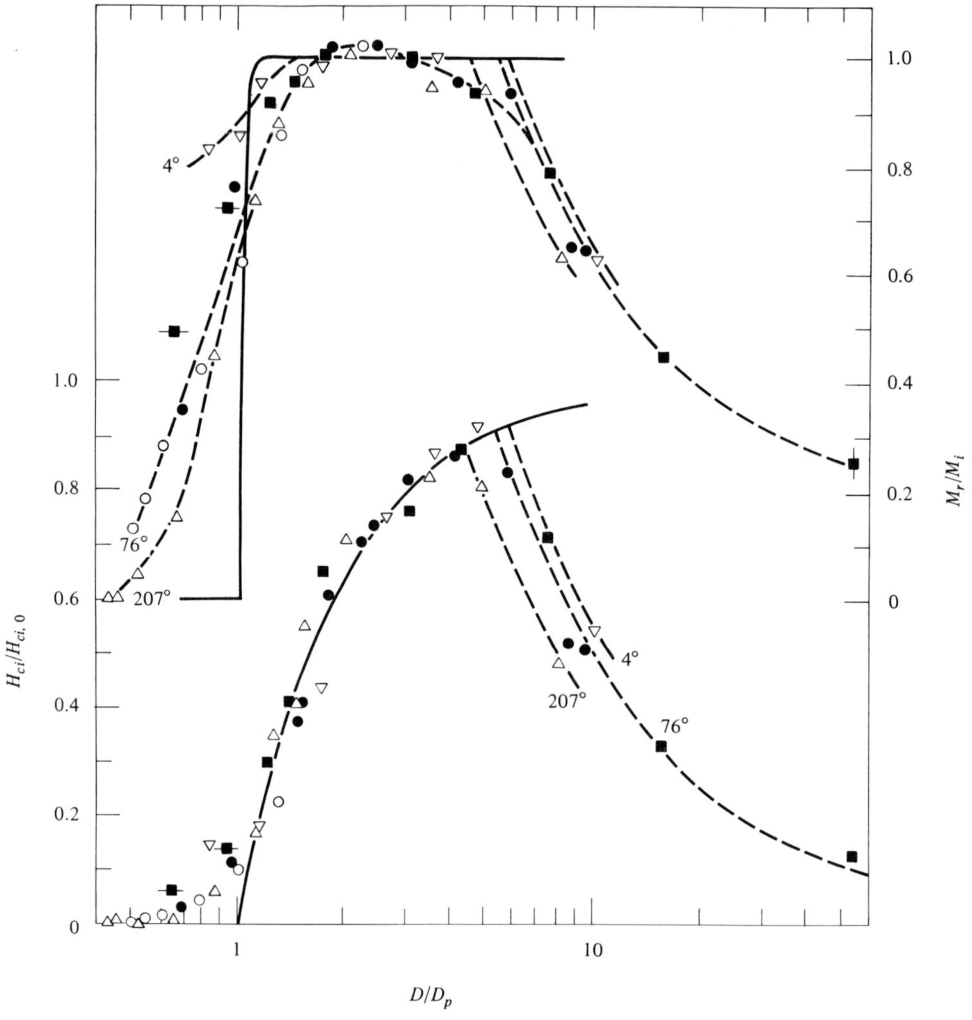

Fig. 11.20 Particle size dependence of the coercivity (lower curve) and retentivity (upper curve) of a Co-Fe alloy. Solid lines are calculated. Kneller and Luborsky [11.25].

agreement is very good. (Measurements were made at three different temperatures, as indicated. The curve is independent of temperature, however, because both coordinates are normalized.) The experimental curve tails off at D/D_p values less than 1; this is due to the particles having a finite range of size, instead of a single size, in each specimen. For these alloy particles D_s equals about $5D_p$; larger particles are therefore multidomain, and the coercivity decreases as the size increases.

Equation (11.36) also predicts the variation of coercivity with temperature, for particles of constant size. Particles of the critical size V_p have zero coercivity at their blocking temperature T_B and above, where T_B is given by Eq. (11.32). Therefore

$$h_{ci} = \frac{H_{ci}}{H_{ci,0}} = 1 - \left(\frac{T}{T_B}\right)^{1/2}. \tag{11.38}$$

This relation is plotted in Fig. 11.21. Similar curves have been found experimentally [11.24]. [Too much generality should not be attached to Eq. (11.38). The value of T_B is so extremely large for the particles in ordinary permanent magnets that any observed variation of H_{ci} with temperature is due, not to the thermal effects discussed here, but to other causes, such as the variation of K and/or M_s with temperature.]

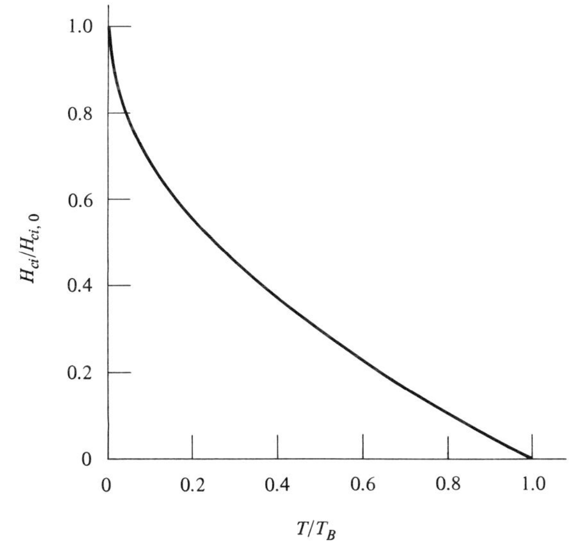

Fig. 11.21 Theoretical temperature dependence of the coercivity of single-domain fine particles. T_B = blocking temperature.

Particles larger than V_p have a finite retentivity, because thermal energy cannot reverse their magnetization in 100 sec. To find the relation between retentivity and size we combine Eqs. (11.28) and (11.29) to obtain

$$\ln \frac{M_r}{M_i} = -\frac{t}{\tau} = -10^9\, t\, e^{-KV/kT}. \tag{11.39}$$

Substituting Eq. (11.31) into this and putting $t = 100$ sec, we obtain

$$\ln \frac{M_r}{M_i} = -10^{11} e^{-25V/V_p}. \tag{11.40}$$

The variation of M_r/M_i with V/V_p, or D/D_p, predicted by this equation, is extremely rapid, as illustrated by Problem 11.11. It is plotted as the upper curve of Fig. 11.20, and its slope is so large that it appears vertical. The experimental retentivities do not agree with theory as well as the coercivities, because the retentivities, which vary so rapidly with size, are more sensitive to the presence of a distribution of sizes. When particles of size V_p are cooled below T_B, the retentivity rises very rapidly to its maximum value (Problem 11.12).

The marked dependence of the magnetic properties of very fine particles on their size means that particle sizes can be measured magnetically. An example is given in the next section.

11.7 SUPERPARAMAGNETISM IN ALLOYS

Fine particles of magnetic metals, alloys, and nonmetals may be prepared by a variety of methods and then dispersed in a nonmagnetic medium, to avoid particle interactions, to produce specimens suitable for magnetic studies. Alternatively, one can heat treat some alloys in such a way that fine particles of a magnetic phase are precipitated out of a nonmagnetic matrix. Magnetic measurements on such a material can furnish valuable information about the very early stages of precipitation, before the precipitate is visible in the optical microscope.

An alloy of copper with 2 weight percent cobalt has received the most attention. The Cu-Co equilibrium diagram is shown in Fig. 11.22. The copper-rich α phase is nonmagnetic, and the β phase, which contains about 90 percent Co, is magnetic. (This diagram is incomplete, in that it does not show the phase change, from HCP to FCC, occurring in the cobalt-rich alloys near 400°C.) In the experiments of Becker [11.26] the alloy was heated to 1010°C, where the solubility of cobalt in copper exceeds 2 percent. The homogeneous α solid solution was then quenched to room temperature. Subsequent aging treatments at 650 and 700°C allowed the magnetic β phase to precipitate in reasonable time periods. The magnetic properties of the alloy were measured after each treatment.

Magnetization curves measured after aging up to 100 min at 650°C had the typical superparamagnetic shape, with zero coercivity. The size of the precipitate particles was determined from the initial susceptibility κ of the alloy and its saturation magnetization M_{sa}. The Langevin law will apply either for particles with no anisotropy or for a random distribution of particles with uniaxial anisotropy. The initial susceptibility will therefore be given by Eq. (3.21):

$$\kappa = \frac{M}{H} = \frac{n\mu^2}{3kT} = \frac{(n\mu)(\mu)}{3kT} = \frac{(M_{sa})(M_s V)}{3kT}, \tag{11.41}$$

$$V = \frac{3kT\kappa}{M_{sa}M_s}. \tag{11.42}$$

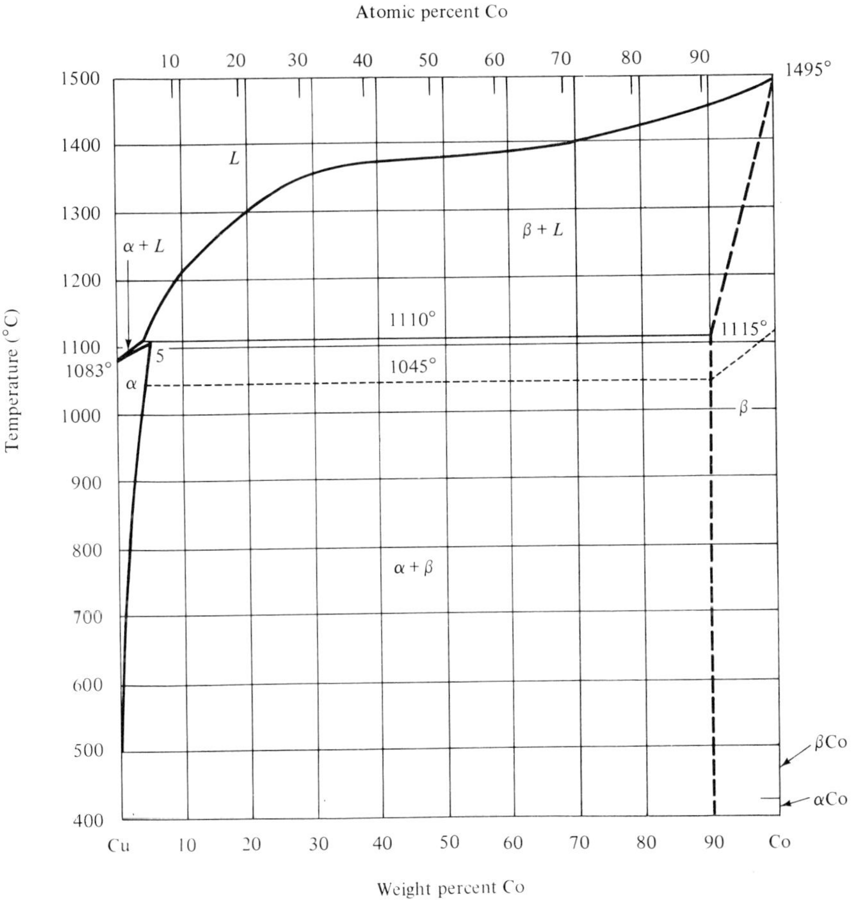

Fig. 11.22 Copper-cobalt equilibrium diagram [4.8].

This equation applies to materials which obey the Curie law [Eq. (3.13)]. However, susceptibility *versus* temperature measurements showed that the Cu-Co alloy obeyed the Curie-Weiss law [Eq. (3.14)] with a small value of the constant θ, about 5 to 10°K. There is, therefore, a small interaction between the precipitate particles. Because of the observed Curie-Weiss behavior, Eq. (11.42) was modified to

$$V = \frac{3k(T - \theta)\kappa}{M_{sa}M_s}. \tag{11.43}$$

The value of M_s in this equation was taken as the value for pure cobalt (1422 emu/cm^3). From the particle volumes V given by Eq. (11.43), the radius R of the particles, assumed spherical, could be calculated. These values are shown

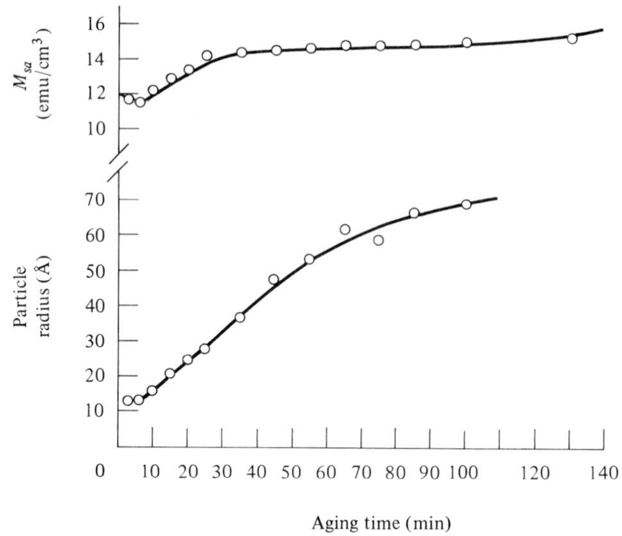

Fig. 11.23 Effect of aging at 650°C on the saturation magnetization M_{sa} of a Cu-Co alloy and on the effective radius R of the precipitate particles. Becker [11.26]

by the lower curve of Fig. 11.23, which shows how the particle size increases with aging time. These particles are very small. They are not only invisible with the light microscope, but can be seen only with great difficulty with the electron microscope. At the shortest aging time of 3 min, their diameter is only about 25 Å, or less than ten unit cells. Probably no other experimental tool could have disclosed this information as readily as these rather simple magnetic measurements.

The observed value of the saturation magnetization M_{sa} of the alloy disclosed the *amount* of the magnetic phase present, independent of its particle size. The upper curve of Fig. 11.23 therefore shows that precipitation of β is essentially complete in only 3 min at 650°C. Thereafter, the changes that occur in the alloy consist solely in an increase in particle size, and a decrease in the number of particles, of a nearly constant amount of precipitate. This process, called coarsening or coagulation, occurs by the re-solution of some particles and the growth of others in order to decrease the total surface energy of the precipitate. This kind of behavior, initial rapid nucleation followed by a decrease in the nucleation rate to zero, is typical of precipitation from a supersaturated solid solution and is technically known as *continuous precipitation* [11.27].

The cobalt-rich precipitate particles are believed to have an FCC crystal structure, rather than HCP, and to owe their uniaxial nature to shape anisotropy. If the particles are actually egg-shaped, then the radius R shown in Fig. 11.23 is to be regarded merely as an effective radius.

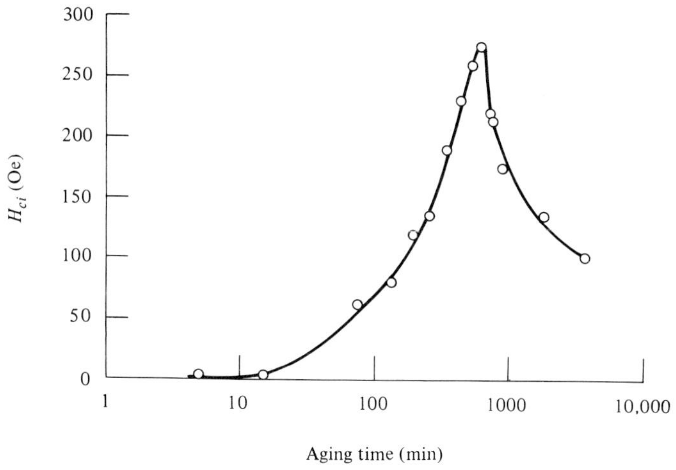

Fig. 11.24 Effect of aging at 700° C on the coercivity H_{ci} of a Cu-Co alloy. Becker [11.26].

After about 100 min at 650°C, the particle volume increased beyond the critical value V_p appropriate to the measuring temperature (room temperature), and hysteresis appeared. The time to grow particles larger than V_p can be shortened from 100 to about 10 min by raising the aging temperature from 650 to 700°C, as shown by Fig. 11.24. Hysteresis, as evidenced by a finite value of the coercivity, begins to appear after about 10 min. Figure 11.24 is interesting because it shows in one single curve, on a time base, all the changes that are schematically displayed in Fig. 11.2 on a particle-size base. Up to about 10 min, the particles behave superparamagnetically. At longer times the particles have grown to such a size that thermal energy alone can no longer rapidly overcome the energy barrier KV for reversal, and a finite coercivity field H_{ci} is needed in addition. With further growth H_{ci} increases to a maximum at about 800 minutes. The particles then become multidomain, and H_{ci} decreases.

Fine magnetic particles dispersed in a nonmagnetic matrix exist, not only in certain alloys, but in many kinds of rocks. Then the magnetic phase is usually a ferrimagnetic oxide. Studies in rock magnetism often involve the problem of determining the direction and magnitude of the earth's field in the remote past when the rock was formed, when the field had a different strength and direction than it has today. This brings up the question of the stability of the original magnetization. Has it been altered by subsequent changes in the earth's field? One must then consider the possibility of magnetization relaxation, not over a period of the order of 100 sec, but over geologic times.

In the copper-cobalt alloy cited in this section and in most rocks, the magnetic particles responsible for the superparamagnetic behavior of the material are discrete particles of a second phase. This condition is not essential, because a single-phase solid solution can also be superparamagnetic if it has local inhomogeneities of the right kind. For example, in a solid solution containing 90 mol percent Zn ferrite and 10 percent Ni ferrite, there may be small clusters containing more than the average number of magnetic ions (Fe and Ni), surrounded by nonmagnetic ions (Zn). These magnetic clusters within the solid solution then act superparamagnetically [6.10].

11.8 EXCHANGE ANISOTROPY

Another interesting small-particle effect was discovered in 1956 by Meiklejohn and Bean [11.28], who called it *exchange anisotropy*. They took fine, single-domain particles of cobalt and partially oxidized them, so that each cobalt particle was covered with a layer of CoO. A compact of these particles was then cooled in a strong field to 77°K, and its hysteresis loop was measured at that temperature. This loop, shown by the full curve of Fig. 11.25, is not symmetrical about the origin but shifted to the left. If the material is not cooled in a field, the loop is symmetrical and entirely normal (dashed curve). The field-cooled material has another unusual characteristic: its rotational hysteresis loss W_r, measured at 77°K, remains high even at fields as large as 16 kOe, whereas W_r decreases to zero at high fields in most materials.

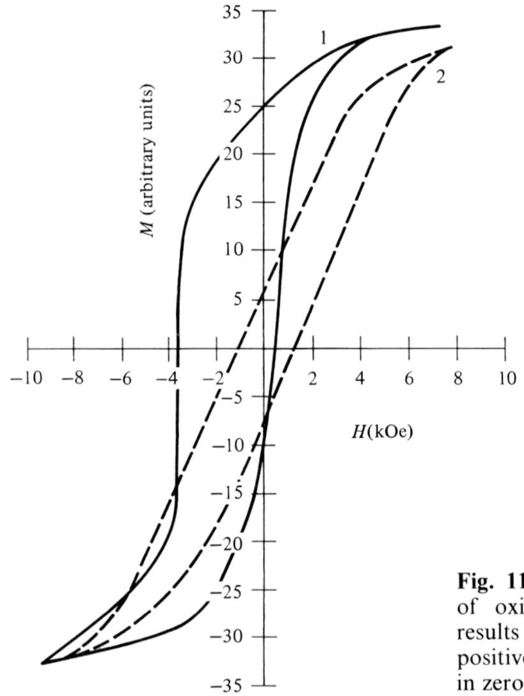

Fig. 11.25 Hysteresis loops measured at 77°K of oxide-coated cobalt particles. Loop (1) results from cooling in a 10 kOe field in the positive direction, and loop (2) from cooling in zero field. Meiklejohn and Bean [11.28]

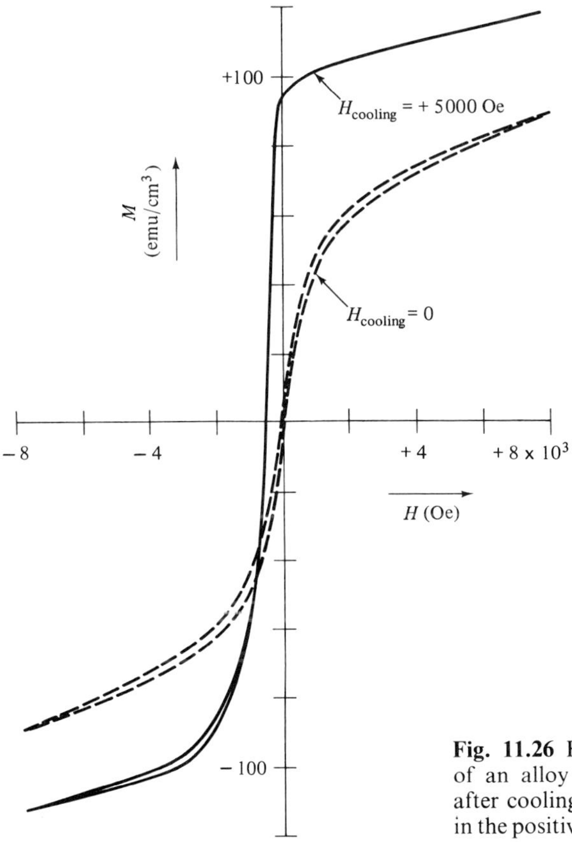

Fig. 11.26 Hysteresis loops measured at 4.2°K of an alloy of Ni + 26.5 atomic percent Mn after cooling with and without a field of 5 kOe in the positive direction. Kouvel *et al.* [11.29]

These two features of exchange anisotropy, a shifted loop and high-field rotational hysteresis, have since been found in other materials, including alloys. For example, disordered nickel-manganese alloys at and near the composition Ni_3Mn are paramagnetic at room temperature but show exchange anisotropy when field-cooled to low temperatures (Fig. 11.26). The hysteresis loop is then shifted so far that the retentivity is positive, after cooling in a field in the positive direction, whether the previous saturating field was applied in the positive or negative direction. (Note that we are dealing here with shifted *major* loops, representing saturation in both directions. A shifted minor loop, in which saturation is not achieved in one or both directions, is no evidence for exchange anisotropy and can be obtained in any material.)

Both the Co-CoO particles and the Ni_3Mn alloy display *unidirectional*, rather than uniaxial, anisotropy. The anisotropy energy is therefore proportional to the first power, rather than the square, of the cosine:

$$E = -K \cos \theta, \tag{11.44}$$

where K is the anisotropy constant and θ is the angle between M_s and the direction of the cooling field. (For oxidized cobalt particles 200 A in diameter, K is 4×10^6 ergs/cm^3 of metallic cobalt.) This dependence of energy on θ is just the same as that of a magnet in a field, Eq. (1.5).

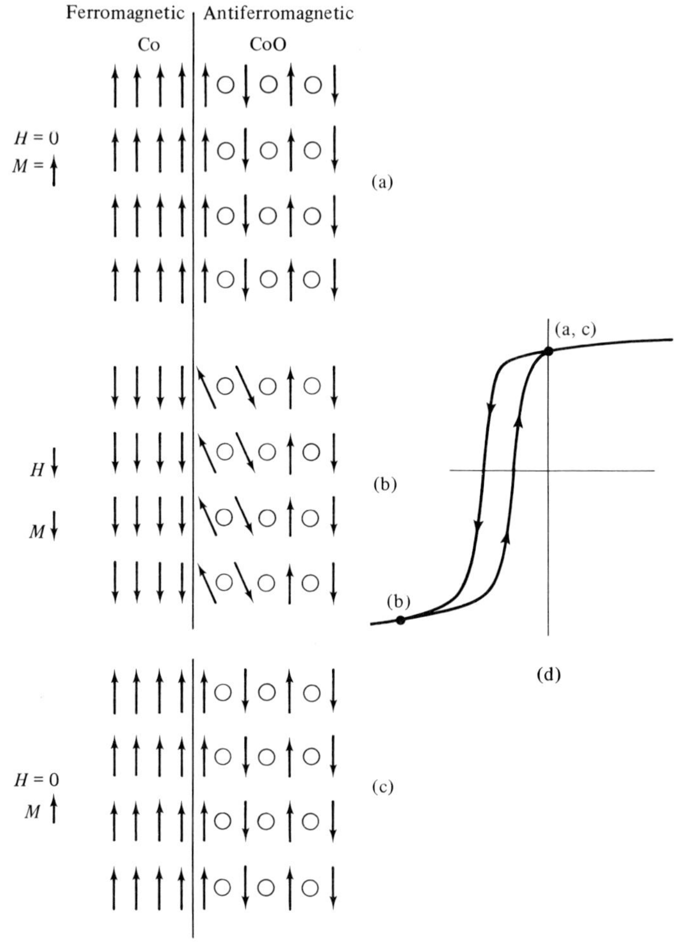

Fig. 11.27 Mechanism of the shifted loop in Co-CoO. Arrows represent spins on cobalt atoms or ions, and open circles are oxygen ions. Points (a), (b), and (c) on the loop correspond to the three states at the left. Graham [10.1]

The unusual properties of field-cooled Co-CoO particles are due to exchange coupling between the spins of ferromagnetic Co and antiferromagnetic CoO at the interface between them. The Néel temperature T_N of CoO is about 20°C. When a strong field is applied at 20°C, the cobalt saturates but the oxide, being paramagnetic, is little affected. However, the spins of the first layer of Co ions in the oxide are forced to be parallel to the adjoining spins in the metal, because of the positive exchange force between the spins of adjacent cobalt atoms. When the particles are cooled, still in the field, far below T_N, the antiferromagnetic ordering that occurs in the oxide will then have to take on the form sketched in Fig. 11.27 (a). Furthermore, this spin arrangement will persist when the field is removed. If now a strong field is applied in the downward (negative) direction, as in (b), the spins in the cobalt will reverse, and the exchange coupling at the interface will try to reverse the spin system of the oxide. This rotation is resisted by the strong crystal anisotropy of the antiferromagnet, with the result that only partial rotation of a few spins near the interface occurs, as indicated. Finally, when the downward field is removed, the "up" spins in the oxide at the interface force the spins in the metal to turn up, restoring the particle to state (a), with a positive retentivity. This ideal behavior is achieved in the observed loop of Ni_3Mn (Fig. 11.26) but not in the Co-CoO loop (Fig. 11.25).

There are evidently three requirements for the establishment of exchange anisotropy: (1) field-cooling through T_N, (2) intimate contact between ferromagnetic and antiferromagnetic, so that exchange coupling can occur across the interface, and (3) strong crystal anisotropy in the antiferromagnetic. Actually, the role of field-cooling is only to give the specimen as a whole a single easy direction. If it is cooled in zero field, the exchange interaction occurs at all interfaces, leading, in the ideal case, to a random distribution of easy directions in space and zero retentivity, as shown in Fig. 11.26.

The exchange anisotropy exhibited by disordered Ni_3Mn is believed to be due to composition fluctuations in the solid solution, leading to the formation of Mn-rich clusters. These clusters would be antiferromagnetic, because the exchange force between Mn-Mn nearest neighbors is negative, as in pure manganese. Just outside the cluster the solution would be richer in nickel than the average composition, and the preponderance of Ni-Ni and Ni-Mn nearest neighbors would cause ferromagnetism there. (Ordered Ni_3Mn, in which all nearest neighbor pairs are Ni-Mn, is ferromagnetic.) We may therefore conclude that the establishment of exchange anisotropy does not require a two-phase system like Co-CoO; it can also occur in a single-phase solid solution having the right kind of inhomogeneity. The latter behavior resembles that described in the last paragraph of Section 11.7.

Exchange anisotropy has been observed in other materials besides the two mentioned here; they are reviewed by Meiklejohn [11.30] and by Jacobs and Bean [11.2]. But because Néel temperatures are usually low, exchange anisotropy has never been observed at room temperature. This is unfortunate, because a permanent magnet which would resist demagnetization by strong fields applied in *either* direction would be of industrial interest.

11.9 PREPARATION AND STRUCTURE OF THIN FILMS

We turn now to thin films. Much research has been done on the magnetic properties of films because of their application as memory elements in digital computers (Section 13.6). Most of this work has been done on films of a single composition: approximately 80 percent Ni, balance Fe (80 Permalloy). At or near this composition, the magnetostriction and crystal anisotropy become zero (Fig. 13.25). Unless otherwise mentioned, this composition is understood in all the following discussion.

One requirement of a magnetic memory element is a square hysteresis loop, i.e., the film must have uniaxial anisotropy. Inasmuch as the film has no anisotropy of its own, this anisotropy must be built into the film as it is formed. This is easily done by depositing the film in the presence of a field, of a few tens of oersteds, parallel to the plane of the film; the induced easy axis is then parallel to the deposition field. The nature of this anisotropy will be discussed later.

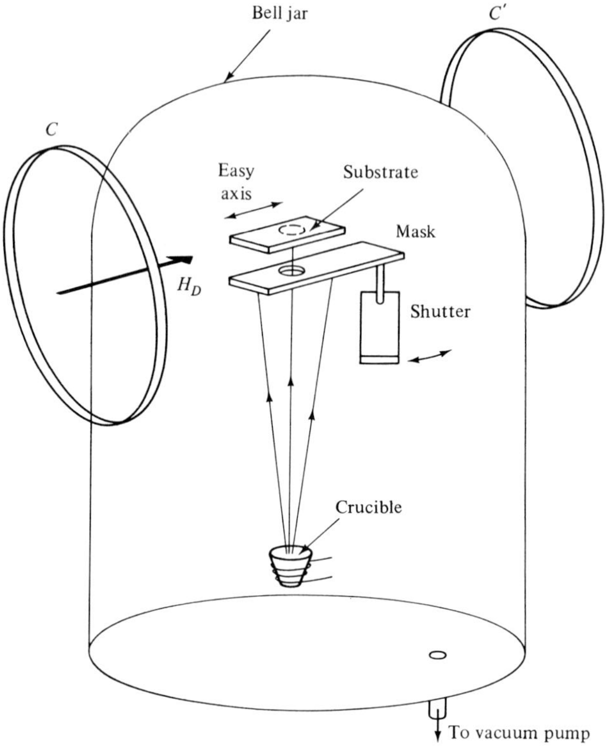

Fig. 11.28 Thin-film preparation by evaporation.

The films of industrial interest are polycrystalline and range in thickness from about 500 to 10,000 A. Although they may be prepared in various ways, only two methods are at all popular: evaporation and electrodeposition.

In the *evaporation method* the nickel and iron atoms evaporate from the surface of a molten alloy in a vacuum and deposit on a flat substrate, usually of glass. The required apparatus is sketched in simplified form in Fig. 11.28. A few grams of the alloy in a crucible are heated to about 1700 C by means of a high-frequency induction coil. The substrate is located about 10 in. above the crucible, and the film thickness, typically 1000 Å, is controlled by the deposition time; the rate of deposition is of the order of 4000 Å/min. The deposition time is in turn controlled by a shutter which can be moved in and out of the stream of atoms from the crucible. The shape of the deposited film, usually circular, is determined by a mask below the substrate. For good adhesion between film and substrate, the latter is maintained at about 300°C by a heater (not shown). The whole apparatus is enclosed in an evacuated bell jar. Outside the jar is a Helmholtz coil system CC′ which sets up the deposition field H_D in the plane of the film. The deposited film is bright and shiny and remains so for months of exposure to the air. The film remains on the substrate on which it was formed during subsequent testing or operation. The nature of this substrate, particularly its degree of cleanliness and surface roughness, can have important effects on the magnetic properties of the film.

Electrodeposited films are made by deposition on a wire cathode from an aqueous electrolyte containing nickel and iron salts; both metals deposited simultaneously. The starting material is usually beryllium copper wire 5×10^{-3} in. in diameter. ("Beryllium copper" is a very strong age-hardenable alloy of Cu + 2 percent Be, with quite high electrical conductivity.) An intermediate layer of copper or gold is plated on the wire, and then the outer layer, typically 8000 Å thick, of Ni-Fe alloy. The purpose of the intermediate layer is to provide a substrate with a controlled degree of surface roughness, which is required for proper magnetic properties in the outer layer. During the electrodeposition of the Ni-Fe alloy a circumferential deposition field H_D is maintained by establishing a current along the wire axis, as indicated in Fig. 11.29. This results in a circumferential, magnetically easy axis in the Ni-Fe film. However, this induced anisotropy is found to deteriorate with time, even at room temperature. It is therefore necessary to "stabilize" the anisotropy by heating the electroplated wire in a circumferential field for 1 min at 350°C. (The induced anisotropy of evaporated films is presumably stabilized by the elevated temperature at which they are deposited.) The whole process is continuous: the wire passes through the Cu (or Au) plating bath, then through the Ni-Fe bath, then through the stabilization furnace. (From a production point of view, plated wire is more attractive than evaporated flat films, because the plated wire can be made by a continuous, rather than a batch, process, and no high vacuum equipment is needed.) The relevant magnetic properties of plated wires have been reviewed by Doyle *et al.* [11.31].

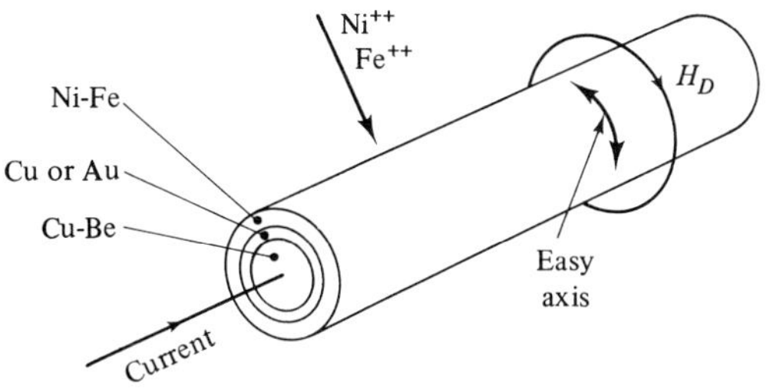

Fig. 11.29 Electroplating Ni-Fe alloy on wire.

The *structure* of thin films is complex and not well known. Writing particularly of evaporated films, Pugh [11.7] refers to "the metallurgical nightmare of imperfections, impurities, and stresses which externally appear as shiny thin films." Similar remarks apply to electrodeposited films, which normally contain an even greater concentration of impurities and defects. The grain size of films is extremely small, of the order of 1000 Å or less, which is less than 10^{-2} times the grain size of most bulk materials. Preferred orientation (texture) may or may not be present, depending on deposition conditions and nature of the substrate. If a texture is present, it is a single or multiple fiber texture, with fiber axes such as ⟨100⟩, ⟨111⟩, etc., normal to the plane of the film [4.9]. Such a texture introduces no crystallographic anisotropy in the plane of the film. Even if it did, little magnetic anisotropy would result, because the crystal anisotropy of 80 Permalloy is close to zero.

The system film-substrate contains residual macrostress, due to the difference in thermal expansion coefficients between film and substrate, and the film itself has microstress due to various imperfections. The existence of these stresses dictates the film composition, which is chosen so that the magnetostriction is as close as possible to zero. Stress anisotropy is therefore avoided because magnetoelastic effects are always proportional to the product of the magnetostriction λ and the stress σ.

The quantity of experimental work on the purely magnetic behavior of films far exceeds the work, which is more difficult, on film structure. As a result, there is insufficient understanding of the relation between structure and magnetic behavior.

11.10 INDUCED ANISOTROPY IN FILMS

The uniaxial anisotropy induced by deposition in a field is described by the energy relation

$$E = K_u \sin^2 \theta, \tag{11.45}$$

where θ is the angle between M_s and the easy axis (axis of deposition field). The value of K_u is generally of the order of 1000 to 3000 ergs/cm^3 for 80 Permalloy, which is about equal to the value observed for magnetically annealed bulk alloys (Fig. 10.5).

The measurement of K_u is difficult, because the specimen size is so very small. Two methods are available:

1. *Measurement of the torque curve.* This is done in the usual way, but the torque magnetometer has to be extremely sensitive. The volume of a circular film specimen 1 cm in diameter and 1000 Å thick is only 10^{-5} cm^3, and the torque on the specimen is proportional to the volume. A similar bulk specimen would have about the same diameter and be about 0.1 cm thick. Thus the film specimen has a volume of only 10^{-4} of the usual bulk specimen. The torsion fiber of the thin-film magnetometer must accordingly be very fine. (Torque magnetometry of electrodeposited films requires that the film be plated on a flat substrate rather than a wire.)

2. *Measurement of the hysteresis loop.* From measurements of the loop in the easy and hard directions, the anisotropy field H_K can be determined. The value of K_u is then calculated from the relation $H_K = 2K_u/M_s$. (Details are given in Section 12.4). The hysteresis loop, measured at a frequency of a few hundred Hz, is displayed on an oscilloscope by means of a loop tracer (Fig. 2.19). The saturation induction $B_s(\approx 4\pi M_s)$ of 80 Permalloy is about 10,000 gauss, so that the total flux

change through a film specimen with a cross section of 1 cm by 1000 Å is only 0.2 maxwell from tip to tip of the hysteresis loop. The search coil (pickup coil) must therefore have a large number of turns (several thousand), and the integrator of the loop tracer must be very sensitive. Details of such measurements have been reviewed by Humphrey [11.32].

The *origin* of the induced anisotropy in 80 Permalloy films has been much investigated and much debated. Although all metal films are known to contain large stresses, stress anisotropy cannot contribute much of the observed anisotropy of 80 Permalloy because this alloy has essentially zero magnetostriction. The induced anisotropy of 80 Permalloy films is generally believed to have two chief causes:

1. *Directional order* of like-atom (Fe-Fe) pairs. The mechanism of this anisotropy is exactly the same as that described in Section 10.2 for bulk Ni-Fe alloys, with one important exception. Ordering is very much faster in films that in bulk alloys; it occurs in minutes rather than hours, and at lower temperatures. Roth [11.33] has shown that the faster ordering is due to the large vacancy concentration in films; vacancies are always more plentiful near surfaces and grain boundaries, and the ratio of surface and grain-boundary area to volume is large in a thin film. Inasmuch as diffusion in a substitutional solid solution takes place by an interchange of atoms and vacancies, the diffusion rate will be high when the vacancy concentration is high.

2. *Oriented imperfections.* As noted earlier, thin films have an abnormally high concentration of defects, such as foreign atoms, dislocations, and vacancies. During formation of the film, when defects are highly mobile, the deposition field H_D keeps the magnetization aligned in one direction. If the defects are somehow coupled to the local direction of M_s, then they will take up a particular orientation in the film and retain that orientation when H_D is removed. For example, a pair of adjacent vacancies, called a divacancy, might have a lower energy when its axis has a particular orientation (parallel, say) to M_s; if so, M_s is thereafter bound to that axis with a certain strength, and a uniaxial anisotropy has been created. There is as yet little experimental evidence for oriented imperfections, and this effect probably makes a much smaller contribution to the observed anisotropy then directional order of Fe-Fe pairs.

11.11 DOMAIN WALLS IN FILMS

The magnetization in thin films lies in the plane of the film, because a huge demagnetizing field $H_d (= 4\pi M_s \approx 10{,}000$ Oe) would act normal to the plane of the film if M_s were turned in that direction. Domains in the film extend completely through the film thickness, and the walls between them are mainly of the 180° kind, roughly parallel to the easy axis of the film.

However, two new *kinds* of domain walls can exist in thin films. The first is called a *Néel wall*, because it was first suggested on theoretical grounds by Néel

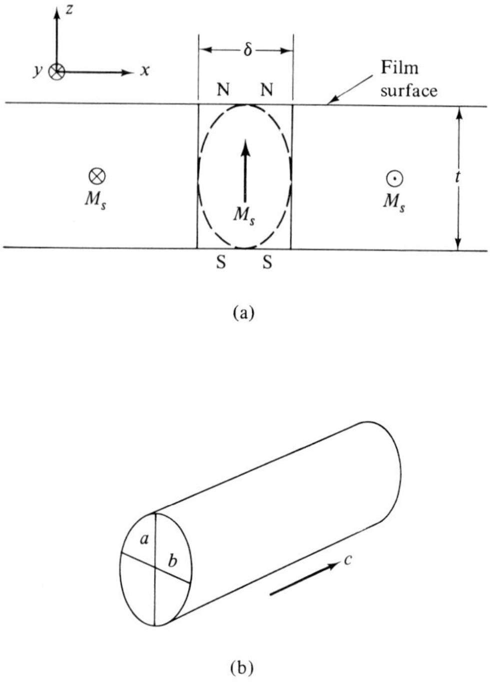

Fig. 11.30 (a) Cross section of 180° Bloch wall in thin film. (b) Elliptic cylinder.

[11.34] in 1955. Ordinary walls such as those found in bulk materials can also exist in thin films; they are then specifically called Bloch walls to distinguish them from Néel walls.

Néel showed that the energy per unit area γ of a Bloch wall is not a constant of the material but depends also on the thickness of the specimen, when the thickness is less than a few thousand angstroms. The magnetostatic energy of the wall them becomes appreciable, relative to the usual exchange and anisotropy energy. Free poles are formed where the wall intersects the surface, as indicated in Fig. 11.30(a) where only the central spin in the wall is shown; when the specimen thickness t is of the same order of magnitude as the wall thickness δ, the field created by these poles constitutes an appreciable magnetostatic energy. To calculate this energy Néel approximated the actual wall, a nonuniformly magnetized rectangular block in which the spins continuously rotate from the direction $+y$ to $-y$ by a uniformly magnetized elliptic cylinder. This is sketched in (b); its major axis c is infinite. When magnetized along the a axis, its demagnetizing coefficient is

$$N_a = \frac{4\pi b}{a + b}, \tag{11.46}$$

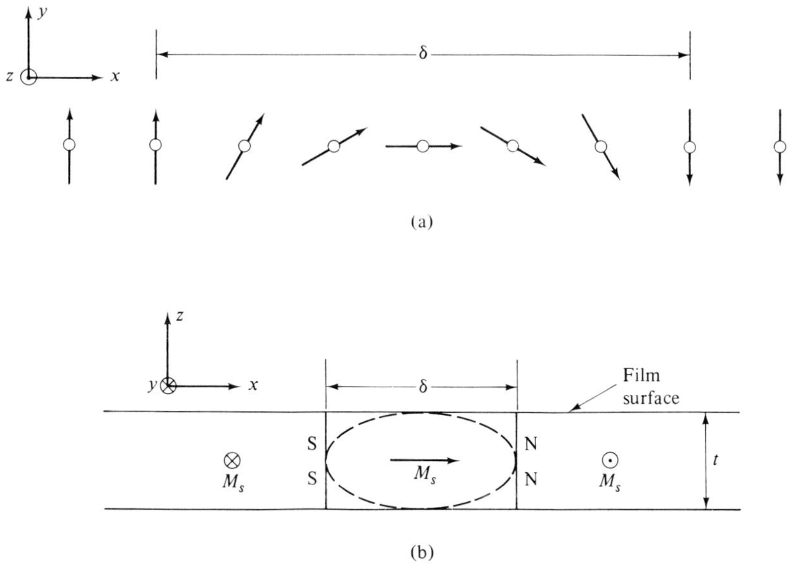

Fig. 11.31 Structure of Néel wall. (a) Section parallel to film surface. (b) Cross section of wall.

and when magnetized along b,

$$N_b = \frac{4\pi a}{a + b} \cdot \tag{11.47}$$

The magnetostatic energy density of the wall is then

$$\begin{aligned} E_{ms} &= \tfrac{1}{2} N_t M_s^2 \text{ erg/cm}^3 \\ &= \tfrac{1}{2}\left(\frac{4\pi\,\delta}{t + \delta}\right) M_s^2. \end{aligned} \tag{11.48}$$

This must be multiplied by δ to obtain the magnetostatic energy per unit area of wall in the yz plane:

$$\gamma_{ms,B} = \frac{2\pi\,\delta^2 M_s^2}{t + \delta} \text{ erg/cm}^2. \tag{11.49}$$

This energy is negligible when t/δ is large, as in bulk specimens, but not when it is of the order of unity or less.

When the film thickness t is small, the magnetostatic energy of the wall can be reduced if the spins in the wall rotate, not about the wall normal x, but about the film normal z. The result is a Néel wall. Free poles are then formed, not on the film surface, but on the wall surface, and spins everywhere in the film, both within the domains and within the walls, are parallel to the film surface (Fig. 11.31).

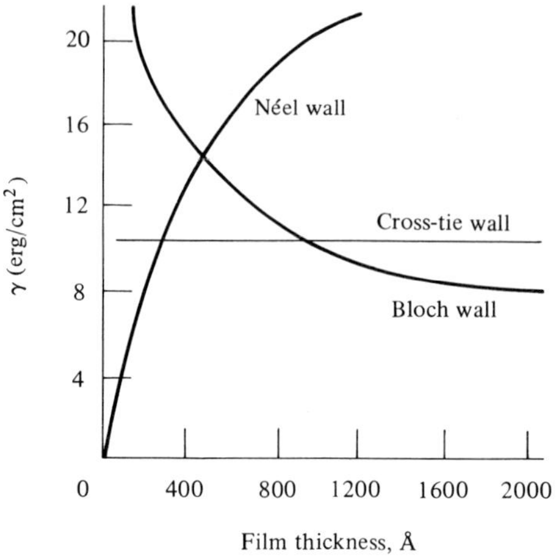

Fig. 11.32 Variation of wall energy with film thickness for various kinds of wall in 80 Permalloy, calculated for exchange constant $A = 10^{-6}$ erg/cm, anisotropy constant $K = 1500$ ergs/cm^3, bulk-wall width = 20,000 Å, and bulk-wall energy = 0.1 erg/cm^2. Prutton [G.24]

We again approximate the wall by an elliptic cylinder and find that its magnetostatic energy is

$$\gamma_{ms,N} = \frac{2\pi t\, \delta M_s^2}{t + \delta} \text{ erg/cm}^2. \tag{11.50}$$

The ratio of the magnetostatic energies of the two kinds of wall is then

$$\frac{\gamma_{ms,B}}{\gamma_{ms,N}} = \frac{\delta}{t}. \tag{11.51}$$

This expression states that the magnetostatic energy of a Néel wall is less than that of a Bloch wall when the film thickness t becomes less than the wall thickness δ. This relation is inexact, however, not only because of the approximations involved in its derivation, but also because δ varies with film thickness. And to really decide which kind of wall is more stable at a given thickness, one must calculate the total wall energy γ, which contains magnetostatic, exchange, and anisotropy terms. Such a calculation is given by Prutton [G.24], Soohoo [G.25], Middelhoek [11.35], and others, and Fig. 11.32 shows the resulting curves, which have been calculated with the constants appropriate to 80 Permalloy. We see that the total energy of a Néel wall, as well as the magnetostatic energy, is less that that of a Bloch wall when the film is very thin, less than about 500 Å. (Actually, for thicknesses of a few hundred angstroms, the total energy of either kind of wall is almost entirely magnetostatic.) The widths of Bloch and Néel walls also vary in different ways with film thickness: the thinner the film, the narrower the Bloch wall and the wider the Néel wall.

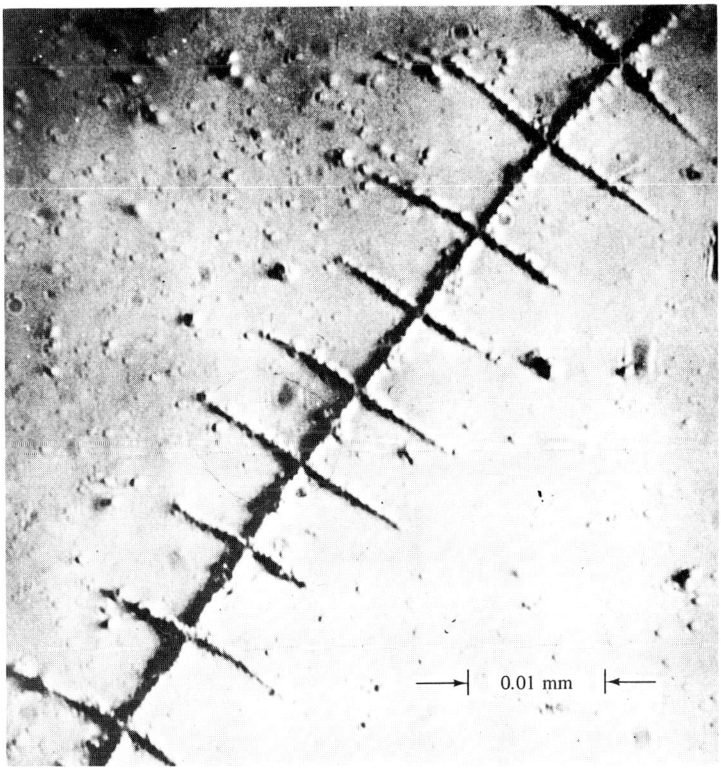

Fig. 11.33 Bitter pattern of a cross-tie wall in an 80 Permalloy film, 300 Å thick. Modified bright-field illumination. Moon [11.37]

The second new kind of wall observed in thin films is the *cross-tie wall*, first seen by Huber *et al.* [11.36]. It consists of a special kind of Néel wall, crossed at regular intervals by Néel wall segments. As shown in Fig. 11.32, its energy is less than that of a Bloch wall or a Néel wall in a certain range of film thickness; the cross-tie wall therefore constitutes a transition form between the Bloch walls of very thick films and the Néel walls of very thin films. Figure 11.33 shows the appearance of a cross-tie wall, and Fig. 11.34 (b) its structure. In Fig. 11.34 (a)

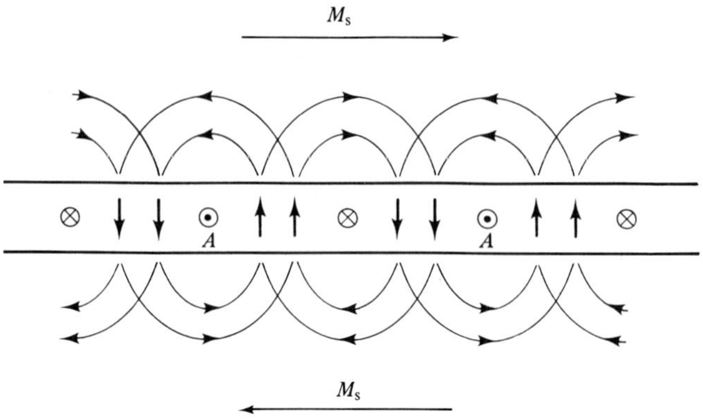

(a)

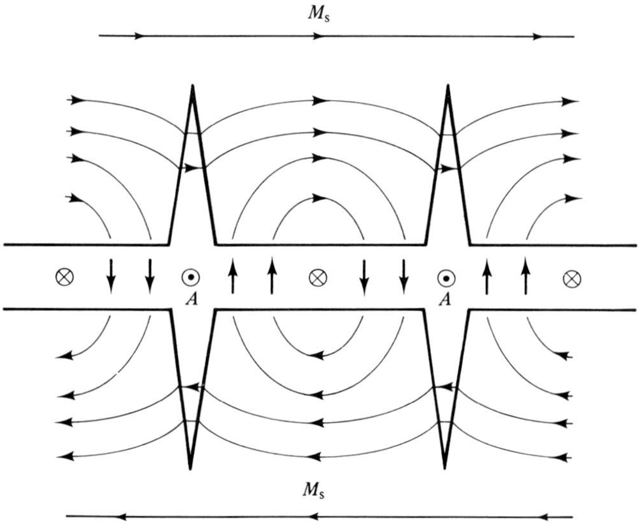

(b)

Fig. 11.34 Sections parallel to film surface of (a) hypothetical Néel wall with sections of opposite polarity, and (b) cross-tie wall. After Craik and Tebble [G.26]

a Néel wall is shown, separating two oppositely magnetized domains. It is not a normal Néel wall because it consists of segments of opposite polarity; these have formed in an attempt to mix the north and south poles on the wall surface more intimately and thus reduce magnetostatic energy. The regions within

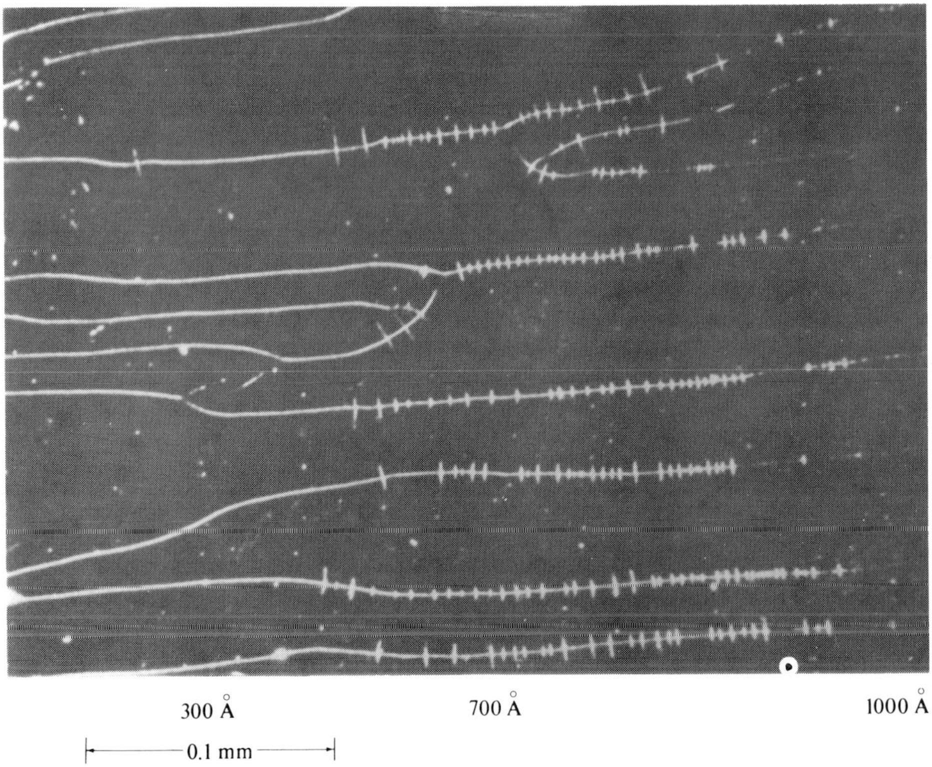

Fig. 11.35 Domain walls in a tapered film of 80 Permalloy. Easy axis horizontal. Dark-field Bitter pattern. Methfessel *et al.* [11.38].

the wall where the polarity changes, marked with small circles, and where the magnetization is normal to the film surface, are called *Bloch lines*. However, this hypothetical wall would have very large energy, because the fields due to the poles on the wall, sometimes called stray fields, are antiparallel to the domain magnetization in the regions opposite the Bloch lines marked A. As a result, spike walls form in these regions, as shown in Fig. 11.34 (b), and the stray fields close in a clockwise direction between the cross ties.

By proper manipulation of a mask during the deposition of an evaporated film, a tapered film can be deposited; its thickness varies continuously from one end to the other. The Bitter pattern of such a film is shown in Fig. 11.35. At the thin end on the left the walls are of the Néel type, and as the thickness increases, cross-tie and then Bloch walls appear. The film thicknesses at which transitions occur from one type to another are in good agreement with the curves of Fig. 11.32. (Néel walls are more easily visible than Bloch walls, which sometimes appear only as a series of dots. This is due to the different modes of flux closure where each type of wall intersects the film surface. At the edge of a Néel wall the stray fields are parallel to the film surface, and these fields apparently attract chains of colloid particles, transverse to the wall, which close the flux. The stray fields at the edge of a Bloch wall are normal to the film surface, have a larger spatial extent, and attract the colloid less strongly.)

While Néel and cross-tie walls are of considerable scientific interest, they are not particularly relevant to the application of thin films in computers, because such films are normally so thick that only Bloch walls are stable.

11.12 DOMAINS IN FILMS

One does not expect a polycrystalline specimen to be a single domain in zero field. It would normally break up into domains arranged in conformity with the easy-axis direction in each grain. A thin film of 80 Permalloy, however, if saturated in the easy direction, can remain a single domain after the saturating field is removed, as shown in Fig. 11.36 (a). The reason is that the demagnetizing field in the plane of the film is nearly zero, because the film is so thin; in addition, the grain size is very small. If each grain became a domain, large numbers of free poles would form at grain boundaries, and the magnetostatic energy would be high.

If a small reverse field is applied, reverse domains are nucleated at the ends of the film, as in (b). It is also common to see small reverse domains in the remanent state ($H = 0$), nucleated by demagnetizing fields due to imperfections at the film edges; these fields need not be large, because this alloy is magnetically so soft that a reverse field of only 2 or 3 oersteds, applied along the easy axis, is enough to reverse the magnetization completely.

When the film is demagnetized by an alternating easy-axis field of decreasing amplitude, the resulting domain structure is shown in (c). This structure is typical. It consists of elongated domains more or less parallel to the easy axis, bounded by gently curved, rather than straight, 180° walls. These walls are normally much wider than the grains. A film can also be demagnetized by saturating it in a hard direction, in the plane of the film and at right angles to the easy axis, and then removing the saturating field. The film then breaks up into the much narrower domains shown in (d). The width of these domains varies from about 2 to 50 μ from one film to another.

This structure (Fig. 11.36 d) is due to *anisotropy dispersion.* By this is meant a variation, from place to place in the film, of the direction of the easy axis and/or the magnitude of the anisotropy constant, because of inhomogeneities in the structure of the film. As a result the direction of the local magnetization M_s varies slightly from one point to another even within a domain. The nonparallelism of the M_s vectors adds exchange energy to the system; in addition, free poles are created within the domain because of the finite divergence of M (Section 2.6), causing stray fields and magnetostatic energy. In order to minimize this exchange and magnetostatic energy, the M_s direction varies in a wavelike manner, called *magnetization ripple*, rather than randomly, even though the film inhomogeneities which cause the effect are themselves randomly distributed. This ripple is illustrated in Fig. 11.37 (a), where the film is shown in the single-domain remanent state after saturation in an easy direction. When a field is now applied

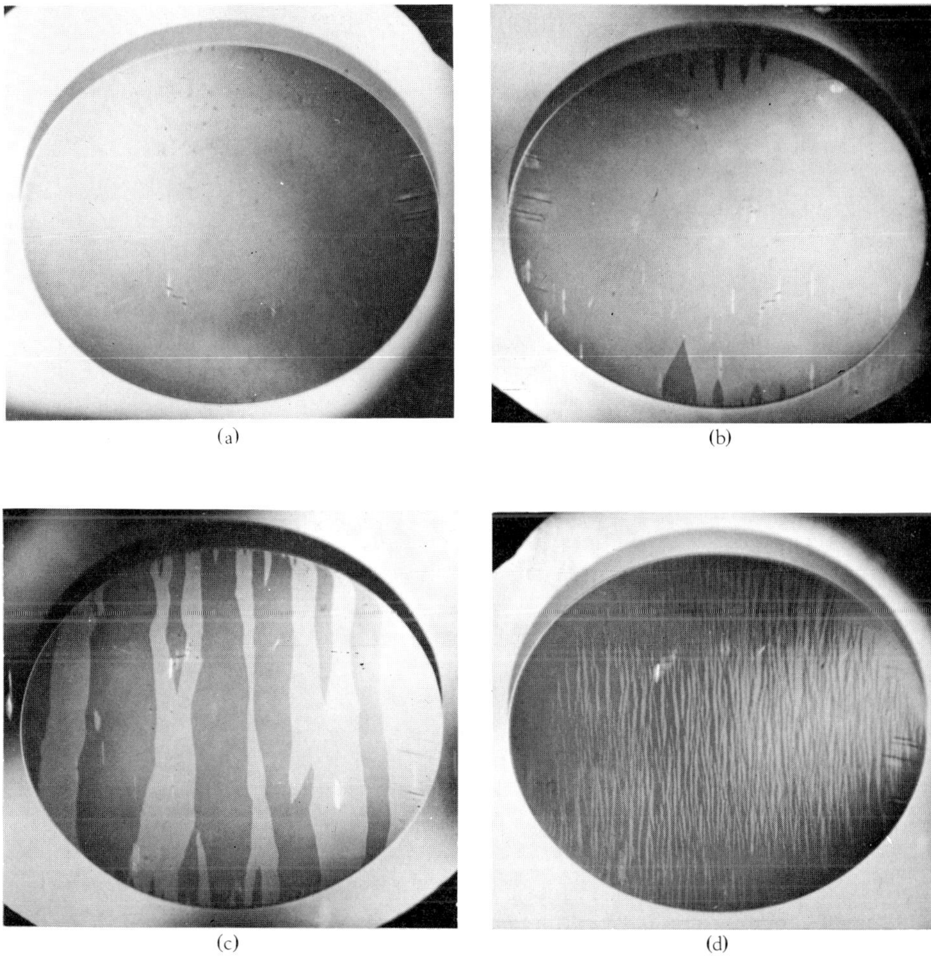

Fig. 11.36 Domains in an 81 Permalloy film observed by the Kerr effect. The film is 8 mm in diameter, 2000 Å thick, and the easy axis is vertical. (a) Single-domain remanent state, $H = 0$. (b) Domains nucleated at edges by a reverse easy-axis field of 0.2 H_c. (c) After demagnetizing in a 60-Hz alternating field of decreasing amplitude. (d) After demagnetizing by a transverse field equal to 1.5 H_K or more. (Courtesy of R. W. Olmen, Sperry Rand Univac Division, St. Paul, Minnesota)

in a hard direction, as in (b), the ripple still persists, and its relation to the local easy-axis directions is indicated. When the hard-axis field is removed, half of the M_s vectors rotate clockwise toward the local easy axis and half counter-clockwise, forming the domain structure of the demagnetized state shown in (c). The vectors indicated in (b) and (c) apply only to a narrow horizontal strip; the ripple persists in the vertical direction in each domain. The easy-axis dispersion is much exaggerated in these drawings; it is typically only 1 or 2°.

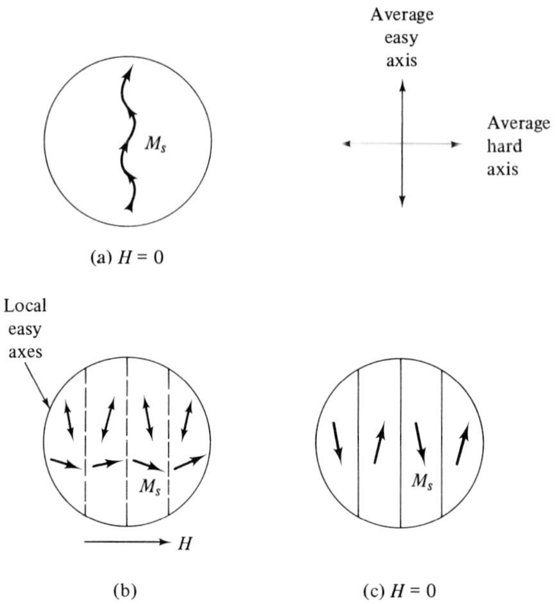

Fig. 11.37 Domain formation in a thin film after magnetization in a hard direction. After Prutton [G.24]

Magnetization ripple is evidenced in transmission electron micrographs by fine striations which are everywhere normal to the local direction of M_s, as in Fig. 11.38. The easy axis is roughly vertical in this photograph, and three 180° walls (two light and one dark) run from top to bottom. Cross ties are also visible.

11.13 FINE WIRES

In the realm of specimens small in one or more dimensions there remain, after fine particles and thin films, fine wires. Relatively little attention has been paid to their magnetic properties. In the early days of magnetic sound recording, the recording medium was stainless steel wire, but this has been entirely replaced by plastic tape containing metal oxide particles. The wire had a two-phase structure of austenite (nonmagnetic) and martensite (magnetic). Studies of the magnetic properties of such wires are reviewed by Wohlfarth [11.3, 11.4].

A wire may be defined as a body with a small diameter and a large length/diameter ratio. Whiskers therefore qualify as "wires," single-crystal wires, in fact, and we have already examined some of their magnetic properties. Wirelike filaments can also be produced by the controlled solidification of certain eutectic alloys. A eutectic is a two-phase alloy in which the two phases form simulta-

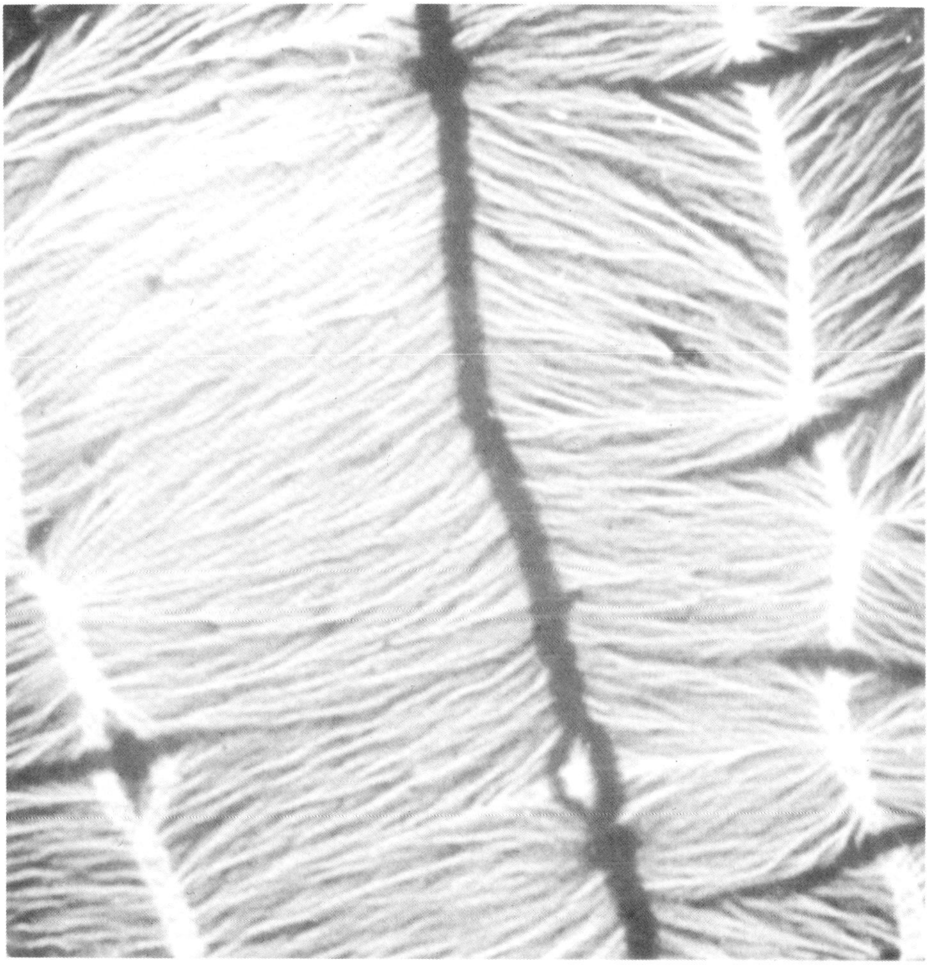

Fig. 11.38 Transmission electron micrograph of a cobalt film, 500 Å thick. (Courtesy of E. J. Torok, Sperry Rand Univac Division, St. Paul, Minnesota.)

neously from the melt as the alloy freezes. If the freezing is made to occur only in one direction, one phase will solidify as long, single-crystal filaments, parallel to the solidification direction, imbedded in a matrix of the second phase. These filaments closely resemble whiskers in size and shape, but, unlike whiskers which are grown in air, the filaments are encased in a solid body. If the alloy system is chosen so that the filament phase is magnetic, the resulting material may have interesting properties.

By directional solidification of the MnSb-Sb eutectic, Jackson *et al.* [11.39] produced a remarkably uniform microstructure. The ferromagnetic MnSb phase had the form of rods about 2μ in diameter and several millimeters in length, parallel to the growth direction. A similar structure has been achieved in the Fe-FeS eutectic [11.40], where iron rods about 2μ in diameter are formed; this alloy shows marked magnetic anisotropy, but the intrinsic coercivity H_{ci} is only 11 Oe.

In the Au-Co eutectic, Livingston [11.41] showed that the size and shape of the elongated phase, which is a cobalt-rich solution in this alloy, can be controlled by changing the rate of the directional freezing. As the freezing rate increases, the cobalt-rich phase changes in form from rods, about 1μ in diameter, to ribbons and then to lamellae, about 0.1μ thick. At the same time H_{ci} increases from about 30 to 330 Oe. Cold drawing the directionally solidified material into wire further reduces the size of the cobalt-rich phase and raises H_{ci} to 925 Oe.

It is too early to tell whether or not directionally solidified eutectics can become useful magnetic materials. One drawback is the rather low packing fraction (volume fraction of magnetic phase), which leads to low magnetization for the material as a whole. The experimenter has no control over this packing fraction, which is a constant of the alloy system involved.

PROBLEMS

11.1 Consider a multidomain single crystal like those making up Specimen B of Section 11.2, with a large demagnetizing factor and domain walls that can be moved by a negligibly small true field H. Assume very low anisotropy. Show that the magnetostatic energy at saturation is equal to the work done by the applied field H_a in bringing the crystal from the demagnetized state to saturation. (Problem 2.14 is relevant here.)

11.2 Derive Eq. (11.8).

11.3 Derive Eq. (11.10).

11.4 Show that the ratio of the energy barriers ΔE_{ms} which must be overcome in reversing the moments of a two-sphere chain from $\theta = 0$ to 180° by fanning and by coherent rotation is

$$\frac{\Delta E_{\text{fanning},\,A}}{\Delta E_{\text{coherent},\,B}} = \frac{\mu^2/a^3}{3\mu^2/a^3} = \frac{1}{3}.$$

11.5 For a prolate spheroid of axial ratio $c/a = 5$, plot h_{ci} versus D/D_0 on linear scales for coherent rotation and curling and obtain the critical value of D/D_0 which separates the two reversal modes.

11.6 a) Will a perfect, spherical, saturated iron crystal break up into domains in zero field? What are the values of H_d and $2K/M_s$?

b) How large must the axial ratio of a prolate spheroid of a perfect iron crystal be in order that it remain saturated in zero field? (The axis of saturation is the easy axis, which is also the long axis of the spheroid.)

11.7 The largest value of H_{ci} observed by DeBlois and Bean in their experiments on iron whiskers (Section 11.5) was 504 Oe at a thickness of 12.4 μ. Assume that the magnetization reversal occurred by curling. The axial ratio of the portion reversed (about 0.8 mm/12.4 μ) is so large that this portion may be considered an infinite cylinder. Calculate the value of H_{ci} due to a curling reversal, over and above that due to crystal anisotropy, in order to show that it forms a negligible fraction of the observed value.

11.8 Calculate the critical diameter D_p of iron particles for superparamagnetic behavior at room temperature, if the particles are (a) spheres and (b) prolate spheroids with an axial ratio of 1.5. (Take the diameter of the spheroids as the length of the minor axis.)

11.9 Calculate the blocking temperature T_B for iron particles 44 Å in diameter for shapes (a) and (b) of the previous problem.

11.10 Derive Eq. (11.34).

11.11 Calculate the values of D/D_p corresponding to the following values of reduced retentivity M_r/M_i: 0.50, 0.90, and 0.99.

11.12 a) Derive a relation between M_r/M_i and T/T_B for particles of volume V_p.
b) When M_r/M_i is 0.90, what is the value of T/T_B?

CHAPTER 12

MAGNETIZATION DYNAMICS

12.1 INTRODUCTION

With a few exceptions, all the phenomena examined so far in this book have been *static* in character. We have asked such questions as: if this material is subjected to a field H, what magnetization M will be reached? But we have largely ignored the *dynamics*, or kinetics, of magnetization. Does M reach its final value quickly or slowly? What controls the rate of change of M? How does M behave when the applied field H is itself varying with time? Questions like these will be examined in this chapter.

We begin with the effect of eddy currents. These not only affect the operation of many kinds of magnetic devices and machines; they can also influence magnetic measurements. It is therefore important that the nature of eddy currents be thoroughly understood.

We will then consider the velocity of magnetization change by domain wall motion and spin rotation, with particular reference to the switching speed of the square-loop materials used in computers.

Other topics include a group of phenomena loosely known as "time effects," and the internal friction (damping) of magnetic materials, because the latter involves the oscillatory motion of domain walls.

In the final section of the chapter, the various forms of magnetic resonance are briefly described.

12.2 EDDY CURRENTS

As Stewart [G.7] remarks, eddy currents are "simple in principle but intricate in detail." In the main, we shall consider only the more qualitative aspects of eddy currents in this chapter and leave the more difficult quantitative problems to the next.

Suppose a rod of magnetic material is wound with a wire, as in Fig. 12.1, connected by a switch to a dc source. When the switch is suddenly closed, a current i_w is established in the wire which creates an applied field H_a along the rod axis and uniform across the rod cross section. This field magnetizes the rod, and the induction B immediately increases from its original value of zero. Therefore, by Faraday's law, Eq. (2.6), an emf e will be induced in the rod proportional to dB/dt.

When the current i_w is increasing, the direction of e is such as to set up an eddy current i_{ec} in the circular path shown. The direction of e and i_{ec} is known from Lenz's law, which states that the direction of the induced emf is such as to oppose the cause producing it. Thus, e and i_{ec} are antiparallel to i_w, when i_w is increasing. Correspondingly, the field H_{ec} due to the eddy current is antiparallel to the field H_a due to i_w. (When the current i_w is in the same direction as shown but *decreasing*, as when the switch is opened, then i_{ec} and H_{ec} reverse directions because they now try to maintain H_a at its former value.) The magnitude of the emf acting around one circular path of radius r is given by

$$e = -10^{-8}\frac{d\phi}{dt} = -10^{-8}A\frac{dB}{dt} \qquad \text{volts,} \tag{12.1}$$

where $A = \pi r^2$ is the cross-sectional area of the rod within the path considered (cm^2), B the induction (gauss), ϕ the flux (maxwells), and t the time (sec). Several points should be noted:

1. This emf will be induced in any material, magnetic or not.
2. For a given dH_a/dt, the induced emf will be larger, the larger the permeability μ, because e depends on dB/dt and $B = \mu H$. Thus the eddy-current effect is much stronger for magnetic materials, with μ values of several hundred or thousand, than for nonmagnetics ($\mu \approx 1$).
3. For a given dB/dt and e, the eddy currents will be larger, the lower the electrical resistivity ρ of the material. In the ferrites, which are practically insulators, the eddy-current effect is virtually absent.

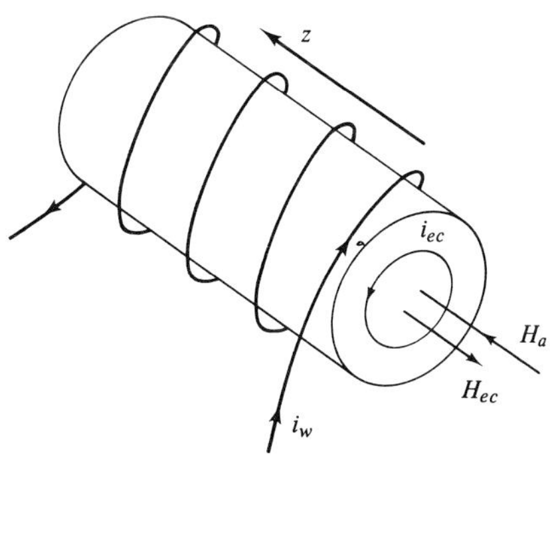

Fig. 12.1 Eddy currents in a rod.

In Fig. 12.1 only one ring of eddy current is shown. Actually, eddy currents are circulating all over the cross section of the rod. *Inside* each ring the eddy current produces a field H_{ec} in the $-z$ direction and outside it in the $+z$ direction. It follows that the eddy-current field is strongest at the center of the rod, where the contributions of all the current rings add, and that it becomes weaker toward

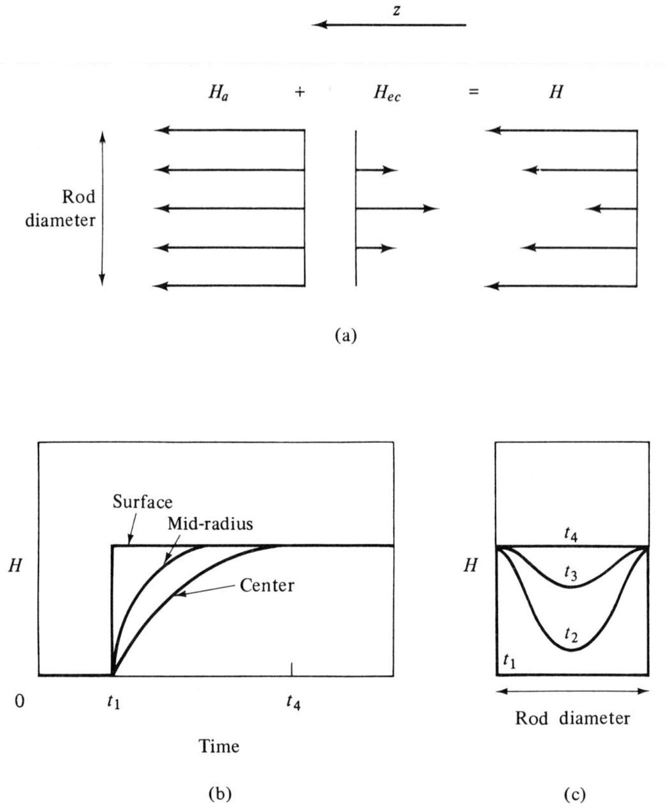

Fig. 12.2 Fields in a rod (schematic).

the edge. The variation of H_{ec} across the bar diameter at one particular instant is sketched in Fig. 12.2 (a). Because the true field H acting in the material is equal to the vector sum of H_a and H_{ec}, we see that the true field decreases below the surface and is a minimum at the center. Thus, as soon as the current i_w is established in the winding, eddy currents are set up which temporarily shield the interior of the rod from the applied field.

This effect is illustrated in another way in Fig. 12.2 (b), which shows how H changes with time at various points along the radius of the rod. The switch is closed at time t_1 and the true field is immediately established at its full value at the rod surface. The field at mid-radius takes some time to reach this value, and the field at the center takes longer still, becoming equal to the surface field at time t_4. The field profile in the rod interior is sketched in (c) for various times. The variation of B with time at any point in the rod is similar to that of H; however the B,t curves do not simply differ from the H,t curves by a constant factor μ, because μ itself is a function of H.

The time required (t_4 sec) for the eddy-current effect to disappear, i.e., for B to reach a stationary value at the center of the rod, depends on μ, ρ, and the rod diameter. In an annealed iron rod about $\frac{1}{4}$ inch in diameter the effect is over in a fraction of a second, but in thick pieces of high-μ, low-ρ material, eddy currents may not die out for several seconds.

[One statement made above is not entirely correct. When the switch is closed, the field does not instantaneously attain its final value at the rod surface. The current i_w in the wire and the field H_a which it creates require a finite time, which depends on the resistance and inductance of the circuit, to reach their final values. However, this time is usually so much smaller than the time required for eddy-current decay in metallic materials that we can neglect it and assume a simple step-function change in H at the bar surface, as in Fig. 12.2(b).]

In all of the above we have assumed a rod so long that demagnetizing effects could be neglected. If the rod is short, a substantial demagnetizing field H_d will exist and introduce additional complications. The true field at the rod surface is given by

$$H = H_a - H_d = H_a - N_d M, \tag{12.2}$$

where N_d is the demagnetizing factor. Rewriting this for the changes which occur after the switch in the magnetizing circuit is closed, we have

$$\Delta H = \Delta H_a - N_d \Delta M. \tag{12.3}$$

But the change ΔM in the rod interior is slowed by eddy currents. Therefore, H_d is initially less than its final value, and H at the rod surface, which is not shielded by eddy currents, is initially *larger* than its final value. We can reach the same conclusion if we write the equation as

$$\Delta H = \Delta H_a - M \Delta N_d. \tag{12.4}$$

The surface layer becomes very quickly magnetized, and N_d has an initial low value characteristic of a hollow tube, rather than a solid rod. H_d is again less than its final value, and H larger at the rod surface.

It follows then that, after a sudden increase in H_a, the rod surface finds itself first on an ascending portion of the magnetization curve (or hysteresis loop) and then on a *descending* loop, whereas the interior is always on an ascending portion. The total change in B, as measured with a ballistic galvanometer, may then not agree with that measured for a slow change in H_a of the same amount. These matters are discussed by Snoek [12.1], who shows how some testing conditions can lead to an initial magnetization curve which lies partly *outside* the hysteresis loop. Snoek recommends inserting a nonmagnetic conducting tube, of brass or copper, for example, between the rod and the coil; the eddy currents induced in the tube will decrease the rate of change of the field applied to the rod.

Whether the specimen rod is long or short, a fluxmeter is a better measuring instrument than a ballistic galvanometer, if the eddy-current effect is large, because a ballistic galvanometer can accurately measure a flux change in the search coil only if that change occurs before the galvanometer coil has moved appreciably from its rest position.

If the field applied to the rod is *alternating*, and at a frequency such that H_a goes through a maximum and begins to decrease before B at the center can attain the same maximum value that was reached by B at the surface, then the maximum value of B at the center will *always* be less than the maximum of B at the surface.

Furthermore, the two will differ in phase, so that, at certain times in the cycle, B at the center can point in the $-z$ direction of the rod axis while B at the surface is in the $+z$ direction. Eddy currents then circulate continuously, in opposite directions during each half cycle, and are a continuous source of heat. This heat is exploited in the induction furnace, where eddy currents are deliberately induced in the charge. But in a transformer core, eddy-current heating represents a power loss, and every attempt is made to minimize it. We will return to this question of power loss in the next chapter.

12.3 DOMAIN WALL VELOCITY

Magnetization change can occur by domain wall motion and/or rotation. In some specimens and in certain ranges of applied field, only one mechanism is operative; in others, both operate. In this section we will examine only the wall-motion mechanism and the factors that influence wall velocity. This subject has been reviewed by Kittel and Galt [9.2].

Let a field H be applied parallel to a 180° wall and to the M_s vector in one of the adjacent domains. Then, as mentioned in Section 9.11, the field exerts a force per unit area, or pressure, on the wall equal to $2HM_s$. Suppose that, when $H = 0$, the wall is initially located at a potential energy minimum, like position 1 of Fig. 9.36. Then the equation of motion of the wall per unit area is

$$m\frac{d^2x}{dt^2} + \beta\frac{dx}{dt} + \alpha x = 2M_sH, \tag{12.5}$$

where x describes the position of the wall. If H is a weak alternating field, the wall will oscillate back and forth about its initial position. In fact, it will behave just like a mass on a spring acted on by an alternating force or like charge in an electric circuit acted on by an alternating emf; the equations describing these mechanical or electrical oscillations are exactly similar in form to Eq. (12.5).

The first term in this equation, the product of the mass per unit area m of the wall and its acceleration, represents the inertia of the wall, or the resistance of the spins to sudden rotation. This term is not usually important; for Fe_3O_4 the value of m has been calculated to be only 10^{-10} g/cm^2. The second term represents a resistance to motion which is proportional to velocity, and β is accordingly called the *viscous damping parameter*. (Any force on a moving body which is proportional to velocity is, by definition, a viscous force.) The physical origin of β will be dealt with later. The third term αx represents a force due to crystal imperfections such as microstress or inclusions, and α is related to the shape of the potential-energy minimum in which the wall is located. [For example, if only microstress is present, then α is equal to $6\delta\lambda_{100}k$ of Eq. (9.36).] The value of α determines the field required to move the wall out of the energy minimum, and the ensemble of α values for the whole specimen determines the coercive force H_c, which is the field required for extensive wall movement.

Suppose the field H is now constant rather than alternating and large enough to cause extensive wall motion at constant velocity. Then the first term of Eq. (12.5) drops out because the acceleration is zero, and the third term must be modified because the wall is now moving large distances and is not affected by the value of α at one particular energy minimum. Instead, the third term will become a constant representing the average resistance to wall motion caused by crystal imperfections. Equation (12.5) becomes

$$\beta v + \text{constant} = 2M_s H, \tag{12.6}$$

where v is the wall velocity. This may be written as

$$\beta v = 2M_s(H - H_0), \tag{12.7}$$

$$v = C(H - H_0), \tag{12.8}$$

where H_0 is a constant and $C(= 2M_s/\beta)$ is another constant called the *domain wall mobility*. The constant H_0 is the field which must be exceeded before extensive wall motion can occur; it is approximately equal to H_c but is measured differently, by extrapolating a v,H plot to zero velocity. The mobility is simply the velocity per unit of excess driving field.

Equation (12.8) is in good agreement with most of the experimental data. Wall velocities have been measured in two ways, and both depend on having a specimen in which magnetization reversal is accomplished by motion of a single wall:

1. In whiskers and in certain polycrystalline wires with an axial easy axis, as in certain Fe-Ni alloys with positive magnetostriction under tension, a single reverse domain can be nucleated at one end of a previously saturated specimen. The resulting wall can be driven along the length of the wire by an axial field from a solenoid, as in Fig. 12.3 (a). Two pickup coils, a known distance apart, also surround the wire and are connected to an oscilloscope. As the wall passes each coil, the flux reversal induces a momentary emf, resulting in a voltage pulse on the oscilloscope screen. The time between the two pulses gives the time required for the wall to move from one coil to the other, from which the velocity can be calculated.

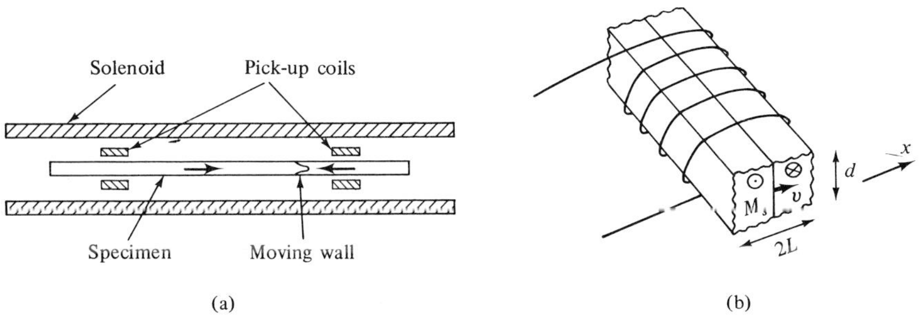

Fig. 12.3 Measurement of domain wall velocity. (a) Wire or whisker specimen. (b) Portion of one leg of a picture-frame specimen wound with secondary coil.

2. Picture-frame specimens, like that of Fig. 9.22, can be cut from large single crystals so that each leg is an easy axis and contains only one domain wall. The specimen is wound with a primary (magnetizing) and a secondary coil. For low drive fields (small wall velocity) the hysteresis loop can be recorded on a hysteresigraph and the time measured with a stop watch for the recording pen to move from top to bottom of the square loop, corresponding to motion of the wall over the distance $2L$ in Fig. 12.3 (b). For high drive fields the size of the voltage pulse induced in the secondary is measured with an oscilloscope, when the wall is driven from one side of the leg to the other. This voltage is

$$e = 10^{-8} N (2Ld) \frac{dB}{dt}, \tag{12.9}$$

where N is the number of terms in the secondary. At the low drive fields involved, dB/dt is essentially equal to $d(4\pi M)/dt$, and $M = (x/L)M_s$, where x is the wall position measured from the center of the leg. Therefore

$$\frac{1}{4\pi} \frac{dB}{dt} = \frac{dM}{dt} = \frac{dx}{dt} \left(\frac{M_s}{L} \right) = v \left(\frac{M_s}{L} \right), \tag{12.10}$$

from which v can be calculated. These relations depend on the assumption that the wall remains flat during its motion; this is not always true, as we shall see below.

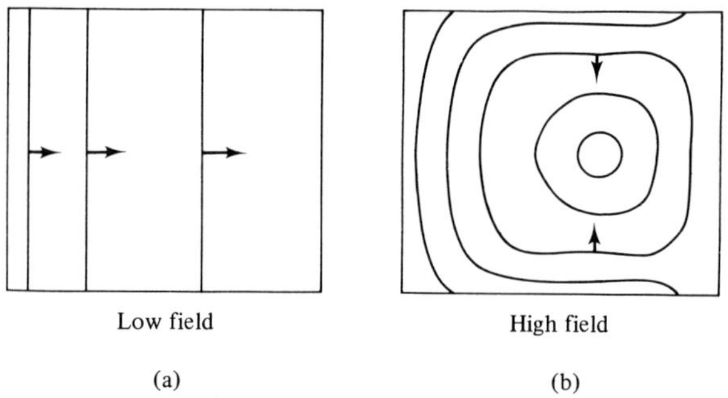

Fig. 12.4 Cross sections of leg of picture-frame crystal, showing wall motion in low and high fields. The magnetization directions are normal to the page. Williams *et al.* [12.2].

Williams, Shockley, and Kittel [12.2] determined wall velocity in the same Fe-Si picture-frame crystal previously studied by Williams and Shockley [9.16] and shown in Fig. 9.22. At very low excess fields, $(H - H_0)$ no larger than 0.003 Oe, they found v equal to about 0.017 cm/sec by timing the hysteresis loop, corresponding to a mobility C of about 5.5 cm/sec Oe. At higher fields of 2 to 80 Oe they found that the domain wall no longer remained flat but assumed the successive positions shown in Fig. 12.4 (b). The wall is retarded more at the center than

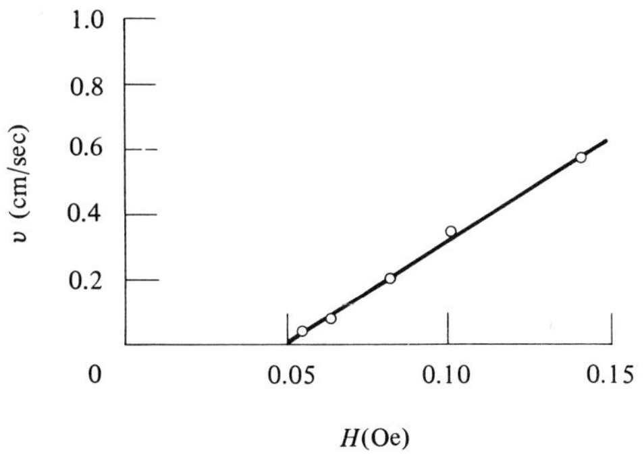

(a)

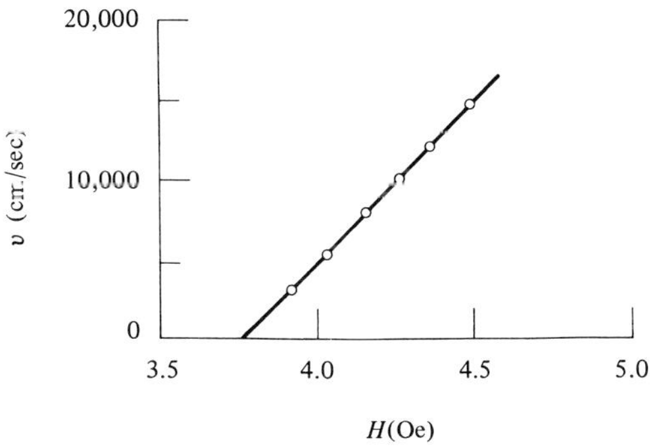

(b)

Fig. 12.5 Domain wall velocity v as a function of applied field H. (a) In an Fe + 3 percent Si single-crystal picture frame. After Stewart [12.3]. (b) In an Fe + 14 percent Ni wire, 0.038 cm diameter, under 92 kg/mm^2 tension. Sixtus and Tonks [12.5].

at the top and bottom, until it curls around on itself to enclose a cylindrical domain which shrinks and finally vanishes. The same effect occurs in a wire but in a less extreme form because of the difference in specimen shape, as suggested in the sketch of Fig. 12.3 (a); there the domain wall, moving from right to left, has the form of a cone, much more elongated than shown, with the tip of the cone lagging behind.

Stewart [12.3] also determined wall velocity in a picture-frame Fe-Si crystal, by measuring the secondary-coil output, with the results shown in Fig. 12.5 (a). The mobility C is 6.3 cm/sec Oe.

In sharp contrast to these data are the results of Galt [12.4] on a single-crystal picture-frame specimen of a mixed ferrite (75 mol percent Ni ferrite, 25 mol percent Fe ferrite). The mobility was about 30,000 cm/sec Oe.

Mobilities of this high order have also been found in metals and alloys, but only in the form of fine wires, whiskers, and thin films. For example, Sixtus and Tonks [12.5], who were the first to measure wall velocity in any material, observed a mobility of 20,000 cm/sec Oe in an Fe-Ni wire under tension; some of their extensive data are shown in Fig. 12.5 (b). For the iron whiskers examined by DeBlois [11.18], the v,H relation was not strictly linear; instead, the points fell approximately on two straight lines, with the higher-field segment having a higher slope. At low fields the mobility was of the order of 10,000 cm/sec Oe. The highest observed velocity was 49 km/sec at 135 Oe. (These experiments on wires and whiskers are complicated by the fact that the greater portion of the moving wall makes a small angle with the specimen axis. As a result the axial wall mobilities or velocities, quoted above, are 100 or more times the true velocity of the wall normal to itself, called the normal velocity. The true shape of the moving wall is not always easy to deduce from the experimental observations.)

The various mobilities C noted above are a measure of the viscous damping parameter β, because $\beta = 2M_s/C$. Two causes of viscous damping have been identified: (1) eddy currents, characterized by a parameter β_e, and (2) an intrinsic or relaxation effect, characterized by β_r. The observed β is the sum of β_e and β_r.

Eddy-current damping. This is due to the field H_{ec} set up locally by eddy currents circulating in and near a moving domain wall, as indicated in Fig. 12.6. When the applied field H_a moves the wall from 1 to 2, the flux change in the region swept out induces an emf which causes eddy currents in the direction shown, namely, the direction which causes an eddy-current field H_{ec} opposed to the applied field. Because the true field actually acting on the wall is now less than H_a, the velocity v of the wall is less than it would be if the eddy currents did not exist, i.e., the wall motion is damped. (Another way of looking at this effect is to note that the heat developed by the eddy currents per unit of time represents a power loss, and the input power required to move the wall past various kinds of obstacles with a particular velocity must be increased to replace this loss.) To distinguish the scale, the large-scale eddy currents in the rod of Fig. 12.1 are usually called *macro eddy currents*, while those associated with a moving domain wall are known as *micro eddy currents*. However, the two are not inherently different. The rod of Fig. 12.1

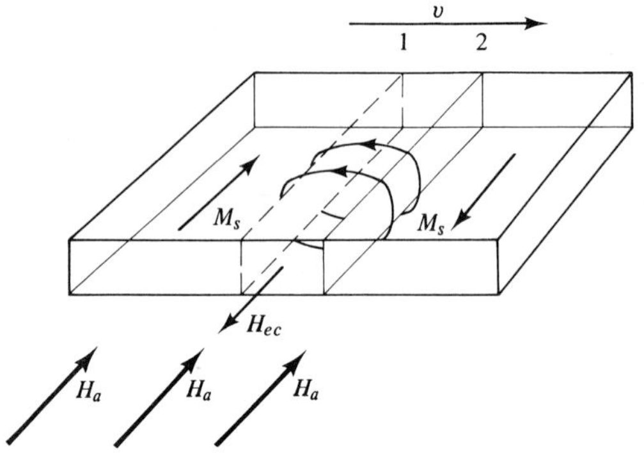

Fig. 12.6 Micro eddy currents associated with a moving domain wall.

is tacitly assumed to be polycrystalline and to contain a great many domains; its macro-eddy-current system is simply the vector sum of all the tiny whirls of micro eddy currents associated with its moving domain walls. *Micro eddy currents are an important aspect of domain wall motion.*

Eddy-current damping can be accurately calculated only in special cases. One is a single wall moving in the specimen of Fig. 12.3 (b), which has a cross section of $2L$ by d. The drive field is assumed to be low enough that the moving wall remains flat. Crystal imperfections are ignored, so that the only force resisting wall motion is eddy-current damping; the constant H_0 in Eq. (12.8) is then zero and the mobility is simply $C = v/H$. The power dissipated in eddy currents can then be equated to the rate at which the applied field does work on the moving wall. With these assumptions and after a complex calculation, Williams *et al.* [12.2] found that

$$C = \frac{v}{H} = \frac{\pi^2 \rho c^2}{32 D B_s d}, \tag{12.11}$$

where ρ is the resistivity, c the velocity of light, and D a numerical constant whose value (Problem 12.2) depends only on the ratio $2L/d$ and equals 0.97 for a square rod ($2L = d$). The units in this equation are Gaussian [12.6]. These are mixed units, in which electrical quantities like ρ are in *cgs* electrostatic units and magnetic quantities like B are in *cgs* electromagnetic units. (A major advantage of the *mks* system, in dealing with electrical problems or mixed electrical-magnetic problems, like eddy currents, is that it replaces this confusing system of mixed units with a single system.) To rephrase this equation in the mixed practical-electromagnetic units in which this book is written, we note that 1 statohm, the electrostatic unit of resistance, equals $1/(9 \times 10^{11})$ ohm. Inserting this factor and the value of c ($= 3.00 \times 10^{10}$ cm/sec) and replacing B_s with $4\pi M_s$, because H is normally negligible in comparison, we have

$$C = \frac{v}{H} = \frac{10^9 \pi \rho}{128 D M_s d}, \tag{12.12}$$

where C is in cm/sec Oe, ρ in ohm-cm, M_s in emu/cm^3, and d in cm. The larger the resistivity ρ and the smaller d, the more restricted are the eddy currents and the larger the mobility. This conclusion agrees with the experimental results previously described: mobility is high even in thick specimens of ferrites because ρ is so very large, and in metallic fine wires and whiskers because d is so small. (Note also that the quantities D and d, characteristic of the specimen shape and size, appear in the above equation. This dependence on dimensions is a general characteristic of eddy currents, whether macro or micro and whether due to a suddenly applied and then constant field or to an alternating field. Eddy currents are always a function of specimen size and shape as well as of the intrinsic properties of the material. This means that eddy-current calculations cannot be done once and for all for a given material but must be repeated for each different specimen.)

When the values appropriate to the Fe-Si specimen of Fig. 12.3 (b) are inserted into Eq. (12.12), the calculated value of C agrees very well with the experimental value of Williams *et al.*, namely, 5.5 cm/sec Oe. This shows that the observed mobility can be explained *entirely* by eddy-current damping; the intrinsic damping is negligible. This is a quite general result, applying even when the drive field is so high that the wall does not remain flat: *wall motion in metallic materials is damped only by eddy currents*, essentially because of the low resistivity of these materials. The only proviso needed is that the specimen be of "normal" size, where this term excludes fine wires, whiskers, and films.

When the drive fields is high, the flat wall becomes distorted, as in Fig. 12.4, because eddy currents slow down the wall more in the interior of the specimen than at the surface, essentially because of the effects sketched in Fig. 12.2. The wall finally becomes a collapsing cylinder, and the eddy currents associated with its motion have also been treated quantitatively by Williams *et al.* [12.2].

Intrinsic damping. This effect, also called relaxation damping, is a retardation of the rotation of electron spins. When an applied field attempts to rotate a spin into the field direction, and that is the elemental act occurring during domain wall motion, this rotational motion is damped. As a result, a wall cannot move infinitely fast, even if eddy currents are absent. The mechanism of intrinsic damping in metals is not understood, and in insulators there appear to be several mechanisms. The theories of intrinsic damping are difficult, and the calculation of the intrinsic damping parameter β_r is inexact. In practice, what is usually done is to subtract the calculated eddy-current damping parameter β_e from the measured parameter β to arrive at the value of β_r.

That some other damping mechanism besides eddy currents must be the major cause of damping in some specimens may be concluded as follows. As noted above, the wall mobility in a metal (Fe-Si) crystal is about 6 cm/sec Oe at low drive fields and is entirely eddy-current limited. According to Eq. (12.12), this mobility is directly proportional to the resistivity ρ. Therefore, a semiconductor such as a ferrite, which has a resistivity of some 10^7 to 10^{11} times that of a metal, should have a wall mobility of the order of 10^8 to 10^{12} cm/sec Oe. Yet the measured mobility of the Ni-Fe ferrite crystal studied by Galt [12.4], which had dimensions of the same order of mafnitude as the Fe-Si crystal, is only 3×10^4 cm/sec Oe. In other words, eddy-current damping is so feeble in this material that it would permit mobilities some 10^4 to 10^8 times those actually observed, if some other damping mechanism did not intervene. This conclusion is also fairly general: *wall motion in ferrites is controlled entirely by intrinsic damping.* Eddy currents play a negligible role.

Examples of specimens in which both kinds of damping are effective together are (1) magnetite crystals of "normal" size, because the resistivity of Fe_3O_4 is very much less than that of most ferrites, and (2) whiskers or fine wires of metals and alloys, because the dimension d of Eq. (12.12) is so small that eddy-current damping is not so large that it obscures intrinsic damping. For example, β_e/β_r is estimated as about 0.4 for certain Ni-Fe wires under tension and as about 3 for certain iron whiskers [11.18], whereas this ratio would approach infinity for metal specimens of "normal" size and zero for most ferrites.

Metallic thin films have a thickness much less than that of any wire, and one would therefore expect that eddy-current damping would be negligible in films. Most investigations have supported this view, but there have been some contrary reports. Hatfield [12.7] concluded that the damping in 80 Permalloy films at room temperature was entirely intrinsic, but that eddy currents began to contribute a substantial portion of the total damping at temperatures below 200°K, because of the decrease in resistivity. The mobility in an 80 Permalloy film, 4000 Å thick, is of the order of 2000–5000 cm/sec Oe.

12.4 SWITCHING SPEED

The memory elements of digital computers (Section 13.6) are tiny bits of magnetic material which must have three characteristics: (1) a square hysteresis loop, (2) the capability of very fast magnetization reversal, and (3) a low coercive force.

A piece of magnetic material which has a square loop and is so shaped that it has a negligible demagnetizing field will have two clearly defined remanent states, shown by the black dots in Fig. 12.7 (a); one of these states signifies the binary digit 1 and the other the binary digit 0. The number 1 can be "written" on the element by applying, and removing, a field somewhat greater than $+H_c$.

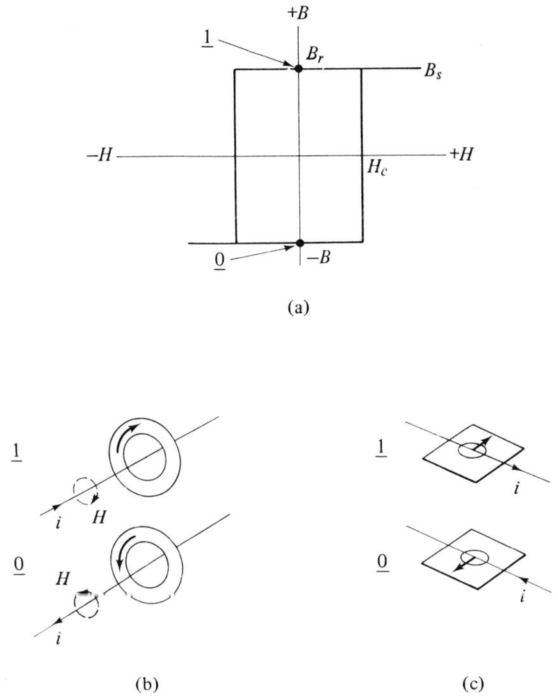

Fig. 12.7 Ideal square hysteresis loop and memory elements.

If we now wish to write 0, we can switch the element to the 0 state by applying and removing a field negatively greater than $-H_c$. The speed at which the computer can operate depends on the speed with which each element can be switched, and this fact has provided the impetus for most of the research on switching speed. The switch can be exceedingly fast. An appropriate time unit is not the millisecond (= msec = 10^{-3} sec) but the microsecond (= μsec = 10^{-6} sec), or even the nanosecond (= nsec = 10^{-9} sec). A light beam travels only about one foot in one nanosecond.

To increase switching speed requires an understanding of the factors limiting this speed, and this in turn requires a knowledge of the various mechanisms of magnetization reversal. Some very fundamental problems in magnetism are involved, and the subject is complex.

In order to reduce power consumption and heat production in the computer, the fields needed to reverse the magnetization of the memory elements, and the currents that produce these fields, must be low. This is the reason for the requirement of low coercive force mentioned above. It is typically of the order of 1 or 2 Oe.

The memory elements in present computers are ferrite toroids and thin films. The *ferrite toroids*, called cores, are less than 0.1 inch in diameter and are made by pressing and sintering a mixture of ferrites. The chief constituent is manganese ferrite. Each core is threaded with one or more wires, as in Fig. 12.7 (b), and a current pulse in the wire will magnetize the core circumferentially, clockwise for 1 or counterclockwise for 0. The reason that ferrite cores have square loops is not clear: it appears to be a matter of composition and sintering conditions. Low magnetostriction is desirable, to minimize the effect of residual microstress. The overall demagnetizing factor, for circumferential magnetization, is zero, but ferrites are usually porous and the free poles on pore surfaces are the source of local demagnetizing fields that may affect loop squareness. *Thin films* are in the form of small circular deposits of 80 Permalloy, evaporated on to a flat substrate. They have a square loop by virtue of the easy axis induced during deposition (Section 11.10). Bits of thin film are shown in Fig. 12.7 (c); magnetization in the positive or negative direction of the easy axis represents 1 or 0. The magnetizing field is provided by a current pulse in a wire or strip near the film and at right angles to the easy axis.

Ferrite cores far outnumber thin films in present computers. Although thin films switch faster, flat evaporated films have encountered production problems that have restricted their application. However, cylindrical films plated on wire are currently thought to have great promise as a replacement for ferrite cores. The switching behavior of cores and films is described separately below.

Ferrite cores

The switching of cores is measured by threading the core with two wires, as in Fig. 12.8 (a), forming a one-turn primary "winding" and a one-turn secondary. A pulse of current i is passed through the primary, producing a pulse of circumferential field H which magnetizes the core. The flux change in the core induces an emf e in the secondary, which is connected to an oscilloscope on which the form of the

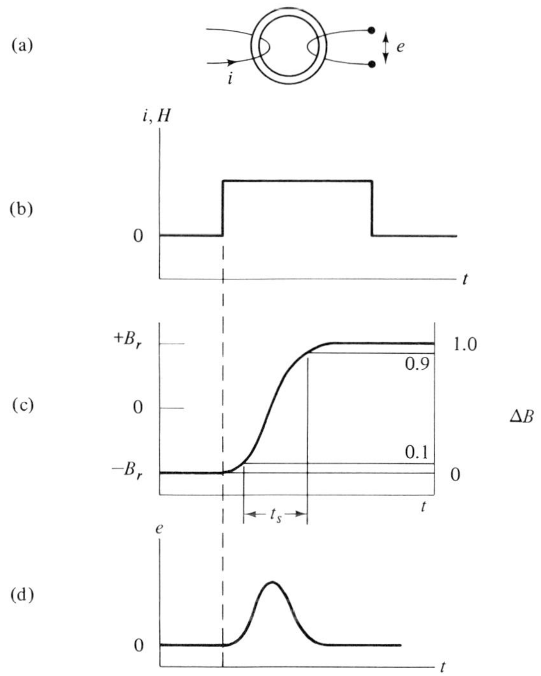

Fig. 12.8 Switching of a ferrite core (schematic).

output voltage pulse can be examined. Suppose the induction in the core is initially $-B_r$ (0 state) and that a positive field pulse is applied, large enough to switch the core to $+B_r$ (1 state). Suppose also that the field pulse is square, as in (b). Then the induction will change with time t approximately as in (c). The *switching time* t_s is usually defined as the time required for B to change from 0.1 to 0.9 of the total change ΔB. The induced emf e, which is proportional to dB/dt, varies with time as in (d); the size and shape of this voltage pulse depends on the magnitude of the drive field H.

It is found experimentally that the larger the drive field H the shorter the switching time t_s, as shown by Fig. 12.9 for a particular ferrite. Most of this curve is a straight line given by

$$S = t_s(H - H_0) \tag{12.13}$$

or

$$\frac{1}{t_s} = \frac{1}{S}(H - H_0), \tag{12.14}$$

where S is called the *switching constant*, or coefficient, and H_0 is the intercept of the straight line on the field axis. These equations, although similar in form to Eq. (12.8) for domain wall velocity, do not imply anything about the mechanism of magnetization reversal. They merely describe measurements of switching time.

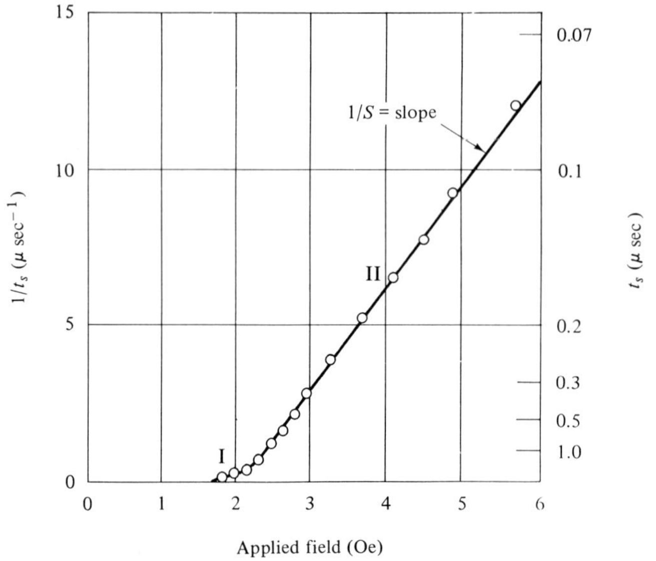

Fig. 12.9 Variation of switching time t_s (right ordinate) and its reciprocal (left) for a ferrite core of composition $Mn_{0.46}Mn_{0.71}Fe_{1.76}O_4$. Gyorgy [12.8].

Figure 12.9 shows that the switching time t_s decreases rapidly as the drive field is increased; for the highest field shown there, t_s for this ferrite is about 80 nsec. However, it is not usual to apply fields of this size. In fact, in one mode of computer operation (coincident-current mode, Section 13.6), the drive field must not exceed about twice the coercive force H_c. Ferrite cores are normally operated at a switching time of about 1 μsec.

By what mechanism does the magnetization reverse? The cross section of a core is shown in Fig. 12.10, and the mechanism illustrated there can be quickly ruled out. It is a rotation of the spins, either coherently about a single axis in the core plane or about the core radii; either leads to the intermediate state of high magnetostatic energy shown in (b). If the dimension b of the core is somewhat larger than a, as shown, the demagnetizing factor parallel to b might be conservatively estimated as about equal to π; circumferentially, normal to the plane of the drawing, the demagnetizing factor is zero. The shape anisotropy constant K is then $\pi M^2/2$, from Eq. (7.54). The field required to produce the rotation would then be $2K/M$ or πM, which is just the demagnetizing field for the intermediate state. Inasmuch as M is about 200 emu/cm^3, this field amounts to some 600 Oe or several hundred times the field actually required for reversal. Coherent rotation in the plane of the core would be even more difficult.

Other reversal mechanisms must therefore operate, and these have been reviewed by Gyorgy [12.10, 12.11]. In the lowest fields at which reversal takes

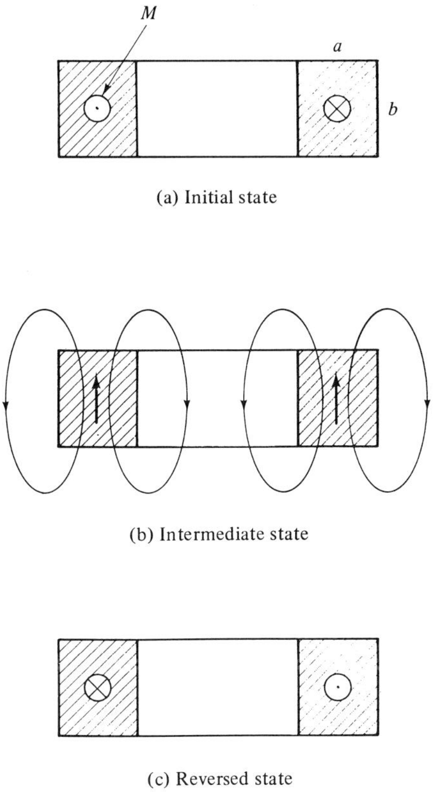

Fig. 12.10 Rotational switching of a ferrite core. After Smith [12.9].

place, and where the $1/t_s, H$ relation is usually a curve rather than a straight line, the reversal mechanism is generally considered to be domain wall motion. The range of fields in which this occurs is called region I by Gyorgy. If we knew the number of walls moving and the distance they moved before reversal was complete, we could calculate the wall velocity from the switching time. The simplest assumption is that a single 180° wall is nucleated or unpinned in each grain and then travels across the grain. The grain size of ferrite cores is typically about 50μ. If t_s is 1 μsec, then the wall velocity would be $50 \times 10^{-4}/10^{-6}$ or 5000 cm/sec. If the excess drive field $(H - H_0)$ is of the order of 1 Oe, this velocity corresponds to a mobility of 5000 cm/sec Oe. This is not unreasonable, in that a wall mobility of 3000 cm/sec Oe in a Mn ferrite single crystal has been observed [9.4]. If the reversal is accomplished by the motion of more than one wall per grain, the average wall velocity will be less. The fact that the $1/t_s, H$ relation is not linear suggests that the number of walls nucleated per grain increases with the drive field.

In region II the $1/t_s,H$ curve becomes a straight line, t_s decreases very rapidly with increasing H, and the reversal mechanism becomes incoherent rotation of the spins, with the reversal velocity limited by intrinsic damping. (Note that reversal by simultaneous spin rotation, coherent or incoherent, is a very much faster process than reversal by wall motion, for the same angular velocity of spin rotation. In the time required for a spin to rotate 180°, the whole specimen can reverse by simultaneous rotation, but in the same time a 180° domain wall will move a distance equal only to its own width δ. If there are many walls and each has to move a distance $k\delta$ to accomplish reversal, then the switching time t_s for wall motion is k times t_s for simultaneous rotation; k is usually a large number.) The rotation mechanism envisaged by Gyorgy [12.10, 12.11] is such that the free poles formed on the core surface, when the rotation has gone half way, are not all of the same sign all around the circumference on a surface such as a of Fig. 12.10 (a). Instead, regions of alternate sign succeed one another. This leads to lower magnetostatic energy than the intermediate state sketched in Fig. 12.10, where north poles would exist all around the top surface of the core.

Thin films

The switching of films is in many ways similar to that of ferrite cores: at low fields reversal occurs by wall motion and at higher fields by rotation. There is, however, one marked geometrical difference. Simply because the film is so thin, the demagnetizing factor is negligible in any direction in the plane of the film. Thus spins can rotate in this plane without going through any high-energy state. It is also easy in a film, but not in a core, to superimpose on the main easy-axis drive field another field at right angles to this axis in the plane of the film and constant in magnitude. This transverse field causes faster switching.

We digress now from the main topic of switching speed to examine the effect of two fields at right angles on the quasi-static hysteresis loop. We assume that the film is a single domain and that its magnetization changes only by coherent rotation. (This is true only in certain circumstances. However, it is useful to calculate this kind of behavior in order to have a standard with which experimental results can be compared.) The uniaxial anisotropy energy of the film is given by

$$E_a = K_u \sin^2 \theta,$$

and its anisotropy field by

$$H_K = \frac{2K_u}{M_s} \cdot$$

We express the easy-axis field H_e and the transverse, hard-axis field H_t in terms of reduced fields h, just as in Section 9.13:

$$h_e = \frac{H_e}{H_K} = \frac{H_e M_s}{2K_u} \quad \text{and} \quad h_t = \frac{H_t}{H_K} = \frac{H_t M_s}{2K_u}.$$

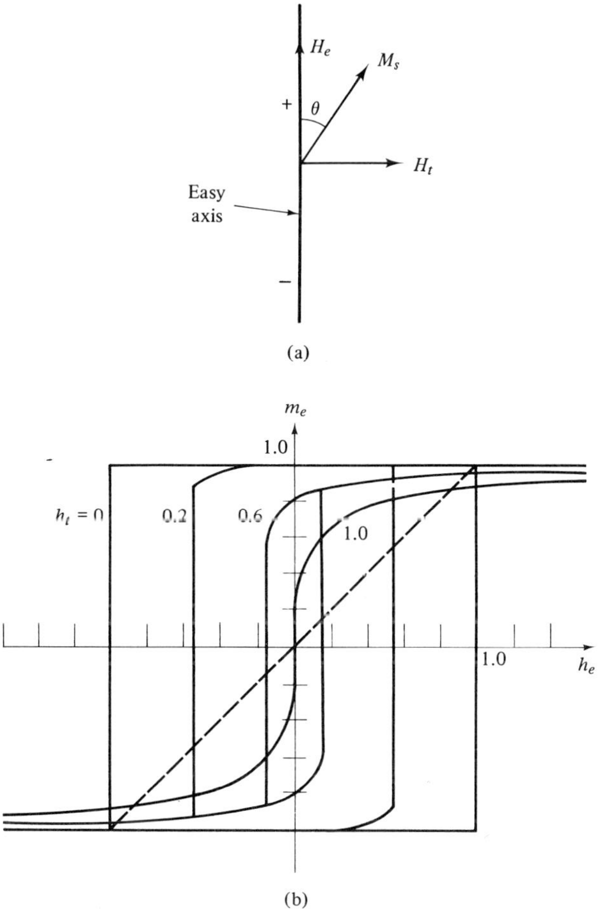

Fig. 12.11 Calculated hysteresis loops as a function of reduced transverse field h_t. Smith [12.9].

Similarly, reduced magnetization $= m = M/M_s = \cos\theta$. When $h_e = 0$, the transverse hysteresis loop is merely a straight line, shown dashed in Fig. 12.11 (b) and as a full line in Fig. 9.38. When $h_t = 0$, the longitudinal loop is square, with a reduced intrinsic coercivity of $h_{ec} = 1$, again as we found in Fig. 9.38. To find the effect of both fields acting together, we write the equation for the total energy:

$$E = K_u \sin^2\theta - H_e M_s \cos\theta - H_t M_s \cos(90° - \theta), \tag{12.15}$$

where the positive direction of H_e is indicated in Fig. 12.11 (a). The first term in this equation is the anisotropy energy and the next two are the potential energy of the magnetization in the easy-axis and transverse fields. Proceeding as in

Section 9.13, we form the first derivative to find the equilibrium position of M_s:

$$\frac{dE}{d\theta} = 2K_u \sin\theta\cos\theta + H_e M_s \sin\theta - H_t M_s \cos\theta = 0, \tag{12.16}$$

$$\sin\theta\cos\theta + h_e \sin\theta - h_t \cos\theta = 0. \tag{12.17}$$

To find the critical field h_{ec} at which the magnetization will irreversibly flip, which is also the coercivity, we set the second derivative equal to zero:

$$\frac{d^2E}{d\theta^2} = \cos^2\theta - \sin^2\theta + h_e\cos\theta + h_t\sin\theta = 0. \tag{12.18}$$

Solving (12.17) and (12.18) together, we find h_{ec} in terms of the parametric equations

$$h_t = \sin^3\theta, \quad h_{ec} = -\cos^3\theta. \tag{12.19}$$

To plot the hysteresis loop for a given value of h_t, we find h_{ec} from Eqs. (12.19), thus locating the vertical sides of the loop, and we find points on the remainder of the loop from Eq. (12.18). Loops calculated in this way are shown in Fig. 12.11 for $h_t = 0.2$, 0.6, and 1.0; the last is merely a line, indicating reversible but non-linear magnetization, with $h_{ec} = 0$. The effect of a transverse field is seen to be a *decrease* in the easy-axis coercivity, because the transverse field supplies some of the energy needed to overcome the anisotropy energy.

Observed hysteresis loops, made with a loop tracer at a few hundred Hz, agree only in part with the calculated loops. When the field is applied along the easy axis ($H_t = 0$), the loop is square as expected (Fig. 12.12 a, b), and M_s and H_c can be obtained directly from the loop. When the field is in the hard direction ($H_e = 0$), the loop is merely a straight line, as expected from the rotation theory, but only at low fields (c). If the drive field is increased, as in (d), the line opens up into a loop of finite area, showing that irreversible processes are occurring. It is customary to determine the anisotropy field H_K by extrapolating the low-field line to intersect the M_s level, as in (a).

The easy-axis coercivity H_c is usually considerably less than H_K, in disagreement with the theoretical loops of Fig. 12.11. The reason is that reversal is occurring by wall motion rather than rotation, and the field necessary to nucleate or unpin walls is less than the field required for rotation. That wall motion must be occurring can be shown by examining the form of the unintegrated output from the secondary coil during reversal; it shows the spikes characteristic of Barkhausen jumps (Fig. 9.21 b).

On the other hand, magnetization along the hard axis produces the linear M,H relation predicted by rotation theory, as long as the field does not exceed about 0.5 H_K. But if the field is large enough to produce saturation in the hard direction, the film will break up into domains when the field is reduced, as shown in Fig. 11.36 (d), because of the magnetization ripple illustrated in Fig. 11.37. Wall motion is accompanied by hysteresis, and the M,H line opens up into a loop (Fig. 12.12 d).

If a constant transverse field is applied, in addition to the main easy-axis field, reversal can be made to occur by rotation, as evidenced by the fact that the unin-

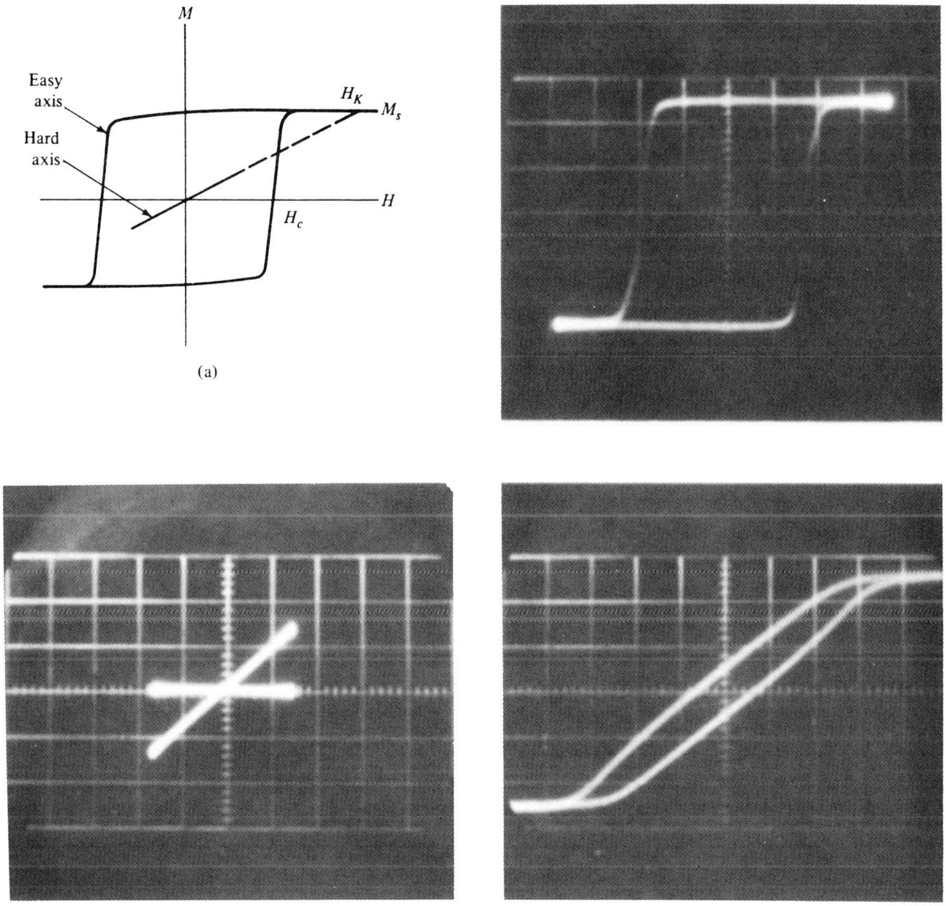

Fig. 12.12 Typical thin-film hysteresis loops. (a) is schematic. (b), (c), and (d) are for a Permalloy film and are made with a loop tracer; 1 horizontal division = 1 Oe. (b) results from a high drive field along the easy axis, (c) and (d) from low and high drive fields, respectively, along the hard axis. The horizontal line in (c) is obtained with the film removed from the search coil. (Courtesy of E. J. Torok, Sperry Rand Univac Division, St. Paul, Minnesota.)

tegrated output of the secondary coil is now a smooth curve without spikes. Even more direct proof of this has been obtained by aligning the secondary-coil axis with the hard axis, as in Fig. 12.13. When H_t is zero, and reversal occurs by motion of 180° walls parallel to the easy axis, no emf is induced in this transverse pick-up coil. But when rotation occurs, in the presence of a transverse field, the flux through this coil changes and a voltage is induced in it.

Further proof of the rotation mechanism has been found by Olson and Pohm [12.12] and Smith [12.9] by determining the various combinations of H_e and H_t necessary for rotational switching. The results are in good agreement with

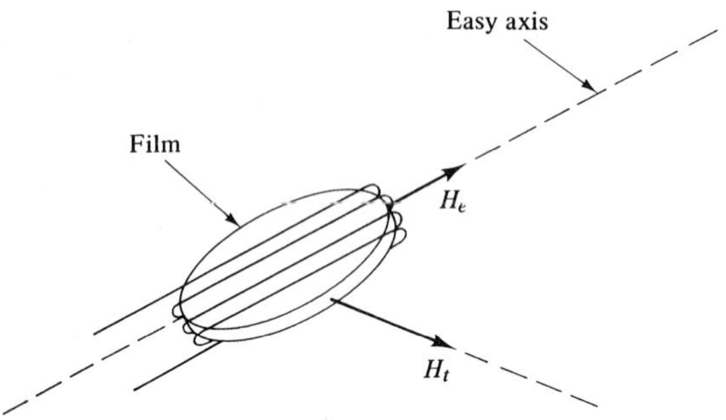

Fig. 12.13 Transverse pick-up coil.

Eqs. (12.19) of the rotation theory, as shown in Fig. 12.14. The main point to note is that, when h_t exceeds about 0.3, the film will switch by rotation when acted on by an easy-axis field h_{ec} smaller than that required for wall-motion reversal in the absence of a transverse field. A region in which reversal by wall motion can occur is also indicated; the lower limit of this region, shown as 0.5 H_K, is not exact, inasmuch as H_c/H_K often varies from one film to another. When h_t is less than about 0.2, h_c for wall motion is less than h_{ec} and rotational switching can occur only when the "rise time" of the easy-axis field is less than the time required for domain walls to form and move; this is the time required for H_e to rise from zero to its maximum value; it was equal to 1.5 nsec in the experiments on which Fig. 12.14 is based.

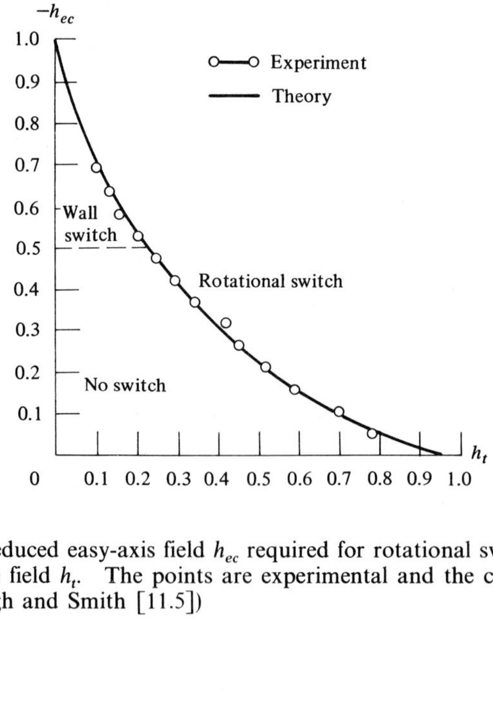

Fig. 12.14 Critical reduced easy-axis field h_{ec} required for rotational switching as a function of reduced transverse field h_t. The points are experimental and the curve is a plot of Eqs. (12.19). (Goodenough and Smith [11.5])

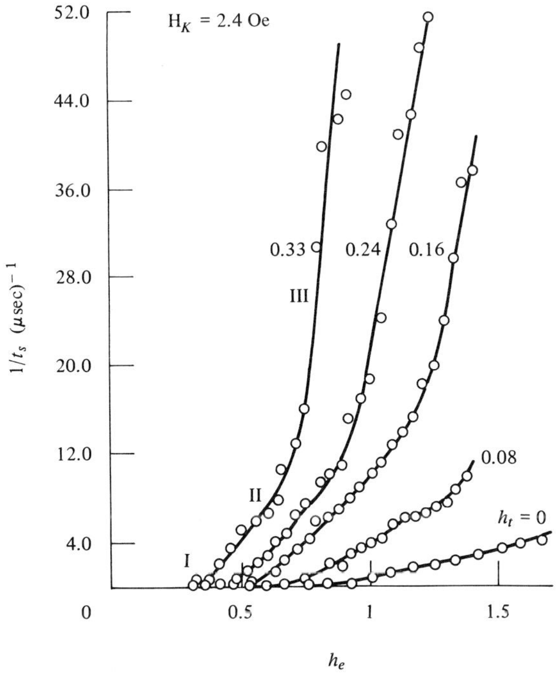

Fig. 12.15 Inverse switching time $1/t_s$ of a 1600 Å film as a function of easy-axis drive field h_e for various transverse fields h_t. The reduced coercive force h_c is 0.56. Measurements of Olson and Pohm [12.12], as presented by Goodenough and Smith [11.5].

Curves of switching speed for films resemble those for ferrite cores, with the additional effect of the tranverse field. Figure 12.15 shows $1/t_s$,H curves for various values of h_t for a particular film. When h_t is zero, the line is first curved, then straight, as for ferrites, comprising regions I and II. When a transverse field is present, the curve has two straight-line segments; the one of larger slope is called region III. In region I reversal occurs by wall motion, in region II by incoherent rotation, and in region III by coherent rotation. The reversal mechanism in region II is not well understood and appears to be called "incoherent rotation" only for convenience; high-speed Kerr photographs by Kryder and Humphrey [12.13] show that the reversal mechanism is a mixture of rotation and the motion of some kind of diffuse boundaries.

As Fig. 12.15 shows, a constant transverse field can radically decrease the switching time by causing reversal to occur by rotation rather than wall motion. Thus, at a drive field h_e of about 0.9, a transverse field h_t of 0.33 reduces t_s a hundredfold, from about 2 μsec to about 20 nsec. Dietrich and Proebster [12.14] observed a switching time of only 1 nsec for a 1500 Å film with $h_e = 2.0$ and $h_t = 0.42$.

12.5 TIME EFFECTS

We turn now to a group of phenomena loosely known as "time effects," which manifest themselves on a time scale ranging from seconds to days. The most important of these effects, which have been given a bewildering variety of names, can be classified as follows:

1. *Time decrease* (or *decay) of the permeability* or *disaccommodation.*
2. *Magnetic after-effect* or *magnetic viscosity.* Two different kinds have been observed:
   a) *Diffusion after-effect* or *reversible after-effect* or *Richter after-effect.*
   b) *Thermal fluctuation after-effect* or *irreversible after-effect* or *Jordan after-effect.*

These effects have been reviewed by Rathenau [12.15] and by Rathenau and DeVries [12.16].

Excluded from this category are any time-dependent magnetic changes caused by metallurgical phase changes. The magnetic properties of certain magnetic alloys will change with time if the alloy is in an unstable state, such that a second phase is being precipitated; an example is the aging of steel, in which carbon is withdrawn from solution in the iron to precipitate as cementite Fe_3C. Or a magnetic phase may precipitate from a nonmagnetic alloy, causing pronounced changes in the magnetic properties; an example is the Cu-Co alloy described in Section 11.7. Such topics are not included in this section, which is restricted to time effects occurring in pure substances or single-phase solid solutions.

Time decrease of the permeability

This effect has been most often studied in iron containing interstitial carbon and/or nitrogen in solution. The effect is dependent on these interstitials and does not exist in pure iron. Suppose a specimen of iron containing some carbon, say, in solution is partially magnetized and then demagnetized in the usual way, i.e., with a gradually decreasing alternating field. Let the initial permeability μ_0 be measured, immediately after demagnetization and at intervals thereafter, with a *small* field, i.e., a field so small that very little change in the domain structure is produced by the measurement of μ_0. Then μ_0 is found to decrease with time after demagnetization. This process is shown schematically in Fig. 12.16 (a) and by actual measurements in (b). The effect is substantial, in that the drop in permeability can amount to more than 50 percent of the initial value.

This effect is due to the preferential distribution of carbon atoms with respect to the local direction of the M_s vector, namely, in a plane at right angles to the M_s vector in iron, as described in Section 10.3. Once this distribution is set up, the domain walls tend to become stabilized (Section 10.2). The kind of potential wells caused by the carbon atoms, and the forces required to free domain walls from these wells, have been sketched in Fig. 10.7. With these facts in mind we can interpret Fig. 12.16. During demagnetization, domain walls are in contin-

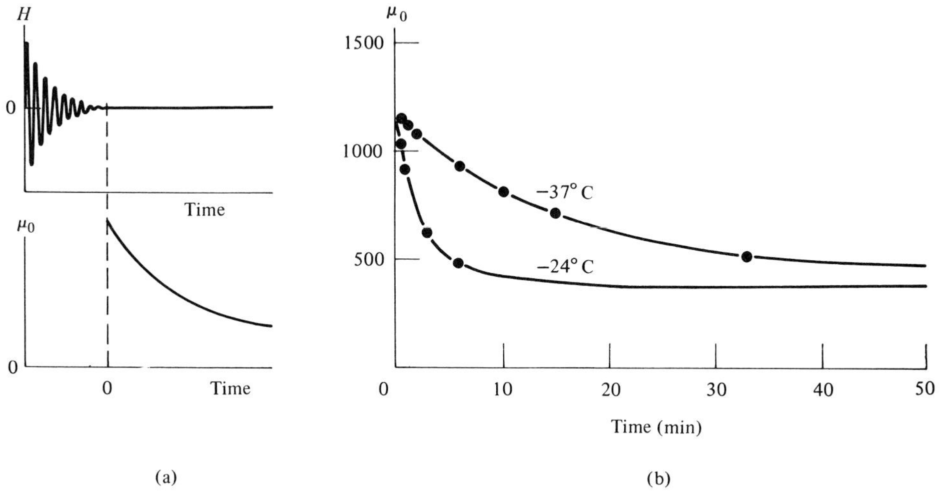

Fig. 12.16 Time decrease of the permeability. (a) Schematic (Rathenau and DeVries [12.16]). (b) Measurements on iron containing carbon. (Snoek [12.17])

ual motion over large distances, and in any local region there is no preferred direction of M_s for any length of time and therefore no preferred distribution of carbon atoms; x, y, and z sites are equally populated. When demagnetization ceases, at time $t = 0$, the permeability is high, because the isotropic carbon distribution cannot stabilize domain walls. But the carbon atoms immediately begin to distribute themselves preferentially in each domain, i.e., the material self-magnetically anneals, and potential wells begin to build up with time at each domain wall, somewhat as sketched in Fig. 12.17. The walls become increasingly difficult to move, which is another way of saying that the permeability μ_0 decreases.

Domain wall stabilization can be demonstrated in another way. Instead of measuring just μ_0, Brissonneau [12.18] determined the initial portion of the

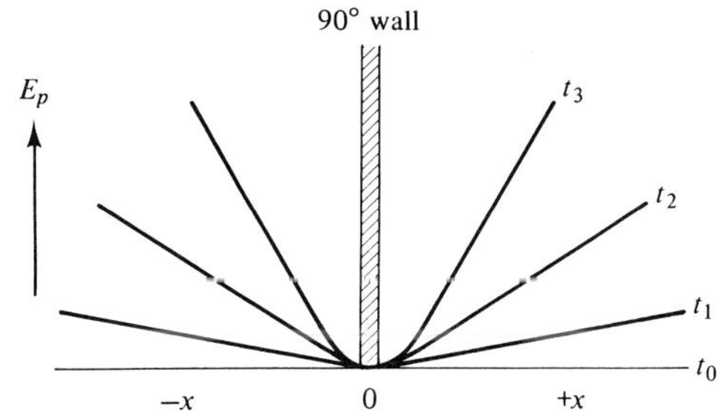

Fig. 12.17 Build up of a potential well at the position of a 90° domain wall as a function of time t. Demagnetization ceased at time t_0. (Schematic)

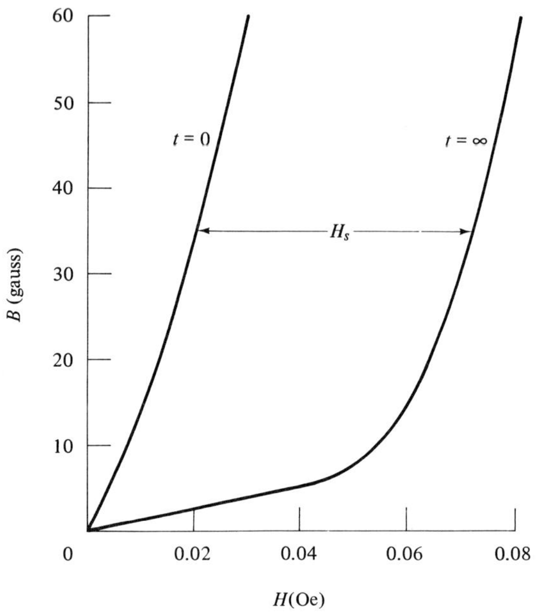

Fig. 12.18 Normal induction curves, at a constant time t after demagnetization, of iron containing 46 ppm C. Temperature = $-$ 27.3°C. (After Brissonneau [12.18])

B,H curve for 17 time periods after demagnetization, ranging from 0.7 to 1000 minutes. Only two of his extreme curves, for $t = 0$ and $t = \infty$ (both obtained by extrapolation), are shown in Fig. 12.18. The horizontal separation of the two curves is called the *stabilization field* H_s; this is the extra field that must be applied a long time after demagnetization to produce the same induction B that a smaller field would produce immediately after demagnetization. (It has already been alluded to in Section 10.2 with reference to Perminvar.) The value of H_s measures the strength of the wall stabilization, and it is itself a function of the induction B. As B increases, Brissonneau found that H_s increases, goes through a maximum at about 100 gauss, and then decreases to a constant value at about 2500 gauss and beyond. The maximum in H_s is associated with the freeing of 180° walls from their potential wells, because the force-distance curve of Fig. 10.7 (b) goes through a maximum for 180° walls; the constant value of H_s at higher inductions is associated with the motion of 90° walls over large distances, in conformity with the corresponding force-distance curve.

The total decrease in permeability due to wall stabilization by carbon, or the value of H_s at a particular level of B, is proportional to the amount of carbon in solution. Therefore magnetic measurements can be used to determine the carbon content of iron [12.19]. The magnetic method has the advantage, for certain applications, that it reveals only the amount of carbon actually in solution; any carbon not in solution will be present as Fe_3C, which cannot cause any time-dependent wall stabilization.

The rate of wall stabilization is strongly dependent on temperature, because it depends on the diffusion rate of carbon; this varies as $e^{-Q/kT}$, where Q is the activation energy for diffusion. The experiments reported in Figs. 2.16 and 2.18 had to be made at subzero temperatures to slow the effect to a pace at which it could be conveniently measured. At room temperature wall stabilization occurs almost instantly.

Time decay of the permeability is also found in substitutional solid solutions like Fe-Si and Fe-Ni. Wall stabilization in such alloys is due to directional order of like-atom pairs, which, when fully established, tends to keep the local M_s vector parallel to a particular direction in the lattice, as described in Section 10.2. Smolinski *et al.* [12.20] have described permeability decay in steel containing 4 percent Si.

Many ferrites show a time decay of permeability. The permeability of Mn-Zn ferrites can decrease in 24 hours at room temperature by amounts ranging up to 2.5 percent, an amount objectionable in certain device applications where long-term stability of ferrite inductors is required [G.35]. The effect is thought to be due to directional order of Mn^{2+}–Fe^{2+} pairs. In ferrites containing cobalt, the preferential occupation of a particular $\langle 111 \rangle$ axis by Co^{2+} ions, as described in Section 10.2, can cause wall stabilization. These time effects are greater, the larger the concentration of metal-ion vacancies, which promote rapid diffusion; the vacancy concentration can be kept low by ensuring that the atmosphere during sintering does not contain excess oxygen. Further details can be found in other sources [G.23, G.26, G.36].

Time decay of the permeability has been discussed by Néel [12.21], who was the first to clarify the nature of this effect [10.14].

Diffusion after-effect

The nature of the diffusion after-effect is sketched in Fig. 12.19. When a field H is suddenly applied to a magnetic material, B (or M) does not always reach its final value instantaneously. First there is an "instantaneous" change B_i, followed by a slower change, the after-effect induction, which reaches a maximum value of B_{ma}; the two together make up the total change B. It is usual to measure B_a as a function of time rather than the after-effect induction itself, which is $(B_{ma} - B_a)$. The effect has also been called magnetic viscosity. An even better name would be *magnetic creep*, because it is entirely analogous to mechanical creep, which is a slow increase in strain (analogous to M) at constant stress (analogous to H).

The converse effect occurs when H is suddenly reduced to zero from some finite value. B then decreases slowly to its final value.

It is important to ensure that eddy currents do not interfere with the effect to be measured. Even in materials which do not show an after-effect, B is delayed by eddy currents in reaching its final value, as described in Section 12.2 and as illustrated in an exaggerated way in Fig. 12.19. A rather unrealistic break in the B,t curve is there indicated. Actually the two portions of the curve run smoothly

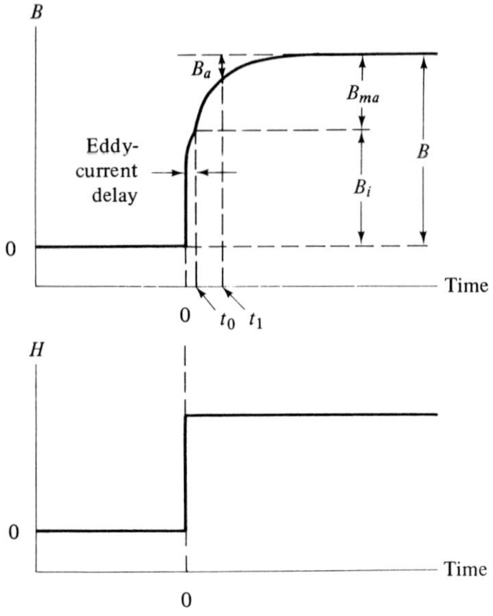

Fig. 12.19 Magnetic after-effect.

together, so that it may be difficult to establish an origin for the measurement of the after-effect. The eddy-current delay should be made as small as possible, for example by reducing the specimen thickness, relative to the time required for B to reach its final value.

The diffusion after-effect was first observed many years ago in iron containing carbon. Like the time decay of permeability, the effect is due to the retardation of domain wall motion by dissolved carbon or other interstitials. An example is shown in Fig. 12.20. The material was commercially pure iron (Armco ingot iron) containing dissolved carbon and nitrogen, in the form of rods 10 in. long and 3/16 in. in diameter. The applied field was 0.20 Oe, the initial state was the demagnetized state, and the temperature was 23°C. Changes in B were measured with a fluxmeter. This could be connected to the secondary coil on the specimen by an interval timer and relay at any selected time after the field H was applied. With a specimen heat treated to remove interstitials and which therefore showed no after-effect, it was found that closure of the fluxmeter circuit 0.1 sec after the application of H produced no fluxmeter deflection. This means that the "instantaneous" flux change B_i was complete in 0.1 sec and eddy currents had died out; this is the time t_0 in Fig. 12.19, and it is from this time that the after-effect was measured. The value of B_a at any time, such as t_1, was determined by closing the fluxmeter circuit at t_1 and observing the total fluxmeter deflection until the light spot ceased to move, which required up to 15 sec in some specimens. The value of B_{ma} was taken as the value of B_a at 0.1 sec. For the annealed specimen of Fig. 12.20, B_{ma} was 26 gauss; this total after-effect change amounted to about 20 percent of the instantaneous change B_i, which was 140 gauss at a field level of 0.2 Oe. The ratio B_{ma}/B_i decreased with increasing H.

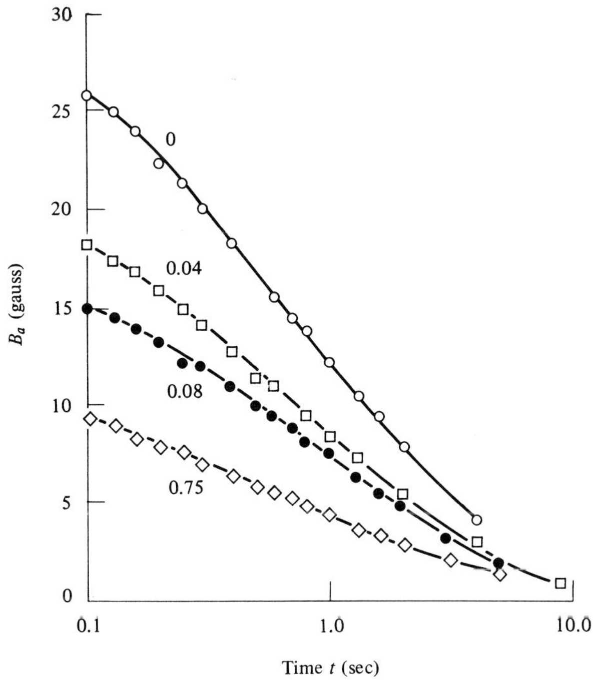

Fig. 12.20 Variation of B_a with time after application of a magnetic field, for iron containing 32 ppm C and 37 ppm N. Numbers on the curves indicate the amount of plastic elongation (in percent) previously applied to annealed specimens. (Rusnak *et al.* [12.22])

In iron containing carbon, permeability decay and the diffusion after-effect have the same basic cause: the tendency of carbon atoms to take up preferred positions and thus stabilize domain walls. But the two effects are not the same. To measure permeability decay, we measure a time-dependent change in the ease of magnetization with a small temporarily applied field; wall motion is very restricted. To measure the after-effect, we measure a time-dependent change in the magnetization itself under a constant applied field; wall motion proceeds as far as it can. The after-effect is the more complex of the two, simply because the domain walls are in constant motion until the effect ceases. When the fast change B_i is over, carbon atoms immediately start to arrange themselves so as to stabilize walls at their new positions. The stabilization is at first weak, and the still-applied field is able to push the walls on at a fairly high velocity. But the slight slowing down allows the carbon atoms more time for stabilization at the later position of the wall, the effect becomes cumulative, and wall motion finally ceases. Despite the very rapid diffusion rate of carbon at room temperature, wall motion can persist for some 10 to 15 sec, in the specimens of Fig. 12.20, simply because of the difficulty of stabilizing a *moving* wall.

In the simplest case the rate of change of B_a is governed by a single time constant τ:

$$B_a = B_{ma}e^{-t/\tau}. \tag{12.20}$$

Then a plot of log B_a against t will be a straight line, and Tomono [12.23] observed such a relation for Armco iron at a field level of 0.015 Oe. But a plot of B_a against log t will be decidedly curved, for a single time constant. The fact that the curves of B_a versus log t in Fig. 12.20 are almost straight means that more than one time constant is involved. It is significant that the field level for the data of Fig. 12.20 is more than ten times that of Tomono. His measurements were carried out in the reversible (constant permeability) range of magnetization, and the single time constant he observed is presumably due to the fact that wall motion is reversible in this range. The data of Fig. 12.20, on the other hand, correspond to Rayleigh-region magnetization, where wall motion has both a reversible and an irreversible component. The multiple time constants made evident by Fig. 12.20 are presumably connected with irreversible wall motion, but it is difficult to say what physical processes in the material are responsible for them. But wall motion is slowed down not only by carbon atoms in particular sites but also by the crystal imperfections which are always present. Presumably the nature and distribution of these imperfections determine the relaxation times. Chikazumi [G.23] has given a formal treatment of the effect of a continuous range of relaxation times between certain limits.

As shown in Fig. 12.20, small amounts of cold work drastically reduce the magnitude of the after-effect. But cold work also reduces B_i by increasing the microstress hindrances to wall motion. A B_i,H curve was determined in the usual way, except that the search coil was disconnected from the fluxmeter 0.1 sec after the field was applied; the after-effect contribution was thus excluded. The differential permeability dB_i/dH was then obtained from this curve at $H = 0.2$ Oe; this permeability is a measure of the magnetic hardness of the material at the same field level as that at which the after-effect was measured. It was found that B_{ma} was directly proportional to dB_i/dH for various degrees of cold work. The reduction in B_{ma} by cold work is therefore due simply to the increase in magnetic hardness caused by cold work and not to any change in the stabilizing power of the interstitials.

The diffusion after-effect is strongly dependent on temperature. The relaxation time or times are proportional to $e^{Q/kT}$, where Q is the activation energy for diffusion of the interstitials.

The quasi-static experiment of Fig 12.19 is one way of demonstrating the effect of dissolved carbon, say, on wall motion. Another quite different way is to apply a small alternating field. If the frequency is so low that the carbon atoms always have time to be in their equilibrium positions with respect to the local M_s vector, B and H will be in phase and there will be no power loss due to the diffusion after-effect. At the other extreme, the frequency may be so high that the carbon atoms never have time to reach equilibrium positions; again B and H will be in phase. At an intermediate frequency, the carbon atoms are mobile enough to make B lag behind H by a phase angle ϕ. There is then an energy loss per cycle proportional to tan ϕ, which is called the *loss factor*. If the after-effect is characterized by a single relaxation time τ and if we put $B_{ma}/B_i = R$, then, as Tomono [12.23] showed,

$$\tan \phi = \frac{\omega\tau R}{\omega^2\tau^2 + (1 + R)}, \tag{12.21}$$

where ω is the angular frequency, equal to 2π times the frequency in Hz. This function has a maximum value of $R/2\sqrt{1 + R}$ when $\omega = \sqrt{1 + R}/\tau$. Because R is of the order of 0.1–0.4, the frequency that maximizes tan ϕ is not much larger than $1/\tau$. Tomono measured the total losses of his iron specimens under alternating magnetization with an inductance bridge, at various frequencies and temperatures. By subtracting the losses due to hysteresis and to eddy currents, he obtained the loss factor due to the after-effect. From the observed variation of tan ϕ with ω, the value of τ and R can be found from the relations given above. At constant frequency, tan ϕ also goes through a maximum as a function of temperature; at 200 Hz, for example, this maximum occurred at about 100°C, while at 4000 Hz it occurred at 150°C. This behavior is due to the temperature dependence of the relaxation time, given by

$$\tau = \tau_0 e^{Q/kT}, \tag{12.22}$$

where Q is the activation energy of the process that controls the rate of the after-effect. At low temperatures, τ is so large that the after-effect essentially does not exist (carbon atoms cannot reach equilibrium positions); B and H are in phase. At high temperatures, the after-effect is so fast that B and H are again always in phase. At intermediate temperatures, losses will occur, and tan ϕ will have its maximum value when

$$T = \frac{2Q}{k \ln\left(\dfrac{1 + R}{\omega^2\tau_0^2}\right)}. \tag{12.23}$$

The larger ω, the higher is this temperature, in agreement with experiment. From values of τ determined at different temperatures, Tomono found that Q was 0.99 eV/atom or 23 kcal/mol, which is in fair agreement with the value of Q for diffusion of carbon in alpha iron (20.1 kcal/mol).

This result gives a satisfying explanation of the observed single relaxation time and establishes carbon diffusion as the rate-controlling process of the after-effect in the *reversible* range of magnetization. In the irreversible range other processes must play an additional role.

Thermal fluctuation after-effect

This effect differs from the diffusion after-effect in three respects: (1) it does not require the diffusion of anything and therefore can occur in pure metals or other pure substances, (2) it is much less sensitive to temperature, and (3) it causes a loss factor tan ϕ in alternating fields which does not depend on frequency. The magnitude of the fluctuation after-effect is larger in magnetically hard than in magnetically soft materials.

Street and Woolley [12.24] studied the effect in an Alnico permanent-magnet alloy containing 10 Al, 18 Ni, 12 Co, 6 Cu, and 54 Fe (weight percentages). The specimens were cast rods 30 cm long and 0.65 cm in diameter; magnetization was measured with a suspended-magnet magnetometer. The after-effect was larger on the hysteresis loop than on the initial magnetization curve and was measured on the descending part of the loop. The abrupt change in field ΔH

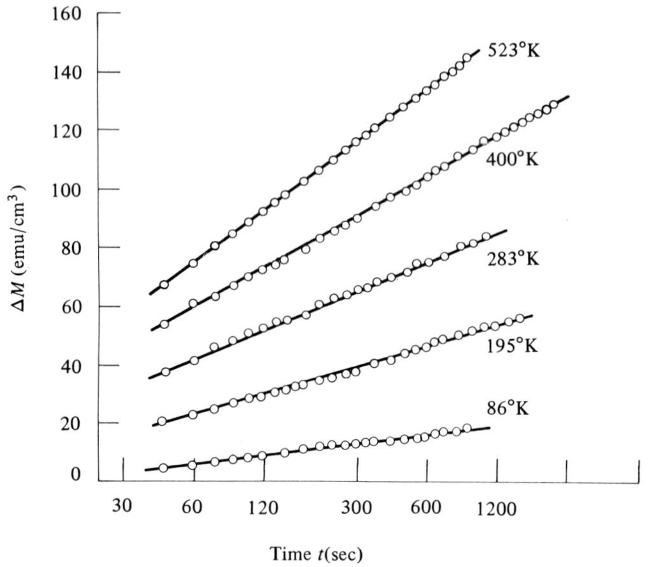

Fig. 12.21 Fluctuation after-effect in Alnico. (Street and Woolley [12.24])

was such that the final value of H was near the coercivity point, and this ΔH produced an instantaneous magnetization change ΔM_i of about 200 emu/cm^3. This was followed by the after-effect change ΔM shown in Fig. 12.21, where the curves have been shifted vertically and drawn from arbitrary origins for greater clarity. Near room temperature, the after-effect change is about 35 emu/cm^3 in 20 minutes or about 17 percent of the instantaneous change. The results of Fig. 12.21, including the temperature dependence, conform to the relation

$$M = a\,T \ln t + b, \tag{12.24}$$

where a and b are constants. Street and Woolley ascribed the after-effect to fluctuations of thermal energy sufficient to cause irreversible rotations in small volumes of the material. (This view is consistent with the current opinion that Alnico contains single-domain regions.) The probability that the energy of a particular volume will fluctuate by an amount Q above the average is proportional to $e^{-Q/kT}$. One would therefore expect the after-effect ΔM to vary in the same way, whereas it is found to vary linearly with T, and the rate of the after-effect $d(\Delta M)/dt$ to vary as $e^{-t/\tau}$, whereas actually it varies as $1/t$. [See Eqs. (11.25) and (11.28).] However, Street and Woolley showed that Eq. (12.24) could be accounted for by assuming that the activation energy Q had not a single value but a continuous range of values from 0 to ∞. This is equivalent, because of Eq. (12.22), to a similar range of relaxation times τ.

Néel [12.25] considered the problem in another way. He assumed that the effect of thermal fluctuations could be represented by a fluctuating internal field H_i which acted in addition to the applied field; H_i is sometimes positive, sometimes negative, and its absolute value varies linearly with ln t after an abrupt change in the applied field H. This theory leads to an equation for the after-effect change ΔM which is, at constant temperature, identical to Eq. (12.24). (This equation, incidentally, cannot be literally true, as it predicts an infinite change ΔM at infinite time. However, it does accurately represent experimental data over the duration of the measurements. Street and Woolley show how it can be modified to keep ΔM finite; this modification consists in allowing the activation energies Q to range from zero up to a finite limit rather than to infinity.)

In the theory of Street and Woolley, irreversible rotations are specifically mentioned. Néel, on the other hand, was concerned with the Rayleigh region and thought of the fluctuating field H_i as a field which aided domain walls to pass obstacles. Actually it is not necessary to be specific about the way the after-effect change occurs. There are energy barriers both to wall motion, as in Fig. 9.36, and to rotation, as in Fig. 9.37. One can therefore regard the after-effect change ΔM as due to energy fluctuations altering the shape and height of the barrier, or as due to a fluctuating internal field aiding the applied field to overcome a fixed barrier.

Barbier [12.26] has reported measurements of the fluctuation after-effect in various materials and on various parts of the hysteresis loop. The effect is very small in commercially pure iron in the Rayleigh region.

The reader may wonder at the absence, in this account of the fluctuation after-effect, of any mention of superparamagnetism, inasmuch as superparamagnetism is due entirely to thermal fluctuations. But we saw in Section 11.6 that the relaxation time of a particle varies extremely rapidly with its volume. Therefore, at a given temperature, an assembly of small particles of the same size will show magnetic viscosity only if their size corresponds to a relaxation time of the same order of magnitude as the time of the experiment. If there is a large range of particle size, then an after-effect can be expected, and Chikazumi [G.23] has shown how the theory of superparamagnetism then leads to Eq. (12.24).

12.6 MAGNETIC DAMPING

Suppose a weight is hung on a helical spring, then pulled downward and released. The spring will contract, then lengthen, then contract, etc. If the displacement of the end of the spring is plotted as a function of time, the result will look like the curves of Fig. 12.22. Frictional forces make the oscillations die out eventually. What frictional forces? Air resistance is one, and quite probably there is some friction at the point of support of the spring. But when the greatest care is taken to reduce these exterior sources to a minimum, the oscillations still die out, although more time now elapses before they stop. Evidently, frictional processes within the material of the spring itself must be the cause. This "internal friction", or damping, occurs in any vibrating material.

There are many measures of internal friction, and the most fundamental is the fractional energy loss per cycle, also called the specific damping capacity:

$$\frac{\Delta W}{W} = \frac{\text{energy loss per cycle}}{\text{total vibrational energy}}. \tag{12.25}$$

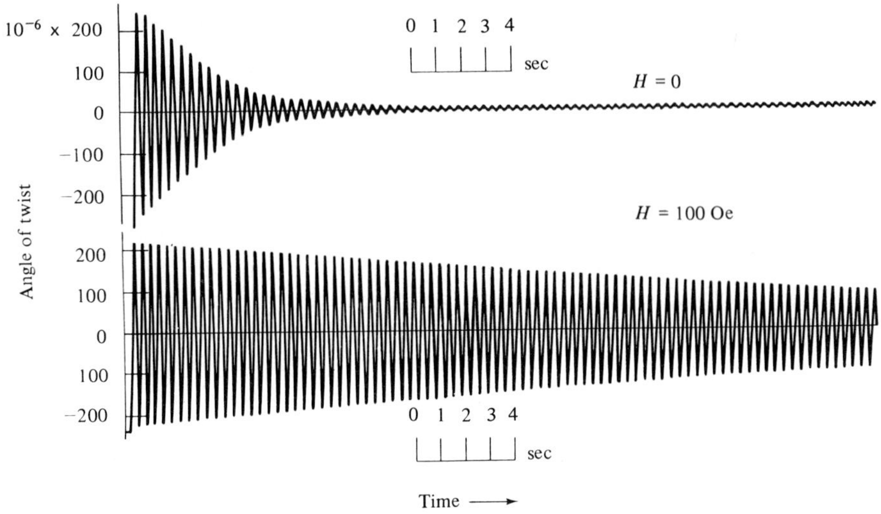

Fig. 12.22 Photographically recorded trace of the angle of twist of an iron wire in a torsion pendulum, when the wire is unmagnetized (above) and magnetized (below). Frequency = 2.6 Hz. (Becker and Kornetski [12.27])

For the weight on the spring, W is the elastic strain energy stored in the spring at either extreme, extension or contraction, of the cycle. (A value of $\Delta W/W$ of 10^{-1} is considered large, while some mechanisms of internal friction make $\Delta W/W$ as low as 10^{-5}.) Another way to measure damping is to determine the rapidity with which the oscillations of the system die out. If A_n is the amplitude of any one oscillation and A_{n+1} that of the next, then the extent of damping is characterized by the logarithmic decrement δ, defined by

$$\delta = \ln \frac{A_n}{A_{n+1}}. \tag{12.26}$$

If δ is small, $\delta = \Delta A/A$, where ΔA is the difference between successive amplitudes and A is their mean. (If $\Delta W/W$ is 0.1, the amplitude will decrease to one-tenth of its original value in about 46 cycles; if $\Delta W/W$ is 10^{-5}, some 460,000 cycles will be required). The vibrational energy W is always proportional to A^2, no matter how complex the stress system. For example, suppose a rod is stretched by a stress σ to a strain, or amplitude, ε. Then the stored energy = work done = $(1/2)(\sigma\varepsilon) = (1/2)(\varepsilon E)(\varepsilon) = (1/2)(E\varepsilon^2)$, where E is the elastic modulus. Therefore, if c is a constant,

$$W = cA^2,$$

$$\Delta W = 2cA\Delta A,$$

$$\frac{\Delta W}{W} = \frac{2cA\Delta A}{cA^2} = 2\frac{\Delta A}{A} = 2\delta. \tag{12.27}$$

The spring-weight system mentioned above is a freely vibrating system. Alternatively, we may choose to drive the system at constant amplitude, with the input power balancing the losses. The strain will then lag behind the applied stress by a phase angle ϕ, and $\tan \phi$, which equals ϕ when ϕ is small, is a measure of the damping. (This is entirely analogous to the loss resulting from B lagging behind H in alternating magnetization, as described in the previous section.) In an electrical circuit, "vibration" consists in the oscillation of charge back and forth, and a quantity Q^{-1} describes the damping of these oscillations. These various measures of damping are related as follows, when the damping is small:

$$\frac{\Delta W}{2W} = \delta, \qquad \tan \phi = \phi = Q^{-1} = \frac{\delta}{\pi}. \tag{12.28}$$

When an alternating stress is applied to a solid, the various imperfections within the solid move in particular ways, each according to its own nature and the kind of stress imposed. Dislocations move back and forth in slip planes in response to a shear stress parallel to those planes. Solute atoms jump from crystal axes that are shortened by the applied stress to nearby positions on elongated axes, where they have more room, and back again when the stress reverses. If the solid is polycrystalline, a kind of rubbing action occurs at the grain boundaries. In addition, if the material is magnetic, domain walls will move when the material is stressed. The work done in moving these imperfections during a quarter cycle of stress is *not* returned to the system during the next quarter cycle, as is the work of purely elastic deformation of a perfect crystal. The work done in moving these imperfections back and forth during a cycle constitutes the internal friction loss ΔW. It is converted into heat.

The relative magnitude of any one of these damping mechanisms (dislocations, solute atoms, grain boundaries, domain walls) depends markedly on the frequency and amplitude of the vibration imposed on the solid. One mechanism can make a far greater contribution to the total damping in a certain frequency or amplitude range than some other mechanism, and yet be practically negligible in some other range.

Internal friction has been studied at frequencies ranging from about one Hz to a few MHz (1MHz = 1 megahertz = 10^6 Hz). The nature of the apparatus required differs from one frequency range to another and, for the lowest range, takes the simple form of a torsion pendulum. The specimen is in the form of a wire suspended vertically from a fixed support. The lower end of the wire is attached to a fairly heavy weight, either a horizontal rod called an *inertia bar* or a disk like that shown in Fig. 7.11 (a). The weight is turned a few degrees, so as to twist the wire, and then released. The resulting torsional oscillations of the wire are indicated by a light beam reflected from a mirror, fixed to the lower end of the wire, on to a graduated scale.

Some adjustment of the torsional frequency is possible by altering the moment of inertia of the suspended weight, for example, by moving weights attached to the inertia bar toward or away from the wire axis, but the range of adjustment is small: from perhaps 0.1 to 2 Hz. In this range the decay of the amplitude can be measured by eye. If the specimen is made thick enough to be called a rod, rather than a

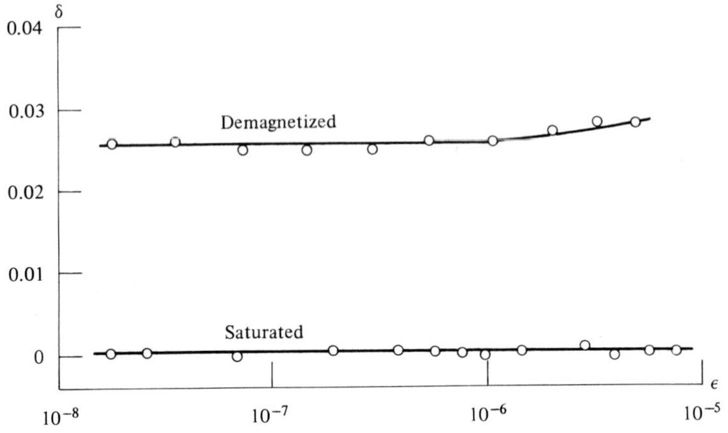

Fig. 12.23 Damping δ (logarithmic decrement) as a function of strain amplitude ε in a nickel single-crystal rod with a ⟨100⟩ axis. Room temperature. Vibrated in the longitudinal mode at 50 kHz. Ganganna *et al.* [12.29].

wire, and other parts of the apparatus are made more robust, frequencies of some tens of Hz can be attained. The motion of the reflected light beam must then be photographically recorded, as in Fig. 12.22. The rather narrow frequency range to which the torsion pendulum is restricted is an unfortunate feature of all internal friction devices; each is best suited to a particular frequency range and that range cannot be greatly extended. Wert [12.28] has surveyed the main experimental techniques and given examples of some common damping mechanisms.

Domain wall damping, often called *magneto-mechanical damping*, is easily separated from other effects. The specimen is enclosed in a solenoid capable of producing a saturating magnetic field. Damping is then measured with the specimen first demagnetized, then saturated to remove domain walls. The difference between the two measurements gives the magnetic contribution. Figure 12.22 vividly demonstrates the large difference in damping between the two states.

At extremely low amplitudes magnetic damping is small and independent of amplitude, as shown by the data of Fig. 12.23 for longitudinal vibration of a rod. In these measurements any one region of the rod is alternately in states of axial tension and compression. This is accomplished by cementing a quartz "driver" crystal to the end of the rod and applying an alternating voltage to opposite faces of the quartz. Because quartz is piezoelectric, this voltage causes the quartz to vibrate, and the vibrations are transmitted to the specimen rod. Another quartz crystal, called the "gage" and cemented to the driver, senses the vibrations and generates a voltage between opposite faces proportional to the vibrational amplitude. The specimen will vibrate substantially (resonate) only at its natural frequency, which is determined by its dimensions, or at integral multiples (har-

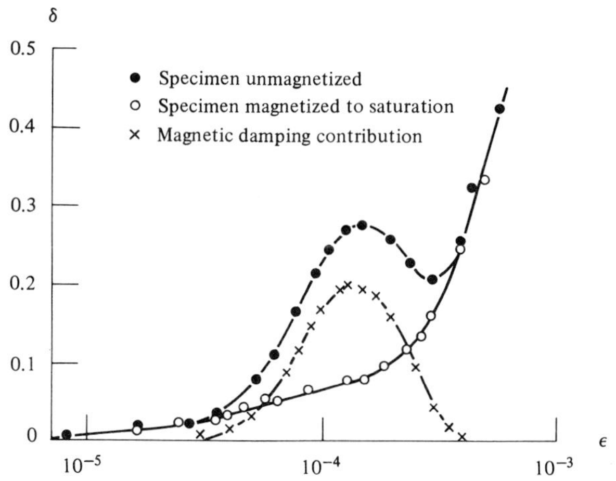

Fig. 12.24 Damping δ (logarithmic decrement) as a function of maximum shear strain amplitude ε at the surface of polycrystalline nickel at -40°C. Torsion pendulum measurements at about 1 Hz. Roberts and Barrand [12.30].

monics) of this frequency. The vibration frequency is determined by the frequency of the applied voltage. At each selected frequency, the damping is found by measuring the shape of the resonance peak, i.e., the curve of vibrational amplitude (proportional to voltage output of the gage crystal) versus frequency (equal to frequency of the applied voltage). The smaller the damping, the sharper is the resonance peak, according to the relation

$$\delta = \frac{\pi}{\sqrt{3}} \frac{\Delta f}{f_0}, \tag{12.29}$$

where f_0 is the frequency of the resonance and Δf the width of the resonance peak at half its maximum height. To make the measurement at various amplitudes requires only a change in the applied voltage.

The nonmagnetic damping exhibited by the saturated crystal of Fig. 12.23 is so nearly zero that the damping of the demagnetized crystal is virtually all magnetic damping. The behavior of a $\langle 111 \rangle$ crystal is quite similar except that its magnetic damping is much smaller, about 30 percent of that of a $\langle 100 \rangle$ crystal, at the same frequency of 50 kHz.

At higher amplitudes, magnetic damping becomes very large and dependent on amplitude, as shown by Fig. 12.24. (In these torsion-pendulum measurements the maximum amplitude is varied simply by varying the initial twist given to the wire.) The nonmagnetic damping shown by open circles in Fig. 12.24, which increases rapidly with strain, is due to dislocation motion. These data were obtained at -40°C, but a similar peak in magnetic damping, somewhat broader and about half as high (maximum $\delta = 0.1$), was observed at room temperature.

Sumner and Entwistle [12.31] observed extremely high magnetic damping in polycrystalline nickel in torsional oscillation at 20 Hz and room temperature: the maximum value of δ was 0.25. They also found that magnetic damping in iron and steel went through a maximum as a function of amplitude, just as in nickel. If an annealed specimen is cold worked, the magnetic damping decreases, usually by a large amount.

Magnetic damping in metals is chiefly due to the micro eddy currents set up by moving domain walls. We can understand the main features of the experimental results, at least qualitatively, if we consider the kind of wall motion an applied alternating stress can produce. We assume that the only hindrance to wall motion is residual microstress or, rather, that inclusion hindrance can be discussed in essentially the same terms, and we consider the two kinds of walls separately.

90° walls

As we saw in Section 9.11, a certain kind of 90° wall is stable where the stress σ_y is zero, provided the stress gradient has the correct sign. The wall sketched at the top of Fig. 12.25 (a) is stable, at zero applied stress, at position A where the residual stress is zero. If a uniform tensile (positive) stress is now applied, the total stress is given by the full line. The point of zero stress is now at B and the wall therefore moves to that position. (The behavior of this stress-driven wall is in sharp contrast to the field-driven wall of Fig. 9.33. When an applied field displaced that wall, the magnetoelastic energy of the system increased. But when the wall of Fig. 12.25 moves, in response to an applied stress, the magnetoelastic energy remains constant.)

If the applied stress σ_{ya} is alternating with time, as in Fig. 12.25 (b), it becomes tedious to make drawings like (a) to show the various positions of the total-stress curve during the cycle. It is simpler to draw just the residual-stress curve, as in (c), and indicate the zero of total stress by a horizontal line. The heavy horizontal line in (c) indicates zero stress when the applied stress in the y direction σ_{ya} is zero. If tension is now applied, the new zero-stress line is drawn at the appropriate distance *below* the heavy line. Thus, at point 2 of the stress cycle, the point of zero total stress is at b, and this applied stress would cause the wall to move to this position on the x axis from its original position at a.

Suppose now that the range of applied stress is small, as indicated by the dashed line in (b). The wall will then merely oscillate about position a over a short distance, *without making any Barkhausen jumps.* Moreover, its average velocity over a half-cycle will be a function of only the frequency f and the amplitude σ_m of the applied stress. If this stress varies sinusoidally with time, so will the wall position x, measured from a:

$$x = x_m \sin 2\pi f t. \tag{12.30}$$

If the microstress gradient $d\sigma/dx$ in the vicinity of a has a constant value g, the amplitude x_m of the wall motion will be σ_m/g. The velocity of the wall at any instant is

$$v = \frac{dx}{dt} = 2\pi f x_m \cos 2\pi f t. \tag{12.31}$$

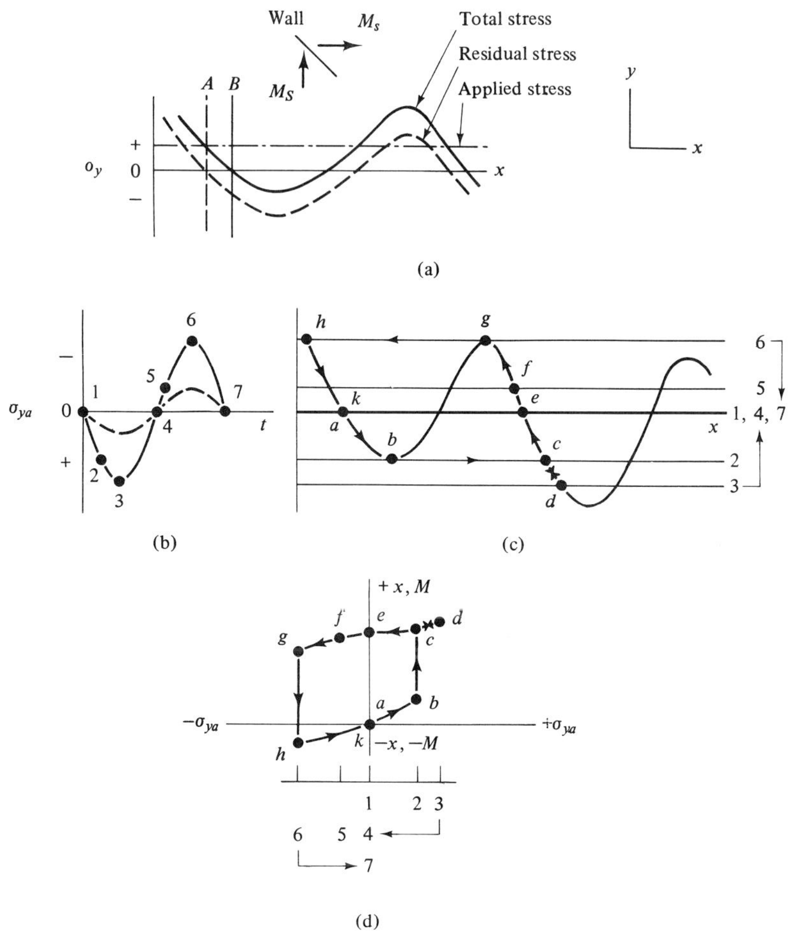

Fig. 12.25 Motion of a 90° wall wall due to an applied alternating stress, for a material with positive magnetostriction in the easy direction.

The average of the absolute value of v over a cycle is $2/\pi$ times the maximum value, or

$$\overline{|v|} = 4fx_m. \tag{12.32}$$

Williams *et al.* [12.2] have shown that the eddy-current *power* loss P due to a moving 180° wall is proportional to the square of its velocity, and this should also be true of a 90° wall. Therefore, the energy loss per cycle is

$$\Delta W = P\left(\frac{1}{f}\right) = (c_1 f^2 x_m^2)\left(\frac{1}{f}\right) = \frac{c_1 f \sigma_m^2}{g^2}, \tag{12.33}$$

where c_1 is a constant. Inasmuch as the magneotelastic energy does not change with wall position, the only stored energy in the cycle is elastic, equal to (1/2) $\sigma_m \varepsilon_m = \frac{1}{2}(\sigma_m^2/E) = W$. Therefore the damping is given by

$$\frac{\Delta W}{W} = \frac{c_2 f E}{g^2}, \tag{12.34}$$

where c_2 is a constant. This result applies only to small strains. The strain amplitude is proportional to x_m, which in turn is proportional to σ_m. Since σ_m does not appear in Eq. (12.34), we conclude that the damping at small strain amplitudes is independent of amplitude, which agrees with the experimental results of Fig. 12.23. [As amplitude increases, at constant frequency, the wall velocity must increase. This increases the eddy currents and therefore ΔW. But the increased amplitude also increases W, and the two change in such a way that the damping $\Delta W/W$ remains constant. For further comment on Eq. (12.34), see Section 13.3.]

The situation is entirely different at higher strains. If the applied stress maxima are now higher, as shown by the full curve in Fig. 12.25 (b), the wall will make a Barkhausen jump from b to c at point 2 in the stress cycle. It will then move to its extreme position at d, then to e, f, and g, whereupon it jumps to h and then moves back to its starting point at k (= a). [Wall motion changes the magnetization M of this small region of the specimen, and the M,σ hysteresis loop is shown in (d).]

During a Barkhausen jump the velocity of the wall is limited only by eddy-current damping and is expected to be considerable larger than that of a wall which is merely driven by the changing stress. Thus the wall velocity between positions h and b, and between d and g, is determined solely by the frequency of the applied stress, but the velocity during jumps is quite unrelated to this frequency. As the amplitude increases, wall motion includes more and more high-velocity jumps, with the result that ΔW increases faster than W, and the damping $\Delta W/W$ itself increases. This increase, which has only just begun in Fig. 12.23, is very marked in Fig. 12.24, which extends to larger strains. In this range of strain the damping goes through a maximum and begins to decrease.

This observed variation of magnetic damping with strain can be understood in terms of Fig. 12.26. The vibrational energy W equals $c_3\varepsilon^2$ over the whole range of strain, where c_3 is a constant. At very low strains, up to ε_1, the losses are given by a similar equation, namely, $\Delta W = c_4\varepsilon^2$, where c_4 is a constant, in accordance with Eq. (12.33). In this range the damping $\Delta W/W$ is a constant given by c_4/c_3. At the strain ε_1, Barkhausen jumps begin, and the losses ΔW increase more rapidly with strain, namely, as ε^n where n is greater than 2, causing an increase in damping.

Up to this point we have considered the moving wall as though it existed in isolation. Actually, it is connected at its ends to other walls, or to grain boundaries or to the specimen surface, in a complex domain structure, and these end connections constitute impediments to wall motion. At small strains (small amplitude of wall motion), the walls are perhaps pinned at their ends, so that the motion of a wall consists in a bowing back and forth between fixed ends rather than a pure translation. But at some larger strain, ε_2 of Fig. 12.26, these restraints must

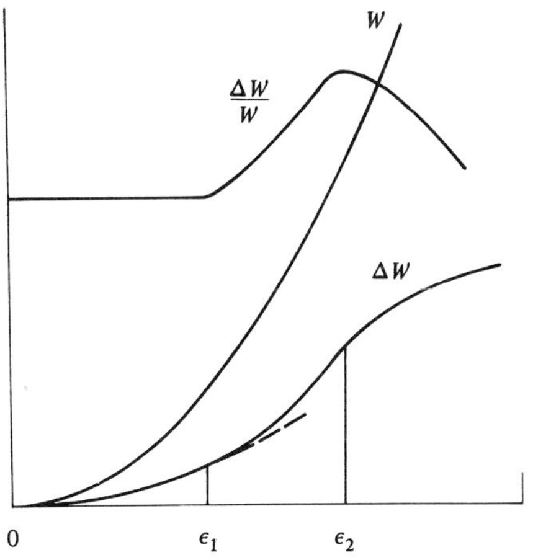

Fig. 12.26 Dependence on strain ε of vibrational energy W, losses ΔW, and damping $\Delta W/W$ (schematic). The $\Delta W/W$ curve is drawn from an arbitrary origin.

become so severe that they begin to decrease the extent of Barkhausen jumps. The $\Delta W, \varepsilon$ curve then becomes concave to the strain axis, and the damping goes through a maximum. [One way in which these restraints might operate is suggested by Fig. 8.20, where an imaginary domain structure consisting entirely of 90° walls is shown in (a). An applied tensile stress changes this structure to (b). But two new energies have been introduced: (1) magnetostatic energy due to the free poles formed on the ends, and (2) wall energy due to the newly created 180° wall, These are only partly balanced by the reduction in 90° wall energy. As a result the 90° walls cannot move in response to the applied stress as freely as they might if isolated and hindered only by microstress.]

Cold working an annealed material causes a large decrease in magnetic damping, because cold work increases the magnitude of both the residual microstress peaks and the average stress gradient g, making wall motion more difficult. The effect is a strong one, because the damping is expected to be inversely proportional to g^2, as shown by Eq. (12.34).

This discussion has focused on the variation of damping with amplitude at constant frequency. The related problem of how damping should vary with frequency at constant amplitude has hardly been touched on, except that Eq. (12.34) predicts that $\Delta W/W$ should be proportional to f. This does not agree with experiment. Instead, it is found that $\Delta W/W$ increases with f to a maximum value and then decreases [12.29].

The wrong frequency dependence of the damping given by Eq. (12.34) is due to ignoring the inertia of the domain wall. In deriving this equation, the real back-and-forth motion of the wall was replaced by an average, constant-velocity motion in order to simplify the analysis of the amplitude dependence. Actually, the equation of motion of the wall is Eq. (12.5), with the term on the right involving a field replaced by one involving the applied stress. The first term md^2x/dt^2 is the inertia force, and it cannot be omitted in any analysis of the frequency dependence of the damping. Whereas damping initially increases with frequency, at constant amplitude, because of increased wall velocity, a frequency is ultimately reached where the wall inertia is so large that the average wall velocity, and the resultant production of eddy currents, begin to decrease.

180° walls

These are stable wherever the stress goes through a minimum, whether the stress at that point be tensile or compressive, as shown in Section 9.11. The effect of an applied stress is merely to change the value of the stress at the minimum; the minimum itself does not change its position, as shown in Fig. 12.25 (a). Therefore applied stress will not move 180° walls, and they cannot contribute to magnetic damping. This conclusion is valid for small strains, when each wall in the specimen can be considered as approximately "isolated." At larger strains, 180° walls will move and thus contribute to the damping, for two reasons:

1. At larger strains, nonmagnetic damping due to dislocation motion becomes important (Fig. 12.24). Dislocation motion means plastic flow, which will alter the residual-microstress distribution. Stress minima will change their positions, and 180° walls will move to follow the minima.
2. In a complex set of interconnected walls, extensive 90° wall motion is expected to cause some motion of the 180° walls to which they are attached, although not in the simple example of Fig. 8.20.

General

In summary, magnetic damping in pure metals is mainly due to the dissipation of heat caused by the micro eddy currents associated with moving domain walls. The greatest contribution is made by 90° walls. The role of 180° walls is negligible at small strains and probably minor at large. In support of these conclusions is the observation [12.29] that magnetic damping in nickel single-crystal rods is much larger for a rod with a ⟨100⟩ axis than for one with a ⟨111⟩ axis. The easy direction of magnetization in nickel is ⟨111⟩. A ⟨100⟩ rod can therefore be expected to contain a great many 90° walls, while a ⟨111⟩ rod should contain a preponderance of 180° walls, inasmuch as its domain structure is expected to consist mainly of long, columnar domains magnetized parallel to the rod axis.

Another possible contribution to magnetic damping in pure metals, at strain amplitudes large enough to cause Barkhausen jumps, might be called the "phonon effect." When a 90° wall makes a Barkhausen jump, the volume swept out is very suddenly strained magnetostrictively, because the M_s vector in this volume is abruptly rotated through 90°. As a result, a strain pulse (elastic wave) is sent through the lattice. This pulse eventually dies out, but not before adding to the phonon energy (atomic vibrational energy) of the crystal; in short, the strain pulse is converted into heat. The relative importance of the phonon effect is not known and appears to be difficult to estimate.

In impure metals and in alloys, solute atoms can make a contribution to magnetic damping entirely independent of the eddy-current and phonon effects. Consider, for example, iron containing interstitial carbon and/or nitrogen. As we saw in Section 10.3, these interstitials have preferred positions in the lattice relative to the direction of the local M_s vector. When a 90° domain wall moves back and forth in response to an applied alternating stress, the region swept out has its M_s vector rotated by 90° once in each half-cycle. The interstitials therefore move back and forth, trying at all times to take up positions dictated by the orientation of the M_s vector at that time. This motion of the interstitials causes a loss and contributes to the damping.

(The situation is complicated by the fact that these interstitials can cause damping even in magnetically *saturated* iron that contains no domain walls. The interstitials then have preferred positions dictated solely by the stress. They prefer to be in sites lying on crystal directions which are elongated by the applied stress, simply because there is more room there. When the stress reverses, they then avoid these sites. This back-and-forth motion of the interstitials causes a peak in the damping versus temperature curve at constant frequency. It is known as the Snoek peak after its discoverer and is well described by Reed-Hill [12.32]. The interstitials in *demagnetized* iron undergoing mechanical vibration therefore have divided loyalties. In a region, for example, where the stress, at a certain instant, is tensile and parallel to M_s, and both are parallel to the x direction, the interstitials would rather be in y or z sites for magnetic reasons but in x sites for mechanical reasons. The *height* of the Snoek peak is proportional to the amount of carbon, or other interstitials, actually in solution in the iron. Damping measurements on magnetically saturated iron can therefore disclose the amount of carbon in solution, just as measurements of the time decrease of permeability can, as described in Section 12.5. The damping measurement is actually the more popular method.)

12.7 MAGNETIC RESONANCE

When an alternating magnetic field of high frequency is applied to a substance, certain resonance effects are observed at particular values of the frequency and magnitude of the field. The effect is one of two kinds: one involves the magnetic moment of the *electron*, and the other that of the *nucleus*. The subject of magnetic resonance is somewhat out of the mainstream of this book, and only brief descriptions of resonance effects are given here. However, magnetic resonance is of great importance in many fields, particularly chemistry, and many books and papers deal with the subject. An extended account has been given by Morrish [G.27].

Electron paramagnetic resonance (EPR)

This effect, also called *electron spin resonance (ESR)*, is a resonance between the applied field and the net magnetic moment of the atom, which is usually due only to electron spin. It can therefore be observed in all substances except those which are diamagnetic. However, as the name EPR implies, the effect was first observed in paramagnetic materials and these have received the most study.

To observe spin resonance experimentally, place the specimen in a constant field H of a few thousand oersteds in the gap of an electromagnet. Doing this causes the spins to precess about the direction of H at a frequency ν proportional to H. At the same time the specimen is subjected to an alternating field at right angles to H in the form of a microwave travelling in a waveguide. When the frequency of the microwave field equals the frequency ν of precession, the system is in resonance, and a sharp drop in transmitted microwave power is indicated by a receiver placed on the other side of the specimen. The condition for resonance can be found as follows. The potential energy of each atomic moment in the field is $-\mu_H H$, and μ_H is given by Eq. (3.32):

$$\mu_H = gM_J\mu_B. \tag{3.32}$$

Whatever the value of J, adjacent values of the quantum number M_J always differ by unity. If one thinks of the magnetic state of the material in the field H as a distribution of the atomic moments among a set of $(2J + 1)$ energy levels, each distinguished by a particular value of μ_H, then the separation between adjacent levels is given by

$$\Delta E = \Delta(\mu_H H) = g\mu_B H. \tag{12.35}$$

The condition for resonance is that the energy per quantum $h\nu$ of the microwave beam be equal to ΔE, because this energy will then be absorbed by the specimen in raising an atom from one energy level to the next higher one. Therefore,

$$h\nu = g\mu_B H. \tag{12.36}$$

The resonant state is usually found by varying H at a fixed frequency ν and plotting power absorbed versus H. The wavelength λ of the microwaves is usually a few centimeters. If $\lambda = 3$ cm, then $\nu = c/\lambda = 3.00 \times 10^{10}/3 = 10^{10}$ Hz $= 10$ GHz (1 GHz $=$ 1 gigahertz $= 10^3$ MHz $= 10^9$ Hz). If g is 2, then the field at resonance is

$$H = \frac{h\nu}{2\mu_B} = \frac{(6.62 \times 10^{-27})(10^{10})}{2(0.927 \times 10^{-20})} = 3570 \text{ Oe}. \tag{12.37}$$

If the field H at resonance is found experimentally, then Eq. (12.36) can be solved for g. Thus EPR measurements have the important function of revealing the g value of the specimen, as mentioned in Section 3.7.

Spin resonance can also be observed in ferro-, antiferro-, and ferrimagnetic substances, where the spins are coupled by exchange forces. Here resonance measurements can reveal not only the g factor but also the crystal anisotropy constant K_1. Suppose the constant field H is directed parallel to the easy axis of a uniaxial crystal. Then two forces act to turn the spins toward the easy axis: the applied field H and the crystal anisotropy, which can be regarded as an anisotropy field H_K (Section 7.6). The resonance condition then becomes

$$h\nu = g\mu_B(H + H_K), \tag{12.38}$$

and H_K is related to K_1, as previously shown, by

$$H_K = \frac{2K_1}{M_s}. \tag{7.46}$$

If resonance measurements are made parallel to several crystal directions in a cubic crystal, the value of both K_1 and K_2 may be found.

Spin resonance in ferromagnetic metals, called simply *ferromagnetic resonance*, is complicated by eddy-current effects. At frequencies of about 10^{10} Hz, eddy-current shielding of the interior of the specimen is so nearly complete that the depth of penetration of the alternating field is only about 1000 Å. The specimen is therefore usually composed of powder particles of about this diameter.

Nuclear magnetic resonance (NMR)

This effect is a resonance between the applied field and the magnetic moment of the nucleus. The moment of the nucleus is due to its spin, depends on the size of the nucleus, and is spatially quantized in a way similar to the quantization of atomic moments. But nuclear moments are much smaller than atomic moments and are measured in terms of the *nuclear magneton* μ_n, defined similarly to the Bohr magneton μ_B:

$$\mu_n = \frac{eh}{4\pi Mc}, \tag{12.39}$$

where M is the mass of the proton. Because the proton mass is 1840 times the electron mass, we have

$$\mu_n = \frac{\mu_B}{1840} = \frac{0.927 \times 10^{-20}}{1840} = 0.505 \times 10^{-23} \text{ erg/Oe}. \tag{12.40}$$

The magnetic moment of a nucleus in the direction of the field is given by

$$\mu_H = gm\mu_n, \tag{12.41}$$

where g is the g factor of the nucleus and m can have the values $I, I - 1, \ldots, -(I - 1), -I$, where I is the quantum number describing the nuclear spin. Because adjacent values of m must differ by unity, the separation ΔE of adjacent energy levels and the condition for resonance are given by

$$h\nu = \Delta E = g\mu_n H. \tag{12.42}$$

The simplest nucleus is the proton, the nucleus of the hydrogen atom. Its spin I is $\frac{1}{2}$ and $g = 5.58$, so that μ_H is 2.79 μ_n. If H is 10,000 Oe, the resonant frequency is

$$\nu = \frac{(5.58)(0.505 \times 10^{-23})(10^4)}{6.62 \times 10^{-27}} = 42.6 \text{ MHz}. \tag{12.43}$$

This frequency is in the radio region of the electromagnetic spectrum, which means that the resonant circuit can be made of coils rather than waveguides. The constant field H is supplied by an electromagnet, as in EPR.

Because hydrogen is an almost universal constituent of organic sustances, observation of the proton resonance has proved so helpful to organic chemists that NMR measurements are now routine in their laboratories. The precise value of the resonant frequency depends slightly on the chemical surroundings of the proton and, with apparatus of high resolution, the resonance peak is seen to split into two or more peaks. By such measurements the chemist can reach certain conclusions about the structure of the molecule being examined.

Proton resonance is also the basis of the *proton precession magnetometer*, an instrument used primarily for the measurement of the earth's field. The sensor is simply a bottle of water, or other hydrogen-bearing substance, wound with a coil whose axis is roughly at right angles to the earth's field. A direct current in this coil produces a field of about 100 Oe; this aligns a certain fraction of the proton moments with the coil axis. After a second or so this field is turned off. The moments then begin to precess around the only field then acting, the earth's field, at a rate proportional to that field. As they precess, they induce a weak alternating emf in the coil around the bottle, and the frequency of this emf, equal to the precession frequency, is measured electronically. The relation between the precession frequency, which is of the order of a few kilohertz, and the field strength is known with much more accuracy than the rough calculation of Equation (13.43) suggests. The field strength in gammas ($1\gamma = 10^{-5}$ Oe) equals 23.4874 times the frequency in Hz. This magnetometer needs no calibration, because it is an absolute instrument, and it measures the earth's field, which is of the order of 50,000 γ, to an accuracy of 1 γ. It is widely used, both in magnetic observatories and as a portable instrument. By towing the sensor behind an aircraft, geologists can make rapid magnetic surveys of large areas in their search for the magnetic "anomalies," slight irregularities in the earth's field, that may disclose ore bodies or oil deposites. The proton magnetometer is also used for the measurement of laboratory magnetic fields.

Although proton resonance is probably the most widely studied, NMR measurements are not confined to the hydrogen nucleus. Resonance in other, more complex nuclei have also been investigated. For example, Rodbell [12.33] has surveyed some of the applications of magnetic resonance, both EPR and NMR, in metallurgical research; these include studies of local order in solid solutions and of the shape of precipitate particles in two-phase alloys.

PROBLEMS

12.1 Domain wall velocity is to be measured on a picture-frame crystal of a mixed Ni-Fe ferrite by the method of Fig. 12.3(b). Take $M_s = 350$ emu/cm^3, and $2L = d = 2$ mm. If the wall velocity at a certain drive field is 3000 cm/sec, how many turns must the secondary coil have to obtain an induced voltage of 1 volt?

12.2 The constant D of Eqs. (12.11) and (12.12) is given by

$$D = \sum_{\text{odd } n} n^{-3} \tanh\left(\frac{n\pi L}{d}\right).$$

This series converges rapidly.

a) Show that $D = 0.97$ for $2L = d$ (square rod).
b) When L/d is very large, as in a thin film, show that $D = 1.05$, independent of the actual values of L and d.

12.3 What would be the value of the domain wall mobility in an 80 Permalloy film, 4000 A thick, if wall motion were damped only by eddy currents? Take $M_s = 800$ emu/cm^3 and $\rho = 15$ microhm-cm.

12.4 Derive Eqs. (12.19).

12.5 Compute and plot the easy-axis hysteresis loop of a thin film reversing by rotation, for a reduced transverse field h_t of 0.20. What is the reduced coercivity h_{ec}?

12.6 The Néel theory of the thermal-fluctuation after-effect predicts that the after-effect change in magnetization is given, in emu/cm^3, by

$$M = cS_v(Q + \ln t),$$

where c is the irreversible differential susceptibility at the field at which ΔM is measured, S_v and Q are constants, and t is the time (sec). Between two times t_1 and t_2 after a variation of the applied field, the after-effect change will be

$$\Delta M = cS_v(\ln t_2 - \ln t_1).$$

For the annealed specimen of Fig. 12.20, the maximum after-effect induction B_{ma}, due to the diffusion after-effect, is 26 gauss between 0.1 and 10 sec after application of a field of 0.2 Oe. The Rayleigh constants of this specimen are $\mu_0 = 322$ and $v = 1830$ Oe^{-1}. The value of S_v for "soft iron" is reported by Barbier [12.26] to be 5×10^{-4}.

Assume that this value of S_v applies to the annealed specimen of Fig. 12.20. What is the ratio of the calculated fluctuation after-effect to the observed diffusion after-effect?

COMMERCIAL MAGNETIC MATERIALS

CHAPTER 13

SOFT MAGNETIC MATERIALS

13.1 INTRODUCTION

In this chapter and the next we will examine the main applications of magnetic materials, the requirements which these applications impose on the materials, and finally the materials themselves. The wide variety of magnetic materials can be rather sharply divided into two groups, the magnetically soft (easy to magnetize and demagnetize) and the magnetically hard (hard to magnetize and demagnetize). The distinguishing characteristic of the first group is high permeability, and it is chiefly this flux-multiplying power of the magnetically soft materials that fits them for their job in machines and devices. Magnetically hard materials, on the other hand, are made into permanent magnets; here a high coercivity is a primary requirement because a permanet magnet, once magnetized, must be able to resist the demagnetizing action of stray fields including its own.

The statistics of Table 13.1 show that the total market for magnetic materials in 1968 was divided almost exactly half and half between the magnetically soft and hard. Traditionally, "electrical steel," which forms the magnetic core of

Table 13.1 Market Value of U.S. Production of Magnetic Materials in 1968 [13.1].

			Millions of dollars	Percent	
Soft magnetic materials	Electrical steel		$ 180	28	
	Soft ferrites		110	17	49
	Nickel-iron alloys		25	4	
Hard magnetic materials	Magnetic recording materials	Tape	180	28	
		Disks and drums	100	15	51
	Permanent magnets		55	8	
	Total		650	100	100

Table 13.2 Approximate 1971 Metal Prices.

Material		Price per lb	Price relative to iron	Material		Price per lb	Price relative to iron
Iron *	Fe	$ 0.09	1	Silver	Ag	$ 24.	270
Cobalt	Co	2.30	25	Gold	Au	577.	6400
Nickel	Ni	1.33	15	Platinum	Pt	1780.	20000
Silicon †	Si	0.15	2	Yttrium	Y	155.	1700
Aluminum	Al	0.29	3	Lanthanum	La	35.	390
Copper	Cu	0.52	6	Cerium	Ce	19.	210
Vanadium	V	1.94	22	Praseodymium	Pr	175.	1900
Molybdenum	Mo	3.73	41	Neodymium	Nd	110.	1200
Chromium	Cr	1.01	11	Samarium	Sm	85.	950
Titanium	Ti	2.81	31	Europium	Eu	3400.	38000
Tungsten	W	4.50	50	Gadolinium	Gd	230.	2600
Manganese	Mn	0.31	3	Terbium	Tb	725.	8000
Tin	Sn	1.67	19	Dysprosium	Dy	150.	1700
Lead	Pb	0.13	1	Holmium	Ho	325.	3600
Zinc	Zn	0.15	2	Erbium	Er	180.	2000
Antimony	Sb	0.79	9	Thulium	Tm	2900.	32000
Bismuth	Bi	6.00	67	Ytterbium	Yb	250.	2800
Cadmium	Cd	2.25	25				
				Mischmetal		3.	33
Magnesium	Mg	0.37	4				
Mercury	Hg	4.61	51				

* Low-carbon sheet steel.
† Per lb of silicon in ferrosilicon.

almost all transformers, generators, and motors, has always occupied the first place among magnetic materials, in terms both of total tonnage and total market value. But in 1968 the rapidly expanding market for magnetic recording tape made the dollar value of this material equal to that of electrical steel for the first time in history. And if one includes with magnetic tape, as it is quite reasonable to do, magnetic disks and drums, then the total value of magnetic recording media exceeds that of electrical steel and has done so since about 1966, according to Jacobs [13.1].

Soft magnetic materials have been reviewed by Lee and Lynch [13.2] and a committee of specialists [13.3]. These materials can be divided into four main groups, according to function:

1. *Heavy-duty flux multipliers.* These are the cores of transformers, generators, and motors. These machines are usually large and heavy, so that the cost per pound of magnetic material is important; their cores are made of electrical steel, which is the cheapest magnetic material. The economic facts of life underlying the magnetic-materials business are displayed in Table 13.2, in the form of current prices for metals which are used, or may be used, either in magnetically soft or hard materials. Such data should be regarded with extreme caution; they apply only to a particular time, and they can vary greatly with metal purity and form (ingot, sheet, or strip, for example). However, the relative prices are roughly constant with time and can suggest, for example, the cost of increasing the permeability of iron by alloying it with nickel.

2. *Light-duty flux multipliers.* These are the cores of small, special-purpose transformers, inductors, etc., used mainly in communications equipment. Here the cost per pound of magnetic material is usually secondary to some particular magnetic requirement. Soft ferrites and nickel-iron alloys fall in this class.

3. *Square-loop materials.* Used in computers and in magnetic amplifiers and other "saturable-core" devices. They include soft ferrites and nickel-iron alloys.

4. *Microwave system components.* These comprise soft ferrites and garnets.

The electrical steels that form the cores of transformers, generators, and motors will be considered first. These cores are subjected to alternating and/or rotating magnetic fields, and the minimization of the energy loss per cycle is, as a design objective, second only to the attainment of high permeability, because some of these machines handle very large amounts of electric power. The energy losses are so important that their nature, size, and control will be examined before the core materials themselves. And because these losses are primarily due to eddy currents, we begin by treating this subject more quantitatively than was done in Section 12.2.

13.2 EDDY CURRENTS

A simple form of transformer is shown in Fig. 13.1 (a). It consists of a rectangular core of magnetic material, with a primary winding on one leg and a secondary on the opposite. An alternating current in the primary magnetizes the core

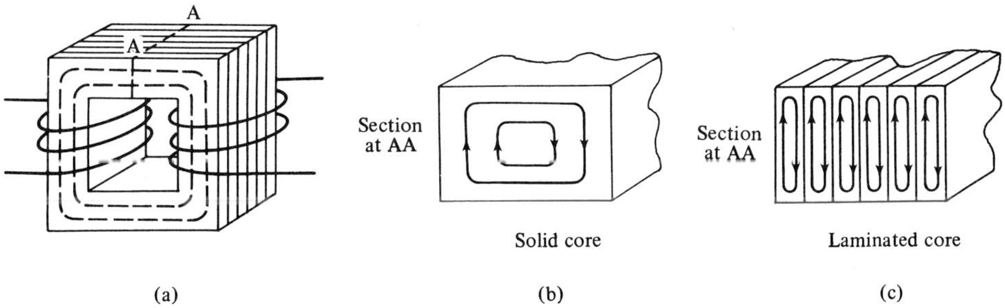

Fig. 13.1 Eddy currents in a transformer core. Sears [1.3].

along the dashed lines, alternately in one direction and the other. The change of flux through the secondary coil induces an emf in it that is proportional at any instant to the rate of change of flux dB/dt through that coil. The ratio of the output voltage to the input depends on the ratio of the number of turns in the secondary coil to the number in the primary.

The alternating flux induces an emf, not only in the secondary, but also in the core itself and, if the core is a good conductor, this emf will cause eddy currents. At a time when the flux is increasing from left to right in the top leg, the eddy currents in the section AA will have the direction shown in (b), i.e., the direction which will produce an eddy-current field from right to left, opposing the main field. Eddy currents are objectionable, not only because they decrease the flux, but also because they produce heat, a direct power loss, proportional to i^2R, where i is the eddy current and R the resistance of its path. One way of decreasing this loss is to form the core of thin sheets, as in (c), rather than from a solid piece. If these sheets are electrically insulated from one another, the eddy currents are forced to circulate within each lamination. The path length in each lamination is now shorter, thus decreasing R, but the cross-sectional area A of the path is also reduced. The induced emf e is therefore reduced, according to Eq. (12.1), and the net effect is a decrease in the current i and in the eddy-current power loss ($= ei = i^2R$). The effect is a strong one (Problem 13.3), and laminated construction is standard for all cores of transformers, motors, or generators made from metallic (conducting) materials.

How thin should the laminations be made? We saw in Section 12.2 that when a body is exposed to an alternating field, eddy currents are generated which partially shield the interior of the body from the field. Thus the value of H and hence B inside the body can be much less than the value at the surface. The reader may have heard of this "skin effect," which is a confinement of the flux to the surface layers of a body at a "high" frequency. However, he may be surprised to learn how strong this effect can be in a material like iron at ordinary power frequencies, which are normally considered "low." Electric power is generated and used in the U.S.A. at a frequency of 60 Hz, and in many other countries at 50 Hz. Figure 13.2 shows that the flux density B at the center of an iron sheet $\frac{1}{16}$ in. thick is only about half the value at the surface at 50 Hz. Visually extrapolating the data of Fig. 13.2 to a sheet $\frac{1}{4}$ in. thick, one can conclude that the flux at its center is virtually zero. In other words, the iron at the center of such a sheet is contributing nothing to the desired multiplication of the applied field and might as well not be there; it is simply wasted material. In the U.S.A. the most popular lamination thickness is 0.012 in. (nominal), for which eddy-current shielding of the center of the sheet is practically negligible at 60 Hz.

(Eddy-current shielding can be convincingly demonstrated by a simple experiment [13.4]. A cylindrical tube was made, 5 in. long, 0.707 in. O.D., and 0.5 in. I.D.; into this could be inserted a solid rod, 0.5 in. O.D. Both pieces were made of annealed steel containing 0.35 percent carbon, with a permeability of about 2000 at the field strength of the experiment. The tube was wound with primary and secondary coils, and an ac voltage was applied to the primary. At frequencies of 25, 60, and 100 Hz, the output voltage was exactly the same, *whether or not the*

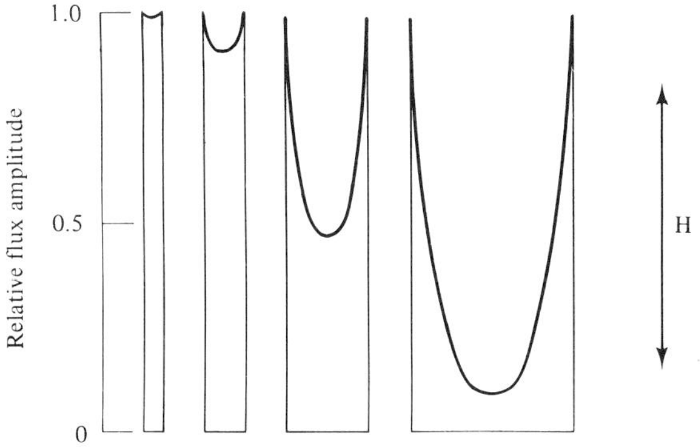

Fig. 13.2 Calculated amplitude of alternating flux in electrical-steel sheets of various thicknesses as a fraction of the value at the surface. Frequency = 50 Hz. The thicknesses are, from left to right, 0.016, 0.032, 0.064, and 0.128 inch. Brailsford [G.12].

inner rod was present in the tube. The tube and rod dimensions are such that the cross-sectional area of the central rod equals just half of the total area of the composite bar, and this central portion carried none of the flux at the frequencies mentioned. In other words, B decreased to zero at a depth below the surface equal to less than 0.207/0.707, or 29 percent of the bar radius.)

The way in which the flux decreases with depth below the surface can be calculated exactly only by making certain simplifying assumptions, namely, that the permeability μ is constant, independent of H, and that it has the same value at all parts of the sheet. The field is applied parallel to the surface of the sheet and is assumed to vary sinusoidally with time t, so that $H = H_0 \cos 2\pi f t$, where f is the frequency in Hz. Then the field amplitude H_x and the induction amplitude B_x ($= \mu H_x$) at a distance x from the center of the sheet are given by

$$\frac{H_x}{H_0} = \frac{B_x}{B_0} = \left[\frac{\cosh(2x/\delta) + \cos(2x/\delta)}{\cosh(d/\delta) + \cos(d/\delta)}\right]^{1/2}, \tag{13.1}$$

where d is the sheet thickness in cm. The quantity δ is given by

$$\delta = 5030 \sqrt{\frac{\rho}{\mu f}} \text{ cm}, \tag{13.2}$$

where ρ is the resistivity in ohm-cm. With the assumptions given, H_x and B_x vary sinusoidally with time, but they lag in time behind H_0 and B_0 at the surface. Equation (13.1) does not give the ratio of the interior value of H or B to the surface value at a given time, but rather the ratio of the amplitude of the interior value to the amplitude of the surface value. These amplitudes (maximum values) are reached at different times. The derivation of Eq. (13.1) is given by Russell [13.5].

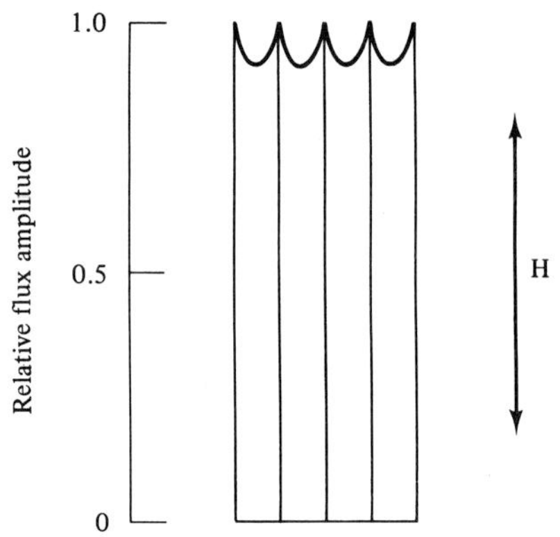

Fig. 13.3 Calculated relative flux amplitude in a stack of four sheets, each 0.032 inch thick, for the same conditions as in Fig. 13.2.

Flux penetration increases as μ and f decrease and as ρ increases, for a constant plate thickness. (Note that μ and f enter only as the product μf. Thus the flux penetration for $\mu = 1000$ and $f = 60$ Hz is the same as for $\mu = 100$ and $f = 600$ Hz.) The quantity δ is called the *skin depth* and has the following physical meaning. In a "thick" sheet H or B falls to $1/e$ (= 37 percent) of its value at the surface at a depth δ below the surface; a sheet is thick if δ is small compared to the sheet thickness d; in a thin sheet δ has no special meaning. (Note that δ is measured from the sheet surface, whereas x is measured from the center of the sheet.)

The curves of Fig. 13.2 were calculated from Eq. (13.1) with $\mu = 2500$ and $\rho = 14 \times 10^{-6}$ ohm-cm, and they apply to a frequency of 50 Hz and the stated thicknesses. Alternatively, they apply to any material, thickness, or frequency for which the dimensionless constant d/δ has the values 0.765, 1.53, 3.06, and 6.12, respectively. The flux penetration in each sheet of a stack of sheets, as in the laminated core of a transformer, is the same as in a single sheet, provided the sheets are electrically insulated from one another, in a manner to be described later. This point is illustrated in Fig. 13.3.

Flux penetration in cylindrical rods is given by the following relation, equivalent to Eq. (13.1):

$$\frac{H_x}{H_0} = \frac{B_x}{B_0} = \left[\frac{ber^2\,(\sqrt{2}\,x/\delta) + bei^2\,(\sqrt{2}\,x/\delta)}{ber^2\,(\sqrt{2}\,r/\delta) + bei^2\,(\sqrt{2}\,r/\delta)}\right]^{1/2}, \tag{13.3}$$

where x is the distance from the rod axis, r is the rod radius, and δ is given by Eq. (13.2) as before. The functions *ber* and *bei* are particular kinds of Bessel functions tabulated by Dwight [13.6] and McLachlan [13.7]. Rods are not used in transformer cores, but Eq. (13.3) is given here because magnetic measurements are often made on rod specimens, and it is sometimes necessary to estimate flux penetration in such specimens in alternating fields.

The *eddy-current power loss* P per unit volume of sheet can be calculated by integrating $I^2\rho$, where I is the current density, over the volume of the sheet and the length of a cycle. The result as derived by Golding [13.8], for example, is

$$P = \frac{10^{-9}\pi^2 d^2 B_0^2 f^2}{6\rho} \text{ erg/sec cm}^3. \tag{13.4}$$

This equation presupposes complete flux penetration, or d/δ less than 1, and constant permeability. The thickness d is in cm, the flux amplitude at the surface B_0 in gauss, the frequency f in Hz, and the resistivity ρ in ohm-cm. To obtain the power loss in watts/cm^3, multiply Eq. (13.4) by 10^{-7}. The strong dependence of the eddy-current loss on d and f should be noted. However, the loss is independent of μ, because complete flux penetration is assumed.

Equations (13.1), (13.3), and (13.4) are *classical*, in that the permeability μ is assumed to be constant not only in time, i.e., independent of the variation of H with time, but also in space, i.e., independent of position in the specimen even on a microscopic scale. These assumptions are valid only for para-. antiferro-, and diamagnetics. They are certainly not true for the strongly magnetic substances, which are made up of self-saturated domains and in which μ varies with H as shown in Fig. 1.14 (b). And μ depends strongly on position, on the scale of the domains. Suppose a field H is applied parallel to a 180° wall in a demagnetized sample, causing the wall to move a certain distance. What is the permeability? On the domain scale the only definition of μ that has much physical meaning is $(\Delta B)/H$. In the two domains adjacent to the displaced wall, but not including the swept-out region, ΔB is simply H, because the magnetization M_s has not changed, and μ is 1. Within the swept-out region, however, ΔB is $(H + 8\pi M_s)$, and μ is $(1 + 8\pi M_s/H)$ or about 43,000 for iron if H is 1 Oe.

The dependence of μ on H and on position in strongly magnetic substances has two effects:

1. If the applied field H varies sinusoidally with time, the variation of B will not be sinusoidal, as we previously assumed. Brailsford [G.12] shows how the B wave form is distorted and how the distortion varies with depth below the surface of a sheet.
2. Equation (13.4) underestimates the eddy-current loss except when the domain size is extremely small.

The effect of domain structure on eddy-current loss has been calculated by Pry and Bean [13.9] and Lee [13.10], following earlier work by Williams, Shockley, and Kittel [12.2]. Pry and Bean assumed the specimen to be a sheet of thickness d, infinite in extent, made up of domains of width $2L$ magnetized antiparallel to one another, as shown in the upper part of Fig. 13.4. Their result for the power loss P_{ed} is complicated and will not be given here, but the quantity of most interest is the ratio P_{ed}/P_{ec}, where P_{ec} is what we will now call the classical eddy-current loss given by Eq. (13.4). This ratio is shown in Fig. 13.4 as a function of $2L/d$. The full line refers to small magnetization, such that the amplitude of the induction

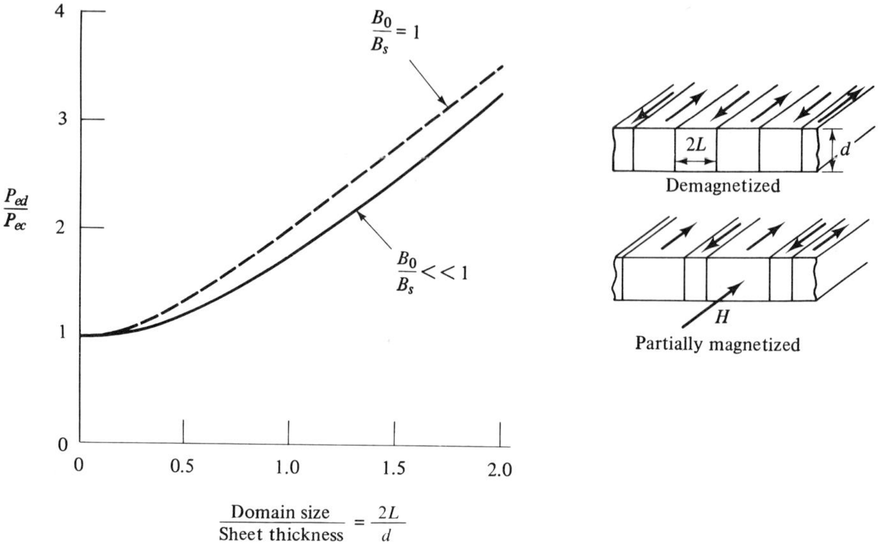

Fig. 13.4 Ratio of the domain-model eddy-current power loss P_{ed} to the classical eddy-current power loss P_{ec} as a function of domain size $2L$ relative to sheet thickness d. Pry and Bean [13.9].

B_0 is small compared to the saturation induction B_s. The dashed line shows the higher losses that occur when the domain walls move so far that the sheet is saturated once in each half cycle. The walls are assumed to remain flat, so that the calculation applies only to low frequencies where the walls do not bow out. (The lower the frequency, the lower is the average wall velocity, for constant B_0.) We conclude from this calculation that P_{ed} becomes equal to the classical value P_{ec} as the domain size approaches zero; this is reasonable in that μ then becomes homogeneous on a microscopic scale. But when the domain size becomes equal to the sheet thickness, as it does in some electrical steels, the eddy-current loss can become almost double the value calculated on a classical model.

The calculation of Pry and Bean was made for a very special arrangement of domain walls, but their qualitative conclusions are believed to be generally true, namely, that the spatial inhomogeneity of μ, which is the inevitable result of a domain structure, must lead to higher eddy-current losses than those classically calculated. In a material containing domains, the eddy currents are localized at the moving domain walls, as indicated in Fig. 12.6, and the eddy-current density I can reach very large values at the walls. The average value of $I^2\rho$ over the specimen is then larger than if the eddy currents were evenly distributed throughout. In other words, the larger the domain size, the fewer the walls and the faster they have to move to accomplish a given total flux change at a given frequency; the result is larger eddy currents.

13.3 LOSSES IN ELECTRICAL MACHINES

The machines to be considered here are transformers, motors, and generators.

Transformers

These are probably the most efficient "machines" ever made. Efficiency generally increases with size, and efficiencies exceeding 98 percent in large transformers are not uncommon. Efficiency is defined as follows:

$$\text{efficiency} = \frac{\text{output}}{\text{input}} = \frac{\text{input} - \text{losses}}{\text{input}}.$$

Transformer losses are mainly composed of:

1. *Core losses*, sometimes called *iron losses*, occurring in the magnetic core and amounting to 20 to 30 percent of the total losses.

2. *Copper losses*, occurring in the windings and amounting to the other 70 to 80 percent of the total. (These should probably be called winding losses, inasmuch as copper is now being replaced by aluminum conductors in some transformers, because of the high price of copper.) Although the copper loss, equal to i^2R, is at first glance purely an electrical loss, it has in fact some dependence on the magnetic properties of the core: the lower the permeability of the core, the larger must be the current in the primary winding to produce the required flux, and the larger the copper loss.

Although core losses in a transformer may appear insignificant, in that they amount to much less than 1 percent of the input to a large transformer, they add up to a very large amount, because virtually all generated electric power passes through transformers before reaching the consumer. In 1965, the electrical energy consumption in the U.S.A. amounted to some 10^{12} kilowatt-hours, and Olson [13.11] estimates that the core loss in transformers cost $300,000,000. It is no wonder then that considerable effort is expended in trying to reduce this loss.

Before attempting any analysis of the sources of core loss, we will digress to consider how this loss is measured. The standard method is the *Epstein test*, which simulates the operation of a transformer on zero load. This test has been standardized in all its details by the American Society for Testing and Materials [13.12] and has been described by various authors [2.1, 2.14, G.5]. The test material is in the form of strips 28 cm long and 3 cm wide, having a total weight of 2 kg. (This amounts to a total of 88 strips, if they are cut from sheet 0.014 inch thick). The strips are inserted into four solenoids arranged in a square, and mounted on a base with the strips interleaved at the corners to form a square core, as shown in Fig. 13.5. The axes of the solenoids form a square 25 cm (10 in.) on each side. Each solenoid has an inner, secondary winding S of 175 turns and an outer, primary (magnetizing) winding P of 175 turns; the four primary windings are connected in series and so are the four secondary windings. The normal oxide skin on the steel strips usually offers enough electrical resistance to prevent eddy-current flow from one strip to another. Moderate pressure

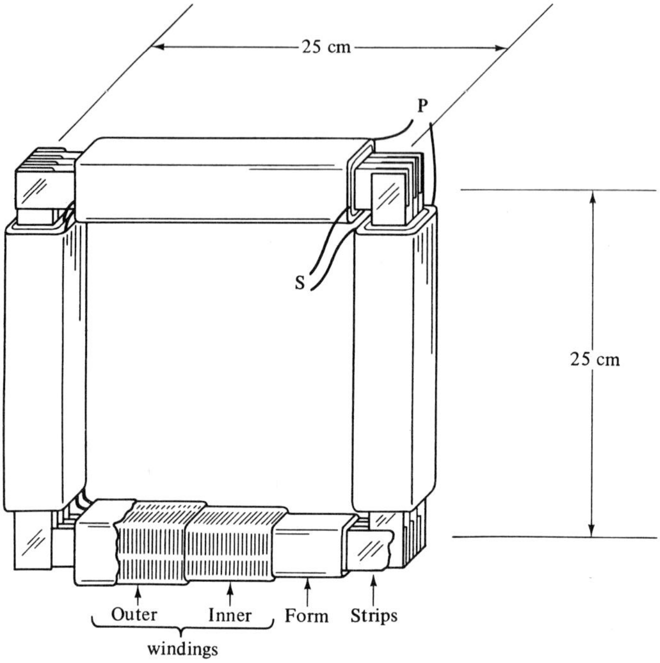

Fig. 13.5 Epstein test frame with strips inserted. Bozorth [G.4].

may be applied at the corners of the assembled stack of strips to improve magnetic continuity at these points where the strips overlap.

The heart of the measuring circuit is a dynamometer type of wattmeter, shown in Fig. 13.6 (a). This is a moving-coil instrument, but the coil moves in the magnetic field due to a current in a second, fixed coil rather than in the field of a permanent magnet. As normally applied to the measurement of power in an ac circuit, the fixed coil (current coil) of the wattmeter is connected in series with the line, and the movable coil (voltage coil) across the line. The magnetic field due to the current coil, which has a low resistance, is proportional to the current i in the line at any instant, while the current in the voltage coil, which has a high resistance, is proportional to the voltage e across the line. The average torque on the moving coil is then proportional to the average value of ei or to the average power. This torque rotates the voltage coil about an axis normal to the plane of the drawing against the resistance of a spiral spring, and a pointer (not shown) indicates the coil's position.

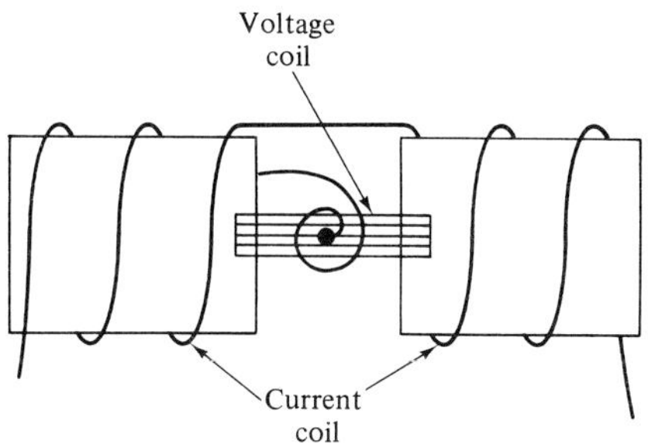

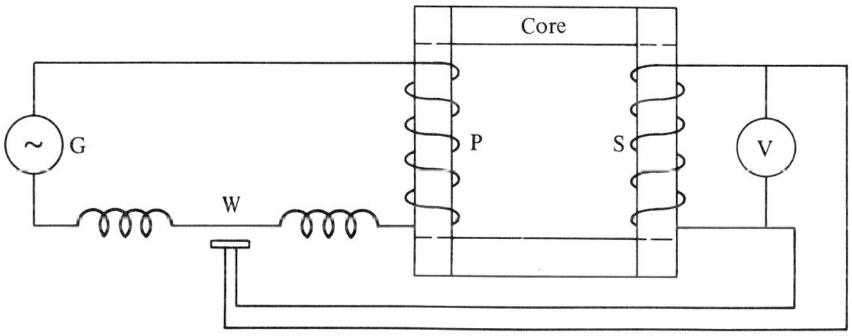

Fig. 13.6 (a) Dynamometer type of wattmeter. (b) Simplified circuit for core-loss measurement on Epstein specimens. G = ac source, P = primary, S = secondary, W = wattmeter, V = voltmeter.

When a transformer is operated with the secondary circuit open (no load), all the input power goes into losses. These losses are (a) core loss, and (b) copper loss in the primary winding. If the wattmeter is connected in the transformer primary circuit in the way described above, it would measure these two sources of loss. Instead, the wattmeter in the Epstein test is connected as in Fig. 13.6 (b), in order to avoid a correction for the copper loss in the primary. The current coil is in series with the primary winding and the voltage coil is connected across the secondary. The voltage applied to the voltage coil of the wattmeter is now due only to the changing magnetization of the core, and the wattmeter measures the power lost in the core alone. The resistance of the transformer secondary circuit is so high that virtually no power is dissipated in this circuit. The voltmeter V indicates the maximum induction B at which the core is being driven. Core-loss measurements are usually made at B values of 10 and 15 kilogauss; the total core loss is divided by the total weight of the strips making up the core and reported in units of watts/lb or watts/kg. Minor corrections to the wattmeter and voltmeter readings are required in practice, and these are described in the references cited above.

The *analysis* of core loss into its constituent parts is difficult and still a matter of some debate. There are two approaches: (1) an older conventional approach that ignores the domain structure of the material, and (2) a more recent approach that tries to identify the losses with actual physical phenomena known to occur in materials undergoing alternating magnetization. The conventional point of view, still maintained in some quarters, will be described first.

Conventionally, the total core losses measured by the Epstein test are divided into two parts, hysteresis loss and eddy-current loss. The hysteresis loss is determined from the area of the so-called "static" or direct-current hysteresis loop, i.e., the B,H loop determined ballistically with a galvanometer or fluxmeter (Section 2.5) or with a slowly operated hysteresigraph. The area of this loop, divided by 4π, equals the hysteresis loss W_h in ergs/cm^3 (Section 7.6). Knowing the density of the material and the conversion 10^7 ergs = 1 joule, we can find the hysteresis loss W_h in joules/lb. This loss is assumed to be independent of frequency. The hysteresis *power* loss P_h is therefore given by $W_h f$ watts/lb, where f is the frequency (1 watt = 1 joule/sec). On the other hand, the classical eddy-current power loss P_{ec} is proportional to f^2, according to Eq. (13.4). Therefore, the total power loss P_t in the core, as measured by the Epstein test, should be given by

$$\text{total power loss} = P_t = P_h + P_{ec}$$

$$= W_h f + kf^2 \text{ watts/lb,} \tag{13.5}$$

$$\text{total energy loss per cycle} = \frac{P_t}{f} = W_h + kf \text{ joules/lb.} \tag{13.6}$$

Here k is a constant. We therefore expect that P_t/f will vary linearly with frequency and equal W_h at zero frequency. Actual loss measurements over the range from 0 to about 100 Hz do show, as indicated in Fig. 13.7, the expected linear variation over a fairly wide range, but the curve becomes concave downward at low frequencies. However, it does extrapolate to the value W_h determined from the static hysteresis loop. But there is a gross discrepancy between the apparent eddy-current loss and the value calculated classically. The difference between the two is called the *anomalous loss*, and an anomaly factor is defined by

$$\eta = \text{anomaly factor} = \frac{\text{apparent eddy-current loss}}{\text{classical eddy-current loss}}. \tag{13.7}$$

If $\eta = 1$, there is no anomaly, but Brailsford [G.30] reports values ranging from 1.5 to 3.5 for various core materials and even values exceeding 8 in exceptional circumstances.

The "anomalous loss" is anomalous in name only. It is now recognized that the classical calculation of eddy-current loss simply does not apply to a material containing domains. Once this is admitted, it follows that it is unreasonable to compare an observed loss with the classical loss and call the difference an anom-

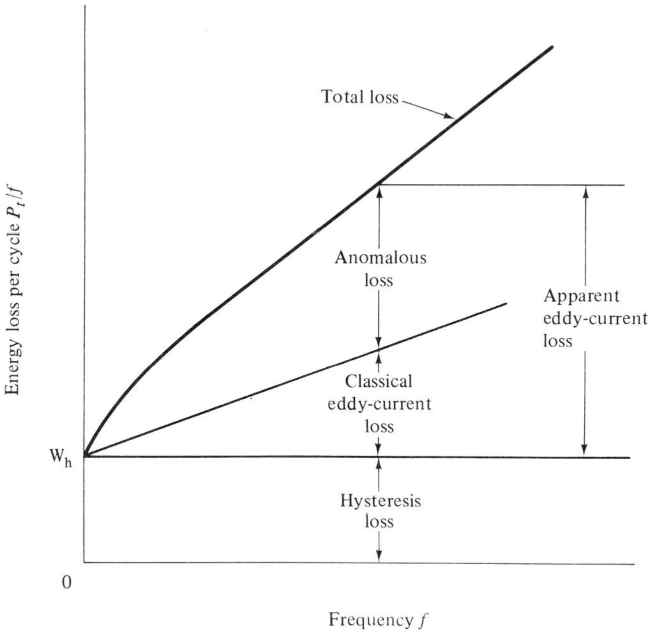

Fig. 13.7 Conventional separation of losses in transformer cores.

aly. When the real domain structure of the material is taken into account, the calculated eddy-current loss exceeds the classical loss, and the difference between the two is larger the larger the domain size, as shown by Fig. 13.4. In fact, the quantity plotted there, P_{ed}/P_{ec}, is simply the anomaly factor η for the particular domain model assumed.

The modern approach to core losses not only rejects the "anomalous loss" as an unreal fiction, but it also regards the conventional separation of the total loss into hysteresis and eddy-current components as artificial. This separation is based on the assumption that there is something fundamental about the static hysteresis loop. This assumption is simply not true, for the following reasons:

1. The loop is really not "static." If measured ballistically, the loop represents the response of the material to a step-function field; each time the field is applied, the induction B changes abruptly. Even when H is smoothly and slowly increased, as in hysteresigraph measurements, Barkhausen jumps occur locally throughout the material.

2. The shape and area of the loop depend on the way in which H changes with time. Figure 9.45 showed that a square loop can be changed into a re-entrant loop, with a consequent reduction of "hysteresis loss," by a feedback circuit which decreases H whenever B begins to change rapidly. This is a rather extreme

example. But if a hysteresis loop is displayed on an oscilloscope at a frequency of, say, 60 Hz by means of a loop-tracer (Section 2.5), the shape and area of the loop are observed to vary with the shape of the drive-field waveform (sinusoidal, triangular, or square, for example).

3. If the drive-field frequency is slowly increased from, say, 10 Hz to 100 Hz, with a constant waveform and constant B_{max}, the width and area of the loop are observed to increase as the frequency increases. Inasmuch as "hysteresis loss" is related to loop area, this behavior is at odds with the conventional view of Figure 13.7, where "hysteresis loss" is regarded as independent of frequency.

The current view is that virtually *all* the observed losses in alternating magnetization are due to micro eddy currents associated with moving domain walls and magnetization rotation. This view implies that the "static hysteresis loss" is also due to the same cause. As long as H is increased monotonically, then no matter how slowly the loop is traversed, irreversible changes in magnetization will occur and these will generate micro eddy currents; these changes will be chiefly Barkhausen jumps of domain walls and, to a lesser degree, irreversible rotations. Barkhausen jumps of 90° walls may, in addition, contribute to losses by means of the "phonon effect" postulated in Section 12.6. There simply do not appear to be any other mechanisms for generating heat. (It follows also that the loss mechanisms involved in magnetic damping experiments are precisely the same as those which operate in a transformer core undergoing alternating magnetization. Whether the domain walls are stress-driven or field-driven makes no difference.)

This view of the hysteresis loop in turn implies that we could reduce the hysteresis loss to zero if we could prevent irreversible changes in magnetization by appropriate control of H. A step in this direction is the re-entrant loop of Fig. 9.45. More generally, one would have to devise a means of nursing a domain wall at infinitesimal velocity over energy hills, such as those pictured in Fig. 9.36 (a). To prevent a Barkhausen jump from position 2 to 3, the positive field would have to be reduced from a maximum at 2 to zero at the top of the hill and then reversed in direction; in this way the wall could be made to move reversibly from one energy valley (1) to another (4). While perhaps feasible for a single wall, this kind of control in a material with a complex domain structure "is not even conceptually possible, as it would require H to be changing in different ways in different parts of the material," as Becker [13.13] points out. Furthermore, one kind of Barkhausen jump could not be prevented, even by sensitive field control of a single wall. This is the jump that occurs when the main wall suddenly jerks away from the tubular domains connecting it to an inclusion, as in Fig. 9.27 (d).

At first glance, the increase in hysteresis-loop width with increasing drive frequency, at constant B_{max}, is easy to understand, at least qualitatively. To accomplish a given change in B each domain wall must move back and forth over a certain distance in each cycle. As the frequency increases, the average velocity of the wall must increase. This higher velocity can be obtained only with a higher drive field, according to the results of Fig. 12.5. Therefore the hysteresis loop must widen along the H axis.

The difficulty with this explanation is that it is based on an assumption which, although plausible, has been shown experimentally to be false. This is the assumption that the number of walls does not change with frequency, or, more explicitly, that the total area of moving walls is the same at any frequency as the wall area of the static, demagnetized specimen.

Actually, as a number of investigators have concluded on the basis of various kinds of evidence, the moving wall area *increases* with frequency. Perhaps the most direct evidence for this is furnished by high-speed, Kerr-effect photographs of alternating magnetization in single crystals of Fe + 3 percent Si. The surface examined is a {110} plane, with an ⟨001⟩ easy axis parallel to the surface. (This is the orientation of most of the grains in the "grain-oriented" silicon steel used in large transformer cores. See Section 13.4.) When the crystal orientation is perfect, the domain structure is a simple one of 180° walls parallel to the easy axis, separating antiparallel domains. An example is shown in Fig. 13.8.

Houze [13.14] has taken motion pictures of such domain structures at speeds up to 5500 frames/sec. The alternating field H was applied parallel to the easy axis with Helmholtz coils and measured with a magnetic potentiometer (Appendix 11); the induction B was measured, not with a search coil, but by comparing the areas of the light and dark domains on the photograph. Measurements on the photographs also yielded the velocities and average spacing d of the domain walls, measured at the midpoint of the cycle ($B = 0$). Some of Houze's data, obtained with a triangular H waveform, are summarized in Table 13.3. The wall spacing d was observed to be 0.14 cm in the static, demagnetized specimen; this distance decreased to 0.07 cm at a drive frequency of 120 Hz, showing that the number of domains has doubled. Comparing the results for 30 and 120 Hz, for which the excitation conditions are nearly alike, we note that quadrupling the frequency does not quadruple the average wall velocity $\bar{v}$, as might have been expected; it only triples it, from 3.5 to 10.3 cm/sec. Thus an increase in frequency does not simply cause the same number of walls to move faster; it increases the number of walls *and* makes them move faster.

Houze also observed that wall motion was frequently jerky rather than smooth, and that part of a wall can be temporarily stopped by an imperfection while another part of the same wall moves ahead. It follows that the actual velocity v of a wall

Table 13.3 Domain-Wall Spacings and Velocities Measured during Alternating Magnetization of a {110} ⟨001⟩ Crystal of Fe + 3 percent Si (Houze [13.14]).

Frequency (Hz)	H_{max} (Oe)	B_{max}/B_s	v (cm/sec)	$\bar{v}$ (cm/sec)	d (cm)
30	2.1	0.43	17.8	3.5	0.128
60	1.7	0.38	24.6	6.4	0.083
120	2.2	0.48	23.7	10.3	0.071

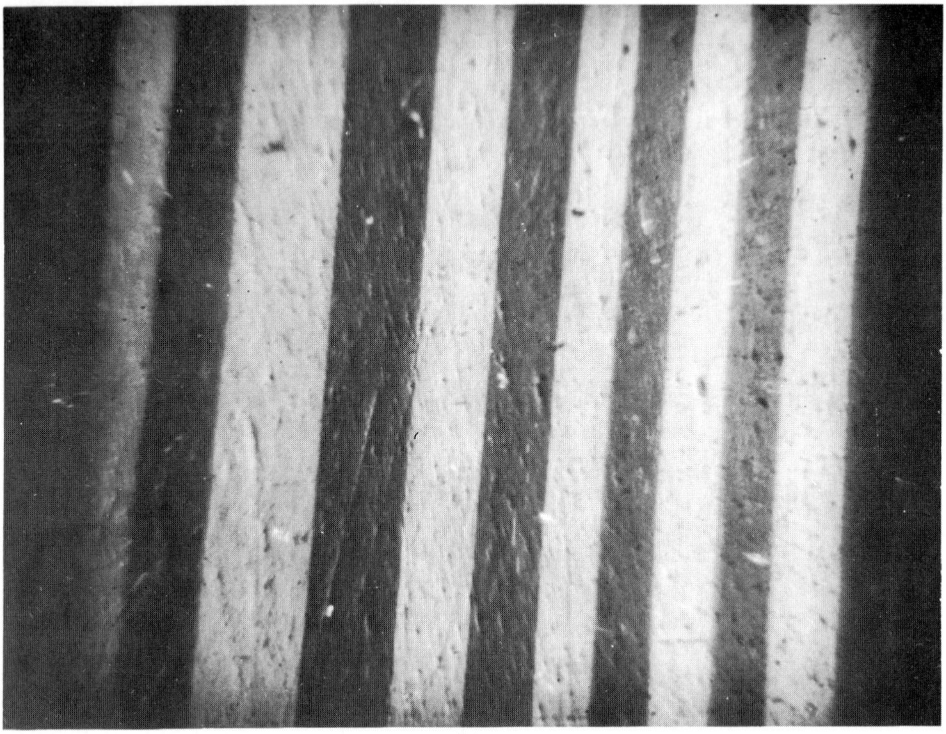

(a)

Fig. 13.8 Static Kerr-effect photographs of domains in Fe + 3 percent Si. The surface observed is {110}, and the ⟨001⟩ direction is approximately vertical on the page. (a) Demagnetized. (b) Partially magnetized, to $B = 7500$ gauss. (Courtesy of J. W. Shilling, Westinghouse Research Laboratories.)

(its velocity while actually in motion) is larger than its average velocity $\bar{v}$ (distance traveled divided by travel time, including the time when it is stopped). The ratio $v/\bar{v}$ was found to lie between 1.7 and 5.1, depending on f and H_{max}.

Haller and Kramer [13.15] made similar measurements on a crystal of the same composition (Fe + 3 percent Si), but with a {100} plane parallel to the surface. For constant B_{max} the static wall spacing d remained constant until a critical frequency of 40 Hz was reached; d then decreased as $1/\sqrt{f}$ (Fig. 13.9). Similarly, at a constant frequency of 90 Hz, d remained constant until B_{max}/B_s reached 0.6 and then decreased as $1/B_{max}$.

These direct observations of domain walls *during motion* at power frequencies are of great value in understanding core losses. The finding that the number of walls increases with frequency accounts for the initial downward curvature of the curve of energy loss per cycle versus frequency (Fig. 13.7); the loss per cycle does not increase with frequency as fast as might be expected, simply because

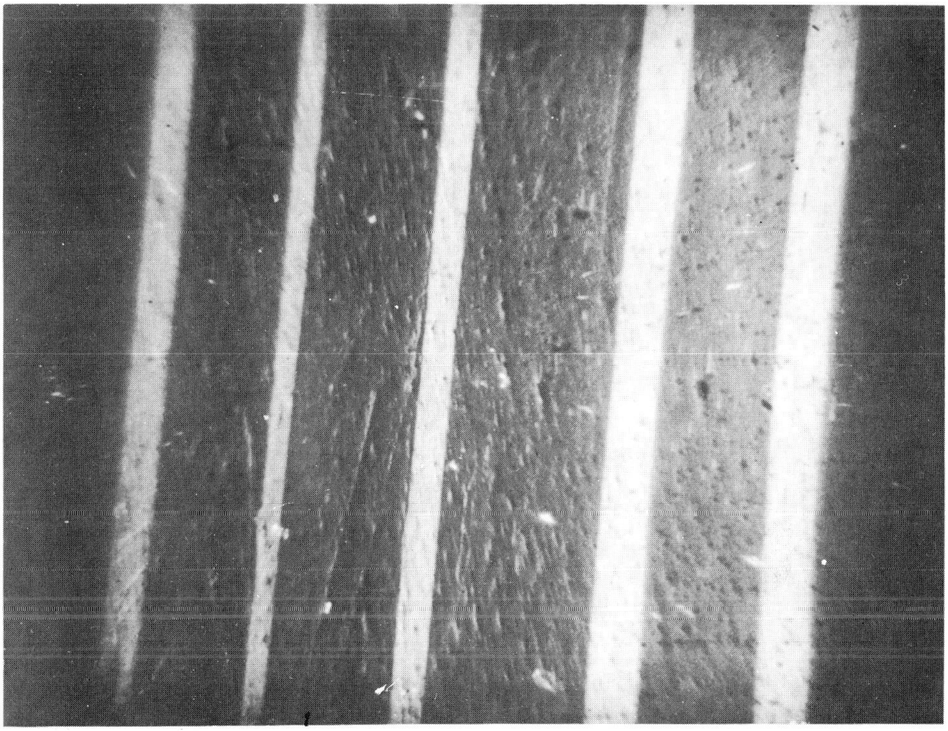

(b)

Fig. 13.8 (continued).

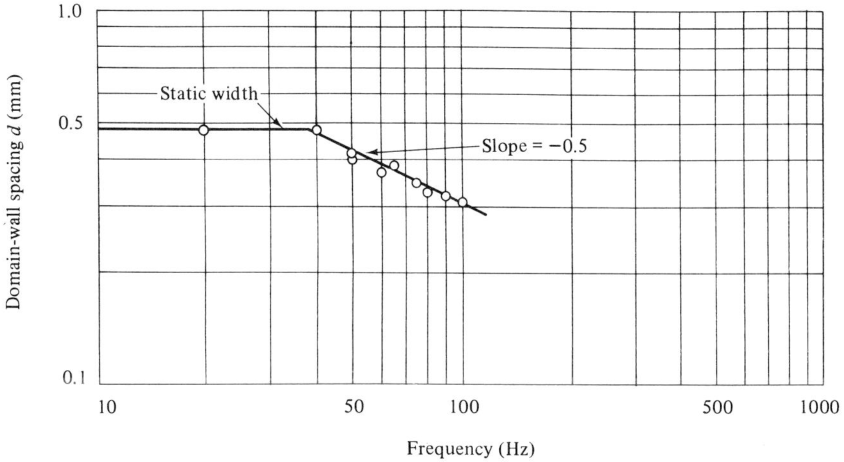

Fig. 13.9 Variation of domain-wall spacing d with frequency for an Fe + 3 percent Si crystal. Sinusoidal drive field. $B_{max} = 0.9\ B_s$. Haller and Kramer [13.15].

more walls are appearing, with the result that the average wall velocity does not have to increase as fast with frequency as it would if the number of walls were constant.

The average value of dB/dt during one half cycle is proportional to the frequency. It is also proportional to $A\bar{v}$, where A is the area per unit volume of moving domain walls and $\bar{v}$ their average velocity. On the other hand, the power loss due to microeddy currents is proportional to $A\bar{v}^2$, according to Williams *et al.* [12.2]. Thus doubling the frequency does not necessarily quadruple the loss. Doubling the frequency doubles $A\bar{v}$, but the change in $\bar{v}$ is less than double because A increases. In short, the material responds in such a way as to minimize its losses, as Haller and Kramer [13.16] put it. [In the analysis of magnetic damping measurements made in Section 12.6, the moving wall area A is implicitly assumed to be constant and independent of amplitude. This is apparently true, because otherwise Eq. (12.34) would not agree with experiment. Correspondingly, Haller and Kramer [13.15] observed, as mentioned above, that the area A of field-driven walls does not vary with amplitude, as long as the amplitude is small.]

An increase in the number of walls with increasing frequency requires that new walls be nucleated. We have already considered the problem of wall nucleation in Section 11.5, but there we were concerned with nucleation in a *saturated* crystal. Here it is a question of nucleatings walls in a specimen that already contains them; the data of Table 13.3, for example, show that more walls are appearing at values of B_{max}/B_s of less than 0.5, i.e., at less than half saturation in each cycle. Presumably new domains with antiparallel magnetization form within existing ones, at the edges of the crystal or at imperfections within it. Each new domain adds two 180° walls to the number already present.

Motors and generators

The subject of core losses in these machines is even more complex than in transformers, because the conditions of magnetization are more complicated. These are rotating machines, which means that the core of magnetic material is subjected to fields which may not only alternate in magnitude and sign along a given axis, as in a transformer, but also change direction in space. Thus in the limit we must examine the behavior of a specimen which is rotated through 360° in a constant field. Actually it is easier to visualize the equivalent behavior of a fixed specimen in a rotating field, as in Fig. 13.10, which is drawn for a field large enough to cause only partial saturation. As the field makes one complete revolution, the domain walls simply move back and forth. Even if the rotation is infinitesimally slow, Barkhausen jumps will occur generating micro eddy currents and heat; the sum of these losses for one revolution is the *rotational hysteresis loss* W_r; in the general case, W_r will also include losses due to irreversible rotations of the magnetization vector.

Rotational hysteresis is measured with a torque magnetometer on disk-shaped specimens, and Fig. 13.11 shows representative curves. If Eq. (7.5) is integrated, we have

$$W_r = -\int_0^{2\pi} L\,d\theta, \tag{13.8}$$

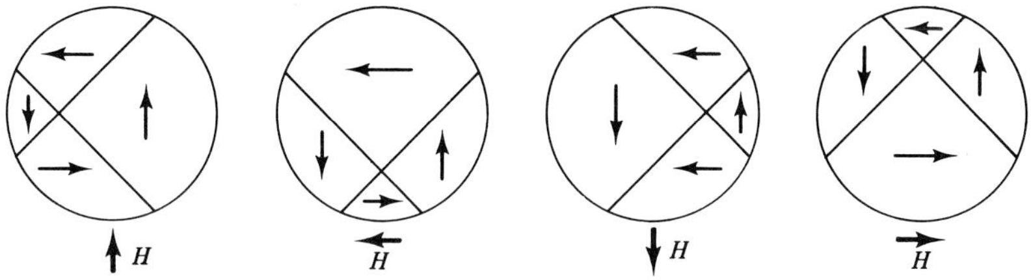

Fig. 13.10 Wall motion in a rotating field (schematic).

where L is the torque on the specimen when M makes an angle θ with some reference line on the specimen. The quantity on the right is just 2π times the average torque required to rotate the specimen in the field; it is also the net (algebraic) area between the L,θ curve and the $L = 0$ axis. In Fig. 13.11 the torque in a field of 50 Oe is practically constant. At 170 Oe the overall anisotropy of the specimen has become evident, but there is still rotational hysteresis because the net area between the curve and axis is still negative. At 1500 Oe, however, the curve is symmetrical about the axis, no net torque is required to rotate the specimen, and W_r is zero. The specimen is then saturated and its M_s vector rotates reversibly

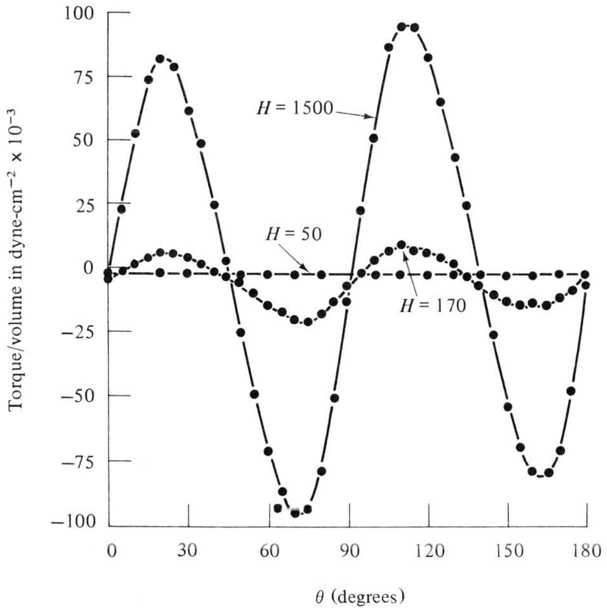

Fig. 13.11 Torque curves for a disk of cold-rolled iron at different values of the applied field. Williams [13.17].

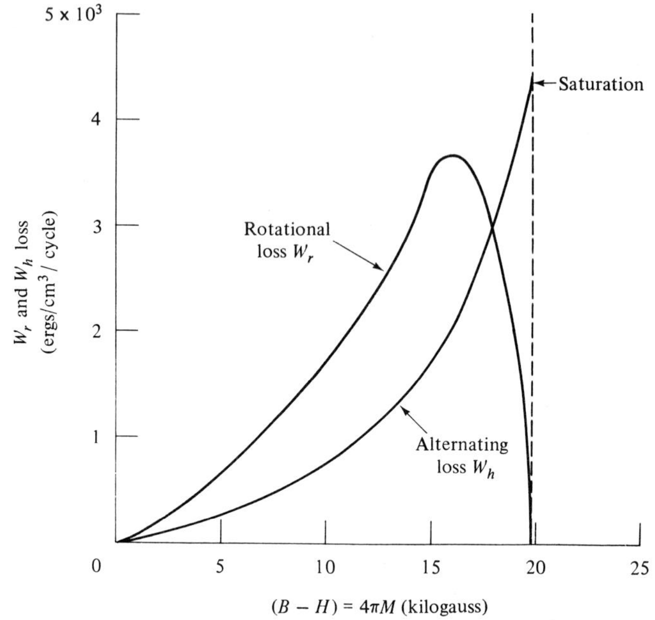

Fig. 13.12 Dependence on $4\pi M$ of the rotational hysteresis loss W_r and the alternating hysteresis loss W_h in the rolling direction for grain-oriented 3.13 percent silicon steel. Brailsford [13.18].

with the field; the torque curve is then simply an expression of the specimen's anisotropy, just as the curve of Fig. 7.15, for example.

In this behavior we see a distinct difference between the alternating hysteresis loss W_h and the rotational hysteresis loss W_r. The value of W_h steadily increases with field until the specimen is saturated, but W_r goes through a maximum and becomes zero at saturation. Figure 13.12 illustrates this point for silicon steel. The reason for the difference is simple: a specimen rotated in a saturating field is always saturated, but in alternating magnetization the specimen goes from saturation in one direction through the demagnetized state to saturation in the other twice each cycle.

Although alternating and rotational hysteresis losses are conceptually different and separately measurable, both are due to the same fundamental processes, irreversible wall movements and irreversible rotations.

13.4 ELECTRICAL STEEL

Three kinds of materials are used for the cores of most electrical machines: (1) low-carbon steel, (2) nonoriented silicon steel, and (3) grain-oriented silicon steel. These materials are usually called *electrical steel* or *transformer steel;* sometimes

Table 13.4 Approximate Price and Annual U.S. Production of Electrical Steel in 1970.

Kind of material	Price per ton	Annual production (tons)	Total value
Low-carbon steel	$ 160	300,000	$ 48,000,000
Nonoriented silicon steel	300	500,000	150,000,000
Grain-oriented silicon steel	480	250,000	120,000,000

all of them are referred to simply as "iron," regardless of composition. The magnetic quality and the price increase in the order listed. Table 13.4 gives price and production statistics.

Before these three grades are examined, some general remarks on texture (preferred orientation) are in order. This subject was introduced in Section 7.18, where we saw that the proper texture could in itself result in a polycrystalline specimen having an easy axis of magnetization. In an iron or iron-silicon crystal the easy axis is $\langle 100 \rangle$ and the hard axis is $\langle 111 \rangle$. In sheet for some transformer cores, the material can be positioned so that the rolling direction in the sheet is parallel to the direction of magnetization (the direction of "flux travel.") The desired texture is therefore $\{hkl\}$ $\langle 100 \rangle$, because this makes the easy direction parallel to the field with a resultant increase in permeability and decrease in losses. This kind of texture has been achieved in silicon steel, and such material is then called "grain-oriented" in the electrical industry.

In cores for rotating machines, on the other hand, the field is in the plane of the sheet, but the angle between the field and rolling direction is variable. Here there is no point in having the easy direction in the rolling direction, and a satisfactory texture would be $\{100\}$ $\langle uvw \rangle$, which keeps the hard $\langle 111 \rangle$ direction out of the plane of the sheet. A $\langle 100 \rangle$ fiber texture would be even better, i.e., a texture in which all grains had a $\langle 100 \rangle$ direction normal to the sheet surface and all possible rotational positions about this normal, because the sheet would then be isotropic in its own plane.

Electrical steel for cores must be in the fully recrystallized, magnetically soft condition. Now there are two kinds of recrystallization, primary and secondary. These will be distinguished later on. It is enough to note here that magnetically desirable textures have been achieved to date only by secondary recrystallization and that this process requires a rather high annealing temperature. After these preliminaries we can turn to the materials themselves.

Low-carbon steel

This material was the original core material for transformers, motors, and generators, but it is limited today mainly to the cores of small motors. This is essentially the same material as that used for automobile bodies, washing machines,

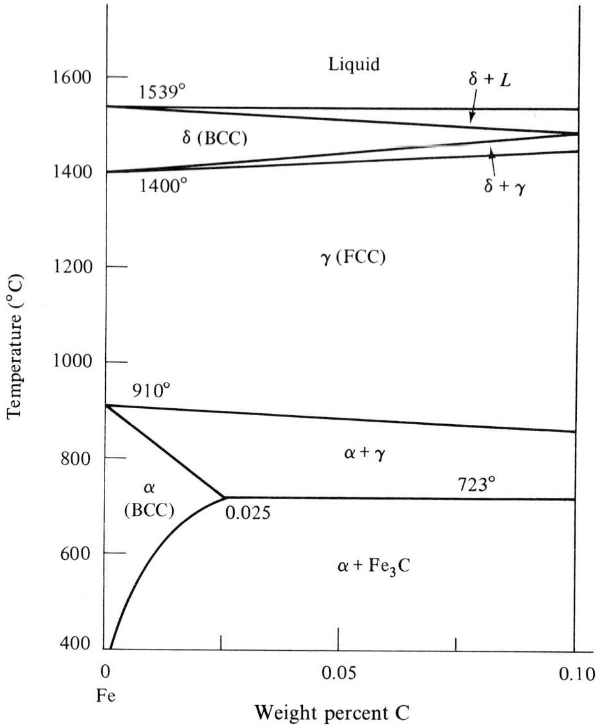

Fig. 13.13 Portion of iron-carbon equilibrium diagram.

refrigerators, and the like. Low-carbon sheet steel is the cheapest steel product made and is produced in large tonnages; the portion destined for motor cores forms only about one percent of the total.

The carbon content is about 0.05 percent. The core loss at 15 kilogauss is 5 to 6 watts/lb, which is some 10 times the loss for grain-oriented silicon steel. But core loss is of small importance to the manufacturer of small motors; low cost rather than efficiency is the chief object. These motors are designed for intermittent operation, chiefly in consumer products. The market is very large. It has been estimated that the average American home now contains some 25 small electric motors; this figure may appear surprising until one makes a careful count, not overlooking such items as clocks, fans, shavers, toys, etc.

The normal (primary) recrystallization texture of sheet steel is complex and magnetically unattractive. Some producers are currently studying ways of improving this texture, not particularly in order to reduce losses but to increase the permeability, which would in turn permit lower core weight or fewer ampere-turns in the winding. Increased permeability can probably be obtained more

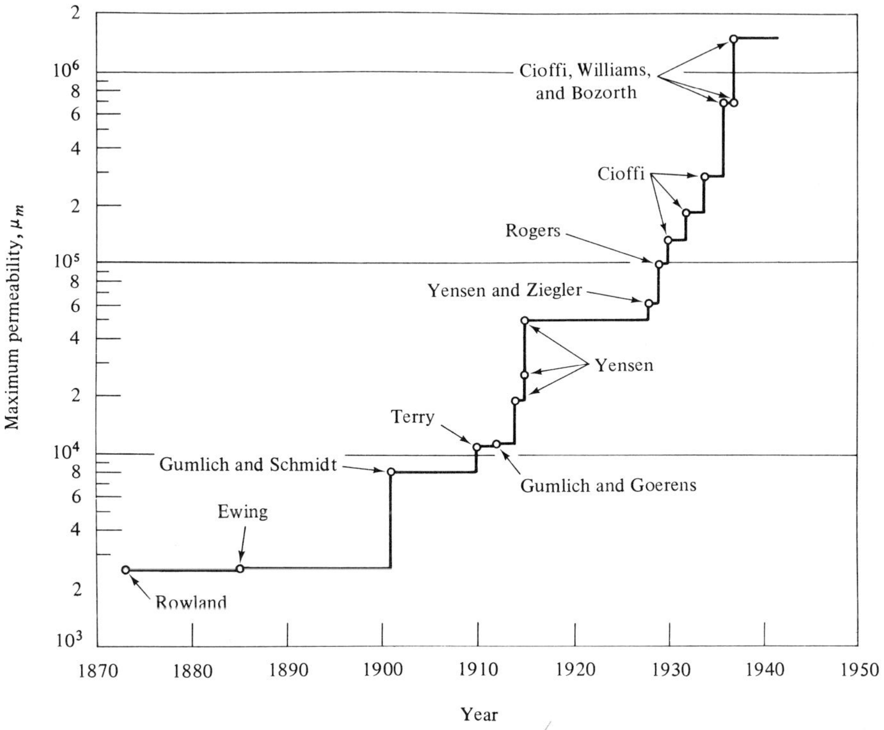

Fig. 13.14 Highest values of maximum permeability of iron reported in various years. Bozorth [G.4].

cheaply by improving the texture than by adding alloying elements, provided the proper combination of rolling and heat treatment to achieve the desired texture can be found. One difficulty is the fact that high annealing temperatures, which might produce a favorable secondary recrystallization texture, are ruled out by the $\alpha \rightarrow \gamma$ phase transformation. This occurs at 910°C in pure iron and at even lower temperatures with increasing carbon content, as shown in Fig. 13.13. (If steel is recrystallized in the γ region, each grain will transform into several α grains of differing orientation on cooling to room temperature. Grain orientations will therefore tend to become random, and the texture formed by high-temperature recrystallization will largely be wiped out.) The lower the carbon content the closer the annealing temperature may approach 910°C, but a decarbutizing operation, to permit a higher annealing temperature, itself adds to the cost of the product.

Low-carbon sheet steel has another application, not involving alternating magnetization, and that is in the core of electromagnets. Here the main requirement is high permeability at minimum cost. The largest electromagnets made are those which supply the constant magnetic field required in synchrotrons, linear accelerators, and other machines of particle physics. Such magnets can weigh several thousand tons per machine.

Low-carbon sheet steel has a maximum permeability of 5,000 to 10,000 and might just as well be called impure iron. Proper purification treatments in the laboratory can lead to remarkable improvement in the permeability of iron, and Figure 13.14 shows the progress that has been made in this direction since 1873. The purest iron is made by vacuum melting and annealing in hydrogen above 1300° C to remove carbon, nitrogen, and oxygen. Despite its high permeability "pure" iron is not of commercial value, because (1) it costs too much to make, and (2) its electrical resistivity is so low that eddy-current losses would be very high in applications involving alternating fields.

Nonoriented silicon steel

This steel was developed by the English metallurgist Robert Hadfield in 1900 and soon became the preferred core material for large transformers, motors, and generators. The history of silicon steel development for electrical applications has been traced by Walter [13.19] and by Bechtold and Wiener [13.20].

The addition of silicon to iron profoundly modifies the phase changes, as shown by Fig. 13.15. The temperature of the $\alpha \rightarrow \gamma$ transformation is raised and that of the $\gamma \rightarrow \delta$ transformation is lowered until the two meet at about 2.5 percent Si, forming a closed "gamma loop." As a result an alloy containing, say, 3 percent Si is body-centered cubic right up to the melting point. This in turns means that (1) such an alloy may be recrystallized at quite high temperatures without interference by phase changes on cooling, and (2) single crystals of such an alloy can be made for research purposes by slow solidification from the liquid. (This cannot be done with pure iron. If a single δ iron crystal is formed from the melt, it would consist of several crystals after it had undergone the $\delta \rightarrow \gamma$ and $\gamma \rightarrow \alpha$ transformations on cooling to room temperature. Instead, α iron crystals must be made by the more difficult *strain-anneal method.* The iron is strained a few percent in tension and then annealed below 910°C in an attempt to recrystallize a single grain which will consume the strained matrix.) The α solid solution of silicon in iron is often called silicon ferrite by metallurgists.

The preceding remarks apply to pure Fe-Si alloys. The presence of carbon widens the $(\alpha + \gamma)$ region, and only 0.07 percent C is enough to shift the nose of the gamma loop over to about 6 percent Si [G.4]. Actually, the carbon content of silicon steel varies from about 0.03 percent (nonoriented) to less than 0.01 percent (oriented), and it would be reduced to even lower levels if it were economically feasible to do so. (Iron carbide and nitride precipitates degrade the magnetic properties by interfering with wall motion, and the slow precipitation of carbides during service, called "aging," can cause a substantial increase in core losses. In fact, this problem was so severe in the plain-carbon steels available before 1900 that transformers often had to be taken apart, after some months or years of service, in order to re-anneal the steel laminations [13.19].) The carbon content of modern silicon steels is so extremely low that it seems they are called "steels" only because

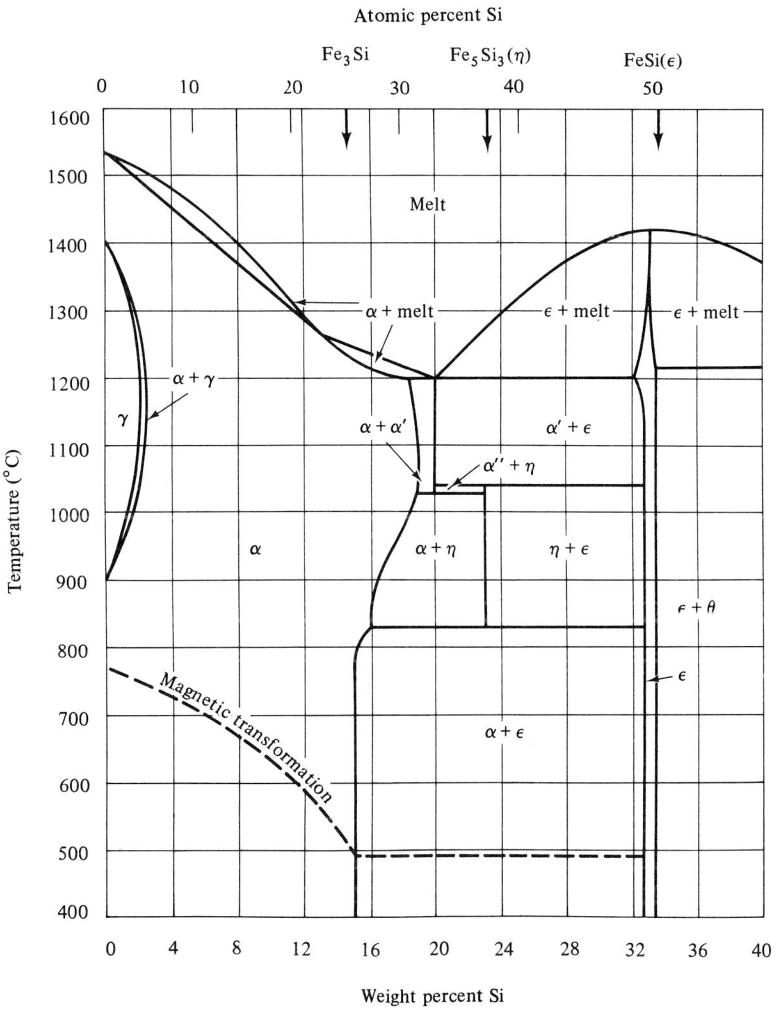

Fig. 13.15 Portion of iron-silicon equilibrium diagram. Bozorth [G.4].

they are made in a steel mill. They are actually Fe-Si alloys and are often called silicon iron.

The addition of silicon to iron has the following beneficial effects on magnetic properties:

1. The electrical resistivity increases, causing a marked reduction in eddy currents and therefore in losses.
2. The crystal anisotropy decreases, causing an increase in permeability.

On the debit side, silicon additions decrease the saturation induction and tend

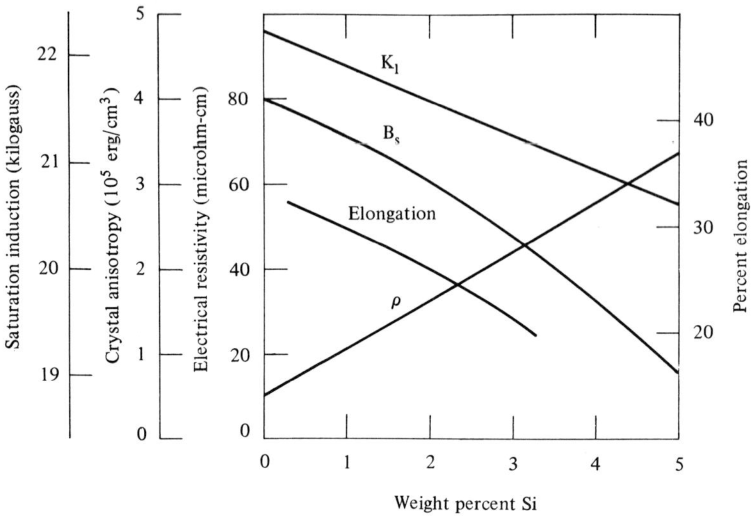

Fig. 13.16 Data on iron-silicon alloys: crystal anisotropy constant K_1, saturation induction B_s, electrical resistivity ρ [13.21], and percent elongation in 2 inches for nonoriented, polycrystalline sheet, 0.0185 inch thick [13.22].

Table 13.5 Some Magnetic Properties of 0.014-inch Silicon Steel [13.22, 13.23].

	Grade	Percent silicon	Core loss at 15 kG (watts/lb)	Permeability at 15 kG*
Nonoriented	M-45	1.05	2.72	1100
	M-43	1.50	2.33	1100
	M-36	2.25	1.98	700
	M-27	2.80	1.81	700
	M-22	3.20	1.67	700
	M-19	3.25	1.60	700
Grain-oriented	M-7	3.25	0.70	16,000
	M-6	3.25	0.64	20,000
	M-5	3.25	0.57	23,000

* ac magnetization.

to make the alloy brittle, so that it becomes difficult to roll into sheet when the silicon content is about 5 percent or higher. Figure 13.16 shows the effect of silicon on these various properties; the percent elongation in the tensile test is a measure of ductility.

To eliminate a confusion of trade names, the American Iron and Steel Institute has assigned AISI type numbers to electrical steels. These consist of the letter "M" (for magnetic material) followed by a number which, when the designations

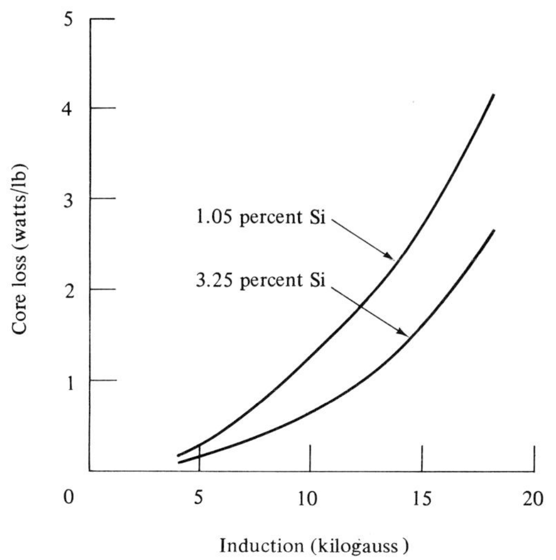

Fig. 13.17 Variation of core loss with maximum induction for nonoriented silicon steel, 0.014 inch thick [13.22].

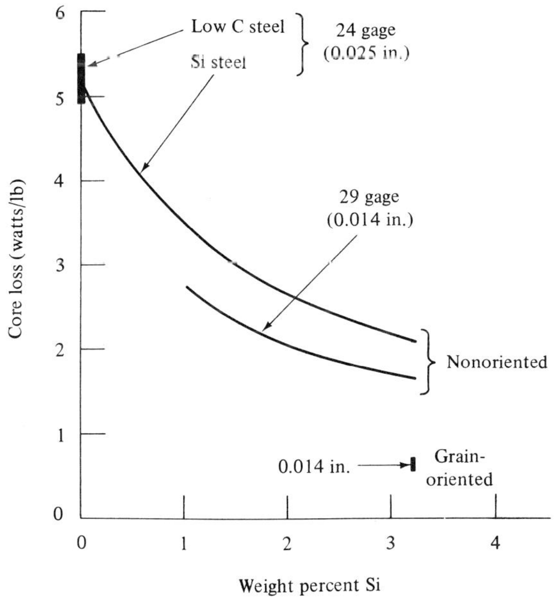

Fig. 13.18 Core loss at 15 kilogauss of low-carbon and silicon steels [13.22, 13.23].

were originally made, was about equal to ten times the core loss in watts/lb at 15 kilogauss and 60 Hz for 29 gage sheet (0.014 inch). Core losses have since been reduced but the type numbers remain. Table 13.5 gives data on several grades for one manufacturer.

Core loss increases rapidly with the maximum induction reached in the cycle, as shown in Fig. 13.17. For a constant induction of 15 kilogauss, Fig. 13.18 shows how silicon content and thickness affect core loss.

Nonoriented silicon steel sheet is made by hot rolling almost to final thickness, pickling in acid to remove the oxide scale, slightly cold rolling to improve flatness, and batch annealing.

Grain-oriented silicon steel

This material was developed by the American metallurgist Norman Goss in 1933. He discovered that *cold rolling* with intermediate anneals, plus a final high-temperature anneal, produced sheet with much better magnetic properties in the rolling direction than hot-rolled sheet. This improvement was due to a magnetically favorable texture produced by secondary recrystallization during high-temperature annealing. Grain-oriented sheet went into commercial production about 1945, and its properties have since been continually improved. Material with a core loss of the order of 0.6 watt/lb is now routinely produced, as shown in Fig. 13.18, and such sheet is the standard core material for large transformers. (Grain-oriented silicon steel was the first metallurgical product of any kind in which preferred orientation was exploited to achieve a desired property. Only in the past ten years have serious efforts been made to improve the *mechanical* properties of steel and other alloys by texture control).

Secondary recrystallization differs markedly from primary. Primary recrystallization occurs when a cold-worked metal is heated to a temperature at which new, strain-free grains can nucleate and grow throughout the cold-worked matrix. It is truly a recrystallization. Secondary recrystallization, on the other hand, is a particular kind of grain growth and is sometimes called discontinuous, exaggerated, or abnormal grain growth. It occurs in some, but not all, materials when (a) normal grain growth is inhibited, and (b) the material is annealed, usually for a long time, at a temperature much higher than that required for primary recrystallization. The result is the preferential growth of a relatively few grains at the expense of the others, leading to extremely large grains. The grains are no longer microscopic in size, with grain diameters of some tens of microns, but they are now visible to the naked eye, having diameters of the order of several millimeters. The grain size has therefore increased by a factor of several hundred (Fig. 13.19), and such grains occupy the entire thickness of 0.014-inch sheet.

Two kinds of normal grain growth inhibition, necessary for secondary recrystallization, have been identified:

1. A sharp primary recrystallization texture, which can be produced in some materials by heavy cold rolling. The driving force for grain growth is the surface energy of the grain boundaries. If the primary texture is sharp, then adjoining grains have nearly the same orientation, and the boundary between them has low surface energy.
2. A small and critical amount of a second phase dispersed in the grain boundaries. These particles impede grain boundary movement.

The point about secondary recrystallization that is of interest here is the fact that a secondary-recrystallization texture is usually quite different from the primary-recrystallization texture. In silicon steel the primary texture is weak and com-

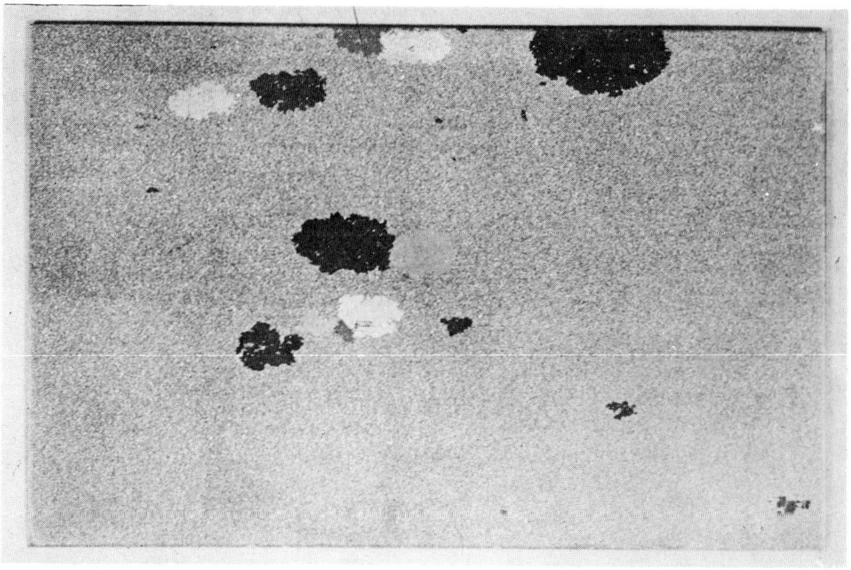

Fig. 13.19 Partial secondary recrystallization in silicon iron. A few large secondary grains are growing in a matrix of primary grains during a 900°C anneal. About actual size. Dunn [13.24].

plex, whereas the secondary texture is a quite sharp single-component texture, namely, {110} ⟨100⟩. This texture is illustrated in Fig. 13.20. In (a) is shown the stereographic projection of the principal directions in a single crystal, with the plane of the sheet taken as the projection plane; (011) is parallel to the sheet surface and [100] parallel to the rolling direction. The sketch in (b) of the orientation of the unit cell of each grain relative to the sheet shows why this is called the *cube-on-edge texture*. The schematic {100} pole figure in (c) shows where the ⟨100⟩ directions are concentrated for a large number of grains in polycrystalline sheet; the texture is sharper, i.e., has less scatter about the ideal orientation shown in (a), the smaller the areas of the high-density regions on the pole figure. Pole figures are the result of x-ray diffraction measurements and are the most direct description of a texture. Useful information about texture in magnetic materials can also be obtained from torque curves, subject to the limitations mentioned in Section 7.8; Fig. 13.21 shows how closely the torque curve for a grain-oriented polycrystalline sheet can approach that of a single crystal.

The cube-on-edge texture in silicon steel is generally held to be due to fine particles of MnS, which inhibit normal grain growth. The exact schedules of rolling, annealing, and atmosphere control required for successful production are trade secrets, but Graham [7.24] has concluded, largely from an examination of the patent literature, that the procedure is essentially as follows:

1. Begin with a steel containing, as important ingredients, about 3.2 percent Si, 0.03 percent C, 0.06 to 0.10 percent Mn, and 0.02 percent S.

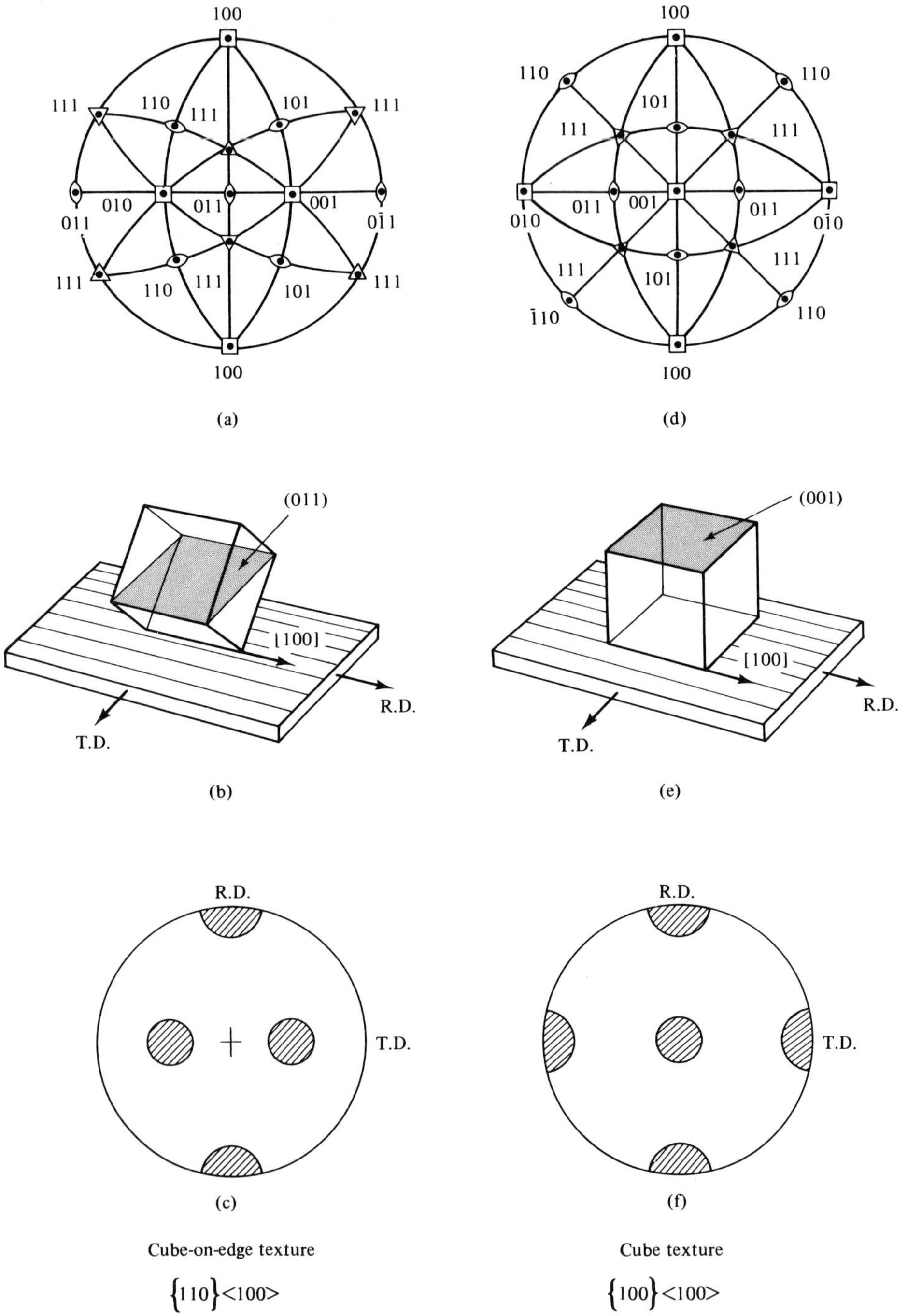

Fig. 13.20 Single-crystal projections (top), unit-cell orientations (center), and {100} pole figures (below). R.D. = rolling direction, T. D. = transverse direction.

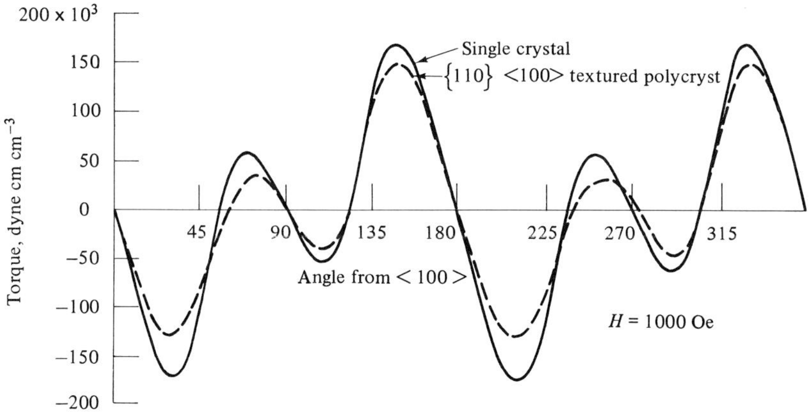

Fig. 13.21 Torque curves of a single crystal in the {110} plane and of a disk cut from a sheet of grain-oriented silicon steel having a strong cube-on-edge, {110} ⟨100⟩ texture. Graham [7.24].

2. Hot roll the cast ingot at about 1300°C to a thickness of 0.06 to 0.10 inch.
3. Remove oxide scale with acid (pickling).
4. Cold roll to a final thickness of 0.010 to 0.014 inch in two approximately equal steps, with an intermediate anneal at 800 to 1000°C to soften the sheet enough for the second stage of cold rolling. (The total cold reduction in thickness is of the order of 85 percent.)
5. Decarburize at about 800°C in moist hydrogen. (This anneal reduces the carbon content to about 0.003 percent and causes primary recrystallization.)
6. Anneal in dry hydrogen at 1100 to 1200°C to form the cube-on-edge texture by secondary recrystallization.

This final anneal is a batch operation requiring several days, and the steel is in the form of large coils about 30 inches wide and weighing 3 to 4 tons. The sulphur, which in the form of MnS particles provided the necessary grain-growth inhibition during the 800°C anneal, is itself reduced (by reaction with hydrogen and removal as H_2S) to levels as low as 0.002 percent in the final high-temperature anneal. The manganese thus liberated goes into solution in the iron. This is important, because any particles that are capable of inhibiting normal grain growth are also capable of inhibiting domain wall motion, as Fiedler [13.25] points out, and such particles must therefore be eliminated in one way or another during the final anneal.

These appear to be the main features of the process. Recrystallization in silicon steel, because of its commercial importance, has received more study than in any other material, and the main mechanisms are well understood. However, certain important features are still obscure. The interested reader can gain entry into the extensive literature on the subject through Graham's review [7.24].

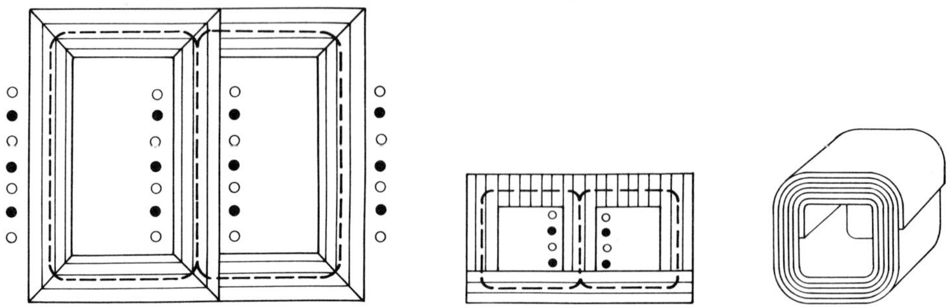

(a) (b) (c)

Fig. 13.22 Transformer cores.

A group of papers published in 1967 is also recommended, both on recrystallization and on the magnetic properties of the grain-oriented product [13.26].

The cube-on-edge texture makes the easy ⟨100⟩ directions in all grains almost parallel to one another and to the rolling direction of the sheet. The magnetic properties in this direction are therefore excellent, and all properties quoted in this section for grain-oriented sheet refer to this direction. For example, Epstein strips for core-loss measurements are all cut parallel to the rolling direction of the sheet. (In nonoriented sheet, on the other hand, properties are an average for the rolling and transverse directions, and the Epstein square is built up of strips cut half in one direction and half in the other.) The magnetic properties in the other directions in grain-oriented sheet are drastically inferior to those in the rolling direction; for example, the permeability at 15 kilogauss in the transverse direction is less than 2 percent of the permeability in the rolling direction. The reason is not far to seek: as Fig. 13.20 (a) shows, both the hard ⟨111⟩ direction and the medium-hard ⟨110⟩ direction lie in the plane of the sheet, the former at 55° to the rolling direction and the latter at 90°. The ductility also differs markedly: the percent elongation in tension is only 8 percent in the rolling direction compared to 28 percent in the transverse.

Because the magnetic properties are so directional, care must be taken in the construction of transformers to make the rolling direction of the sheet parallel, as far as possible, to the direction of flux travel. This is no problem in the largest transformers, because each leg is formed of separate pieces of sheet. This kind of construction is illustrated in Fig. 13.22 (a), where the striations indicate the rolling direction of each piece. (The largest transformers are called *power transformers.* They are located at the generating plant and at large substations. They have cores weighing up to 250 tons, and their size is limited only by transportation problems. *Distribution transformers* are the next smaller size; they are often seen mounted on telephone poles.) In smaller transformers this construction is not economical; one solution is to make E-I laminations, as in (b), where the core is made of two sets of sheets, cut in the form of these letters. Flux travel is in the easy direction except in the top leg, where it is in the transverse direction of the sheet; this leg may be widened to decrease the flux density and thus compensate for the lower permeability in this direction. (Both primary and secondary windings are placed on the central leg. These are only schematically indicated; the windings actually occupy most of the volume between the legs.) Still another solution is to roll up a long length of strip on itself to form the *tape-wound core* shown in (c), with windings (not shown) placed on opposite legs. Here the entire flux path is in the easy (rolling) direction.

Grain-oriented steel is more strain-sensitive than other grades, and laminations must be given a stress-relief anneal, at 800°C in dry nitrogen, after any shearing or punching operations, or the magnetic properties will be degraded. Wound cores must be similarly annealed. During actual construction of a transformer, care must be exercised so that strain is not introduced into the sheets by handling or assembly.

In 1957 three industrial laboratories succeeded in marking silicon steel sheet with the *cube texture*, i.e., {100} ⟨100⟩ as illustrated at the right of Fig. 13.20. This has the advantage over the cube-on-edge texture that two ⟨100⟩ easy directions lie in the plane of the sheet, parallel to the rolling and transverse directions. If E-I laminations were cut from such sheet, flux travel would be in an easy direction in all legs of the core. However, in large transformers such as the one shown in Fig. 13.22(a), cube-texture sheet would offer no advantage over cube-on-edge sheet.

Research on the cube texture has shown that it forms during secondary recrystallization and is due to differences in the gas/metal surface energy of grains having different {*hkl*} planes parallel to the sheet surface [7.24]. If the primary recrystallization texture has a {100} ⟨100⟩ component and if the annealing atmosphere or the metal has a small concentration of the proper impurity, then {100} grains will at the expense of all others during the secondary-recrystallization anneal because {100} surfaces have lower energy. The ⟨100⟩ directions of these grains will then be parallel to the rolling direction, because that was their orientation in the primary texture. Close control of the annealing atmosphere is critical.

Achievement of the cube texture in silicon iron appears to be, at least so far, a metallurgical success and a commercial failure. Its superior magnetic properties are outweighed by high production costs, so that it is not now competitive with cube-on-edge sheet for general use in the cores of electrical machines. There has been some limited production of very thin, cube-texture strip for tape-wound cores in magnetic amplifiers. (The thinner the material, the more effective is the surface-energy mechanism in producing the cube texture.)

General

We consider here several points of general importance relative to the application of electrical sheet steel.

1. *Lamination insulation.* The sheets that make up a core must be insulated from one another to prevent gross eddy-current circulation in the core. Occasionally, the sheets are coated with an organic varnish, but this will not withstand stress-relief annealing. More usually the insulation is simple a film of tightly adherent iron oxide, formed by means of a slightly oxidizing annealing atmosphere.

Grain-oriented steel is often coated with MgO before the high-temperature anneal. This combines with SiO_2, from the silicon in the steel, to form a glassy magnesium silicate. This not only acts as an insulator but, having a smaller coefficient of thermal contraction, tends to put the steel in tension when the coated steel has cooled to room temperature. As a result, core losses are found to decrease. Presumably, the residual tensile stress is predominantly longitudinal, tending to form 180° domain walls parallel to the rolling direction.

2. *Cooling.* Both the core and winding losses show up as heat in the material. Core loss alone can lead to a temperature rise of the order of 0.2 to 1.5°C/min, if all generated heat remains in the core (Problem 13.5). Transformers are more difficult to cool than motors or generators; the construction of the latter is somewhat less compact to begin with, rotation of part of the machine naturally draws in air, and this flow can be augmented by mounting fan blades on the shaft. The smallest transformers, as in communication equipment, need no artificial cooling. A common cooling method for larger sizes is to mount the transformer in a tank full of oil and allow natural convection currents in the oil to bring heat out to the tank surface. In some transformers, core losses are required to be low, not because of any consideration of electrical efficiency, but simply to minimize heat generation.

3. *Noise.* Transformer "hum" is due to magnetostrictive vibration of the core and to the attractive and repulsive forces between laminations. This noise becomes objectionable when large power and distribution transformers are placed in or near residential areas, and transformer manufacturers have gone to considerable effort to reduce noise. The magnetostriction of grain-oriented sheet in the rolling direction is remarkably low, sometimes positive and sometimes negative, as shown in Fig. 13.23. Inasmuch as λ_{100} is about 23×10^{-6} for Fe + 3.2 percent Si [13.27], the low values indicated in Fig. 13.23 show that the demagnetized state of this material is far from ideal, in the special sense of "ideal" used in Section 8.2; it has a preferred domain arrangement, with a preponderance of 180° walls parallel to the rolling direction. This arrangement has two causes:

a) The other two $\langle 100 \rangle$ easy directions make angles of 45° with the sheet surface [Fig. 13.20 (a)]. Domains so oriented are unlikely because of the large demagnetizing field in this direction. If they do exist, their volume is reduced by closure domains which form to reduce free poles at the sheet surface.

b) Residual tensile stress in the rolling direction, due to the coating on the sheet, favors the formation of $\langle 100 \rangle$ domains aligned with the rolling direction.

4. *Loss reduction.* How can silicon steel be improved? Reduction in the cost of cube-texture material is one approach, and further reduction in the losses of cube-on-edge material is another. With respect to the latter, considerable attention has been paid to the effect of grain size, which is usually 1 to 5 mm. Losses are then much lower than if the grain size were a hundred times smaller, 0.05 mm, say. This suggests that the larger the grain size, the lower the losses. But the problem is complicated by the fact that increasing the grain size generally improves the texture. For the grain size range of 1 to 5 mm, Littmann [13.28] has shown that losses slowly *increase* with grain size, when care is taken to compare laboratory

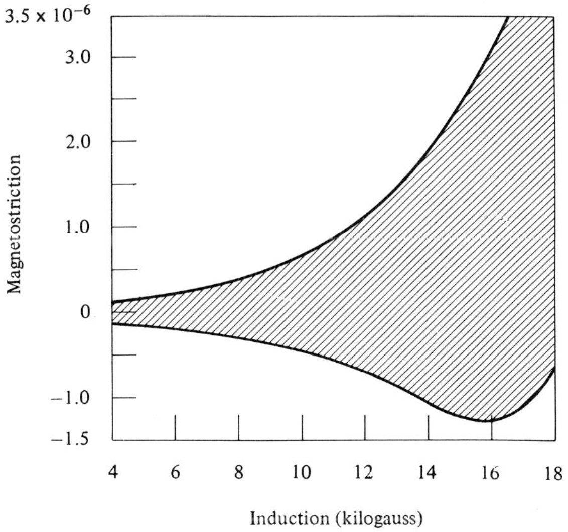

Fig. 13.23 Limits of static longitudinal magnetostriction as a function of induction for several grades and thicknesses of grain-oriented silicon steel sheet [13.23].

specimens with the same texture, and that the curve of loss versus grain size goes through a minimum at about 1 mm. When the grain size is large, the domains also tend to be large, causing high losses even in a material free from imperfections (Fig. 13.4). When the grain size is small, there are more domain walls, but the grain boundaries themselves impede wall motion. What is needed is a way of introducing more walls without at the same time introducing impedances to their motion. As Becker [13.29] points out, dynamic behavior depends on the dynamic domain structure, but not enough is now known about the relation of this structure to the metallurgical structure of the material.

13.5 SPECIAL ALLOYS

These are mainly iron-nickel alloys containing about 50 to 80 percent Ni. They form a very versatile series, usually called Permalloys, chiefly characterized by very high permeability at low applied fields. They were developed by Elmen and his associates, mainly in the period 1913–1921; Bozorth [G.4] has recounted the history of these investigations and Chin [13.44] has reviewed the properties of the alloys. Although Permalloy is a trade name of the Bell Telephone Laboratories, it has passed into the common language of magnetism and now denotes almost any alloy of nickel and iron; the number preceding the name is the nickel content in percent. Special trade names abound in this area, and essentially similar alloys are given different names by different manufacturers.

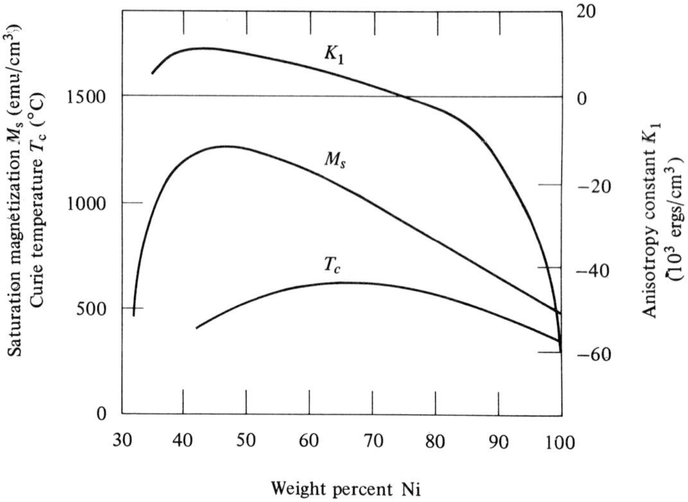

Fig. 13.24 Iron-nickel alloys. Variation with nickel content of saturation magnetization M_s, Curie temperature T_c, and crystal anisotropy K_1 (quenched alloys). After Bozorth [G.4].

The equilibrium diagram of the iron-nickel system has been given in Fig. 10.1 and discussed in Section 10.2. In the range of 50 to 80 percent Ni the alloys are all face-centered cubic. At and near the composition $FeNi_3$ the alloys can undergo long-range ordering below a temperature of 503°C; the ordered unit cell is shown in Fig. 10.6. The magnetic properties of the ordered alloys are inferior to those of the disordered. The disorder-to-order transformation is fortunately sluggish and can be avoided simply by fairly rapid cooling through the 500 to 400°C range; water quenching is unnecessary.

The manner in which the saturation magnetization and the Curie temperature vary with composition is shown in Fig. 13.24. The exact course of the M_s curve between 30 and 40 percent Ni is uncertain.

Figure 13.24 also shows the variation of the crystal anisotropy, and K_1 is seen to pass through zero at about 75 percent Ni. The magnetostriction constants are given in Fig. 13.25. The value of λ_{100} is zero at two compositions, near 46 and 83 percent Ni, while λ_{111} is zero at about 80 percent. (The solid lines denote alloys given a "normal furnace cool" and correspond to "the more-nearly disordered state," according to Hall. The dashed line denotes alloys cooled at 1.2°C/hr, corresponding to "the more-nearly ordered state." The course of these two branches of the λ_{111} curve suggest that both cooling rates have produced essentially the same state, the disordered state, because the results are in good agreement with those of Bozorth [13.30] for quenched alloys.) Note also that the magnetostriction becomes isotropic near 60 and 86 percent Ni.

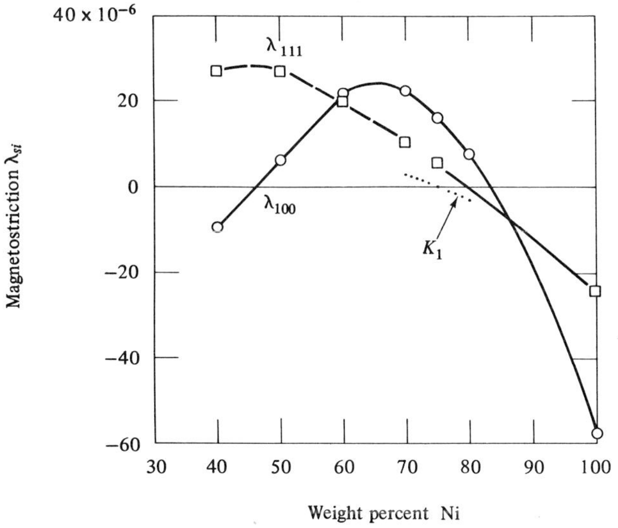

Fig. 13.25 Iron-nickel alloys. Variation of saturation magnetostriction with nickel content. See text for details. The dotted line shows K_1 from Fig. 13.24. After Hall [13.27].

We expect high permeability in any material when K and λ are small. A low value of K decreases domain-wall energy, and inclusions then become less effective hindrances to wall motion. A low value of λ means that microstress becomes similarly less effective. Figure 13.25 shows that both K_1 and λ_{111} are near zero just below 80 percent Ni and λ_{100} is not very large. This is the composition of 78 Permalloy, and Fig. 13.26 shows the remarkable effect of heat treatment on the maximum permeability of this alloy. (A similar curve showing the variation of initial permeability is almost identical with the one shown, when plotted on a permeability scale one-tenth as large.)

The maxima in μ near 50 and 80 percent Ni are due to near-zero values of K_1 and/or λ. Note that it is the value of λ in the easy direction of magnetization that determines the effect of microstress on wall motion; near 50 percent Ni, the value of K_1 is positive, $\langle 100 \rangle$ are easy directions, and λ_{100} is zero; at 80 percent Ni the value of K_1 is negative, $\langle 111 \rangle$ are easy directions, and λ_{111} is zero.

The great sensitivity of 78 Permalloy to heat treatment is due to the variation of K_1, which has a value of about -2000 ergs/cm^3 after quenching and $-18{,}000$ ergs/cm^3 after slow cooling; the values of λ_{100} and λ_{111} change hardly at all with heat treatment. X-ray and neutron diffraction show that slow cooling produces partial long-range order, which for some reason has a larger effect on K_1 than on λ.

Ordinarily 50 Permalloy has a much lower permeability than 78 Permalloy, because K_1 is larger. Yensen, however, found that prolonged annealing at 1000°C

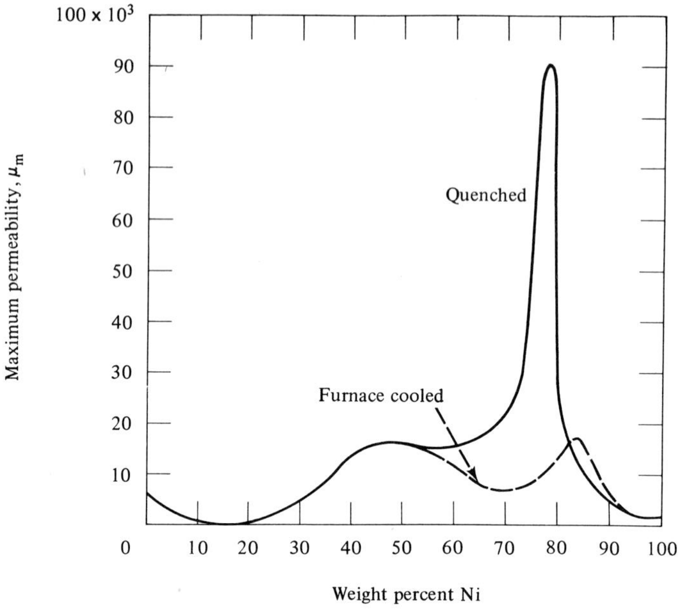

Fig. 13.26 Maximum permeabilities of iron-nickel alloys. Bozorth [13.30].

in hydrogen so purified it that its maximum permeability approached that of 78 Permalloy. The alloy so treated was called Hipernik. It has a higher maximum induction than 78 Permalloy; this is an advantage in applications where a core is driven to saturation or nearly so, because a smaller volume of core material is then required.

The properties of these and several other alloys are listed in Table 13.6. They have important applications in communication equipment (radio, telephone, and television) and other devices which will be described later.

The cube texture is fairly easy to produce in Fe-Ni alloys at or near the 50-50 composition. The texture forms by primary recrystallization after heavy cold rolling. The result is a square hysteresis loop, and such materials are produced under a variety of trade names.

An alloy called Isoperm was developed in Germany by making a cube-textured 50 Fe–50 Ni alloy and then cold rolling it to a 50 percent reduction in thickness. The resulting material has an easy axis in the transverse direction, as described in Section 10.5. Along the hard-axis rolling direction the permeability is low (100 or less) and constant. Constancy of permeability is a requirement for the cores of loading coils, as explained later in this section, and Isoperm has been used in the form of tape-wound cores for this purpose.

When minor amounts of one or more other elements are added to an Fe-Ni base, other useful alloys result. The addition of 4 to 5 percent molybdenum

Table 13.6 Magnetically Soft Alloys. Adapted from [13.20]

Name	Approximate composition (weight percent) *Ni*	*Fe*	*Other*	Initial permeability	Maximum permeability	Coercivity H_c (Oe)	B_s (gauss)	T_c (°C)	Resistivity (microhm-cm)
				Low-Cost Alloys					
Iron	–	100	–	150	5,000	1.0	21,500	770	10
Silicon iron	–	96	4 Si	500	7,000	0.5	19,700	690	60
Grain-oriented silicon iron	–	97	3 Si	1,500	40,000	0.1	20,000	740	47
				High-Permeability Alloys					
78 Permalloy	78	22	–	8,000	100,000	0.05	10,800	580	16
Hipernik	50	50	–	4,000	70,000	0.05	16,000	500	45
4-79 Permalloy	79	17	4 Mo	20,000	100,000	0.05	8,700	460	55
Mumetal	77	16	5 Cu, 2 Cr	20,000	100,000	0.05	6,500		62
Supermalloy	79	16	5 Mo	100,000	1,000,000	0.002	7,900	400	60
				High-Saturation Alloys					
Permendur	–	50	50 Co	800	5,000	2.0	24,500	980	7
2V-Permendur	–	49	49 Co, 2 V	800	4,000	2.0	24,500	980	27
Hiperco	–	64	35 Co, 0.5 Cr	650	10,000	1.0	24,200	970	28
Supermendur	–	49	49 Co, 2 V		60,000	0.2	24,000	980	27

to 78 Permalloy increases the initial permeability and more than triples the electrical resistivity, so that core losses are lower. These alloys are called "4-79 Mo Permalloy" and Supermalloy. The latter is given special care in melting, casting, and heat treatment and it has the largest permeability of any polycrystalline material.

The addition of copper gives Mumetal, which originally contained 5 percent Cu, 75 percent Ni, balance Fe, but 2 percent Cr is now often added. It is easier to roll into thin sheets and is often used in that form for magnetic shielding, chiefly to shield certain components of electrical apparatus from stray magnetic fields. (For example, the electron beam of a cathode ray tube can be deflected by the field of a transformer in the circuit, unless the tube is wrapped in shielding material. Or the signal recorded on magnetic tape can be distorted by the field from the motor of the tape recorder, unless the recording head is shielded.)

The addition of cobalt gives the Perminvar alloys, already described in Section 10.2. As a result of domain wall stabilization by self-magnetic-annealing, they have constant permeability at low fields. The original alloy (25 Co, 45 Ni, 30 Fe) has fewer applications than a cheaper one, made by reducing the cobalt content, namely 7-70 Perminvar (7 Co, 70 Ni, 23 Fe). Its initial permeability of 850 remains constant within 1 percent up to an induction of 600 gauss. (Incidentally, the simplest conceivable domain structure in a "macroscopic" specimen was observed by Williams and Goertz [13.31] in a tiny *polycrystalline* ring, 3 mm in outside diameter, of a Perminvar containing 23 percent Co, 43 percent Ni, and 34 percent Fe. It had been heat treated in a circumferential field to give it an easy circumferential axis. In the demagnetized state, the ring consisted of only two domains, one magnetized clockwise and the other counterclockwise. The single 180° domain wall separating them ran right around the ring, with little regard for grain boundaries. This alloy has low crystal anisotropy.)

Cobalt is the only alloying element that substantially increases the Curie temperature and the saturation magnetization of iron, and the 30 Co–70 Fe alloy has the largest value of M_s at room temperature of any material (Fig. 4.21). The 50 percent Co alloy, invented by Elmen and called Permendur, has almost as large a value of M_s but a much larger permeability, although not as large as the Fe-Ni alloys. The Fe-Co alloys are brittle, but the addition of vanadium makes them ductile enough to be hot and cold rolled down to thin sheet. The result is 2V-Permendur (2 V, 49 Co, 49 Fe), which is used for the diaphragms of telephone receivers and, to a limited extent, as a magnetostrictive transducer for sonar. Supermendur has the same composition but is vacuum melted and magnetically annealed,

Further details of the structure and magnetic properties of the alloys decribed in this section can be found in the books by Bozorth [G.4] and Tebble and Craik [G.36]. A few representative magnetization curves, on log-log scales, are shown in Fig. 13.27. The scale of H may have more meaning if it is realized that the horizontal component of the earth's field is only about 0.3 Oe in northern latitudes. (Of the alloys shown in Fig. 13.27, Hymu 80 contains 4 percent Mo, 79 percent Ni, and 17 percent Fe. Deltamax is a cube-textured 50 Fe–50 Ni alloy, and 4750 alloy contains 47 to 50 percent Ni, balance Fe.)

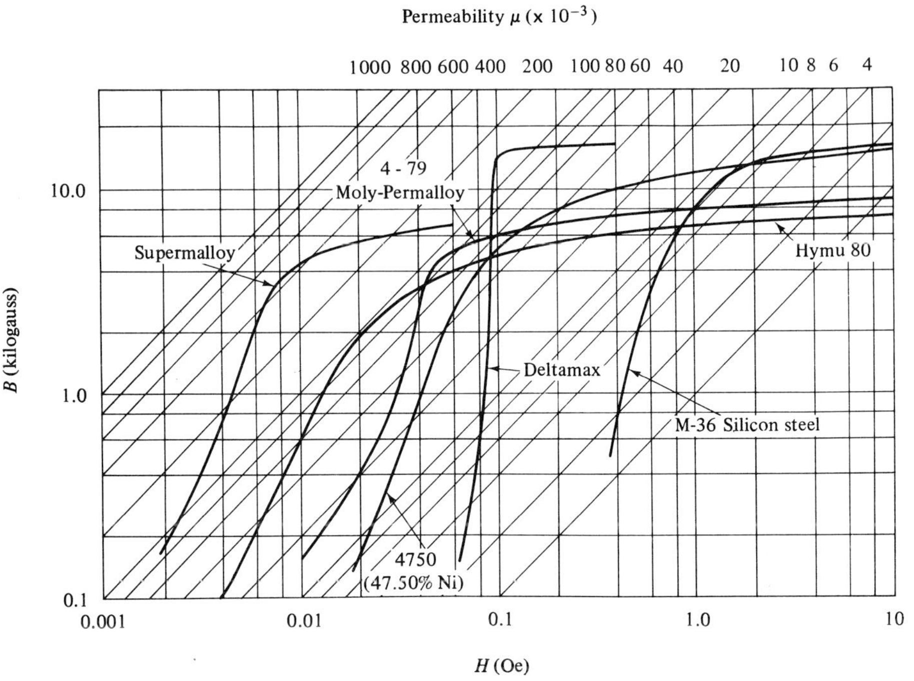

Fig. 13.27 Static magnetization curves for some magnetically soft alloys. The light diagonal lines indicate the permeability, in units of 1000, at any point on the graph. [13.3].

The magnetically-soft special alloys are used mainly in the form of cores, of varying shape and with external diameters from several inches to less than $\frac{1}{4}$ inch. They can be of three forms (Fig. 13.28). In stacked laminations the flux travels in all possible directions in relation to the original rolling direction of the sheet from which the laminations were cut; only the average magnetic properties in the plane of the sheet are of interest. In contrast, the flux in a tape-wound core travels only in the rolling direction of the strip (tape). To minimize core losses, the tape should be thinner the higher the frequency; tape-wound cores are available made

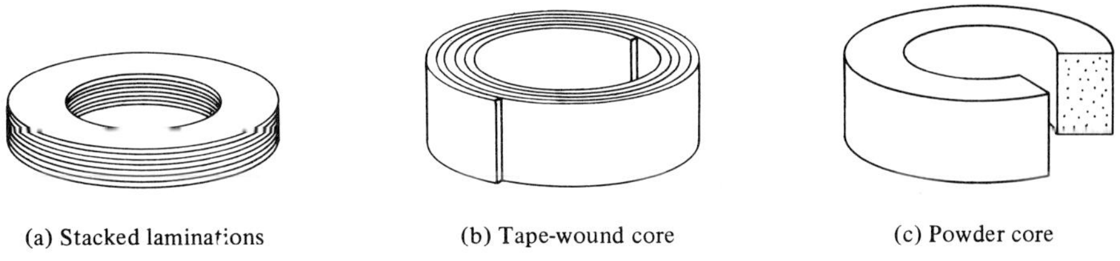

Fig. 13.28 Types of cores (schematic). The powder core has been sectioned to indicate its internal structure.

from material ranging from 14 to $\frac{1}{8}$ mils thick. Powder (dust) cores are made of iron powder or iron-nickel alloy powder, some 50 to 100 μ in diameter and therefore multidomain, each particle electrically insulated from the other by a suitable coating, and the whole pressed into ring shape. Powder cores are intended for high-frequency applications, up to about 100 kHz. At these frequencies, losses would be intolerable in bulk metallic materials, but subdivision of the metal into fine powder essentially eliminates the eddy currents that cause the loss. This result is achieved at the expense of a very large decrease in permeability, to values of the order of 10 to 100, even when the powder is made of high-permeability alloy particles. The reason is simply that each particle is effectively surrounded by an "air gap" of insulating material, and the internal demagnetizing fields are therefore large. As a result the hysteresis loop is sheared over, resembling Fig. 9.44 (b), and the permeability becomes constant over a considerable range of field. (At still higher frequencies, in the megahertz range, ferrite cores have better properties than any metal powder cores because of their inherently high electrical resistivity.)

Some applications of the soft magnetic alloys will now be described; many of them are treated in detail by Bardell [G.11]:

1. *Loading coils.* The transmission of a signal through a telephone or telegraph line with minimum distortion and attenuation requires that the circuit have the proper balance of capacitance and inductance. Capacitance is always present, between the conductors of the cable, and it is necessary to add inductance, an operation called "loading." Submarine cables are continuously loaded by wrapping a thin Permalloy tape helically around the central copper conductor. Cables on land are loaded by connecting them to loading coils at regular intervals, generally of the order of a mile. These coils are wound on cores of Mo Permalloy powder, which have the required properties of low loss and constant inductance. (The inductance L of any coil, in henries, is equal to 10^{-8} times the flux linkages $N\phi$, in maxwell turns, per ampere of current in the coil. The flux $\phi = BA = \mu HA$, where H is the field produced by the current in the coil. The only function of the magnetic material is to increase the permeability μ above the value of unity which an air core would have. But if the inductance is also to be constant, independent of the size of the current in the coil, then μ must be constant and independent of H. Powder cores of Mo Permalloy have permeabilities of the order of 50 to 100, which vary less than 0.2 percent for fields as large as 100 Oe.)

2. *Special transformers*, mainly for communication equipment. One requirement for the core is high permeability, to make best use of the low currents available. Another is low hysteresis loss, not for the sake of efficiency or low heat generation, but because large hysteresis leads to distortion of the current waveform.

3. *Magnetic amplifiers.* These involve square-loop cores and are an example of a group of devices called *saturable reactors.* The principle is illustrated in Fig. 13.29, which shows a core with two windings, one connected to a dc source (the control winding) and the other through a load to an ac source. When the control current is zero, the current through the load is low because of the large

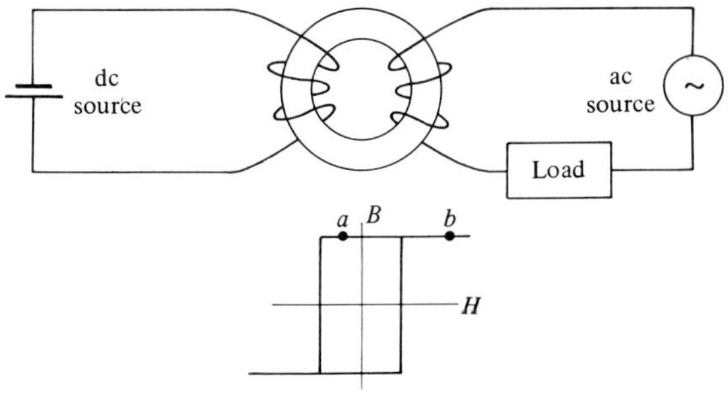

Fig. 13.29 Magnetic amplifier.

inductance of the core. (Alternatively stated, the voltage drop across the right-hand winding is large, being proportional to $d\phi/dt$, and the flux in the core is alternating through a large range.) Suppose now the control current is large enough not only to saturate the core but to keep it saturated through a complete cycle, so that the operating point moves back and forth between a and b on the loop top. Because this involves little change in ϕ, the voltage drop across the winding is small and the load current large. Intermediate values of the control current produce intermediate values of load current; in short, the device acts like an amplifier. The larger the permeability of the core, the smaller is the required control current.

4. *Sensitive relays.* A relay is a control device consisting of a small electromagnet which, when energized by a current in its winding, attracts a piece of iron, thus operating a switch in another circuit. Most relay cores are made of low-carbon steel, but sensitive, fast-acting types have high permeability cores that can be activated by a weak control current.

A variant of the usual relay is the *reed switch*, widely used in telephone systems. It consists of two narrow strips (reeds) of 50 Permalloy sealed into opposite ends of a glass tube, with the ends of the strips within the tube overlapping by a small amount. The strips form part of an electrical circuit, which is closed when the overlapping ends of the strips touch and open when they are apart. The ends of the strips are made to attract or repel one another by means of magnetic fields produced by currents in coils wound on the tube and by an external magnet.

5. *Twistor memory.* This memory does not switch as fast as the ferrite cores to be described in the next section, and so it is not suitable for computers, but its speed is adequate for the switching systems of telephone exchanges, for which it was developed. There are two variants of the Twistor memory and the basic material in each is 4-79 Mo Permalloy in the form of thin tape. Chin [13.44] has described how to adjust the composition of this alloy and how to treat it mechanically and thermally to achieve the required properties.

6. *Saturable-core (fluxgate) magnetometers.* These are instruments designed to measure the earth's field and its variations and are used, along with other types

of magnetometers, for such purposes as magnetic prospecting and submarine detection. Suppose a thin rod, some 4 inches long and made of a square-loop material, is enclosed in a solenoidal winding energized with alternating current strong enough to drive the core into saturation in each direction during each cycle. If the core is now exposed to an external field H_e, the field to be measured, along its axis, then the total alternating field will not be symmetrical about zero, and the output of a secondary coil will be modified. The usual form of the instrument has two primary coils, wound so that their fields are equal and opposed. When H_e is zero, so is the secondary voltage. When H_e is not zero, a secondary voltage is generated, proportional to H_e and of twice the frequency of the drive voltage. This and other kinds of magnetometer are fully described by Hine [13.32].

13.6 DIGITAL COMPUTER APPLICATIONS

Computers are of two types:

1. *Analog.* In this kind numbers are represented by the magnitude of some physical quantity, generally an electric current or voltage.
2. *Digital.* Here numbers are represented by the state of a bistable physical system.

We consider only digital computers in this section. First, consider that in the decimal system of numbers a number such as 237 can be written as

$$2 \times 10^2 + 3 \times 10^1 + 7 \times 10^0 = 237$$

or, in general,

$$\cdots d_2 \times 10^2 + d_1 \times 10^1 + d_0 \times 10^0.$$

Here the coefficients d_i can range from 0 to 9. In the binary system a number such as 13 is written, in a corresponding way, as

$$1 \times 2^3 + 1 \times 2^2 + 0 \times 2^1 + 1 \times 2^0 = 1101$$

or, in general,

$$\cdots b_3 \times 2^3 + b_2 \times 2^2 + b_1 \times 2^1 + b_0 \times 2^0.$$

In this system only limited demands are made on the coefficients b_i. They have only two possible values, 0 and 1. Thus any cœfficient can be represented by *any* bistable system and any number by a row of such systems, one for each coefficient. In practically all digital computers, the bistable system is a small bit of magnetic material, which can be definitely and positively magnetized in one direction or its opposite because it has a square hysteresis loop. Magnetization in one direction makes a particular coefficient equal to 1 and in the opposite direction makes it 0, as indicated in Fig. 13.30 for ferrite cores.

A digital computer is composed of four main parts: (1) *input devices*, such as punched cards, punched paper tape, and magnetic tape, by which the data to be

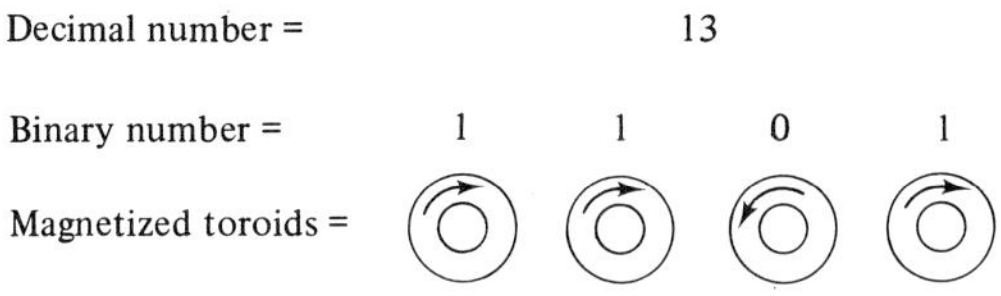

Fig. 13.30 Representation of a number by the directions of magnetization of a group of toroids (ferrite cores).

processed and the instructions (program) for its processing are fed into the computer, (2) the *central processing unit*, which includes the internal memory and the arithmetic and control units, (3) *auxiliary memory units*, such as magnetic tape, disks, or drums, and (4) *output devices*, such as printers, plotters, or cathode-ray tubes for visual display. All of this adds up to a complex mass of electrical and magnetic equipment. The magnetic elements with which we are concerned make up one or the other of the various kinds of *memory:*

1. *Fast memories*, such as the internal memory in the central processing unit, where operating speed is more important than storage capacity. It is in this memory and its associated circuitry that the arithmetical operations on the input data are performed. This kind of memory is usually made of ferrite cores, flat films, or plated wire, listed in order of decreasing usage in present computers. It is with these elements that we are chiefly concerned in this section.

2. *Auxiliary memories* with large storage capacity but lower operating speeds. These are required to store all the data and programs that the internal memory cannot hold because of its restricted size. These memories consist of magnetic tape, disks, or drums in which the magnetic material is usually gamma ferric oxide (γ-Fe_2O_3). This material is magnetically hard and will be described in Section 14.9.

A digital computer can manipulate only binary digits. Thus not only must the data to be processed be put into digital form, but also the instructions that make up the program, such as "add this to that." When this has been done, the input data and program must be *written* into the memory and then selected portions must be *read* from the memory at particular times, sent to the arithmetic unit, then back to the same or another memory. Writing and reading are therefore the basic operations that must be performed on a memory.

Ferrite cores

Figure 13.31 shows how writing and reading are accomplished with a typical kind of ferrite-core memory. A small portion of a rectangular array of cores is shown in (a), and a number of these planar arrays are stacked to form a three-dimensional array. Each core is threaded with three wires: the X and Y wires parallel to these axes are called *drive wires*, and the wire passing through all the cores of the array is called a *sense wire.* If counterclockwise magnetization

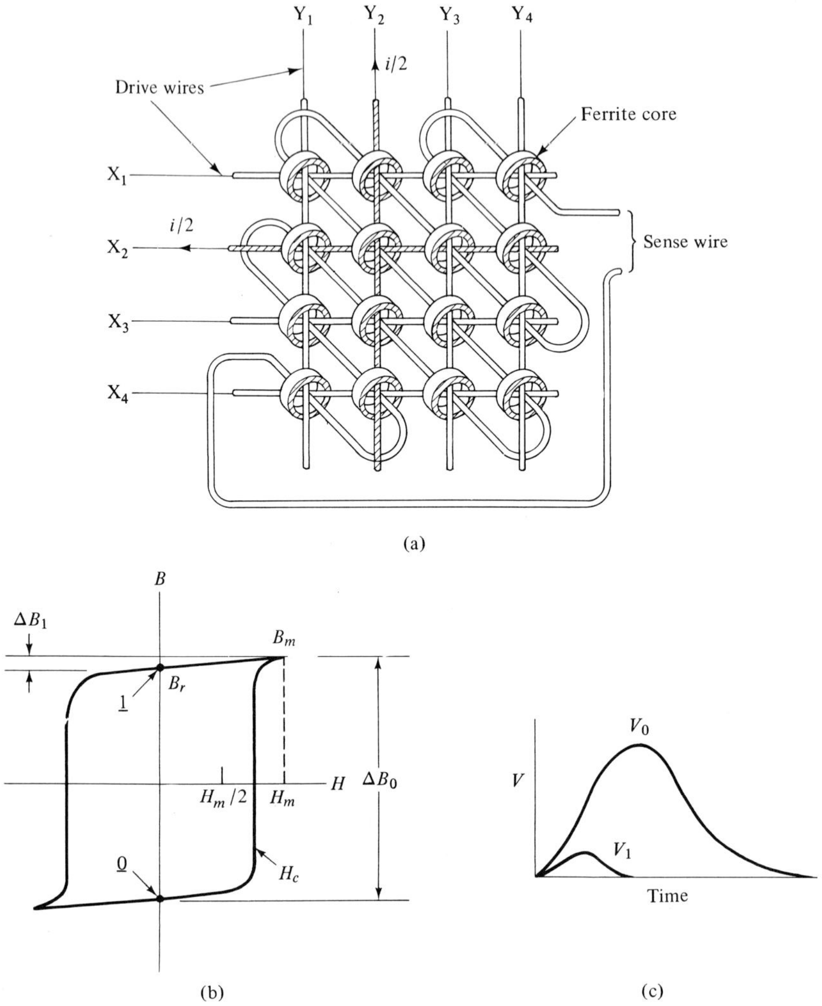

Fig. 13.31 Operation of a ferrite-core memory. (a) Memory array, after Chikazumi [G.23]. (b) Hysteresis loop. (c) Output voltage curves (schematic).

stands for 0, then the whole array can be set in the 0 state, whatever the original state of any core, by sending current pulses i through all the X wires from left to right. [These pulses must produce a field H_m large enough to switch the cores, as indicated in (b); this field must be somewhat larger than the coercive force H_c but not necessarily large enough to saturate the core.] Suppose we now wish to write the binary number 1101 on the second row of cores from the top, so that three of them must be switched to the 1 state. To switch the X_2Y_2 core, for example, we send half-pulses $i/2$ simultaneously through the X_2 and Y_2 wires in the directions indicated. This subjects six cores to fields of magnitude $H_m/2$, which are not strong enough to switch them, but at the selected X_2Y_2 core these half-fields add to H_m and the core switches. This is called the *coincident-current mode* of operation.

To read a selected core, i.e., to determine whether it is in a 0 or 1 state, we proceed in a similar way. To read core X_2Y_2, pulses $i/2$ are sent in the directions shown in wires X_2 and Y_2. If this core is in the 1 state, a small flux density change $\Delta B_1 = B_m - B_r$ will occur in the core, inducing a voltage V_1 in the sense wire, as indicated in (c). But if the core is in the 0 state, the coincident pulses will switch it to the 1 state, the flux density change $\Delta B_0 = B_m - (-B_r)$ is large, and so is the output pulse V_0 produced in the sense wire; V_0 is of the order of tens of millivolts. The electronics connected to the sense wire can easily distinguish between V_0 and V_1 and thus decide the original state of the core. Whatever this state was, it is a 1 state after the core is read. This reading operation is therefore *destructive*, and if we wish the information in the core to remain there after reading, that information must be written back in. *Nondestructive read-out* (NDRO) can be achieved by more complex kinds of operation [13.33].

The ratio V_0/V_1 should be as large as possible, which requires that the top of the hysteresis loop be as flat as possible, to minimize ΔB_1. Note also that when core X_2Y_2 is being read, all the other cores threaded by the X_2 and Y_2 wires are successively receiving half-pulses $H_m/2$, causing flux density changes of the order of $\Delta B_1/2$. These induce small voltage pulses in the sense wire and constitute unwanted noise in the circuit.

As we saw in Section 12.4, the switching speed of a core increases rapidly with applied field (Fig. 12.9), which is proportional to the size of the current pulse in the drive wire. However, coincident-current operation places an upper limit on the drive field H_m, because if this exceeds $2H_c$, half-pulses of size $H_m/2$ will cause unwanted switching. Thus large rather than small values of H_c are desirable, other factors being equal, in order to take advantage of the high switching speeds available at high drive fields. (One method of increasing H_c is to underfire the ferrite during manufacture, so that it is incompletely sintered and retains some porosity. The pores impede wall motion and raise H_c, thus allowing higher drive fields and faster switching by rotational processes, i.e., operation in region II of Fig. 12.9.) On the other hand, increasing the value of H_m increases the power required to operate the memory and the amount of heat generated in it. But H_m can be increased with the same drive current by making the core smaller, because the field increases as the distance to the drive-wire axis decreases [Eq. (1.10)]. Drive currents are of the order of 0.4 to 0.8 ampere. A current of 0.5 ampere for example, produces a field of 2.6 Oe at a distance of 15 mils from the wire axis. Cores with outside diameters of 20 to 50 mils (0.5 to 1.3 mm) are commonplace, and still smaller cores, having holes scarcely visible to the naked eye, have been made. The smallest cores made by the standard method (dry pressing) have an O.D. of 11 mils, but cores with an O.D. of 7 mils have been made by another technique [13.34]. Reduced core size not only yields the benefits noted above but also reduces the size of the memory, permitting denser storage of information. The planar array of cores shown in Fig. 13.31 (a) is only a small portion of a complete "memory plane," which contains many thousands of cores.

Ferrite cores are made of mixed ferrites, of which the chief constituent is manganese ferrite, by methods described in the next section. Manufacturers rarely disclose compositions, nor do they state magnetic characteristics such as H_c and B_r. Instead, they make cores to specified values of drive current, output voltage when switched, etc., because only these quantities are of direct interest to the designer of a memory.

Russel *et al.* [13.35] and Greifer [13.45] have reviewed many aspects of ferrite-core memories, including methods of core manufacture.

Flat films

Flat-film memories are composed of rectangular arrays of circular spots a few mm in diameter of 80 Permalloy film, about 1000 Å thick, deposited by evaporation on a glass or metal substrate. Lying under and over each bit of film are flat conductors, made by printed-circuit techniques, which act as drive and sense lines. (Alternatively, a continuous sheet of film may be deposited; only the areas under the intersections of the drive lines are magnetically effective.)

We recall from Section 12.4 that (1) these films have an easy axis built into them during deposition, (2) they switch relatively slowly, by wall motion, when an easy-axis field H_e is applied, and (3) they switch faster, mainly by rotation, when a transverse field H_t is applied at the same time as H_e. The switch is then much faster than that of a ferrite core.

One element of a flat-film array is shown in Fig. 13.32 (a), with X and Y conductors crossing over it. The easy axis of the film is vertical, and we take upward magnetization to mean 1 and downward 0. A current pulse i_e in the X line applies an easy-axis field H_e, and a pulse i_t in the Y line applies a transverse field H_t. Writing is accomplished by simultaneous i_e and i_t pulses. If the film is in the 0 state, it can be switched to the 1 state by an i_e pulse in the direction shown and a coincident i_t pulse in either direction. (The i_e pulse alone is not enough to cause a switch,

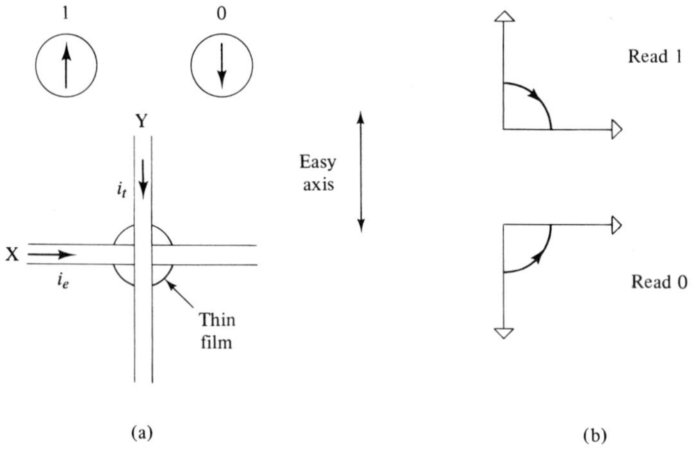

Fig. 13.32 Operation of a flat-film memory.

because the corresponding field H_e is less than H_c. See Fig. 12.14. Hence the i_e pulse will not disturb other film elements crossed by the X line.) A film is read by a single i_t pulse. If it is in the 1 state, the M_s vector rotates clockwise, as in (b), and a voltage pulse is induced in the X line. If it is in the 0 state, the rotation is counterclockwise, and the induced voltage has opposite polarity. Reading is therefore merely a matter of detecting a polarity difference rather than, as in ferrite cores, of detecting a voltage difference between two pulses of the same polarity. After passage of the i_t pulse the film demagnetizes itself by breaking up into domains, as we saw in Fig. 11.36 (d). Reading is therefore destructive, just as in ferrite cores.

The status of flat-film memories as of 1968 has been surveyed by Overn [13.36].

Plated wires

Plated-wire memories are the newest entry into the magnetic-memory competition, and they differ from flat films in at least two respects, although both are made of 80 Permalloy. First, they are about 8000 Å thick, or about 8 times as thick as flat films; as a result, switching operations involve a much larger flux change and induce larger voltages. Second, the easy axis in the cylindrical film on a wire is circumferential, as we saw in Fig. 11.29. The magnetic circuit is therefore closed and there is no demagnetizing field. (The objection to making flat films thicker than about 1000 Å is the increase of demagnetizing field with thickness. If this field becomes too large, the film will break up into domains, and the single-domain remanent state and the attendant square loop will be lost.) Plated-wire arrays are expected to be cheaper to produce than flat-film arrays.

Figure 13.33 shows the method of operation. A section of wire is shown in (a), with a conductor called a *word strap* passing above and below it. This conductor carries the i_t pulses, going in opposite directions, that produce the transverse field H_t parallel to the wire axis. A current pulse i_e in the wire itself produces the easy-axis field H_e. That portion of wire which is beneath a word strap comprises the single element of an array where a single binary digit is stored. A digit is written by coincident i_e and i_t pulses; for the directions shown, 1 would be written. The single i_e pulse is not enough to disturb information under other word straps not carrying current at that time. Information is read *nondestructively* by i_t pulses. These cause the M_s vector, indicated in (b), to turn toward the hard axis and then fall back to the original position, 1 or 0, inducing a voltage pulse in the plated wire of one polarity or the other. A short section of wire is sketched in (c), showing the magnetization directions after the binary number 1101 has been written. Adjacent nonidentical digits are represented by portions of the wire separated by a circumferential 180° domain wall. This wall is not a normal one but a complex transition region; it develops after continuous operation of the memory, i.e., after repeated rotations of the magnetization below each word strap.

The reason that the read-out is nondestructive is not entirely clear, although NDRO is known to depend on having a controlled surface roughness of the plated film. This roughness may cause a variation in the induced anisotropy constant K

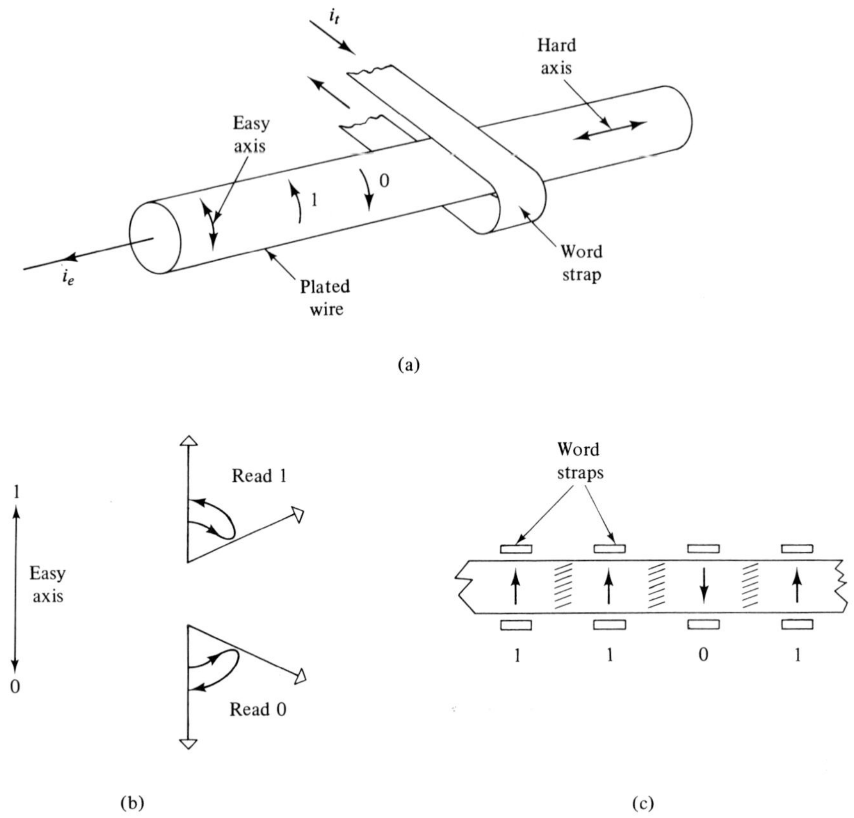

Fig. 13.33 Operation of a plated-wire memory.

such that the M_s rotation in high-K regions is less than that in low-K regions, for an i_t pulse of given size. The couping between these regions then pulls the magnetization back to the initial state, the mean rotation being less than 90°. If the i_t read pulse is too large, this does not occur and the read-out becomes destructive.

One problem of some consequence in plated wires is that of *creep* of domain walls. Here the word creep means wall motion, not because of the pressure of a constant field, but as a result of many repetitions of a pulsed field. For example, a wall will creep when subjected to a small, constant, easy-axis field H_e parallel to the wall and a repetitive, pulsed, transverse field H_t, even though H_e is well below the value H_c which would cause motion if H_t were zero, and even though a field at right angles to a 180° wall should ideally cause no motion of the wall. Creep also occurs when H_e and H_t are both pulsed. Information stored in a wire, as in Fig. 13.33 (c), can be lost if the wall separating adjacent digits becomes unstable and creeps one way or the other under the onslaught of billions of current pulses in the wire and in adjacent word straps. One solution is to keep the word straps well separated, but this decreases the density of information storage. Wall creeping is presumably related to imperfections in the film, but the nature of the relation is not clear. Doyle *et al.* [11.31] have reviewed the problem.

The general characteristics of the plated-wire memory and its present and future status have been surveyed by Fedde and Chong [13.37] and Mathias and Fedde [13.46].

Other applications

At least two other kinds of magnetic memory, both currently in a stage of active development, are worth noting. They may be the wave of the future.

Magneto-optic devices

The Kerr and Faraday effects were described in Section 9.4 merely as means for studying domain behavior by direct observation. However, these effects may also be exploited as means of reading digital information stored in a magnetic medium, because two oppositely magnetized regions can be distinguished by the way in which they reflect or transmit a polarized light beam.

Consider the thin film of Fig. 13.34, which has a remanent magnetization upwards, normal to its surface, as indicated by the gray tone. Writing a binary digit (a *bit*) on this film can be accomplished by striking it for a very short time (a microsec) with a very intense and very narrow beam (a laser or electron beam). This raises the temperature of a small region above the Curie point. When it cools, it becomes a small cylindrical domain magnetized in the downward direction, because it is exposed to the downward demagnetizing field of the surrounding material. This reversely magnetized bit can then be detected by scanning the film with a transmitted light beam which goes through a polarizer and analyser (not shown) as well as the film, before entering a photomultiplier tube which measures its intensity. The bias field H_B shown in the sketch, which is the field that reverses the magnetization, may be either the demagnetizing field of the specimen itself or an externally applied field. The presence or absence of a reverse domain at a particular location in the film can then stand for 1 or 0. Information on the film can be erased by saturating the film in the direction of its original magnetization.

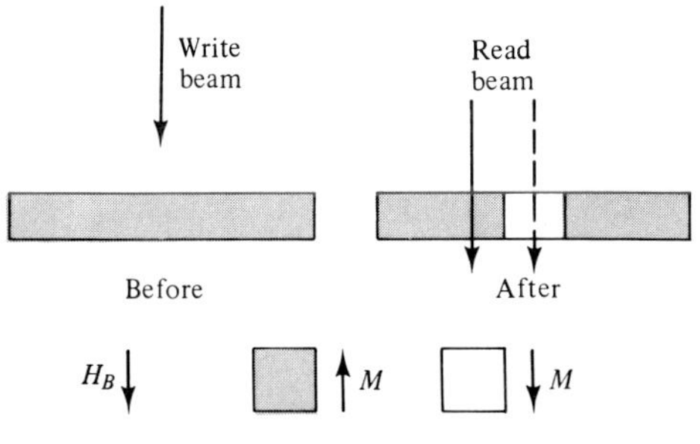

Fig. 13.34 Thermomagnetic recording.

The method just described is called *Curie-point writing*. It is just a special case of what is more generally called *thermomagnetic* (or *thermoremanent*) *writing*. *Compensation-point writing* can be done on a ferrite that has a compensation point. Actually the only requirement is that the material have a coercivity which decreases fairly rapidly with rise in temperature, not necessarily to zero as at the Curie point, but to a value less than the bias field.

The specimen must be in thin-film form in order to have the low heat capacity that will result in rapid heating. The materials successfully investigated to date include Co, MnBi, EuO, and GdIG (gadolinium iron garnet). A spot (bit) diameter of 1μ has been achieved on a film of MnBi [13.38].

Read and write speeds are apparently comparable with ferrite cores, but the density of information storage by thermomagnetic writing is very much larger. A bit size of 1μ suggests potential densities of some 10^7 to 10^8 bits/cm^2. This is to be compared with densities of about 300 bits/cm^2 in planar arrays of ferrite cores or plated wires.

MacDonald and Beck [13.39] have described the general problems involved in thermomagnetic writing and magneto-optic reading and have reported on their own experiments with GdIG. More extensive reviews have also appeared [13.47, 13.48, 13.49].

Bubble domains

If a small cylindrical domain like that sketched in Fig. 13.34 can be formed in a thin specimen by any means, and if in addition the domain is highly mobile and can be moved quickly from one point to another, then the presence or absence of a domain at any point can stand for 1 or 0, and the basis for a new kind of memory is established. Bobeck and his colleagues [13.40, 13.41] have done just this by showing that such domains can be formed in thin single-crystal plates of the rare-earth orthoferrites.

These substances are canted antiferromagnets (Section 5.3) with the general formula $RFeO_3$, where R is yttrium or a rare earth. The saturation magnetization M_s varies with R but is always very small, typically about 8 emu/cm^3, which is only 0.5 percent of that of iron. The easy axis is the c axis of the orthorhombic unit cell except when R is samarium. The plate-shaped crystals examined by Bobeck *et al.* are 1 to 2 mils (25 to 50 μ) thick and have the easy axis normal to the plate surface. The crystals are transparent to red light and domains are easily seen by the Faraday effect. The Néel temperature varies from about 350° to 450° C.

The magnetic properties are remarkable. The crystal anisotropy constant K is about normal, some 10^5 to 10^6 ergs/cm^3, but because M_s is so small, the anisotropy field H_K $(= 2K/M_s)$ is of the order of 10^5 Oe. This is the field that must be overcome to nucleate reversed domains in a perfect crystal. The demagnetizing field $4\pi M_s$ normal to the plate, on the other hand, is only about 100 Oe. As a result, once these crystals are saturated normal to the plate, they are quite stable; fields of several thousand oersteds are needed to reverse the magnetization. The high nucleation field is associated with an extremely small value of the domain-wall coercivity H_c, because walls can be moved by fields as low as 0.01 Oe. Thus

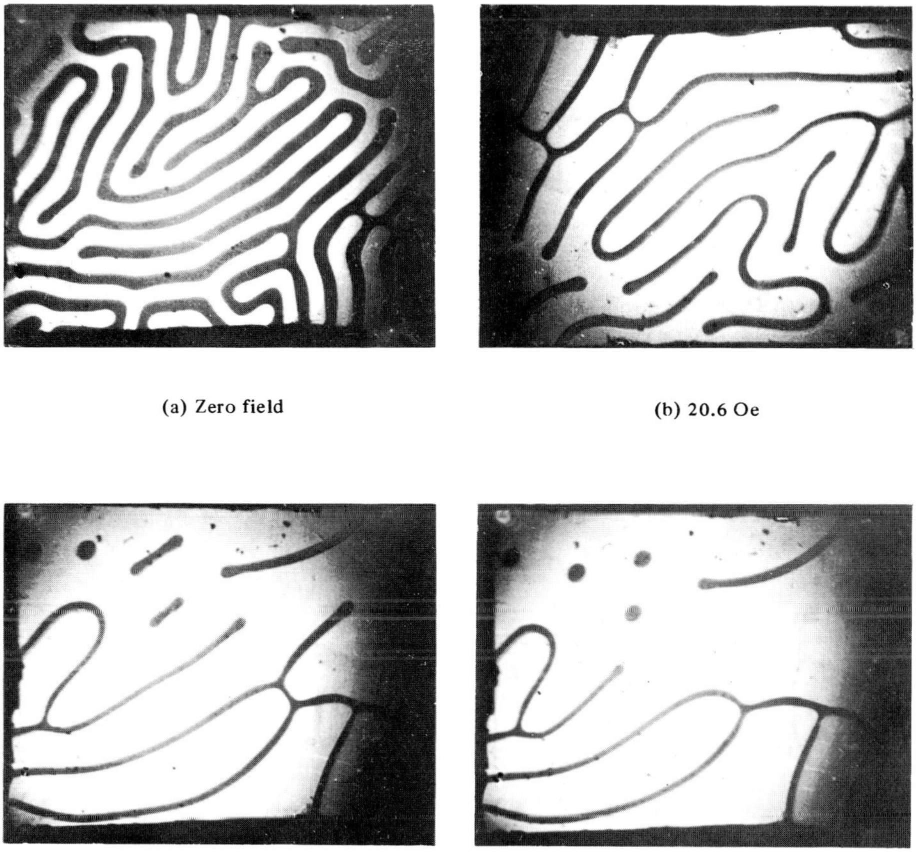

(a) Zero field

(b) 20.6 Oe

(c) 24.6 Oe

(d) 26.5 Oe

Fig. 13.35 Faraday-effect photographs of domains in a crystal of $TmFe0_3$, 2.3 mils thick. The cylindrical domains in (d) have a diameter of 5 mils. Bobeck [13.40]. (Copyright 1967, The American Telephone and Telegraph Co. Reprinted by permission.)

these crystals are at once both magnetically hard and magnetically soft and, like whiskers, their true hysteresis loops are re-entrant loops of the most extreme kind.

Figure 13.35 (a) shows the domain structure of a demagnetized crystal of $TmFeO_3$, consisting of a stripe pattern of "black" and "white" domains magnetized into and out of the paper. Application of a field normal to the crystal plate causes the white domains to grow at the expense of the black, and at 24.6 Oe one domain has shrunk to a cylinder 5 mils wide. Nearby are two strip domains, and a slight further increase in field, to 26.5 Oe, shrinks these also into cylinders. If the field is now reduced to zero, the cylindrical domains will run out into strips, existing strips will lengthen, and both will grow to fill the entire crystal. The resulting

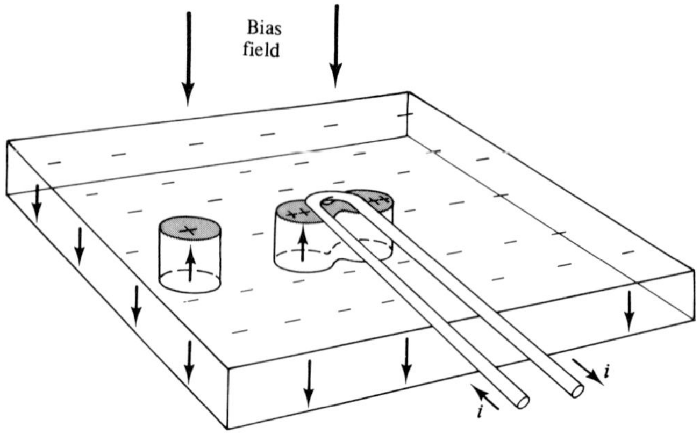

Fig. 13.36 Division of a strip domain into two bubble domains by current in a wire.

structure will resemble (a) but will not be identical with it. The small cylindrical domains have been nicknamed "bubbles" and this name is likely to stick. It is by manipulation of these bubble domains that digital information can be stored and processed.

All the strip domains of Fig. 13.35 (b) can be converted into bubble domains, with the applied field maintained at 20.6 Oe, because the strips can be cut into parts that will contract into bubbles. One way of doing this is to bring the tip of a fine, permanent-magnet wire near a strip domain. Suppose the bias field and M_s in the white domains are both directed into the paper. Then any locally intense downward field, such as that near the north end of the wire magnet, applied to the center of a strip domain, will cause it to neck down and then divide into two parts. (These two parts will then repel each other and move apart in the plate, in order to reduce magnetostatic energy. They repel one another like similar charges, and much of the behavior of these bubbles can be understood in these terms.) Another way of producing the local field is by an electric current in a hairpin loop of wire, as in Fig. 13.36, which shows one bubble domain and a strip domain in the process of being cut into two bubbles.

Bubble domains are stable only within a particular range (about 10 Oe) of the bias field H_B. They shrink and disappear if H_B is too large and run out into strips if H_B is too small. The size of a bubble results from a balance of three energies: domain wall energy and bias field energy act to shrink the bubble, while magnetostatic energy acts to enlarge it. (The theory of bubble stability has been given by Thiele [13.50].) The minimum diameter of a stable bubble depends on the crystal thickness; it is smallest when the crystal is about 1 or 2 mils thick. The smallest bubble so far observed was produced in the mixed orthoferrite $Sm_{0.55}Tb_{0.45}FeO_3$; the crystal was 2.0 mils thick and the bubble was 0.75 mil wide at

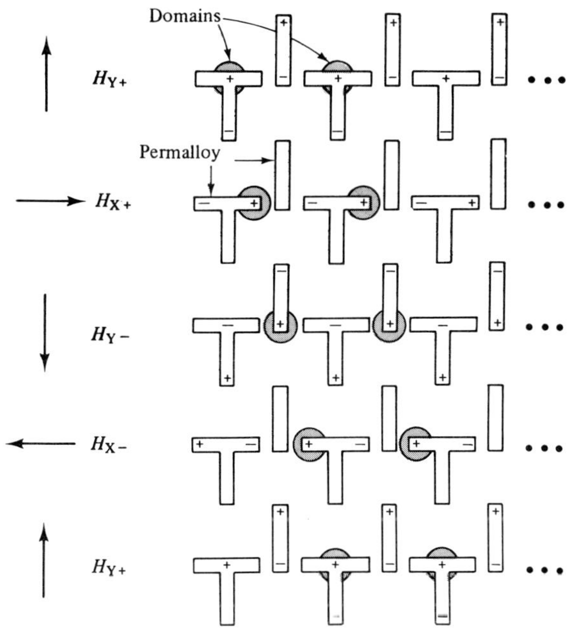

Fig. 13.37 Motion of a bubble domain in a T-BAR channel. Perneski [13.42].

a bias field of 61 Oe. Stable bubbles this small suggest potential storage densities of some 10^4 bits/cm^2.

In order that bubble domains be useful in computers, it must be possible to move them fast from one point to another and to generate new ones at will. These operations can be done as follows:

1. *Motion.* One method consists in laying down a grid of conductors on the crystal surface by printed-circuit techniques. Current pulses in appropriate conductors then create local fields which move the bubbles; details are given by Bobeck *et al.* [13.41]. In another method [13.42] a pattern of spots of Permalloy, about 5000 Å thick, is evaporated onto the crystal surface. These spots are alternately in the shape of the letter T and a bar, as shown in Fig. 13.37. When a magnetic field is applied in the plane of the crystal, north (+) and south (−) poles will appear on the Permalloy spots as indicated; when the field is transverse to the narrow portions of a T or bar, no poles are formed because of the large demagnetizing field in this direction. This in-plane field has no effect on the bubble domains themselves because of the large anisotropy field of the orthoferrite. But the bubble, whose upper end is charged negatively in Fig. 13.37, is attracted by and moves toward the nearest positive pole on the Permalloy and is repelled by a negative pole. (Actually, it is the gradient of the field normal to the crystal surface that moves the bubbles, but it is simpler to think of the forces involved in terms of poles.) If now the in-plane field is rotated clockwise, the poles on the Permalloy

spots will shift their positions, and study of the diagram will show that positive poles always appear immediately to the right of the bubble domains, causing them to move to the right. In one complete revolution of the in-plane field the bubbles move a distance equal to one period of the T-BAR pattern. Note that no current-carrying conductors on the crystal surface are needed in this method of domain drive; all that is required is the Permalloy pattern itself and a rotating field, and this field can be generated by two sets of coils, external to the crystal, carrying alternating currents with the proper phase relation. In a typical experiment of this kind, the period of the pattern was 8 mils, the contant bias field about 40 Oe, and the minimum in-plane field required was 10 Oe. Bubble domains can be moved very fast, whether by currents in a conductor grid or by a rotating magnetic field. Specifically, a bubble can be moved a distance equal to its own diameter in about a microsecond. This means that information can be transmitted along a channel at the rate of 10^6 bits/sec.

2. *Generation.* Even a small cylindrical domain can be cut in two by the method of Fig. 13.36, because the small scale of printed circuits permits two parallel conductors, close together, to be positioned exactly over the center of a cylindrical domain. Currents in the conductors then divide the domain in two, and the children of this reproduction by fission are of exactly the same size as their parent. Another, most ingenious method generates domains right at the entrance of a T-BAR channel, along which they are subsequently moved. The generator is a rotating in-plane field and a Permalloy disk with a tab on it (Fig. 13.38). The field rotates clockwise and the upper face of the domain has negative polarity. The generator disk maintains a domain which stays in contact with the positive poles formed on the disk by the rotating field. In (a) the domain is stretched out along the tab by the positive pole at the end. In (b) it has become attached to the first bar of the pattern. Further field rotation causes further stretching of the domain until it becomes unstable and breaks in two at (d), aided by repulsion by the negative poles then opposite it on the disk. Finally, there is one domain on the disk, another in the channel, and the process is ready to repeat. In one experiment of this kind it was found that a 20-Oe rotating field was required to generate domains, but 10 Oe was enough to move them through the channel.

Suppose that information is represented by a certain sequence of binary digits like 1011010, represented, for example, by bubble domains at certain positions in a channel, and that it is required to move this sequence, as a unit either to right or left in the channel. A device or circuit that will do this is called a *shift register*, which is widely used in communication equipment and computers to store binary digits temporarily. It is much like a delay line, for example, the magnetostrictive delay line of Fig. 8.27 (a). In fact, a shift register may be called a digital delay line. It is believed that one of the future applications of bubble domains will be in large-capacity shift registers.

The digital information represented by the positions and movements of bubble domains in a crystal plate can be *read* by at least two methods. The position of a static domain can be determined magneto-optically, and the motion of a domain can be detected by the voltage induced by this motion in a nearby conductor.

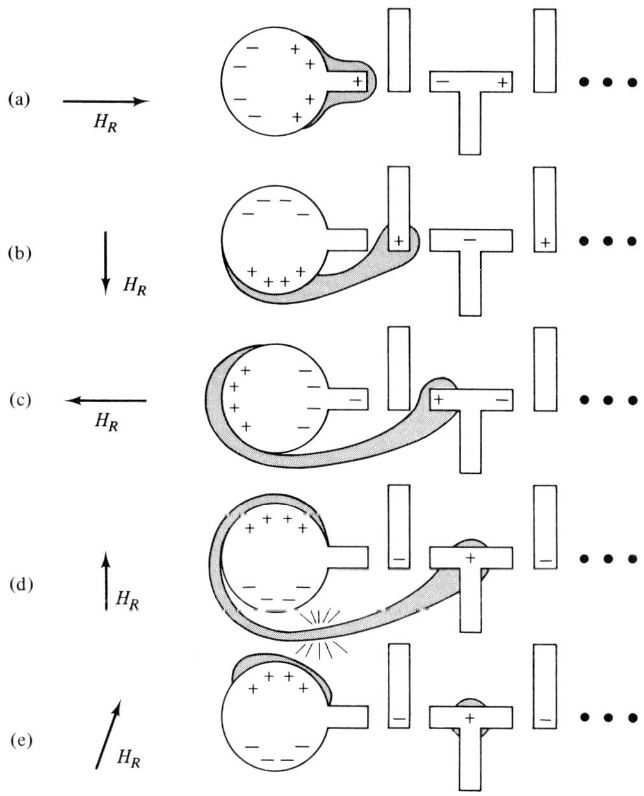

Fig. 13.38 Bubble domain generator. Perneski ⌊13.42⌋.

Complex rare-earth garnets are currently being investigated as substitutes for orthoferrites, with the object of getting smaller bubbles and larger crystals. Although the crystal structure of garnet is cubic, thin slices cut from certain portions of garnet crystals are found to have uniaxial anisotropy, for reasons that are not understood.

In concluding this section on computer applications it is worth pointing out again that a set of *any* bistable systems can store binary digits, and that magnetic devices are not the only kind possible. In particular, certain semiconductor circuits called "flip-flops" are bistable and can be switched in a few nanoseconds. With the development of large-scale integrated circuits these devices are expected to become competitive with magnetic memories.

13.7 SOFT FERRITES

The magnetically soft ferrites first came into commercial production in 1948. They have many applications other than memory cores, and these will be examined here, as well as the methods of making ferrites and the effects of such variables

Table 13.7 Market Value of U.S. Production of Soft Ferrites in 1968 [13.1].

	Millions of dollars	Percent of market
Computer memory cores	$ 55	50
Television receivers	20	18
Communications and radio components, recording heads, magnetostriction transducers	20	18
Telephone communications	12	11
Microwave ferrite components	3	3
Total	$110	100

as porosity and grain size on their magnetic properties. Detailed information is given by Smit and Wijn [G.10], Standley [G.19], Snelling [G.35], and Tebble and Craik [G.36].

The soft-ferrite market in the U.S.A. accounted for some 17 percent of all magnetic materials in 1968 and is itself divided into the groups listed in Table 13.7, as estimated by Jacob [13.1]. Ferrite cores form the largest group, and the figure shown corresponds to some 10 to 15 billion cores.

The soft ferrites have a cubic crystal structure and the general formula $MO \cdot Fe_2O_3$, where M is a divalent metal such as Mg, Mn, Ni, etc. Nonmagnetic Zn ferrite is often added to increase M_s (Section 6.3) and all the commercial ferrites are mixed ferrites (solid solutions of one ferrite in another). The intrinsic properties of the pure ferrites have been given in Table 6.4. Their densities are a little over 5 g/cm^3, Curie points range from about 300° to 600° C, and M_s from about 100 to 500 emu/cm^3. All of them have ⟨111⟩ easy directions of magnetization, low crystal anisotropy, and low to moderate magnetostriction.

The ferrites are distinctly inferior to magnetic metals and alloys for applications involving static or moderate-frequency (power-frequency) fields, because they have M_s values of a third or less of those of iron and its alloys and far lower dc permeabilities. But the outstanding fact about the ferrites is that they combine extremely high electrical resistivity with reasonable good magnetic properties. This means that they can operate with virtually no eddy-current loss at high frequencies, where metal cores, even those made of extremely thin tape or fine particles, would be useless. This fact accounts for virtually all the applications of soft ferrites. (There are, however, frequency ranges where cores of metal tape or powder have lower losses than some ferrites and where metal-powder cores have lower permeabilities than some ferrites. Then the proper choice of magnetic material is not clear-cut.)

Ferrites are made in the following way:

1. *Starting material.* This is a powder of ferric oxide Fe_2O_3 and whatever oxides MO are required. Metal carbonates may also be used; during the later firing, CO_2 will be given off and they will be converted to oxide.

2. *Grinding.* Prolonged wet grinding of the powder mixture in steel ball mills produces good mixing and a smaller particle size, which in turn decreases the porosity of the final product. After grinding, the water is removed in a filter press, and the ferrite is loosely pressed into blocks and dried.

3. *Presintering.* This is done in air, and the temperature goes up to about 1000° C and down to 200° C in about 20 hours. In this step at least partial formation of the ferrite takes place:

$$MO + Fe_2O_3 \rightarrow MO \cdot Fe_2O_3.$$

This step produces a more uniform final product and reduces the shrinkage that occurs during final sintering.

4. *Grinding.* The material is ground again to promote mixing of any unreacted oxides and to reduce the particle size.

5. *Pressing or extrusion.* The dry powder is mixed with an organic binder and formed into its final shape. Most shapes, such as toroidal cores, are pressed (at 1 to 10 tons/in^2), but rods and tubes are extruded.

6. *Sintering.* This is the final and critical step. The heating and cooling cycle typically extends over 8 hours, during which the temperature reaches 1200 to 1400° C. Any unreacted oxides form ferrite, interdiffusion occurs between adjacent particles so that they stick (sinter) together, and porosity is reduced by the diffusion of vacancies to the surface of the part. Strict control of the furnace temperature and atmosphere is very important, because these variables have marked effects on the magnetic properties of the product. Ideally, the partial pressure of oxygen in the furnace atmosphere should equal the equilibrium oxygen pressure, which changes with temperature, of the ferrite. Iron and some other ions can exist in more than one valency state; too little oxygen, then, will change Fe^{3+} to Fe^{2+} and too much will change Mn^{2+}, for example, to Mn^{3+}.

The final product is a hard and brittle ceramic, so hard, in fact, that grinding of certain faces to make them smooth and flat is the only practical finishing operation. During sintering, all linear dimensions of a part shrink from 10 to 25 percent, and allowance for this must be made in designing the pressing mold or extrusion die. Further details of ferrite manufacture are given in the books noted above and in a review by Kriesman and Goldberg [13.43], with particular reference to the preparation of specimens for laboratory studies.

The grain size of commercial ferrites ranges from about 5 to 40 μ, and they are never completely dense. (This is true of most sintered products, metallic or nonmetallic.) Percent porosity is given by $100(\rho_x - \rho_a)/\rho_x$, where ρ_a is the aparent density and ρ_x the "x-ray density"; the latter is the mass of all the atoms in the unit cell divided by the volume of the unit cell; the cell volume is in turn

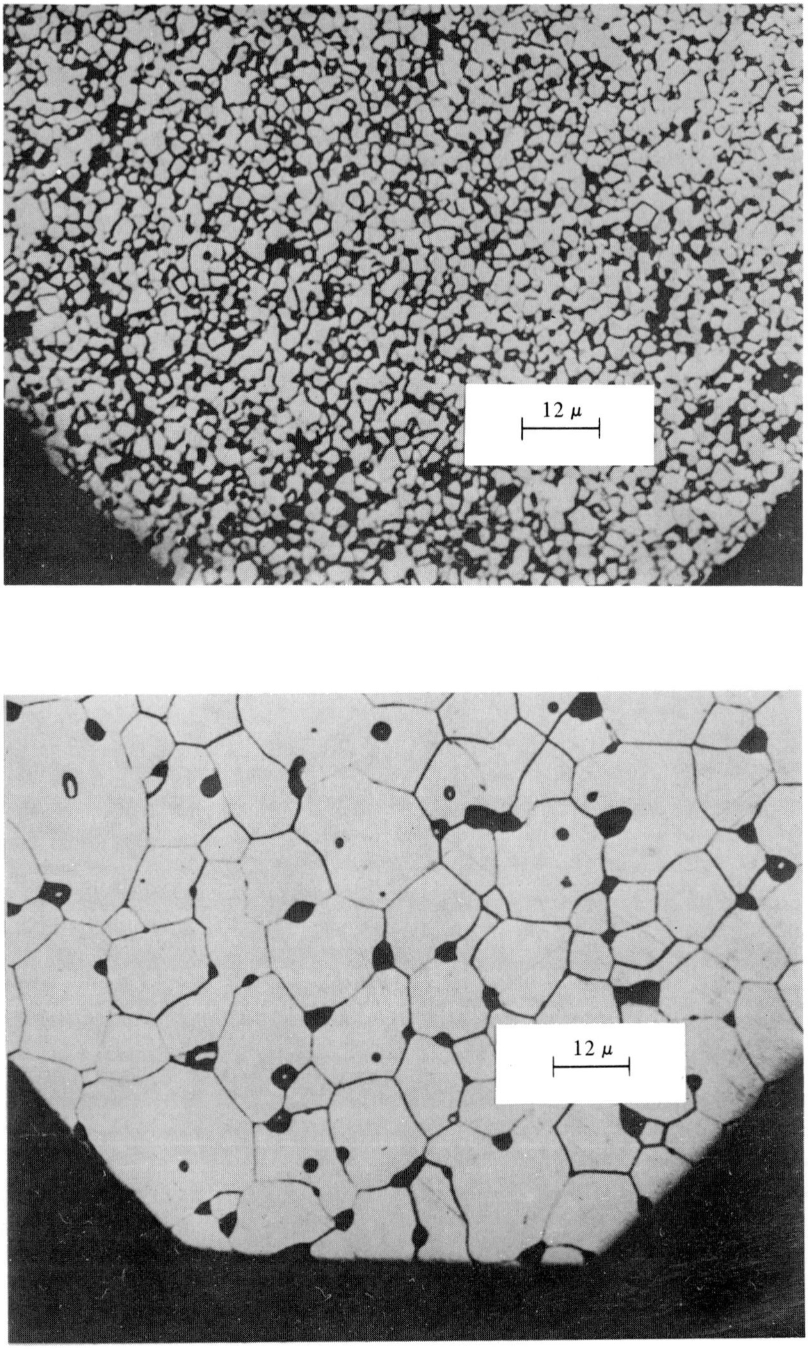

Fig. 13.39 Microstructure of a Mn-Zn ferrite. (a) After sintering 5 min at 1375° C (10 percent porosity). (b) After sintering 1 min at 1435° C (5 percent porosity). Smit and Wijn [G.10].

found from the cell dimensions disclosed by x-ray diffraction. The extreme range of porosity in ferrites is about 1 to 50 percent; a more typical range is 5 to 25 percent. Figure 13.39 shows how an increase in sintering temperature increases the grain size and decreases the porosity; the black areas are voids. Permeability increases as the grains become larger and as the porosity decreases. The grain-size effect is the stronger of the two. Porosity at the grain boundaries is less damaging to the permeability, because it causes less hindrance to domain wall motion than porosity within the grains. Both types are evident in Fig. 13.39.

(It is not easy, incidentally, to perform clear-cut experiments on ferrites to isolate the effect of a single variable. A mixed ferrite like (Mn, Zn)O · Fe_2O_3 is a complex system; it contains, or can contain, six kinds of ions: Mn^{2+}, Mn^{3+}, Zn^{2+}, Fe^{2+}, Fe^{3+}, and O^{2-}. The oxygen content of the sintering atmosphere can not only change the valence of some of these ions but also alter the stoichiometry, by creating oxygen or metal-ion vacancies in the lattice. And variations in sintering temperature or time normally cause simultaneous changes in both grain size and porosity.)

Commercial ferrites are sold under many trade names. The Dutch, who originally developed them, called them "Ferroxcube," and they are still known by that name in the Netherlands and other places. Manufacturers rarely give the chemical composition of their products, preferring instead to specify the magnetic properties; these are adjustable over a rather wide range by varying the composition and the sintering conditions. Two broad classes of ferrites are produced:

1. *Mn-Zn ferrites.* These have initial permeabilities of the order of 1000 to 2000, coercivities of less than 1 Oe, and are usable without serious losses up to frequencies of about 500 kHz. Resistivity is about 20 to 100 ohm-cm.

2. *Ni-Zn ferrites.* These are designed for very high frequency operation, to more than 100 MHz. Initial permeabilities are about 10 to 1000, and coercivities are several oersteds. At the highest frequencies, losses are found to be lower if domain wall motion is inhibited and the magnetization forced to change by rotation. For this reason some grades of Ni-Zn ferrites are deliberately underfired. The resulting porosity interferes with wall motion and decreases both losses and permeability. The Ni-Zn ferrites have very high resistivity, about 10^5 ohm-cm.

Both of these main classes of materials contain zinc ferrite. As we saw in Fig. 6.4, the addition of zinc ferrite increases the value of M_s at 0°K. It also weakens the exchange interaction between ions on A and B sites, with the result that the Curie point decreases. The curves of M_s (or σ_s) versus temperature must therefore cross over, as shown in Fig. 13.40 for a series of Mn-Zn ferrites. In any magnetic material it is usual for the initial permeability μ_0 to increase with temperature to a maximum just below the Curie point, because the crystal anisotropy and magnetostriction normally decrease with rising temperature. This effect is shown for the Mn-Zn ferrites in Fig. 13.41. Not only do zinc ferrite additions shift T_c and the accompanying peak in μ_0 to lower temperatures, but

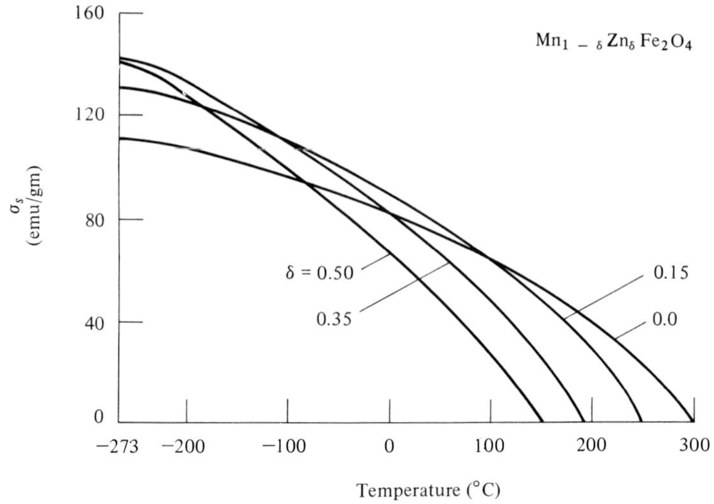

Fig. 13.40 Variation with temperature of the saturation magnetization σ_s of Mn-Zn ferrites. Smit and Wijn [G.10].

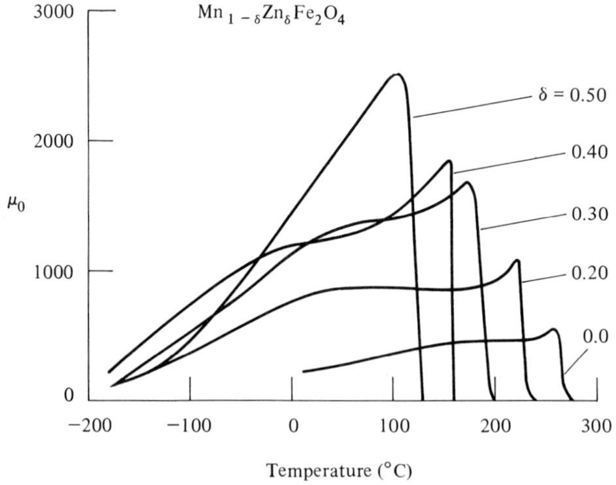

Fig. 13.41 Variation with temperature of the initial permeability μ_0 of Mn-Zn ferrites. These specimens also contain a small ferrous (Fe^{2+}) ion content. Smit and Wijn [G.10].

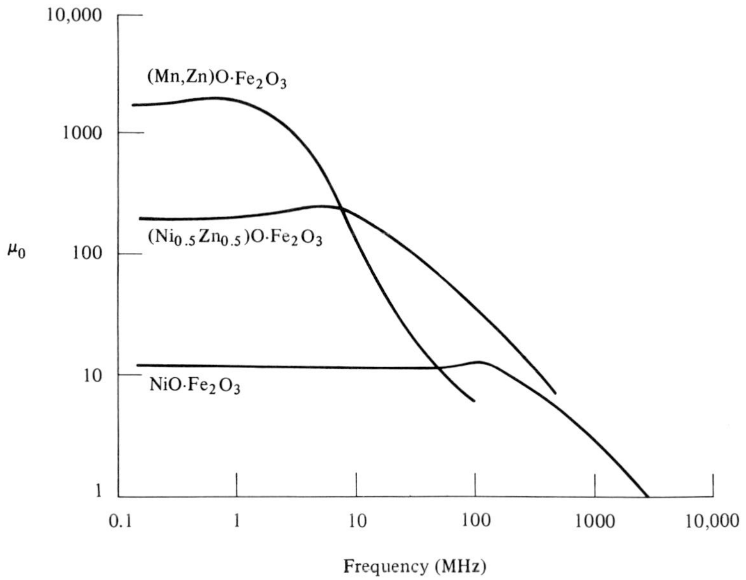

Fig. 13.42 Variation of initial permeability μ_0 with frequency for three ferrites.

they increase the height of the peak; as a result the room-temperature value of μ_0 also increases with the zinc content. Ni-Zn ferrites behave in the same way.

Figure 13.42 shows how the initial permeability μ_0 varies with frequency in the megahertz range for three different ferrites. The first is a Mn-Zn ferrite of unspecified composition [G.35], the second a 50-50 mixture of Ni and Zn ferrites [G.10], and the third is pure Ni ferrite [G.10]. The shape of these curves is typical. As the frequency increases, μ_0 remains equal to its static (dc) value until a critical frequency is reached and then decreases rapidly, as may be seen more clearly if the data are plotted on linear rather than logarithmic paper. The larger the static value of μ_0, the lower is the frequency at which this decrease occurs. Therefore, if a particular ferrite core must have constant inductance, which requires constant permeability, at all frequencies up to several hundred megahertz, there is no option but to choose a material with low permeability.

The decrease in μ_0 at a particular frequency is due to the onset of ferrimagnetic resonance. As mentioned in Section 12.7, electron spin resonance can occur in ferrimagnetics and is normally studied by applying a strong field to wipe out the domain structure and align the magnetization throughout the specimen with the easy axis. The spins then precess about this axis. But in a multidomain ferrite with no applied field, the spins are still precessing about the easy axis, the M_s direction, in each domain. The frequency of this precession depends on how strongly the magnetization is bound to the easy axis; the stronger the coupling, the higher is the natural frequency of precession. The strength of this coupling is described by the value of the crystal anisotropy K or by the anisotropy field H_K, which is proportional to K.

Suppose that the natural frequency of this spin precession is 1 MHz for a particular ferrite. Suppose also that the material is now exposed to a weak alternating field H of frequency, say, 60 Hz. Then the induction B developed in each cycle is governed by H and the initial permeability μ_0. (High-frequency ferrites are usually operated at low fields, which is why we are mainly interested in the initial permeability.) Let the drive frequency now be increased to, say, 1 kHz. The value of μ_0 will remain the same, because of the great disparity between the drive frequency and the precession frequency But when the drive frequency reaches 1 MHz, the two frequencies are matched and the precessing spins absorb power from the drive field. As a result the induction B produced in each cycle decreases, which means that μ_0 decreases. Now the static value of μ_0 is generally higher, the lower the value of K, or H_K. A low value of H_K means a low precession frequency. We therefore expect that μ_0 will begin to decrease at a lower frequency if μ_0 is large than if μ_0 is small, in agreement with experiment (Fig. 13.42). This argument suggests that spin resonance is the only loss mechanism in ferrites. This is not quite true, but the theory outlined does explain the general features of ferrite behavior.

[When losses occur, B lags behind H in time, and the permeability becomes a complex number and should be written $\mu_0 = \mu' - i\mu''$, where μ' is the real part (B in phase with H) and μ'' the imaginary part (B 90° out of phase). It is the real part μ' which is plotted in Fig. 13.42. The imaginary part μ'' is very small until μ' begins to decrease. Then μ'' and the losses increase. Curves that show the variation of both μ' and μ'' with frequency are called the *magnetic spectrum* or *permeability spectrum* of the material. The variation of permeability with frequency is referred to as *dispersion*.]

The high-frequency applications of soft ferrites are mainly as cores for special transformers or inductors. Certain communication equipment requires broadband transformers; as the name implies, they must have cores that show the same behavior over a broad range of frequencies. Another application is in pulse transformers used in certain data-handling circuits, because the Fourier components of a square pulse extend over a wide range of frequencies; excessive distortion of the pulse shape occurs if the permeability of the core varies with frequency. Built-in ferrite antennas are much used in modern radio receivers, particularly portable ones. Such an antenna is merely a ferrite rod wound with a coil. It responds to the magnetic component of the incident electromagnetic wave, and the alternating flux in the ferrite induces an emf in the coil; the ferrite core in effect multiplies the area enclosed by the coil by a factor of at least one hundred.

Finally, there are important microwave applications of ferrites, both in communications and in radar circuits. These involve frequencies of some 10^{10} Hz and fundamental effects not even mentioned in this book. The reader is referred to the book by Lax and Button [G.16].

PROBLEMS

13.1 Verify the point at the center of the third curve from the left of Fig. 13.2 by showing that $B_x/B_0 = 0.46$ at the center of this sheet.

13.2 Calculate the field penetration ratio H_x/H_0 at the center of a nickel rod 0.5 inch in diameter for frequencies of 20 and 60 Hz. Take $\mu = 100$ and $\rho = 9.5$ microhm-cm.

13.3 Suppose a solid transformer core is divided into n laminations of equal thickness. Show that the laminated core has a classical eddy-current power loss, which can be approached if the domain size is small, equal to $1/n^2$ times the loss of the solid core.

13.4 Nonoriented electrical steel sheet, 0.014 inch thick and containing 3.25 percent Si, has a total core loss of 1.60 watts/lb at 60 Hz and 15 kilogauss. Its density is 7.65 g/cm^3 and its resistivity is 50 microhm-cm. The static hysteresis loss is 1.10 watts/lb. Calculate the anomaly factor.

13.5 If the core loss of plain-carbon sheet steel is 5.3 watts/lb at 15 kilogauss and 60 Hz, what is the temperature rise in °C/min if no heat is lost from the material? The specific heat is 0.113 cal/g °C.

CHAPTER 14

HARD MAGNETIC MATERIALS

14.1 INTRODUCTION

The purpose of a permanent magnet is to provide a magnetic field in a particular volume of space. A magnetic field can be produced by current in a conductor or by poles in a magnet. For a great many applications a permanent magnet is the better choice, because it provides a constant field without the continuous expenditure of electric power and without the generation of heat. A magnet is fundamentally an energy-storage device. This energy is put into it when it is first magnetized and it remains in the magnet, if properly made and properly handled, indefinitely. In short, the magnetism is *permanent*. Moreover, the energy of a magnet, which is chiefly the energy of its external field, is always available for use and is not drained away by repeated use, like the energy of a battery, because a magnet does no net work on its surroundings.

The earliest, and for a long time the only, application of the permanent magnet was the compass needle. Today the applications of the magnet, in industry, in the home, in the automobile, form a list so long, if set down in detail, as to astonish the uninitiated. The main divisions are shown in Table 14.1, and they will be described at the end of the chapter. So pervasive are the radio, television and record player that loudspeaker magnets form the largest single part of the market. And so perennial is the fascination of magnetic forces for the young, and not so young, that there is always a small but steady sale of magnetic toys and novelties. Recording tape, which exceeds in dollar value all the items of Table 14.1, is not listed there; it appears as a separate item in Table 13.1.

Table 14.1 shows that the permanent-magnet market, with magnetic recording media excluded, is dominated by two materials, Alnico and barium ferrite. Alnico is actually a series of alloys, named after the chemical symbols of three essential components, Al, Ni, and Co. Barium ferrite is frequently known simply as the "ceramic magnet"; its share of the market has grown rapidly in the last ten years, as evidenced by an earlier estimate [14.1].

The compositions and magnetic properties of a number of magnetically hard materials are given in Table 14.2, which is part of a much longer list [14.2]. The materials are listed more or less in historical order. Not all are in actual production. The stated properties should be regarded as representative rather than "accurate." Different sources rarely give the same property values for the same alloy, and it is hardly ever clear whether the stated properties are minimum

Table 14.1 Permanent-Magnet Material and Applications

Application	Percent of U.S. production by value		
	Alnico	Barium ferrite	Total
Motors, generators, magnetos, and other rotating devices	8	13	21
Loudspeakers	10	6	16
Telephone ringers and receivers, hearing aids, etc.	7	–	7
Meters, switches, controls, etc.	7	–	7
Microwave tubes, computer-related devices, etc.	7	–	7
Separators for ore beneficiation, tramp iron removal, etc.	3	3	6
Holding magnets, coin selectors, toys, novelties, etc.	8	8	16
	50	30	80
Other materials (magnet steels, flexible barium ferrite, Remalloy, Cunife, etc.)			20
			100

guaranteed values, average values, or best values. Properties obtained in regular, commercial production are always inferior to those achieved under closely controlled laboratory conditions.

The most thorough reference work on permanent magnets is the book edited by Hadfield [G.17]. The subject it also treated in books by Bozorth [G.4], Parker and Studders [G.18], Tebble and Graik [G.35], and Bardell [G.11]. The principal review papers are by Wohlfarth [11.3, 11.4], Luborsky [14.3], Becker *et al.* [14.4], McCaig [14.5], Becker [11.14], and a committee of specialists [14.6]. The reader's attention is again called to Andrade's interesting account [1.2] of the early history of the magnet.

Magnetic measurements on permanent magnets are not inherently different from those made on magnetically soft materials, except in the measurement of H. This point is discussed in Appendix 11.

14.2 OPERATION OF PERMANENT MAGNETS

Before considering the materials of which magnets are made, we must examine the conditions under which a magnet operates, in order to determine what material

Table 14.2 Properties of Permanent-Magnet Materials [14.2]

Material	Composition (weight percent)	Remanence B_r (gauss)	Coercivity H_c (oersted)	Maximum energy product $(BH)_{max}$ (MGOe)
Magnet steels				
1 C steel	1 C	9,000	51	0.20
6 W steel	6 W, 0.7 C	9,500	74	0.33
3.5 Cr steel	3.5 Cr, 1 C	9,500	66	0.29
36 Co steel	36 Co, 3.75 W, 5.75 Cr, 0.8 C	9,600	228	0.93
Alnico alloys (cast)				
	Al Ni Co Cu Fe			
Alnico 1	12 23 5 – 60	6,600	540	1.4
Alnico 2	10 18 13 6 63	7,000	650	1.7
Alnico 3	12 26 – 3 59	6,400	560	1.4
Alnico 4	12 28 5 – 55	5,500	730	1.4
Alnico 5*	8 15 24 3 50	12,000	720	5.0
Alnico 5 DG*	8 15 24 3 50	13,100	700	6.5
Alnico 5-7*	8 14 24 3 51	13,400	740	7.5
Alnico 6*	8 17 23 3 45 +4 Ti	7,500	975	2.8
Alnico 8*	8 14 38 3 29 +8 Ti	7,100	2,000	5.5
Alnico 9*	7 15 35 4 34 +5 Ti	10,400	1,600	8.5
Alnico alloys (sintered)				
Alnico 2	10 17 13 6 54	7,200	550	1.5
Alnico 4	12 28 5 – 55	5,500	730	1.3
Alnico 5*	8 14 24 3 50 +1 Ti	10,500	600	3.8
Ferrites				
Ceramic 1	$BaO \cdot 6\ Fe_2O_3$	2,250	1,850 4,000 (H_{ci})	1.2
Ceramic 5*	$BaO \cdot 6\ Fe_2O_3$	3,950	2,400 2,470 (H_{ci})	3.5
Ceramic 6*	$BaO \cdot 6\ Fe_2O_3$	3,600	2,900 3,100 (H_{ci})	3.1
Ceramic 7*	$SrO \cdot 6\ Fe_2O_3$	3,425	3,330 4,100 (H_{ci})	2.9
Special alloys				
Cunife 1*	60 Cu, 20 Ni, 20 Fe	5,700	590	1.9
Cunife 2*	50 Cu, 20 Ni, 2.5 Co, 27.5 Fe	7,300	260	0.8
Remalloy	12 Co, 17 Mo, 71 Fe	10,000	230	1.1

Table 14.2 (cont'd)

Material	Composition (weight percent)	Remanence B_r (gauss)	Coercivity H_c (oersted)	Maximum energy product $(BH)_{max}$ (MGOe)
Vicalloy 1	10 V, 52 Co. 38 Fe	9,000	300	1.0
Vicalloy 2*	13 V, 52 Co, 35 Fe	10,000	450	3.0
PtCo	77 Pt, 24 Co	6,450	4,300	9.5
MnBi*	21 Mn, 79 Bi	4,800	3,650	5.3
MnAl*	70 Mn 30 Al	4,280	2,700	3.5
Iron-powder magnets				
ESD 32*	72 Pb, 10 Co, 18 Fe	6,800	960	3.0
ESD 42	72 Pb, 10 Co, 18 Fe	4,800	830	1.3

* Anisotropic properties.

properties are important. Because the only function of a magnet is to provide an external field, it must have free poles. A circumferentially magnetized ring, forming a closed magnetic circuit, is of no use whatever. A magnet *always* operates on open circuit. The resulting free poles create a demagnetizing field H_d which makes the induction lower than the remanence value B_r found in a closed ring, a point that was illustrated in Fig. 2.32.

After a magnet is manufactured, a strong field H_1 is applied to it and removed, causing the induction B to follow the path shown in Fig. 14.1. The operating point P of the magnet is determined by the intersection of the line OC with the second quadrant of the hysteresis loop. This quadrant is called the *demagnetization curve* of the material. The shape and placement of this curve, and not only the values of H_c and B_r, determine the suitability of a material as a permanent magnet.

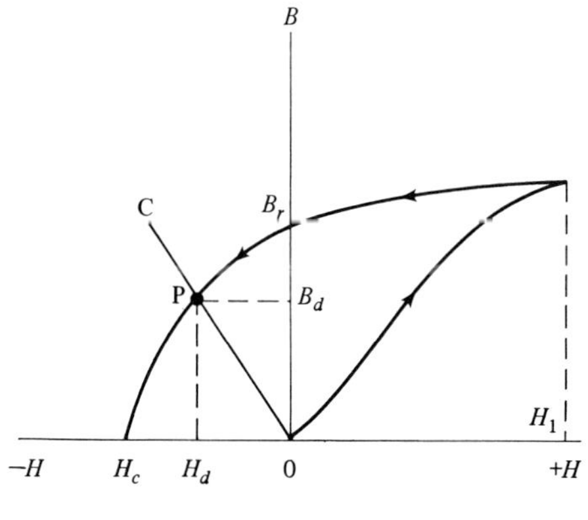

Fig. 14.1 Residual induction B_d in an open magnetic circuit.

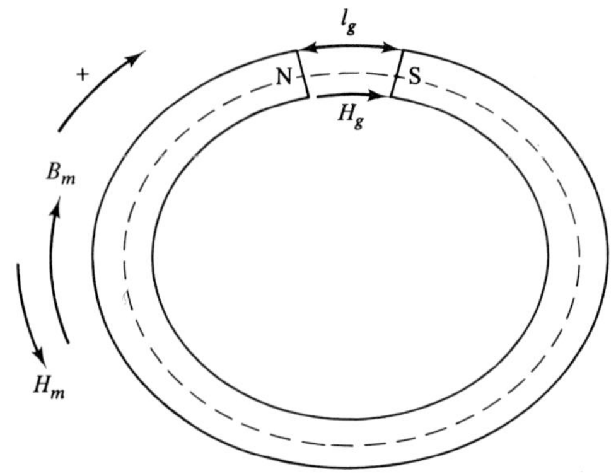

Fig. 14.2 Open magnetic circuit.

The slope of the line OC is, from Fig. 2.32, equal to $-(4\pi - N_d)/N_d$, where N_d is the demagnetizing factor of the magnet. Because N_d depends on the length/diameter ratio of the magnet, it can be altered at will by the magnet designer, who has the freedom to choose the slope of OC, often called the *load line*, and therefore put the operating point P anywhere on the demagnetization curve. The question then arises: is one point P better than any other?

We might pose this question for the specific case of the gapped ring of Fig. 14.2, which simulates many actual magnetic circuits. This could be the magnet for a moving-coil ammeter or galvanometer, with the moving coil located in the air gap. The magnet must provide a field H_g of constant strength in the air gap. The induction in the magnet is B_d, which we will now call B_m, and the field is H_d, which we will now call H_m. According to Ampere's law, Eq. (2.46), the line integral of H around the dashed curve of Fig. 14.2 must be zero, because there is no current:

$$\oint H dl = 0, \tag{14.1}$$

$$H_g l_g - H_m l_m = 0, \tag{14.2}$$

where l_m is the length of the magnet. (Note that concept of magnetomotive force, as defined in Section 2.8, here loses most of its meaning, of a force that "drives" magnetic flux in a circuit. In the circuit of Fig. 14.2 the flux ϕ is everywhere clockwise and positive, but the *net* magnetomotive force is zero.) The continuity of the lines of B furnishes us with a second equation:

$$\phi = B_g A_g = H_g A_g = B_m A_m, \tag{14.3}$$

because $B = H$ in the air gap. Here A_g and A_m are the cross-sectional areas of the air gap and the magnet. In Fig. 14.2 these are equal, because fringing (widening) of the flux in the gap is ignored. Generally, however, A_g and A_m are

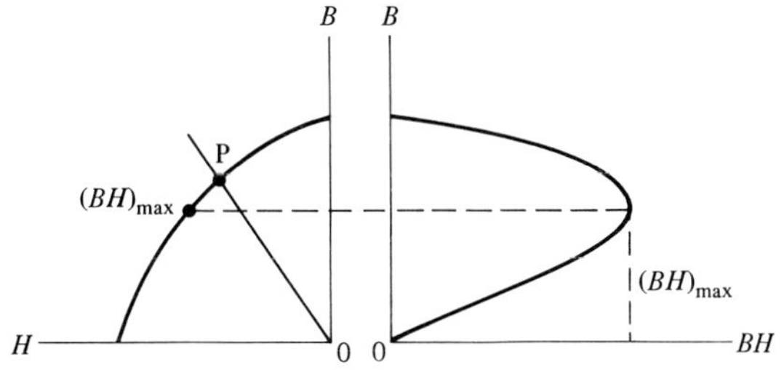

Fig. 14.3 Variation with B of the energy product BH.

unequal, if only because soft iron (high-permeability, annealed iron) pole caps are often fitted to the ends of the magnet; these then define the gap area A_g. The problem of the magnet designer is to choose l_m and A_m so as to best use the material.

If Eqs. (14.2) and (14.3) are each solved for H_g and multiplied together, we find

$$H_g^2 = \frac{B_m H_m l_m A_m}{l_g A_g}, \qquad H_g^2 V_g = (B_m H_m) V_m, \tag{14.4}$$

where V stands for volume. This result shows that the volume V_m of magnet that is required to produce a given field in a given gap is a minimum when the product BH in the magnet is a maximum. Furthermore, Eq. (7.38) shows that the energy stored in the field in the air gap is $H_g^2 V_g/8\pi$ ergs. For a magnet of any volume, this energy is directly proportional to BH, which is accordingly called the *energy product*. Figure 14.3 shows how BH varies with B over the demagnetization curve, going through a maximum value $(BH)_{\max}$ for a particular value of B. For most efficient use of material, the magnet should be so shaped that it operates at its $(BH)_{\max}$ point. Evidently, the magnet of Fig. 14.3 should be made thicker, or shorter, to bring its operating point P down to the $(BH)_{\max}$ point.

The demagnetization curves of permanent-magnet materials are often presented on graphs on which lines of constant BH are lightly drawn, as in Fig. 14.4. The reader can then tell at a glance which of two materials has the higher $(BH)_{\max}$ and approximately where that point lies on the demagnetization curve. Thus material 1 has a higher maximum energy product (about 3 MGOe) than material 2 (less than 2 MGOe). The units of $(BH)_{\max}$ are 10^6 gauss-oersteds or megagauss-oersteds (MGOe).

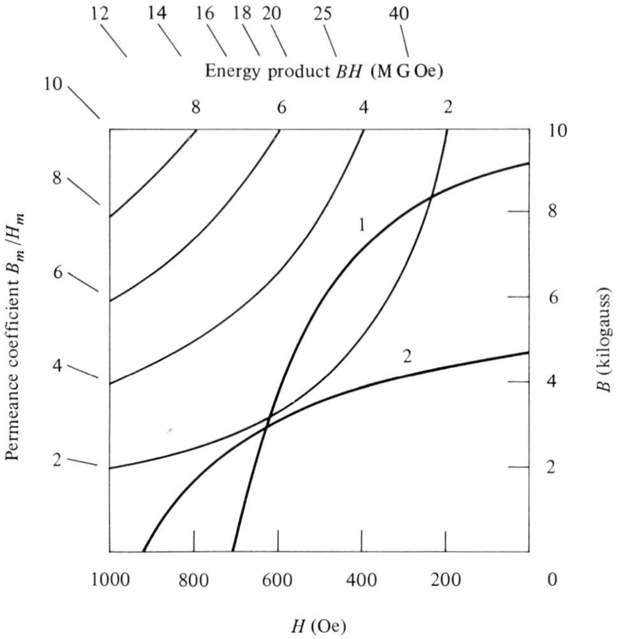

Fig. 14.4 Demagnetization curves (schematic).

The operating point of a magnet depends on the slope of the load line, which in turn depends on the demagnetizing factor N_d. But the magnet designer or, to give him his proper name, the mágnetic-circuit designer prefers to work, not with the demagnetizing factor, but with the *permeance coefficient*. This is defined as the slope of the load line or B_m/H_m, with the negative sign understood, and it is given, from Eqs. (14.2) and (14.3) by

$$\frac{B_m}{H_m} = \frac{A_g l_m}{A_m l_g}. \tag{14.5}$$

This is equivalent to the expression previously given in terms of N_d (Problem 14.1). To aid the designer, values of B_m/H_m, which depend only on the dimensions of the circuit, are often indicated at the left and top of the demagnetization curve, as in Fig. 14.4. A straight line drawn from the origin to the appropriate value of B_m/H_m will then intersect the demagnetization curve at the operating point.

The gapped ring discussed above is a highly idealized example. Not only has fringing at the air gap been neglected, but so has flux leakage from the sides of the magnet itself. These two are illustrated separately in Fig. 14.5. Leakage is a large effect: of the total flux provided by the magnet, less than 50 percent usually appears in the air gap. The chief design problem is to allow for the leakage flux; it is difficult to calculate and the designer usually relies on empirical rules derived

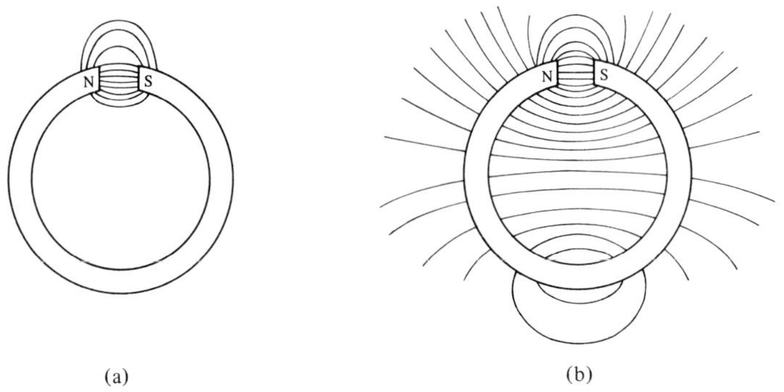

Fig. 14.5 Ring magnet with (a) fringing and (b) leakage. Parker and Studders [G.18].

from experience. Formally, fringing and leakage are allowed for by introducing leakage factors C_1 and C_2 into the design equations:

$$C_1 H_g A_g = B_m A_m, \tag{14.6}$$

$$C_2 H_g l_g = H_m l_m. \tag{14.7}$$

Proper design of permanent-magnet circuits is not an easy matter, and the reader should consult [G.17] and [G.18] for detailed explanations.

For a magnet of given volume the maximum attainable gap energy $H_g^2 V_g/8\pi$ is larger, the larger the energy product $(BH)_{max}$ of the magnet material. For this reason, the maximum energy product is considered to be the best all-around, single index of quality, and it is always included, as in Table 14.2, in any list of magnetic properties.

Although $(BH)_{max}$ is an index of material quality, it is not the only index or even the most suitable one for certain applications. All magnets operate under either static or dynamic conditions. A magnet supplying a field for a moving-coil ammeter exemplifies a static application. But the field in the gap of a holding or lifting magnet changes during operation, and the magnet "works" in a quite different way.

In Fig. 14.6 (a) the iron armature represents the piece being lifted. When this is far away, the magnet is open-circuited and at point P on the demagnetization curve. As the iron approaches, part of the flux is diverted from the leakage gap to the useful gap. When contact is made, the demagnetizing field is reduced and the operating point moves to D. If the armature is pulled away, the point returns to P. The minor hysteresis loop thus traced out is very thin, and is often approximated by a straight line through the tips of the loop; the slope of this line, which is about equal to the slope of the major hysteresis loop at the B_r point, is called the *recoil permeability* μ_r. Under these dynamic conditions the best location for P does not correspond to $(BH)_{max}$ but to a more complex condition given by Edwards [14.7].

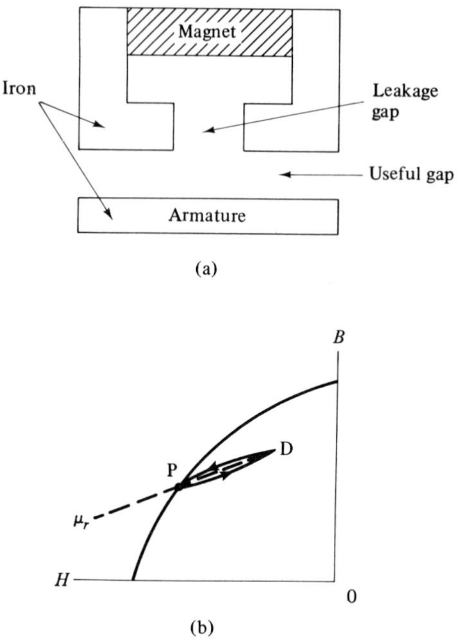

Fig. 14.6 (a) Lifting magnet. (b) Path followed on B,H curve. After Edwards [14.7].

14.3 MAGNET STEELS

We turn now to the magnet materials themselves. The earliest of these was the lodestone, generally "armed" by attaching soft iron at each end to concentrate the flux. Next came high-carbon steel magnets, hardened by quenching, and these were used for centuries as compass needles.

With the advent in the 19th century of alloy steel for structural purposes came interest in alloy steel for magnets, and by 1885 a steel containing about 5 percent tungsten was in use for magnets. This was supplanted by the cheaper chromium steel during World War I, but both types had coercivities of less than 100 Oe.

The next advance was made in 1917 by the Japanese investigators Honda and Takagi, who showed that a steel containing 30 to 40 percent cobalt, plus tungsten and chromium, had a coercivity of 230 Oe. This is still the best magnet steel. Some properties of this and other steels are shown in Fig. 14.7.

The magnetic hardness of magnet steels is presumably due to the hindrances offered to domain wall motion by microstress and inclusions. The main constituent of quenched steel is martensite, and undissolved carbides are often present. Martensite is a metastable phase in a state of high residual stress, and the tempering treatment that follows quenching only partially relieves this stress.

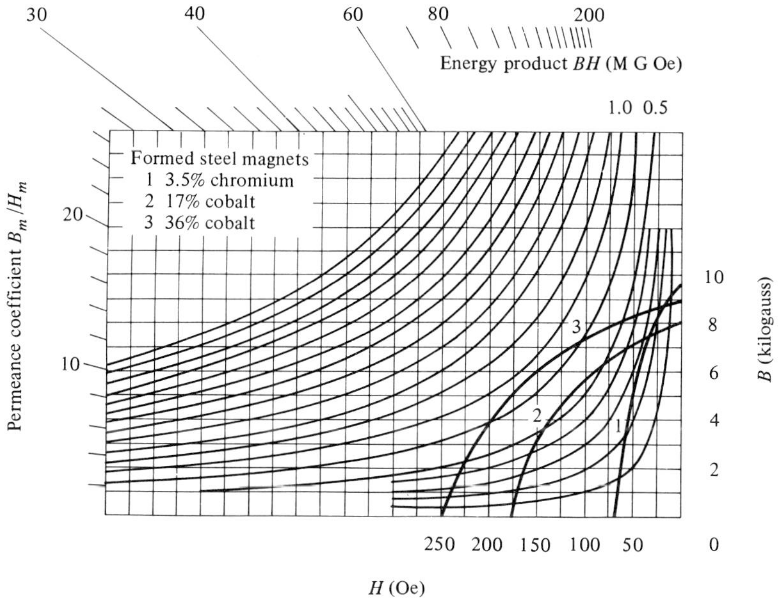

Fig. 14.7 Properties of some magnet steels. Parker and Studders [G.18].

Magnet steels, although cheap, find little application today, because other materials are so much better. Of those that are used, 3.5 percent chromium steel is the most popular.

14.4 ALNICO

Alnico is actually a family of alloys, known by many trade names, containing substantial amounts of all three of the ferromagnetic metals, Fe, Co, and Ni, plus smaller amounts of Al, Cu, and sometimes other elements. They are the most widely used permanent-magnet materials.

The development of Alnico dates from 1931, when Mishima in Japan discovered that an alloy of 58 percent Fe, 30 percent Ni, and 12 percent Al had a coercivity of over 400 Oe, or about double that of the best magnet steel. This alloy and the Alnicos are essentially carbon-free. No martensite formation is involved, and the mechanism of their magnetic hardness is quite different from that of steel.

It was soon discovered that the addition of Co and Cu improved the properties of the Mishima alloy, and Fig. 14.8 shows the various composition changes that have been made.

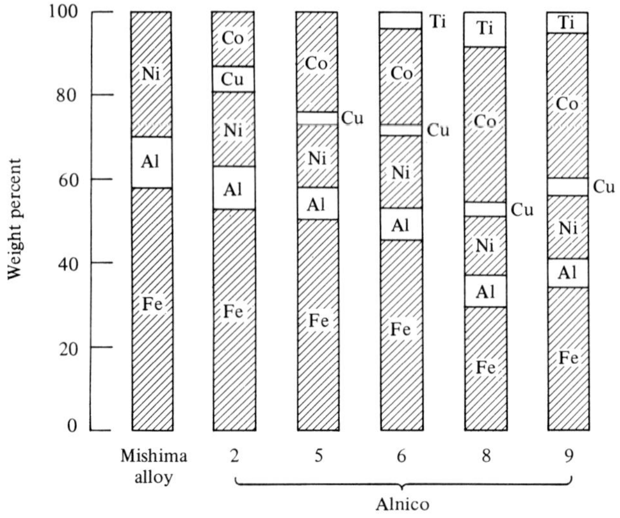

Fig. 14.8 Development of Alnico alloys.

A dramatic improvement in properties was achieved in 1938–1940 by English and Dutch workers, who (a) increased the Co content and (b) heat treated the alloy in a magnetic field. (That field annealing could improve the properties of *soft* magnetic materials had previously been demonstrated, but this was the first application of the technique to hard materials.) Only a modest increase in coercivity was obtained but the remanence increased markedly, leading to a value of $(BH)_{max}$ about triple that of Alnico 2, which is not treated in a field. The resulting material is called Alnico 5 in the U.S.A., Ticonal and Alcomax in Europe, and TK7 in Japan. Figure 14.9 shows demagnetization curves of several Alnico alloys. The effect of titanium, added to Alnico 6, 8, and 9, is to increase the coercivity and decrease the remanence. Alnico 5 is the most widely used of the whole series.

All the Alnicos are hard and brittle, much too brittle to be cold worked. (Hot working is possible but is not done commercially.) All production is therefore by casting of the liquid alloy or by pressing and sintering metal powders. The cast alloys have very coarse grains, of the order of 1 mm diameter. The sintered alloys are finer grained and mechanically stronger but with inferior magnetic properties; they are usually limited to small, intricately shaped magnets, for which the pressing operation is well suited. Surface grinding is the only finishing operation possible on either type of Alnico.

The magnetic properties of as-cast or as-sintered Alnico are not particularly good. A special, three-stage heat treatment is necessary to produce optimum properties:

1. Heat to 1250°C for a time sufficient to produce a homogeneous solid solution.

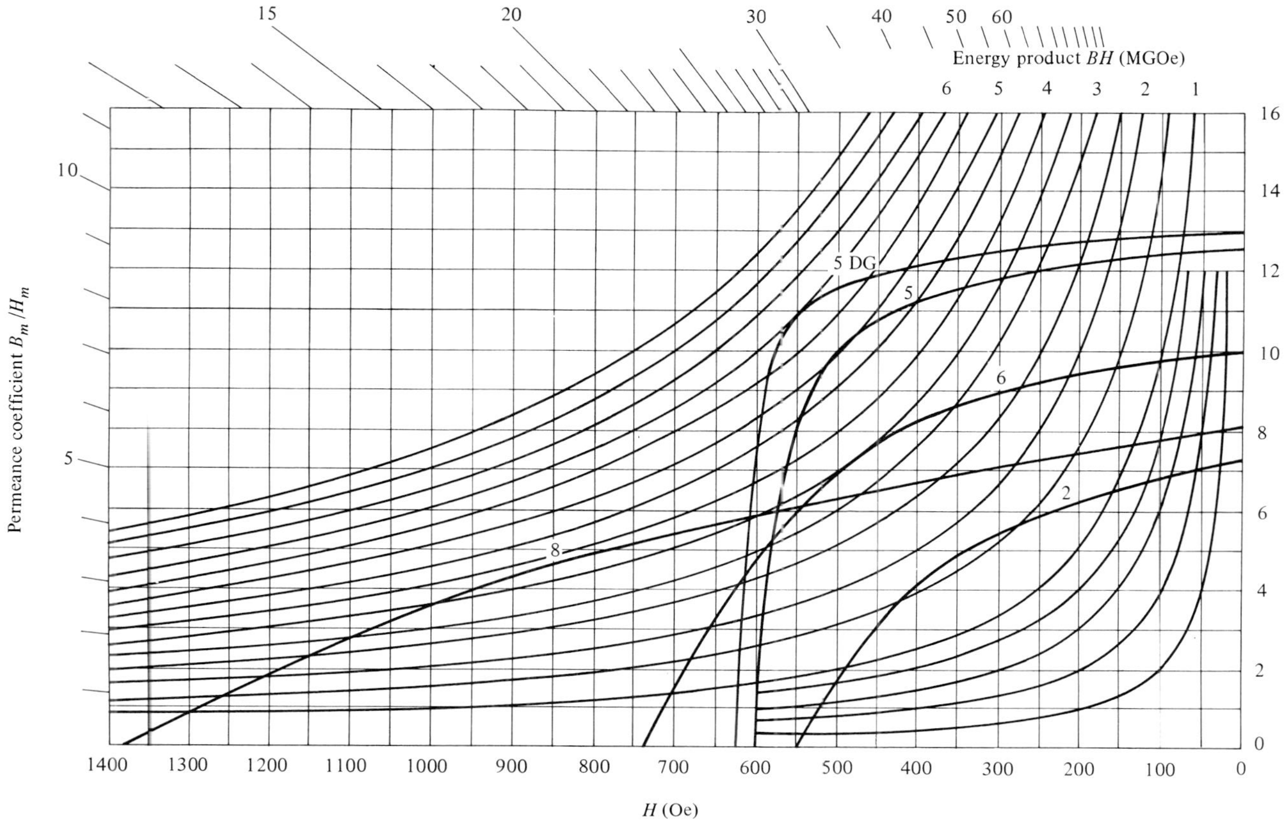

Fig. 14.9 Properties of some Alnico alloys. All of these except Alnico 2 are heat treated in a magnetic field. Parker and Studders [G.18].

2. Cool at a rate of the order of 1°C/sec to about 500°C or lower.

3. Reheat (temper) at 600°C for a few hours.

This is the heat treatment applied to Alnicos 1 to 4.

The field-treated Alnicos 5 to 7 are not heat treated in a field at constant temperature, as might be supposed, but the cooling step (2) above is done in a field. The white-hot alloy is removed from the furnace and placed in the gap of an electromagnet or inside a water-cooled solenoid, which produces a field of about 1000 Oe, while the alloy cools to about 500°C. It is then ready for tempering, which is done in an ordinary furnace (no field). The properties are then good along the direction of the cooling field and inferior at right angles. (The field treatment of Alnicos 8 and 9 is done isothermally, for 10 to 20 min at 815°C, because these alloys have to be held at a particular temperature for a longer time than Alnicos 5 to 7.)

Because of the commercial importance of Alnico, much research has been devoted to the problem of understanding its magnetic hardness. The earlier research was hampered by the fact that an alloy treated to have optimum magnetic properties had a microstructure too fine to be resolved by the light microscope. Later the electron microscope provided the necessary resolution, and the alloy was shown to have an extremely fine, two-phase structure. The most recent review of research on phase transformations in Alnico is by de Vos [14.8], who summarizes his own work and that of others. Although some details remain obscure, the main conclusions are as follows:

1. At the solution-treatment temperature of 1250°C, the alloy is a single-phase solid solution with a BCC crystal structure.

2. During cooling, a rodlike precipitate forms parallel to ⟨100⟩ directions of the matrix. This precipitate is called α'; it is rich in Fe ane Co and is strongly magnetic. The matrix is called α; it is rich in Ni and Al and is weakly magnetic. Both α and α' have BCC structures, with not much difference in lattice parameter. (Actually, both the high-temperature phase and the α matrix formed during cooling have an ordered CsCl structure, i.e., one in which the atom at the cell center is on the average different from the atom at the cell corner. However, it is usual to refer to both of these phases as BCC.)

3. During tempering, the difference between the spontaneous magnetizations of the α and α' phases becomes greater, probably by diffusion of Fe and Co from α to α', to the extent that the α phase may even become paramagnetic.

4. The rods of α' precipitate are very small, of the order of $300 \times 300 \times 1200$ Å.

5. The cause of the magnetic hardness is *shape anisotropy of single-domain particles* of the α' phase. (The larger the difference between the M_s values of precipitate and matrix, the larger is the effective shape anisotropy of the precipitate particles; this explains the fact that tempering increases the coercivity above the value which cooling alone can produce. Another supporting fact is that Ti additions are known to increase the length of the precipitate rods and therefore their shape anisotropy, which accounts for the fact that the Ti-bearing alloys have larger coercivity.)

When a polycrystalline Alnico alloy composed of grains oriented at random is cooled from 1250°C in a strong field, the precipitate rods form preferentially along the ⟨100⟩ axis of each grain that is closest to the field direction, in order to minimize magnetostatic energy. (The smaller the angle between M_s and the axis of a rod-shaped particle, the smaller is the magnetostatic energy.) As a result, the alloy as a whole acquires an easy axis of magnetization parallel to the direction of the cooling field. The partial alignment of the α' rod axes with the cooling field increases not only the coercivity H_c but also the remanence B_r. This follows from the calculated hysteresis loops of Fig. 9.38, and it is still qualitatively true whether or not magnetization reversal in each particle occurs coherently. The resulting greater squareness of the hysteresis loop is evident in Fig. 14.9, when the curve for Alnico 5 (field-cooled) is compared with the curve for Alnico 2 (non-field-cooled).

The effect of field cooling on the microstructure is beautifully shown by the electron micrographs of de Vos and others. Such experiments are usually done on single crystals, so that the annealing field can be directed along a single ⟨100⟩ axis of the crystal; after heat treatment, the alloy is sectioned parallel and at right angles to this axis, and both sections are examined with the microscope. (In the laboratory, it is more convenient to anneal in a field at constant temperature, in order to study isothermal precipitation as a function of time; in the plant, cooling in a field is a more practical operation; both yield essentially the same results, as long as the field-cooled alloy is not taken too quickly through the precipitation-temperature range.) Figure 14.10 shows the microstructure of a single crystal of Alnico 8 after an initial solution heat treatment at high temperature, followed by 9 min at 800°C in a field. The section in (a) is parallel to the annealing field and shows the α' rods (dark) in the α matric (light); cross sections of these rods are shown in (b). The alignment is remarkably good.

A magnetic field can have no effect on precipitation unless the precipitation takes place below the Curie point. It was for this reason that the Co content of the early Alnicos was raised to the 24 percent level of Alnico 5. The additional cobalt increased the Curie point to a temperature (about 850°C) such that precipitation at, say, 700 to 800°C could take place both reasonably fast and still below T_c.

Although the heat treatment of Alnico 5 involves annealing or cooling in a magnetic field, the term *magnetic annealing* has been purposely avoided in the above discussion. It is better, in the interest of overall clarity, to restrict this term to the meaning defined in Chapter 10. There a *single-phase* solid solution was said to magnetically anneal if a field caused directional order to occur within the solution. The phenomena that occur in Alnico are quite different. There a single phase decomposes into two phases, and the field only controls the orientation of the *precipitate particles*. Incidentally, in all the research on Alnico, there has been no suggestion that the field might cause directional order in the parent phase before precipitation or in the product phases during precipitation.

The properties of Alnico 5 can be further improved if we begin with an alloy which has a particular preferred, rather than random, orientation of its grains, namely, one in which ⟨100⟩ directions in all the grains are parallel, or nearly

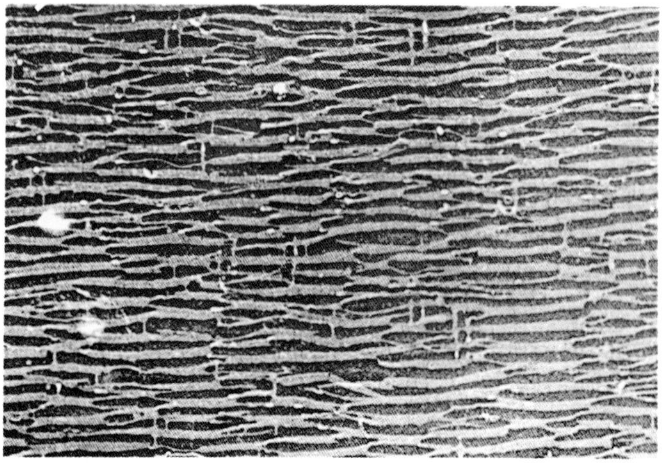

(a)

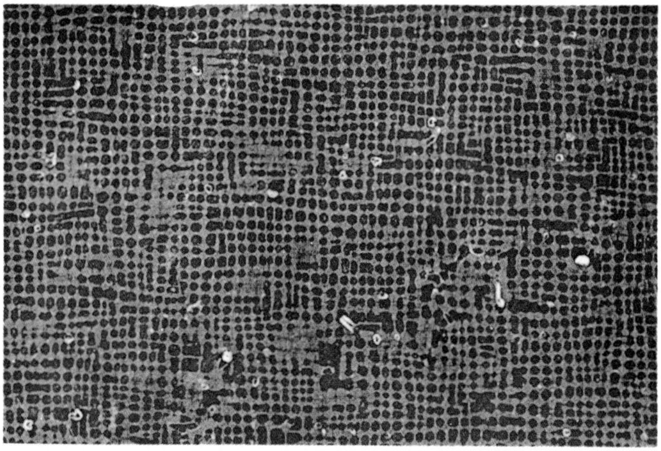

(b)

Fig. 14.10 Electron micrographs of oxide replicas from the surface of field-annealed Alnico 8. (a) Section parallel to the annealing field, (b) section at right angles. 50,000 ×. De Vos [14.8].

so. If the cooling field is then made parallel to this direction, all the precipitate rods will have their axes closely aligned with this direction, and H_c and B_r will both increase. It is a fortunate fact that this texture, a $\langle 100 \rangle$ fiber texture, forms in *all* cubic metals and alloys which are directionally solidified. Figure 14.11 shows how such a casting is made. The object is to achieve unidirectional heat flow from the liquid metal and, with this in mind, the bottom of the mold is made of a thick piece of metal, a good heat conductor, which may also be water-cooled, or of a thick piece of graphite: the sides of the mold are made of a refractory ma-

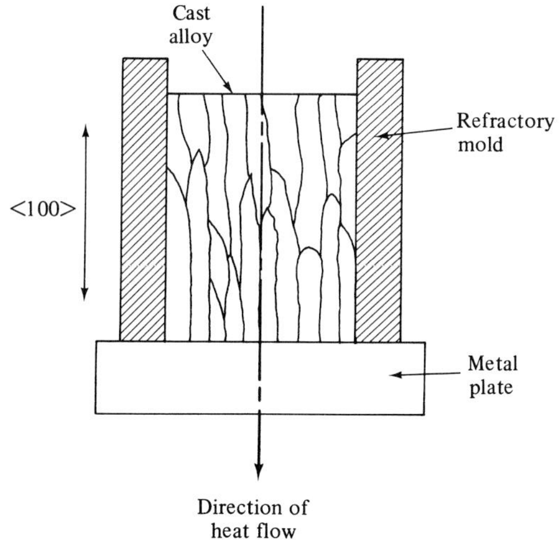

Fig. 14.11 Casting with columnar ⟨100⟩ grains.

terial, poorly conducting, which may also be heated to further restrict lateral heat transfer. As a result, freezing begins, ideally, at the bottom plate and proceeds vertically, forming long columnar grains with ⟨100⟩ axes, as sketched. When Alnico 5 is cast in this way and then heat treated in the usual way but with the cooling field parallel to the axis of the casting, the result is Alnico 5-DG, where DG stands for "directional grain." Its demagnetization curve is compared with that of Alnico 5 in Fig. 14.9. A completely columnar product is not always obtained, as shown by Fig. 14.12 (b), but $(BH)_{max}$ is substantially larger

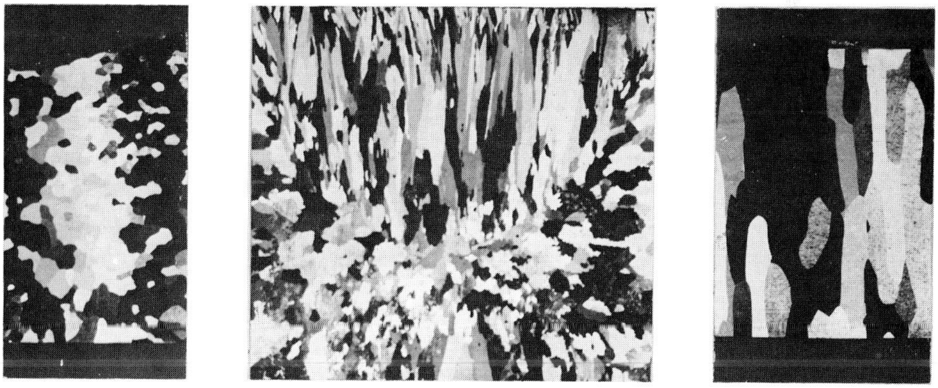

(a) (b) (c)

Fig. 14.12 Approximately full-size macrostructures of (a) Alnico 5, (b) Alnico 5-DG, and (c) UNI-80 supercolumnar Alnico 5 (Japanese manufacture). Graham [7.24].

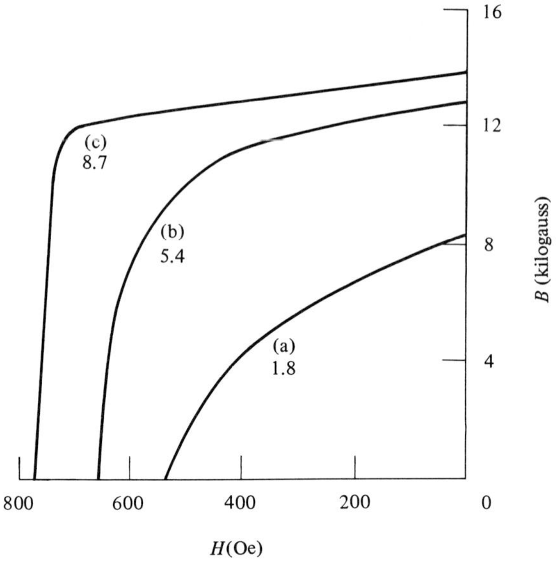

Fig. 14.13 Properties of an Alnico 5 alloy to which 0.7 percent Nb has been added. See text for details. Gould [14.9].

than that of Alnico 5. The normal, equiaxed grain structure of Alnico 5 appears in (a) and a very highly columnar structure appears in (c). Alnico 5–7 also has a columnar structure, in fact, a more highly developed one than Alnico 5-DG. It is called Columax in England. Alnico 9 is a Ti-bearing alloy with a columnar structure.

We now see that three essentially different varieties of Alnico can be made, and all are actually produced:

1. Magnetically isotropic, no preferred grain orientation, not field-cooled (Alnico 2).
2. Magneticaly anisotropic, no preferred grain orientation, field-cooled (Alnico 5).
3. Magnetically anisotropic, preferred grain orientation, field-cooled (Alnico 5-DG).

How these three structures, in an alloy of the *same* composition, result in three progressively better materials is shown in Fig. 14.13. Curves (a), (b), (c) refer to structures (1), (2), (3) above. Numbers on the curves are values of $(BH)_{max}$ in MGOe.

The actual mechanism of "precipitation" in Alnico has been much investigated, and it is now generally agreed that it takes place by *spinodal decomposition* rather than by the more usual process of nucleation and growth. The two are contrasted in Fig. 14.14. Although Alnico contains five components, we will assume that the phase relations in Alnico can be discussed in terms of the hypo-

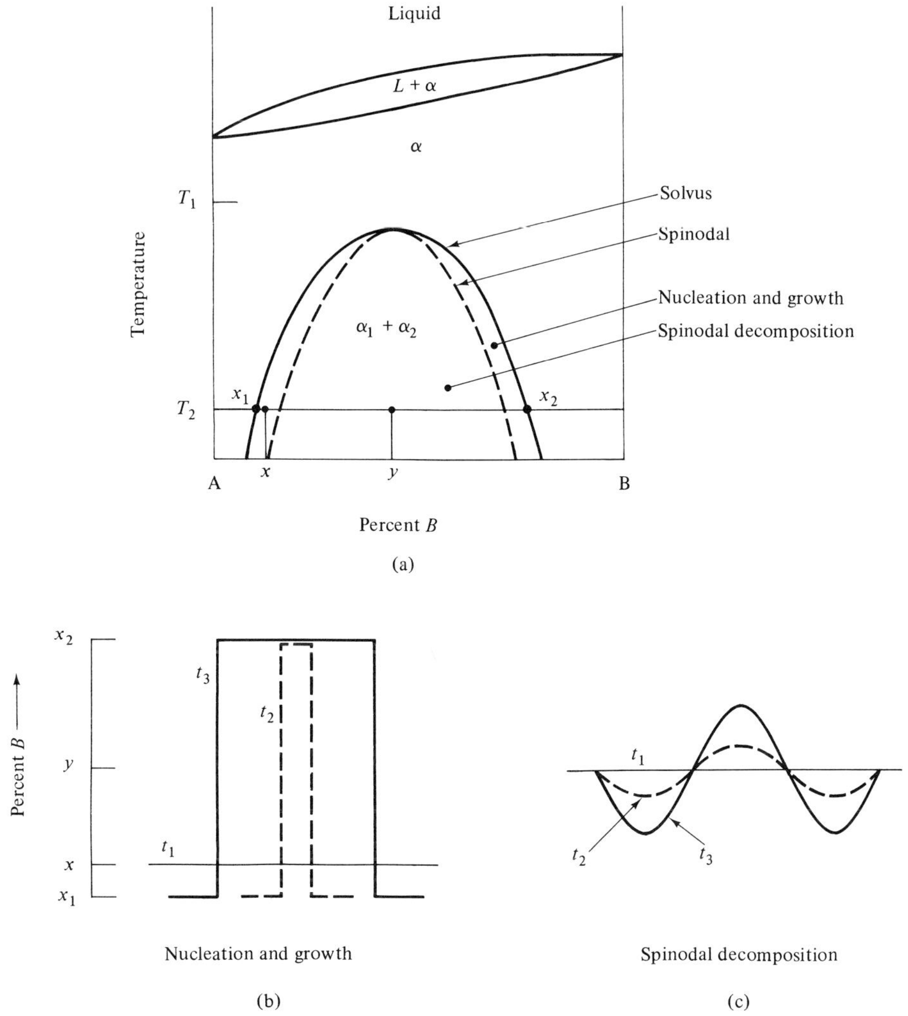

Fig. 14.14 Spinodal decomposition.

thetical binary system of A and B shown in (a). The high-temperature α phase breaks down at lower temperatures into the two stable phases α_1 and α_2 whenever its composition and temperature are inside the solvus curve. (These two phases represent the α and α' phases of Alnico.) Inside the solvus curve is another curve, shown dashed, called the *spinodal*. If a homogeneous α alloy is cooled quickly from the solution treatment temperature T_1 to T_2, it will transform eventually into α_1 of composition x_1 and α_2 of composition x_2. How this transformation takes place depends on the composition of α. If it has composition x, between the solvus and the spinodal, it transforms by nucleation and growth, i.e., a tiny nucleus of α_2 forms and grows. From the very beginning, this nucleus gas composition x_2, rich in B, and its formation leaves the solution around the nucleus depleted in B and of composition x_1. There is therefore a sharp interface between nucleus α_2 and matrix α_1, and a concentration profile (plot of percent B versus distance in the alloy) would appear as in (b) at time t_2. At some later time t_3 the α_2 nucleus will have grown thicker, but the abrupt concentration change across the interface will remain the same.

On the other hand, if the alloy has composition y inside the spinodal, it prefers, for cogent thermodynamic reasons [11.27], to decompose by developing composition fluctuations which are periodic in space. These fluctuations are initially gentle, as shown by the dashed curve for time t_2 in Fig. 14.14 (c), but become progressively more marked. As time goes on, the amplitude of the composition fluctuation increases, but not its extent, whereas the reverse is true for nucleation and growth.

The hallmark of a spinodal decomposition is the periodicity of the "precipitated" phase (Fig. 14.10), whereas particles formed by nucleation and growth are irregularly distributed throughout the matrix.

What is the mechanism of magnetization reversal in the rodlike α' particles of Alnico? It is almost certainly not coherent rotation. From the estimates of critical diameters given for iron in Section 11.4, it follows that a reasonably elongated prolate spheroid of iron would have to be thinner than about 150 Å for coherent rotation to occur, and this dimension would be about the same for an Fe-Co particle. Inasmuch as the α' particles in Alnico are at least twice this thick, the reversal must occur by some incoherent mode like curling or fanning. The sides of the rods in Fig. 14.10 (a) are fairly smooth, but not perfectly so. Any bulging or other irregularity of shape could lead to fanning reversals, which are associated with a lower coercivity than curling (Fig. 11.9).

Can Alnico be further improved? According to de Vos [14.8], the best properties so far obtained, presumably in the laboratory, are a coercivity of 2300 Oe and an energy product of 13 MGOe. He points out that the maximum intrinsic coercivity attainable by shape anisotropy, if the magnetization can be made to reverse coherently, is given by

$$H_{ci} = (1 - p)(N_a - N_c)M_s, \tag{14.8}$$

where p is the packing fraction and $(N_a - N_c)$ depends on the axial ratio of the particles. [This equation follows from Eqs. (11.3) and (11.4).] He estimates M_s as 1700 to 1900 emu/cm^3 for the Fe-Co precipitates; if $(N_a - N_c)$ is then near 2π and p is in the range 0.4 to 0.6, the calculated coercivities are 4000 to 6000 Oe. Since B_s ($\approx 4\pi M_s$) for the alloy is as high as 11,000 to 12,000 gauss, de Vos further estimates that $(BH)_{max}$ could reach ultimate values of 30 to 35 MGOe. These properties can only be approached by somehow making the precipitate rods thinner and smoother.

Finally, something should be said of the Bitter patterns seen on the polished surface of some demagnetized Alnico alloys. The colloid particles collect in broad, hazy lines, seeming to separate different "domains." This was a surprising observation in an alloy supposed to consist of single-domain particles. It was

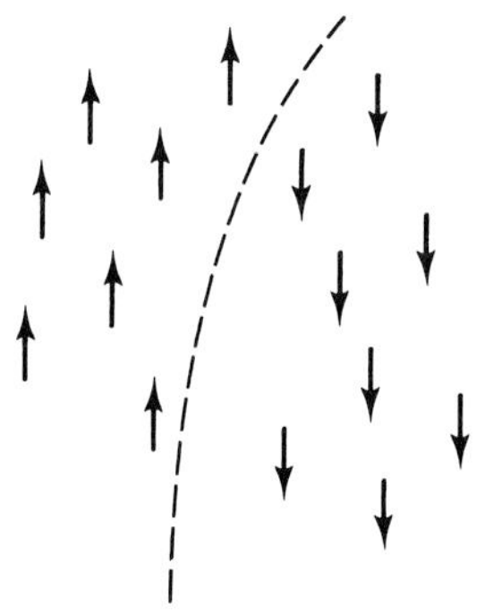

Fig. 14.15 Interaction domains.

then realized that the lines of colloid were delineating what are now called *interaction domains.* In Fig. 14.15, each vector represents the magnetization of an individual particle, and parts of two interaction domains are shown. These are regions in which all the particles are magnetized in the same direction. The dashed line shown, separating two such regions, is the locus of a large stray field, and the colloid accordingly collects along it. But this is in no sense a normal domain wall, which is an atom-by-atom transition in spin direction. Interaction domains have also been seen in the single-domain iron-powder magnets to be described in Section 14.7. This subject is discussed by Craik and Tebble [G.26] and Craik and Lane [14.10].

14.5 BARIUM FERRITE

This substance has the formula $BaO \cdot 6\,Fe_2O_3$, a hexagonal crystal structure, and a fairly large crystal anisotropy. Its intrinsic properties were described in Section 6.5. The hexagonal *c* axis is the easy axis and the crystal anisotropy constant K is 3.3×10^6 ergs/cm^3. The value of M_s is low, 380 emu/cm^3. The Curie point is 450°C. Strontium ferrite $SrO \cdot 6\,Fe_2O_3$ has almost identical properties except that K is somewhat larger.

Barium ferrite was developed into a commercial magnet material in 1952 in the Netherlands by the Philips Company, which called it Ferroxdure. Some American trade names are Indox, Ferrimag, etc. In recent years, it has been replaced to some extent by strontium ferrite [14.11]. In the following account, the term "barium ferrite" can be understood to mean either material. (Sometimes the general term *hexaferrite* is used for both materials.)

Barium ferrite is made by practically the same method as the soft ferrites. Barium carbonate is mixed with Fe_2O_3 and fired at about 1200°C to form the ferrite. This material is then ball milled to reduce the particle size, pressed dry in a die, and sintered at about 1200°C. The resulting magnet has a grain size of about 1μ and is very brittle. Anisotropic grades are made by wet pressing in

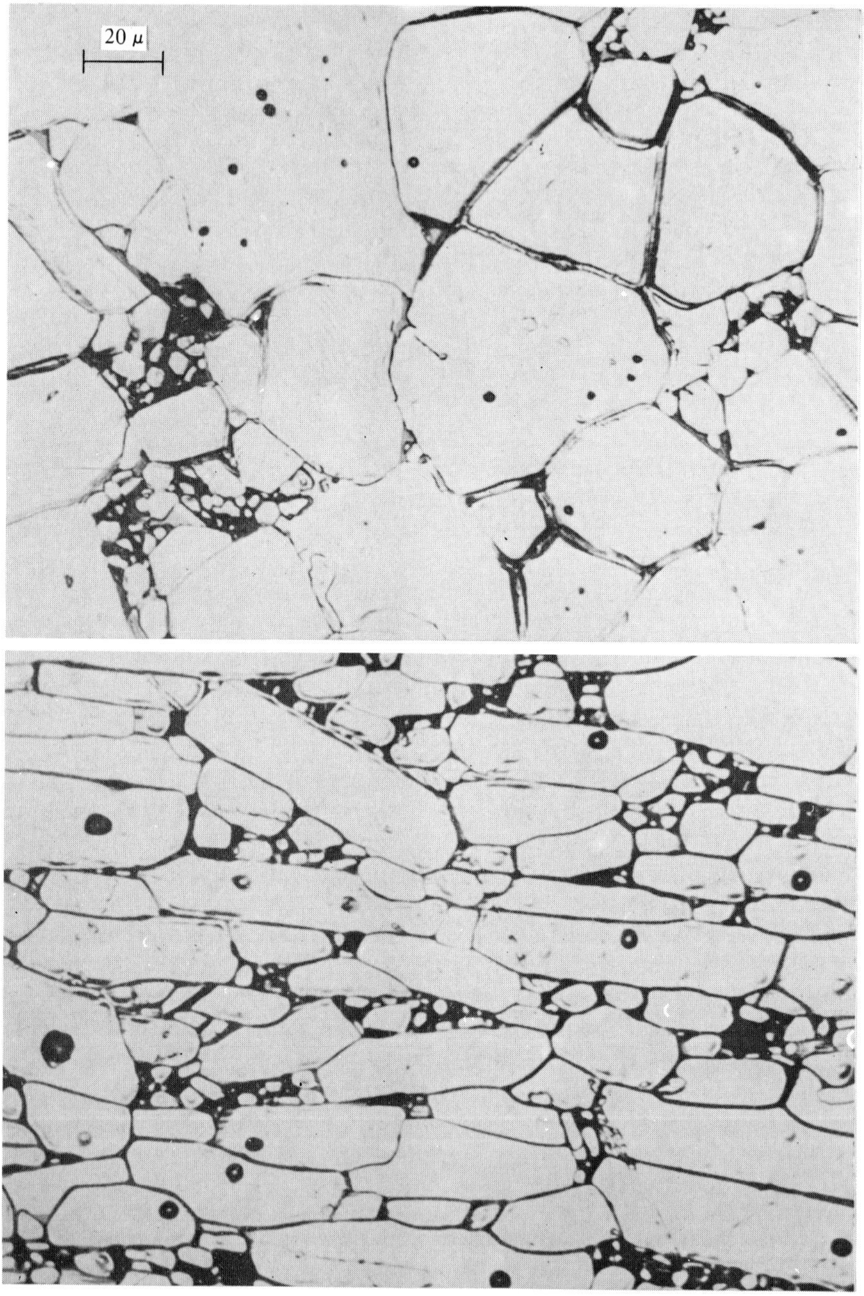

(a)

(b)

Fig. 14.16 Photomicrographs of sintered barium ferrite with oriented grains. (a) Section normal to c axes, (b) section parallel to c axes. Smit and Wijn [G.10].

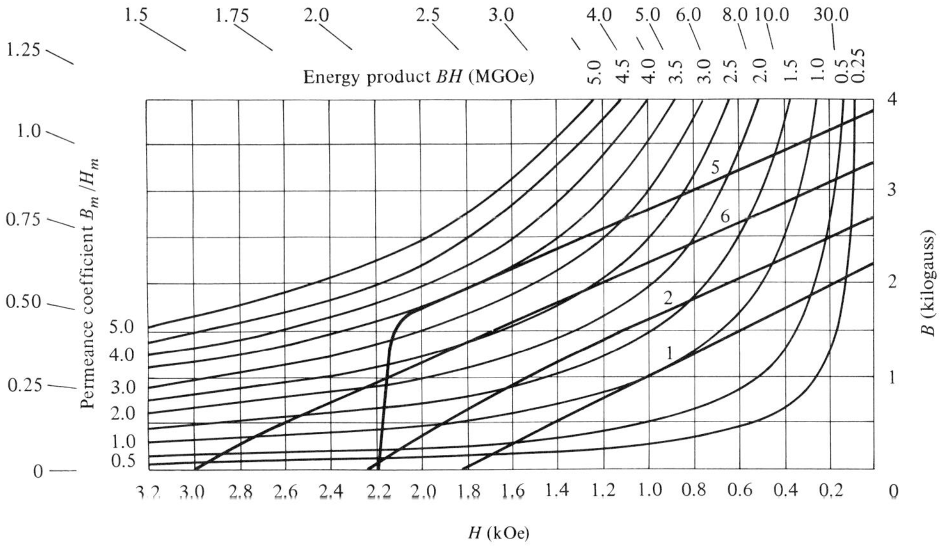

Fig. 14.17 Properties of Indox 1 (isotropic) and Indox 2, 5, 6 (anisotropic) barium-ferrite magnets. (Indiana General)

a magnetic field to align the *c* axes of the particles with the field, which is along the press axis; the usual product is a cylindrical magnet with the easy axis parallel to the cylinder axis. Figure 14.16 shows the microstructure of a coarse-grained specimen with field-oriented grains. Barium ferrite has a tabular "habit," as a mineralogist would say; i.e., it habitually crystallizes in the form of flat plates with the basal plane of the unit cell parallel to, and the *c* axis at right angles to, the plate surface. In (a) these plates are parallel to the surface examined; in (b) the plates are in profile, with the *c* axes more or less vertical and in the plane of the page. Even when a field is not applied during pressing, some preferred orientation will result because of the tendency for the particles to pack together with their flat surfaces parallel to one another and at right angles to the pressing direction.

Flexible rubber magnets and plastic magnets are made simply by mixing ferrite powder with the base material when compounding the rubber or plastic. A familiar application of rubber magnets is the gasket on refrigerator doors.

In Table 14.2, the terms Ceramic 1, 5, 6, and 7 are designations of the Magnetic Materials Producers Association; the last three are anisotropic, field oriented grades. Figure 14.17 shows the properties of the materials made by the largest U.S. manufacturer.

The properties of an oriented Alnico (Alnico 5) and an oriented barium ferrite (Indox 5) are contrasted in Fig. 14.18. Compared to Alnico, barium ferrite has a much lower remanence and a much higher coercivity. Barium ferrite is therefore best used in the form of a disk or plate magnetized through the thickness, because

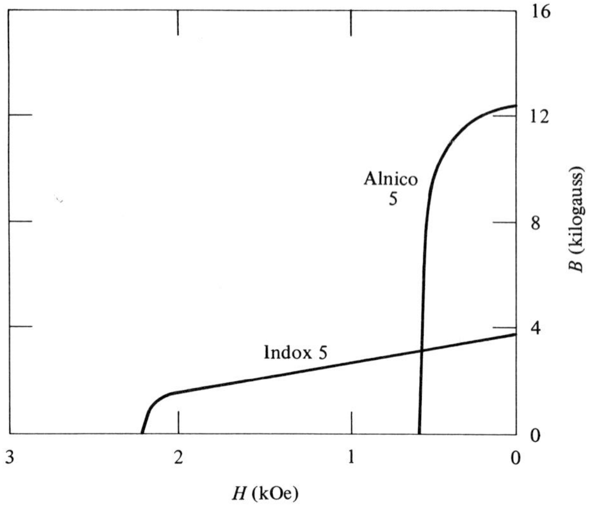

Fig. 14.18 Comparison of Alnico 5 and Indox 5 (barium ferrite).

its coercivity is large enough to resist the large demagnetizing fields associated with such shapes. Stated in another way, the permeance coefficient B_m/H_m at the $(BH)_{max}$ point is about 1 for barium ferrite, compared to about 20 for Alnico 5. In order to deliver the same flux in a gap, the cross-sectional area of a ferrite magnet has to be much larger than that of Alnico because the flux density B_m is so much smaller. The cost of barium ferrite is much less than that of Alnico, because of cheaper raw materials. The density of barium ferrite is 5.3 g/cm^3 compared to 8.7 for Alnico. Generally speaking, the two materials do not compete for the same application; each is used where its special characteristics best fit it for the job.

Barium ferrite owes its magnetic hardness to *crystal anisotropy*. If it were in the form of aligned, spherical, single-domain particles, its intrinsic coercivity should be

$$H_{ci} = H_K = \frac{2K}{M_s} = \frac{2(3.3 \times 10^6)}{380} = 17{,}000 \text{ Oe}. \tag{14.9}$$

(Particle interactions do not affect crystal anisotropy, so this value should apply both to an isolated particle and to a pressed and sintered compact with 100 percent density.) Actually, the particles are plate-shaped, which introduces shape anisotropy. And the shape is "wrong" with respect to best permanent-magnet properties, because the easy axis due to shape is at right angles to, rather than parallel to, the easy axis due to crystal anisotropy. It is difficult to allow for this effect exactly. For an isolated particle, the shape effect reduces H_{ci} by 4800 Oe to a value of 12,000 Oe, as mentioned in Section 11.5; for commercial magnets with 5 to 10 percent porosity, the reduction would not be nearly as much, although the pores are the source of internal demagnetizing fields; for a compact of 100 percent porosity, particle shape has no effect.

The observed values of H_{ci} given in Table 14.2 are one-third or less of the lower theoretical limit of 12,000 Oe and less than one-quarter of the upper theoretical limit of 17,000 Oe. It follows that barium ferrite magnets are not composed of single-domain particles reversing coherently. The typical grain size of 1μ is simply too large, inasmuch as we estimated (Problem 9.2) the critical size for single-domain behavior to be of the order of 1000 Å ($= 0.1\ \mu$).

Magnetization reversal in barium ferrite magnets must take place by wall nucleation and motion (Section 11.5). The coercivity could be increased by making the particles (1) smaller, and/or (2) smoother and with fewer crystal imperfections, in order to decrease the number of sites for wall nucleation. This may require a process other than ball milling to reduce the particle size. The work of Mee and Jeschke [14.12] is an indication of what can be done in this direction: by an undisclosed method of chemical precipitation they prepared barium ferrite plate-shaped particles about 0.1 μ thick; the coercivity of these particles in a dilute, nonaligned assembly was 93 percent of the theoretical value.

The maximum value of $(BH)_{max}$ for commercial magnets is about 3.5 MGOe. Becker [11.14] estimates that this could be raised to 5.0 MGOe if the particles were completely aligned and single-domain, even allowing for the porosity and impurities present in the commercial product.

14.6 SPECIAL ALLOYS

Most permanent-magnet materials are so extremely brittle that they can be put into usable form only by casting or by grinding into powder, pressing, and sintering. But there are a few exceptional alloys which are magnetically hard and yet ductile enough to be hot or cold worked into wire, sheet, and other forms. These are Cunife, Remalloy, Vicalloy, and CoPt. The first three have moderate coercivities, less than Alnico and barium ferrite but larger than the magnet steels, while CoPt has very high coercivity.

Cunife. This contains 60 percent Cu, 20 percent Ni, and 20 percent Fe, and it resembles Alnico in its metallurgy and heat treatment. When heated to 1000°C it consists of a single-phase, FCC, solid solution. It is then quenched and tempered at 650°C, whereupon it decomposes spinodally into two FCC phases, one rich in Fe and Ni and magnetic, the other rich in Cu and nonmagnetic. Its magnetic hardness is due to the shape anisotropy of the single-domain particles of the Fe-Ni-rich phase. It is often heavily cold worked after quenching and before tempering.

Remalloy. This Fe-Co-Mo alloy is also called Comol. It resembles some of the high-alloy magnet steels but contains less cobalt and no carbon; in fact, it was the first carbon-free permanent-magnet material. It is heated to 1200°C, quenched, and tempered at 650° to 700°C. Its magnetic hardness appears to be due to hindered wall motion associated with the precipitation of a second phase. It has only one major application, as a bias magnet in telephone receivers (Section 14.12), but large numbers are required.

Vicalloy. This is an Fe-Co-V alloy, which may be thought of as a modification of the magnetically *soft* alloy 2 V-Permendur (Section 13.5), the modification being an increase in the V content to 10 to 13 percent at the expense of the iron. The result is a ductile, magnetically hard alloy. The Fe-Co phase diagram has been given in Fig. 4.21 (a). The γ phase (FCC) is nonmagnetic, and the magnetic phase exists in two forms near the equiatomic composition FeCo: disordered α (BCC) and ordered α' (CsCl structure). The addition of a few percent V to the 50-50 Fe-Co alloy lowers the temperature of the $\alpha/(\alpha + \gamma)$ phase boundary to such an extent [G.4] that both α' and γ may become stable at room temperature, although the ternary phase diagram does not appear to have been fully worked out.

The heat treatment of Vicalloy 1 consists in quenching from 1200 C and tempering (aging) at 600°C for 8 hours. Vicalloy 2 is cast, hot worked to a convenient intermediate size, cold reduced by 90 percent or more, and aged at 600°C for 8 hours; it then has an easy axis in the direction of elongation during cold working. The origin of the magnetic hardness of these alloys is not clear. Presumably Vicalloy 1 is all γ after quenching, and subsequent aging causes the precipitation of ordered α'. In Vicalloy 2 the heavy cold work transforms any γ originally present to α, and aging then causes ordering to α'. The coercivity of both alloys is thought to be due to shape or crystal anisotropy of small particles. In Vicalloy 1 these particles are probably α' in a matrix of γ, and in Vicalloy 2 they may be ordered α' regions in a matrix of disordered α.

Cobalt platinum. The equiatomic alloy CoPt is the most expensive magnetic material in commercial production. Until quite recently, its high cost was justified by its high coercivity (4300 Oe), which suited it to applications like electric wrist watches and traveling-wave tubes for space satellites (Section 14.12). It is currently being displaced by the cheaper rare-earth alloys (Section 14.8), which have even higher coercivity. CoPt is a single-phase FCC solid solution above 820°C, and below this temperature it undergoes long-range ordering to a structure known as the CuAu-I type [4.9]. This structure is tetragonal, with the (002) planes, normal to the c axis, occupied alternately by Co and Pt atoms; the c/a ratio is about 0.97. Both the ordered and disordered phases are magnetic. CoPt magnets are made by disordering at 1000°C, cooling to room temperature at a controlled rate, and aging at 600°C for 5 hours to achieve a *partially ordered* structure. The extreme magnetic hardness is due to the high crystal anisotropy of the single-domain "particles" of ordered material dispersed in a disordered matrix.

14.7 IRON-POWDER MAGNETS

The composition and processing of most permanent-magnet materials evolved through trial and error without the guide of adequate theory. For example, field-treated Alnico with excellent properties had been made some ten years before publication of the classic papers of Stoner and Wohlfarth in 1948 on domain rotation [7.23] and of Williams and Shockley on domain wall motion [9.16], i.e., years before any one had any clear idea of how the magnetization of a body actually changed. In fact, far from theory guiding practice, it was the other way around, because it is fair to say that the *existence* of Alnico stimulated theorists to search for an explanation of its high coercivity.

By the late 1940's, however, there was sufficient understanding of fine-particle theory to allow magneticians to proceed, for the first time, with the development of a *synthetic* magnet material. What was required was an assembly of elongated single-domain particles of, for example, iron, dispersed in a suitable matrix. The particles themselves could have negligible crystal anisotropy, because their shape alone would lead to high coercivity (Table 9.2). It is not easy to make such particles. The initial development work was done in France but did not lead to a commercially successful material, chiefly because the particles produced were not sufficiently elongated. Finally, in 1955, Mendelsohn, Luborsky, and Paine [14.13] of the General Electric Company in the United States announced the successful development of fine-particle magnets. These are produced under the trade name of Lodex. They are also called ESD (elongated single-domain) magnets and two grades are so listed in Table 14.2. Luborsky [11.9] has reviewed the development of these materials, and Falk [14.14] has described production methods and applications as of 1966.

The production of any fine-powder magnet is complicated by the fact that fine metal powders are pyrophoric, i.e., they spontaneously burst into flame or explode when exposed to air. Great care must therefore be taken to keep the powder out of contact with air until it is pressed into a compact form.

The original Lodex was made of iron particles, but the iron was later alloyed with cobalt in order to increase B_r and H_c by increasing M_s. Magnets are made by the following operations, illustrated in Fig. 14.19:

1. Iron and cobalt are simultaneously electrodeposited from an aqueous solution of their salts on a liquid mercury cathode. The deposited particles are dendritic, as indicated in the sketch. (Plating conditions are critical. The mercury surface, for example, must be absolutely still, undisturbed by ripples caused by vibration of the apparatus. The larger the current density, the thicker are the particles.)

2. The slurry of Fe-Co particles and mercury is heated to a temperature of about 200°C. This treatment removes the branches of the dendrites and thickens the particles to their optimum diameter of 150 to 300 Å. Their shape at this point is shown in Fig. 11.5 (a).

3. A small amount of antimony is added to the slurry. (The antimony adsorbs on the particle surfaces and prevents their further growth when they are subjected, in later operations, to temperatures up to 400°C. The antimony also tends to disperse the particles so that they do not clump together.)

4. Lead is added to the slurry at an elevated temperature. (The lead dissolves in the mercury and protects the Fe-Co particles from oxidation when the mercury is subsequently removed. The lead is the matrix of the final magnet, and the amount of lead determines the packing fraction p, the volume fraction of magnetic material in the magnet.)

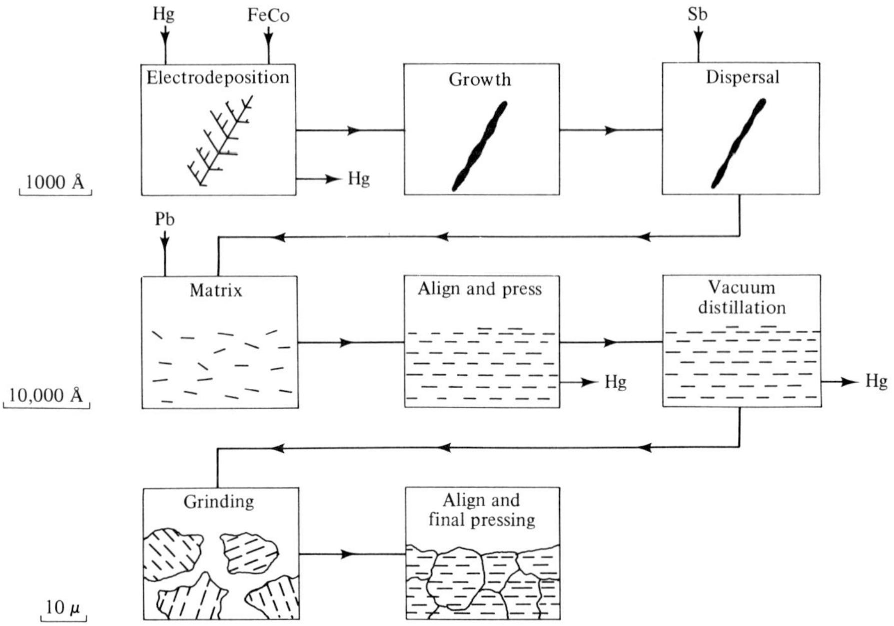

Fig. 14.19 Manufacture of Lodex fine-particle magnets. Falk [14.14].

5. The slurry is lightly pressed in a magnetic field to a solid ingot. (This step removes some of the mercury and locks the particles into the aligned state.)

6. The compact is heated under vacuum to remove the rest of the mercury.

7. The compact is demagnetized and ground into a coarse powder (−20 mesh). The Fe-Co particles within each particle of powder are still aligned with one another.

8. The powder is pressed into final form, with or without an aligning magnetic field, depending on whether an anisotropic or isotropic magnet is to be made. (The lead matrix gives excellent pressing characteristics, and virtually 100 percent density is easily achieved.)

The microstructure of a finished magnet is shown in Fig. 14.20; its similarity to the Alnico structure of Fig. 14.10 (b) is evident. The peanutlike shape of the Fe-Co particles in Lodex was discussed in Section 11.4; this shape causes their magnetization to reverse by fanning rather than curling. As a result the particles have a much smaller coercivity than they would have if their sides could be made smooth and parallel.

The packing fraction p in Lodex magnets can be easily varied, with the result that magnets with a graded set of properties can be produced at will. This is not possible with other materials. If p is increased, M_s and hence B_r also increase, but H_c decreases because of particle interactions, as we saw in Fig. 11.3. As a

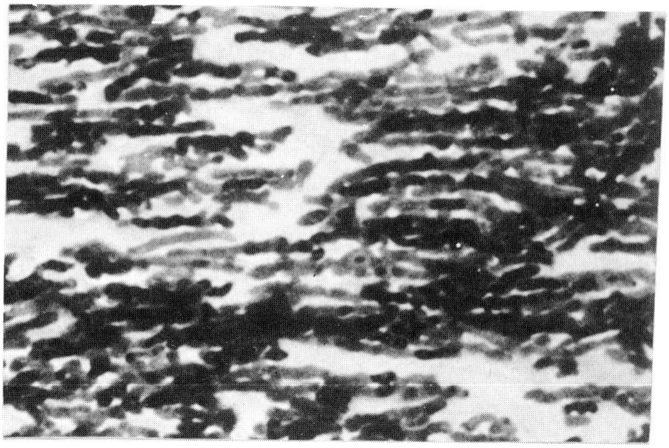

Fig. 14.20 Electron micrograph (extraction replica) of a Lodex magnet. Section parallel to the long axis of the aligned particles. (Courtesy of Fred E. Luborsky, Research and Development Center, General Electric Company.)

result, $(BH)_{max}$ increases with p only up to a certain point, and then it decreases. Table 14.3 shows these effects.

Although Lodex has a smaller value of $(BH)_{max}$ than Alnico or barium ferrite, it has particular advantages over these materials for certain applications. Pressing is the final operation; no sintering or finish grinding is needed. This means that small, intricately shaped magnets can be made to very stringent dimensional and magnetic specifications at low cost. Such magnets are used in devices like hearing aids, small motors, and electrical meters.

The large amount of effort expended on the development of Lodex has yielded a dividend beyond that of a new material useful for certain applications. Lodex is a synthetic analog to Alnico or, for that matter, to any other material owing its coercivity to the shape anisotropy of fine particles. Furthermore, in Lodex

Table 14.3 Magnetic Properties of Lodex Magnets as a Function of Packing Fraction. Parker and Studders [G.18]

	Isotropic			Anisotropic		
Packing fraction	B_r (gauss)	H_c (Oe)	$(BH)_{max}$ (MGOe)	B_r (gauss)	H_c (Oe)	$(BH)_{max}$ (MGOe)
0.25	3350	1140	1.2	4500	1325	2.6
0.30	4100	1050	1.4	5700	1250	3.1
0.35	4850	970	1.6	6900	1150	3.4
0.40	5620	820	1.6	7900	1000	3.5
0.45	6500	680	1.4	9000	850	3.6

important parameters such as particle diameter, length/diameter ratio, and packing fraction can be easily varied, whereas these parameters can be varied with great difficulty or not at all in Alnico and similar materials. Studies on the synthetic material have thrown much light on the relation between these structural parameters and the magnetic properties of the finished magnet and have thus led, indirectly, to a better understanding of Alnico itself. In addition, the experimental work aimed at understanding the magnetization-reversal mechanism in these electrodeposited particles stimulated a great deal of theoretical work, and the two together have greatly clarified the whole problem of magnetization reversal in small particles generally, whether these be particles in a synthetic mixture or precipitated particles in an alloy. Much of this work was described in Section 11.4.

14.8 RARE-EARTH AND OTHER ALLOYS

Whether called alloys or intermetallic compounds, combinations of metals that can be represented by simple chemical formulae such as A_xB_y are often magnetically hard. The ones described in this section are MnBi, MnAl, and certain rare-earth alloys. These materials owe their high coercivity mainly to high crystal anisotropy [11.14].

The equiatomic alloy MnBi has figured prominently in the literature of magnetism ever since 1943, when Guillaud in France reported on its very high crystal anisotropy ($K_1 = 9 \times 10^6$ ergs/cm^3). Although very high coercivities can be obtained with fine-particle MnBi, work on this material in bulk has been abandoned because of two properties which make it unsuitable for a practical magnet. It corrodes in damp air, and K_1 varies too rapidly with temperature. (This variation, incidentally, is abnormal: K_1 *decreases* as the temperature decreases.) Thin films of MnBi on a suitable substrate can be made stable in air, either by a natural oxide layer or by a thin coating of another oxide; these films may have applications in computer memories, as mentioned in Section 13.6.

In Mn-Al alloys near the equiatomic composition MnAl, a metastable phase can be formed by appropriate heat treatment. This phase has the composition $Mn_{1.11}Al_{0.89}$, a partially ordered body-centered tetragonal structure, and an anisotropy constant of about 10^7 ergs/cm^3. High coercivities can be obtained either by grinding a powder to small sizes or by deforming a cast bar by a special swaging technique. The magnetic hardness of this material appears to be due, not to the difficulty of magnetization rotation, but to interference with domain-wall motion by stacking faults (irregularities in the crystal structure) produced by the deformation. A swaged bar has a coercivity H_c of 2750 Oe and a $(BH)_{max}$ value of 3.5 MGOe.

But the materials which are exciting by far the greatest current interest are compounds of R (a rare-earth metal or yttrium) and M (a transition metal such as Mn, Fe, Co, or Ni) with compositions such as RM, RM_2, RM_5, and R_2M_{17}. Their properties have been reviewed by Becker [14.15]. The RCo_5 compounds are magnetically the most attractive, and of these $SmCo_5$ seems to offer the best promise as a practical magnet material.

Widespread interest in RCo compounds dates from 1966, when Hoffer and Strnat of the U.S. Air Force Materials Laboratory reported that YCo_5 had an anisotropy constant of 5.5×10^7 ergs/cm^3, by far the largest value for any material then known. Since then it has been found that SmCo has an even larger value, about 7.7×10^7 ergs/cm^3, which is more than 20 times that of barium ferrite. These and other properties of some RCo_5 compounds are given in Table 14.4, which was compiled by Strnat *et al* [14.16]. The value of K for YCo_5 was found from measurements on a single crystal, while K for the other compounds was estimated from magnetization curves made on samples of aligned powder particles. The extremely high values of $2K/M_s$ should be noted, because these are the upper limits for the intrinsic coercivities of well-aligned powders. The values of $(BH)_{max}$ are theoretical upper limits given by $4\pi^2 M_s^2$ (see Section 14.10). In the last two compounds, Y-MM and Ce-MM stand for yttrium-rich and cerium-rich misch-metal, respectively. *Mischmetal* (literally, mixed metal) is a solid solution of a number of rare earths; it is relatively cheap.

All the RCo_5 compounds have the hexagonal $CaCu_5$ structure. The c axis is the easy axis of magnetization at room temperature for most of these compounds, including all those listed in Table 14.4.

Table 14.4 Properties of RCo_5 Compounds [14.16]

Compound	K (10^7 ergs/cm^3)	$2K/M_s$ (k Oe)	T_c (°C)	$4\pi M_s$ (gauss)	Calculated $(BH)_{max}$ (MGOe)
YCo_5	5.5	129	630–700	10,600	28.1
$CeCo_5$	~7.2	~200	464	8,700	18.9
$PrCo_5$	~7.7	~167	639	11,200	31.3
$SmCo_5$	~7.7	~200	747	9,500	22.5
(Y-MM)Co_5	~5.4	~142	~700	9,500	22.5
(Ce-MM)Co_5	~6.0	~173	~500	8,500	18.

High coercivities have been observed in RCo_5 powders when the particles are much too large to be single domains. The current view is that magnetization reverses by wall motion and that the high coercivity is due to the difficulty of nucleating or unpinning domain walls, as discussed in Section 11.5.

Although many of the R-Co compounds are attractive as magnet materials, most of the development work has been concentrated on $SmCo_5$, because it has given consistently better results than other, quite similar compounds. The reasons for this are not clear.

The earliest $SmCo_5$ magnets [14.16] were made simply by pressing with a binder; $(BH)_{max}$ was 5.1 MGOe and the porosity must have been rather high.

Various investigators are now approaching the problem of making practical, high-density magnets in various ways. The main approaches and the resulting properties are listed below [14.15]:

1. Press fine particles in the usual way and sinter. One group has obtained $H_{ci} = 25$ kOe, $H_c = 8$ to 9 kOe, and $(BH)_{max} = 16$ to 20 MGOe. Another group has obtained $H_{ci} = 10$ to more than 30 kOe and $(BH)_{max}$ consistently exceeding 15 MGOe.

2. Press fine particles hydrostatically with simultaneous uniaxial compression in a special die. No sintering. $H_{ci} = 15.8$ kOe, $H_c = 8.4$ KOe, $(BH)_{max} =$ 18.5 MGOe.

3. Make a two-phase Sm-Co-Cu alloy by casting and annealing; one phase is rich in $SmCo_5$ and the other is rich in $SmCu_5$. The composition Co_3Cu_2Sm has $H_{ci} = 28$ kOe. But an increase in the cobalt content and the addition of some iron results in better overall properties. The composition $Co_{3.5}Fe_{0.4}Cu_{1.35}Sm$ has $H_{ci} = 6$ kOe, $H_c = 4$ kOe, $(BH)_{max} = 8.8$ kOe.

All these properties are impressive, and workers in this area are confident that still better properties can be obtained. Improvement is expected to come, not so much from making the particles smaller, but from making them more nearly perfect.

Finally, a word about the cost of rare-earth alloys is in order. The rare earths are not particularly rare, and the high prices of the pure metals shown in Table 13.2 are due, not to their scarcity, but to the cost of separating them from one another; they usually occur together in their ores and they are chemically very similar. It is for this reason that mischmetal is so cheap compared to the pure metals. The prices given in Table 13.2 are for 1-lb lots and become substantially less for larger quantities. In fact, one manufacturer states that cerium could be supplied at the rate of one million pounds per year for less than $5.00 per lb. All of which means that demand will fix the price; if a large market for RCo_5 magnets can be created, the cost of the rare earths involved will fall considerably. Even at present small-lot prices, $SmCo_5$ magnets are far cheaper than CoPt and are replacing CoPt in applications like travelling-wave tubes. Many other applications are currently being explored.

14.9 MAGNETIC RECORDING

The magnetic medium in recording tape, disks, and drums is not usually thought of as a permanent magnet, possibly because it is not a single magnet but an assembly of a huge number of microscopic magnets. Wire and thin metal tape were used in the earlier recorders, but since 1947 the magnetic material in virtually all recording tape, disks, and drums is a dispersion of fine γ-Fe_2O_3 particles in a plastic binder. That the vast magnetic recording industry (audio, video, and digital) rests on a firm foundation of what is essentially iron *rust* is a thought that must bemuse a corrosion engineer.

A book by Mee [14.17] gives the most thorough coverage of the magnetic problems involved in recording. Useful review papers have been written by Speliotis [14.18], Wohlfarth [14.19, 11.3, 11.4], and Bate and Alstad [14.33]. The magnetic processes that occur in tapes, disks, and drums are all essentially similar and are described below solely in terms of tape.

Magnetic tape is made in widths of $\frac{1}{4}$, $\frac{1}{2}$, and 1 inch. It consists of a plastic film 1 mil thick coated on one side with a suspension 0.2 mil (5μ) thick of γ-Fe_2O_3 particles in plastic. The packing fraction of the particles in the coating is about 0.4. The crystal and spin structure of γ-Fe_2O_3 have already been described (Section 6.6); it is a ferrimagnetic spinel with $\sigma_s = 76$ emu/g ($M_s = 390$ emu/cm^3). The oxide particles are elongated single-domain particles, about 0.1μ in diameter and with a length/diameter ratio of about 6. They owe their coercivity, about 250 to 300 Oe, to shape anisotropy. Since this value is only about a tenth of that expected for coherent rotation ($2\pi M_s = 2450$ Oe), the magnetization of these particles must reverse incoherently. The particles in the tape coating are aligned by a field during the coating process so that their long axes tend to be parallel with the length of the tape; this is done in order to make the remanence, after magnetization in that direction, as large as possible. (It may seem curious that, of all the permanent-magnet materials available for use in recording tape, γ-Fe_2O_3 was chosen instead. There are at least two reasons for this choice. Most permanent-magnet materials have coercivities which are too *large*: values greater than about 700 Oe are undesirable in recording. In addition, γ-Fe_2O_3 is chemically very stable, inasmuch as it is the highest oxide of iron, and very cheap.)

The elongated oxide particles are made from hydrated ferric oxide (α-FeO · OH) particles, which are already elongated, in the following steps: (1) dehydration to hematite (α-Fe_2O_3), (2) reduction by hydrogen at 400°C to magnetite (Fe_3O_4), and (3) careful oxidation at about 250°C to γ-Fe_2O_3.

To appreciate the particular magnetic properties that are important in tape requires some knowledge of the recording and reproducing (playback) processes. Figure 14.21 (a) shows the tape transport system of a recorder. No matter how much tape is on reel A, it is moved at constant linear velocity by the drive E, consisting of a rotating capstan and a pinch roller. Between A and E the tape is made to contact the three transducers (heads) B, C, and D. (Less elaborate recorders may have only a single head.) B is an erase head which demagnetizes the coating, C is a recording head which magnetizes the coating in a direction more or less parallel to the length of the tape, and D is a playback head which senses the degree of magnetization of the coating. The enlargement of the recording head in (b) shows that it is an electromagnet with a tiny air gap; the gap width is typically 60 microinches (1.5μ) or only about a third of the thickness of the tape coating. (The electromagnet is usually made of a stack of thin Permalloy sheets, but solid heads of a soft ferrite are also used.) The further enlargement in (c) shows the tape in the vicinity of the gap; it is drawn out of proportion for clarity.

The recording and playback of digital information is the easiest to understand, because regions of the tape are simply magnetized in one direction or the other to the same extent. To record such information, current pulses are fed

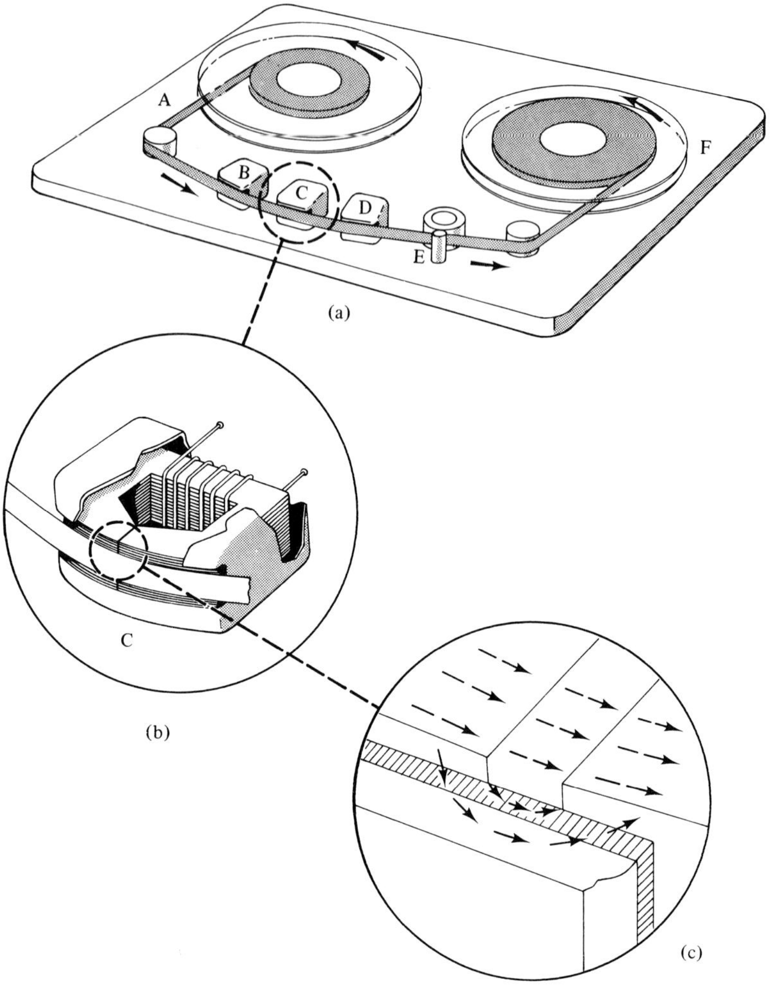

Fig. 14.21 Tape transport in a tape recorder. After Mee [14.17].

to the winding on the recording-head core. Each pulse magnetizes the core, thus creating a field in the air gap. As the moving tape passes the gap, the fringing field of the gap magnetizes the tape. Not all this magnetization is retained, as this particular section of tape leaves the gap area, but a certain *remanent* portion is retained and this portion constitutes the recording (Fig. 14.22a). Reading the tape depends on the ability of the playback head to sense the exterior field created by the magnetization of each section of the tape. When a 1 is below the gap, as in (b), the field at the tape surface prefers the low-reluctance path through the high-permeability core to the air-gap path, and the flux in the core is counterclockwise, as indicated. When a 0 is below the gap, as in (c), the flux is reversed.

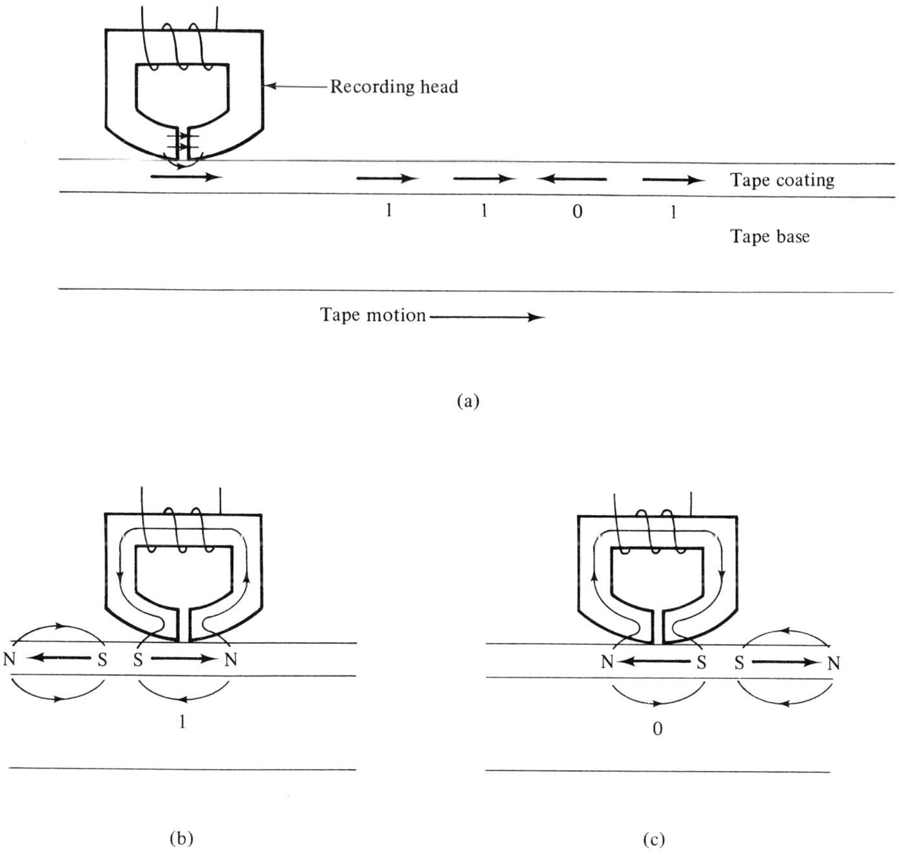

Fig. 14.22 (a) Recording digital information, such as 1101. Heavy vectors show magnetization direction. (b) and (c) Reading digital information. Not to scale. After Ragosine [14.20].

Therefore, as the tape moves past the gap, the transition from a 1 to a 0 will reverse the flux in the playback core and induce a voltage pulse in its winding.

The main problem in digital recording is to pack as much information into each inch of tape as possible. Densities up to 1600 bits/inch have been obtained on one track, and there are several parallel tracks running lengthwise on the tape. High density requires high remanence, high coercivity, and, especially, small coating thickness [14.18].

Digital information is not hard to record, because only two levels of magnetization are involved. The recording of speech and music introduces new problems. A microphone converts sound waves into a varying electric current which is

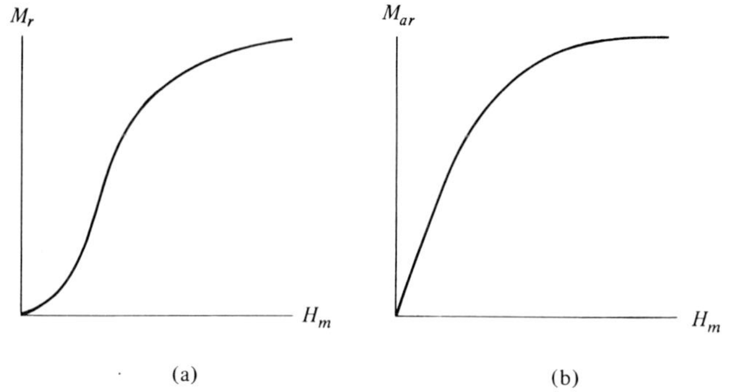

Fig. 14.23 Remanence curves: (a) normal, (b) anhysteretic.

amplified and fed to the winding on the recording head. The tape leaving this head then has impressed on it an intensity of magnetization which varies *continuously* along the tape, at a rate depending on the frequency of the recorded sound. When this tape passes in front of the playback head, a varying voltage is induced in the head winding. This voltage is then amplified and fed to a loudspeaker. The extent to which the sound coming from the loudspeaker resembles the sound entering the microphone determines the fidelity of the whole system, and the tape is one link in the fidelity, or lack of it, of the system.

The prime requirement for high fidelity is linearity of response. If the sound entering the microphone doubles in volume, the voltage output on playback should double in volume. As far as the tape is concerned, this means that the remanent magnetization should be proportional to the maximum field H_m to which the tape has been exposed in the gap. (It is the remanent magnetization M_r which is all-important in tape recording, and not the actual magnetization M when H_m is acting, because the external field of the tape, which actuates the playback head, is proportional to M_r.) However, all magnetic materials have a nonlinear relation between M_r and H_m, as indicated in Fig. 14.23 (a); in fact, such curves usually resemble magnetization (M,H) curves. Under these circumstances the output sound of the recorder is distorted.

However, it is found that this distortion can be almost eliminated by recording with "ac bias." This means that an alternating voltage is applied to the recording-head winding simultaneously with the input-signal voltage (the voltage from the recording microphone). This ac voltage has a frequency and amplitude much higher than the input signal. As a result the field in the gap varies with time as sketched in Fig. 14.24. When a particular section of tape is in the gap for a brief period, it finds itself subjected to an alternating field of amplitude sufficient to saturate it in either direction and with a mean value proportional to the input-signal voltage at that time. As that section of tape leaves the gap, typically at a speed of 7.5 in./sec, the amplitude of the rapidly varying field decreases to zero

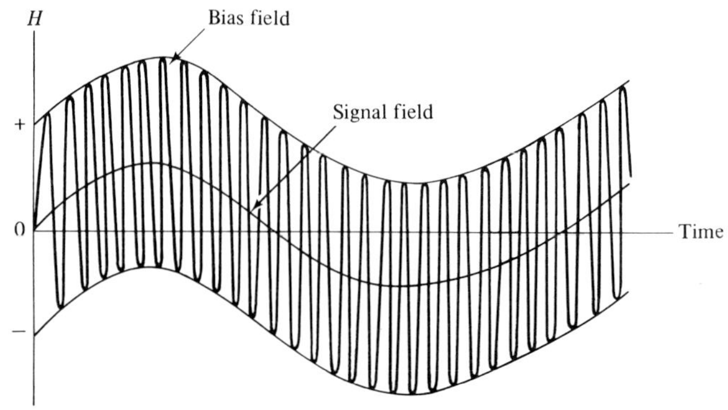

Fig. 14.24 Variation of gap field with time when recording with a,c, bias.

and so does its mean value, the signal field. In short, the tape is subject to conditions closely resembling those of the anhysteretic magnetization described in Section 9.16. There is one difference. The anhysteretic magnetization M_1 of Fig. 9.47 is the value attained when the steady field H_1 remains on. But when a particular section of recording tape has moved far from the gap, that section is subject to no applied field. This corresponds, in Fig. 9.47, to the removal of the *steady* field H_1; the magnetization remaining is called the anhysteretic *remanent* magnetization M_{ar}; it is somewhat smaller than M_1. (The steady field H_1 of Fig. 9.47 is analogous to the signal field of Fig. 14.24; the frequency of the bias field, typically 100 kHz for audio recording, is so much higher than the frequency of the signal field that the latter may be regarded as roughly constant during many alternations of the bias field.) The all-important property of the tape coating is then the shape of its M_{ar},H_m curve. This curve is shown in Fig. 14.23 (b). It resembles the anhysteretic curve of Fig. 9.47 but is somewhat lower. It is also practically linear up to about half-saturation. If the signal field does not go beyond this linear portion, essentially distortion-free recording will result. The initial slope s of the M_{ar},H_m curve is a special kind of susceptibility that measures the sensitivity of the tape coating; the larger s, the larger is the output signal induced in the playback-head winding, under constant recording conditions.

The measurements required to evaluate magnetic particles for tapes are fully described by Mee [14.17]. These are of two kinds:

1. *Measurements on the powder alone.* The powder sample is packed tightly into a small glass tube, and the hysteresis loop is measured by standard techniques. The properties required are the saturation magnetization M_s and the coercivity H_{ci}.

2. *Measurements on finished tapes.* To increase the amount of magnetic material, the specimen is a bundle of short parallel sections cut from the tape. The

properties of most interest are the anhysteretic remanence curve and the ordinary remanence ratio M_r/M_s. This quantity is a measure of the degree of alignment of the particles in the tape coating. For perfectly aligned single-domain particles, M_r/M_s is 1; values of about 0.6 to 0.7 are typical of actual tapes. (The presence of low-coercivity, multidomain particles would also reduce M_r/M_s below 1.)

The anhysteretic remanence curve is strongly dependent on *particle interactions.* This knotty problem was described in Section 11.3, and it is the central magnetic problem in tape recording. If there were no interactions between particles, the initial slope s of the anhysteretic remanence curve would be infinite, because a vanishingly small steady applied field in a particular direction would be enough to turn the M_s vectors of all the particles toward that direction as the large-amplitude ac field is reduced toward zero. But a particular particle in the tape does not feel only the applied field; it feels this field plus the vector sum of the external field of all the particles in the tape. These interaction fields reduce s to a low value, typically about 2 (in units of emu/cm^3 of tape coating per oersted). The value of s depends, in ways not completely understood, on the *assembly properties* of the coating, i.e., on packing fraction, on degree of particle alignment, and on how uniformly the particles are dispersed in the plastic matrix.

The dominant position of γ-Fe_2O_3 particles in the magnetic tape industry is now being challenged by CrO_2 particles. This material is one of nature's rarities, a ferromagnetic oxide (Section 4.5). It has σ_s equal to 90 to 100 emu/g, depending on how it is made, and fine particles can be prepared with greater elongations and smoother sides than γ-Fe_2O_3. Its coercivity is therefore higher, of the order of 400 Oe.

This abbreviated account of magnetic recording only skims the surface of an intricate subject involving very fundamental magnetic problems. The interested reader can gain entry to the literature of this field through the references cited earlier.

14.10 SUMMARY OF MAGNETICALLY HARD MATERIALS

It may be useful at this point to look back over all the magnetically hard materials and attempt to classify them into broad groups and make some generalizations.

The materials we have examined are classified in Table 14.5 according to the mechanism responsible for their magnetic hardness.

a) *Fine particles with shape anisotropy.* These particles are all small enough to be single domains.

b) *Fine particles with crystal anisotropy.* Of the four materials listed only one (CoPt) is composed of single-domain particles. The others owe their magnetic hardness to the difficulty of nucleating domains. These materials are made in small-particle form simply to lower the probability of having wall-nucleating defects on the particle surfaces.

Table 14.5 Classification of Magnetically Hard Materials

Magnetic hardness due to

- Fine particles having
  - Shape anisotropy
    1) **Alnico**
    2) **Lodex**
    3) **γ-Fe_2O_3**
    4) **Cunife**
    5) **Vicalloy ?**
  - Crystal anisotropy
    1) **Barium ferrite**
    2) **CoPt**
    3) **MnBi**
    4) **RCo_5**
- Mechanical hardness
  1) **Magnet steels**
  2) **Remalloy**
  3) **MnAl**

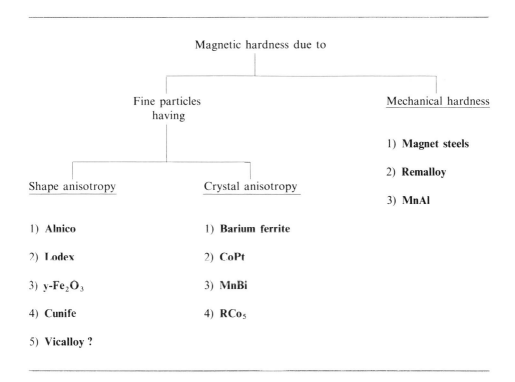

c) *Mechanical hardness.* In these the magnetization changes by wall motion, which is strongly hindered by precipitates, strain, and crystal imperfections, all of which also cause the material to be mechanically hard.

Table 14.5 shows that all the important magnets are "fine-particle magnets" of one sort or another. These could also be divided another way into two groups, depending on whether the fine particles are made (a) "naturally," by precipitation out of a matrix (as in Alnico) or (b) "artificially," by grinding and pressing powders (barium ferrite) or mixing particles and matrix (Lodex).

Shape and crystal anisotropy may be compared as follows:

1. *Shape anisotropy.* The intrinsic coercivity is

$$H_{ci} = (N_a - N_c)M_s, \tag{9.49}$$

directly proportional to M_s. Increasing M_s has two desirable effects, an increase in H_{ci} and an increase in the remanence B_r. But the magnetic particles must of necessity be diluted with a nonmagnetic phase, or the whole shape-anisotropy effect disappears. Then the value of B_r for the *magnet* becomes typically 50 percent or less of the value of B_r for the magnetic phase.

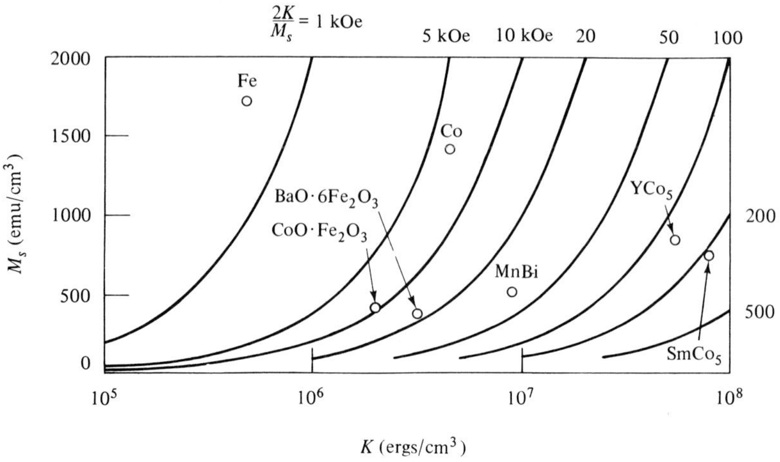

Fig. 14.25 Values of K, M_s, and $2K/M_s$ for several materials.

2. *Crystal anisotropy.* The coercivity is now

$$H_{ci} = 2K/M_s. \qquad (9.51)$$

No dilution is involved; in fact, the less porosity the better; but an increase in M_s alone decreases H_{ci}. It is therefore necessary to search for materials which combine large K with a reasonably large value of M_s. It is not enough to look simply for large values of $2K/M_s$. (For example, the rare-earth orthoferrites described in Section 13.6 have $2K/M_s$ values of the order of 10^5 Oe, larger than that of MnBi. But they would be useless as permanent magnets because M_s is only about 8 emu/cm^3). Figure 14.25 shows lines of constant $2K/M_s$ on a plot of K versus M_s.

In the search for better materials, many investigators feel that crystal anisotropy holds far more promise than shape anisotropy, despite the present dominance of Alnico. Unfortunately, the search for high-K materials in hampered by the almost total lack of an adequate theory of crystal anisotropy. The investigator is aided only by the empirical knowledge that large values of K are to be found, not in cubic substances, but in crystallographically uniaxial substances (hexagonal, tetragonal, etc.).

Becker [11.14] has suggested that there are two main types of materials, characterized by the shapes of their hysteresis loops, which can in turn be described by the relative magnitudes of B_r and the two kinds of coercivity:

1. $H_{ci} \approx H_c \ll B_r$. Figure 14.26 (a). Example: Alnico.
2. $H_{ci} \gg H_c \approx B_r$. Figure 14.26 (b). Example: barium ferrite and RCo_5.

The Type 1 material illustrated has $H_{ci} = H_c = 700$ Oe. The $(BH)_{max}$ value is 5 MGOe, occurring at $B_d = 10{,}000$ gauss and $H_d = 500$ Oe. The Type 2 material has a square hysteresis loop because of crystal anisotropy, and H_{ci} is 10,000 Oe.

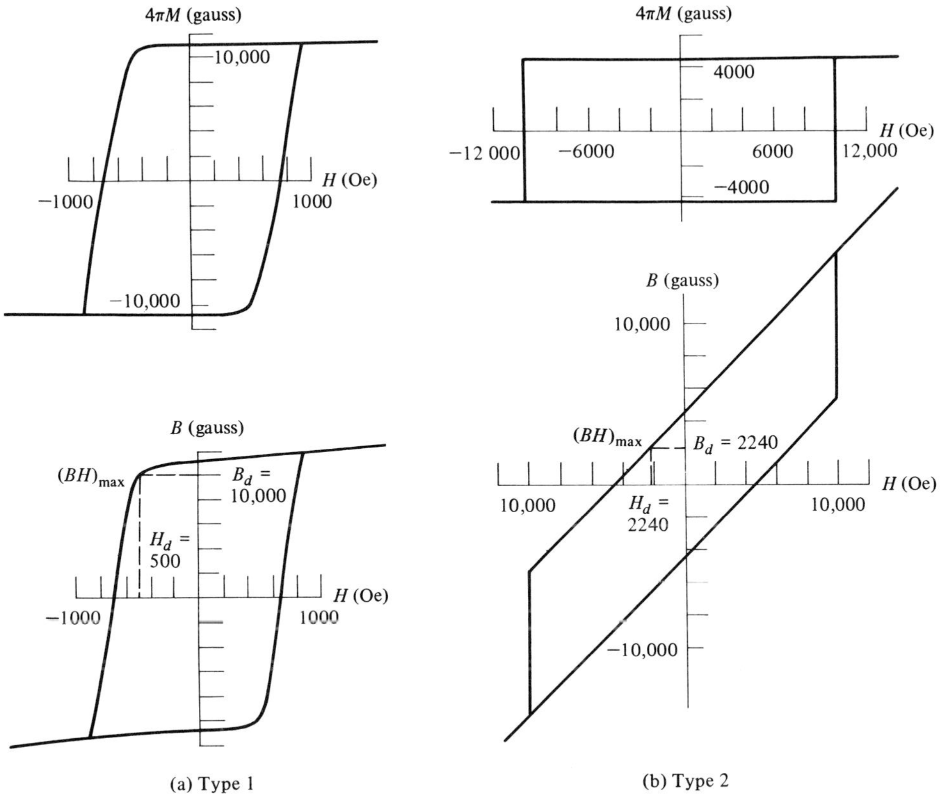

Fig. 14.26 Schematic $4\pi M,H$ and B,H hysteresis loops for two types of magnetic material. Becker [11.14].

Because M in the Type 2 material is constant and equal to $+M_s$ for all fields between plus and minus 10 kOe, the equation for the B,H curve is that of a straight line:

$$B = H + 4\pi M_s. \tag{14.10}$$

Therefore, $H_c = B_r = 4\pi M_s$ and

$$(BH)_{\max} = \left(\frac{H_c}{2}\right)\left(\frac{B_r}{2}\right) = 4\pi^2 M_s^2. \tag{14.11}$$

The value of M_s has been adjusted in Fig. 14.26 (b) to make $(BH)_{\max}$ equal to 5 MGOe, as in (a). The value of H_c for Type 2 is 4480 Oe, and the $(BH)_{\max}$ point occurs at $B_d = H_d = 2240$ Oe.

These two materials require quite different designs. To work at its $(BH)_{\max}$ point, Type 1 has to be in the shape of a rod with a length/diameter ratio of about 5/1. Type 2 has to be in the form of a disk with length (thickness)/diameter ratio of about 1/2. Type 1 delivers 10,000/2240 or 4.5 times the flux density of Type 2, but Type 2 resists demagnetizing fields much better. For example, a 700-Oe field would demagnetize Type 1, but a reverse field of up to 10,000 Oe could be applied to Type 2 and, on its removal, it would return to the $(BH)_{max}$ point, because the reverse field did not change M. As we will see later, considerations like these are important in applications like permanent-magnet motors.

14.11 MAGNET STABILITY

If a permanent magnet is to live up to its name, it must be stable: it must deliver the same flux in the air gap at all times. This condition is not difficult to meet if the magnet operates in a controlled environment. More often the magnet is exposed to such disturbing influences as external magnetic fields and temperature changes. We will examine these effects below. They are discussed in detail by Gould [14.34].

External fields

Suppose that the magnet of Fig. 14.27 (a) has a load line such that it operates at point a. What is the result of external fields that may increase or decrease the field acting on the magnet by an amount ΔH? If the true field changes from H_d to H_1 and back again, the operating point moves from a to b and back to a along the recoil line defined in Fig. 14.6 (b). The change in induction B during this reversible excursion is small, and the magnet is back at its original operating point. But if the true field changes from H_d to H_2 and back, the operating point moves from a to c to d, causing a large permanent drop in induction. However, once the magnet is at d, temporary external fields in either direction merely move it reversibly back and forth along the recoil line through d. The resulting changes in B are small and temporary.

This suggests a method for *stabilizing* a magnet against the effects of external fields. After magnetization the magnet is subjected to a temporary demagne-

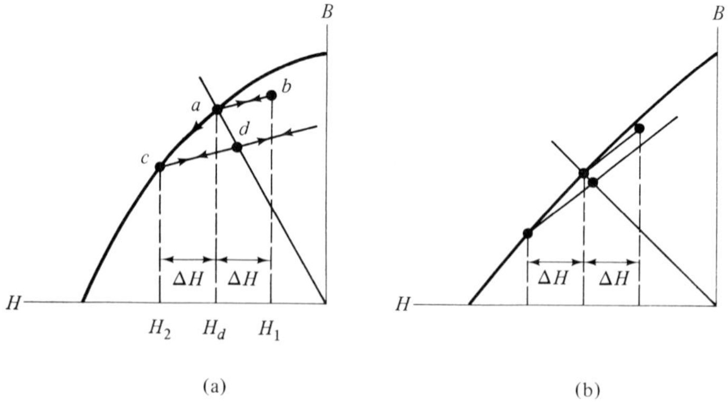

Fig. 14.27 Stabilization of a magnet.

tizing field that moves its operating point from a to d. This treatment, which is quite common, increases stability at the cost of decreased induction and is called "knockdown." An ac field is usually preferred over dc for stabilization, because it will cycle the material many times over the recoil line.

A person ignorant of these effects can innocently degrade a magnetic circuit merely by removing the magnet from the circuit and replacing it. In the circuit, which consists of the magnet and pieces of soft iron, the magnet is normally exposed to a smaller demagnetizing field than it is if removed. If an unstabilized magnet is operating at point a in the circuit, then removal would shift it to c, if the change in true field were ΔH, and replacing it would shift it to d.

A material with a demagnetization curve like that in Fig. 14.27 (b) is inherently more stable than the material in (a), because its recoil line has a slope almost as large as the demagnetization curve. Temporary external fields therefore cause only a small permanent flux change in an unstabilized magnet. [The idealized material of Fig. 14.26 (b) would suffer *no* permanent change, because its demagnetization curve and recoil line coincide.]

Temperature changes

These can produce three kinds of effects. One is reversible, and the other two irreversible, in special senses of this word. The property of interest is the flux B_d at the operating point, because this determines the flux in the air gap.

Reversible changes

If B_d returns to its original value after a temperature cycle, as in Fig. 14.28 (a), the change is called reversible. Reversible changes occur because the relevant structure-insensitive properties M_s and K are both temperature dependent. Both normally decrease as the temperature increases, but $2K/M_s$ can vary in peculiar ways. In a shape-anisotropy material like Alnico, the coercivity decreases as the temperature increases, because M_s decreases. The demagnetization curve then shifts toward the origin, as in (b), leading to a decrease in B_d. In barium ferrite, a crystal-anisotropy material, the coercivity depends on $2K/M_s$ and increases with temperature from quite low temperatures to about 300°C and then decreases rapidly to zero at the Curie point (450°C). The value of B_d decreases as the temperature increases above room temperature.

In Alnico the temperature coefficient of B_d is quite small, of the order of -0.01 to -0.02 percent/°C, and it depends on the slope of the load line. In barium ferrite this coefficient is some ten times as large, about -0.2 percent/°C, and independent of the load line.

Irreversible changes

If B_d does not return to its original value after a temperature cycle, as in Fig. 14.28 (c), the change is called irreversible; remagnetization is required to restore the original value. This irreversible change is due to the thermal fluctuation aftereffect described in Section 12.5. The demagnetizing field of the magnet, aided by fluctuations

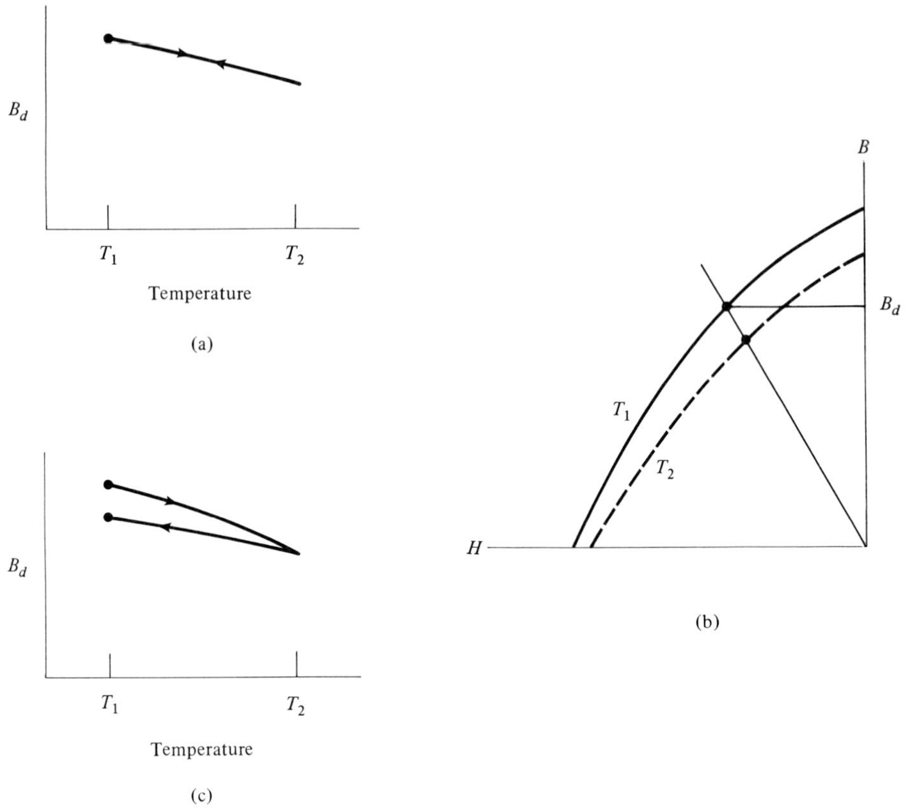

Fig. 14.28 Some effects of temperature changes on the induction B_d at the operating point.

of thermal energy, causes a small, slow decrease in magnetization, even at constant temperature, by irreversible rotations or wall movements. If the temperature is changing, the effect is integrated over the temperature cycle but is complicated by the change of coercivity with temperature. The irreversible change in B_d becomes less for each successive temperature cycle; in practice, a magnet can be stabilized against such changes by cycling it three times over the temperature range involved.

Because certain critical applications require exceptional stability over long periods, Kronenberg and Bohlmann [14.21] measured the value of B_d for Alnico and barium ferrite magnets at 24°C for times up to 10,000 hours (about 1 year) after magnetization. They found that B_d decreased logarithmically with time in accordance with Eq. (12.24). For various grades of material and various length/diameter ratios, they observed decreases in B_d in 10,000 hours of 0 to 0.4 percent for barium ferrite and 0.05 to 3.0 percent for Alnico. (The rather large changes in Alnico shown in Fig. 12.21 are not relevant here. They refer to changes observed immediately after a large change in *applied* field.) The logarithmic time dependence of this effect means (1) that stability improves with time after demagnetization (half of the total change in 10,000 hours occurs in the first 100 hours), and (2) that the measured data can be extrapolated to fairly long times (20 years) with reasonable confidence.

This isothermal effect can be made negligible by stabilizing the magnet with a reverse field, as explained above. It was found [14.21] that an 8 percent reduction in B_d (8 percent knockdown) caused "complete" stability. More sensitive measurements by Zingery *et al.* [14.22] show that a properly stabilized Alnico 6 magnet produces a gap flux which is constant to within $\pm$ 5 ppm over a period of one year.

Structural changes

If the temperature change is sufficient to cause permanent structural changes in the material, such as precipitation of a new phase or a change in the composition of existing phases, then another kind of irreversible change will result. Not only will B_d be different after the temperature cycle, but the original value cannot be restored simply by remagnetization. The material must be re-heat-treated to its original condition and then remagnetized. Years ago this kind of instability was a serious problem in magnet steels, chiefly because of improper heat treatment. It is not a problem with modern materials.

14.12 APPLICATIONS

The applications of magnetically hard materials in recording have already been dealt with in Section 14.9. Other applications can be classified in various ways, but Tyack's [14.23] system is perhaps the most straightforward: (1) electrical-to-mechanical energy conversion, (2) mechanical-to-electrical energy conversion, (3) electron control, and (4) force applications. Figure 14.29 suggests the wide variety of shapes and sizes in which permanent magnets are made.

Electrical-to-mechanical

If a current flows in a conductor located in the magnetic field of a permanent magnet, a force will be exerted on that conductor. This is the operating principle of such devices as the loudspeaker, permanent-magnet motor, and moving-coil electrical instruments.

Loudspeakers. These form the largest single application of permanent magnets. One design suitable for an Alnico magnet is shown in Fig. 14.30. The base of the speaker cone carries a coil which fits in the annular gap of the magnet circuit. Flux travels from the magnet into the iron pole piece, radially across the gap, and back through the outer return path of iron. A varying current in the coil causes a varying axial force on the coil, which therefore vibrates axially, as does the cone attached to it, thus generating sound. An efficient design involving a barium-ferrite magnet would look quite different. The cross section of the magnet would have to be much larger, to allow for the smaller B_d of the material, but its thickness in the axial direction could be much less, because its H_c is larger

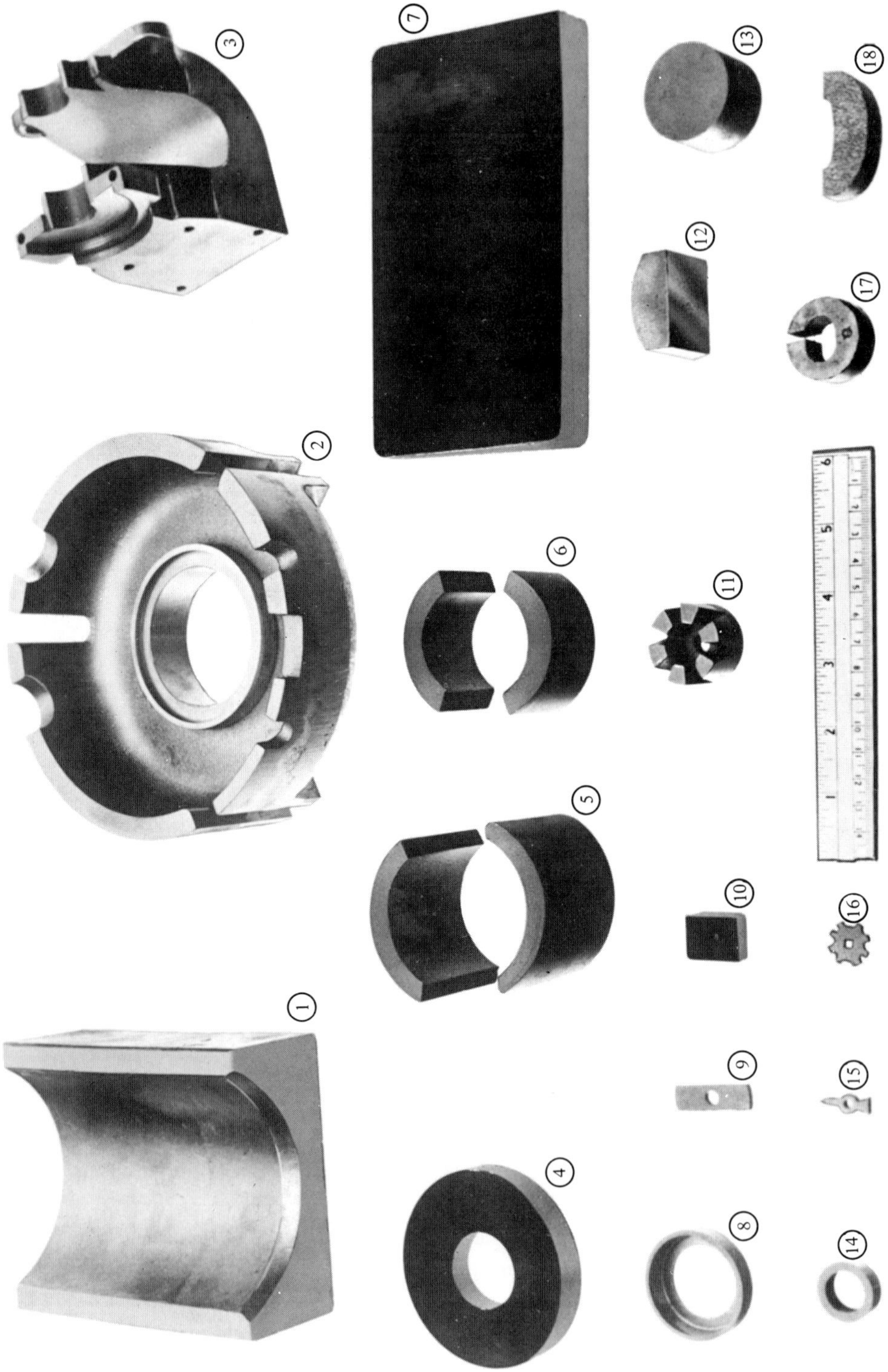
1
2
3
4
5
6
7
8
9
10
11
12
13
14
15
16
17
18

Fig. 14.29 Miscellaneous permanent magnets. 1. Alnico magnet used in linear actuators for disk drives (peripheral computer equipment). 2 and 3. Alnico magnets for magnetron tubes. 4. Ceramic loudspeaker magnet. 5 and 6. Ceramic magnets for small electric motors. 7. Ceramic magnet used for beneficiation of taconite iron ore. 8. Alnico magnet used in telephone receivers. 9. Cunife magnet used in automobile speedometer. 10. Ceramic magnet for a cabinet latch. 11. Alnico magnet used in a drive. 12. Alnico magnet used in a flywheel magneto. 13. Alnico loudspeaker magnet. 14. Alnico magnet used in a generator. 15. Sintered Alnico compass magnet. 16. Cunife magnet used in an appliance timing motor. 17. Alnico magnet used in a watt-hour meter. 18. Alnico magnet used in a panel meter. (Courtesy of Indiana General.)

than that of Alnico. Actually, barium-ferrite speaker magnets are axially magnetized rings rather than solid cylinders (Fig. 14.29, item 4).

Permanent-magnet motors. The rotor (armature) of an electric motor is wound with copper wire carrying a current. The turns of the winding are at the same time located in a magnetic field. The result is a force on the conductors and

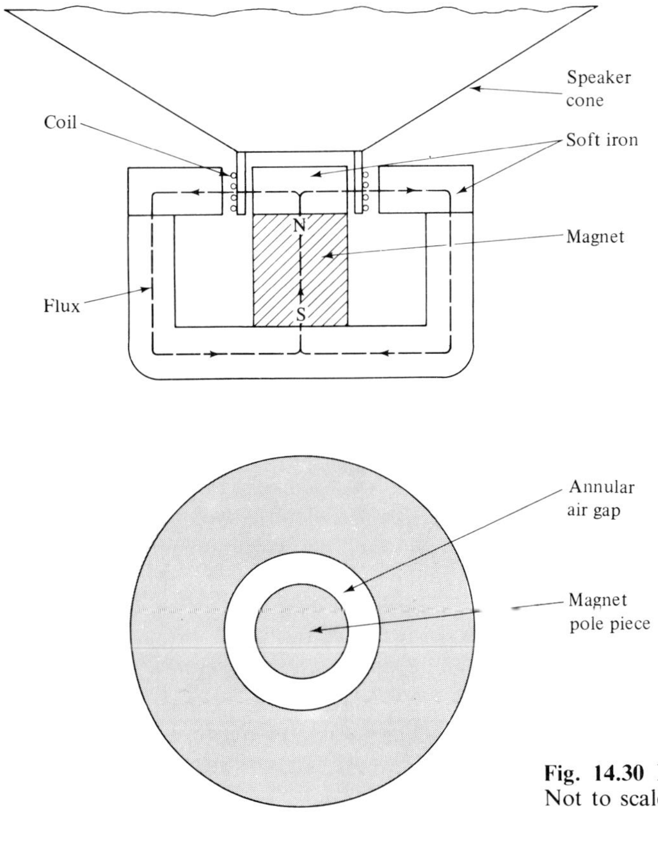

Fig. 14.30 Loudspeaker construction. Not to scale.

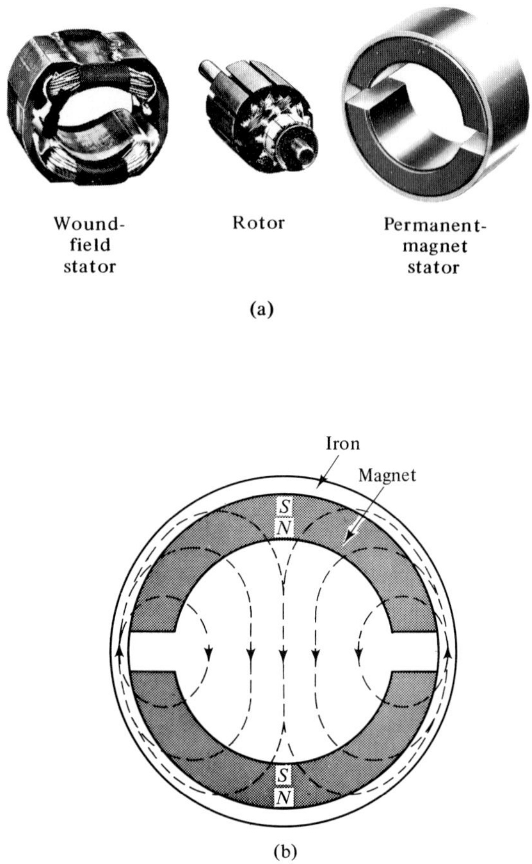

Fig. 14.31 Permanent-magnet motor. (a) Permanent-magnet stator at right, instead of wound-field stator at left. (b) Section through permanent magnet stator. Dashed lines indicate flux. (Courtesy of Indiana General.)

rotation of the rotor. The necessary magnetic field is supplied, in most motors, by an electromagnet; this is the stator, an iron (i.e., low-carbon steel or silicon steel) core wound with another winding. For small dc motors, there has been an increasing preference for a permanent magnet as a field source rather than an electromagnet. This means replacement of the wound-field stator at the left of Fig. 14.31 (a) by the permanent-magnet stator at the right. The result is a cheaper, more compact, and lighter motor. Such motors have many applications in the automobile (windshield-wipers, heater fans, etc.) and in the home (electric carving knives, electric toothbrushes, etc.). The magnetic circuit is shown in (b). The preferred material is barium ferrite, because the fairly thin sections involved can be magnetized in the thickness direction.

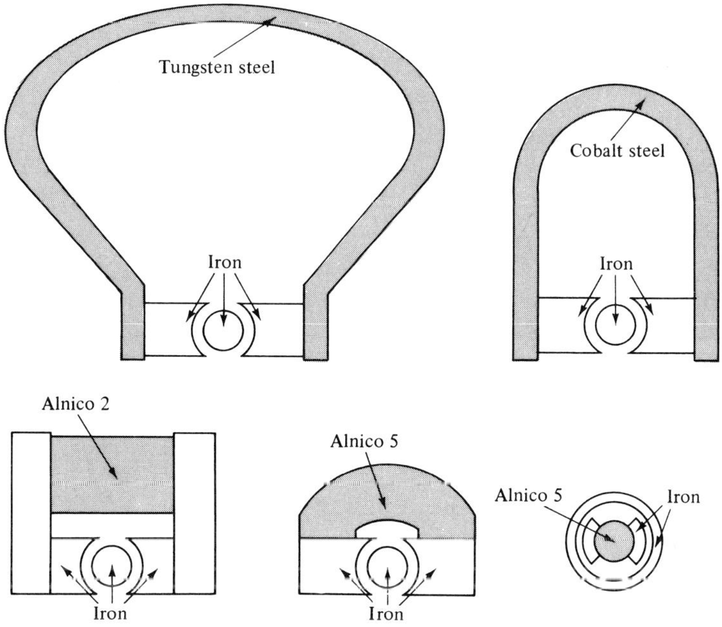

Fig. 14.32 Development of the moving-coil instrument magnet. After Clegg [14.1].

Ireland [14.24, 14.25] has pointed out that $(BH)_{max}$, while an adequate index of quality for a static application like a loudspeaker, is not the correct index for a motor magnet. The magnet in a motor is exposed not only to its own demagnetizing field but also to the field due to current in the rotor winding, and the strength and direction of this armature field vary with the rotational position of the armature. Ireland shows how these effects lead to a different figure of merit, based on the M,H curve of the material.

Moving-coil instruments, such as galvanometers, ammeters, voltmeters, etc., operate under conditions of static flux, like loudspeakers, but magnet stability with respect to time and temperature is a much more critical requirement. How the size of the magnet, and therefore the size of the instrument, is controlled by the material available is shown in Fig. 14.32. Tungsten steel had to be used in the earliest instruments, and its coercivity was so low that it had to be made very long or it would practically demagnetize itself. As materials improved, magnets could be made shorter and shorter, but the central iron core was still retained, as in the galvanometer of Fig. 2.11. In the last design of Fig. 14.32, the horseshoe magnetic circuit is replaced by a hollow iron cylinder and a permanent-magnet core. The magnet is magnetized along a diameter, and the iron ring acts both as a return path for the flux and as a shield against exterior fields.

Telephone receiver. This has undergone a similar shrinkage in size over the years, as magnets have improved. A thin diaphragm of 2V-Permendur is attracted by a Remalloy or Alnico magnet. The bias field due to this magnet is modified by a varying field from the coils carrying the incoming voice current; the resultant field causes the diaphragm to vibrate and emit sound.

Mechanical-to-electrical

If the flux through a winding is changed, an emf will be induced in that winding. The flux change can be effected by any mechanically produced relative motion of a magnet and the winding. This principle is the basis for devices like the magneto and the microphone.

The *magneto* consists of a magnet rotating within an iron core carrying a winding. It is used to produce the sparks for ignition in small gasoline engines, such as those used in power lawn mowers and outboard motors.

The magnetic *microphone* is simply the inverse of the loudspeaker. It is rarely used, except in some intercom systems in which the loudspeaker also serves as the microphone.

Electron control

A moving charge like an electron experiences a force when it moves in a magnetic field. A magnet or an array of magnets can provide this field, and proper design of the magnetic circuit can result in close control of the electron's path. This kind of control is required in many kinds of electrical equipment; the most important applications are in microwave equipment and television receivers.

Microwave equipment. Included here is equipment both for communications and radar. The devices involved bear such names as magnetron, travelling wave tube, and crossed-field amplifier, and the details of their operation are matters for a specialist. Harrold and Reid [14.26] have reviewed the requirements for magnets in such devices. In some travelling wave tubes for military applications, the requirements are so severe that only materials of very high coercivity, like PtCo and $SmCo_5$, are suitable.

Television picture tubes and the cathode ray tubes of oscilloscopes require magnets for two purposes. Near the cathode, where the electrons are emitted, an ion-trap magnet maintains a transverse field which removes ions from the electron beam. Further down the tube, a ring magnet around the tube produces an axial field to focus the electron beam.

Force applications

If either pole of a permanent magnet is brought near a piece of iron, the field of the magnet will magnetize the iron in such a direction that an opposite pole will be, to use the ancient term, "induced" in the iron, and *attraction* will occur. On the other hand, if like poles of two permanent magnets are brought together, *repulsion* will result.

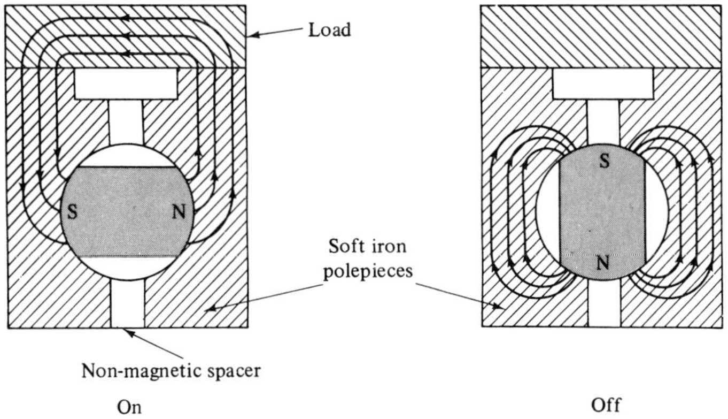

Fig. 14.33 Permanent-magnet chuck. Edwards [14.7].

In dealing with these forces we come full circle from the first page of this book. Offhand, one would think that Coulomb's law, with which we began our subject [Eq. (1.1)], says all there is to say about magnetic forces. However, that law is limited to "point" poles, which can only be approximated by the poles of long, needlelike magnets. It says nothing of the forces between thick magnets, which have poles spread over considerable volumes, or about the effect of the near approach of two magnets in changing the initial pole distribution of one or both of them. The calculation of these forces is not easy. They are dealt with separately below, under the two application categories of attraction and repulsion.

Attraction. If the flat end of a magnet makes good contact with a material of high permeability or with the flat end of a magnet of opposite polarity, the force of attraction between the two is

$$F = \frac{AB^2}{8\pi} \quad \text{dynes,} \tag{14.12}$$

where A is the area of contact (cm^2) and B is the induction in the magnet (gauss). This equation can be derived by summing the Coulomb forces between all the poles assumed to lie on the two surfaces. But it can be understood more simply in another way. Suppose the two pieces are pulled apart to a distance d, creating a very thin gap. If there is no fringing, the field in this gap is B, and the energy density of the field is $B^2/8\pi$, from Eq. (7.38). The work done is Fd and the field energy is $(B^2/8\pi)(Ad)$. If the two are equated, Eq. (14.12) follows. This equation holds only when the two pieces are in contact, and it is found experimentally that the attractive force falls off very rapidly as the two are separated. Edwards [14.7] derives the rather complicated relation between the force and the distance of separation; this force also depends on the shape and dimensions of the two pieces.

Attractive forces are exploited in permanent-magnet chucks, which are gripping devices that can be used, for example, to hold steel objects in place for a machining or grinding operation. Naturally, there must be some means of "turning off" the magnetism, so to speak, and many ingenious ways of doing so have been contrived. The simplest is shown in Fig. 14.33. When the magnet is rotated into the vertical position, the flux is short-circuited by the pole pieces, which have a lower reluctance than the path through the load.

Another holding application is the magnetic latch, widely used on kitchen cupboard doors. These can be cheaply made in large volume, out of barium ferrite.

Magnets are used as separators. Such applications vary from separating certain minerals from others in an ore on the basis of differing magnetic properties to the removal of bits of scrap iron from a variety of materials. Certainly the most bizarre application in this or any other category occurs on the farm. Cattle can pick up along with their food bits of baling wire, nails, and other pieces of iron. This sharp-edged material lodges in the cow's second stomach, which it can irritate or puncture, leading to infection and other complications, a malady known as "hardware disease." The remedy is a cylindrical Alnico magnet, 2 to 3 inches long and $\frac{1}{2}$ inch in diameter, with rounded ends. It is simply dropped down the cow's throat; it lodges in the second stomach and remains there for the life of the cow, attracting and holding any iron that enters. Magnets removed from slaughtered animals have been found to be surrounded with bits of iron, chiefly baling wire, held in a compact mass with no sharp protruding points. This has proved to be a simple and effective means of combating hardware disease, estimated to affect 3 percent of all dairy cows [14.27].

Repulsion. No major applications of repulsive forces exist, but there are some interesting problems. Equation (14.12) is not symmetrical like Coulomb's law and therefore is useless for predicting repulsive forces, as McCaig points out [14.5]. For if opposite poles of two identical magnets are forced close together, the flux density B is zero at the interface; a repulsive force certainly exists, although Eq. (14.12) predicts it to be zero. And if the magnets are not of identical strength, the polarity of the weaker magnet may be reversed, and repulsion will be turned into attraction.

By considering the force on the poles of one magnet by the field of an identical second magnet, McCaig [14.28] derives this equation for the repulsive force at zero separation:

$$F = \frac{AH_c^2}{8\pi} \text{ dynes,} \tag{14.13}$$

where H_c is the coercivity (in Oe) of either magnet. He finds experimentally that the force is usually less than half the predicted value, either because the magnets are not uniformly magnetized or because one or both become partly demagnetized and work on an inner loop.

For centuries there have been men fascinated with the idea of *magnetic levitation*, i.e., the stable suspension in the air of a body made of iron or other magnetic material, without contact above or below, by an artful arrangement of magnets exerting the proper attraction and repulsion. Writing in 1600, Gilbert touches on this subject in Book 1, Chapter 1 and in Book 2, Chapter 24 of his book *On The Magnet* [1.1]. He relates the legend, and calls it false, of Mahomet's shrine where the iron coffin of the prophet was said to float in the air under a vaulted roof of lodestones. However, there probably remained people who thought the trick could be done if only enough ingenuity were brought to bear, because the idea has the same kind of fascination as perpetual motion. But such hopes were dashed in 1839 when Earnshaw [14.29] proved that it could not be done. His theorem relates to both electrostatic and magnetostatic forces, in fact, to any system of particles which exert forces on one another varying inversely as the square of the distance. For magnetic systems, Earnshaw's theorem may be stated as follows: stable levitation of one body by one or more other bodies is impossible, if all bodies in the system have a permeability μ greater than 1. Theory and experiment are reviewed by Boerdijk [14.30].

Metastable levitation can be temporarily achieved with bodies having $\mu > 1$, but if the body is slightly displaced from its point of metastability, it does not return but moves off in one direction or the other. Stable levitation can be achieved ($\mu > 1$) if power is fed into the system. Thus a steel shaft can be supported metastably in space by magnets; as soon as instability occurs, it is detected by a sensor, and a servo system causes currents to flow in nearby conductors in such a way as to correct the field and force the shaft back to its stable point [14.28].

If diamagnetic materials are included in the system, stable levitation becomes possible without power input. Levitation of this kind is easiest with a perfect diamagnet, namely, a superconductor (Section 3.5). Figure 14.34 shows a bar

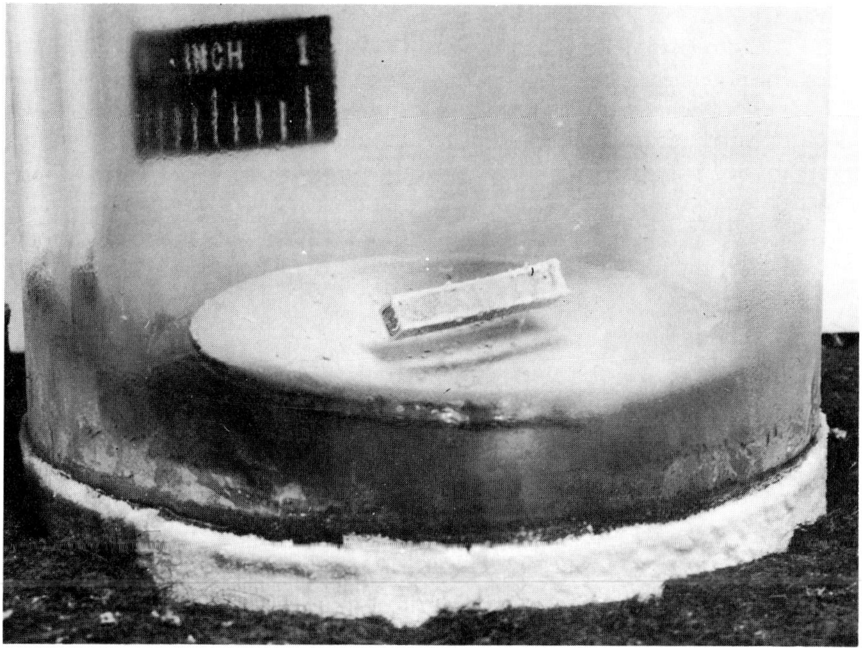

Fig. 14.34 Bar magnet floating over superconducting lead. (Courtesy of Lewis Research Center, National Aeronautics and Space Administration.)

magnet floating over the slightly concave surface of superconducting lead at 4°K. The magnet is in effect supported by its own field, which cannot penetrate into the lead.

The permeability of a superconductor is zero. Values of μ then jump to values only slightly less than 1; unfortunately, no substances with intermediate values of μ, like 0.5, are known. This means that only very light bodies, a few grams in weight, can be stably levitated at room temperature with the help of ordinary diamagnetic materials, like graphite or bismuth, because their flux-repelling powers are so feeble. Nevertheless, such levitation is possible and is exploited in the construction of some very sensitive, frictionless measuring instruments [14.31].

Quite large weights can be supported by the repulsion between permanent magnets, the lower one fixed to a base and the upper one to the underside of the load. Some lateral constraint is needed to make the load stable, but it need not be strong. The magnets must have high coercivity. Barium ferrite is well suited to this application; it can be made in flat pieces of fairly large area, magnetized in the thickness direction. Polgreen [14.32] believes that engineers are not sufficiently aware of the potentialities of this relatively new material, with magnetic properties so different from older materials like Alnico. He has proposed that railway cars be supported by magnetic repulsion and driven by linear electric motors; the track would be made of two rows of ferrite bricks, with poles opposed to those on similar bricks on the underside of the car structure. A later suggestion [14.35] is a combined permanent magnet and electromagnet. It is made of a ferrite-strip magnet sandwiched between two iron strips, whose projecting ends (the pole pieces) are wound with a coil. Current through this coil provides a field opposed to that of the magnet. A momentary current in the coil can release a load carried by the magnet, and a continuous current of the right magnitude can support a load, such as a vehicle from an overhead track, without contact but with lateral constraint.

PROBLEMS

14.1 Show that the slope of the load line of a permanent-magnet circuit is given by

$$\frac{B_m}{H_m} = \frac{4\pi - N_d}{N_d}.$$

Assume that $A_m = A_g$, in which case N_d can be found from Eq. (2.44).

14.2 You are given two iron rods, identical in all respects except that one is magnetized lengthwise and the other is demagnetized. How would you tell them apart without any apparatus whatsoever, not even a string to hang them by?

APPENDIXES

APPENDIX 1

SOME DATES IN THE MODERN HISTORY OF MAGNETISM

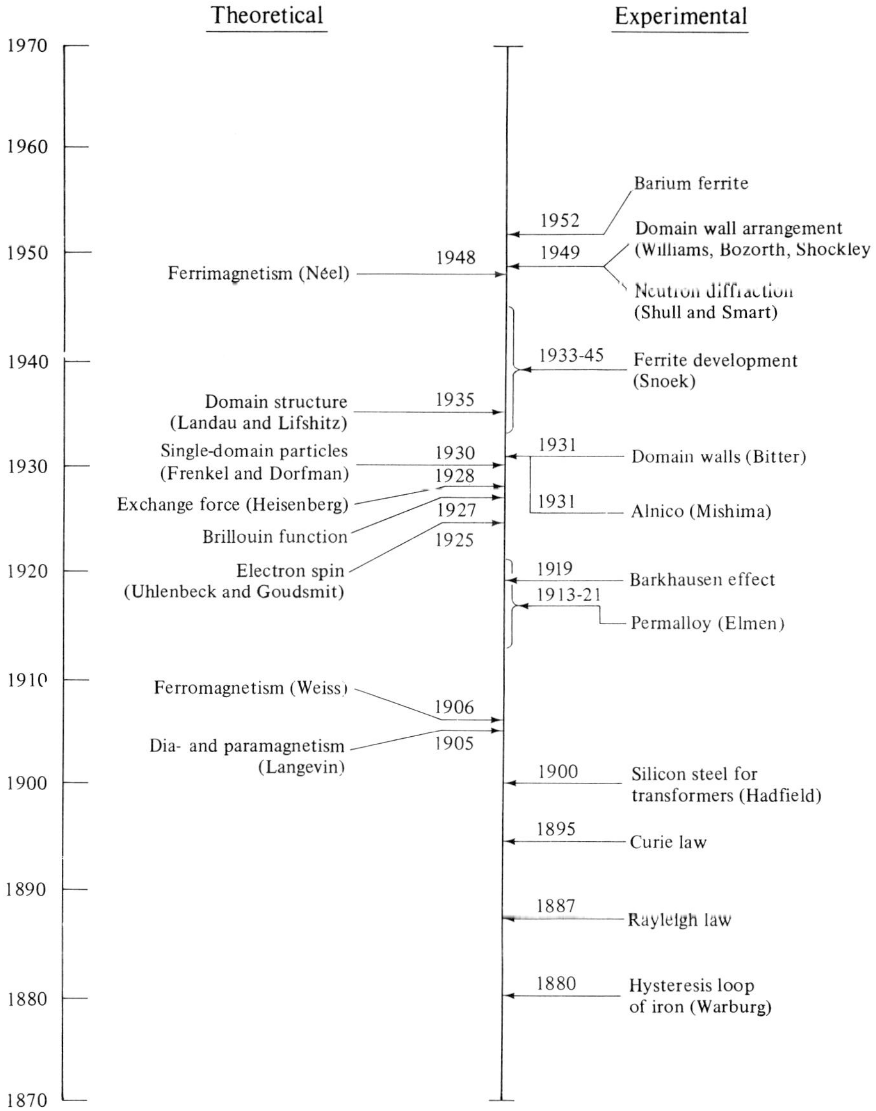

APPENDIX 2

MAGNETIC STATE OF THE ELEMENTS AT ROOM TEMPERATURE

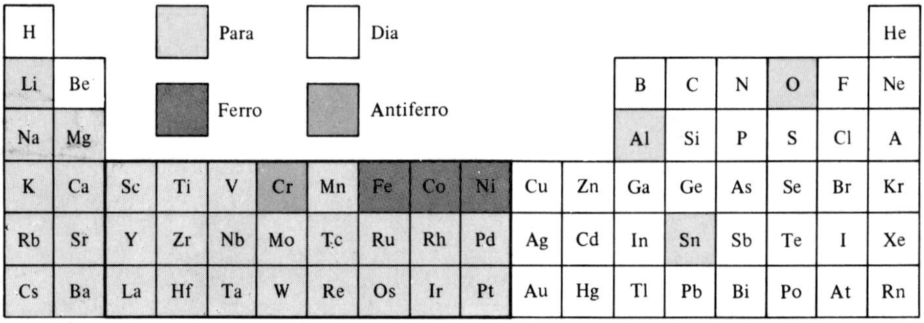
Para
Dia
Ferro
Antiferro
H He
Li Be B C N O F Ne
Na Mg Al Si P S Cl A
K Ca Sc Ti V Cr Mn Fe Co Ni Cu Zn Ga Ge As Se Br Kr
Rb Sr Y Zr Nb Mo Tc Ru Rh Pd Ag Cd In Sn Sb Te I Xe
Cs Ba La Hf Ta W Re Os Ir Pt Au Hg Tl Pb Bi Po At Rn

The transition elements are enclosed by a heavy line. See Appendix 3 (opposite page) for data on the rare earths.

APPENDIX 3

MAGNETIC STATES OF THE RARE EARTHS

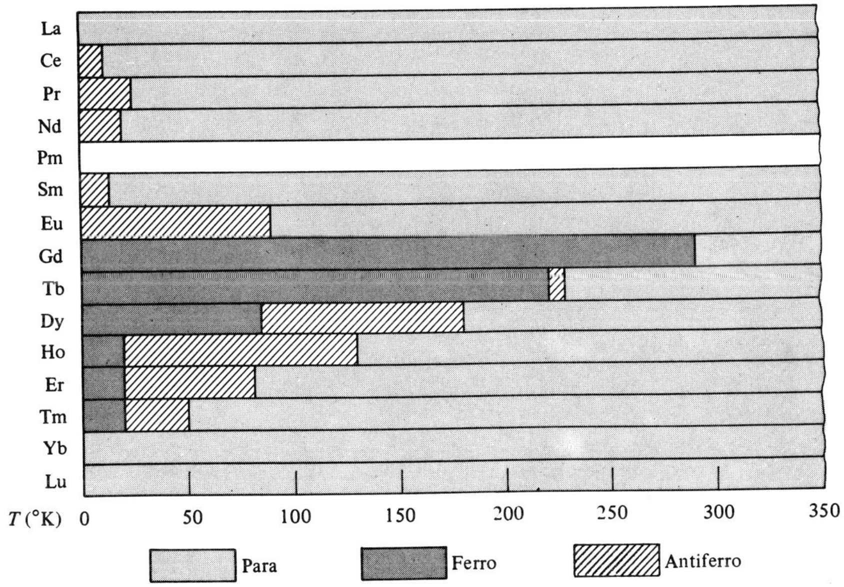

APPENDIX 4

DIPOLE FIELDS AND ENERGIES

We consider a magnet, or dipole, consisting of two point poles of strength p, interpolar distance l, and magnetic moment $m = pl$.

The field H_1 of the magnet at a point P distant r from the magnet center and in line with the magnet (Fig. A4.1 a) is given by, from Eq. (1.3),

$$H_1 = \frac{p}{[r - (l/2)]^2} - \frac{p}{[r + (l/2)]^2} = \frac{2prl}{[r^2 - (l^2/4)]^2}.$$

If r is large compared to l, this expression becomes

$$H_1 = \frac{2pl}{r^3} = \frac{2m}{r^3}. \tag{1}$$

Similarly, the field H_2 at a point P abreast of the magnet center (Fig. A4.1 b) is the sum of the two fields $H(+)$ and $H(-)$, equal in magnitude:

$$\begin{aligned} H_2 &= 2H(+) \cos \alpha \\ &= 2\left[\frac{p}{r^2 + (l^2/4)}\right]\left[\frac{l/2}{\{r^2 + (l^2/4)\}^{1/2}}\right] \\ &= \frac{pl}{[r^2 + (l^2/4)]^{3/2}}. \end{aligned}$$

If r is large compared to l, this expression becomes

$$H_2 = \frac{pl}{r^3} = \frac{m}{r^3}. \tag{2}$$

In Fig. A4-1 (c) we wish to know the field H at P, where the line from P to the magnet makes an angle θ with the magnet axis. The moment of the magnet can be resolved into components parallel and normal to the line to P, so that

$$H_r = \frac{2(m \cos \theta)}{r^3}, \tag{3}$$

$$H_\theta = \frac{m \sin \theta}{r^3}, \tag{4}$$

$$\begin{aligned} H &= (H_r^2 + H_\theta^2)^{1/2} \\ &= \frac{m}{r^3}(3 \cos^2 \theta + 1)^{1/2}. \end{aligned} \tag{5}$$

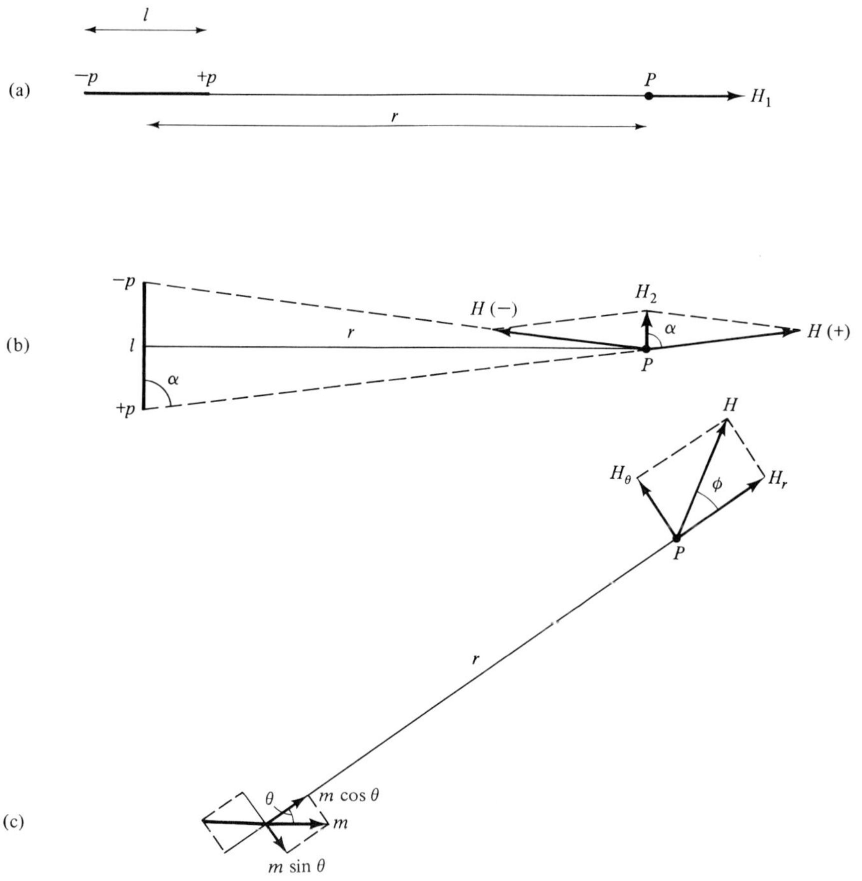

Fig. A4.1 Fields of dipoles.

The resultant field H is inclined at an angle $(\theta + \phi)$ to the magnet axis, where

$$\tan \phi = \frac{H_\theta}{H_r} = \frac{\tan \theta}{2}.$$

We now want an expression for the mutual potential energy of two magnets. Figure A4.2 shows two magnets of moment m_1 and m_2 at a distance r apart and making angles θ_1 and θ_2 with the line joining them. The field at m_2, parallel to m_2 and due to m_1, is

$$H_p = H_r \cos \theta_2 - H_\theta \cos (90^\circ - \theta_2).$$

The potential energy of m_2 in the field of m_1 is, from Eq. (1.5),

$$E_p = - m_2 (H_r \cos \theta_2 - H_\theta \sin \theta_2).$$

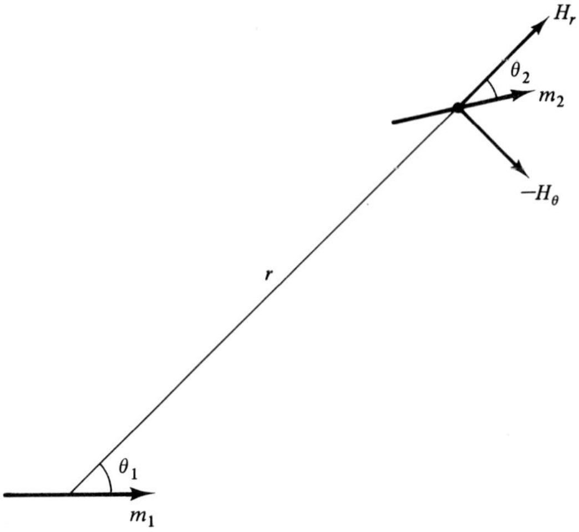

Fig. A4.2 Interacting dipoles.

Combining this with Eqs. (3) and (4) gives

$$E_p = -\frac{m_1 m_2}{r^3}(2\cos\theta_1\cos\theta_2 - \sin\theta_1\sin\theta_2) \tag{6}$$

$$= \frac{m_1 m_2}{r^3}[\cos(\theta_1 - \theta_2) - 3\cos\theta_1\cos\theta_2]. \tag{7}$$

Similarly, it can be shown that the potential energy of m_1 in the field of m_2 is given by the same expression. Thus E_p is the mutual potential energy of the two dipoles. It is also called the dipole-dipole energy or the interaction energy between the two dipoles. It is fundamentally a magnetostatic energy.

APPENDIX 5

DATA ON THE FERROMAGNETIC ELEMENTS

Element	20° C			0° K			
	σ_s (emu/g)	M_s (emu/cm^3)	$4\pi M_s$ (G)	σ_0 (emu/g)	μ_H (μ_β)	T_c (° C)	$\frac{\sigma_s}{\sigma_0}$
Fe	218.0	1,714.	21,580	221.9	2.219	770	0.982
Co	161.	1,422.	17,900	162.5	1.715	1131	0.991
Ni	54.39	484.1	6,084	57.50	0.604	358	0.946

Relative Saturation Magnetization

$\frac{T}{T_c}$	σ_s/σ_0		
	Fe	Co, Ni	$J = \frac{1}{2}$ (theory)
0	1	1	1
0.1	0.996	0.996*	1.000
0.2	0.99	0.99	1.000
0.3	0.975	0.98	0.997
0.4	0.95	0.96	0.983
0.5	0.93	0.94	0.958
0.6	0.90	0.90	0.907
0.7	0.85	0.83	0.829
0.8	0.77	0.73	0.710
0.85	0.70	0.66	0.630
0.9	0.61	0.56	0.525
0.95	0.46	0.40	0.380
1.0	0	0	0

* Value for Ni only.

(From *American Institute of Physics Handbook*, 2d ed. (New York: McGraw-Hill, 1963))

APPENDIX 6

DEMAGNETIZING FACTORS FOR ELLIPSOIDS OF REVOLUTION

These tables are from Stoner [2.12].

TABLE 1

This table is most useful for *problate* spheroids. See Fig. 2.27 for drawings of ellipsoids. For oblate spheroids, a is the polar axis (axis of revolution). For prolate spheroids, c is the polar axis. $N_a + N_b + N_c = 4\pi = 12.566\ 371$.

Oblate

a/c	$N_a/4\pi$	a/c	$N_a/4\pi$
0.0	1.000 000	0.5	0.527 200
0.1	0.860 804	0.6	475 826
0.2	750 484	0.7	432 065
0.3	661 350	0.8	394 440
0.4	588 154	0.9	361 822

Prolate

c/a	$N_c/4\pi$	c/a	$N_c/4\pi$	c/a	$N_c/4\pi$
1.0	0.333 333	3.5	0.089 651	10	0.020 286
1.1	308 285	3.6	86 477	11	17 515
1.2	286 128	3.7	83 478	12	15 297
1.3	266 420	3.8	80 641	13	13 490
1.4	248 803	3.9	77 954	14	11 997
1.5	0.232 981	4.0	0.075 407	15	0.010 749
1.6	218 713	4.1	72 990	16	09 692
1.7	205 794	4.2	70 693	17	08 790
1.8	194 056	4.3	68 509	18	08 013
1.9	183 353	4.4	66 431	19	07 339
2.0	0.173 564	4.5	0.064 450	20	0.006 749
2.1	164 585	4.6	62 562	21	6 230
2.2	156 326	4.7	60 760	22	5 771
2.3	148 710	4.8	59 039	23	5 363
2.4	141 669	4.9	57 394	24	4 998
2.5	0.135 146	5.0	0.055 821	25	0.004 671
2.6	129 090	5.5	48 890	30	3 444
2.7	123 455	6.0	43 230	35	2 655
2.8	118 203	6.5	38 541	40	2 116
2.9	113 298	7.0	34 609	45	1 730
3.0	0.108 709	7.5	0.031 275	50	0.001 443
3.1	104 410	8.0	28 421	60	1 053
3.2	100 376	8.5	25 958	70	0 805
3.3	096 584	9.0	23 816	80	0 637
3.4	093 015	9.5	21 939	90	0 518

c/a	$N_c/4\pi$	c/a	$N_c/4\pi$
100	0.000 430	400	0.000 036
110	363	500	24
120	311	600	17
130	270	700	13
140	236	800	10
150	0.000 209	900	0.000 008
200	125	1000	7
250	083	1100	6
300	060	1200	5
350	045	1300	4

TABLE 2

This table is most useful for *oblate* spheroids. See Fig. 2.27 for drawings of ellipsoids. For prolate spheroids, c is the polar axis (axis of revolution). For oblate spheroids, a is the polar axis. $N_a + N_b + N_c = 4\pi = 12.566\,371$.

Prolate

a/c	$N_c/4\pi$	a/c	$N_c/4\pi$
0.0	0.000 000	0.5	0.173 564
0.1	020 286	0.6	209 962
0.2	055 821	0.7	244 110
0.3	095 370	0.8	275 992
0.4	135 146	0.9	305 689

Oblate

c/a	$N_a/4\pi$	c/a	$N_a/4\pi$	c/a	$N_a/4\pi$
1.0	0.333 333	2.5	0.588 154	4.0	0.703 641
1.1	359 073	2.6	598 539	4.1	709 097
1.2	383 059	2.7	608 422	4.2	714 357
1.3	405 437	2.8	617 837	4.3	719 432
1.4	426 344	2.9	626 817	4.4	724 330
1.5	0.445 906	3.0	0.635 389	4.5	0.729 061
1.6	464 237	3.1	643 581	4.6	733 633
1.7	481 442	3.2	651 417	4.7	738 055
1.8	497 615	3.3	658 920	4.8	742 332
1.9	512 843	3.4	666 110	4.9	746 473
2.0	0.527 200	3.5	0.673 006	5.0	0.750 484
2.1	540 758	3.6	679 625	5.5	768 780
2.2	553 578	3.7	685 984	6.0	784 585
2.3	565 717	3.8	692 097	6.5	798 373
2.4	577 227	3.9	697 979	7.0	810 506

Oblate

c/a	$N_a/4\pi$	c/a	$N_a/4\pi$	c/a	$N_a/4\pi$
7.5	0.821 265	20	0.926 181	100	0.984 490
8.0	830 870	21	929 494	110	985 885
8.5	839 497	22	932 522	120	987 048
9.0	847 288	23	935 301	130	988 034
9.5	854 359	24	937 860	140	988 881
10	0.860 804	25	0.940 224	150	0.989 616
11	872 125	30	949 778	200	992 196
12	881 743	35	956 700	250	993 749
13	890 017	40	961 944	300	994 786
14	897 210	45	966 056	350	995 528
15	0.903 520	50	0.969 366	400	0.996 085
16	909 101	60	974 359	500	996 866
17	914 071	70	977 961	600	997 388
18	918 526	80	980 673	700	997 760
19	922 542	90	982 790	800	998 040
				900	0.998 257
				1000	998 431
				1100	998 574
				1200	998 692
				1300	998 793

APPENDIX 7

MAJOR SYMBOLS AND DEFINING EQUATIONS

Pole strength	$= p\ (g^{1/2}\ cm^{3/2}\ sec^{-1})$
Interpole distance	$= l$ (cm)
Magnetic moment	$= m$ (erg/Oe) $= pl$
Magnetic field	$= H$ (Oe)
Magnetization	= magnetic moment per unit volume $= M$ (erg/Oe cm^3 *or* emu/cm^3) $= m/v$
Specific magnetization	= magnetic moment per unit mass $= \sigma$ (erg/Oe g *or* emu/g) $= m/w$
Magnetic induction	$= B$ (gauss) $= H + 4\pi M$
Permeability	$= \mu = B/H$
Permeability of a vacuum	= 1
Permeability of air	= 1.000 000 37
Volume susceptibility	$= \kappa$ (emu/cm^3 Oe) $= M/H$
Mass susceptibility	$= \chi$ (emu/g Oe) $= \kappa/\rho = \sigma/H$
Atomic susceptibility	$= \chi_A$ (emu/g atom Oe) $= \chi A$
Molecular susceptibility	$= \chi_M$ (emu/g mol Oe) $= \chi M'$

ρ = density
A = atomic weight
M' = molecular weight

APPENDIX 8

UNITS AND CONVERSIONS

Magnetism

cgs units	*mks* units
$B = H + 4\pi M$	$B = \mu_0 H + M$
B in gauss	B in webers/meter^2 (tesla)
H in oersteds	H in amperes/meter
M in emu/cm^3	M in webers/meter^2
μ (vacuum) = 1	μ_0 (vacuum) = $4\pi \times 10^{-7}$ weber/ampere meter

cgs to *mks*	*mks* to *cgs*
B: 1 gauss = 10^{-4} weber/meter^2	1 weber/meter^2 = 10^4 gauss
H: 1 oersted = 79.6 amperes/meter	1 ampere/meter = 12.57×10^{-3} Oe
M: 1 emu/cm^3 = 12.57×10^{-4} weber/meter^2	1 weber/meter^2 = 796 emu/cm^3
ϕ: 1 maxwell = 10^{-8} weber	1 weber = 10^8 maxwells

Length

1 angstrom (Å) = 10^{-8} cm	1 cm = 0.394 in.
1 micron (μ) = 10^{-6} meter	1 in. = 2.54 cm
= 10^{-4} cm	10^{-3} in. = 1 mil = 25.4 μ
= 10^4 Å	1μ = 39.4×10^{-6} in.

Mass

1 kg = 2.205 lb	1 lb = 0.454 kg

Energy

1 cal = 4.19×10^7 ergs	1 erg = 2.39×10^{-8} cal
1 erg = 10^{-7} joule	1 joule = 10^7 ergs
1 eV = 1.602×10^{-12} erg	1 erg = 6.25×10^{11} eV

Stress and energy density

1 dyne/cm^2	$= 1.02 \times 10^{-8}$ kg/mm^2	1 lb/in^2	$= 6.90 \times 10^{4}$ dynes/cm^2
	$= 1.45 \times 10^{-5}$ lb/in^2		$= 7.03 \times 10^{-4}$ kg/mm^2
1 kg/mm^2	$= 9.80 \times 10^{7}$ dynes/cm^2		
	$= 1420$ lb/in^2		
1 erg/cm^3	$= 10^{-1}$ joule/meter^3	1 joule/meter^3	$= 10$ ergs/cm^3

Power

1 watt = 1 joule/sec
= 10^7 ergs/sec
= 0.239 cal/sec

Core loss per unit weight

1 watt/kg = 0.454 watt/lb	1 watt/lb = 2.205 watt/kg

Metric prefixes

10^{12}	tera	T	10^{1}	deka	da	10^{-9}	nano	n
10^{9}	giga	G	10^{-1}	deci	d	10^{-12}	pico	p
10^{6}	mega	M	10^{-2}	centi	c	10^{-15}	femto	f
10^{3}	kilo	k	10^{-3}	milli	m	10^{-18}	atto	a
10^{2}	hecto	h	10^{-6}	micro	μ			

APPENDIX 9

PHYSICAL CONSTANTS

Planck's constant	$= h$	$= 6.62 \times 10^{-27}$ erg sec
Boltzmann's constant	$= k$	$= 1.38 \times 10^{-16}$ erg/°K
		$= 0.862 \times 10^{-4}$ eV/°K
Avogadro's number	$= N$	$= 6.02 \times 10^{23}$/g mol
Gas constant	$= R$	$= 1.99$ cal/°K g mol
		$= 8.32 \times 10^{7}$ ergs/°K g mol
Velocity of light	$= c$	$= 3.00 \times 10^{10}$ cm/sec
Charge on the electron	$= e$	$= 4.80 \times 10^{-10}$ esu
Charge on the electron	$= e/c$	$= 1.60 \times 10^{-20}$ emu
Mass of electron	$= m$	$= 9.11 \times 10^{-28}$ g
Bohr magneton	$= \mu_B$	$= 0.927 \times 10^{-20}$ erg/Oe

APPENDIX 10 ATOMIC WEIGHTS OF THE ELEMENTS

Based on the assigned relative atomic mass of $^{12}C = 12$

Element	Symbol	Atomic number	Atomic weight
Actinium	Ac	89	(227)
Aluminium	Al	13	26.9815
Americium	Am	95	(243)
Antimony	Sb	51	121.75
Argon	Ar	18	39.948
Arsenic	As	33	74.9216
Astatine	At	85	(210)
Barium	Ba	56	137.34
Berkelium	Bk	97	(247)
Beryllium	Be	4	9.01218
Bismuth	Bi	83	208.9806
Boron	B	5	10.81
Bromine	Br	35	79.904
Cadmium	Cd	48	112.40
Calcium	Ca	20	40.08
Californium	Cf	98	(249)
Carbon	C	6	12.011
Cerium	Ce	58	140.12
Cesium	Cs	55	132.9055
Chlorine	Cl	17	35.453
Chromium	Cr	24	51.996
Cobalt	Co	27	58.9332
Copper	Cu	29	63.546
Curium	Cm	96	(247)
Dysprosium	Dy	66	162.50
Einsteinium	Es	99	(254)
Erbium	Er	68	167.26
Europium	Eu	63	151.96
Fermium	Fm	100	(253)
Fluorine	F	9	18.9984
Francium	Fr	87	(223)
Gadolinium	Gd	64	157.25
Gallium	Ga	31	69.72
Germanium	Ge	32	72.59
Gold	Au	79	196.9665
Hafnium	Hf	72	178.49
Helium	He	2	4.00260
Holmium	Ho	67	164.9303
Hydrogen	H	1	1.0080
Indium	In	49	114.82
Iodine	I	53	126.9045
Iridium	Ir	77	192.22
Iron	Fe	26	55.847
Krypton	Kr	36	83.80
Lanthanum	La	57	138.9055
Lawrencium	Lr	103	(257)
Lead	Pb	82	207.2
Lithium	Li	3	6.941
Lutetium	Lu	71	174.97
Magnesium	Mg	12	24.305
Manganese	Mn	25	54.9380
Mendelevium	Md	101	(256)
Mercury	Hg	80	200.59
Molybdenum	Mo	42	95.94
Neodymium	Nd	60	144.24
Neon	Ne	10	20.179
Neptunium	Np	93	237.0482
Nickel	Ni	28	58.71
Niobium	Nb	41	92.9064
Nitrogen	N	7	14.0067
Nobelium	No	102	(254)
Osmium	Os	76	190.2
Oxygen	O	8	15.9994
Palladium	Pd	46	106.4
Phosphorus	P	15	30.9738
Platinum	Pt	78	195.09
Plutonium	Pu	94	(242)
Polonium	Po	84	(210)
Potassium	K	19	39.102
Praseodymium	Pr	59	140.9077
Promethium	Pm	61	(147)
Protactinium	Pa	91	231.0359
Radium	Ra	88	226.0254
Radon	Rn	86	(222)
Rhenium	Re	75	186.2
Rhodium	Rh	45	102.9055
Rubidium	Rb	37	85.4678
Ruthenium	Ru	44	101.07
Samarium	Sm	62	150.4
Scandium	Sc	21	44.9559
Selenium	Se	34	78.96
Silicon	Si	14	28.086
Silver	Ag	47	107.868
Sodium	Na	11	22.9898
Strontium	Sr	38	87.62
Sulfur	S	16	32.06
Tantalum	Ta	73	180.9479
Technetium	Tc	43	98.9062
Tellurium	Te	52	127.60
Terbium	Tb	65	158.9254
Thallium	Tl	81	204.37
Thorium	Th	90	232.0381
Thulium	Tm	69	168.9342
Tin	Sn	50	118.69
Titanium	Ti	22	47.90
Tungsten	W	74	183.85
Uranium	U	92	238.029
Vanadium	V	23	50.9414
Xenon	Xe	54	131.30
Ytterbium	Yb	70	173.04
Yttrium	Y	39	88.9059
Zinc	Zn	30	65.37
Zirconium	Zr	40	91.22

Values in parentheses represent the most stable known isotopes.

APPENDIX 11

MEASUREMENT OF INTERNAL FIELDS

In the determination of a B,H curve or loop, the measurement of the induction B poses no particular problem. The specimen is simply wound with a B-coil, which is connected to a ballistic galvanometer, ordinary fluxmeter, or electronic fluxmeter; any one of these instruments can measure changes in B, from which the actual values of B at particular fields can be determined.

If the specimen is in the form of a ring, the H field is easily found from the current in the magnetizing winding, as described in Section 2.5. But if the specimen has any other shape, the determination of H can present some difficulty. By H is meant the true, or internal, field inside the specimen; it is the difference between the applied field H_a and the demagnetizing field H_d. For a rod specimen tested in a solenoid, the approach adopted in the ballistic method of Section 2.7 was to estimate H_d from tabulated values of the demagnetizing factor N_d and then calculate H. The alternative is to *measure* H directly by one of the methods described below. (If H is measured, then $H_a - H$ yields H_d, from which N_d can be found. These methods can therefore be used to obtain experimental values of N_d.)

It is impossible to measure H inside a solid specimen. But, according to a theorem given in Section 2.6, the tangential component of H is continuous across an interface. This means that H just inside the surface of the specimen is equal to H just outside, if H inside is parallel to the surface, as it normally is in experimental work. We can therefore measure H just outside the specimen, with the assurance that it will equal H inside. We will examine this technique with reference to a permanent magnet tested in an electromagnet, but it is equally applicable to a rod specimen tested in a solenoid.

Permanent-magnet materials are usually tested in the form of a short cylinder placed between, and in contact with, the pole faces of an electromagnet, as in Fig. A11.1 (a). The induction is measured by means of a B-coil wound around the center of the specimen. It is easy to measure the H field in the air gap of the electromagnet before the specimen is inserted, but this field is not equal to the H field inside the specimen when the latter is in place, for two reasons:

1. The joints between specimen and pole faces are never mechanically perfect. Tiny air gaps therefore exist and these create free poles, according to Eq. (2.28).
2. Even if the joints were perfect, the magnetization of the specimen is not equal to that of the pole caps. Therefore M is discontinuous across the ends of the specimen and more free poles are formed, according to Eq. (2.29).

These new poles produce an H field which alters the field that was present before the specimen was inserted. The true field inside the specimen must therefore be measured, and any one of the following methods will serve.

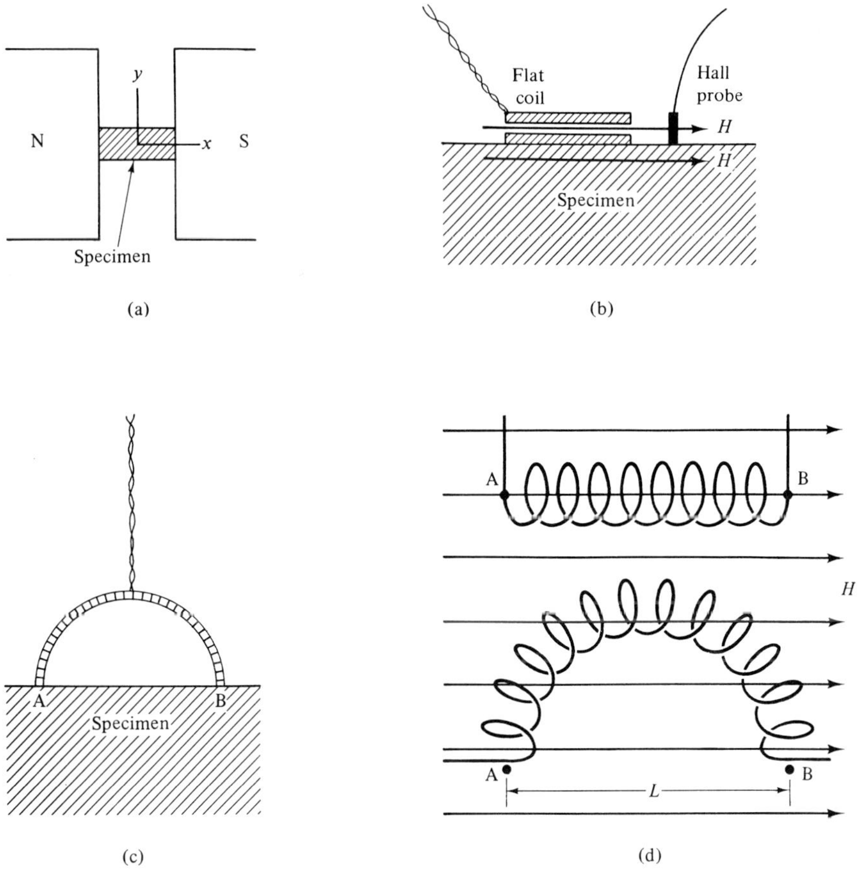

Fig. A11.1. Measurement of internal fields.

Hall probe

The sensing element of a Hall probe (Section 2.4) is placed in contact with the specimen surface at or near its center ($x = 0$), as shown in Fig. A11.1 (b), with the flat face of the probe normal to the field to be measured. The read-out instrument connected to the probe then gives the value of H. The H vectors shown in (b) are drawn to suggest the equality of H just inside and outside the specimen.

If the area of the electromagnet pole faces is large relative to the cross-sectional area of the specimen, H should be uniform over the cross section of the specimen and for a considerable distance outside it. This uniformity can be tested by measuring the field with the Hall probe at various distances from the specimen surface, along the y axis. If a marked gradient of H with y is observed, it can be concluded that the field inside the specimen is *not* uniform. In this case, the average field over the specimen cross section can be estimated by extrapolating the fields measured outside the specimen to $y = 0$.

Flat H coil

A coil is wound on a thin strip of a nonmagnetic material and placed in contact with the specimen surface, with the coil axis (long axis of the strip) parallel to the field to be measured, as shown in Fig. A11.1 (b). The coil is connected to a fluxmeter, and H is measured by reversing the direction of H or by pulling the coil away from the surface to a point where H is essentially zero.

The size of the coil depends on the size of the specimen, but a typical coil might be 20 mm long, 10 mm wide, and 2 mm thick. It will normally have to have a large number of turns of fine wire. Sensitivity can be increased by placing two or more flat coils, connected in series, at points around the specimen.

Magnetic potentiometer

This device, also called a *Chattock coil*, is simply a curved coil with coplanar ends. It is usually, but not necessarily, semicircular in shape. It is connected to a fluxmeter, and its ends are placed in contact with the specimen surface, as in Fig. A 11.1 (c). The field H is measured by reversing its direction or by pulling the coil away to a position of zero field.

The principle on which the magnetic potentiometer operates is illustrated in Fig. A 11.1 (d). A uniform H field is depicted there and, at the top, a straight coil of N turns, cross-sectional area A, and length L. Use of such a coil to measure a field has already been described in Section 2.4, but we will now regard this measurement in a slightly different way. The flux linkage with the coil is

$$N\phi = NHA = NAHL/L = N'AHL \quad \text{maxwell-turns,} \tag{1}$$

where N' is the number of turns per cm. But HL is the value of the line integral $\int_B^A H dl$ from one end of the coil to the other. And, according to Eq. (2.51), this also equals the difference in magnetic potential V_{AB} between points A and B. If the coil is connected to a fluxmeter and removed from the field H to a point of zero field, the fluxmeter deflection, according to Eq. (2.10), is

$$d = \frac{N\Delta\phi}{K} = \frac{N\phi}{K} = \frac{NHA}{K} = \frac{N'AV_{AB}}{K} \quad \text{div,} \tag{2}$$

where K is the fluxmeter constant in maxwell-turns/div. The fluxmeter deflection d is therefore a measure both of the field H and the magnetic potential V_{AB}.

We turn now to the curved coil shown in Fig. A 11.1 (d). Let it have N' turns/cm, a length along its axis of l cm, and a cross-sectional area of A cm^2. Then the line integral of H along the curved axis of the coil from B to A is

$$\int_B^A H dl = V_{AB} = HL \quad \text{Oe cm.} \tag{3}$$

The line integral on the left is the summation of products $(H \cos\theta)(dl)$, where θ is the angle between H and dl. The identities in Eq. (3) are an expression of the fact that the potential difference between two points is independent of the path taken to evaluate it; i.e., the work done on a unit pole in moving it from B to A against the field is independent of the path followed. The flux linkage with the curved coil is

$$\int_B^A HAN'dl = AN' \int_B^A H dl = AN'HL \quad \text{maxwell-turns.} \tag{4}$$

Therefore, when the coil is removed from the field H to a region of zero field, the fluxmeter deflection is

$$d = \frac{AN'HL}{K} \quad \text{div,} \tag{5}$$

from which H can be calculated. Note that the length l of the coil does not enter; the coil may be of any length and bent into any shape. The area A should be small so that the measurement can be made between two points. The coil must be uniformly wound in order to make N' constant.

Both the straight coil and the curved coil are magnetic potentiometers. The only purpose in curving the coil is to make its ends coplanar. It can then measure the potential difference between two points on a flat surface, as in Fig. A 11.1 (c), and therefore measure H essentially in the plane of that surface.

Although the method of the magnetic potentiometer (curved coil) has been illustrated for a uniform field, it can be used equally well in a nonuniform field. Whether the field is uniform or not in the y direction has no effect whatever on the fluxmeter deflection. If the field has a gradient in the x direction, the measured value of H will be an average over the straight line between A and B.

Magnetic-potentiometer coils have been wound on rigid forms and, in order to vary L, on flexible forms. Constructional details are given by Bates on pp. 68 and 85 of [G.13], and Edwards has shown, on p. 267 of [G.17], how two straight, interconnected, rigid coils, used in the form of "chopsticks," can replace a curved coil.

CHAPTER REFERENCES

CHAPTER 1. DEFINITIONS AND UNITS

1.1 Gilbert, William, *De Magnete.* English translation by P. Fleury Mottelay (New York: Dover Publications, 1958). Another translation, done mainly by Silvanus P. Thompson, was published by Basic Books, Inc. (New York, 1958). Edited by Derek J. Price. It is a facsimile of the original 1900 edition.

1.2 Andrade, E. N. da C., "The Eearly History of the Permanent Magnet," *Endeavour*, pp. 22-30, January, 1958. (This paper is also reprinted as Chapter 1 of reference [G.18], and an expanded version as Chapter 1 of [G.17]).

1.3 Sears, Francis Weston, *Electricity and Magnetism.* 434 pp. (Reading, Mass.: Addison-Wesley, 1953).

1.4 Shoenberg, David, *Magnetism.* 216 pp. (London: Sigma Books, 1949).

CHAPTER 2. EXPERIMENTAL METHODS

2.1 Sanford, Raymond L., and Cooter, Irvin L., *Basic Magnetic Quantities and the Measurement of the Magnetic Properties of Materials.* 36 pp. National Bureau of Standard Monograph 47 (1962). Available for 30 cents from the Superintendent of Documents, U. S. Gov. Printing Office, Washington 25, D. C.

2.2 Scott, G. G., "Compensation of the Earth's Magnetic Field," *J. Appl. Phys.*, **28**, 270-272 (1957). A drawing of the field lines within the total volume of a Helmholtz coil system has been given by Maxwell. See Fig. XIX, Vol. 2, of reference [2.11] below.

2.3 Bitter, F., "The Design of Powerful Electromagnets. Part I. The Use of Iron," *Rev. Sci. Inst.*, **7**, 479-481 (1936). "Part II. The Magnetizing Coil," *ibid.*, **7**, 482-488 (1936). "Part III. The Use of Iron," *ibid.*, **8**, 318-319 (1937). "Part IV. The New Magnet Laboratory at M.I.T.," *ibid.*, **10**, 373-381 (1939).

2.4 Kolm, Henry; Lax, Benjamin; Bitter, Francis; and Mills, Robert, editors, *High Magnetic Fields.* 751 pp. (Cambridge, Mass.: M.I.T. Press; New York and London: Wiley, 1962).

2.5 Jacobs, I. S., and Lawrence, P. E., "Measurement of Magnetization Curves in High Pulsed Magnetic Fields," *Rev. Sci. Instr.*, **29**, 713-714 (1958).

2.6 Kunzler, J. E., Buehler, E., Hsu, F. S. L., and Wernick, J. H., "Super-conductivity in Nb_3Sn at High Current Densities in a Magnetic Field of 88 kgauss," *Phys. Rev. Letters*, **6**, 89-91 (1961).

2.7 Sampson, W. B., "Superconducting Magnets," *IEEE Trans. Magnetics*, **4**, 99-107 (1968).

2.8 Foner, S., McNiff, E. J., Jr., Matthias, B. T., Geballe, T. H., Willens, R. H., and Corenzwit, E., "Upper Critical Fields of High-Temperature Superconducting $Nb_{1-y}(Al_{1-x}Ge_x)_y$ and Nb_3Al: Measurements of $H_{c2} > 400$ kG at 4.2°K," *Physics Letters*, **31A**, 349-350 (1970).

2.9 For one method of doing this, see Kinnard, Isaac F., *Applied Electrical Measurements*, p. 275. 600 pp. (New York: Wiley, 1956).

2.10 For an account of the methods adopted in World War II to counteract the magnetic mine, see a series of papers on pp. 430-455 and pp. 488-524 of the *J. Institution of Elec. Engs.*, Vol. 93, Part I (1946).

2.11 Maxwell, James Clerk, *A Treatise On Electricity and Magnetism*, Articles 437 and 438, Vol. 2. 1006 pp., 2 vols., 3d (1891) ed. (New York: Dover Publications, 1954).

2.12 Stoner, E. C., "The Demagnetizing Factors for Ellipsoids," *Phil. Mag.* [7], **36**, 803-821 (1945).

2.13 Osborn, J. A., "Demagnetizing Factors of the General Ellipsoid," *Phys. Rev.*, **67**, 351-357 (1945).

2.14 Vigoureux, P., and Webb, C. E., *Principles of Electric and Magnetic Measurements*, 392 pp. (New York: Prentice-Hall, 1937).

2.15 An example of numerical data obtained by this method is given by Weiss, Pierre and Forrer, R., "Aimantation et Phénomène Magnétocalorique du Nickel," *Annales de Physique* [10], **5**, 153–213 (1926). The units of Column 3 of their Table 1 are parts per thousand.

2.16 Foner, Simon, "Versatile and Sensitive Vibrating-Sample Magnetometer," *Rev. Sci. Instr.*, **30**, 548-557 (1959).

2.17 Rayleigh, Lord, "On the Behaviour of Iron and Steel under the Operation of Feeble Magnetic Forces," *Phil. Mag.* [5], **23**, 225–248 (1887).

2.18 Hazard, Daniel L., *Directions for Magnetic Measurements*, Serial No. 166, Coast and Geodetic Survey, U. S. Department of Commerce. 129 pp. (Washington: U. S. Gov. Printing Office, 1957). This publication describes the older methods of measuring the earth's field. Information on the newer methods, including the proton-precession magnetometer, a good description of the earth's field, and a history of geomagnetic studies are given by Nelson, James H., Hurwitz, Louis, and Knapp, David G., *Magnetism of the Earth*, Publication 40-1, Coast and Geodetic Survey, U. S. Dept. of Commerce. 79 pp. (Washington: U. S. Govt. Printing Office, 1962).

2.19 Bozorth, Richard M., "A Null-Reading Astatic Magnetometer of Novel Design," *J. Opt. Soc. Amer.*, **10**, 591-598 (1925). An abstract appears in [G.4], p. 857.

2.20 Laufer, Arthur R., "The Orientation of Paramagnetic and Diamagnetic Rods in Magnetic Fields," *Amer. J. Physics*, **19**, 275-279 (1951).

2.21 Heyding, R. D., Taylor, J. B., and Hair, M. L., "Four-Inch Shaped Pole Caps for Susceptibility Measurements by the Curie Method," *Rev. Sci. Instr.*, **32**, 161-163 (1961).

2.22 See description by Bates [G.13] and Sucksmith, W., Clark, C. A., Oliver, D. J., and Thompson, J. E., "Spontaneous Magnetization–Techniques and Measurements," *Rev. Mod. Phys.*, **25**, 34-41 (1953).

2.23 McGuire, T. R., and Flanders, P. J., "Direct Current Magnetic Measurements," on pp. 123-188 of Vol. 1 of [G.37].

2.24 Oguey, H. J., "Alternating Current Magnetic Measurements," on pp. 189-245 of Vol. 1 of [G.37].

2.25 Fleischer, R. L., Hart, H. R., Jr., Jacobs, I. S., Price, P. B., Schwarz, W. M., and Woods, R. T., "Magnetic Monopoles: Where Are They and Where Aren't They?" *J. Appl. Phys.*, **41,** 958-965 (1970).

2.26 *Direct-Current Magnetic Measurements for Soft Magnetic Materials.* 68 pp. Special Technical Publication 371S1 (Philadelphia: American Society for Testing and Materials, 1970).

2.27 Cohen, David, "A Shielded Facility for Low-Level Magnetic Measurements," *J. Appl. Phys.*, **38,** 1295-1296 (1967).

2.28 Cohen, David, Edelsack, Edgar A., and Zimmerman, James E., "Magnetocardiograms Taken Inside a Shielded Room with a Superconducting Point-Contact Magnetometer," *Appl. Phys. Letters*, **16,** 278-280 (1970).

CHAPTER 3. DIAMAGNETISM AND PARAMAGNETISM

3.1 Langevin, P., "Magnétisme et Théorie des Electrons," *Annales de Chemie et de Physique*, [8], **5,** 70-127 (1905).

3.2 Cioffi, P. P., "Approach to the Ideal Magnetic Circuit Concept through Superconductivity," *J. Appl. Phys.*, **33,** 875-879 (1962).

3.3 Weiss, Pierre, "L'Hypothèse du Champ Moléculaire et de la Propriété Ferromagnétique," *J. de Physique*, **6,** 661-690 (1907).

3.4 Ishiwara, T., "On the Thermomagnetic Properties of Various Compounds at Low Temperatures," *Sci. Reports Tohoku Univ.*, [1], **3,** 303-319 (1914).

3.5 Honda, K., and Ishiwara, T., "On the Thermomagnetic Properties of Various Compounds and the Weiss Theory of Magnetons," *Sci. Reports Tohoku Univ.*, [1], **4,** 215-260 (1915).

3.6 See, for example, Bates [G.13], p. 43.

3.7 Brillouin, L., "Les Moments de Rotation et le Magnétisme dans la Mécanique Ondulatoire," *J. de Physique*, [6], **8,** 74-81 (1927).

3.8 Henry, Warren E., "Spin Paramagnetism of Cr^{3+}, Fe^{3+}, and Gd^{3+} at Liquid Helium Temperatures and in Strong Magnetic Fields," *Phys. Rev.*, **88,** 559-562 (1952).

3.9 Foex, G., *Constantes Sélectionnées, Diamagnétisme et Paramagnétisme.* 307 pp. (Paris: Masson et Cie., 1957).

3.10 Pugh, Emerson W., "Completely Non-Magnetic Alloy for Instrumentation in Magnetic Fields," *Rev. Sci. Instr.*, **29,** 1118 (1958).

3.11 *Metals Handbook.* 1444 pp. (Cleveland: American Society for Metals, 1948).

3.12 Bitter, F., and Kaufman, A. R., "Magnetic Studies of Solid Solutions," *Phys. Rev.*, **56,** 1044-1051 (1939).

CHAPTER 4. FERROMAGNETISM

4.1 Weiss, Pierre, "La Variation du Ferromagnétisme avec la Temperature," *Compt. Rend.*, **143,** 1136-1139 (1906). This publication is a brief note. Weiss's ideas were developed more fully in his 1907 paper [3.3], in which the molecular field hypothesis was applied both to ferro- and paramagnetics. In 1906, he supposed that domains are identical with the crystal grains that make up a polycrystalline substance. It was not until several years later that it was realized that domains could be, and normally are, much smaller than the grains.

4.2 Kittel, Charles, *Introduction to Solid State Physics.* 617 pp., 2d ed. (New York: Wiley, 1956).

4.3 Stoner, E. C., "Collective Electron Ferromagnetism," *Proc. Roy. Soc.*, **165**, 372-414 (1938). "Collective Electron Ferromagnetism II. Energy and Specific Heat," *Proc. Roy. Soc*,, **169**, 339-371 (1939).

4.4 Mader, S., and Nowick, A. S., "Metastable Co-Au Alloys: Example of an Amorphous Ferromagnet," *Appl. Phys. Letters*, **7**, 57-59 (1965).

4.5 Koster, G. F., "Density of States Curve for Nickel," *Phys. Rev.*, **98**, 901-902(1955).

4.6 Crangle, J., "The Intrinsic Magnetic Properties of Transition Metals and Their Alloys," pp. 51-68 of *Electronic Structure and Alloy Chemistry of the Transition Elements*, edited by Paul A. Beck. 251 pp. (New York: Interscience, 1963). Other papers in this book deal with portions of the theory of ferromagnetic alloys.

4.7 Kouvel, J. S., "Magnetic Properties," pp. 529-568 of *Intermetallic Compounds*, edited by J. H. Westbrook. 663 pp. (New York: Wiley, 1967).

4.8 *Metals Handbook*. 1444 pp. (Cleveland: American Society for Metals, 1948).

4.9 Barrett, Charles S., and Massalski, T. B., *Structure of Metals*. 654 pp., 3d ed. (New York: McGraw-Hill, 1966). This book is an excellent general reference on order-disorder transformations, phase transformations, preferred orientation, crystal structure, x-ray diffraction, etc.

4.10 Crangle, J., "Ferromagnetism and Antiferromagnetism in Non-Ferrous Metals and Alloys," *Metallurgical Reviews*, **7**, 133-174 (1962).

4.11 Potter, H. H., "The Magneto-Caloric Effect and Other Magnetic Phenomena in Iron," *Proc. Roy. Soc.*, **146**, 362-387 (1934).

4.12 See, for example, Bates [G.13] or Bozorth [G.4].

4.13 Herring, Conyers, "The State of *d* Electrons in Transition Metals," *J. Appl. Phys.*, Supplement to Vol. 31, 3S-11S (1960). This review paper, although nonmathematical, is on a rather advanced level.

4.14 Berkowitz, A. E., "Constitution of Multiphase Alloys," pp. 331-363 of Vol. 1 of [G.37].

CHAPTER 5. ANTIFERROMAGNETISM

5.1 Néel, Louis, "Influence des Fluctuations du Champ Moléculaire sur les Propriétés Magnétiques des Corps," *Annales de Physique*, **18**, 5-105 (1932).

5.2 Lidiard, A. B., "Antiferromagnetism," *Reports on Progress in Physics*, **17**, 201-244 (1954).

5.3 Nagamiya, T., Yosida, K., and Kubo, R., "Antiferromagnetism," *Advances in Physics*, **4**, 1-112 (1955).

5.4 Bizette, H., Squire, C. F., and Tsai, B., "Le point de transition λ de la susceptibilité magnétique du protoxyde de manganèse MnO," *Compt. Rend.*, **207**, 449-450 (1938).

5.5 Bizette, H., and Tsai, B., "Susceptibilités magnétiques principals d'un cristal de siderose et du fluorure manganeux," *Compt. Rend.*, **238**, 1575-1576 (1954).

5.6 Shull, C. G., and Smart, J. Samuel, "Detection of Antiferromagnetism by Neutron Diffraction," *Phys. Rev.*, **76,** 1256 (1949).

5.7 Bacon, G. E., *Neutron Diffraction.* 426 pp., 2d ed. (Oxford: Clarendon Press, 1962).

5.8 Erickson, R. A., "Neutron Diffraction Studies of Antiferromagnetism in Manganous Fluoride and some Isomorphous Compounds," *Phys. Rev.*, **90,** 779-785 (1953).

5.9 Nathans, R., and Pickart, S. J., "Spin Arrangements in Metals," pp. 211-269 of Vol. 3 of [G.22].

5.10 Shull, C. G., "Spin Density Distribution in Fe, Co, and Ni," pp. 15-30 of *Magnetic and Inelastic Scattering of Neutrons by Metals*, T. J. Rowland and Paul A. Beck, editors (New York: Gordon and Breach, 1968).

5.11 Koehler, W. C., "Magnetic Properties of Rare-Earth Metals and Alloys," *J. Appl. Phys.*, **36,** 1078-1087 (1965).

5.12 Palmberg, P. W., DeWames, R. E., and Vredevoe, L. A., "Direct Observation of Coherent Exchange Scattering by Low-Energy Electron Diffraction from Antiferromagnetic NiO," *Phys. Rev. Letters*, **21,** 682-685 (1968).

CHAPTER 6. FERRIMAGNETISM

6.1 Néel, Louis, "Propriétés Magnétique des Ferrites; Ferrimagnétisme et Antiferromagnétisme," *Annales de Physique*, **3,** 137-198 (1948).

6.2 Smart, J. Samuel, "The Néel Theory of Ferrimagnetism," *Amer. J. Physics*, **23,** 356-370 (1955).

6.3 Wolf, W. P., "Ferrimagnetism," *Reports on Prog. in Phys.*, **24,** 212-303 (1961).

6.4 Gorter, E. W., "Some Properties of Ferrites in Connection with Their Chemistry," *Proc. I.R.E.*, **43,** 1945-1973 (1955).

6.5 Pauthenet, R., "Aimantation Spontanée de Ferrites," *Annales de Physique*, **7,** 710-745 (1952).

6.6 Serres, A., "Recherches sur les Moments Atomiques," *Annales de Physique*, **17,** 5-95 (1932).

6.7 See p. 342 of [5.7].

6.8 Shull, C. G., Wollan, E. O., and Koehler, W. C., "Neutron Scattering and Polarization by Ferromagnetic Materials," *Phys. Rev.*, **84,** 912-921 (1951).

6.9 Guillaud, Charles, "Propriétés Magnétiques des Ferrites," *J. de Phys. et Radium*, **12,** 239-248 (1951).

6.10 Ishikawa, Yoshikazu, "Superparamagnetism in Magnetically Dilute Systems," *J. Appl. Phys.*, **35,** 1054-1059 (1964).

CHAPTER 7. MAGNETIC ANISOTROPY

7.1 Honda, K., and Kaya, S., "On the Magnetization of Single Crystals of Iron," *Sci. Reports Tohoku Univ.*, **15,** 721-753 (1926).

7.2 Kaya, S., "On the Magnetization of Single Crystals of Nickel," *Sci. Reports Tohoku Univ.*, **17,** 639-663 (1928).

7.3 Kaya, S., "On the Magnetization of Single Crystals of Cobalt," *Sci. Reports Tohoku Univ.*, **17,** 1157-1177 (1928).

7.4 Luborsky, F. E., and Morelock, C. R., "Magnetization Reversal of Almost Perfect Whiskers," *J. Appl. Physics*, **35,** 2055-2066 (1964).

7.5 Penoyer, R. F., "An Automatic Torque Balance for the Determination of Magnetocrystalline Anisotropy." Papers presented at the Conference of Magnetism and Magnetic Materials, Boston, 1956. 365 pp. (New York: American Institute of Electrical Engineers, 1957).

7.6 Pearson, R. F., "The Magnetocrystalline Anisotropy of Gallium and Aluminum Substituted Magnetite," *J. de Phys. et Radium*, **20,** 409-413 (1959).

7.7 Kouvel, J. S., and Graham, C. D., Jr., "On the Determination of Magnetocrystalline Anisotropy Constants from Torque Measurements," *J. Appl. Physics.*, **28,** 340-343 (1957).

7.8 Williams, H. J., "Magnetic Properties of Single Crystals of Silicon Iron," *Phys. Rev.*, **52,** 747-751 (1937).

7.9 Rathenau, G. W., and Snoek, J. L., "Magnetic Anisotropy Phenomena in Cold Rolled Nickel-Iron," *Physica*, **8,** 555-575 (1941).

7.10 Zijlstra, H., "Device for the Rapid Measurement of Magnetic Anisotropy at Elevated Temperatures," *Rev. Sci. Instr.*, **32,** 634-638 (1961).

7.11 Lawton, H., and Stewart, K. H., "Magnetization Curves for Ferromagnetic Single Crystals," *Proc. Roy. Soc.*, **193,** 72-88 (1948). Or see Stewart [G.7, pp. 25-31].

7.12 Barnier, Y., Pauthenet, R., and Rimet, G., "Thermomagnetic Study of a Hexagonal Cobalt Single Crystal," *Cobalt*, No. 15, pp. 1-7, June, 1962. (Published by the "Centre D'information Du Cobalt," Brussels.) Or see "Etude Magnétique d'un Monocristal de Cobalt dans la Phase Hexagonale," *J. Phys. Soc. Japan*, **17** (Supplement B-1), 309-313 (1962), Proceedings of International Conference on Magnetism and Crystallography, 1961, Vol. 1.

7.13 Trevena, D. H., *Static Fields in Electricity and Magnetism.* p. 225 (London: Butterworths, 1961).

7.14 Bozorth, R. M., "Directional Ferromagnetic Properties of Metals," *J. Appl. Phys.*, **8,** 575-588 (1937).

7.15 Honda, K., Masumoto, H., and Kaya, S., "Magnetization of Single Crystal of Fe at High Temperatures," *Sci. Reports Tohoku Univ.*, **17,** 111-130 (1928).

7.16 Aubert, G., "Torque Measurements of the Anisotropy Energy and Magnetization of Nickel," *J. Appl. Phys.*, **39,** 504-510 (1968).

7.17 Bhandary, Vittal S., and Cullity, B. D., "The Texture and Mechanical Properties of Iron Wire Recrystallized in a Magnetic Field," *Trans. Met. Soc. A.I.M.E.*, **224,** 1194-1200 (1962).

7.18 Becker, J. J., "A Recording Torque Magnetometer," General Electric Research Laboratory Report No. RL-887 (1953).

7.19 Dunn, C. G., and Walter, J. L., "Synthesis of a (110) [001] Type Torque Curve in Silicon Iron," *J. Appl. Phys.*, **30,** 1067-1072 (1959).

7.20 Jacobs, I. S., "Spin-Flopping in MnF_2 by High Magnetic Fields," *J. Appl. Phys.*, **32,** 61S-62S (1961).

7.21 Starr, C., Bitter, F., and Kaufmann, A. R., "The Magnetic Properties of the Iron Group Anhydrous Chlorides at Low Temperatures," *Phys. Rev.*, **58,** 977-983 (1940).

7.22 Williams, D. E. G., *The Magnetic Properties of Matter.* 232 pp. (New York: American Elsevier, 1966).

7.23 Stoner, E. C., and Wohlfarth, E. P., "A Mechanism of Magnetic Hysteresis in Heterogeneous Alloys," *Phil. Trans. Roy. Soc.* **A-240,** 599-642 (1948). But note that their *a* and *b* axes, for a prolate spheroid, correspond to the *c* and *a* axes, respectively, of this book.

7.24 Graham, C. D., Jr., "Textured Magnetic Materials," pp. 723-748 of Vol. 2 of [G.37].

7.25 Nesbitt, E. A., Williams, H. J., and Bozorth, R. M., "Factors Determining the Permanent Magnet Properties of Single Crystals of Fe_2NiAl," *J. Appl. Phys.*, **25,** 1014-1020 (1954).

CHAPTER 8. MAGNETOSTRICTION AND THE EFFECTS OF STRESS

8.1 Lee, E. W., "Magnetostriction and Magnetomechanical Effects," *Reports on Prog. in Phys.*, **18,** 184-229 (1955).

8.2 Birss, R. R., "The Saturation Magnetostriction of Ferromagnetics," *Advances in Physics.*, **8,** 252-291 (1959).

8.3 Carr, W. J. Jr., "Magnetostriction," pp. 200-250 of [G.9].

8.4 Kittel, Charles, "Physical Theory of Ferromagnetic Domains," *Rev. Mod. Phys.*, **21,** 541-583 (1949).

8.5 Bozorth, R. M., "Magnetostriction and Crystal Anisotropy of Single Crystals of Hexagonal Cobalt," *Phys. Rev.*, **96,** 311-315 (1954).

8.6 Callen, Herbert B., and Goldberg, Norman, "Magnetostriction of Polycrystalline Aggregates," *J. Appl. Phys.*, **36,** 976-977 (1965).

8.7 Lee, E. W., "Magnetostriction Curves of Polycrystalline Ferromagnetics," *Proc. Phys. Soc.*, **72,** 249-258 (1958).

8.8 Bagchi, D. K., Unpublished research, University of Notre Dame.

8.9 Kuruzar, Michael E., Unpublished research, University of Notre Dame.

8.10 Kirchner, H., "Uber den Einfluss von Zug, Druck und Torsion auf die Langsmagnetostriktion," *Ann. Physik*, [5], **27,** 49-69 (1936).

8.11 Kuruzar, Michael E., and Cullity, B. D., "The Magnetostriction of Iron under Tensile and Compressive Stress," *Inter. J. Magnetism* **1,** 323-325 (1971).

8.12 These measurements are reported by McKeehan, L. W., "Magnetostriction," *J. Franklin Institute*, **202,** 737-773 (1926).

8.13 Bagchi, D. K., and Cullity, B. D., "Effects of Applied and Residual Stress on the Magnetoresistance of Nickel," *J. Appl. Phys.*, **38,** 999-1000 (1967).

8.14 Webster, W. L., "Magneto-Striction in Iron Crystals," *Proc. Roy. Soc. A*, **109,** 570-584 (1925).

8.15 Masiyama, Yoshio, "On the Magnetostriction of a Single Crystal of Nickel," *Sci. Reports Tohoku Univ.*, **17,** 947-961 (1928).

8.16 Tatsumoto, E., and Okamoto, T., "Temperature Dependence of the Magnetostriction Constants of Iron and Silicon Iron," *J. Phys. Soc. Japan*, **14,** 1588-1594 (1959).

8.17 Cullity, B. D., "Fundamentals of Magnetostriction," *J. Metals* **23,** 35-41 (1971).

CHAPTER 9. DOMAINS AND THE MAGNETIZATION PROCESS

9.1 Williams, H. J., Bozorth, R. M., and Shockley, W., "Magnetic Domain Patterns on Single Crystals of Silicon Iron," *Phys. Rev.*, **75,** 155-178 (1949).

9.2 Kittel, C., and Galt, J. K., "Ferromagnetic Domain Theory," *Solid State Physics*, **3,** 437-564 (1956).

9.3 Craik, D. J., and Tebble, R. S., "Magnetic Domains," *Reports on Prog. in Phys.*, **24,** 116-166 (1961).

9.4 Dillon, J. F., Jr., "Domains and Domain Walls," pp. 415-464 of Vol. 3 of [G.22].

9.4a Dillon, J. F., Jr., "Observation of Domains in the Ferrimagnetic Garnets by Transmitted Light," *J. Appl. Phys.*, **29,** 1286-1291 (1958).

9.5 Bitter, F., "On Inhomogeneities in the Magnetization of Ferromagnetic Materials," *Phys. Rev.*, **38,** 1903-1905 (1931).

9.6 Bozorth, R. M., "Magnetic Domain Patterns," *J. de Phys. et Radium*, **12,** 308-321 (1951).

9.7 Michalak, J. T., and Glenn, R. C., "Transmission Electron Microscope Observations of Magnetic Domain Walls," *J. Appl. Phys.*. **32,** 1261-1265 (1961).

9.8 Landau, L., and Lifshitz, E., "On the Theory of the Dispersion of Magnetic Permeability in Ferromagnetic Bodies," *Physik. Z. Sowjetunion*, **8,** 153-169 (1935).

9.9 For other photographs of domains in whiskers see DeBlois, R. W., and Graham, C. D., Jr., "Domain Observations in Iron Whiskers," *J. Appl. Phys.*, **29,** 931-939 (1958).

9.10 DeBlois, R. W., personal communication.

9.11 This photograph appears in Becker, J. J., "Metallurgical Structure and Magnetic Properties," pp. 68-92 of [G.9].

9.12 Frenkel, J., and Dorfman, J., "Spontaneous and Induced Magnetization in Ferromagnetic Bodies," *Nature*, **126,** 274-275 (1930).

9.13 Kittel, C., "Theory of the Structure of Ferromagnetic Domains in Films and Small Particles," *Phys. Rev.*, **70,** 965-971 (1946).

9.14 Néel, L., "Propriétés d'un Ferromagnétique Cubiques en Grains Fins," *Compt. Rend.*, **224,** 1488-1490 (1947).

9.15 Shtrikman, S., and Treves, D., "Micromagnetics," pp. 395-414 of Vol. 3 of [G.22].

9.16 Williams, H. J., and Shockley, W., "A Simple Domain Structure in an Iron Crystal Showing a Direct Correlation with the Magnetization," *Phys. Rev.*, **75,** 178-183 (1949).

9.17 Néel, L., "Effet des Cavités et des Inclusions sur le Champ Coercitif," *Cahiers de Physique*, **25,** 21-44 (1944).

9.18 Williams, H. J., "Direction of Domain Magnetization in Powder Patterns," *Phys. Rev.*, **71,** 646-647 (1947).

9.19 Baldwin, William Marsh, Jr., "Residual Stress in Metals," *Proc. A.S.T.M.*, **49,** 1-45 (1949).

9.20 Richards, D. G., "Relief and Redistribution of Residual Stress in Metals," pp. 129-204 of *Residual Stress Measurements*. 210 pp. (Cleveland: Amer. Soc. for Metals, 1952).

9.21 An introductory treatment of these matters is given by Cullity, B. D., *Elements of X-Ray Diffraction*. 514 pp. (Reading, Mass.: Addison-Wesley, 1956). See also [9.9].

9.22 Warren, B. E., "X-Ray Studies of Deformed Metals," *Progress in Metal Physics*, **8,** 147-202 (1959).

9.23 Warren, B. E., *X-Ray Diffraction*. 381 pp. (Reading, Mass.: Addison-Wesley, 1969).

9.24 For a good, relatively brief account of dislocation theory, see Hull, Derek, *Introduction to Dislocations*. 259 pp. (Pergamon Press, 1965).

9.25 Polcarova, Milena, and Lang, A. R., "X-Ray Topographic Studies of Magnetic Domain Configurations and Movements," *Appl. Phys. Letters*, **1,** 13-15 (1962).

9.25a Polcarova, Milena, "Applications of X-Ray Diffraction Topography to the Study of Magnetic Domains," *I.E.E.E. Trans. Magnetics*, **5**, 536-544 (1969).

9.26 Cullity, B. D., "Residual Stress after Plastic Elongation and Magnetic Losses in Silicon Steel," *Trans. Met. Soc. A.I.M.E.*, **227**, 356-358 (1963).

9.27 Krause, R. F., and Cullity, B. D., "Formation of Uniaxial Magnetic Anisotropy in Nickel by Plastic Deformation," *J. Appl. Phys.*, **39**, 5532-5537 (1968).

9.28 Kondorski, E., "On the Nature of Coercive Force and Irreversible Changes in Magnetization," *Phys. Z. Sowjetunion*, **11**, 597-620 (1937).

9.29 Vicena, Frantisek, "The Effect of Dislocations on the Coercive Force of Ferromagnetics," *Czech. J. Phys.*, **5**, 480-499 (1955). (Associated Technical Services, Inc. translation RJ-1725 from the Russian.)

9.30 Seeger, A., Kronmuller, H., Reiger, H., and Trauble, H., "Effect of Lattice Defects on the Magnetization Curve of Ferromagnets," *J. Appl. Phys.*, **35**, 740-748 (1964).

9.30a Scherpereel, D. E., Kazmerski, L. L., and Allen, C. W., "The Magnetoelastic Interaction of Dislocations and Ferromagnetic Domain Walls in Iron and Nickel," *Met. Trans.*, **1**, 517-524 (1970).

9.31 Rhodes, P., "Thermal Changes in Irreversible Magnetization," *Proc. Leeds. Phil. and Lit. Soc. (Sci. Sect.)*, **5**, 116-127 (1949).

9.32 Rusnak, R. M., Ph.D. Thesis, University of Notre Dame, 1967.

9.33 Rusnak, R. M., and Cullity, B. D., "Correlation of Magnetic Permeability and X-Ray Diffraction Line Broadening in Cold-Worked Iron," *J. Appl. Phys.*, **40**, 1581-1582 (1969).

9.34 Elwood, W. B., "A New Ballistic Galvanometer Operating in High Vacuum," *Rev. Sci. Instr.*, **5**, 300-305 (1934).

9.35 Stoner, Edmund C., "Ferromagnetism: Magnetization Curves," *Reports on Prog. in Phys.*, **13**, 83-183 (1950).

9.36 Chou, H., and Cullity, B. D., unpublished research, University of Notre Dame.

9.37 Hansen, H. J., Jr., "Continuous Nondestructive Magnetic Testing of Steel Strip," *J. of Metals*, 1131-1136 (October, 1966).

9.38 Damiano, V. V., Domenicali, C., and Collings, R. W., "Recovery and Recrystallization," pp. 689-721 of Vol. 2 of [G.37].

CHAPTER 10. INDUCED MAGNETIC ANISOTROPY

10.1 Graham, C. D. Jr., "Magnetic Annealing," pp. 288-329 of [G.9].

10.2 Slonczewski, John C., "Magnetic Annealing," pp. 205-242 of Vol. 1 of [G.22].

10.3 Chikazumi, S., and Graham, C. D., Jr., "Directional Order," pp. 577-619 of [G.37].

10.4 Pender, H., and Jones, R. L., "The Annealing of Steel in an Alternating Magnetic Field," *Phys. Rev.*, **1**, 259 (1913).

10.5 Goertz, Matilda, "Iron-Silicon Alloys Heat Treated in a Magnetic Field," *J. Appl. Phys.*, **22**, 964-965 (1951).

10.6 Hansen, Max, and Anderko, Kurt, *Constitution of Binary Alloys*, 2d ed. 1305 pp. (New York: McGraw-Hill, 1958).

10.7 Williams, H. J., Bozorth, R. M., and Christensen, H., "Magnetostriction, Young's Modulus and Damping of 68 Permalloy as Dependent on Magnetization and Heat Treatment," *Phys. Rev.*, **59**, 1005-1012 (1941).

10.8 Chikazumi, Soshin, "Ferromagnetic Properties and Superlattice Formation of Iron-Nickel Alloys," *J. Phys. Soc. Japan*, **5**, 327-333, 333-338 (1950).

10.9 Néel, Louis, "Anisotropie Magnétique Superficielle et Surstructures D'Orientation," *J. de Phys. et Radium*, **15**, 225-239 (1954).

10.10 Taniguchi, S., and Yamamoto, M., "A Note on a Theory of the Uniaxial Ferromagnetic Anisotropy Induced by Cold Work or by Magnetic Annealing in Cubic Solid Solutions," *Sci. Reports Tohoku Univ.*, **A6**, 330-332 (1954).

10.11 Taniguchi, S., "A Theory of the Uniaxial Ferromagnetic Anisotropy Induced by Magnetic Annealing in Cubic Solid Solutions," *Sci. Reports Tohoku Univ.*, **A7**, 269-281 (1955).

10.12 Chikazumi, S., and Oomura, T., "On the Origin of Magnetic Anisotropy Induced by Magnetic Annealing," *J. Phys. Soc. Japan*, **10**, 842-849 (1955).

10.13 Néel, Louis, "Le Trainage Magnétique," *J. de Phys. et Radium*, **12**, 339-351 (1951).

10.14 Néel, Louis, "Théorie du Trainage Magnétique de Diffusion," *J. de Phys. et Radium*, **13**, 249-264 (1952).

10.15 Snoek, J. L., "Effect of Small Quantities of Carbon and Nitrogen on the Elastic and Plastic Properties of Iron," *Physica*, **8**, 711-733 (1941).

10.16 De Vries, G., Van Geest, D. W., Gersdorf, R., and Rathenau, G. W., "Determination of the Magnetic Anisotropy Energy Caused by Interstitial Carbon or Nitrogen in Iron," *Physica*, **25**, 1131-1138 (1959).

10.17 Birkenbeil, H. J., and Cahn, R. W., "Induced Magnetic Anisotropy Created by Magnetic or Stress Annealing of Iron-Aluminum Alloys," *J. Appl. Phys.*, **32**, 362S-363S (1961).

10.18 Chin, G. Y., "Slip-Induced Directional Order in Fe-Ni Alloys. I. Extension of the Chikazumi-Suzuki-Iwata Theory," *J. Appl. Phys.*, **36**, 2915-2924 (1965).

10.19 Chin, G. Y., Nesbitt, E. A., Wernick, J. H., and Vanskike, L. L., "Slip-Induced Directional Order in Fe-Ni Alloys. II. Experimental Observations," *J. Appl. Phys.*, **38**, 2623-2629 (1967).

10.20 Schindler, A. I., and Salkovitz, E. I., "Effect of Applying a Magnetic Field During Neutron Irradiation on the Magnetic Properties of Fe-Ni Alloys," *J. Appl. Phys.*, **31**, 245S-246S (1960).

10.21 Schindler, A. I., Kernohan, R. H., and Weertman, J., "Effect of Irradiation on Magnetic Properties of Fe-Ni Alloys," *J. Appl. Phys.*, **35**, 2640-2646 (1964).

10.22 Néel, L., Pauleve, J., Pauthenet, R., Laughier, J., and Dautreppe, D., "Magnetic Properties of an Iron-Nickel Single Crystal Ordered by Neutron Bombardment." *J. Appl. Phys.*, **35**, 873-876 (1964).

10.23 Chin, G. Y., "Effect of Plastic Deformation on the Magnetic Properties of Metals and Alloys," in Vol. 5 of *Advances in Materials Research*, H. Herman, ed. (New York: Interscience, 1970).

CHAPTER 11. FINE PARTICLES AND THIN FILMS

11.1 Paine, Thomas O., "Magnetic Properties of Fine Particles." pp. 146-167 of [G.9].

11.2 Jacobs. I. S., and Bean, C. P., "Fine Particles, Thin Films and Exchange Anisotropy," pp. 271-350 of Vol. 3 of [G.22].

11.3 Wohlfarth, E. P., "Hard Magnetic Materials," *Advances in Physics*, **8**, 87-224 (1959).

11.4 Wohlfarth, E. P., "Permanent Magnet Materials," pp. 351-393 of Vol. 3 of [G.22].

11.5 Goodenough, John B., and Smith, D. O., "Magnetic Properties of Thin Films." pp. 112-145 of [G.9].

11.6 Smith, Donald O., "The Structure and Switching of Permalloy Films," pp. 465-523 of Vol. 3 of [G.22].

11.7 Pugh, Emerson W., "Magnetic Films of Nickel-Iron," pp. 277-334 of *Physics of Thin Films*, Vol. 1., Georg Hass, ed. (New York: Academic Press, 1963).

11.8 Kittel, C., Galt, J. K., and Campbell, W. E., "Crucial Experiment Demonstrating Single Domain Property of Fine Ferromagnetic Powders," *Phys. Rev.*, **77**, 725 (1950).

11.9 Luborsky, Fred E., "Development of Elongated Particle Magnets," *J. Appl. Phys.*, **32**, 171S-183S (1961).

11.10 Jacobs, I. S., and Bean, C. P., "An Approach to Elongated Fine-Particle Magnets," *Phys. Rev.*, **100**, 1060-1067 (1955).

11.11 Frei, E. H., Shtrikman, S., and Treves, D., "Critical Size and Nucleation Field of Ideal Ferromagnetic Particles," *Phys. Rev.*, **106**, 446-455 (1957).

11.12 Aharoni, Amikam, "Some Recent Developments in Micromagnetics at the Weizmann Institute of Science," *J. Appl. Phys.*, **30**, 70S-78S (1959).

11.13 Jacobs, I. S., and Luborsky, F. E., "Magnetic Anisotropy and Rotational Hysteresis in Elongated Fine-Particle Magnets," *J. Appl. Phys.*, **28**, 467-473 (1957).

11.14 Becker, J. J., "Permanent Magnets Based on Materials with High Crystal Anisotropy," *IEEE Trans. Magnetics*, **4**, 239-249 (1968).

11.15 Shtrikman, S., and Treves, D., "On the Resolution of Brown's Paradox," *J. Appl. Phys.*, **31**, 72S-73S (1960).

11.16 Fowler, C. A., Jr., Fryer, E. M., and Treves, D., "Domain Structures in Iron Whiskers as Observed by the Kerr Effect," *J. Appl. Phys.*, **32**, 296S-297S (1961).

11.17 DeBlois, R. W., and Bean, C. P., "Nucleation of Ferromagnetic Domains in Iron Whiskers," *J. Appl. Phys.*, **30**, 225S-226S (1959).

11.18 DeBlois, R. W., "Domain Wall Motion in Metals," *J. Appl. Phys.*, **29**, 459-467 (1958).

11.19 DeBlois, R. W., "Ferromagnetic Nucleation Sources on Iron Whiskers," *J. Appl. Phys.*, **32**, 1561-1563 (1961).

11.20 Becker, J. J., "Observations of Magnetization Reversal in Cobalt-Rare Earth Particles," *IEEE Trans. Magnetics*, **5**, 211-214 (1969).

11.21 Becker, J. J., "A Domain-Boundary Model for a High Coercive Force Material," *J. Appl. Phys.*, **39**, 1270-1271 (1968).

11.22 Kooy, C., and Enz, U., "Experimental and Theoretical Study of the Domain Configuration in Thin Layers of $BaFe_{12}O_{19}$," *Philips Research Reports*, **15**, 7-29 (1960).

11.23 Néel, Louis, "Influence des fluctuations thermiques sur l'aimantation de grains ferromagnétiques trés fins," *Compt. Rend.*, **228**, 664-666 (1949).

11.24 Bean, C. P., and Livingston, J. D., "Superparamagnetism," *J. Appl. Phys.*, **30**, 120S-129S (1959),

11.25 Kneller, E. F., and Luborsky, F. E., "Particle Size Dependence of Coercivity and Remanence of Single-Domain Particles," *J. Appl. Phys.*, **34**, 656-658 (1963).

11.26 Becker, J. J. "Magnetic Method for the Measurement of Precipitate Particle Sizes in a Cu-Co Alloy," *Trans. A.I.M.E.*, **209**, 59-63 (1957).

11.27 Burke, J., *The Kinetics of Phase Transformations in Metals*. 226 pp. (New York: Pergamon Press, 1965).

11.28 Meiklejohn, W. H., and Bean, C. P., "New Magnetic Anisotropy," *Phys. Rev.*, **105**, 904-913 (1957).

11.29 Kouvel, J. S., Graham, C. D., Jr., and Jacobs, I. S., "Ferromagnetism and Antiferromagnetism in Disordered Ni-Mn Alloys," International Conference on Magnetism, Grenoble, 1958. *J. de Phys. et Radium*, **20**, 198-202 (1959).

11.30 Meiklejohn, W. H., "Exchange Anisotropy–A Review," *J. Appl. Phys.*, **33**, 1328-1335 (1962).

11.31 Doyle, W. D., Josephs, R. M., and Baltz, A., "Electrodeposited Cylindrical Magnetic Films," *J. Appl. Phys.*, **40**, 1172-1181 (1969).

11.32 Humphrey, Floyd B., "Magnetic Measurement Techniques for Thin Films and Small Particles," *J. Appl. Phys.*, **38**, 1520-1527 (1967).

11.33 Roth, Michel, "Annealing Kinetics of the Induced Anisotropy in Ni-Fe Films: The Role of Atom Self-Diffusion in Grain Surface Regions," *J. Appl. Phys.*, **41**, 1286-1294 (1970).

11.34 Néel, Louis, "Energie des parois de Bloch dans les couches minces," *Compt. Rend.*, **241**, 533-536 (1955).

11.35 Middelhoek, S., "Domain Walls in Thin Ni-Fe Films," *J. Appl. Phys.*, **34**, 1054-1059 (1963).

11.36 Huber, E. E., Smith, D. O., and Goodenough, J. B., "Domain-Wall Structure in Permalloy Films," *J. Appl. Phys.*, **29**, 294-295 (1958).

11.37 Moon, Ralph M., "Internal Structure of Cross-Tie Walls in Thin Permalloy Films through High-Resolution Bitter Techniques," *J. Appl. Phys.*, **30**, 82S-83S (1959).

11.38 Methfessel, S., Middelhoek, S., and Thomas, H., "Domain Walls in Thin Magnetic Ni-Fe Films," *J. Appl. Phys.*, **31**, 302S-304S (1960).

11.39 Jackson, M. R., Tauber, R. N., and Kraft, R. W., "Magnetic Anisotropy of MnSb-Sb Eutectic," *J. Appl. Phys.*, **39**, 4452-4457 (1968).

11.40 Albright, D. L., Conard, G. P., II, and Kraft, R. W., "Magnetic Behavior of a Eutectic Alloy Featuring an Aligned Array of Iron Rods," *J. Appl. Phys.*, **38**, 2919-2923 (1967).

11.41 Livingston, J. D., "Structure and Magnetic Properties of Au-Co Aligned Eutectic," *J. Appl. Phys.*, **41**, 197 (1970).

CHAPTER 12. MAGNETIZATION DYNAMICS

12.1 Snoek, J. L., "The Influence of Eddy Currents on the Apparent Hysteresis Loop of Ferromagnetic Bars," *Physica*, **8**, No. 4, 426 (1941). This also appears as Appendix II of [G.3].

12.2 Williams, H. J., Shockley, W., and Kittel, C., "Studies of the Propagation Velocity of a Ferromagnetic Domain Boundary," *Phys. Rev.*, **80**, 1090-1094 (1950).

12.3 Stewart, K. H., "Experiments on a Specimen with Large Domains," *J. de Phys. et Radium*, **12**, 325-331 (1951).

12.4 Galt, J. K., "Motion of Individual Domain Walls in a Nickel-Iron Ferrite," *Bell System Tech. Jour.*, **33,** 1023-1054 (1954).

12.5 Sixtus, K. J., and Tonks, L., "Propagation of Large Barkhausen Discontinuities," *Phys. Rev.*, **37,** 930-958 (1931).

12.6 Bleaney, B. I., and Bleaney. B., *Electricity and Magnetism.* 676 pp. (Oxford: Clarendon Press, 1957). The last chapter of this book contains a good description and comparison of units.

12.7 Hatfield, W. B., "Domain Wall Velocities in Permalloy Films," *J. Appl. Phys.*, **37,** 1934-1935 (1966).

12.8 Gyorgy, E. M., "Rotational Model of Flux Reversal in Square Loop Ferrites," *J. Appl. Phys.*, **28,** 1011-1015 (1957).

12.9 Smith, D. O., "Static and Dynamic Behavior of Thin Permalloy Fiims," *J. Appl. Phys.*, **29,** 264-273 (1958).

12.10 Gyorgy, E. M., "Flux Reversal in Soft Ferromagnetics," *J. Appl. Phys.*, **31,** 110S-117S (1960).

12.11 Gyorgy, E. M., "Magnetization Reversal in Nonmetallic Ferromagnets," pp. 525-552 of Vol. 3 of [G.22].

12.12 Olson, C. D., and Pohm, A. V., "Flux Reversal in Thin Films of 82% Ni, 18 % Fe," *J. Appl. Phys.*, **29,** 274-282 (1958).

12.13 Kryder, M. H., and Humphrey, F. B., "Dynamic Kerr Observations of High-Speed Flux Reversal and Relaxation Processes in Permalloy Thin Films," *J. Appl. Phys.*, **40,** 2469-2474 (1969).

12.14 Dietrich, W., and Proebster, W. E , "Millimicrosecond Magnetization Reversal in Thin Magnetic Films," *J. Appl. Phys.*, **31,** 281S-282S (1960).

12.15 Rathenau, G. W., "Time Effects in Magnetization," pp. 168-199 of [G.9].

12.16 Rathenau, G. W., and De Vries, G., "Diffusion," pp. 748-814 of Vol. 2 of [G.37].

12.17 Snoek, J. L.,"Time Effects in Magnetization," *Physica*, **5,** 663-688 (1938).

12.18 Brissonneau, Pierre, "Contribution à L'Etude Quantitative du Traînage Magnétique de Diffusion du Carbone dans la Fer α," *J. Phys. Chem. Solids*, **7,** 22-51 (1958).

12.19 Singer, Joseph, and Anolick, E. S., "The Solubility of Carbon in Alpha-Fe as Determined by the Time Decay of Permeability," *Trans. Met. Soc. A.I.M.E.*, **218,** 405-409 (1960).

12.20 Smolinski, A. K., Kaczkowski, Z., and Zbikowski, M., "Influence of Plastic Deformation on the Time Decrease of Permeability in Transformer Steel," *J. Appl. Phys.*, **30,** 195S-199S (1959).

12.21 Néel, Louis, "Directional Order and Diffusion Aftereffect," *J. Appl. Phys.*, **30,** 3S-8S (1959).

12.22 Rusnak, R. M., and Cullity, B. D., "The Effect of Plastic Deformation on the Magnetic Aftereffect in Iron," *J. Appl. Phys.*, **39,** 984-986 (1968).

12.23 Tomono, Yuso, "Magnetic After Effect of Cold Rolled Iron (I)," *J. Phys. Soc. Japan*, **7,** 174-179 (1952).

12.24 Street, R., and Woolley, J. C., "A Study of Magnetic Viscosity," *Proc. Phys. Soc.*, **A 62,** 562-572 (1949).

12.25 Néel, Louis, "Théorie du Traînage Magnétique des Substances Massives dans le Domaine de Rayleigh," *J. de Phys. et Radium*, **11,** 49-61 (1950).

12.26 Barbier, J. C., "The Thermal-Agitation After-Effect," pp. 130-134 of *Soft Magnetic Materials for Telecommunications.* 346 pp. C. E. Richards and A. C. Lynch, editors (London: Pergamon Press, 1953).

12.27 Becker, R., and Kornetski, M., "Einige Magnetoelastiche Torsionsversuche," *Zeit. fur Physik*, **88**, 634-646 (1934).

12.28 Wert, C., "The Metallurgical Use of Anelasticity," pp. 225-250 of *Modern Research Techniques in Physical Metallurgy.* 335 pp. (Cleveland: American Society for Metals, 1953).

12.29 Ganganna, H. V., Fiore, N. F., and Cullity, B. D., "Micro Eddy Current Damping in Nickel," *J. Appl. Phys.* **42**, 5792 (1971).

12.30 Roberts, J. T. A., and Barrand, P., "Magnetomechanical Damping Behavior in Pure Nickel and a 20 Wt.% Copper-Nickel Alloy," *Acta Met.*, **15**, 1685-1692 (1967).

12.31 Summer, G., and Entwistle, K. M., "The Stress-Dependent Damping Capacity of Ferromagnetic Metals," *J. Iron and Steel Institute*, **192**, 238-245 (1959).

12.32 Reed-Hill, Robert E., *Physical Metallurgy Principles.* 630 pp. (Princeton: Van Nostrand, 1964).

12.33 Rodbell, D. S., "Interpretation of Magnetic Resonance Measurements in Metals," pp. 815-838 of Vol. 2 of [G.37].

CHAPTER 13. SOFT MAGNETIC MATERIALS

13.1 Jacobs, I. S., "Role of Magnetism in Technology," *J. Appl. Phys.*, **40**, 917-928 (1969).

13.2 Lee, E. W., and Lynch, A. C., "Soft Magnetic Materials," *Advances in Physics*, **8**, 292-348 (1959).

13.3 "Magnetically Soft Materials," pp. 785-797 of Vol. I, 8th ed., *Metals Handbook* (Novelty, Ohio: American Society for Metals, 1961).

13.4 Geiss, R., Unpublished research, University of Notre Dame.

13.5 Russell, Alexander, *A Treatise on the Theory of Alternating Currents*, 2d ed. Vol. 1, 534 pp. Vol. 2, 488 pp. (Cambridge: University Press, 1914).

13.6 Dwight, Herbert Bristol, *Tables of Integrals and Other Mathematical Tables*, 3d ed. 288 pp. (New York: Macmillan, 1957).

13.7 McLachlan, N. W., *Bessel Functions for Engineers*, 2d ed. 239 pp. (Oxford: Clarendon Press, 1955).

13.8 Golding, E. W., *Electrical Measurements and Measuring Instruments*, 4th ed. 913 pp. (London: Pitman and Sons, 1961).

13.9 Pry, R. H., and Bean, C. P., "Calculation of the Energy Loss in Magnetic Sheet Materials Using a Domain Model," *J. Appl. Phys.*, **29**, 532-533 (1958).

13.10 Lee, E. W., "Eddy-Current Losses in Thin Ferromagnetic Sheets," *Proc. Instn. Elec. Engrs.*, **C105**, 337-342 (1958).

13.11 Olson, R. D., "Application of Soft Magnetic Materials and Specialty Alloys," *J. Appl. Phys.*, **37**, 1197-1201 (1966).

13.12 "Alternating Current Magnetic Properties of Materials Using Epstein Specimens (A343-60)." pp. 29-65 of a booklet entitled "ASTM Standards Relating to Magnetic Properties." 1966. American Society for Testing and Materials, 1916 Race St., Philadelphia, Pa. 19103.

13.13 Becker, J. J., "Metallurgical Structure and Magnetic Properties," pp. 68-92 of [G.9].

13.14 Houze, G. L., Jr., "Domain-Wall Motion in Grain-Oriented Silicon Steel in Cyclic Magnetic Fields," *J. Appl. Phys.*, **38,** 1089-1096 (1967).

13.15 Haller, T. R., and Kramer, J. J., "Observation of Dynamic Domain Size Variation in a Silicon-Iron Alloy," *J. Appl. Phys.*, **41,** 1034-1035 (1970).

13.16 Haller, T. R., and Kramer, J. J., "A Model for Reverse-Domain Nucleation in Ferromagnetic Conductors," *J. Appl. Phys.*, **41,** 1036-1037 (1970).

13.17 Williams, H. J., "Some Uses of the Torque Magnetometer," *Rev. Sci. Instr.* **8,** 56-60 (1937).

13.18 Brailsford, F., "Alternating Hysteresis Loss in Electrical Sheet Steels," *J. Instn. Elec. Engrs.*, **84,** 399-407 (1939).

13.19 Walter, J. L., "History of Silicon-Iron," pp. 519-540 of *The Sorby Centennial Symposium on the History of Metallurgy.* 558 pp. (New York: Gordon and Breach, 1965).

13.20 Bechtold, J. H., and Wiener, G. W., "The History of Soft Magnetic Materials," pp. 501-518 of [13.19].

13.21 Data from pp. 76, 77, and 572 of Bozorth [G.4], who gives original sources.

13.22 "Non-Oriented Electrical Steel Sheets," U. S. Steel Corporation, Pittsburgh, Pa.

13.23 "Oriented Electrical Steel Sheets," United States Steel Corporation, Pittsburgh, Pa.

13.24 Dunn, C. G., "Recrystallization Textures," pp. 113-130 of *Cold Working of Metals* (Cleveland: American Society for Metals, 1949).

13.25 Fiedler, H. C., "A Comparison of the Use of Aluminum and Vanadium Nitrides for Making Grain-Oriented Silicon-Iron," *J. Appl. Phys.*, **38,** 1098-1099 (1967).

13.26 *J. Appl. Phys.* **38,** 1074-1108 (1967).

13.27 Hall, R. C., "Single Crystal Anisotropy and Magnetostriction Constants of Several Ferromagnetic Materials Including Alloys of NiFe, SiFe, AlFe, CoNi, and CoFe." *J. Appl. Phys.*, **30,** 816-819 (1959).

13.28 Littmann, M. F., "Structures and Magnetic Properties of Grain-Oriented 3.2% Silicon-Iron." *J. Appl. Phys.*, **38,** 1104-1108 (1967).

13.29 Becker, J. J., "Magnetization Changes and Losses in Conducting Ferromagnetic Materials," *J. Appl. Phys.*, **34,** 1327-1332 (1963).

13.30 Bozorth, R. M., "The Permalloy Problem," *Rev. Mod. Phys.*, **25,** 42-48 (1953).

13.31 Williams, H. J., and Goertz, Matilda, "Domain Structure of Perminvar Having a Rectangular Hysteresis Loop," *J. Appl. Phys.*, **23,** 316-323 (1952). See also "Ferromagnetic Domains," by H. J. Williams on pp. 251-277 of [G.9].

13.32 Hine, Alfred, *Magnetic Compasses and Magnetometers.* 385 pp. (Toronto: University of Toronto Press, 1968).

13.33 Meyerhoff, Albert J., ed., *Digital Applications of Magnetic Devices.* (N. Y.: Wiley, 1960).

13.34 Brownlow, J. M., and Grebe, K. R., "Miniature Ferrite Cores," *J. Appl. Phys.*, **38,** 1190-1191 (1967).

13.35 Russel, Louis A., Whalen, Robert M., and Leilich, Hans O., "Ferrite Memory Systems," *IEEE Trans. Magnetics*, **4,** 134-145 (1968).

13.36 Overn, William M., "Status of Planar Film Memory," *IEEE Trans. Magnetics*, **4,** 308-312 (1968).

13.37 Fedde, George A., and Chong, Carlos F., "Plated Wire Memory–Present and Future," *IEEE Trans. Magnetics*, **4,** 313-318 (1968).

13.38 Tchernev, Dimiter I., and Lewicki, George, "Extremely High Density Magnetic Information Storage," *IEEE Trans. Magnetics*, **4**, 75 (1968).

13.39 MacDonald, R. E., and Beck, J. W., "Magneto-Optical Recording," *J. Appl. Phys.*, **40**, 1429-1435 (1969).

13.40 Bobeck, A. H., "Properties and Device Applications of Magnetic Domains in Orthoferrites," *Bell System Tech. Jour.*, **46**, 1901-1925 (1967).

13.41 Bobeck, Andrew H., Fischer, Robert F., Perneski, Anthony J., Remeika, J. P., and Van Uitert, L. G., "Application of Orthoferrites to Domain-Wall Devices," *IEEE Trans. Magnetics*, **5**, 544-553 (1969).

13.42 Perneski, Anthony J., "Propagation of Cylindrical Magnetic Domains in Orthoferrites," *IEEE Trans. Magnetics*, **5**, 554-557 (1969).

13.43 Kriesman, C. J., and Goldberg, N., "Preparation and Crystal Synthesis of Magnetic Oxides," pp. 553-597 of Vol. 3 of [G.22].

13.44 Chin, G. Y., "Review of Magnetic Properties of Iron-Nickel Alloys." To be published in *IEEE Trans. Magnetics.*

13.45 Greifer, A. P., "Ferrite Memory Materials," *IEEE Trans. Magnetics*, **5**, 774-811 (1969).

13.46 Mathias, J. S., and Fedde, G. A., "Plated-Wire Technology: A Critical Review," *IEEE Trans. Magnetics*, **5**, 728-751 (1969).

13.47 Hunt, Robert P., "Magnetooptics, Lasers, and Memory Systems," *IEEE Trans. Magnetics*, **5**, 700-716 (1969).

13.48 Eschenfelder, A. H., "Promise of Magneto-Optic Storage Systems Compared to Conventional Magnetic Technology," *J. Appl. Phys.*, **41**, 1372-1376 (1970).

13.49 Rajchman, Jan A., "Promise of Optical Memories," *J. Appl. Phys.*, **41**, 1376-1383 (1970).

13.50 Thiele, A. A., "Theory of the Static Stability of Cylindrical Domains in Uniaxial Platelets," *J. Appl. Phys.*, **41**, 1139-1145 (1970).

CHAPTER 14. HARD MAGNETIC MATERIALS

14.1 Clegg, A. G., "The Development of the Permanent Magnet Industry," pp. 473-502 of [G.17].

14.2 Data sheet, *Metal Progress*, pp. 82, 83, 85, and 87. January, 1968.

14.3 Luborsky, F. E., "Permanent Magnets in Use Today," *J. Appl. Phys.*, **37**, 1091-1094 (1966).

14.4 Becker, J. J., Luborsky, F. E., and Martin, D. L., "Permanent Magnet Materials," *IEEE Trans. Magnetics*, **4**, 84-99 (1968).

14.5 McCaig, M., "Present and Future Technological Applications of Permanent Magnets," *IEEE Trans. Magnetics*, **4**, 221-228 (1968).

14.6 "Permanent Magnet Materials," pp. 779-785 of Vol. 1, 8th ed. *Metals Handbook* (Novelty, Ohio: American Society for Metals, 1961).

14.7 Edwards, Alun, "Magnet Design and Selection of Material," pp. 191-296 of [G.17].

14.8 De Vos, K. J., "Alnico Permanent Magnet Alloys," pp. 473-512 of Vol. 1 of [G.37].

14.9 Gould, J. E., "Progress in Permanent Magnet Materials," *Proc. Instn. Elec. Engrs.*, **106 A**, 493-500 (1959).

14.10 Craik, D. J., and Lane, R., "Magnetostatic and Exchange Interactions between Particles in Permanent Magnet Materials," *Brit. J. Appl. Phys.*, **2**, 33-45 (1969).

14.11 Cochardt, A., "Recent Ferrite Magnet Developments," *J. Appl. Phys.*, **37**, 1112-1115 (1966).

14.12 Mee, C. D., and Jeschke, J. C., "Single-Domain Properties in Hexagonal Ferrites," *J. Appl. Phys.*, **37**, 1271-1272 (1963).

14.13 Mendelsohn, L. I., Luborsky, F. E., and Paine, T. O., "Permanent-Magnet Properties of Elongated Single-Domain Iron Particles," *J. Appl. Phys.*, **26**, 1274-1280 (1955).

14.14 Falk, R. B., "A Current Review of Lodex Permanent Magnet Technology," *J. Appl. Phys.*, **37**, 1108-1112 (1966).

14.15 Becker, J. J., "Rare-Earth-Compound Magnets," *J. Appl. Phys*,, **41**, 1055-1064 (1970).

14.16 Strnat, K., Hoffer, G., Olson, J., Ostertag, W., and Becker, J. J., "A Family of New Cobalt-Base Permanent Magnet Materials," *J. Appl. Phys.*, **38**, 1001-1002 (1967).

14.17 Mee, C. D., *The Physics of Magnetic Recording.* 270 pp. (Amsterdam: North-Holland, 1964).

14.18 Speliotis, D. E., "Magnetic Recording Materials," *J. Appl. Phys.*, **38**, 1207-1214 (1967).

14.19 Wohlfarth, E. P., "A Review of the Problem of Fine-Particle Interactions with Special Reference to Magnetic Recording," *J. Appl. Phys.*, **35**, 783-790 (1964).

14.20 Ragosine, Victor E., "Magnetic Recording," *Sci. American*, pp. 71-82, Nov., 1969.

14.21 Kronenberg, K. J., and Bohlmann, M. A., "Long Term Magnetic Stability of Alnico and Barium Ferrite Magnets," *J. Appl. Phys.*, **31**, 82S-84S (1960).

14.22 Zingery, W. L., Whalley, W. B., Romberg, E. B., and Wheeler, F. W., "Evaluation of Long-Term Magnet Stability," *J. Appl. Phys.*, **37**, 1101-1103 (1966).

14.23 Tyack, F. G., "Permanent Magnet Applications," pp. 297-372 of [G.17].

14.24 Ireland, J. R., "New Figure of Merit for Ceramic Permanent Magnet Material Intended for dc Motor Applications," *J. Appl. Phys.*, **38**, 1011-1012 (1967).

14.25 Ireland, J. R., *Ceramic Permanent-Magnet Motors.* 188 pp. (New York: McGraw-Hill, 1968).

14.26 Harrold, William, and Reid, William R., "Permanent Magnets for Microwave Devices," *IEEE Trans. Magnetics*, **4**, 229-239 (1968).

14.27 Eckard, Mel, "Hardware Disease Can Be Prevented," *Hoard's Dairyman*, Nov. 10, 1957.

14.28 McCaig, M., "Permanent Magnets for Repulsion Devices," 9 pp. 1966. A technical bulletin of the Permanent Magnet Association, 301 Glossop Road, Sheffield 10, England.

14.29 Earnshaw, S., "On the Nature of the Molecular Forces which regulate the Constitution of the Luminiferous Ether," *Trans. Camb. Phil. Soc.*, **7**, 97-112 (1837-1842).

14.30 Boerdijk, A. H., "Technical Aspects of Levitation," *Philips Research Reports*, **11**, 45-56 (1956).

14.31 Simon, Ivan, Emslie, Alfred G., Strong, Peter F., and McConnell, Robert K., Jr., "Sensitive Tiltmeter Utilizing a Diamagnetic Suspension," *Rev. Sci. Instr.*, **39**, 1166-1171 (1968).

14.32 Polgreen, G. R., *New Applications of Permanent Magnets.* 330 pp. (London: Macdonald, 1966).

14.33 Bate, G., and Alstad, J. K., "A Critical Review of Magnetic Recording Materials," *IEEE Trans. Magnetics*, **5**, 821-839 (1969).

14.34 Gould, J. E., "Magnet Stability," pp. 443-471 of [G.17].

14.35 Polgreen, G. R., "The ideal magnet - fully controllable permanent magnets for power and transport," *Electronics and Power* (J. Inst. Elec. Engrs.) **17**, 31-34 (1971).

GENERAL REFERENCES

The following books in English on magnetic materials and magnetic measurements are listed in order of publication.

[G.1] Ewing, J. A., *Magnetic Induction in Iron and Other Metals*, 3d ed. 393 pp. (London: "The Electrician" Printing and Publishing Co., 1900). Although the theoretical portions of this book are of only historical interest, the clearly written accounts of experimental work are still well worth reading, particularly those dealing with the effects of stress. *cgs units.*

[G.2] Bitter, Francis, *Introduction to Ferromagnetism.* 314 pp. (New York: McGraw-Hill, 1937). Contains a detailed formal treatment of the effect of elastic crystal deformation on magnetic properties. *cgs units.*

[G.3] Snoek, J. L., *New Developments in Ferromagnetism.* 139 pp. (Elsevier, 1949). Primarily a report on research done in Holland during World War II, particularly on hysteresis, magnetic aftereffects, and ferrites. *cgs units.*

[G.4] Bozorth, Richard M., *Ferromagnetism.* 968 pp. (New York: Van Nostrand, 1951). Probably the most widely used reference book in the field. A treatise containing a very large amount of experimental data on a number of magnetic phenomena. Particularly valuable for its survey of the properties of ferromagnetic alloys. Includes a large number of bibliographic references, chronologically arranged from 1842 to 1951. *cgs units.*

[G.5] Astbury, N. F., *Industrial Magnetic Testing.* 132 pp. (London: The Institute of Physics, 1952). A critical examination of methods of measuring permeability, hysteresis loops, power loss, and permanent magnet properties. Alternating-current bridge methods are also described. *cgs units.*

[G.6] Hoselitz, K., *Ferromagnetic Properties of Metals and Alloys.* 317 pp. (Oxford: Clarendon Press, 1952). Contains a valuable section on phase determination by magnetic analysis. *cgs units.*

[G.7] Stewart, K. H., *Ferromagnetic Domains.* 176 pp. (Cambridge: University Press, 1954). A very well-written description of the phenomena responsible for the formation and arrangement of domains, and the properties and motion of domains walls. *cgs units.*

[G.8] Selwood, Pierce W., *Magnetochemistry*, 2d ed. 435 pp. (New York and London: Interscience, 1956). Concerned with the relation between chemical bonding and magnetic properties, and the problem of obtaining chemical information by means of magnetic measurements. *cgs units.*

[G.9] *Magnetic Properties of Metals and Alloys.* 349 pp. (Cleveland: American Society for Metals, 1959). Contains 13 papers, some of which are lengthy reviews, presented at an A.S.M. seminar.

[G.10] Smit, J., and Wijn, H. P. J., *Ferrites.* 369 pp. (New York: Wiley, 1959. This book is published in the Philips Technical Library series.) The first full-length treatment of the subject, written by two men associated with the Philips Research Laboratories in Holland, where the ferrites were first developed into materials. *cgs units.*

[G.11] Bardell. P. R., *Magnetic Materials in the Electrical Industry*, 2d ed. 320 pp. (London: Macdonald, 1960). Applications are stressed, rather than theory. Includes discussions of the applications of magnetic materials in transformers, relays, motors, generators, telephones, recording tapes, transducers, etc. *cgs units.*

[G.12] Brailsford, F., *Magnetic Materials*, 3d ed. 188 pp. (London: Methuen, and New York: Wiley, 1960). Contains a great deal of information presented in a very compact form. *mks units.*

[G.13] Bates, L. F., *Modern Magnetism*, 4th ed. 514 pp. (Cambridge: University Press, 1961). An unusual book, in that it combines a considerable amount of theory with detailed descriptions of many fundamental experiments. Full descriptions of gyromagnetic and magnetothermal effects are included. *cgs units.*

[G.14] Belov, K. P., *Magnetic Transitions.* W. H. Furry, translator. 242 pp. (New York: Consultants Bureau, 1961). An examination of the changes which occur in magnetic and other properties of ferro-, antiferro-, and ferrimagnetics in the neighborhood of their Curie or Néel temperatures. *cgs units.*

[G.15] Brown, William F., Jr., *Magnetostatic Principles in Ferromagnetism.* 202 pp. (Amsterdam: North-Holland, 1962). Concerned with the application of magnetostatic principles to single-domain particle theory, micromagnetics, domain theory, resonance, and spin-wave calculations. *Generalized gaussian units.*

[G.16] Lax, Benjamin, and Button, Kenneth J., *Microwave Ferrites and Ferrimagnetics.* 752 pp. (New York: McGraw-Hill, 1962). A thorough treatment of the theory and practice of ferrites at high frequencies. It also contains, in the first three chapters, a good review of the various kinds of magnetism. *cgs units.*

[G.17] Hadfield, D., ed., *Permanent Magnets and Magnetism.* 556 pp. (London: Iliffe Books, and New York: Wiley, 1962). A comprehensive book, by twelve British experts, on the theory, properties, design, applications, manufacture, and testing of permanent magnets. Interesting historical accounts are also included. *Mainly mks units, although some equations are given both in mks and cgs units.*

[G.18] Parker, Rollin J., and Studders, Robert J., *Permanent Magnets and Their Applications.* 406 pp. (New York: Wiley, 1962). Chiefly concerned with the design, application, and testing of permanent magnets. *cgs units.*

[G.19] Standley, K. J., *Oxide Magnetic Materials.* 204 pp. (Oxford: Clarendon Press, 1962). Theory and properties of ferrites, garnets, and other magnetic oxides. Applications are not described. *cgs units.*

[G.20] Goodenough, John B., *Magnetism and the Chemical Bond.* 393 pp. (New York and London: Interscience, 1963). A fairly advanced treatment, written in the form of an expanded review paper, of magnetochemistry, particularly of compounds of transition metals. *cgs units.*

[G.21] Brown, William Fuller, Jr., *Micromagnetics.* 143 pp. (New York and London: Interscience, 1963). A critical examination, on the phenomenological level, of the theory of magnetism, on a scale intermediate between that of the individual atomic moment and that of the domain. *cgs units.*

[G.22] Rado, George T., and Suhl, Harry, eds., *Magnetism.* (New York and London: Academic Press). An advanced treatise composed of chapters written by individual specialists on various topics in ferro-, antiferro-, and ferrimagnetism. *cgs units.*
Vol. I. *Magnetic Ions in Insulators, Their Interactions, Resonances, and Optical Properties.* 688 pp. (1963).
Vol. IIA. *Statistical Models, Magnetic Symmetry, Hyperfine Interactions, and Metals.* 443 pp. (1965).
Vol. IIB. *Interactions and Metals.* 428 pp. (1966).

Vol. III. *Spin Arrangements and Crystal Structure, Domains, and Micromagnetics.* 623 pp. (1963).
Vol IV. *Exchange Interactions among Itinerant Electrons.* 407 pp. (1966).

[G.23] Chikazumi, Soshin, *Physics of Magnetism.* 554 pp. (New York: Wiley, 1964). As the title indicates, theory is stressed, rather than applications. Contains a very thorough analysis of domain theory and of the ways in which magnetization can change by wall motion and by magnetization rotation, and more material on magnetostatic energies than most books. Special topics include magnetic annealing, thin films, and rare earths. *mks units.*

[G.24] Prutton, M., *Thin Ferromagnetic Films.* 269 pp. (London: Butterworths, 1964). Describes how thin films are made and their magnetic properties measured. Contains a detailed treatment of the magnetization reversal process. *cgs units.*

[G.25] Soohoo, Ronald F., *Magnetic Thin Films.* 316 pp. (New York: Harper and Row, 1965). A detailed examination of the magnetic properties of thin ferromagnetic films, of magnetization reversal, and of magnetic resonance in thin films. Experimental methods are described. *cgs units.*

[G.26] Craik, D. J., and Tebble, R. S., *Ferromagnetism and Ferromagnetic Domains.* 337 pp. (Amsterdam: North-Holland, and New York: Wiley, 1965). An examination of ferro- and ferrimagnetic behavior on the basis of domain structure, both in the massive and in the thin-film state. Experimental results are emphasized, particularly the direct observation of domains. *cgs units.*

[G.27] Morrish, Allan H., *The Physical Principles of Magnetism.* 680 pp. (New York: Wiley, 1965). Covers the whole range of magnetic phenomena, with particular emphasis on relaxation and resonance effects. *cgs units.*

[G.28] Smart, J. Samuel, *Effective Field Theories of Magnetism.* 188 pp. (Philadelphia: Saunders, 1966). An exposition and critical examination of the molecular field theories of ferro-, antiferro-, and ferrimagnetism. Includes extensive tables of Brillouin and related functions. *cgs units.*

[G.29] Carey, R., and Isaac, E. D., *Magnetic Domains and Techniques for Their Observation.* 168 pp. (New York and London: Academic Press, 1966). Contains an introductory review of domain theory, followed by descriptions and illustrations of the following methods of domain observation: colloid, magneto-optical, electron beam, and probe. *cgs units.*

[G.30] Brailsford, F., *Physical Principles of Magnetism.* 274 pp. (New York and London: Van Nostrand, 1966). An enlarged version of [G.12]. The main addition is a section on atomic theory, and other chapters are expanded. Commercial magnetic materials are described. *mks units.*

[G.31] Zijlstra, H., *Experimental Methods in Magnetism.* (Amsterdam: North-Holland, and New York: Wiley, 1967). *mks units.*

Vol. 1. *Generation and Computation of Magnetic Fields.* 236 pp. Detailed treatment of solenoid design, including normal and superconducting solenoids, and continuous and pulsed operation. Brief treatment of electromagnets.
Vol 2. *Measurements of Magnetic Quantities.* 296 pp. Covers the measurement of magnetic fields, magnetization, magnetostriction, and magnetic anisotropy. Resonance methods and the Mossbauer effect are discussed. Neutron diffraction and methods for domain observation are not included.

[G.32] Schieber, Michael M., *Experimental Magnetochemistry.* 572 pp. (Amsterdam: North-Holland, and New York: Wiley, 1967). Restricted to nonmetallic materials, namely, oxides, halides, chalcogenides, and miscellaneous compounds. Includes magnetic theory, specimen preparation (polycrystalline and single crystal), measurement methods, and 328 pages of crystallographic and magnetic data. *cgs units.*

[G.33] Anderson, J. C., *Magnetism and Magnetic Materials.* 248 pp. (London: Chapman and Hall, 1968). Theory is stressed, rather than applications. *mks units.*

[G.34] Thompson, John E., *The Magnetic Properties of Materials.* 173 pp. (Cleveland: CRC Press, 1968). Materials are stressed, rather than theory. Contains a fairly detailed account of the production and properties of silicon steel for magnetic applications. *cgs units.*

[G.35] Snelling, E. C., *Soft Ferrites, Properties and Applications.* 390 pp. (London: Iliffe, 1969). Contains a wealth of data on Ni-Zn and Mn-Zn ferrites and on their applications to inductors, transformers, and related devices. Microwave ferrites and ferrites for computer cores are not included. *mks and cgs units.*

[G.36] Tebble, R. S., and Craik, D. J., *Magnetic Materials.* 726 pp. (London and New York: Wiley-Interscience, 1969). Dubbed a "handbook" by the authors, it is more than that. The first half contains extensive data, systematically arranged, on the intrinsic magnetic properties of metallic and nonmetallic materials, together with many phase diagrams. The second half deals with structure-sensitive properties and how these affect magnetic phenomena in permanent magnets, magnetically soft materials (metals, alloys, and ferrites), and square-loop switching and memory elements. *cgs units.*

[G.37] Berkowitz, Ami E., and Kneller, Eckart, eds., *Magnetism and Metallurgy.* (New York and London: Academic Press, 1969). Composed of chapters written by individual specialists on correlations between magnetic properties and metallurgical structure. *Mainly cgs units.*

Vol. 1. Magnetic moments in solids, principles of ferromagnetic behavior, magnetic resonance, experimental methods (ac and dc), magnetic moments and transition temperatures, constitution of multiphase alloys, fine particle theory, Alnico permanent magnet alloys. 551 pp.

Vol. 2. Nonferromagnetic precipitate in a ferromagnetic matrix, effects of atomic order-disorder on magnetic properties, directional order, influence of crystal defects on magnetization processes in ferromagnetic single crystals, recovery and recrystallization, textured magnetic materials, diffusion, interpretation of magnetic resonance measurements in metals. 358 pp.

CONFERENCE PROCEEDINGS

The published proceedings of the various conferences on magnetism and magnetic materials are good sources of further information.

International

These are generally held every three years. The most important of these since 1950 are as follows.

Date	Place	Published
1950	Grenoble	*J. de Phys. et Radium,* **12**, 153-254 (1951)
1952	Washington	*Rev. Mod. Phys.,* **25,** 1-352 (1953)
1956	London	*Proc. Instn. Elec. Engrs.,* **B 104,** Supp., 127-570 (1957)
1958	Grenoble	*J. de Phys. et Radium,* **20,** 70-442 (1959)
1961	Kyoto	*J. Phys. Soc. Japan,* **17,** Supp. B.1, 1-718 (1962)
1964	Nottingham	*Proc. of the Inter. Conf. on Magnetism.* 878 pp. (London: The Institute of Physics and the Physical Society)
1967	Boston	*J. Appl. Phys.,* **39,** 363-1390 (1968)
1970	Grenoble	*J. de Physique* (1971)

U.S.A.

Two national conferences are held each year, generally with some contributions from abroad:
1. A conference is held each November, and the proceedings are published in a special issue of the *Journal of Applied Physics* in the following spring.
2. The *Intermag* conference on applied magnetism is held each April, and the proceedings are published in the September issue of the *IEEE (Institute of Electrical and Electronics Engineers) Transactions on Magnetics.* This conference is occasionally held abroad. The papers were formerly heavily oriented toward devices but lately are more balanced between devices and materials.

ANSWERS TO SELECTED PROBLEMS

Chapter 1

1.1 $g^{1/2}$ $cm^{3/2}$ sec^{-1}
1.2 $g^{1/2}$ $cm^{-1/2}$ sec^{-1}
1.3 a) 181 emu/cm^3 b) 276 amperes
1.4 30.8 ergs/Oe
1.5 19,800 gauss, 1580 emu/cm^3, 2010

Chapter 2

2.1 a) 39.6 Oe, 8.04 volts, 16.08 watts
2.2 a) 0.9987 b) 0.5002
2.3 $(4i\rho DL/E)^{1/3}$
2.4 H is $\frac{1}{2}$, E is $\frac{1}{8}$, and W is $\frac{1}{8}$ of its former value.
2.5 0.0179
2.6 a) 0.0469×10^{-6} coulomb/mm b) 1510 maxwell-turns/mm
2.7 $B_{\max} = 6950$ gauss, $B_r = 3160$ gauss, $H_c = 1.3$ Oe
$\mu_{\max} = 2590$, $M_{\max} = 553$ emu/cm^3
2.9 $2cx$, $3cl^2/4$
2.10 500 gauss
2.11 a) 3.0 b) 18 c) 1600 d) 5000
2.12 b) $\mu = 1700$ (approx)
2.13 500 turns (for $\Delta B = B_s/10$)
2.14 c) 4.24
2.15 $-$ 0.077 g

Chapter 3

3.1 a) $5\mu_B/6$ b) $\mu_B/2$
3.3 a) $5.39\mu_B$ b) $J = 2.24$, $\mu_H = 4.48\mu_B$ c) 1960 Oe; 6400 Oe
3.5 a) $\sqrt{3}\,\mu_B$ b) 1.28×10^{-3} emu/g atom-Oe c) 2.3 percent d) 54,500 Oe
3.6 a) 1.28×10^{-3} emu/g atom-Oe b) 1.3 percent
3.7 a) $3\,\mu_B$, μ_B, $-\mu_B$, $-3\,\mu_B$ b) 39.2°, 75.0°, 105.0°, 140.8°
c) 0.868, 0.115, 0.015, 0.002
3.8 3.8 percent

Chapter 4

4.3 b) 1.7×10^{-9} per oersted at 20°C, 1.8×10^{-5} per oersted at 750°C

Chapter 5

5.1 a) 2.41 b) 145 emu/g MnF_2, 5.61×10^5 Oe c) 0.51 degree
5.3 Between $\sqrt{2}/\sqrt{3}$ (= 0.816) and $\sqrt{2}$ (= 1.414)

Chapter 6

6.2 2.35 μ_B/molecule
6.3 4.29 μ_B/molecule

Chapter 7

7.2 a) 97.5 dyne-cm/degree b) 4.1 degree

7.5 a) $\langle 100 \rangle$ $H = -\dfrac{K_1}{M_s}\left(\dfrac{M}{M_s}\right)\left[3\left(\dfrac{M}{M_s}\right)^2 - 1\right]$

$\langle 110 \rangle$ $H = -\dfrac{K_1}{M_s}\left(\dfrac{M}{M_s}\right)\left[3\left(\dfrac{M}{M_s}\right)^2 - 2\right]$

c) $\langle 100 \rangle$ 206 Oe, $\langle 110 \rangle$ 103 Oe

7.8 a) Biaxial b) $L = -(3K_1/8)\sin 4\alpha$
7.9 Oblate
7.10 Saturation is easier normal to disk.
a) $K_s = 37 \times 10^5$ ergs/cm^3, $K_1 = 45 \times 10^5$ ergs/cm^3
b) 9430 Oe normal to disk, 10,550 Oe in plane of disk
7.11 Saturation easier normal to disk if $(N_a - N_c) < 2K_1/M_s^2$.
7.12 2.3

Chapter 8

8.1 $\Delta l/l = \frac{3}{2}\lambda_{100}\sin^2\delta$
8.2 $\Delta l/l = \frac{3}{4}\lambda_{111}\sin^2 2\gamma$
8.3 $\Delta l/l = -\frac{3}{2}\lambda_{111}\sin^2\theta$
8.4 $\lambda_{110} = \frac{1}{4}(\lambda_{100} + 3\lambda_{111})$
8.5 [001] gage; $\Delta l/l = -\frac{3}{2}\lambda_{100}\sin^2\theta$
[110] gage; $\Delta l/l = \frac{3}{4}(\lambda_{100} + \lambda_{111})\sin^2\theta$
8.7 M_s points in [010] until σ exceeds $(2K_1/3\,\lambda_{100})$.
It then jumps to the [100] or $[\bar{1}00]$ orientation.
8.9 b) $\Delta E/E = 3.5$ percent

Chapter 9

9.1 a) $\gamma = 7.6$ ergs/cm^2, $\delta = 84$ Å
b) $L_c = 65$ Å for cobalt
9.2 $\gamma = 6.1$ ergs/cm^2, $\delta = 93$ Å, and $L_c = 730$ Å for barium ferrite
9.6 a) 0.574, 35.5°

Chapter 11

11.5 $D/D_0 = 1.13$
11.6 a) Yes. $H_d = 7180$ Oe, $2K/M_s = 560$ Oe
b) c/a larger than about 8.5

11.7 H_{ci} (curling) $\approx$ 0.01 Oe
11.8 a) 250 Å b) 77 Å
11.9 a) 1.6°K b) 54°K
11.11 $D/D_p = 1.00_4, 1.03, 1.06$
11.12 a) $\ln\frac{M_r}{M_i} = -10^{11}e^{-25T_B/T}$ b) $T/T_B = 0.92$

Chapter 12

12.1 19 turns
12.3 11,000 cm/sec Oe
12.5 -0.534
12.6 $1.7/26 = 0.065$

Chapter 13

13.2 0.63 and 0.22
13.4 2.51
13.5 1.48°C/min

INDEX

INDEX